Optical Electronics in Modern Communications

The Oxford Series in Electrical and Computer Engineering

M. E. Van Valkenburg, Senior Consulting Editor
Adel S. Sedra, Series Editor, Electrical Engineering
Michael R. Lightner, Series Editor, Computer Engineering

ALLEN AND HOLBERG
CMOS Analog Circuit Design

BOBROW
Elementary Linear Circuit Analysis, 2/e

BOBROW
Fundamentals of Electrical Engineering, 2/e

CAMPBELL
The Science and Engineering of Microelectronic Fabrication

CHEN
Linear System Theory and Design

CHEN
System and Signal Analysis, 2/e

COMER
Digital Logic and State Machine Design, 3/e

COMER
Microprocessor-Based System Design

COOPER AND MCGILLEM
Probabilistic Methods of Signal and System Analysis, 2/e

GHAUSI
Electronic Devices and Circuits: Discrete and Integrated

HOUTS
Signal Analysis in Linear Systems

JONES
Introduction to Optical Fiber Communication Systems

KENNEDY
Operational Amplifier Circuits: Theory and Application

KUO
Digital Control Systems, 3/e

LEVENTHAL
Microcomputer Experimentation with the IBM PC

LEVENTHAL
Microcomputer Experimentation with the Intel SDK-86

LEVENTHAL
Microcomputer Experimentation with the Motorola MC6800 ECB

MCGILLEM AND COOPER
Continuous and Discrete Signal and System Analysis, 3/e

MINER
Lines and Electromagnetic Fields for Engineers

NAVON
Semiconductor Microdevices and Materials

PAPOULIS
Circuits and Systems: A Modern Approach

RAMSHAW AND VAN HEESWIJK
Energy Conversion

SCHWARZ
Electromagnetics for Engineers

SCHWARZ AND OLDHAM
Electrical Engineering: An Introduction, 2/e

SEDRA AND SMITH
Microelectronic Circuits, 3/e

STEFANI, SAVANT, AND HOSTETTER
Design of Feedback Control Systems, 3/e

VAN VALKENBURG
Analog Filter Design

VRANESIC AND ZAKY
Microcomputer Structures

WARNER AND GRUNG
Semiconductor Device Electronics

YARIV
Optical Electronics in Modern Communications, 5/e

Optical Electronics in Modern Communications

Fifth Edition

Amnon Yariv
California Institute of Technology

New York Oxford
Oxford University Press
1997

Oxford University Press

Oxford New York
Athens Auckland Bangkok Bogotá Bombay Buenos Aires
Calcutta Cape Town Dar es Salaam Delhi Florence Hong Kong
Istanbul Karachi Kuala Lumpur Madras Madrid Melbourne
Mexico City Nairobi Paris Singapore Taipei Tokyo Toronto

and associated companies in
Berlin Ibadan

Published by Oxford University Press, Inc.,
198 Madison Avenue, New York, New York, 10016-4314

Library of Congress Cataloging-in-Publication Data
Yariv, Amnon.
Optical electronics in modern communications / Amnon Yariv.—5th ed.
p. cm.—(The Oxford series in electrical engineering)
Rev. ed of: Optical electronics. 4th ed., c1991.
Includes bibliographical references and index.
ISBN 0-19-510626-1 (cl)
1. Lasers. 2. Fiber optics. 3. Electrooptical devices.
4. Acoustooptical devices. 5. Photoelectronic devices. I. Title.
II. Series.
TA 1675.Y37 1996
621.382′7—dc20 95-52302

9 8 7 6 5 4 3

Printed in the United States of America
on acid-free paper

Preface to the Fifth Edition

"If it is beautiful, wear it around your neck. If it serves a useful purpose, carry it on your back. It if is neither, get rid of it."

In the process of deciding which material to include in this new edition and which to discard, I attempted to follow the above-quoted advice of my sergeant major (whose name I forgot) in the Israeli army as we were preparing to go into the field. I thus limited the material to what I consider, subjectively, beautiful, or important, or, on more than one occasion, both. The user of this book will have his own candidates for each of these categories, which, I hope, are not different from mine. In the process, most of the topics and chapters of the fourth edition survived the transition to the fifth. A considerable amount of new material, however, has been added. The changes reflect the continuous ascendance of optical communication as the foremost communication technology. With the new additions, the center of gravity of the book has swung clearly to the side of low-power, communication-related topics that made it appropriate to change the title to *Optical Electronics in Modern Communications*.

The main new features of this edition are:

1. Use of the transfer function Fourier transform formalism to treat pulse propagation in fibers.
2. The temporal-spatial equivalence of pulse and beam propagation including temporal lenses.
3. Compensation of dispersive pulse spread in fibers.

4. New treatment of the optical susceptibility ($\chi'(\nu)$ and $\chi''(\nu)$), using the Kramers–Kroning relations. Derivation of Kramers–Kroning relations.
5. A major overhaul of the discussion on distributed feedback lasers, including a treatment of gain-coupled lasers.
6. Dynamic chirp in semiconductor lasers.
7. Vertical-cavity semiconductor lasers.
8. A new chapter on solitons, with a first-principles derivation of the propagation equation in nonlinear fibers.
9. A new chapter consisting of a classical treatment of quantum optics and quantum noise, consequences for optical measurements, shot noise, and "squeezing" of amplitude fluctuations, below the classical limit, by degenerate parametric amplification.

The academic requirements for the use of this book are unchanged from those stated in the preface to the fourth edition, repeated here.

I am indebted to Mrs. Jana Mercado and Mrs. Mary Eleanor Johnson for typing and editing under fire. I also benefited from specific technical inputs by John Kitching, William Marshall, John O'Brien, and Matt McAdams.

To Mr. Ali Adibi my deep appreciation for the countless hours spent rederiving all the major results. The errors and inconsistencies that he corrected will go a long way toward making this a rigorous and relatively error-free text.

Pasadena, California **Amnon Yariv**
June 1996

Preface to the Fourth Edition

The five years that have intervened since the appearance of the third edition of *OPTICAL ELECTRONICS* witnessed significant technical developments in the field and the emergence of some major trends. A few of the important developments are

1. Optical fiber communication has established itself as the key communication technology.
2. The semiconductor laser and especially the longer wavelength GaInAsP/InP version has emerged as the main light source for high-data-rate optical fiber communication systems.
3. Quantum well semiconductor lasers started replacing their conventional counterparts for high-data-rate long distance communication and most other sophisticated applications including ultra-low threshold and mode-locked lasers.
4. Optical fiber amplifiers are causing a minor revolution in fiber communication due to their impact on very long distance transmission and on large scale optical distribution systems.

The accumulated weight of the new developments was such that when I last taught the course at Caltech in 1989 I found myself using a substantial fraction of course material that was not included in the text. The fourth edition brings this material into the fold. The main additions to the third edition, include major revisions and new chapters dealing with

1. Jones calculus and its extension to Faraday effect elements.
2. Radiometry and infrared detection.
3. Optical fiber amplifiers and their impact on fiber communication links.
4. Laser arrays.
5. Distributed feedback lasers, including multi-element lasers with phase shift sections.
6. Quantum well and ultra-low threshold semiconductor lasers.
7. Photorefractive crystals and two-beam coupling in dynamic holography and image processing.
8. Two-beam coupling and phase conjugation in stimulated Brillouin scattering.
9. Intensity fluctuations and coherence in semiconductor lasers and their impact on fiber communication systems.

The book continues to be aimed at the student interested in learning how to generate and manipulate optical radiation and how to use it to transmit information. At Caltech the course is taken, almost in equal proportions, by electrical engineering, physics, and applied physics students. About half the students tend to be seniors and the rest graduate students.

The prerequisites for taking the course at Caltech are a sound undergraduate background in electromagnetic theory—usually a one year course in this area—and an introcution to atomic physics.

The hands-on and research flavor of the book owes greatly to the exciting mix of visitors, talented students, and postdocs who bombard me continually with their newest findings and thoughts.

This edition includes acknowledged and unacknowledged contributions from Chris Harder, Kerry Vahala, Eli Kapon, Kam Lau, Pamela Derry, Israel Ury, Nadav Bar-Chaim, Hank Blauvelt, Michael Mittelstein, Lars Eng, Norman Kwong, Shu Wu Wu, Bin Zhao, and Rudy Hoffmeister. The Caltech Applied Physics 130 and 131 classes during 1987 and 1989, helped ferret out inconsistencies and insisted on clearer presentations.

My wife Fran and my administrative assistant Jana Mercado are responsible for the typing and editing. To them and to all of the above, my gratitude.

Pasadena, California
January 1991

Amnon Yariv

Contents

Optical Electronics in Modern Communications

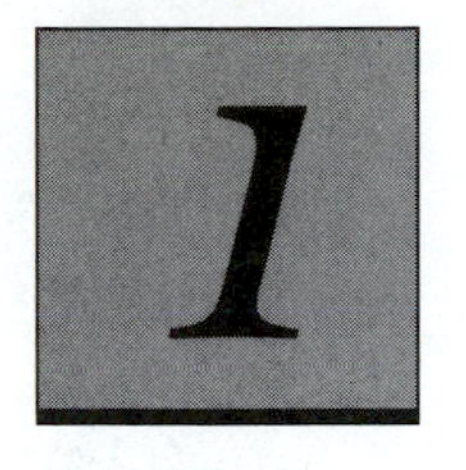

Electromagnetic Theory

1.0 INTRODUCTION

In this chapter we derive some of the basic results concerning the propagation of plane, single-frequency, electromagnetic waves in homogeneous isotropic media, as well as in anisotropic crystal media. Starting with Maxwell's equations we obtain expressions for the dissipation, storage, and transport of energy resulting from the propagation of waves in material media. We consider in some detail the phenomenon of birefringence, in which the phase velocity of a plane wave in a crystal depends on its direction of polarization. The two allowed modes of propagation in uniaxial crystals—the ''ordinary'' and ''extraordinary'' rays—are discussed using the formalism of the index ellipsoid.

We also derive the Fresnel-Kirchhoff diffraction integral. This integral, the key for work in coherent and Fourier optics, will be used extensively throughout this book.

1.1 COMPLEX-FUNCTION FORMALISM

In problems that involve sinusoidally varying time functions, we can save a great deal of manipulation and space by using the complex-function formalism. As an example consider the function

$$a(t) = |A| \cos(\omega t + \phi_a) \tag{1.1-1}$$

where ω is the circular (radian) frequency[1] and ϕ_a is the phase. Defining the complex amplitude of $a(t)$ by

$$A = |A|\, e^{i\phi_a} \tag{1.1-2}$$

we can rewrite (1.1-1) as

$$a(t) = \mathrm{Re}[Ae^{i\omega t}] \tag{1.1-3}$$

We will often represent $a(t)$ by

$$a(t) = Ae^{i\omega t} \tag{1.1-4}$$

instead of by (1.1-1) or (1.1-3). This, of course, is not strictly correct so that when this happens *it is always understood* that what is meant by (1.1-4) is the *real part of* A$\exp(i\omega t)$. In most situations the replacement of (1.1-3) by the complex form (1.1-4) poses no problems. The exceptions are cases that involve the product (or powers) of sinusoidal functions. In these cases we must use the real form of the function (1.1-3). To illustrate the case where the distinction between the real and complex form is not necessary, consider the problem of taking the derivative of $a(t)$. Using (1.1-1) we obtain

$$\frac{da(t)}{dt} = \frac{d}{dt}[|A| \cos(\omega t + \phi_a)] = -\omega|A| \sin(\omega t + \phi_a) \tag{1.1-5}$$

If we use instead the complex form (1.1-4), we get

$$\frac{da(t)}{dt} = \frac{d}{dt}(Ae^{i\omega t}) = i\omega Ae^{i\omega t}$$

Taking, as agreed, the real part of the last expression and using (1.1-2), we obtain (1.1-5).

As an example of a case in which we have to use the real form of the function, consider the product of two sinusoidal functions $a(t)$ and $b(t)$, where

$$\begin{aligned} a(t) &= |A| \cos(\omega t + \phi_a) \\ &= \frac{|A|}{2}[e^{i(\omega t+\phi_a)} + e^{-i(\omega t+\phi_a)}] \\ &= \mathrm{Re}[Ae^{i\omega t}] \end{aligned} \tag{1.1-6}$$

and

$$\begin{aligned} b(t) &= |B| \cos(\omega t + \phi_b) \\ &= \frac{|B|}{2}[e^{i(\omega t+\phi_b)} + e^{-i(\omega t+\phi_b)}] \\ &= \mathrm{Re}[Be^{i\omega t}] \end{aligned} \tag{1.1-7}$$

[1]The radian frequency ω is to be distinguished from the real frequency $\nu = \omega/2\pi$.

with $A = |A| \exp(i\phi_a)$ and $B = |B| \exp(i\phi_b)$. Using the real functions, we get

$$a(t)b(t) = \frac{|A|\,|B|}{2}\,[\cos(2\omega t + \phi_a + \phi_b) + \cos(\phi_a - \phi_b)] \tag{1.1-8}$$

Were we to evaluate the product $a(t)b(t)$ using the complex form of the functions, we would get

$$a(t)b(t) = ABe^{i2\omega t} = |A|\,|B|\,e^{i(2\omega t + \phi_a + \phi_b)} \tag{1.1-9}$$

Comparing the last result to (1.1-8) shows that the time-independent (dc) term $\frac{1}{2}|A|\,|B| \cos(\phi_a - \phi_b)$ is missing, and thus the use of the complex form led to an error.

Time-Averaging of Sinusoidal Products[2]

Another problem often encountered is that of finding the time average of the product of two sinusoidal functions of the same frequency

$$\overline{a(t)b(t)} = \frac{1}{T}\int_0^T |A| \cos(\omega t + \phi_a)|B| \cos(\omega t + \phi_b)\,dt \tag{1.1-10}$$

where $a(t)$ and $b(t)$ are given by (1.1-6) and (1.1-7) and the horizontal bar denotes time-averaging. $T = 2\pi/\omega$ is the period of the oscillation. Since the integrand in (1.1-10) is periodic in T, the averaging can be performed over a time T. Using (1.1-8) we obtain directly

$$\overline{a(t)b(t)} = \frac{|A|\,|B|}{2} \cos(\phi_a - \phi_b) \tag{1.1-11}$$

This last result can be written in terms of the complex amplitudes A and B, defined immediately following (1.1-7), as

$$\overline{a(t)b(t)} = \tfrac{1}{2}\mathrm{Re}(AB^*) \tag{1.1-12}$$

This important result will find frequent use throughout the book.

1.2 CONSIDERATIONS OF ENERGY AND POWER IN ELECTROMAGNETIC FIELDS

In this section we derive the formal expressions for the power transport, power dissipation, and energy storage that accompany the propagation of electromagnetic radiation in material media. The starting point is Maxwell's curl equations (in MKS units)

$$\nabla \times \mathbf{h} = \mathbf{i} + \frac{\partial \mathbf{d}}{\partial t} \tag{1.2-1}$$

$$\nabla \times \mathbf{e} = -\frac{\partial \mathbf{b}}{\partial t} \tag{1.2-2}$$

---

[2]The problem of the time average of the product of two nearly sinusoidal functions is considered in Problems 1.1 and 1.2.

and the constitutive equations relating the polarization of the medium to the displacement vectors

$$\mathbf{d} = \varepsilon_0 \mathbf{e} + \mathbf{p} \tag{1.2-3}$$

$$\mathbf{b} = \mu_0(\mathbf{h} + \mathbf{m}) \tag{1.2-4}$$

where **i** is the current density (amperes per square meter); $\mathbf{e}(\mathbf{r}, t)$ and $\mathbf{h}(\mathbf{r}, t)$ are the electric and magnetic field vectors, respectively; $\mathbf{d}(\mathbf{r}, t)$ and $\mathbf{b}(\mathbf{r}, t)$ are the electric and magnetic displacement vectors; $\mathbf{p}(\mathbf{r}, t)$ and $\mathbf{m}(\mathbf{r}, t)$ are the electric and magnetic polarizations (dipole moment per unit volume) of the medium; and ε_0 and μ_0 are the electric and magnetic permeabilities of vacuum, respectively. We adopt the convention of using lowercase letters to denote the time-varying functions, reserving capital letters for the amplitudes of the sinusoidal time functions. For a detailed discussion of Maxwell's equations, the reader is referred to any standard text on electromagnetic theory such as Reference [1].

Using (1.2-3) and (1.2-4) in (1.2-1) and (1.2-2) leads to

$$\nabla \times \mathbf{h} = \mathbf{i} + \frac{\partial}{\partial t}(\varepsilon_0 \mathbf{e} + \mathbf{p}) \tag{1.2-5}$$

$$\nabla \times \mathbf{e} = -\frac{\partial}{\partial t}\mu_0(\mathbf{h} + \mathbf{m}) \tag{1.2-6}$$

Taking the scalar (dot) product of (1.2-5) and **e** gives

$$\mathbf{e} \cdot \nabla \times \mathbf{h} = \mathbf{e} \cdot \mathbf{i} + \frac{\varepsilon_0}{2}\frac{\partial}{\partial t}(\mathbf{e} \cdot \mathbf{e}) + \mathbf{e} \cdot \frac{\partial \mathbf{p}}{\partial t} \tag{1.2-7}$$

where we used the relation

$$\frac{1}{2}\frac{\partial}{\partial t}(\mathbf{e} \cdot \mathbf{e}) = \mathbf{e} \cdot \frac{\partial \mathbf{e}}{\partial t}$$

Next we take the scalar product of (1.2-6) and **h**:

$$\mathbf{h} \cdot \nabla \times \mathbf{e} = -\frac{\mu_0}{2}\frac{\partial}{\partial t}(\mathbf{h} \cdot \mathbf{h}) - \mu_0 \mathbf{h} \cdot \frac{\partial \mathbf{m}}{\partial t} \tag{1.2-8}$$

Subtracting (1.2-8) from (1.2-7) and using the vector identity

$$\nabla \cdot (\mathbf{A} \times \mathbf{B}) = \mathbf{B} \cdot \nabla \times \mathbf{A} - \mathbf{A} \cdot \nabla \times \mathbf{B} \tag{1.2-9}$$

results in

$$-\nabla \cdot (\mathbf{e} \times \mathbf{h}) = \mathbf{e} \cdot \mathbf{i} + \frac{\partial}{\partial t}\left(\frac{\varepsilon_0}{2}\mathbf{e} \cdot \mathbf{e} + \frac{\mu_0}{2}\mathbf{h} \cdot \mathbf{h}\right) + \mathbf{e} \cdot \frac{\partial \mathbf{p}}{\partial t} + \mu_0 \mathbf{h} \cdot \frac{\partial \mathbf{m}}{\partial t} \tag{1.2-10}$$

We integrate the last equation over an arbitrary volume V and use the Gauss theorem [1]

$$\int_V (\nabla \cdot \mathbf{A})\, dv = \int_S \mathbf{A} \cdot \mathbf{n}\, da \tag{1.2-10a}$$

where $\mathbf{A}$ is any vector function, $\mathbf{n}$ is the unit vector normal to the surface S enclosing V, and dv and da are the differential volume and surface elements, respectively. The result is

$$\begin{aligned} -\int_V \nabla \cdot (\mathbf{e} \times \mathbf{h})\, dv &= -\int_S (\mathbf{e} \times \mathbf{h}) \cdot \mathbf{n}\, da \\ &= \int_V \left[\mathbf{e} \cdot \mathbf{i} + \frac{\partial}{\partial t}\left(\frac{\varepsilon_0}{2}\mathbf{e} \cdot \mathbf{e}\right) + \frac{\partial}{\partial t}\left(\frac{\mu_0}{2}\mathbf{h} \cdot \mathbf{h}\right) + \mathbf{e} \cdot \frac{\partial \mathbf{p}}{\partial t} + \mu_0 \mathbf{h} \cdot \frac{\partial \mathbf{m}}{\partial t}\right] dv \end{aligned} \tag{1.2-11}$$

According to the conventional interpretation of electromagnetic theory, the left side of (1.2-11), that is,

$$-\int_S (\mathbf{e} \times \mathbf{h}) \cdot \mathbf{n}\, da$$

gives the total power flowing *into* the volume bounded by S. The first term on the right side is the power expended by the field on the moving charges; the sum of the second and third terms corresponds to the rate of increase of the vacuum electromagnetic stored energy $\mathscr{E}_{\text{vac}}$ where

$$\mathscr{E}_{\text{vac}} = \int_V \left[\frac{\varepsilon_0}{2}\mathbf{e} \cdot \mathbf{e} + \frac{\mu_0}{2}\mathbf{h} \cdot \mathbf{h}\right] dv \tag{1.2-12}$$

Of special interest in this book is the next-to-last term

$$\mathbf{e} \cdot \frac{\partial \mathbf{p}}{\partial t} \tag{1.2-13}$$

which represents the power per unit volume expended by the field *on* the electric dipoles. This power goes into an increase in the potential energy stored by the dipoles as well as into supplying the dissipation that may accompany the change in $\mathbf{p}$. We will return to it again in Chapter 5, where we treat the interaction of radiation and atomic systems.

Dipolar Dissipation in Harmonic Fields

According to the discussion in the preceding paragraph, the average power per unit volume expended by the field on the medium electric polarization is

$$\overline{\frac{\text{Power}}{\text{Volume}}} = \overline{\mathbf{e} \cdot \frac{\partial \mathbf{p}}{\partial t}} \tag{1.2-14}$$

where the horizontal bar denotes time-averaging. Let us assume for the sake of simplicity that $\mathbf{e}(t)$ and $\mathbf{p}(t)$ are parallel to each other and take their time dependence to be

$$e(t) = \mathrm{Re}[Ee^{i\omega t}] \tag{1.2-15}$$

$$p(t) = \mathrm{Re}[Pe^{i\omega t}] \tag{1.2-16}$$

where E and P are the complex amplitudes. The electric susceptibility χ_e of the medium is defined by

$$P = \varepsilon_0 \chi_e E \tag{1.2-17}$$

and is thus a complex number, in general a function of the frequency ω. Substituting (1.2-15) and (1.2-16) in (1.2-14) and using (1.2-17) gives

$$\begin{aligned}\overline{\frac{\text{Power}}{\text{Volume}}} &= \overline{\mathrm{Re}[Ee^{i\omega t}]\,\mathrm{Re}[i\omega Pe^{i\omega t}]} \\ &= \tfrac{1}{2}\,\mathrm{Re}[i\omega\varepsilon_0\chi_e EE^*] \\ &= \frac{\omega}{2}\varepsilon_0|E|^2\,\mathrm{Re}(i\chi_e)\end{aligned} \tag{1.2-18}$$

where in going from the first to the second equality we used (1.1-12). Since χ_e is complex, we can write it in terms of its real and imaginary parts as

$$\chi_e = \chi_e' - i\chi_e'' \tag{1.2-19}$$

which, when used in (1.2-17), gives

$$\overline{\frac{\text{Power}}{\text{Volume}}} = \frac{\omega\varepsilon_0\chi_e''}{2}|E|^2 \tag{1.2-20}$$

which is the desired result.

We leave it as an exercise (Problem 1.3) to show that in anisotropic media in which the complex field components are related by

$$P_i = \varepsilon_0 \sum_j \chi_{ij}E_j \tag{1.2-21}$$

the application of (1.2-14) yields

$$\overline{\frac{\text{Power}}{\text{Volume}}} = \frac{\omega}{2}\varepsilon_0 \sum_{i,j} \mathrm{Re}(i\chi_{ij}E_i^*E_j) \tag{1.2-22}$$

The study of power exchange between electrons (bound or free) and electromagnetic fields is central to this book. It is thus instructive to rederive Equation (1.2-18), obtained here formally from Maxwell's equations, using another, and possibly more familiar, point of view.

Consider the case of a single localized electric dipole $\boldsymbol{\mu}$. In this case, the power flow from *the dipole to the field* is obtained by replacing $\mathbf{p}$ by $\boldsymbol{\mu}$ in (1.2-13).

$$\frac{\text{Power}}{\text{dipole} \rightarrow \text{field}} = -\mathbf{e} \cdot \frac{\partial \boldsymbol{\mu}}{\partial t} \tag{1.2-23}$$

The simplest oscillating dipole imaginable is arguably that of an electron whose position is given by

$$x = x_0 \cos(\omega t + \phi_e) \tag{1.2-24}$$

which is subject to an electric field

$$e_x = E_0 \cos \omega t \tag{1.2-25}$$

The dipole moment of the oscillating electron (whose charge is $-e$) is

$$\mu_x = -ex = -ex_0 \cos(\omega t + \phi_e)$$

which leads to

$$\frac{\text{Power}}{\text{elect.} \rightarrow \text{field}} = e\, e_x \frac{\partial x}{\partial t} = e\, e_x(t) v(t) = -Fv \tag{1.2-26}$$

where $F = -e\, e_x$ is the force on the electron, while $v = \dfrac{\partial x}{\partial t}$ is its velocity. We have thus shown that the dipolar result for power exchange, Equation (1.2-23) or equivalently (1.2-13) is equivalent to Equation (1.2-27), well familiar from classical dynamics. It is now easy to understand why the power flow from a field to the electron (polarization) depends on their relative phase. The case $\phi_e = -\dfrac{\pi}{2}$, for example, is one where

$$\frac{\text{Power}}{\text{elect.} \rightarrow \text{field}} = \omega e x_0 E_0 \cos^2 \omega t \tag{1.2-27}$$

The electron *is subject to a braking force* at all times and continually loses power to the field. If $\phi_e = \dfrac{\pi}{2}$, the reverse is true:; the electron is always accelerated and the power flow is given by (1.2-28) with a minus sign.

The reader is encouraged to make plots of the power flow vs. t during one optical cycle for, say, $\phi_e = 0, \pi$, the power flow reverses sign four times per (optical) period, thus averaging out to zero.

1.3 WAVE PROPAGATION IN ISOTROPIC MEDIA

Here we consider the propagation of electromagnetic plane waves in homogeneous and isotropic media so that ε and μ are scalar constants. Vacuum is, of course, the best example of such a "medium." Liquids and glasses are material media that, to a first approximation, can be treated as homogeneous and isotropic.[3] We choose the direction of propagation as z and, taking the plane wave to be uniform in the x-y plane, put $\partial/\partial x = \partial/\partial y = 0$ in (1.2-1) and (1.2-2). Assuming a lossless ($\sigma = 0$) medium, (1.2-1) and (1.2-2) become

$$\nabla \times \mathbf{e} = -\mu \frac{\partial \mathbf{h}}{\partial t} \tag{1.3-1}$$

$$\nabla \times \mathbf{h} = \varepsilon \frac{\partial \mathbf{e}}{\partial t} \tag{1.3-2}$$

$$\frac{\partial e_y}{\partial z} = \mu \frac{\partial h_x}{\partial t} \tag{1.3-3}$$

$$\frac{\partial h_y}{\partial z} = -\varepsilon \frac{\partial e_x}{\partial t} \tag{1.3-4}$$

$$\frac{\partial e_x}{\partial z} = -\mu \frac{\partial h_y}{\partial t} \tag{1.3-5}$$

$$\frac{\partial h_x}{\partial z} = \varepsilon \frac{\partial e_y}{\partial t} \tag{1.3-6}$$

$$0 = \mu \frac{\partial h_z}{\partial t} \tag{1.3-7}$$

$$0 = \varepsilon \frac{\partial e_z}{\partial t} \tag{1.3-8}$$

From (1.3-7) and (1.3-8) it follows that the time dependent parts of h_z and e_z are both zero; therefore, a uniform plane wave in a homogeneous isotropic medium can have no longitudinal field components. We can obtain a self-consistent set of equations from (1.3-3) through (1.3-8) by taking e_y and h_x (or e_x and h_y) to be zero.[4] In this case the last set of equations reduces to Equations (1.3-4) and (1.3-5). Taking the derivative of (1.3-5) with respect to z and using (1.3-4), we obtain

$$\frac{\partial^2 e_x}{\partial z^2} = \mu\varepsilon \frac{\partial^2 e_x}{\partial t^2} \tag{1.3-9}$$

---

[3]The individual molecules making up the liquid or glass are, of course, anisotropic. This anisotropy, however, is averaged out because of the very large number of molecules with random orientations present inside a volume $\sim \lambda^3$.

[4]More fundamentally it can be easily shown from (1.3-1) and (1.3-2) (see Problem 1.4) that, for uniform plane harmonic waves, **e** and **h** are normal to each other as well as to the direction of propagation. Thus, **x** and **y** can simply be chosen to coincide with the directions of **e** and **h**.

A reversal of the procedure will yield a similar equation for h_y. Since our main interest is in harmonic (sinusoidal) time variation, we postulate a solution in the form of

$$e_x^{\pm} = E_x^{\pm}\, e^{i(\omega t \mp kz)} \tag{1.3-10}$$

where $E_x^{\pm}\exp(\mp ikz)$ are the complex field amplitudes at z. Before substituting (1.3-10) into the wave equation (1.3-9), we may consider the nature of the two functions $e_x^{\pm}$. Taking first e_x^{+}: If an observer were to travel in such a way as to always exercise the same field value, he would have to satisfy the condition

$$\omega t - kz = \text{constant}$$

where the constant is arbitrary and determines the field value ''seen'' by the observer. By differentiation of the last result, it follows that the observer must travel in the $+z$ direction with a velocity

$$c = \frac{dz}{dt} = \frac{\omega}{k} \tag{1.3-11}$$

This is the *phase velocity* of the wave. If the wave were frozen in time, the separation between two neighboring field peaks—that is, the wavelength—would be

$$\lambda = \frac{2\pi}{k} = 2\pi\frac{c}{\omega} \tag{1.3-12}$$

The e_x^{-} solution differs only in the sign of k, and thus, according to (1.3-11), it corresponds to a wave traveling with a phase velocity c in the $-z$ direction.

The value of c can be obtained by substituting the assumed solution (1.3-10) into (1.3-9), which results in

$$c = \frac{\omega}{k} = \frac{1}{\sqrt{\mu\varepsilon}} \tag{1.3-13}$$

or

$$k = \omega\sqrt{\mu\varepsilon}$$

The phase velocity in vacuum is

$$c_0 = \frac{1}{\sqrt{\mu_0\varepsilon_0}} = 3 \times 10^8 \text{ m/s}$$

whereas in material media it has the value

$$c = \frac{c_0}{n}$$

where $n = \sqrt{\varepsilon/\varepsilon_0}$ is the *index of refraction.*

Turning our attention next to the magnetic field h_y, we can express it, in a manner similar to (1.3-10), in the form of

$$h_y^{\pm} = H_y^{\pm} e^{i(\omega t \mp kz)} \tag{1.3-14}$$

Substitution of this equation into (1.3-4) and using (1.3-10) gives

$$-ikH_y^+ e^{i(\omega t - kz)} = -i\omega\varepsilon E_x^+ e^{i(\omega t - kz)}$$

Therefore, from (1.3-13),

$$H_y^+ = \frac{E_x^+}{\eta} \qquad \eta = \sqrt{\frac{\mu}{\varepsilon}} \tag{1.3-15}$$

In vacuum $\eta_0 = \sqrt{\mu_0/\varepsilon_0} \simeq 377$ ohms. Repeating the same steps with H_y^- and E_x^- gives

$$H_y^- = -\frac{E_x^-}{\eta} \tag{1.3-16}$$

so that for negative ($-z$) traveling waves the relative phase of the electric and magnetic fields is reversed with respect to the wave traveling in the $+z$ direction. Since the wave equation (1.3-9) is a linear differential equation, we can take the solution for the harmonic case as a linear superposition of e_x^+ and e_x^-

$$e_x(z, t) = E_x^+ e^{i(\omega t - kz)} + E_x^- e^{i(\omega t + kz)} \tag{1.3-17}$$

and, similarly,

$$h_y(z, t) = \frac{1}{\eta} [E_x^+ e^{i(\omega t - kz)} - E_x^- e^{i(\omega t + kz)}]$$

where E_x^+ and E_x^- are arbitrary complex constants.

Power Flow in Harmonic Fields

The average power per unit area—that is, the intensity (W/m^2)—carried in the direction of propagation by a uniform plane wave is given by (1.2-11) as

$$|I| = |\overline{\mathbf{e} \times \mathbf{h}}| \tag{1.3-18}$$

where the horizontal bar denotes time averaging. Since $\mathbf{e} \parallel x$ and $\mathbf{h} \parallel y$, we can obtain from (1.3-18) for the power flow in the z direction

$$I = \overline{e_x h_y}$$

Taking advantage of the harmonic nature of e_x and h_y, we use (1.3-17) and (1.1-12) to obtain

$$\begin{aligned} I = \tfrac{1}{2}\,\mathrm{Re}[E_x H_y^*] &= \frac{1}{2\eta}\,\mathrm{Re}\{[E_x^+ e^{-ikz} + E_x^- e^{ikz}] \\ &\quad \times [(E_x^+)^* e^{ikz} - (E_x^-)^* e^{-ikz}]\} \\ &= \frac{|E_x^+|^2}{2\eta} - \frac{|E_x^-|^2}{2\eta} \end{aligned} \tag{1.3-19}$$

The first term on the right side of (1.3-19) gives the intensity associated with the positive ($+z$) traveling wave, whereas the second term represents the negative traveling wave, with the minus sign accounting for the opposite direction of power flow.

An important relation that will be used in a number of later chapters relates the intensity of the plane wave to the stored electromagnetic energy density. We start by considering the second and fourth terms on the right of (1.2-11)

$$\frac{\partial}{\partial t}\left(\frac{\varepsilon_0}{2}\mathbf{e}\cdot\mathbf{e}\right) + \mathbf{e}\cdot\frac{\partial \mathbf{p}}{\partial t}$$

Using the relations

$$\mathbf{p} = \varepsilon_0 \chi_e \mathbf{e}$$

$$\varepsilon = \varepsilon_0(1 + \chi_e) \tag{1.3-20}$$

we obtain

$$\frac{\partial}{\partial t}\left(\frac{\varepsilon_0}{2}\mathbf{e}\cdot\mathbf{e}\right) + \mathbf{e}\cdot\frac{\partial \mathbf{p}}{\partial t} = \frac{\partial}{\partial t}\left(\frac{\varepsilon}{2}\mathbf{e}\cdot\mathbf{e}\right) \tag{1.3-21}$$

Since we assumed the medium to be lossless, the last term must represent the rate of change of electric energy density stored in the vacuum as well as in the electric dipoles; that is,

$$\frac{\mathscr{E}_{\text{electric}}}{\text{Volume}} = \frac{\varepsilon}{2}\mathbf{e}\cdot\mathbf{e} \tag{1.3-22}$$

The magnetic energy density is derived in a similar fashion using the relations

$$\mathbf{m} = \chi_m \mathbf{h}$$

$$\mu = \mu_0(1 + \chi_m)$$

resulting in

$$\frac{\mathscr{E}_{\text{magnetic}}}{\text{Volume}} = \frac{\mu}{2}\mathbf{h}\cdot\mathbf{h} \tag{1.3-23}$$

Considering only the positive traveling wave in (1.3-17), we obtain from (1.3-22) and (1.3-23)

$$\begin{aligned}\frac{\overline{\mathscr{E}}}{\text{Volume}} = \frac{\overline{\mathscr{E}_{\text{magnetic}}} + \overline{\mathscr{E}_{\text{electric}}}}{\text{Volume}} &= \left(\frac{\varepsilon}{2}\right)\overline{(e_x^+)^2} + \left(\frac{\mu}{2}\right)\overline{(h_y^+)^2} \\ &= \frac{\varepsilon}{4}|E_x^+|^2 + \frac{\mu}{4}|H_y^+|^2 \\ &= \frac{\varepsilon}{4}|E_x^+|^2 + \frac{\mu}{4}\frac{|E_x^+|^2}{\eta^2} \\ &= \tfrac{1}{2}\varepsilon|E_x^+|^2\end{aligned} \tag{1.3-24}$$

where the second equality is based on (1.1-12), and the third and fourth use (1.3-15). Comparing (1.3-24) to (1.3-19), we get

$$\frac{I}{\overline{\mathscr{E}}/\text{Volume}} = \frac{1}{\sqrt{\mu\varepsilon}} = c \tag{1.3-25}$$

where $\overline{\mathscr{E}} = \overline{\mathscr{E}_{\text{magnetic}}} + \overline{\mathscr{E}_{\text{electric}}}$ is the electromagnetic field energy and c is the phase velocity of light in the medium. In terms of the electric field we get, putting $|E_x^+| \equiv E$,

$$I = \frac{c\varepsilon|E|^2}{2} = \frac{|E|^2}{2\eta}$$

$$\eta = \sqrt{\frac{\mu}{\varepsilon}} = 377 \text{ ohms in free space} \tag{1.3-26}$$

1.4 WAVE PROPAGATION IN CRYSTALS—THE INDEX ELLIPSOID

In the discussion of electromagnetic wave propagation up to this point, we have assumed that the medium was isotropic. This causes the induced polarization to be parallel to the electric field and to be related to it by a (scalar) factor that is independent of the direction along which the field is applied. This situation does not apply in the case of dielectric crystals. Since the crystal is made up of a regular periodic array of atoms (or ions), we may expect that the induced polarization will depend in its magnitude and direction, on the direction of the applied field. Instead of the simple relation (1.3-20) linking **p** and **e**, we have

$$\begin{aligned} P_x &= \varepsilon_0(\chi_{11}E_x + \chi_{12}E_y + \chi_{13}E_z) \\ P_y &= \varepsilon_0(\chi_{21}E_x + \chi_{22}E_y + \chi_{23}E_z) \\ P_z &= \varepsilon_0(\chi_{31}E_x + \chi_{32}E_y + \chi_{33}E_z) \end{aligned} \tag{1.4-1}$$

where the capital letters denote the complex amplitudes of the corresponding time-harmonic quantities. The 3×3 array of the χ_{ij} coefficients is called the electric susceptibility tensor. The magnitude of the χ_{ij} coefficients depends, of course, on the choice of the x, y, and z axes relative to that of the crystal structure. It is always possible to choose x, y, and z in such a way that the off-diagonal elements vanish, leaving

$$\begin{aligned} P_x &= \varepsilon_0\chi_{11}E_x \\ P_y &= \varepsilon_0\chi_{22}E_y \\ P_z &= \varepsilon_0\chi_{33}E_z \end{aligned} \tag{1.4-2}$$

These directions are called the *principal dielectric axes of the crystal.* In this book we will use only the principal coordinate system. We can, instead of using (1.4-2), describe the dielectric response of the crystal by means of the electric permeability tensor ε_{ij}, defined by

$$\begin{aligned} D_x &= \varepsilon_{11}E_x \\ D_y &= \varepsilon_{22}E_y \\ D_z &= \varepsilon_{33}E_z \end{aligned} \tag{1.4-3}$$

From (1.4-2) and the relation

$$\mathbf{D} = \varepsilon_0 \mathbf{E} + \mathbf{P}$$

we have

$$\begin{aligned} \varepsilon_{11} &= \varepsilon_0(1 + \chi_{11}) \\ \varepsilon_{22} &= \varepsilon_0(1 + \chi_{22}) \\ \varepsilon_{33} &= \varepsilon_0(1 + \chi_{33}) \end{aligned} \tag{1.4-4}$$

Birefringence

One of the most important consequences of the dielectric anisotropy of crystals is the phenomenon of birefringence in which the phase velocity of an optical beam propagating in the crystal depends on the direction of polarization of its **e** vector. Before treating this problem mathematically, we may pause and ponder its physical origin. In an isotropic medium the induced polarization is independent of the field direction so that $\chi_{11} = \chi_{22} = \chi_{33}$, and, using (1.4-4), $\varepsilon_{11} = \varepsilon_{22} = \varepsilon_{33} = \varepsilon$. Since $c = (\mu\varepsilon)^{-1/2}$, the phase velocity is independent of the direction of polarization. In an anisotropic medium the situation is different. Consider, for example, a wave propagating along z. If its electric field is parallel to x, it will induce, according to (1.4-2), only P_x and will consequently "see" an electric permeability ε_{11}. Its phase velocity will thus be $c_x = (\mu\varepsilon_{11})^{-1/2}$. If, on the other hand, the wave is polarized parallel to y, it will propagate with a phase velocity $c_y = (\mu\varepsilon_{22})^{-1/2}$.

Birefringence has some interesting consequences. Consider, as an example, a wave propagating along the crystal z direction and having at some plane, say $z = 0$, a linearly polarized field with equal components along x and y. Since $k_x \neq k_y$, as the wave propagates into the crystal the x and y components get out of phase and the wave becomes elliptically polarized. This phenomenon is discussed in detail in Section 9.2 and forms the basis of the electrooptic modulation of light.

Returning to the example of a wave propagating along the crystal z direction, let us assume, as in Section 1.3, that the only nonvanishing field components are e_x and h_y. Maxwell's curl equations (1.3-5) and (1.3-4) reduce, in a self-consistent manner, to

$$\begin{aligned} \frac{\partial e_x}{\partial z} &= -\mu \frac{\partial h_y}{\partial t} \\ \frac{\partial h_y}{\partial z} &= -\varepsilon_{11} \frac{\partial e_x}{\partial t} \end{aligned} \tag{1.4-5}$$

Taking the derivative of the first of Equations (1.4-5) with respect to z and then substituting the second equation for $\partial h_y/\partial z$ gives

$$\frac{\partial^2 e_x}{\partial z^2} = \mu\varepsilon_{11} \frac{\partial^2 e_x}{\partial t^2} \tag{1.4-6}$$

If we postulate, as in (1.3-10), a solution in the form

$$e_x = E_x e^{i(\omega t - k_x z)} \tag{1.4-7}$$

then Equation (1.4-6) becomes

$$k_x^2 E_x = \omega^2 \mu \varepsilon_{11} E_x$$

Therefore, the propagation constant of a wave polarized along x and traveling along z is

$$k_x = \omega\sqrt{\mu\varepsilon_{11}} \tag{1.4-8}$$

Repeating the derivation but with a wave polarized along the y axis, instead of the x axis, yields $k_y = \omega\sqrt{\mu\varepsilon_{22}}$.

Index Ellipsoid

As shown above, in a crystal the phase velocity of a wave propagating along a given direction depends on the direction of its polarization. For propagation along z, as an example, we found that Maxwell's equations admitted two solutions: one with its linear polarization along x and the second along y. If we consider the propagation along some arbitrary direction in the crystal, the problem becomes more difficult. We have to determine the directions of polarization of the two allowed waves, as well as their phase velocities. This is done most conveniently using the so-called index ellipsoid

$$\frac{x^2}{\varepsilon_{11}/\varepsilon_0} + \frac{y^2}{\varepsilon_{22}/\varepsilon_0} + \frac{z^2}{\varepsilon_{33}/\varepsilon_0} = 1 \tag{1.4-9}$$

This is the equation of a generalized ellipsoid with major axes parallel to x, y, and z whose respective lengths are $2\sqrt{\varepsilon_{11}/\varepsilon_0}$, $2\sqrt{\varepsilon_{22}/\varepsilon_0}$, and $2\sqrt{\varepsilon_{33}/\varepsilon_0}$. The procedure for finding the polarization directions and the corresponding phase velocities for a *given* direction of propagation is as follows: Determine the ellipse formed by the intersection of a plane through the origin and normal to the direction of propagation and the index ellipsoid (1.4-9). The directions of the major and minor axes of this ellipse are those of the two allowed polarizations,[5] and the lengths of these axes are $2n_1$ and $2n_2$, where n_1 and n_2 are the indices of refraction of the two allowed solutions. The two waves propagate, thus, with phase velocities c_0/n_1 and c_0/n_2, respectively, where $c_0 = (\mu_0\varepsilon_0)^{-1/2}$ is the phase velocity in vacuum. A formal proof of this procedure is given in References [2–4].

To illustrate the use of the index ellipsoid, consider the case of a uniaxial crystal (that is, a crystal with a single axis of threefold, fourfold, or sixfold symmetry). Taking the direction of this axis as z, symmetry considerations dictate that $\varepsilon_{11} =$

[5]These are actually the directions of the **D**, not of the **E**, vector. In a crystal these two are separated, in general, by a small angle; see References [2] and [3].

ε_{22}.[6] Defining the principal indices of refraction n_o and n_e by

$$n_o^2 \equiv \frac{\varepsilon_{11}}{\varepsilon_0} = \frac{\varepsilon_{22}}{\varepsilon_0} \qquad n_e^2 \equiv \frac{\varepsilon_{33}}{\varepsilon_0} \tag{1.4-10}$$

the equation of the index ellipsoid (1.4-9) becomes

$$\frac{x^2}{n_o^2} + \frac{y^2}{n_o^2} + \frac{z^2}{n_e^2} = 1 \tag{1.4-11}$$

This is an ellipsoid of revolution with the circular symmetry axis parallel to z. The z major axis of the ellipsoid is of length $2n_e$, whereas that of the x and y axes is $2n_o$. The procedure of using the index ellipsoid is illustrated by Figure 1-1.

The direction of propagation is along **s** and is at an angle θ to the (optic) z axis. Because of the circular symmetry of (1.4-11) about z, we can choose, without any loss of generality, the y axis to coincide with the projection of **s** on the x-y plane. The intersection ellipse of the plane normal to **s** with the ellipsoid is shaded in the figure. The two allowed polarization directions are parallel to the axes of the ellipse and thus correspond to the line segments OA and OB. They are consequently perpendicular to **s** as well as to each other. The two waves polarized along these direc-

[6]See, for example, J. F. Nye, *Physical Properties of Crystals.* New York: Oxford University Press, 1957.

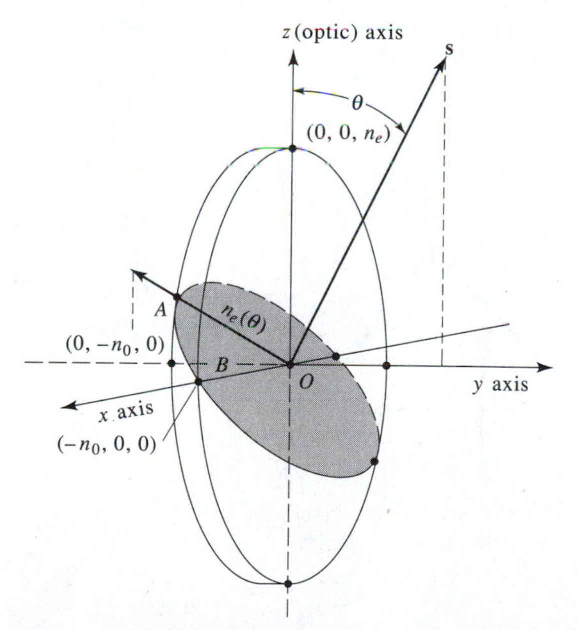

Figure 1-1 Construction for finding indices of refraction and allowed polarization for a given direction of propagation **s**. The figure shown is for a uniaxial crystal with $n_x = n_y = n_o$.

tions have, respectively, indices of refraction given by $n_e(\theta) = |OA|$ and $n_o = |OB|$. The first of these two waves, which is polarized along OA, is called the *extraordinary wave.* Its direction of polarization varies with θ following the intersection point A. Its index of refraction is given by the length of OA. It can be determined using Figure 1-2, which shows the intersection of the index ellipsoid with the y-z plane.

Using the relations

$$n_e^2(\theta) = z^2 + y^2$$

$$\frac{z}{n_e(\theta)} = \sin\theta$$

and the equation of the ellipse

$$\frac{y^2}{n_o^2} + \frac{z^2}{n_e^2} = 1$$

we obtain

$$\frac{1}{n_e^2(\theta)} = \frac{\cos^2\theta}{n_o^2} + \frac{\sin^2\theta}{n_e^2} \tag{1.4-12}$$

Thus, for $\theta = 0°$, $n_e(0°) = n_o$, and for $\theta = 90°$, $n_e(90°) = n_e$.

The ordinary wave remains, according to Figure 1-1, polarized along the same direction OB independent of θ. It has an index of refraction n_o. The amount of birefringence $n_e(\theta) - n_o$ thus varies from zero for $\theta = 0°$ (that is, propagation along the optic axis) to $n_e - n_o$ for $\theta = 90°$.

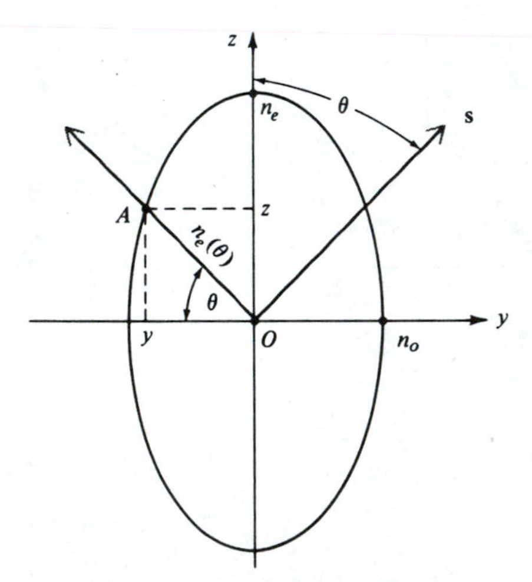

Figure 1-2 Intersection of the index ellipsoid with the z-y plane. $|OA| = n_e(\theta)$ is the index of refraction of the extraordinary wave propagating in the direction **s**.

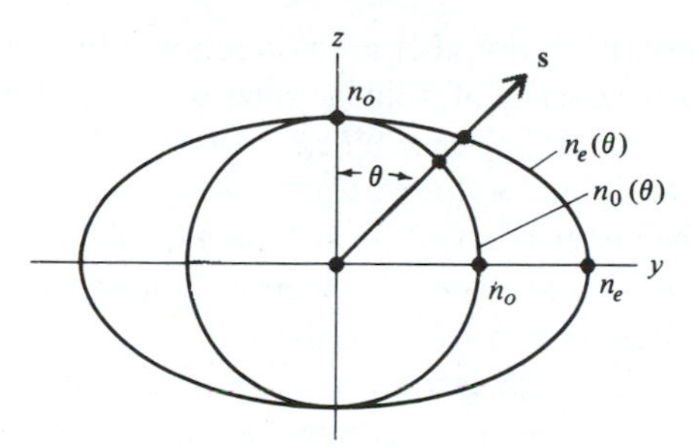

Figure 1-3 Intersection of s-z plane with normal surfaces of a positive uniaxial crystal ($n_e > n_o$).

Normal (index) Surfaces

Consider the surface in which the distance of a given point from the origin is equal to the index of refraction of a wave propagating along this direction. This surface, not to be confused with the index ellipsoid, is called the normal surface. It is constructed using the index ellipsoid (Figure 1-1). The normal surface of the extraordinary ray is constructed by measuring along each direction $\mathbf{s}(\theta, \phi)$ the corresponding index $n_e\ (\theta, \phi)$, which is the distance OA in Figure 1-1. For a uniaxial crystal, this results in an ellipsoid of revolution about the z axis as illustrated by the outer line in Figure 1-3. For the ordinary ray we plot the distance $OB = n_0$ (which is independent of θ, ϕ), resulting in the inner sphere of Figure 1-3.

1.5 JONES CALCULUS AND ITS APPLICATION TO PROPAGATION IN OPTICAL SYSTEMS WITH BIREFRINGENT CRYSTALS

Many sophisticated optical systems, such as electrooptic modulators (to be discussed in Chapter 9) involve the passage of light through a train of polarizers and birefringent (retardation) plates. The effect of each individual element, either polarizer or retardation plate, on the polarization state of the transmitted light can be described by simple means. However, when an optical system consists of many such elements, each oriented at a different azimuthal angle, the calculation of the overall transmission becomes complicated and is greatly facilitated by a systematic approach. The Jones calculus, invented in 1940 by R. C. Jones [5], is a powerful matrix method in which the state of polarization is represented by a two-component vector, while each optical element is represented by a 2×2 matrix. The overall transfer matrix for the whole system is obtained by multiplying all the individual element matrices, and the polarization state of the transmitted light is computed by multiplying the vector representing the input beam by the overall matrix. We will first develop the mathematical formulation of the Jones matrix method and then apply it to some cases of practical interest.

We have shown in the previous section that a unidirectional light propagation in a birefringent crystal generally consists of a linear superposition of two orthogonally polarized waves—the eigenwaves. These eigenwaves, for a given direction of propagation, have well-defined phase velocities and directions of polarization. The birefringent crystals may be either uniaxial ($n_x = n_y, n_z$) or biaxial ($n_x \neq n_y \neq n_z$). However, the most commonly used materials, such as calcite and quartz, are uniaxial. In a uniaxial crystal, these eigenwaves are the so-called *ordinary* and *extraordinary* waves, whose properties were derived in Section 1.4. The directions of polarization for these eigenwaves are mutually orthogonal and are called the *slow* and *fast* axes of the crystal for the given direction of propagation. Retardation plates are usually cut in such a way that the c axis lies in the plane of the plate surfaces. Thus the propagation direction of normally incident light is perpendicular to the c axis.

Retardation plates (also called wave plates) are polarization-state converters, or transformers. The polarization state of a light beam can be converted to any other polarization state by using a suitable retardation plate. In formulating the Jones matrix method, we assume that there is no reflection of light from either surface of the plate and the light is totally transmitted through the plate surfaces. In practice, there is some reflection, though most plates are coated with "antireflection" coatings to greatly reduce such reflection. Referring to Figure 1.4, we consider a light beam

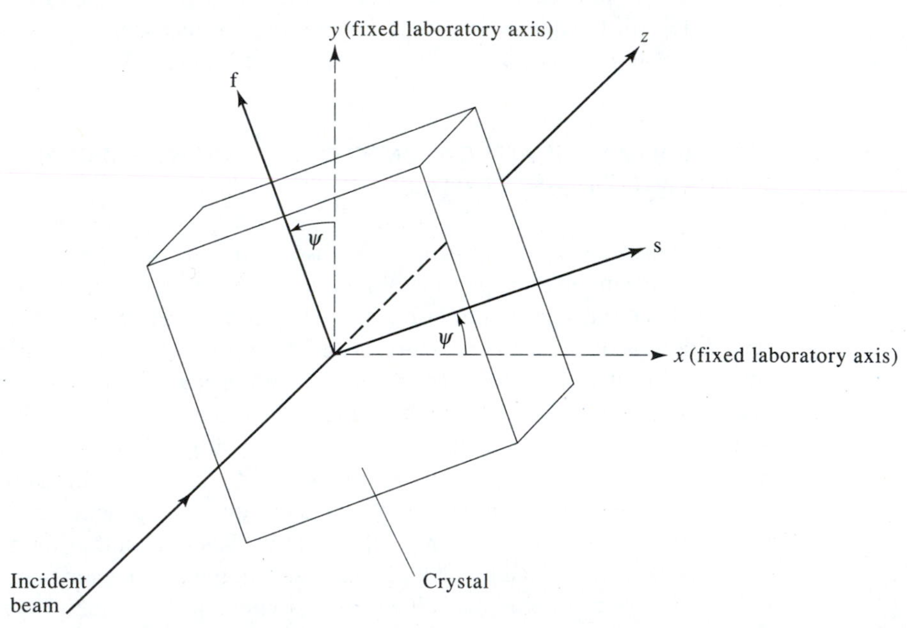

Figure 1-4 A retardation plate rotated at an angle ψ about the z axis. f("fast") and s("slow") are the two principal dielectric axes of the crystal for light propagating along z (see Section 1.4). The x and y axes are fixed in the laboratory frame.

that is incident normally on a retardation plate along the z axis with a polarization state described by the Jones column vector

$$\mathbf{V} = \begin{pmatrix} V_x \\ V_y \end{pmatrix} \tag{1.5-1}$$

where V_x and V_y are two complex numbers representing the complex field amplitudes along x and y. The x, y and z axes are *fixed* laboratory axes. To determine how the light propagates in the retardation plate, we need to resolve it into a linear combination of the fast and slow eigenwaves of the crystal. This is done by the coordinate transformation

$$\begin{pmatrix} V_s \\ V_f \end{pmatrix} = \begin{pmatrix} \cos\psi & \sin\psi \\ -\sin\psi & \cos\psi \end{pmatrix} \begin{pmatrix} V_x \\ V_y \end{pmatrix} \equiv R(\psi) \begin{pmatrix} V_x \\ V_y \end{pmatrix} \tag{1.5-2}$$

V_s is the slow component of the polarization vector $\mathbf{V}$, whereas V_f is the fast component. The slow and fast axes are fixed in the crystal. The angle between the fast axis and the y direction is ψ. These two components are eigenwaves of the retardation plate and will propagate with their own phase velocities and polarizations as discussed in Section 1.4. Because of the difference in phase velocity, the two components undergo a different phase delay in passage through the crystal. This retardation changes the polarization state of the emerging beam.

Let n_s and n_f be the refractive indices of the slow and fast eigenwaves, respectively. The polarization state of the emerging beam in the crystal coordinate system is thus given by

$$\begin{pmatrix} V_s' \\ V_f' \end{pmatrix} = \begin{pmatrix} \exp\left(-in_s \dfrac{\omega}{c} l\right) & 0 \\ 0 & \exp\left(-in_f \dfrac{\omega}{c} l\right) \end{pmatrix} \begin{pmatrix} V_s \\ V_f \end{pmatrix} \tag{1.5-3}$$

where l is the thickness of the plate and ω is the radian frequency of the light beam. The phase retardation is defined as the difference of the phase delays (exponents) in (1.5-3)

$$\Gamma = (n_s - n_f) \frac{\omega l}{c} \tag{1.5-4}$$

Notice that the phase retardation Γ is a measure of the relative change in phase, not the absolute change. The birefringence of a typical crystal retardation plate is small, that is, $|n_s - n_f| << n_s, n_f$. Consequently, the absolute change in phase caused by the plate may be hundreds of times greater than the phase retardation. Let ϕ be the mean absolute phase change

$$\phi = \tfrac{1}{2}(n_s + n_f) \frac{\omega l}{c} \tag{1.5-5}$$

Then Equation (1.5-3) can be written in terms of ϕ and Γ as

$$\begin{pmatrix} V_s' \\ V_f' \end{pmatrix} = e^{-i\phi} \begin{pmatrix} e^{-i\frac{\Gamma}{2}} & 0 \\ 0 & e^{i\frac{\Gamma}{2}} \end{pmatrix} \begin{pmatrix} V_s \\ V_f \end{pmatrix} \tag{1.5-6}$$

The Jones vector of the polarization state of the emerging beam in the xy coordinate system is given by transforming back from the crystal to the laboratory coordinate system

$$\begin{pmatrix} V_x' \\ V_y' \end{pmatrix} = \begin{pmatrix} \cos\psi & -\sin\psi \\ \sin\psi & \cos\psi \end{pmatrix} \begin{pmatrix} V_s' \\ V_f' \end{pmatrix} \tag{1.5-7}$$

By combining Equations (1.5-2), (1.5-6), and (1.5-7), we can write the transformation due to the retardation plate as

$$\begin{pmatrix} V_x' \\ V_y' \end{pmatrix} = R(-\psi)\, W_0 R(\psi) \begin{pmatrix} V_x \\ V_y \end{pmatrix} \tag{1.5-8}$$

where $R(\psi)$ is the rotation matrix of (1.5-2) and W_0 is the Jones matrix of (1.5-6) for the retardation plate. These are given, respectively, by

$$R(\psi) = \begin{pmatrix} \cos\psi & \sin\psi \\ -\sin\psi & \cos\psi \end{pmatrix} \tag{1.5-9}$$

and

$$W_0 = e^{-i\phi} \begin{pmatrix} e^{-i\Gamma/2} & 0 \\ 0 & e^{i\Gamma/2} \end{pmatrix} \tag{1.5-10}$$

The phase factor $e^{-i\phi}$ can usually be left out.[7] A retardation plate, characterized by its phase retardation Γ and its azimuth angle ψ, is represented by the product of three matrices

$$W(\psi, \Gamma) \equiv W = R(-\psi) W_0 R(\psi)$$

$$= \begin{vmatrix} e^{-i(\Gamma/2)} \cos^2\psi + e^{i(\Gamma/2)} \sin^2\psi & -i \sin\frac{\Gamma}{2} \sin(2\psi) \\ -i \sin\frac{\Gamma}{2} \sin(2\psi) & e^{-i(\Gamma/2)} \sin^2\psi + e^{i(\Gamma/2)} \cos^2\psi \end{vmatrix} \tag{1.5-11}$$

Note that the Jones matrix of a wave plate is a unitary matrix, that is,

$$W^{\dagger} W = 1$$

where the dagger † signifies the Hermitian conjugate ($W^*_{ij} = (W^{\dagger})_{ji}$). The passage of a polarized light beam through a wave plate is described mathematically as a unitary transformation. Many physical properties are invariant under unitary trans-

[7]The overall phase factor $\exp(-i\phi)$ is only important when the output field $\mathbf{V}'$ is combined coherently with another field.

formations; these include the orthogonal relation between the Jones vectors and the magnitude of the Jones vectors. Thus, if the polarization states of two beams are mutually orthogonal, they will remain orthogonal after passing through an arbitrary wave plate.

The Jones matrix of an ideal, lossless homogeneous and linear, thin plate polarizer oriented with its transmission axis parallel to the laboratory x axis is

$$P_0 = e^{-i\theta}\begin{pmatrix}1 & 0\\ 0 & 0\end{pmatrix} \tag{1.5-12}$$

where θ is the absolute phase accumulated due to the finite optical thickness of the polarizer. The Jones matrix of a polarizer rotated by an angle ψ from the x axis about z is given by

$$P = R(-\psi)P_0R(\psi) \tag{1.5-13}$$

Thus, if we neglect the (in this case unimportant) absolute phase θ, the Jones matrix representations of the polarizers oriented so as to transmit light with electric field vectors parallel to the x and y laboratory axes, respectively, are given by

$$P_x = \begin{pmatrix}1 & 0\\ 0 & 0\end{pmatrix} \quad \text{and} \quad P_y = \begin{pmatrix}0 & 0\\ 0 & 1\end{pmatrix} \tag{1.5-14}$$

To find the effect of an arbitrary train of retardation plates and polarizers on the polarization state of polarized light, we multiply the Jones vector of the incident beam by the ordered product of the matrices of the various elements.

Example: A Half-Wave Retardation Plate

A half-wave plate has a phase retardation of $\Gamma = \pi$. According to Equation (1.5-4), an x-cut[8] (or y-cut) uniaxial crystal will act as a half-wave plate, provided the thickness is $l = \lambda/2(n_e - n_0)$. We will determine the effect of a half-wave plate on the polarization state of a transmitted light beam. The azimuth angle of the wave plate is taken as 45° and the incident beam as vertically (y) polarized. The Jones vector for the incident beam can be written as

$$\mathbf{V} = \begin{pmatrix}0\\ 1\end{pmatrix} \tag{1.5-15}$$

and the Jones matrix for the half-wave plate is obtained by using Equation (1.5-11) with $\Gamma = \pi$, $\psi = \pi/4$

$$W = \frac{1}{\sqrt{2}}\begin{pmatrix}1 & -1\\ 1 & 1\end{pmatrix}\begin{pmatrix}-i & 0\\ 0 & i\end{pmatrix}\frac{1}{\sqrt{2}}\begin{pmatrix}1 & 1\\ -1 & 1\end{pmatrix} = \begin{pmatrix}0 & -i\\ -i & 0\end{pmatrix} \tag{1.5-16}$$

[8] A crystal plate is called *x-cut* if its facets are perpendicular to the principal x axis.

The Jones vector for the emerging beam is obtained by multiplying Equations (1.5-16) and (1.5-15); the result is

$$\mathbf{V}' = \begin{pmatrix} -i \\ 0 \end{pmatrix} = -i \begin{pmatrix} 1 \\ 0 \end{pmatrix} \qquad (1.5\text{-}17)$$

which corresponds to horizontally (x) polarized light. The effect of the half-wave plate is thus to rotate the input polarization by 90°. It can be shown that for a general azimuth angle ψ, the half-wave plate will rotate the polarization by an angle 2ψ (see Problem 1.7a). In other words, linearly polarized light remains linearly polarized, except that the plane of polarization is rotated by an angle of 2ψ.

When the incident light is circularly polarized, a half-wave plate will convert right-hand circularly polarized light into left-hand circularly polarized light and vice versa, regardless of the azimuth angle. The proof is left as an exercise (see Problem 1.7). Figure 1.5 illustrates the effect of a half-wave plate.

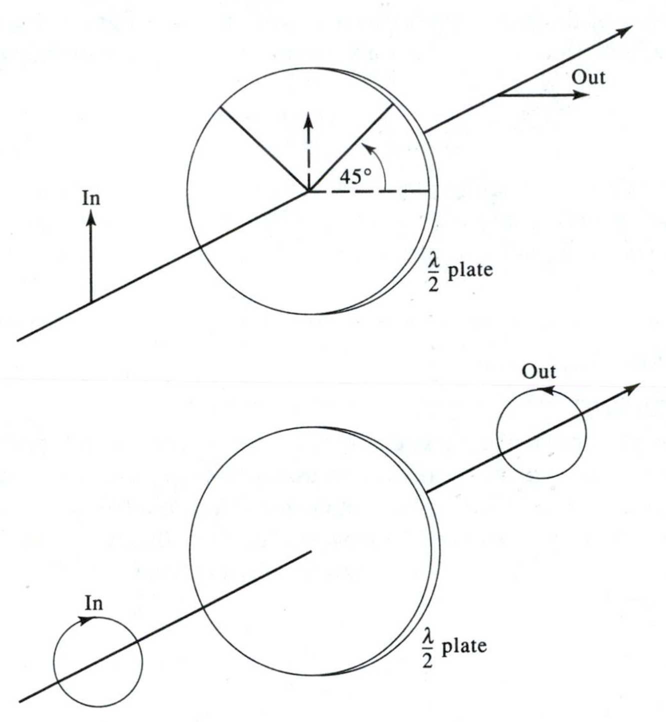

Figure 1-5 The effect of a half-wave plate on the polarization state of a beam.

Example: A Quarter-Wave Plate

A quarter-wave plate has a phase retardation of $\Gamma = \pi/2$. If the plate is made of an x-cut (or y-cut) uniaxially anistropic crystal, the thickness is $l = \lambda/4\ (n_e - n_o)$ (or odd multiples thereof). Suppose again that the azimuth angle of the plate is $\psi = 45°$ and the incident beam is vertically polarized. The Jones vector for the incident beam is given by Equation (1.5-15). The Jones matrix for this quarter-wave plate is

$$W = \frac{1}{\sqrt{2}}\begin{pmatrix} 1 & -1 \\ 1 & 1 \end{pmatrix}\begin{pmatrix} e^{-i\pi/4} & 0 \\ 0 & e^{i\pi/4} \end{pmatrix}\frac{1}{\sqrt{2}}\begin{pmatrix} 1 & 1 \\ -1 & 1 \end{pmatrix}$$

$$= \frac{1}{\sqrt{2}}\begin{pmatrix} 1 & -i \\ -i & 1 \end{pmatrix} \tag{1.5-18}$$

The Jones vector of the emerging beam is obtained by multiplying Equations (1.5-18) and (1.5-15) and is given by

$$\mathbf{V}' = -\frac{i}{\sqrt{2}}\begin{pmatrix} 1 \\ i \end{pmatrix} \tag{1.5-19}$$

To an observer facing the z direction (direction of propagation), this is a clockwise circularly polarized light. The effect of a 45°-oriented quarter-wave plate is thus to convert vertically polarized light into circularly polarized light. If the incident beam is horizontally polarized, the emerging beam will be circularly polarized in a counterclockwise sense. The effect of this quarter-wave plate is illustrated in Figure 1-6.

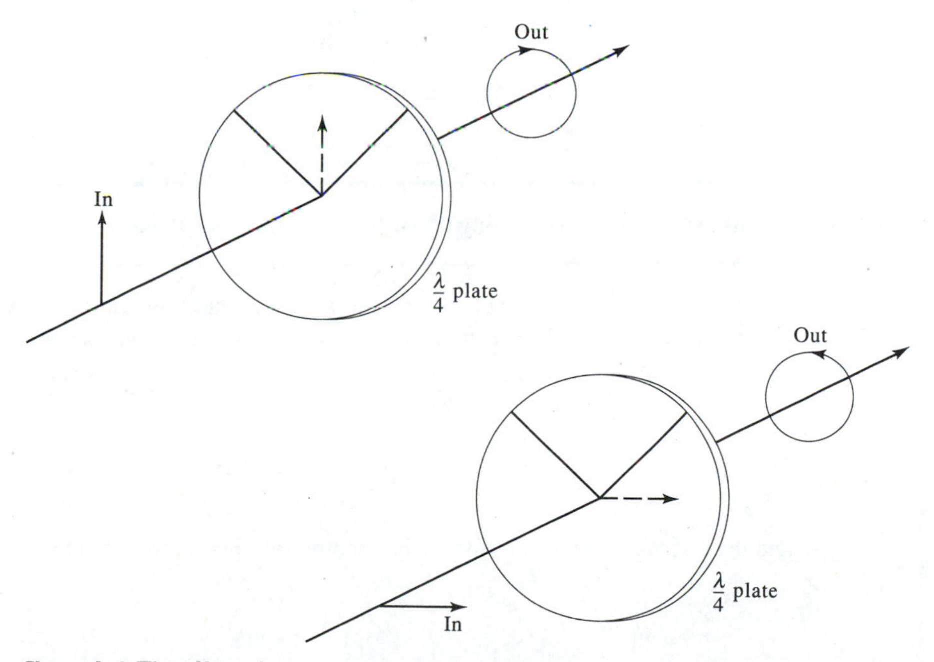

Figure 1-6 The effect of a quarter-wave plate on the polarization state of a linearly polarized input wave.

Intensity Transmission

Up to this point our development of the Jones calculus was concerned with the polarization state of the light beam. In many cases, we need to determine the transmitted intensity. The combination of retardation plates and polarizers is often used to control or modulate the transmitted optical intensity. Because the phase retardation of each wave plate is wavelength-dependent, the polarization state of the emerging beam and its intensity (when polarizers are present) depend on the wavelength of the light. Let us represent the field as a Jones vector

$$\mathbf{V} = \begin{pmatrix} V_x \\ V_y \end{pmatrix} \tag{1.5-20}$$

The intensity is taken using (1.1-12) and (1.3-24) as proportional to:

$$I = \mathbf{V} \cdot \mathbf{V}^* = |V_x|^2 + |V_y|^2 \tag{1.5-21}$$

If the output beam is given by

$$\mathbf{V}' = \begin{pmatrix} V_x' \\ V_y' \end{pmatrix} \tag{1.5-22}$$

the transmissivity of the optical system is calculated as

$$\frac{|V_x'|^2 + |V_y'|^2}{|V_x|^2 + |V_y|^2} \tag{1.5-23}$$

Example: A Birefringent Plate Sandwiched between Parallel Polarizers

Referring to Figure 1-7, we consider a birefringent plate sandwiched between a pair of parallel polarizers. The plate is oriented so that the slow and fast axes are at 45° with respect to the polarizer. Let the birefringence be $n_e - n_0$ and the plate thickness be d. The phase retardation is then given by

$$\Gamma = 2\pi(n_e - n_0)\,\frac{d}{\lambda} \tag{1.5-24}$$

and the corresponding Jones matrix is, according to Equation (1.5-11), with $\psi = 45°$

$$W = \begin{pmatrix} \cos\frac{1}{2}\Gamma & -i\sin\frac{1}{2}\Gamma \\ -i\sin\frac{1}{2}\Gamma & \cos\frac{1}{2}\Gamma \end{pmatrix} \tag{1.5-25}$$

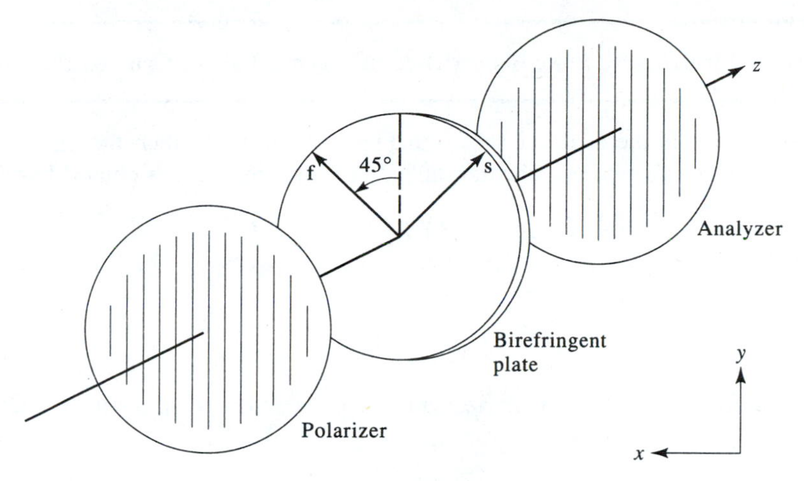

Figure 1-7 A birefringent plate sandwiched between a pair of parallel polarizers.

The incident beam, after it passes through the front polarizer, is polarized parallel to y and can be represented by

$$\mathbf{V} = \begin{pmatrix} 0 \\ 1 \end{pmatrix} \tag{1.5-26}$$

we shall take, arbitrarily, the intensity corresponding to (1.5-26) as unity. The Jones vector representation of the electric field vector of the transmitted beam is obtained as follows:

$$\begin{aligned} \mathbf{V}' &= \begin{pmatrix} 0 & 0 \\ 0 & 1 \end{pmatrix} \begin{pmatrix} \cos\frac{1}{2}\Gamma & -i\sin\frac{1}{2}\Gamma \\ -i\sin\frac{1}{2}\Gamma & \cos\frac{1}{2}\Gamma \end{pmatrix} \begin{pmatrix} 0 \\ 1 \end{pmatrix} \\ &= \begin{pmatrix} 0 \\ \cos\frac{1}{2}\Gamma \end{pmatrix} \end{aligned} \tag{1.5-27}$$

The transmitted beam is y polarized with an intensity given by

$$I = \cos^2\tfrac{1}{2}\Gamma = \cos^2\left[\frac{\pi(n_e - n_0)d}{\lambda}\right] \tag{1.5-28}$$

It can be seen from Equation (1.5-28) that the transmitted intensity is a sinusoidal function of the wave number (λ^{-1}) and peaks at $\lambda = (n_e - n_0)d$, $(n_e - n_0)d/2$, $(n_e - n_0)d/3, \ldots$. The wave-number separation between transmission maxima increases with decreasing plate thickness.

Example: A Birefringent Plate Sandwiched between a Pair of Crossed Polarizers

If we rotate the analyzer shown in Figure 1-7 by 90°, then the input and output polarizers are crossed. The transmitted beam for this case is obtained as follows:

$$\mathbf{V}' = \begin{pmatrix} 1 & 0 \\ 0 & 0 \end{pmatrix} \begin{pmatrix} \cos\frac{1}{2}\Gamma & -i\sin\frac{1}{2}\Gamma \\ -i\sin\frac{1}{2}\Gamma & \cos\frac{1}{2}\Gamma \end{pmatrix} \begin{pmatrix} 0 \\ 1 \end{pmatrix} = -i \begin{pmatrix} \sin\frac{\Gamma}{2} \\ 0 \end{pmatrix} \tag{1.5-29}$$

The transmitted beam is horizontally (x) polarized with an intensity relative to the input value given by

$$\frac{I_{\text{out}}}{I_{\text{in}}} = \sin^2\tfrac{1}{2}\Gamma = \sin^2\left[\frac{\pi(n_e - n_0)\,d}{\lambda}\right] \tag{1.5-30}$$

This is again a sinusoidal function of the wave number λ^{-1}. The transmission spectrum consists of a series of maxima at $\lambda = 2(n_e - n_0)d$, $2(n_e - n_0)d/3, \ldots$. These wavelengths correspond to phase retardations of π, 3π, $5\pi, \ldots$, that is, when the wave plate becomes a "half-wave" plate or odd integral multiples of a half-wave plate.

Circular Polarization Representation

Up to this point we represented the state of the propagating field as a vector $\mathbf{V}$ [Equation (1.5-1)] with components V_x and V_y.

$$\mathbf{V} = \begin{pmatrix} V_x \\ V_y \end{pmatrix} \tag{1.5-31}$$

The orthogonal unit vectors (*basis* vector set) in this representation are

$$\mathbf{V}_x = \begin{pmatrix} 1 \\ 0 \end{pmatrix} \qquad \mathbf{V}_y = \begin{pmatrix} 0 \\ 1 \end{pmatrix} \tag{1.5-32}$$

The above choice is most convenient when dealing with birefringent crystals, since the propagating eigenmodes in this case are linearly and orthogonally polarized. It is often more convenient to express the field in terms of "basis" vectors that are circularly polarized [6]. This is the case, for example, when we propagate through a magnetic medium. We define a wave of unit amplitude seen rotating in the CCW sense by an observer gazing along the $+z$ axis as $\begin{Bmatrix} 1 \\ 0 \end{Bmatrix}$, while $\begin{Bmatrix} 0 \\ 1 \end{Bmatrix}$ denotes a CW rotating wave. As in the case of the linearly polarized basis vectors $\begin{pmatrix} 1 \\ 0 \end{pmatrix}$ and $\begin{pmatrix} 0 \\ 1 \end{pmatrix}$,

$\begin{Bmatrix}1\\0\end{Bmatrix}$ and $\begin{Bmatrix}0\\1\end{Bmatrix}$ constitute a complete set that can be used to describe a transverse field of arbitrary polarization. Let **V** be some such field. We can write

$$\mathbf{V} = V_x\begin{pmatrix}1\\0\end{pmatrix} + V_y\begin{pmatrix}0\\1\end{pmatrix} \equiv \begin{pmatrix}V_x\\V_y\end{pmatrix} \tag{1.5-33}$$

or alternatively

$$\mathbf{V} = V_+\begin{Bmatrix}1\\0\end{Bmatrix} + V_-\begin{Bmatrix}0\\1\end{Bmatrix} \equiv \begin{Bmatrix}V_+\\V_-\end{Bmatrix} \tag{1.5-34}$$

The $\begin{pmatrix}V_x\\V_y\end{pmatrix}$ and $\begin{Bmatrix}V_+\\V_-\end{Bmatrix}$ representations of a given vector can be derived from each other by a 2×2 matrix[9]

$$\begin{Bmatrix}V_+\\V_-\end{Bmatrix} = \frac{1}{2}\begin{vmatrix}1 & i\\1 & -i\end{vmatrix}\begin{pmatrix}V_x\\V_y\end{pmatrix} \equiv T\begin{pmatrix}V_x\\V_y\end{pmatrix} \tag{1.5-35}$$

$$\begin{pmatrix}V_x\\V_y\end{pmatrix} = \begin{vmatrix}1 & 1\\-i & i\end{vmatrix}\begin{Bmatrix}V_+\\V_-\end{Bmatrix} \equiv S\begin{Bmatrix}V_+\\V_-\end{Bmatrix} \tag{1.5-36}$$

so that $T = S^{-1}$. As an example, consider a (unit) field polarized along x. Its rectangular representation is $\begin{pmatrix}1\\0\end{pmatrix}$, while its rotating representation is

$$\begin{Bmatrix}V_+\\V_-\end{Bmatrix} = \begin{vmatrix}1 & i\\1 & -i\end{vmatrix}\begin{pmatrix}1\\0\end{pmatrix} = \begin{Bmatrix}1\\1\end{Bmatrix} \tag{1.5-37}$$

i.e., equal and in-phase admixture of the two counter-rotating eigenmodes. Conversely, a clockwise, circularly polarized unit wave $\begin{Bmatrix}0\\1\end{Bmatrix}$, for example, is expressed in the rectangular representation by

$$\begin{pmatrix}V_x\\V_y\end{pmatrix} = \begin{vmatrix}1 & 1\\-i & i\end{vmatrix}\begin{Bmatrix}0\\1\end{Bmatrix} = \begin{pmatrix}1\\i\end{pmatrix} \tag{1.5-38}$$

Faraday Rotation

In certain optical materials containing magnetic atoms or ions, the natural modes of propagation are the two counter-rotating, circularly-polarized (CP) waves described above. The z direction is usually that of an applied magnetic field or that of the spontaneous magnetization. As in the case of a birefringent crystal, the two CP modes propagate with different phase velocities or, equivalently, have different indices of refraction. This difference is due to the fact that the individual atomic magnetic

---

[9]The form of T implies that at $t=0$ the rotating waves $\begin{Bmatrix}1\\0\end{Bmatrix}$ and $\begin{Bmatrix}0\\1\end{Bmatrix}$ are parallel to the x axis.

moments precess in a unique sense about the z axis and thus interact differently (have slightly displaced resonances) with the two CP waves. Using the notation of (1.5-34), we can describe the propagation of a wave with arbitrary transverse polarization by first resolving it, at $z = 0$, into its components $\begin{Bmatrix} V_+(0) \\ 0 \end{Bmatrix}$ and $\begin{Bmatrix} 0 \\ V_-(0) \end{Bmatrix}$ and propagating each component with its appropriate phase delay through the magnetic medium

$$\begin{Bmatrix} V_+(z) \\ V_-(z) \end{Bmatrix} = \begin{Bmatrix} V_+(0) \\ 0 \end{Bmatrix} e^{-i(\omega/c)n_+z} + \begin{Bmatrix} 0 \\ V_-(0) \end{Bmatrix} e^{-i(\omega/c)n_-z}$$

$$= e^{-(i/2)(\theta_+ + \theta_-)} \begin{vmatrix} e^{(i/2)(\theta_- - \theta_+)} & 0 \\ 0 & e^{-(i/2)(\theta_- - \theta_+)} \end{vmatrix} \begin{Bmatrix} V_+(0) \\ V_-(0) \end{Bmatrix} \tag{1.5-39}$$

where $\theta_\pm \equiv (\omega/c)\, n_\pm z$ is the phase delay for the (+) or (−) circularly polarized wave. Ignoring the prefactor $\exp[-(i/2)(\theta_+ + \theta_-)]$ (it is only relative phase delays that are of interest here) we rewrite (1.5-39) as

$$\begin{Bmatrix} V_+\,(z) \\ V_-\,(z) \end{Bmatrix} = \begin{vmatrix} e^{i\theta_F(z)} & 0 \\ 0 & e^{-i\theta_F(z)} \end{vmatrix} \begin{Bmatrix} V_+(0) \\ V_-(0) \end{Bmatrix} \tag{1.5-40}$$

$$\theta_F(z) \equiv \frac{1}{2}(\theta_- - \theta_+) = \frac{\omega}{2c}(n_- - n_+)z \tag{1.5-41}$$

$$\equiv \text{Faraday rotation angle}$$

The reason for calling θ_F the *Faraday rotation angle* becomes clear if we consider the effect of a magnetic medium on an incident wave that is described in the rectangular component representation

$$\begin{pmatrix} V_x(z) \\ V_y(z) \end{pmatrix} = T^{-1} \begin{vmatrix} e^{i\theta_F(z)} & 0 \\ 0 & e^{-i\theta_F(z)} \end{vmatrix} T \begin{pmatrix} V_x(0) \\ V_y(0) \end{pmatrix}$$

$$= \begin{vmatrix} \cos\theta_F & -\sin\theta_F \\ \sin\theta_F & \cos\theta_F \end{vmatrix} \begin{pmatrix} V_x(0) \\ V_y(0) \end{pmatrix} \tag{1.5-42}$$

$$= R(-\theta_F) \begin{pmatrix} V_x(0) \\ V_y(0) \end{pmatrix} \tag{1.5-43}$$

where $R(-\theta_F)$ is, according to (1.5-2), the matrix representing a *rotation* by $-\theta_F$ about the z axis. The output field is thus rotated by $-\theta_F$ with respect to the input field.

There exists a basic difference between propagation in a magnetic medium and in a dielectric birefringent medium. Consider the latter case first. An x'-polarized eigenwave, for instance, propagating along the z direction in a birefringent crystal has a phase velocity c/n_x, where x' is a principal dielectric axis. The same applies to the wave propagating in the reverse direction. The medium is *reciprocal.* In a magnetic medium the story is quite different. Let a linearly polarized wave traveling from left to right a distance L (in the $+z$ direction) undergo a (Faraday) rotation of its plane of polarization of $+\theta$ (the sign signifies the sense of the rotation about the *direction of propagation*). A wave traveling in the $-z$ direction in the crystal will experience a rotation of $-\theta(L)$ *about the new direction* $(-z)$ *of propagation.* This is because the magnetic field or, equivalently, the magnetic polarization now points in the opposite direction relative to the direction of propagation. (The wave can differentiate between $+z$ and $-z$—something that it cannot do in a birefringent crystal). The medium is termed nonreciprocal. The net effect of a round trip through the medium of length L is that the plane of polarization of the beam returning to the starting, $z = 0$ plane, is rotated by $2\theta_F(L)$. This Faraday rotation is used to make optical isolators to block off back-reflected radiation. The basic configuration of a Faraday isolator is illustrated in Figure 1-8 (a) and (b). A linearly polarized incident wave is rotated by 45° in passage through the Faraday medium and then passed fully by the output polarizer. A reflected wave is rotated an additional 45° in the return trip and is thus blocked off by the input polarizer. Faraday isolators now form an integral part of most optical communication systems employing semiconductor diode lasers since such lasers are extremely sensitive to even small amounts of reflected light that cause instabilities in their power and frequency characteristics.

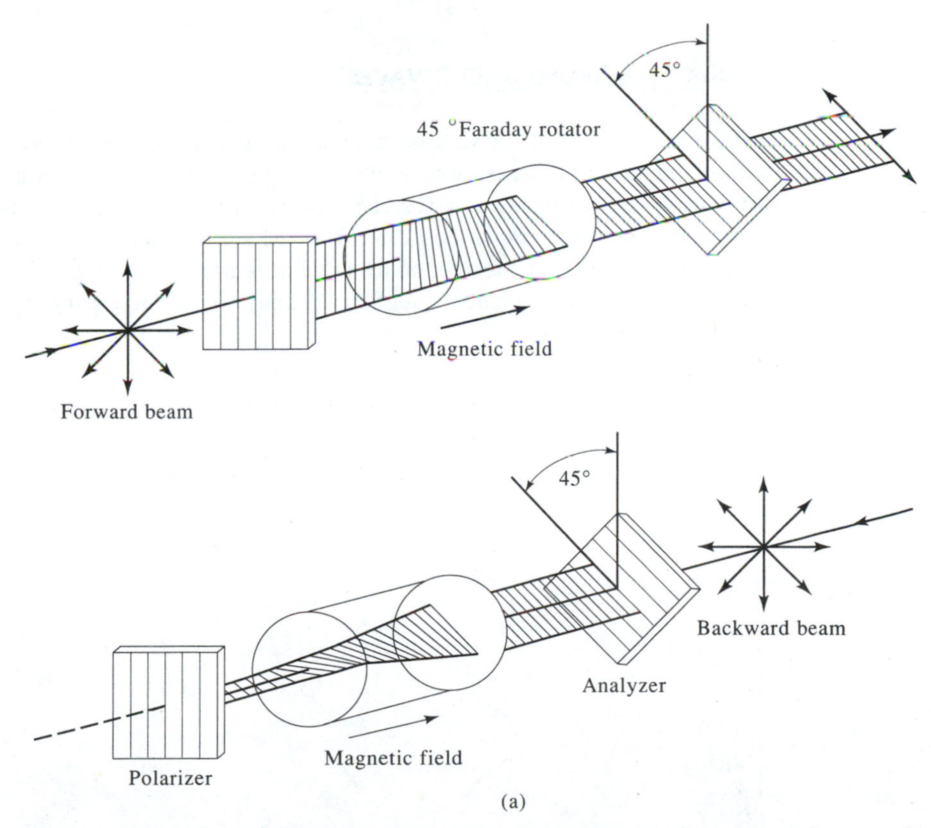

Figure 1-8 (a) A Faraday isolator comprised of two polarizers rotated by 45° relative to each other on either side of a magnetic medium with $\theta_F = 45°$.

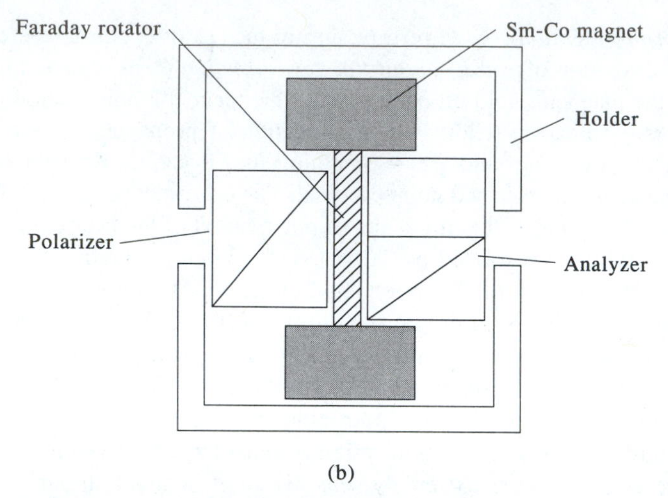

Figure 1-8 (*continued*) (b) A cross-sectional view of a practical commercial isolator. (Courtesy of Namiki Precision Jewel Company.)

1.6 DIFFRACTION OF ELECTROMAGNETIC WAVES

In this section, we will derive a most important result, the Fresnel-Kirchhoff Diffraction Integral, to describe how an electromagnetic field propagates between any two planes, say, the planes $z = 0$ and $z = L$ of an isotropic medium, as shown in Figure 1-9. The result, Equation (1.6-13), is the starting point to many of the important developments in coherent optics and image processing [9] and is used in this book to treat optical resonators (Section 4.9) and image processing by four-wave mixing (Section 17.10).

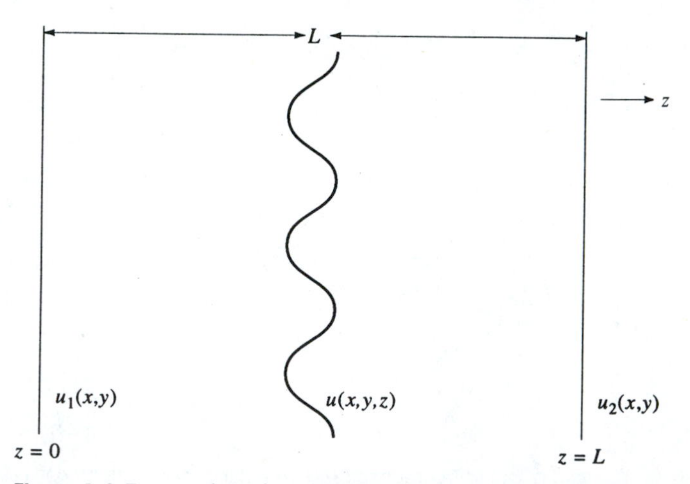

Figure 1-9 Propagation of an optical wave from $z = 0$ to $z = L$.

We will depart from the conventional, Green function, derivation [10] and employ a "linear system" approach that is formally identical to that used to analyze electrical and mechanical systems. As a bonus, we will find the mathematics and resulting formulae *identical* to those we will employ in Chapter 3 to describe how narrow, information-bearing, optical pulses propagate in fibers. This analogy will prove both interesting and useful.

In this section we will make ample use of the Fourier transform (FT) relations

$$f(x,y) = \iint \tilde{f}(k_x,k_y) \exp[i(k_x x + k_y y)]dk_x\, dk_y$$
$$\tilde{f}(k_x,k_y) = \frac{1}{4\pi^2} \iint f(x,y) \exp[-i(k_x x + k_y y)]dx\, dy \tag{1.6-1}$$

between a function $f(x,y)$ and its *FT* $\tilde{f}(k_x,k_y)$. Another important result that follows directly from (1.6-1) is that of the convolution integral

$$\iint \tilde{f}_1(k_x,k_y)\, \tilde{f}_2(k_x,k_y) \exp[i(k_x x + k_y y)]dk_x\, dk_y$$
$$= \frac{1}{4\pi^2} \iint f_1(x',y')\, f_2(x - x', y - y')dx'\, dy' \tag{1.6-2}$$
$$\equiv \frac{1}{4\pi^2} f_1 * f_2$$

where the * symbol represents the convolution integral.

We consider an electromagnetic field of (radian) frequency ω and a scalar complex amplitude $u(x,y,z)$

$$E(x,y,z,t) = u(x,y,z) \exp(i\omega t) \tag{1.6-3}$$

where E is some Cartesian coordinate of the vector field. The complex amplitude $u(x, y, z)$ obeys

$$\nabla^2 u + k^2 u = 0 \qquad k^2 = \omega^2 \mu\varepsilon \tag{1.6-4}$$

which equation results when we substitute (1.6-3) in the equation $\nabla^2 u = \mu\varepsilon \dfrac{\partial^2 u}{\partial t^2}$, which is the three-dimensional extension of Equation (1.3-9) to the case of an isotropic medium.

The scenario considered next is one where we are given an "input" optical field whose complex amplitude at $z = 0$ is

$$u_1(x,y) \equiv u(x,y,0) \tag{1.6-5}$$

Our task is to find the "output" field u_2 at $z = L$:

$$u_2(x,y) \equiv u(x,y,L) \tag{1.6-6}$$

We have already shown that a simple solution of (1.6-4) is of the form $u = e^{\pm ikz}$, which represents a plane wave propagating along $\mp z$. It can be verified, by direct substitution, that a plane wave propagating along any arbitrary direction, say that of **k** can be taken as

$$u(x,y,z) = u(\mathbf{k}) \exp(-i\, \mathbf{k}\cdot\mathbf{r}), \quad k = \omega\sqrt{\mu\varepsilon} = \frac{2\pi n}{\lambda} \tag{1.6-7}$$

since this expression also satisfies (1.6-4). It follows that an arbitrary superposition of plane waves, each of the form of (1.6-7), propagating along all conceivable directions, i.e.,

$$u(x,y,z) = \iint F(k_x, k_y) \exp[i(k_x x + k_y y) - i\sqrt{k^2 - k_x^2 - k_y^2}\, z]dk_x\, dk_y \tag{1.6-8}$$

where F is an arbitrary function, also satisfies Equation (1.6-4). The integrand of (1.6-8) represents the amplitude of a plane wave propagating along the direction

$$\mathbf{k} = -\hat{x}k_x - \hat{y}k_y + \hat{z}\sqrt{k^2 - k_x^2 - k_y^2} \tag{1.6-9}$$

so that integration over k_x and k_y takes in waves propagating along all possible directions. The form of k_z in (1.6-9) is dictated by the requirement that $u(x, y, z)$ satisfy the wave equation (1.6-4) or, physically, to ensure that each plane wave component in (1.6-8) has the same wavelength $\lambda/n = 2\pi/k$ as appropriate to a wave propagating in an isotropic medium.

It follows from (1.6-8) that

$$u(x,y,0) \equiv u_1(x,y) = \iint F(k_x,k_y) \exp[i(k_x x + k_y y)]dk_x\, dk_y \tag{1.6-10}$$

Equation (1.6-10) is of the form of the Fourier integral transform (FT) of (1.6-1). $F(k_x,k_y)$ is thus the Fourier transform of the input field $u_1(x,y)$.

$$F(k_x,k_y) = \frac{1}{4\pi^2}\iint u_1(x,y) \exp[-i(k_x x + k_y y)]dk_x\, dk_y \equiv \tilde{u}_1(k_x,k_y) \tag{1.6-11}$$

We can thus rewrite (1.6-8) as

$$u(x,y,z) = \iint \tilde{u}_1(k_x,k_y) \exp[i(k_x x + k_y y - \sqrt{k^2 - k_x^2 - k_y^2}\, z)]dk_x\, dk_y \tag{1.6-12}$$

Equation (1.6-12) constitutes a powerful algorithm for the propagation of a monochromatic wave. It states that given an "input" field with an arbitrary complex amplitude $u_1(x,y)$ at some plane, which without loss of generality we take as $z = 0$, we can write down the field amplitude at any other plane z by first deriving the Fourier integral transform $\tilde{u}_1(k_x,k_y)$ of $u_1(x,y)$ and then using it in the integral of Equation (1.6-12).

Our main preoccupation in this book is with beamlike optical waves. By this we mean optical beams whose plane wave components propagate either along the z axis of at small angles to it. Mathematically, this is equivalent to stating that in Equation (1.6-12) $\tilde{u}(k_x,k_y)$ is appreciable only in a region where $k_x,k_y << k$. This is the, so-called, paraxial condition. When it applies, we can approximate

$\sqrt{k^2 - k_x^2 - k_y^2} \approx k\left(1 - \frac{k_x^2 + k_y^2}{2k^2}\right)$ and re-express (1.6-8), taking $z = L$, as

$$u_2(x,y) \equiv u(x,y,L)$$

$$= \exp(-ikL) \iint \Bigg[\overbrace{\tilde{u}_1(k_x,k_y)}^{①} \overbrace{\exp\left(i \frac{k_x^2 + k_y^2}{2k} L\right)}^{②}\Bigg] \exp[i(k_x x + k_y y)]dk_x\, dk_y \tag{1.6-13}$$

We recognize the integral of (1.6-13) as the inverse Fourier transform (IFT) of the product of the two functions, designated as 1 and 2. Using the convolution theorem (1.6-2), we write the integral as the convolution of the two functions

$$u_1(x,y) = \text{IFT}\{\tilde{u}_1(k_x,k_y)\} \tag{1.6-14}$$

$$p(x,y) \equiv \text{IFT}\left\{\exp\left(i \frac{k_x^2 + k_y^2}{2k} L\right)\right\} = \frac{2\pi ik}{L} \exp\left(-ik \frac{x^2 + y^2}{2L}\right) \tag{1.6-15}$$

which results in

$$u_2(x,y) = \frac{1}{4\pi^2} \iint u_1(x', y')p(x - x', y - y')dx'\, dy' \equiv \frac{1}{4\pi^2} u_1 * p \tag{1.6-16}$$

$$= \frac{i}{\lambda L} \exp(-ikL) \iint u_1(x', y') \exp\left\{\frac{-ik}{2L} [(x - x')^2 + (y - y')^2]\right\}dx'\, dy' \tag{1.6-17}$$

Equation (1.6-17) is the celebrated Fresnel-Kirchhoff diffraction integral, which we will use on a number of occasions throughout the book.

Another useful relation results when we compare Equation (1.6-13) to the first of Equations (1.6-1). It follows directly that

$$\tilde{u}_2(k_x,k_y) = \tilde{u}_1(k_x,k_y) \exp\left(i \frac{k_x^2 + k_y^2}{2k} L\right) \tag{1.6-18}$$

where we left out the constant delay factor $\exp(-ikL)$ since it does not depend on k_x,k_y. It can always be restored, on the rare occasions when needed, by multiplying the right side through by $\exp(-ikL)$.

The effect of propagation of (1.6-18) a distance L by a monochromatic beam in a homogeneous and isotropic medium can thus be represented by multiplying the Fourier transform $\tilde{u}_1(k_x,k_y)$ of the input amplitude $\mu_1(x,y)$ by the "transfer function,"

$$T(k_x,k_y) = \exp\left(i \frac{k_x^2 + k_y^2}{2k} L\right) \tag{1.6-19}$$

This point of view, which embodies the spirit of system theory [8], is completely equivalent to the spatial relationship (1.6-17). It is often more convenient to use one or the other of these relations depending on the problem at hand.

Problems

1.1 Consider the problem of finding the time average

$$\overline{a^2(t)} = \frac{1}{T}\int_0^T a^2(t)\,dt$$

of

$$\begin{aligned} a(t) &= |A_1|\cos(\omega_1 t + \phi_1) + |A_2|\cos(\omega_2 t + \phi_2) \\ &= \mathrm{Re}[V_a(t)] \end{aligned}$$

where

$$V_a(t) = A_1 e^{i\omega_1 t} + A_2 e^{i\omega_2 t}$$

and $A_{1,2} = |A_{1,2}|e^{i\phi_{1,2}}$. $V_a(t)$ is called the *analytical signal* of $a(t)$. Assume that $(\omega_1 - \omega_2) \ll \omega_1$ and integrate over a time T, which is long compared to the period $2\pi/\omega_{1,2}$ but short compared to the beat period $2\pi/(\omega_1 - \omega_2)$.[10] Show that

$$\overline{a^2(t)} = \tfrac{1}{2}[V_a(t)V_a^*(t)]$$

1.2 Show how we can use the analytic functions as defined by Problem 1.1 to find the time average

$$\overline{a(t)b(t)} = \frac{1}{T}\int_0^T a(t)b(t)dt$$

where $a(t)$ is the same as in Problem 1.1, and the analytic function of $b(t)$ is

$$V_b(t) = [A_3 e^{i\omega_3 t} + A_4 e^{i\omega_4 t}]$$

so that $b(t) = \mathrm{Re}[V_b(t)]$. Assume that the difference between any two of the frequencies ω_1, ω_2, ω_3, and ω_4 is small compared to the frequencies themselves. (*Answer:* $\overline{a(t)b(t)} = \frac{1}{2}\,\mathrm{Re}[V_a(t)V_b^*(t)]$.)

1.3 Derive Equation (1.2-22).

1.4 Starting with Maxwell's curl equations [(1.2-1), (1.2-2)] and taking $\mathbf{i} = 0$, show that in the case of a harmonic (sinusoidal) uniform plane wave, the field vectors $\mathbf{e}$ and $\mathbf{h}$ are normal to each other as well as to the direction of propagation. [*Hint:* Assume the wave to have the form $e^{i(\omega t - \mathbf{k}\cdot\mathbf{r})}$ and show by actual differentiation that we can formally replace the operator ∇ in Maxwell's equations by $-i\mathbf{k}$].

1.5 Derive Equation (1.3-19).

1.6 A linearly polarized electromagnetic wave is incident normally at $z = 0$ on the x-y face of a crystal so that it propagates along its z axis. The crystal electric per-

[10] When this condition is fulfilled, $a(t)$ consists of a sinusoidal function with a "slowly" varying amplitude and is often called a *quasi-sinusoid.*

meability tensor referred to x, y, and z is diagonal with elements ε_{11}, ε_{22}, and ε_{33}. If the wave is polarized initially so that it has equal components along x and y, what is the state of its polarization at the plane z, where

$$(k_x - k_y)z = \frac{\pi}{2}$$

Plot the position of the electric field vector in this plane at times $t = 0$, $\pi/6\omega$, $\pi/3\omega$, $\pi/2\omega$, $2\pi/3\omega$, $5\pi/6\omega$.

1.7 *Half-wave plate.* A half-wave plate has a phase retardation of $\Gamma = \pi$. Assume that the plate is oriented so that the azimuth angle (i.e., the angle between the x axis and the slow axis of the plate) is ψ.

a. Find the polarization state of the transmitted beam, assuming that the incident beam is linearly polarized in the y direction.
b. Show that a half-wave plate will convert right-hand circularly polarized light into left-hand circularly polarized light, and vice versa, regardless of the azimuth angle of the plate.
c. Lithium tantalate ($LiTaO_3$) is a uniaxial crystal with $n_0 = 2.1391$ and $n_e = 2.1432$ at $\lambda = 1$ μm. Find the half-wave-plate thickness at this wavelength, assuming the plate is cut in such a way that the surfaces are perpendicular to the x axis of the principal coordinate (i.e., x-cut).

1.8 *Quarter-wave plate.* A quarter-wave plate has a phase retardation of $\Gamma = \pi/2$. Assume that the plate is oriented in a direction with azimuth angle ψ.

a. Find the polarization state of the transmitted beam, assuming that the incident beam is polarized in the y direction.
b. If the polarization state resulting from (a) is represented by a complex number on the complex plane, show that the locus of these points as ψ varies from 0 to $\frac{1}{2}\pi$ is a branch of a hyperbola. Obtain the equation of the hyperbola.
c. Quartz ($\alpha = SiO_2$) is a uniaxial crystal with $n_0 = 1.53283$ and $n_e = 1.54152$ at $\lambda = 1.1592$ μm. Find the thickness of an x-cut quartz quarter-wave plate at this wavelength.

1.9 A matrix **A** is called unitary if

$$\mathbf{A}^\dagger\mathbf{A} = \mathbf{A}\mathbf{A}^\dagger = \mathbf{1}$$

where **1** is the unity matrix and the Hermitian conjugate $A^\dagger$ of matrix **A** is defined by $(A^\dagger)_{ij} = A^*_{ji}$. Show that if **A** is unitary

$$\sum_j A^*_{ji}A_{jk} = \delta_{ik}$$

This property will be needed in Problem 1.10d.

1.10 *Polarization transformation by a wave plate.* A wave plate is characterized by its phase retardation Γ and azimuth angle ψ.

a. Find the polarization state of the emerging beam, assuming that the incident beam is polarized in the x direction.
b. Use a complex number to represent the resulting polarization state obtained in (a).
c. The polarization state of the incident x-polarized beam is represented by a point at the origin of the complex plane. Show that the transformed polarization state can be anywhere on the complex plane, provided Γ can be varied from 0 to 2π and ψ can be varied from 0 to $\frac{1}{2}\pi$. Physically, this means that any polarization state can be produced from linearly polarized light, provided a proper wave plate is available.
d. Show that the Jones matrix $\mathbf{W}$ of a wave plate is unitary, that is,

$$W^{\dagger}W = 1, \; (W^{\dagger})_{ij} \equiv W^{*}_{ji}$$

where the dagger indicates Hermitian conjugation [see Equation (1.5-11)].
e. Let $\mathbf{V}_1'$ and $\mathbf{V}_2'$ be the transformed Jones vectors of $\mathbf{V}_1$ and $\mathbf{V}_2$, respectively. Show that if $\mathbf{V}_1$ and $\mathbf{V}_2$ are orthogonal, so are $\mathbf{V}_1'$ and $\mathbf{V}_2'$. ($\mathbf{A}$ and $\mathbf{B}$ are orthogonal if $\mathbf{A} \cdot \mathbf{B}^* = 0$.)

1.11 Show that the (Jones) matrix (in the rectangular eigenwave representation) of a birefringent plate with a retardation Γ that is rotated by an angle ψ from the x axis is

$$W(\Gamma,\psi) = \begin{vmatrix} \cos^2\psi \exp(-i\Gamma/2) + \sin^2\psi \exp(+i\Gamma/2) & -i\sin 2\psi \sin\Gamma/2 \\ -i\sin 2\psi \sin\Gamma/2 & \sin^2\psi \exp(-i\Gamma/2) + \cos^2\psi \exp(+i\Gamma/2) \end{vmatrix}$$

Derive an expression for the intensity transmission through a system consistent of a polarizer $\parallel$ to $\hat{x}$, a Faraday rotator θ, wave plate with retardation Γ rotated an angle ψ from the x axis, and a crossed output polarizer ($\parallel$ to $\hat{y}$).

1.12

a. Show that $(AB)^{\dagger} = B^{\dagger}A^{\dagger}$.
b. Show that if an optical element is represented by a unitary matrix, the intensity of an incident wave of arbitrary polarization is preserved in passage through the element.
c. Show that the matrix representing a train of arbitrary retardation plates is unitary.

1.13

a. Show that in an isotropic medium we can take the general solution of the wave equation of a monochromatic field

$$\nabla^2\mathbf{E} + k^2\mathbf{E} = 0$$

as

$$\mathbf{E}(\mathbf{r}) = \int_{-\infty}^{\infty}\int_{\infty}^{\infty} dk_x dk_y \mathbf{A}(\mathbf{k}) e^{-i(k_x x + k_y y + \sqrt{k^2 - k_x^2 - k_y^2} z)}$$

where $\mathbf{A}(\mathbf{k})$ is an arbitrary vector lying on a plane normal to $\mathbf{k}$ and $k^2 = \omega^2\mu\epsilon$.

b. Show that if $\mathbf{E}(\mathbf{r})$ is specified at some plane S, say the plane $z = 0$, as $\mathbf{E}(x,y,0)$, then

$$\mathbf{A}(\mathbf{k}) = \left(\frac{1}{2\pi}\right)^2 \iint_S dxdy\mathbf{E}(x,y,0)e^{i(k_x x + k_y y)}$$

where $\mathbf{k} = \hat{\mathbf{x}}k_x + \hat{\mathbf{y}}k_y + \hat{\mathbf{z}}\sqrt{k^2 - k_x^2 - k_y^2}$. [*Hint:* Compare Equations (1) and (2) with $z = 0$ to the integral Fourier transform relationships.]

c. Assume that at $z = 0$ the field is given by

$$E(x,y,0) = E_0 \begin{cases} -a/2 \leq x \leq a/2 \\ -a/2 \leq y \leq a/2 \end{cases}$$

and is zero everywhere else. Find the spreading angle of the beam far away to the right of the aperture. [*Hint:* Each $\mathbf{k}$ signifies a direction of propagation so that $|A(\mathbf{k})|^2$ can be viewed as the distribution function of directions $\mathbf{k}$ of the beam to the right of the aperture.]

1.14 Consider light propagating through a sequence of $\lambda/2$ retardation plates ($\Gamma = \pi$) as shown:

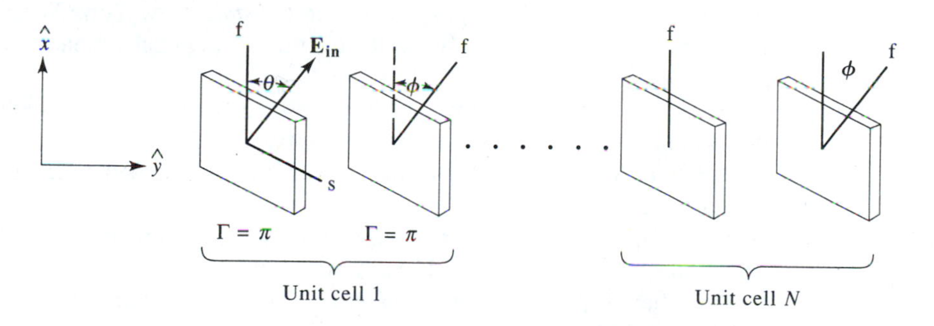

Each unit cell consists of two plates whose surfaces are normal to $\hat{y}$—one with its f (fast) axis ∥ to $\hat{z}$ and one rotated by ϕ about $\hat{x}$. Find the effect of propagation through N cells on a beam initially polarized as shown. Solve the problem first by simple considerations, if possible, then formally.

1.15 Show that if we define

$$\mathbf{v}_g \equiv \nabla_{\mathbf{k}}\omega(\mathbf{k})$$

$$\mathbf{v}_e \equiv \frac{\mathbf{E} \times \mathbf{H}}{\frac{1}{2}[\mathbf{E} \cdot \varepsilon\mathbf{E} + \mathbf{H} \cdot \mu\mathbf{H}]}$$

in a crystal, then

$$\mathbf{v}_g = \mathbf{v}_e$$

a. Recall that ε is a tensor.

b. After giving the problem a real try, you may consult *Optical Waves in Crystals,* A. Yariv and P. Yeh, New York: Wiley, p. 79, 1983.

1.16 Derive the transfer matrix of a polarizer whose transmission direction is rotated by α from the laboratory x axis.

1.17 Prove relation (1.5-35).

1.18 The electric field at some point in a medium and the position x of an electron are given by

$$e_x(\mathbf{r},t) = R_e[E_x\, e^{i(\omega t+\phi_E)}]$$

$$x(\mathbf{r},t) = R_e[X\, e^{i(\omega t+\phi_e)}]$$

Plot the instantaneous power flow ee_xv_x during one complete oscillation cycle for the case $\phi E - \phi e = 0, -\pi/2, +\pi/2$. What is the (cycle) average power flow for each case ($v_x = dx/dt$)?

References

1. Ramo, S., J. R. Whinnery, and T. Van Duzer, *Fields and Waves in Communication Electronics.* New York: Wiley, 1965.

2. Born, M., and E. Wolf, *Principles of Optics.* New York: Macmillan, 1964.

3. Yariv, A., *Quantum Electronics,* 2d ed. New York: Wiley, 1975.

4. Yariv, A., and P. Yeh, *Optical Waves in Crystals.* New York: Wiley, 1983.

5. Jones, R. C., ''New calculus for the treatment of optical systems,'' *J. Opt. Soc. Am.* 31:488, 1941.

6. Yariv, Amnon, ''Operator algebra for propagation problems involving phase conjugation and nonreciprocal elements,'' *Appl. Opt.* 26:4538, 1987.

7. Lohman, A. W., and D. Medlovic, ''Temporal filtering with time lenses,'' *Appl. Opt.* 31:6212, 1992.

8. Papoulis, A. ''Pulse compression, fiber communication, and diffraction: a unified approach, *J. Opt. Soc. Am.* 11:3, 1994.

9. Goodman, J. W., *Introduction to Fourier Optics*, 2d ed. San Francisco: McGraw-Hill, 1995, Ch. 3.

10. See, for example, M. Born and E. Wolf, *Principles of Optics*, 6th ed. Oxford: New York: Pergamon, 1986.

The Propagation of Rays and Beams

2.0 INTRODUCTION

In this chapter we take up the subject of optical ray propagation through a variety of optical media. These include homogeneous and isotropic materials, thin lenses, dielectric interfaces, and curved mirrors. Since a ray is, by definition, normal to the optical wavefront, an understanding of the ray behavior makes it possible to trace the evolution of optical waves when they are passing through various optical elements. We find that the passage of a ray (or its reflection) through these elements can be described by simple 2×2 matrices. Furthermore, these matrices will be found to describe the propagation of spherical waves and of Gaussian beams such as those characteristic of the output of lasers. Gaussian beam propagation is analyzed in second half of the chapter.

2.1 LENS WAVEGUIDE

Consider a paraxial ray[1] passing through a thin lens of focal length f as shown in Figure 2-1. Taking the cylindrical axis of symmetry as z, denoting the ray distance from the axis by r and its slope dr/dz as r', we can relate the output ray $(r_{\text{out}}, r'_{\text{out}})$ to the input ray $(r_{\text{in}}, r'_{\text{in}})$ by means of

$$
\begin{aligned}
r_{\text{out}} &= r_{\text{in}} \\
r'_{\text{out}} &= r'_{\text{in}} - \frac{r_{\text{in}}}{f}
\end{aligned}
\tag{2.1-1}
$$

[1]By paraxial ray we mean a ray whose angular deviation from the cylindrical (z) axis is small enough that the sine and tangent of the angle can be approximated by the angle itself.

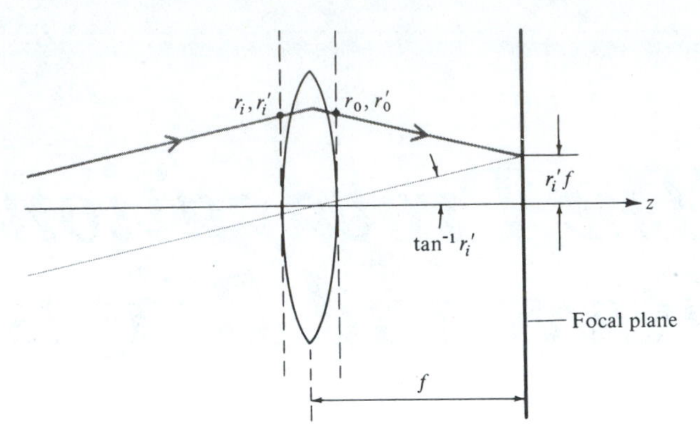

Figure 2-1 Deflection of a ray by a thin lens.

where the first of (2.1-1) follows from the definition of a thin lens and the second can be derived from a consideration of the behavior of the undeflected central ray with a slope equal to r'_{in}, as shown in Figure 2-1.

Representing a ray at any position z as a column matrix,

$$\mathbf{r}(z) = \begin{vmatrix} r(z) \\ r'(z) \end{vmatrix}$$

We can rewrite (2.1-1) using the rules for matrix multiplication (see References [1]–[3]) as

$$\begin{vmatrix} r_{\text{out}} \\ r'_{\text{out}} \end{vmatrix} = \begin{vmatrix} 1 & 0 \\ -\dfrac{1}{f} & 1 \end{vmatrix} \begin{vmatrix} r_{\text{in}} \\ r'_{\text{in}} \end{vmatrix} \tag{2.1-2}$$

where $f > 0$ for a converging lens and is negative for a diverging one.

The ray matrices for a number of other optical elements are shown in Table 2-1.

Consider as an example the propagation of a ray through a straight section of a homogeneous medium of length d followed by a thin lens of focal length f. This corresponds to propagation between planes a and b in Figure 2-2. Since the effect of the straight section is merely that of increasing r by dr', using (2.1-1) we can relate the output b and input (at a) rays by:

$$\begin{vmatrix} r_{\text{out}} \\ r'_{\text{out}} \end{vmatrix} = \begin{vmatrix} 1 & d \\ -\dfrac{1}{f} & \left(1 - \dfrac{d}{f}\right) \end{vmatrix} \begin{vmatrix} r_{\text{in}} \\ r'_{\text{in}} \end{vmatrix} \tag{2.1-3}$$

Notice also that the matrix corresponds to the product of the thin lens matrix times the straight section matrix as given in Table 2-1.

We are now in a position to consider the propagation of a ray through a biperiodic lens system made up of lenses of focal lengths f_1 and f_2 separated by d as

Table 2-1 Ray Matrices for Some Common Optical Elements and Media

(1) Straight Section: Length d	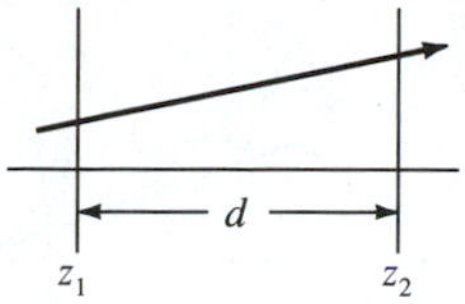 d; z_1, z_2	$\begin{bmatrix} 1 & d \\ 0 & 1 \end{bmatrix}$
(2) Thin Lens: Focal length f ($f > 0$, converging; $f < 0$, diverging)	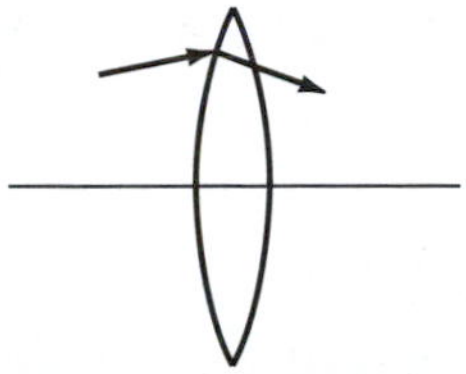	$\begin{bmatrix} 1 & 0 \\ \frac{-1}{f} & 1 \end{bmatrix}$
(3) Dielectric Interface: Refractive indices n_1, n_2	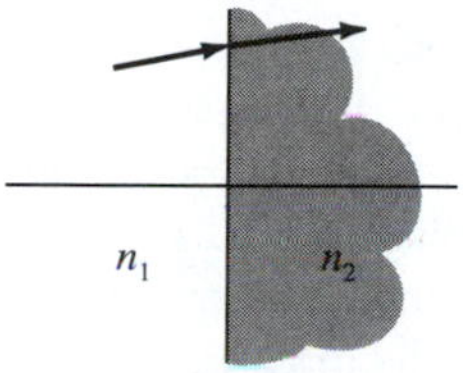 n_1; n_2	$\begin{bmatrix} 1 & 0 \\ 0 & \frac{n_1}{n_2} \end{bmatrix}$
(4) Spherical Dielectric Interface: Radius R	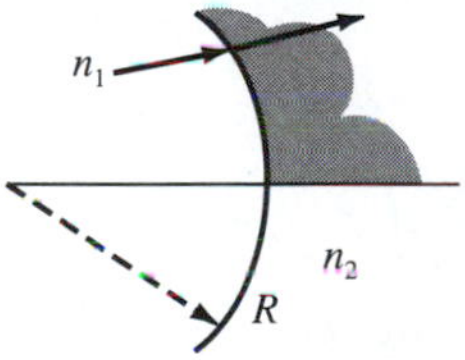 n_1; n_2; R	$\begin{bmatrix} 1 & 0 \\ \frac{n_2 - n_1}{n_2 R} & \frac{n_1}{n_2} \end{bmatrix}$
(5) Spherical Mirror: Radius of curvature R	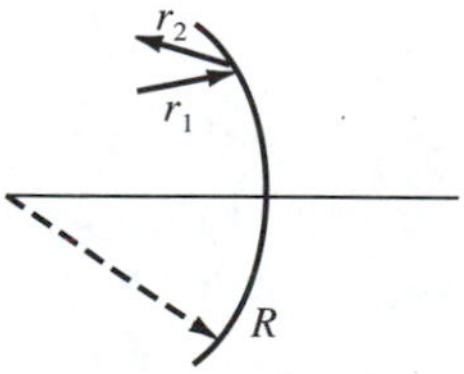 r_2; r_1; R	$\begin{bmatrix} 1 & 0 \\ \frac{-2}{R} & 1 \end{bmatrix}$
(6) A medium with a quadratic index profile	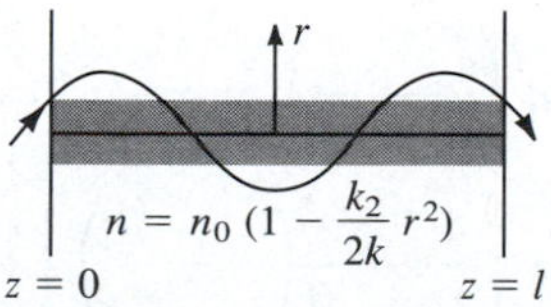 r; $n = n_0 (1 - \frac{k_2}{2k} r^2)$; $z = 0$; $z = l$	$\begin{bmatrix} \cos\left(\sqrt{\frac{k_2}{k}}\, l\right) & \sqrt{\frac{k}{k_2}} \sin\left(\sqrt{\frac{k_2}{k}}\, l\right) \\ -\sqrt{\frac{k_2}{k}} \sin\left(\sqrt{\frac{k_2}{k}}\, l\right) & \cos\left(\sqrt{\frac{k_2}{k}}\, l\right) \end{bmatrix}$

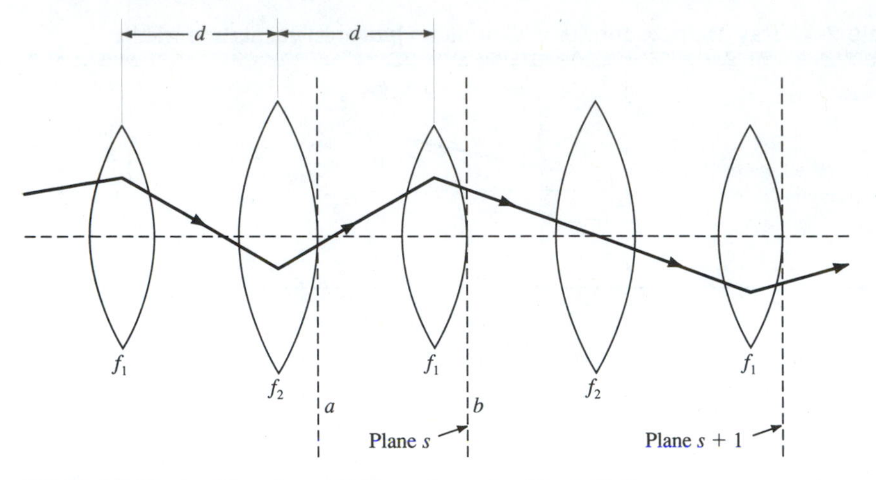

Figure 2-2 Propagation of an optical ray through a biperiodic lens sequence.

shown in Figure 2-2. This will be shown in the next chapter to be formally equivalent to the problem of Gaussian-beam propagation inside an optical resonator with mirrors of radii $R_1 = 2f_1$ and $R_2 = 2f_2$ that are separated by d.

The section between the planes s and $s + 1$ can be considered as the basic unit cell of the periodic lens sequence. The matrix relating the ray parameters at the output of a unit cell to those at the input is the product of two matrices, one for each lens, each of which is of the form of the matrix in (2.1-3).

$$\begin{vmatrix} r_{s+1} \\ r'_{s+1} \end{vmatrix} = \begin{vmatrix} 1 & d \\ -\dfrac{1}{f_1} & \left(1 - \dfrac{d}{f_1}\right) \end{vmatrix} \begin{vmatrix} 1 & d \\ -\dfrac{1}{f_2} & \left(1 - \dfrac{d}{f_2}\right) \end{vmatrix} \begin{vmatrix} r_s \\ r'_s \end{vmatrix} \tag{2.1-4}$$

or, in equation form,

$$\begin{aligned} r_{s+1} &= Ar_s + Br'_s \\ r'_{s+1} &= Cr_s + Dr'_s \end{aligned} \tag{2.1-5}$$

where A, B, C, and D are the elements of the matrix resulting from multiplying the two square matrices in (2.1-4) and are given by

$$\begin{aligned} A &= 1 - \frac{d}{f_2} \\ B &= d\left(2 - \frac{d}{f_2}\right) \\ C &= -\left[\frac{1}{f_1} + \frac{1}{f_2}\left(1 - \frac{d}{f_1}\right)\right] \\ D &= -\left[\frac{d}{f_1} - \left(1 - \frac{d}{f_1}\right)\left(1 - \frac{d}{f_2}\right)\right] \end{aligned} \tag{2.1-6}$$

From the first of (2.1-5) we get

$$r'_s = \frac{1}{B}(r_{s+1} - Ar_s) \tag{2.1-7}$$

and thus

$$r'_{s+1} = \frac{1}{B}(r_{s+2} - Ar_{s+1}) \tag{2.1-8}$$

Using the second of (2.1-5) in (2.1-8) and substituting for r'_s from (2.1-7) gives

$$r_{s+2} - (A + D)r_{s+1} + (AD - BC)r_s = 0 \tag{2.1-9}$$

for the difference equation governing the evolution through the lens waveguide. Using (2.1-6) we can show that $AD - BC = 1$. We can consequently rewrite (2.1-9) as

$$r_{s+2} - 2br_{s+1} + r_s = 0 \tag{2.1-10}$$

where

$$b = \tfrac{1}{2}(A + D) = \left(1 - \frac{d}{f_2} - \frac{d}{f_1} + \frac{d^2}{2f_1f_2}\right) \tag{2.1-11}$$

Equation (2.1-10) is the equivalent, in terms of difference equations, of the differential equation $r'' + Gr = 0$, whose solution is $r(z) = r(0)\exp[\pm i\sqrt{G}z]$. We are thus led to try a solution in the form of

$$r_s = r_0 e^{isq}$$

which, when substituted in (2.1-10), leads to

$$e^{2iq} - 2be^{iq} + 1 = 0 \tag{2.1-12}$$

and therefore

$$e^{iq} = b \pm i\sqrt{1 - b^2} = e^{\pm i\theta} \tag{2.1-13}$$

where $\cos\theta = b$.

The general solution can be taken as a linear superposition of $\exp(is\theta)$ and $\exp(-is\theta)$ solutions or equivalently as

$$r_s = r_{\max}\sin(s\theta + \alpha) \tag{2.1-14}$$

where $r_{\max} = r_0/\sin\alpha$ and α can be expressed using (2.1-7) in terms of r_0 and r'_0.

The condition for a stable, that is, confined, ray is that θ be a real number, since in this case the ray radius r_s oscillates as a function of the cell number s between $r_{\max}$ and $-r_{\max}$. According to (2.1-13), the necessary and sufficient condition for θ to be real is that [5]

$$|b| \leq 1 \tag{2.1-15}$$

In terms of the system parameters, we can use (2.1-11) to reexpress (2.1-15)

$$-1 \leq 1 - \frac{d}{f_2} - \frac{d}{f_1} + \frac{d^2}{2f_1f_2} \leq 1$$

or

$$0 \leq \left(1 - \frac{d}{2f_1}\right)\left(1 - \frac{d}{2f_2}\right) \leq 1 \tag{2.1-16}$$

If, on the other hand, the stability condition $|b| \leq 1$ is violated, we obtain, according to (2.1-10), a solution in the form of

$$r_s = C_1 e^{(\alpha+)s} + C_2 e^{(\alpha-)s} \tag{2.1-17}$$

where $e^{\alpha\pm} = b \pm \sqrt{b^2 - 1}$, and, since the magnitude of either $\exp(\alpha_+)$ or $\exp(\alpha_-)$ exceeds unity, the beam radius will increase without limit as a function of (distance) s.

Identical-Lens Waveguide

The simplest case of a lens waveguide is one in which $f_1 = f_2 = f$; that is, all lenses are identical.

The analysis of this situation is considerably simpler than that used for a biperiodic lens sequence. The reason is that the periodic unit cell (the smallest part of the sequence that can, upon translation, recreate the whole sequence) contains a single lens only. The (A, B, C, D) matrix for the unit cell is given by the square matrix in (2.1-3). Following exactly the steps leading to (2.1-11) through (2.1-14), the stability condition becomes

$$0 \leq d \leq 4f \tag{2.1-18}$$

and the beam radius at the nth lens is given by

$$r_n = r_{\max} \sin(n\theta + \alpha)$$

$$\cos\theta = \left(1 - \frac{d}{2f}\right) \tag{2.1-19}$$

Because of the algebraic simplicity of this problem we can easily express $r_{\max}$ and α in (2.1-19) in terms of the initial conditions r_0 and r_0', obtaining

$$(r_{\max})^2 = \frac{4f}{4f - d}(r_0^2 + dr_0 r_0' + dfr_0'^2) \tag{2.1-20}$$

$$\tan\alpha = \sqrt{\frac{4f}{d} - 1} \Bigg/ \left(1 + 2f\frac{r_0'}{r_0}\right) \tag{2.1-21}$$

where n corresponds to the plane immediately to the right of the nth lens. The derivation of the last two equations is left as an exercise (Problem 2.1).

The stability criteria can be demonstrated experimentally by tracing the behavior of a laser beam as it propagates down a sequence of lenses spaced uniformly. One can easily notice the rapid escape of the beam once condition (2.1-18) is violated.

2.2 PROPAGATION OF RAYS BETWEEN MIRRORS (6)

Another important application of the formalism just developed concerns the bouncing of a ray between two curved mirrors. Since the reflection at a mirror with a radius of curvature R is equivalent, except for the folding of the path, to passage through a lens with a focal length $f = R/2$, we can use the formalism of the preceding section to describe the propagation of a ray between two curved reflectors with radii of curvature R_1 and R_2, which are separated by d. Let us consider the simple case of a ray that is injected into a symmetric two-mirror system as shown in Figure 2-3(a). Since the x and y coordinates of the ray are independent variables, we can take them according to (2.1-19) in the form of

$$\begin{aligned} x_n &= x_{\max} \sin(n\theta + \alpha_x) \\ y_n &= y_{\max} \sin(n\theta + \alpha_y) \end{aligned} \tag{2.2-1}$$

where n refers to the ray parameter immediately following the nth reflection. According to (2.2-1), the locus of the points x_n, y_n on a given mirror lies on an ellipse.

Reentrant Rays

If θ in (2.2-1) satisfies the condition

$$2\nu\theta = 2l\pi \tag{2.2-2}$$

where ν and l are any two integers, a ray will return to its starting point following ν round trips and will thus continuously retrace the same pattern on the mirrors. If we consider as an example the simple case of $l = 1$, $\nu = 2$, so that $\theta = \pi/2$, from (2.1-19) we obtain $d = 2f = R$; that is, if the mirrors are separated by a distance equal to their radius of curvature R, the trapped ray will retrace its pattern after two round trips ($\nu = 2$). This situation ($d = R$) is referred to as symmetric confocal, since the two mirrors have a common focal point $f = R/2$. It will be discussed in detail in Chapter 4. The ray pattern corresponding to $\nu = 2$ is illustrated in Figure 2-3(b).

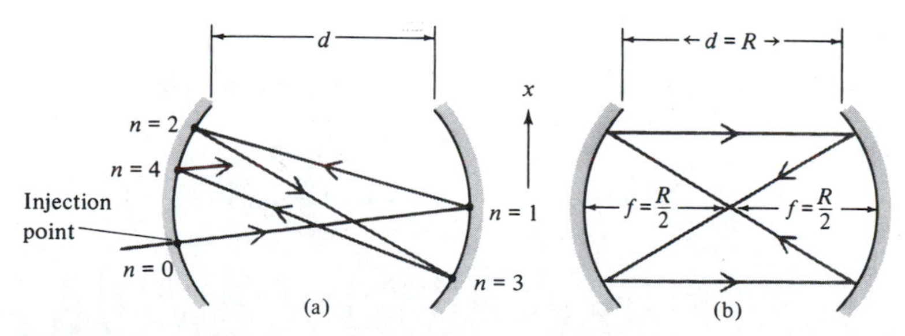

Figure 2-3 (a) Path of a ray injected in plane of figure into the space between two mirrors. (b) Reentrant ray in confocal ($d = R$) mirror configuration repeating its pattern after two round trips.

2.3 RAYS IN LENSLIKE MEDIA (7)

The basic physical property of lenses that is responsible for their focusing action is the fact that the optical path across them $\int n(r, z)\, dz$ (where n is the index of refraction of the medium) is a quadratic function of the distance r from the z axis. Using ray optics, we account for this fact by a change in the ray's slope as in (2.1-1). This same property can be represented by relating the complex field amplitude of the incident optical field $E_R(x, y)$ immediately to the right of an ideal thin lens to that immediately to the left $E_L(x, y)$ by

$$E_R(x, y) = E_L(x, y) \exp\left[+ik \frac{x^2 + y^2}{2f}\right] \tag{2.3-1}$$

where f is the focal length and $k = 2\pi n/\lambda_0$.

The effect of the lens, therefore, is to cause a phase shift $k(x^2 + y^2)/2f$, which increases quadratically with the distance from the axis. We consider next the closely related case of a medium whose index of refraction n varies according to[2]

$$n(x, y) = n_0 \left[1 - \frac{k_2}{2k}(x^2 + y^2)\right] \tag{2.3-2}$$

where k_2 is a constant. Since the phase delay of a wave propagating through a section dz of a medium with an index of refraction n is $(2\pi\, dz/\lambda_0)n$, it follows directly that a thin slab of the medium described by (2.3-2) will act as a thin lens, introducing [as in (2.3-1)] a phase shift proportional to $(x^2 + y^2)$. The behavior of a ray in this case is described by the differential equation that applies to ray propagation in an optically inhomogeneous medium [8],

$$\frac{d}{ds}\left(n \frac{d\mathbf{r}}{ds}\right) = \nabla n \tag{2.3-3}$$

where s is the distance along the ray measured from some fixed position on it and $\mathbf{r}$ is the position vector of the point at s. For paraxial rays we may replace d/ds by d/dz and, using (2.3-2), obtain

$$\frac{d^2 r}{dz^2} + \left(\frac{k_2}{k}\right) r = 0 \tag{2.3-4}$$

If at the input plane $z = 0$ the ray has a radius r_0 and slope r'_0, we can write the solution of (2.3-4) directly as

$$\begin{aligned} r(z) &= \cos\left(\sqrt{\frac{k_2}{k}}\, z\right) r_0 + \sqrt{\frac{k}{k_2}} \sin\left(\sqrt{\frac{k_2}{k}}\, z\right) r'_0 \\ r'(z) &= -\sqrt{\frac{k_2}{k}} \sin\left(\sqrt{\frac{k_2}{k}}\, z\right) r_0 + \cos\left(\sqrt{\frac{k_2}{k}}\, z\right) r'_0 \end{aligned} \tag{2.3-5}$$

---

[2]Equation (2.3-2) can be viewed as consisting of the first two terms in the Taylor series expansion of $n(x, y)$ for the radial symmetric case.

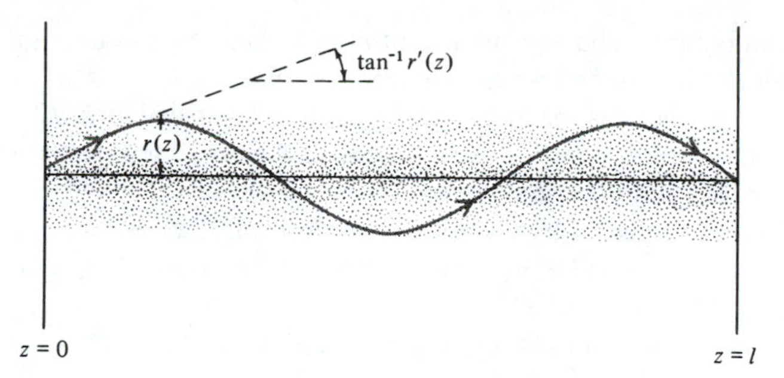

Figure 2-4 Path of a ray in a medium with a quadratic index variation.

That is, the ray oscillates back and forth across the axis, as shown in Figure 2-4. A section of the quadratic index medium acts as a lens. This can be proved by showing, using (2.3-5), that a family of parallel rays entering at $z = 0$ at different radii will converge upon emerging at $z = l$ to a common focus at a distance

$$h = \frac{1}{n_0}\sqrt{\frac{k}{k_2}} \cot\left(\sqrt{\frac{k_2}{k}}\, l\right) \tag{2.3-6}$$

from the exit plane. The factor n_0 accounts for the refraction at the boundary, assuming the medium at $z > l$ to possess an index $n = 1$ and a small angle of incidence. The derivation of (2.3-6) is left as an exercise (Problem 2.3).

Equations (2.3-5) apply to a focusing medium with $k_2 > 0$. In a medium where $k_2 < 0$—that is, where the index increases with the distance from the axis—the solutions for $r(z)$ and $r'(z)$ become

$$\begin{aligned} r(z) &= \cosh\left(\sqrt{\frac{|k_2|}{k}}\, z\right) r_0 + \sqrt{\frac{k}{|k_2|}} \sinh\left(\sqrt{\frac{|k_2|}{k}}\, z\right) r_0' \\ r'(z) &= \sqrt{\frac{|k_2|}{k}} \sinh\left(\sqrt{\frac{|k_2|}{k}}\, z\right) r_0 + \cosh\left(\sqrt{\frac{|k_2|}{k}}\, z\right) r_0' \end{aligned} \tag{2.3-7}$$

so that $r(z)$ increases with distance and eventually escapes. A section of such a medium acts as a negative lens.

Physical situations giving rise to quadratic index variation include:

1. Propagation of laser beams with Gaussian-like intensity profile in a slightly absorbing medium. The absorption heating gives rise, because of the dependence of n on the temperature T, to an index profile [9]. If $dn/dT < 0$, as is the case for most materials, the index is smallest on the axis where the absorption heating is highest. This corresponds to a $k_2 < 0$ in (2.3-2) and the beam spreads with the distance z. If $dn/dT > 0$, as in certain lead glasses [10], the beams are focused.
2. In optically pumped solid-state laser rods, the portion of the absorbed pump power that is not converted to laser radiation is conducted as heat to the rod surface. This heat conduction requires a temperature gradient in which T is

maximum on axis. The dependence of n on T then gives rise to a positive lens effect for $dn/dT > 0$ and a negative lens for $dn/dT < 0$.

3. Dielectric waveguides made by sandwiching a layer of index n_1 between two layers with index $n_2 < n_1$. This situation will be discussed further in Chapter 7 in connection with injection lasers.
4. Optical fibers produced by cladding a thin optical fiber (whose radius is comparable to λ) of an index n_1 with a sheath of index $n_2 < n_1$. Such fibers are used as light pipes.
5. Optical waveguides consisting of glasslike rods or filaments, with radii large compared to λ, whose index decreases with increasing r. Such waveguides can be used for the simultaneous transmission of a number of laser beams, which are injected into the waveguide at different angles. It follows from (2.3-5) that the beams will emerge, each along a unique direction, and consequently can be easily separated. Furthermore, in view of its previously discussed lens properties, the waveguide can be used to transmit optical image information in much the same way as images are transmitted by a multielement lens system to the image plane of a camera [11].

2.4 WAVE EQUATION IN QUADRATIC INDEX MEDIA

The most widely encountered optical beam in quantum electronics is one where the intensity distribution at planes normal to the propagation direction is Gaussian. To derive its characteristics we start with the Maxwell's equations in an isotropic charge-free medium.

$$\nabla \times \mathbf{H} = \varepsilon \frac{\partial \mathbf{E}}{\partial t}$$

$$\nabla \times \mathbf{E} = -\mu \frac{\partial \mathbf{H}}{\partial t}$$

$$\nabla \cdot (\varepsilon \mathbf{E}) = 0 \tag{2.4-1}$$

Taking the curl of the second of (2.4-1) and substituting the first results in

$$\nabla^2 \mathbf{E} - \mu\varepsilon \frac{\partial^2 \mathbf{E}}{\partial t^2} = -\nabla \left(\frac{1}{\varepsilon} \mathbf{E} \cdot \nabla \varepsilon \right) \tag{2.4-2}$$

where we used $\nabla \times \nabla \times \mathbf{E} \equiv \nabla(\nabla \cdot \mathbf{E}) - \nabla^2 \mathbf{E}$. If we assume the field quantities to vary as $\mathbf{E}(x, y, z, t) = \mathrm{Re}[\mathbf{E}(x, y, z)e^{i\omega t}]$ and neglect the right side of (2.4-2),[3] we obtain

$$\nabla^2 \mathbf{E} + k^2(\mathbf{r})\mathbf{E} = 0 \tag{2.4-3}$$

[3]This neglect is justified if the fractional change of ε in one optical wavelength is $\ll 1$.

where

$$k^2(\mathbf{r}) = \omega^2 \mu \varepsilon(\mathbf{r}) \left[1 - \frac{i\sigma(\mathbf{r})}{\omega\varepsilon} \right] \tag{2.4-4}$$

thus allowing for a possible dependence of ε on position $\mathbf{r}$. We have also taken k as a complex number to allow for the possibility of losses ($\sigma > 0$) or gain ($\sigma < 0$) in the medium.[4]

We limit our derivation to the case in which $k^2(\mathbf{r})$ is given by

$$k^2(r, \phi, z) = k^2 - kk_2 r^2 \qquad r = \sqrt{x^2 + y^2} \tag{2.4-5}$$

where, according to (2.4-4),

$$k^2 = k^2(0) = \omega^2 \mu \varepsilon(0) \left(1 - i \frac{\sigma(0)}{\omega\varepsilon(0)} \right)$$

so that k_2 is some constant characteristic of the medium. Furthermore, we assume a solution whose transverse dependence is on $r = \sqrt{x^2 + y^2}$ only, so that in (2.4-3) we can replace ∇^2 by

$$\nabla^2 = \nabla_t^2 + \frac{\partial^2}{\partial z^2} = \frac{\partial^2}{\partial r^2} + \frac{1}{r}\frac{\partial}{\partial r} + \frac{\partial^2}{\partial z^2} + \frac{1}{r^2}\frac{\partial^2}{\partial \phi^2} \tag{2.4-6}$$

The kind of propagation we are considering is that of a nearly plane wave in which the flow of energy is predominantly along a single (for example, $\mathbf{z}$) direction so that we may limit our derivation to a single transverse field component E. Taking E as

$$E = \psi(x, y, z)e^{-ikz} \tag{2.4-7}$$

we obtain from (2.4-3) and (2.4-5) in a few simple steps,

$$\nabla_t^2 \psi - 2ik\psi' - kk_2 r^2 \psi = 0 \tag{2.4-8}$$

where $\psi' \equiv \partial\psi/\partial z$ and where we assume that the variation is slow enough that $k\psi' \gg \psi'' \ll k^2\psi$.

Next we take ψ in the form of

$$\psi = \exp\left\{ -i \left[P(z) + \frac{k}{2q(z)} r^2 \right] \right\} \tag{2.4-9}$$

that, when substituted into (2.4-8) and after using (2.4-6), gives

$$-\left(\frac{k}{q}\right)^2 r^2 - 2i\left(\frac{k}{q}\right) - k^2 r^2 \left(\frac{1}{q}\right)' - 2kP' - kk_2 r^2 = 0 \tag{2.4-10}$$

---

[4]If k is complex (for example, $k_r + ik_i$), then a traveling electromagnetic plane wave has the form of $\exp[i(\omega t - kz)] = \exp[+k_i z + i(\omega t - k_r z)]$.

If (2.4-10) is to hold for all r, the coefficients of the different powers of r must be equal to zero. This leads to [7]

$$\left(\frac{1}{q}\right)^2 + \left(\frac{1}{q}\right)' + \frac{k_2}{k} = 0 \qquad P' = -\frac{i}{q} \tag{2.4-11}$$

The wave equation (2.4-3) is thus reduced to (2.4-11).

2.5 GAUSSIAN BEAMS IN A HOMOGENEOUS MEDIUM

We start with Equation (2.4-11)

$$\frac{1}{q^2} + \frac{d}{dz}\left(\frac{1}{q}\right) + \frac{k_2}{k} = 0 \tag{2.5-1}$$

In a homogeneous medium the quadratic coefficient k_2 of (2.4-5) is zero so that

$$\frac{1}{q^2} + \frac{d}{dz}\left(\frac{1}{q}\right) = 0 \tag{2.5-2}$$

or

$$\frac{dq}{dz} = 1 \tag{2.5-3}$$

$$q = z + q_0 \tag{2.5-4}$$

where q_0 is an arbitrary integration constant. From (2.4-11) and (2.5-4) we have

$$P' = -\frac{i}{q} = -\frac{i}{z + q_0} \tag{2.5-5}$$

so that

$$P(z) = -i \ln\left(1 + \frac{z}{q_0}\right) \tag{2.5-6}$$

where the new constant of integration was chosen as $c = i \ln q_0$.

Combining (2.5-5) and (2.5-6) in (2.4-9), we obtain

$$\psi = \exp\left\{-i\left[-i \ln\left(1 + \frac{z}{q_0}\right) + \frac{k}{2(q_0 + z)} r^2\right]\right\} \tag{2.5-7}$$

We take q_0 to be purely imaginary and reexpress it in terms of a new constant ω_0 as

$$q_0 = i\frac{\pi \omega_0^2 n}{\lambda} \qquad \lambda = \frac{2\pi n}{k} = \text{vacuum wave length} \tag{2.5-8}$$

The choice of imaginary q_0 will be found to lead to physically meaningful waves whose energy density is confined near the z axis. With this last substitution, let us consider, one at a time, the two factors in (2.5-7). The first one becomes

$$\exp\left[-\ln\left(1 - i\frac{\lambda z}{\pi\omega_0^2 n}\right)\right] = \frac{1}{\sqrt{1 + (\lambda^2 z^2/\pi^2\omega_0^4 n^2)}} \exp\left[i\tan^{-1}\left(\frac{\lambda z}{\pi\omega_0^2 n}\right)\right] \quad (2.5\text{-}9)$$

where we used $\ln(a + ib) = \ln\sqrt{a^2 + b^2} + i\tan^{-1}(b/a)$. Substituting (2.5-8) in the second term of (2.5-7) and separating the exponent into its real and imaginary parts, we obtain

$$\exp\left[\frac{-ikr^2}{2(q_0 + z)}\right] = \exp\left\{\frac{-r^2}{\omega_0^2[1 + (\lambda z/\pi\omega_0^2 n)^2]} - \frac{ikr^2}{2z[1 + (\pi\omega_0^2 n/\lambda z)^2]}\right\} \quad (2.5\text{-}10)$$

If we define the following parameters

$$\omega^2(z) = \omega_0^2\left[1 + \left(\frac{\lambda z}{\pi\omega_0^2 n}\right)^2\right] = \omega_0^2\left(1 + \frac{z^2}{z_0^2}\right) \quad (2.5\text{-}11)$$

$$R = z\left[1 + \left(\frac{\pi\omega_0^2 n}{\lambda z}\right)^2\right] = z\left(1 + \frac{z_0^2}{z^2}\right) \quad (2.5\text{-}12)$$

$$\eta(z) = \tan^{-1}\left(\frac{\lambda z}{\pi\omega_0^2 n}\right) = \tan^{-1}\left(\frac{z}{z_0}\right) \quad (2.5\text{-}13)$$

$$z_0 \equiv \frac{\pi\omega_0^2 n}{\lambda}$$

we can combine (2.5-9) and (2.5-10) in (2.5-7) and, recalling that $E(x, y, z) = \psi(x, y, z)\exp(-ikz)$, obtain

$$E(x, y, z) = E_0\frac{\omega_0}{\omega(z)}\exp\left\{-i[kz - \eta(z)] - i\frac{kr^2}{2q(z)}\right\}$$

$$= E_0\frac{\omega_0}{\omega(z)}\exp\left\{-i[kz - \eta(z)] - r^2\left(\frac{1}{\omega^2(z)} + \frac{ik}{2R(z)}\right)\right\}$$

$$k = \frac{2\pi n}{\lambda} \quad (2.5\text{-}14)$$

This is our basic result. We refer to it as the fundamental Gaussian-beam solution, since we have excluded the more complicated solutions of (2.4-3) (that is, those with azimuthal variation) by limiting ourselves to transverse dependence involving $r = (x^2 + y^2)^{1/2}$ only. These higher-order modes will be discussed separately.

From (2.5-14) the parameter $\omega(z)$, which evolves according to (2.5-11), is the distance r at which the field amplitude is down by a factor $1/e$ compared to its value on the axis. We will consequently refer to it as the beam *spot size*. The parameter ω_0 is the minimum spot size. It is the beam spot size at the plane $z = 0$. The parameter R in (2.5-14) is the radius of curvature of the very nearly spherical wavefronts at z.[5] We can verify this statement by deriving the radius of curvature of the constant phase surfaces (wavefronts) or, more simply, by considering the form of a spherical wave emitted by a point radiator placed at $z = 0$. It is given by

$$E \propto \frac{1}{R} e^{-ikR} = \frac{1}{R} \exp\left(-ik\sqrt{x^2 + y^2 + z^2}\right)$$
$$\simeq \frac{1}{R} \exp\left(-ikz - ik\frac{x^2 + y^2}{2R}\right) \qquad x^2 + y^2 \ll z^2 \qquad \textbf{(2.5-15)}$$

since z is equal to R, the radius of curvature of the spherical wave. Comparing (2.5-15) with (2.5-14), we identify R as the radius of curvature of the Gaussian beam. The convention regarding the sign of $R(z)$ is that it is negative if the center of curvature occurs at $z' > z$ and vice versa.

The form of the fundamental Gaussian beam is, according to (2.5-14), uniquely determined once its minimum spot size ω_0 and its location—that is, the plane $z = 0$—are specified. The spot size ω and radius of curvature R at any plane z are then found from (2.5-11) and (2.5-12). Some of these characteristics are displayed in Figure 2-5. The hyperbolas shown in this figure correspond to the ray direction and are intersections of planes that include the z axis and the hyperboloids

$$x^2 + y^2 = \text{const. } \omega^2(z) \qquad \textbf{(2.5-16)}$$

These hyperbolas correspond to the local direction of energy propagation. The spherical surfaces shown have radii of curvature given by (2.5-12). For large z the hyperboloids $x^2 + y^2 = \omega^2$ are asymptotic to the cone

$$r = \sqrt{x^2 + y^2} = \frac{\lambda}{\pi\omega_0 n} z \qquad \textbf{(2.5-17)}$$

[5]Actually, it follows from (2.5-14) that, with the exception of the immediate vicinity of the plane $z = 0$, the wavefronts are parabolic since they are defined by $k[z + (r^2/2R)] = $ const. For $r^2 \ll z^2$, the distinction between parabolic and spherical surfaces is not important.

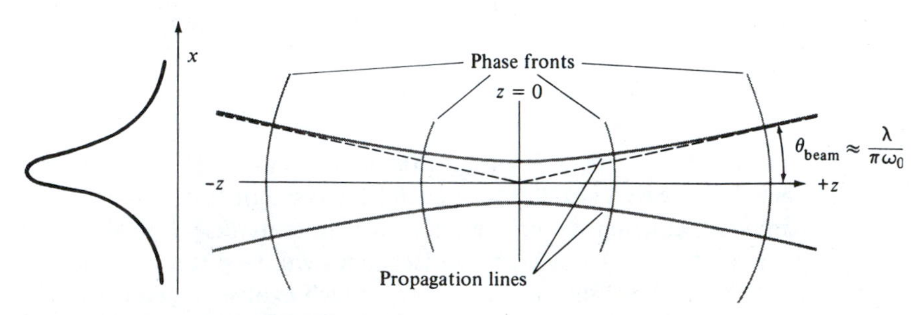

Figure 2-5 Propagating Gaussian beam.

whose half-apex angle, which we take as a measure of the angular beam spread, is

$$\theta_{\text{beam}} = \tan^{-1}\left(\frac{\lambda}{\pi\omega_0 n}\right) \simeq \frac{\lambda}{\pi\omega_0 n} \qquad \text{for } \theta_{\text{beam}} \ll \pi \tag{2.5-18}$$

This last result is a rigorous manifestation of wave diffraction according to which a wave that is confined in the transverse direction to an aperture of radius ω_0 will spread (diffract) in the far field ($z \gg \pi\omega_0^2 n/\lambda$) according to (2.5-18).

2.6 FUNDAMENTAL GAUSSIAN BEAM IN A LENSLIKE MEDIUM—THE ABCD LAW

We now return to the general case of a lenslike medium so that $k_2 \neq 0$. The P and q functions of (2.4-9) obey, according to (2.4-11)

$$\left(\frac{1}{q}\right)^2 + \left(\frac{1}{q}\right)' + \frac{k_2}{k} = 0$$

$$P' = -\frac{i}{q} \tag{2.6-1}$$

If we introduce the function s defined by

$$\frac{1}{q} = \frac{s'}{s} \tag{2.6-2}$$

we obtain from (2.6-1)

$$s'' + s\frac{k_2}{k} = 0$$

so that

$$s(z) = a \sin\sqrt{\frac{k_2}{k}}\,z + b\cos\sqrt{\frac{k_2}{k}}\,z$$

$$s'(z) = a\sqrt{\frac{k_2}{k}}\cos\sqrt{\frac{k_2}{k}}\,z - b\sqrt{\frac{k_2}{k}}\sin\sqrt{\frac{k_2}{k}}\,z \tag{2.6-3}$$

where a and b are arbitrary constants.

Using (2.6-3) in (2.6-2) and expressing the result in terms of an input value q_0 gives the following result for the complex beam radius $q(z)$

$$q(z) = \frac{\cos[(\sqrt{k_2/k})z]q_0 + \sqrt{k/k_2}\sin[(\sqrt{k_2/k})z]}{-\sin[(\sqrt{k_2/k})z]\sqrt{k_2/k}\,q_0 + \cos[(\sqrt{k_2/k})z]} \tag{2.6-4}$$

The physical significance of $q(z)$ in this case can be extracted from (2.4-9). We expand the part of $\psi(r, z)$ that involves r. The result is

$$\psi \propto e^{-ikr^2/2q(z)}$$

If we express the real and imaginary parts of $q(z)$ by means of

$$\frac{1}{q(z)} = \frac{1}{R(z)} - i\frac{\lambda}{\pi n\omega^2(z)} \tag{2.6-5}$$

we obtain

$$\psi \propto \exp\left[\frac{-r^2}{\omega^2(z)} - i\frac{kr^2}{2R(z)}\right]$$

so that $\omega(z)$ is the beam spot size and R its radius of curvature, as in the case of a homogeneous medium, which is described by (2.5-14). For the special case of a homogeneous medium ($k_2 = 0$), (2.6-4) reduces to (2.5-4).

Transformation of the Gaussian Beam—the ABCD Law

We have derived above the transformation law of a Gaussian beam (2.6-4) propagating through a lenslike medium that is characterized by k_2. We note first by comparing (2.6-4) to Table 2-1(6) and to (2.3-5) that the transformation can be described by

$$q_2 = \frac{Aq_1 + B}{Cq_1 + D} \tag{2.6-6}$$

where A, B, C, D are the elements of the ray matrix that relates the ray (r, r') at a plane 2 to the ray at plane 1. It follows immediately that the propagation through, or reflection from, any of the elements shown in Table 2-1 also obeys (2.6-6), since these elements can all be viewed as special cases of a lenslike medium. For future reference we note that by applying (2.6-6) to a thin lens of focal length f we obtain from (2.6-6) and Table 2-1(2)

$$\frac{1}{q_2} = \frac{1}{q_1} - \frac{1}{f} \tag{2.6-7}$$

so that using (2.6-5)

$$\omega_2 = \omega_1$$

$$\frac{1}{R_2} = \frac{1}{R_1} - \frac{1}{f} \tag{2.6-8}$$

These results apply, as well, to reflection from a mirror with a radius of curvature R if we replace f by $R/2$.

Consider next the propagation of a Gaussian beam through two lenslike media that are adjacent to each other. The ray matrix describing the first one is (A_1, B_1, C_1, D_1) while that of the second one is (A_2, B_2, C_2, D_2). Taking the input beam parameter as q_1 and the output beam parameter as q_3, we have from (2.6-6)

$$q_2 = \frac{A_1q_1 + B_1}{C_1q_1 + D_1}$$

for the beam parameter at the output of medium 1 and

$$q_3 = \frac{A_2 q_2 + B_2}{C_2 q_2 + D_2}$$

and after combining the last two equations,

$$q_3 = \frac{A_T q_1 + B_T}{C_T q_1 + D_T} \tag{2.6-9}$$

where (A_T, B_T, C_T, D_T) are the elements of the ray matrix relating the output plane (3) to the input plane (1), that is,

$$\begin{vmatrix} A_T & B_T \\ C_T & D_T \end{vmatrix} = \begin{vmatrix} A_2 & B_2 \\ C_2 & D_2 \end{vmatrix} \begin{vmatrix} A_1 & B_1 \\ C_1 & D_1 \end{vmatrix} \tag{2.6-10}$$

It follows by induction that (2.6-9) applies to the propagation of a Gaussian beam through any arbitrary number of lenslike media and elements. The matrix (A_T, B_T, C_T, D_T) is the ordered product of the matrices characterizing the individual members of the chain.

The great power of the ABCD law is that it enables us to trace the Gaussian beam parameter $q(z)$ through a complicated sequence of lenslike elements. The beam radius $R(z)$ and spot size $\omega(z)$ at any plane z can be recovered through the use of (2.6-5). The application of this method will be made clear by the following example.

Example: Gaussian Beam Focusing

As an illustration of the application of the ABCD law, we consider the case of a Gaussian beam that is incident at its waist on a thin lens of focal length f, as shown in Figure 2-6. We will find the location of the waist of the output beam and the beam spot size at that point.

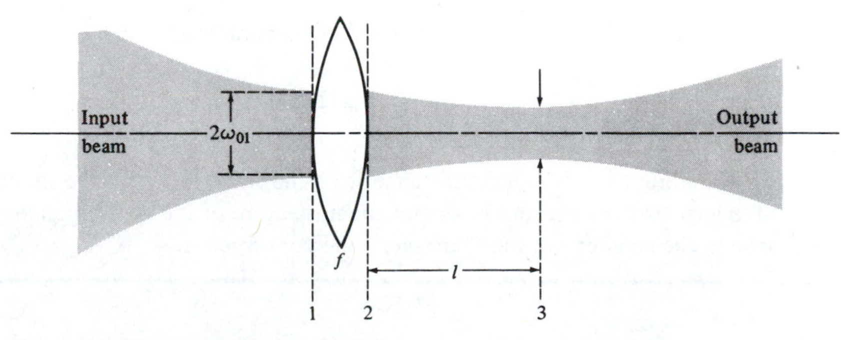

Figure 2-6 Focusing of a Gaussian beam.

At the input plane 1 $\omega = \omega_{01}$, $R_1 = \infty$ so that

$$\frac{1}{q_1} = \frac{1}{R_1} - i\frac{\lambda}{\pi\omega_{01}^2 n} = -i\frac{\lambda}{\pi\omega_{01}^2 n}$$

using (2.6-8) leads to

$$\frac{1}{q_2} = \frac{1}{q_1} - \frac{1}{f} = -\frac{1}{f} - i\frac{\lambda}{\pi\omega_{01}^2 n}$$

$$q_2 = \frac{1}{-1/f - i(\lambda/\pi\omega_{01}^2 n)} = \frac{-a + ib}{a^2 + b^2}$$

$$a \equiv \frac{1}{f} \qquad b \equiv \frac{\lambda}{\pi\omega_{01}^2 n}$$

At plane 3 we obtain, using (2.5-4),

$$q_3 = q_2 + l = \frac{-a}{a^2 + b^2} + \frac{ib}{a^2 + b^2} + l$$

$$\frac{1}{q_3} = \frac{1}{R_3} - i\frac{\lambda}{\pi\omega_3^2 n}$$

$$= \frac{[-a/(a^2 + b^2) + l] - ib/(a^2 + b^2)}{[-a/(a^2 + b^2) + l]^2 + b^2/(a^2 + b^2)^2}$$

Since plane 3 is, according to the statement of the problem, to correspond to the output beam waist, $R_3 = \infty$. Using this fact in the last equation leads to

$$l = \frac{a}{a^2 + b^2} = \frac{f}{1 + (f/\pi\omega_{01}^2 n/\lambda)^2} = \frac{f}{1 + (f/z_{01})^2} \tag{2.6-11}$$

as the location of the new waist, and to

$$\frac{\omega_3}{\omega_{01}} = \frac{f\lambda/\pi\omega_{01}^2 n}{\sqrt{1 + (f\lambda/\pi\omega_{01}^2 n)^2}} = \frac{f/z_{01}}{\sqrt{1 + (f/z_{01})^2}} \tag{2.6-12}$$

for the output beam waist. The confocal beam parameter

$$z_{01} \equiv \frac{\pi\omega_{01}^2 n}{\lambda}$$

is, according to (2.5-11), the distance from the waist in which the input beam spot size increases by $\sqrt{2}$ and is a convenient measure of the convergence of the input beam. The smaller z_{01}, the "stronger" the convergence.

2.7 A GAUSSIAN BEAM IN LENS WAVEGUIDE

As another example of the application of the ABCD law, we consider the propagation of a Gaussian beam through a sequence of thin lenses, as shown in Figure 2-2. The matrix, relating a ray in plane $s + 1$ to the plane $s = 1$ is

$$\begin{vmatrix} A_T & B_T \\ C_T & D_T \end{vmatrix} = \begin{vmatrix} A & B \\ C & D \end{vmatrix}^s \tag{2.7-1}$$

where (A, B, C, D) is the matrix for propagation through a single two-lens, unit cell $(\Delta s = 1)$ and is given by (2.1-6). We can use a well-known formula for the sth power of a matrix with a unity determinant (unimodular) to obtain

$$A_T = \frac{A\sin(s\theta) - \sin[(s-1)\theta]}{\sin\theta}$$

$$B_T = \frac{B\sin(s\theta)}{\sin\theta}$$

$$C_T = \frac{C\sin(s\theta)}{\sin\theta}$$

$$D_T = \frac{D\sin(s\theta) - \sin[(s-1)\theta]}{\sin\theta} \tag{2.7-2}$$

where

$$\cos\theta = \tfrac{1}{2}(A + D) = \left(1 - \frac{d}{f_2} - \frac{d}{f_1} + \frac{d^2}{2f_1 f_2}\right) \tag{2.7-3}$$

and then use (2.7-2) in (2.6-9) with the result

$$q_{s+1} = \frac{\{A\sin(s\theta) - \sin[(s-1)\theta]\}q_1 + B\sin(s\theta)}{C\sin(s\theta)q_1 + D\sin(s\theta) - \sin[(s-1)\theta]} \tag{2.7-4}$$

The condition for the confinement of the Gaussian beam by the lens sequence is, from (2.7-4), that θ be real; otherwise, the sine functions will yield growing exponentials. From (2.7-3), this condition becomes $|\cos\theta| \leq 1$, or

$$0 \leq \left(1 - \frac{d}{2f_1}\right)\left(1 - \frac{d}{2f_2}\right) \leq 1 \tag{2.7-5}$$

that is, the same as condition (2.1-16) for stable-ray propagation.

2.8 HIGH-ORDER GAUSSIAN BEAM MODES IN A HOMOGENEOUS MEDIUM

The Gaussian mode treated up to this point has a field variation that depends only on the axial distance z and the distance r from the axis. If we do not impose the condition $\partial/\partial\phi = 0$ (where ϕ is the azimuthal angle in a cylindrical coordinate system

(r, ϕ, z)) and take $k_2 = 0$, the wave equation (2.4-3) has solutions in the form of [12]

$$\begin{aligned} E_{l,m}(x, y, z) &= E_0 \frac{\omega_0}{\omega(z)} H_l\left(\sqrt{2}\frac{x}{\omega(z)}\right) H_m\left(\sqrt{2}\frac{y}{\omega(z)}\right) \\ &\quad \times \exp\left[-ik\frac{x^2+y^2}{2q(z)} - ikz + i(l+m+1)\eta\right] \\ &= E_0 \frac{\omega_0}{\omega(z)} H_l\left(\sqrt{2}\frac{x}{\omega(z)}\right) H_m\left(\sqrt{2}\frac{y}{\omega(z)}\right) \\ &\quad \times \exp\left[-\frac{x^2+y^2}{\omega^2(z)} - \frac{ik(x^2+y^2)}{2R(z)} - ikz + i(l+m+1)\eta\right] \end{aligned} \tag{2.8-1}$$

where H_l is the Hermite polynomial of order l, and $\omega(z)$, $R(z)$, $q(z)$, and η are defined as in (2.5-11) through (2.5-13).

We note for future reference that the phase shift on the axis is

$$\theta = kz - (l+m+1)\tan^{-1}\left(\frac{z}{z_0}\right)$$

$$z_0 = \frac{\pi\omega_0^2 n}{\lambda} \tag{2.8-2}$$

The transverse variation of the electric field along x (or y) is seen to be of the form $H_l(\xi)\exp(-\xi^2/2)$ where $\xi = \sqrt{2}x/\omega$. This function has been studied extensively, since it corresponds, also, to the quantum mechanical wavefunction $u_l(\xi)$ of the harmonic oscillator [13]. Some low-order functions normalized to represent the same amount of total beam power are shown in Figure 2-7. Photographs of actual field patterns are shown in Figure 2-8. Note that the first four correspond to the intensity $|u_l(\xi)|^2$ plots ($l = 0, 1, 2, 3$) of Figure 2-7.

2.9 HIGH-ORDER GAUSSIAN BEAM MODES IN QUADRATIC INDEX MEDIA

In Section 2.6 we treated the propagation of a circularly symmetric Gaussian beam in lenslike media. Here we extend the treatment to higher-order modes and limit our attention to steady-state (that is, $q(z)$ = const.) solutions in media whose index of refraction can be described by

$$n^2(\mathbf{r}) = n^2\left(1 - \frac{n_2}{n}r^2\right) \qquad r^2 = x^2 + y^2 \tag{2.9-1a}$$

that is consistent with (2.4-5) if we put $k_2 = 2\pi n_2/\lambda$.

The vector-wave equation (2.4-3) takes the form

$$\nabla^2\mathbf{E} + k^2\left(1 - \frac{n_2}{n}r^2\right)\mathbf{E} = 0 \tag{2.9-1b}$$

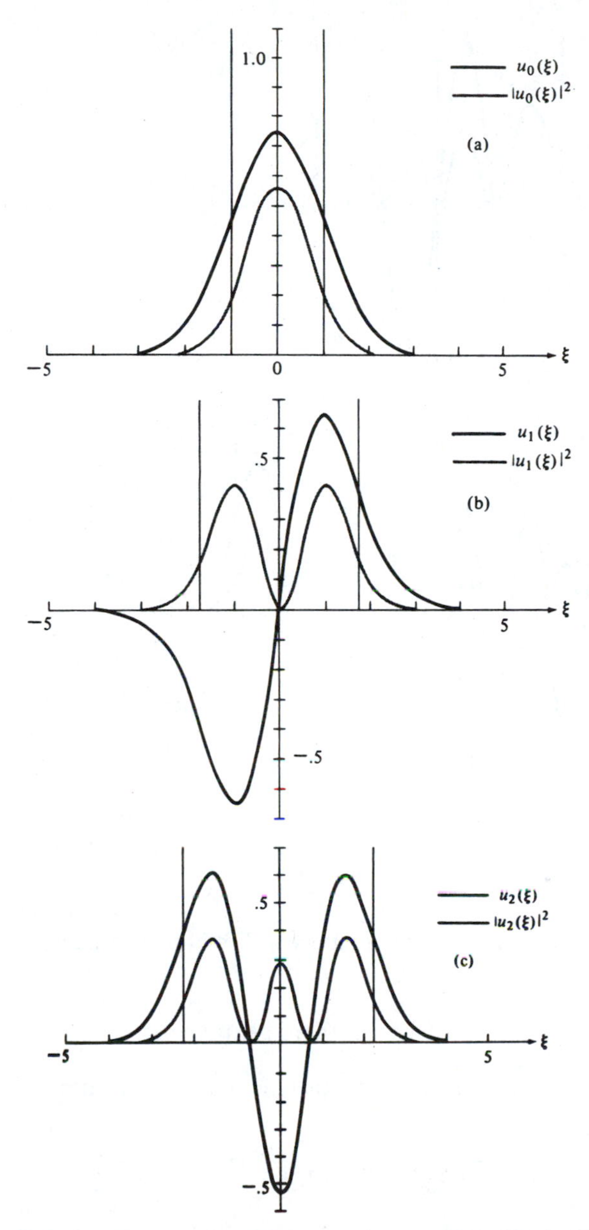

Figure 2-7 Hermite–Gaussian functions $u_l(\xi) = (\pi^{1/2}l!2^l)^{-1/2}H_l(\xi)e^{-\xi^2/2}$ corresponding to higher-order beam solutions (Equation 2.8-1). The curves are normalized so as to represent a fixed amount of total beam power in all the modes

$$\left(\int_{-\infty}^{\infty} u_l^2(\xi)d\xi = 1\right)$$

The solid curves are the functions $u_l(\xi)$ for $l = 0, 1, 2, 3$, and 10. The curves which are limited to the upper half space are $u_l^2(\xi)$.

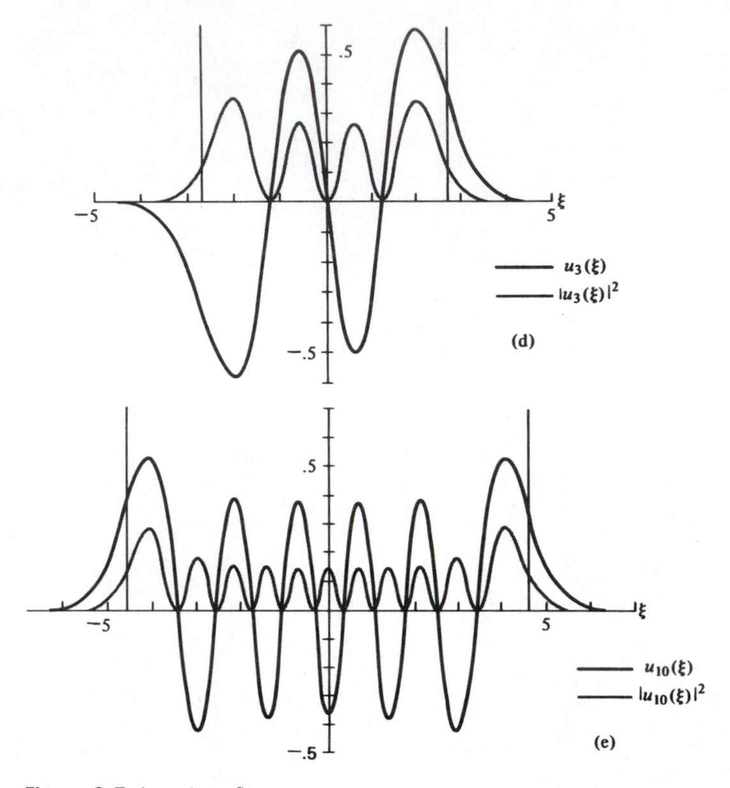

Figure 2-7 *(continued)*

We consider some (scalar) component E of the last equation and assume a solution in the form

$$E(x, y) = \psi(x, y) \exp(-i\beta z)$$

Taking $\psi(x, y) = f(x)g(y)$ the wave equation (2.9-1b) becomes

$$\frac{1}{f}\frac{\partial^2 f}{\partial x^2} + \frac{1}{g}\frac{\partial^2 g}{\partial y^2} + k^2 - k^2\frac{n_2}{n}(x^2 + y^2) - \beta^2 = 0 \qquad \textbf{(2.9-2)}$$

Since (2.9-2) is the sum of a y dependent part and an x dependent part, it follows that

$$\frac{1}{f}\frac{d^2 f}{dx^2} + \left(k^2 - \beta^2 - k^2\frac{n_2}{n}x^2\right) = C \qquad \textbf{(2.9-3)}$$

$$\frac{1}{g}\frac{d^2 g}{dy^2} - k^2\frac{n_2}{n}y^2 = -C \qquad \textbf{(2.9-4)}$$

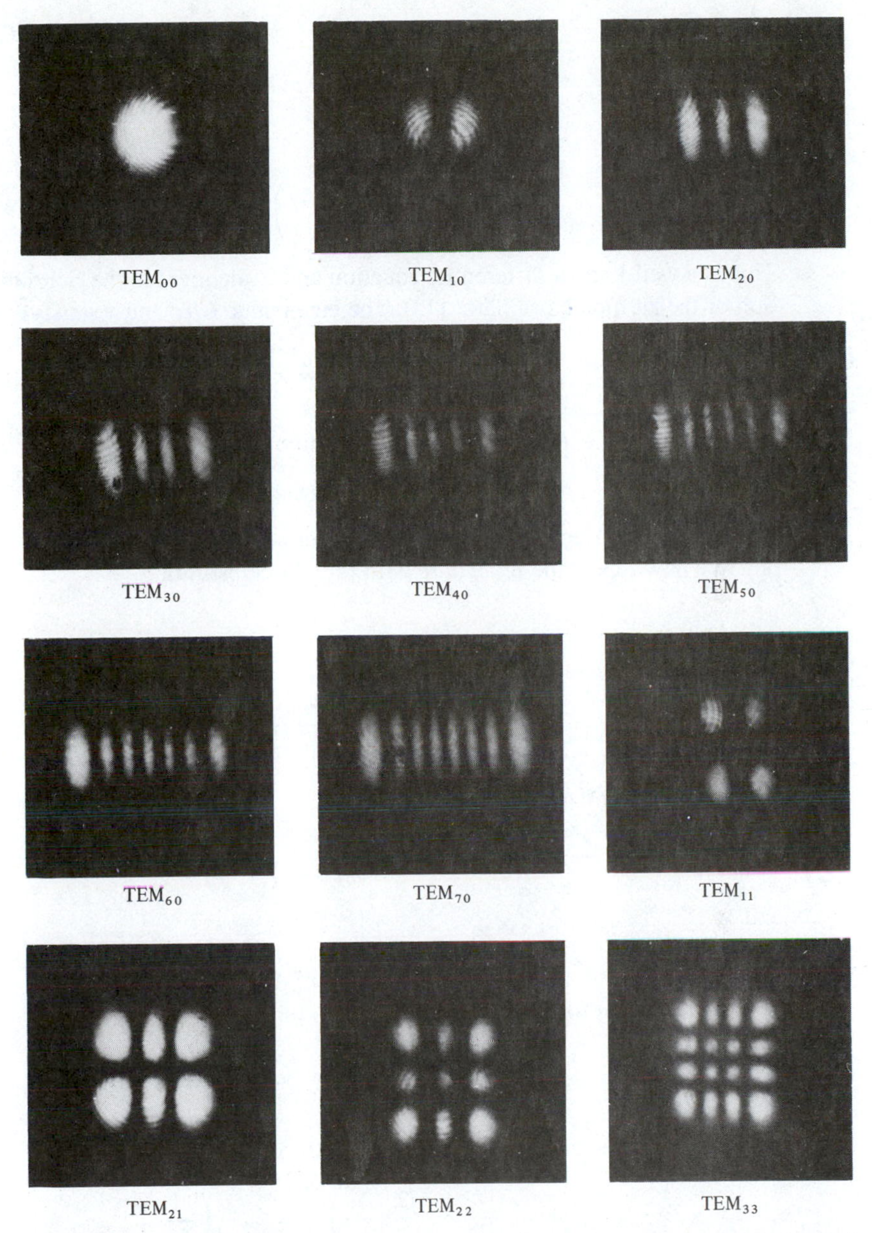

Figure 2-8 Intensity photographs of some low-order Gaussian beam modes. (After Reference [14].)

where C is some constant. Consider first (2.9-4). Defining a variable ξ by

$$\xi = \alpha y \qquad \alpha \equiv k^{1/2}\left(\frac{n_2}{n}\right)^{1/4} \tag{2.9-5}$$

(2.9-4) becomes

$$\frac{d^2 g}{d\xi^2} + \left(\frac{C}{\alpha^2} - \xi^2\right) g = 0 \tag{2.9-6}$$

This is a well-known differential equation and is identical to the Schrödinger equation of the harmonic oscillator [13]. The eigenvalue C/α^2 must satisfy

$$\frac{C}{\alpha^2} = (2m + 1) \qquad m = 1, 2, 3 \ldots \tag{2.9-7}$$

and corresponding to an integer m, the solution is

$$g_m(\xi) = H_m(\xi)e^{-\xi^2/2} \tag{2.9-8}$$

where H_m is the Hermite polynomial of order m.

We now repeat the procedure with (2.9-3). Substituting

$$\zeta = \alpha x$$

it becomes

$$\frac{\partial^2 f}{\partial \zeta^2} + \left[\frac{k^2 - \beta^2 - C}{\alpha^2} - \zeta^2\right] f = 0$$

so that, as in (2.9-7),

$$\frac{k^2 - \beta^2 - C}{\alpha^2} = (2l + 1) \qquad l = 1, 2, 3, \ldots \tag{2.9-9}$$

and

$$f_l(\zeta) = H_l(\zeta)e^{-\zeta^2/2} \tag{2.9-10}$$

The total solution for ψ is thus

$$\psi(x, y) = H_l\left(\frac{\sqrt{2}x}{\omega}\right) H_m\left(\frac{\sqrt{2}y}{\omega}\right) e^{-(x^2+y^2)/\omega^2}$$

where the "spot size" ω is, according to (2.9-5),

$$\omega = \frac{\sqrt{2}}{\alpha} = \sqrt{\frac{2}{k}}\left(\frac{n}{n_2}\right)^{1/4} = \sqrt{\frac{\lambda}{\pi}}\left(\frac{1}{nn_2}\right)^{1/4} \tag{2.9-11}$$

The total (complex) field is

$$\begin{aligned} E_{l,m}(x, y, z) &= \psi_{l,m}(x, y)e^{-i\beta_{l,m}z} \\ &= E_0 H_l\left(\sqrt{2}\,\frac{x}{\omega}\right) H_m\left(\sqrt{2}\,\frac{y}{\omega}\right) \exp\left(-\frac{x^2 + y^2}{\omega^2}\right) \exp\left(-i\beta_{l,m}z\right) \end{aligned} \tag{2.9-12}$$

The propagation constant $\beta_{l,m}$ of the l, m mode is obtained from (2.9-7) and (2.9-9)

$$\beta_{l,m} = k\left[1 - \frac{2}{k}\sqrt{\frac{n_2}{n}}\,(l + m + 1)\right]^{1/2} \tag{2.9-13}$$

Two features of the mode solutions are noteworthy. (1) Unlike the homogeneous medium solution ($n_2 = 0$), the mode spot size ω is independent of z. This can be explained by the focusing action of the index variation ($n_2 > 0$), which counteracts the natural tendency of a confined beam to diffract (spread). In the case of an index of refraction which increases with $r(n_2 < 0)$, it follows from (2.9-11) and (2.9-12) that $\omega^2 < 0$ and no confined solutions exist. The index profile in this case leads to defocusing, thus reinforcing the diffraction of the beam. (2) The dependence of β on the mode indices l, m causes the different modes to have phase velocities $v_{l,m} = \omega/\beta_{l,m}$ as well as group velocities $(v_g)_{l,m} = d\omega/d\beta_{l,m}$ that depend on l and m.

Let us consider the modal dispersion (that is, the dependence on l and m) of the group velocity of mode l, m

$$(v_g)_{l,m} = \frac{d\omega}{d\beta_{l,m}} \tag{2.9-14}$$

If the index variation is small so that

$$\frac{1}{k}\sqrt{\frac{n_2}{n}}\,(l + m + 1) \ll 1 \tag{2.9-15}$$

we can approximate (2.9-13) as

$$\beta_{l,m} \cong k - \sqrt{\frac{n_2}{n}}\,(l + m + 1) - \frac{n_2}{2kn}\,(l + m + 1)^2 \tag{2.9-16}$$

so that, according to (2.9-14),

$$(v_g)_{l,m} = \frac{c/n}{\left[1 + \dfrac{(n_2/n)}{2k^2}\,(l + m + 1)^2\right]} \tag{2.9-17}$$

The effect of the group velocity dispersion on pulse propagation is considered next.

We will show in Chapter 3 that optical pulses propagate at the group velocity v_g. We will also treat the effect of group velocity dispersion $\left(\dfrac{dv_g}{dw} \neq 0\right)$ on pulse broadening.

Pulse Spreading in Quadratic Index Glass Fibers

Glass fibers with quadratic index profiles (2.9-1a) are excellent channels for optical communication systems [15, 16]. The information is coded onto trains of optical pulses and the channel information capacity is thus fundamentally limited by the number of pulses that can be transmitted per unit time [17, 18].

There are two ways in which the group velocity dispersion limits the pulse repetition rate of the quadratic index channel.

1. Modal Dispersion If the optical pulses fed into the input end of the fiber excite a large number of modes (this will be the case if the input light is strongly focused so that the "rays" subtend a large angle), then each mode will travel with a group velocity $(v_g)_{l,m}$, as given by (2.9-17). If all the modes from (0, 0) to $(l_{\max}, m_{\max})$ are excited, the output pulse at $z = L$ will broaden to

$$\Delta\tau \cong L\left[\frac{1}{(v_g)_{l_{\max},m_{\max}}} - \frac{1}{(v_g)_{0,0}}\right] \tag{2.9-18}$$

We can use (2.9-17) to obtain

$$\Delta\tau \cong \frac{n_2 L}{2ck^2}\left[(l_{\max} + m_{\max} + 1)^2 - 1\right] \tag{2.9-19}$$

The maximum number of pulses per second that can be transmitted without serious overlap of adjacent output pulses is thus $f_{\max} \sim 1/\Delta\tau$. High data rate transmission will thus require the use of single mode excitation, which can be achieved by the use of coherent single mode laser excitation [16, 17, 18, 24].

Example: Numerical Example

Consider a 1-km-long quadratic index fiber with $n = 1.5$, $n_2 = 5.1 \times 10^3\ \text{cm}^{-2}$. Let the input optical pulses at $\lambda = 1\ \mu\text{m}$ excite the modes up to $l_{\max} = m_{\max} = 30$. Substitution in (2.9-19) gives

$$\Delta\tau = 8 \times 10^{-9}\ \text{s}$$

and $f_{\max} \sim (\Delta\tau)^{-1} = 1.25 \times 10^8$ pulses per second for the maximum pulse rate.

2. Group Velocity Dispersion The pulse spreading (2.9-19) due to multimode excitation can be eliminated if one were to excite a single mode, say l, m only. In this case pulse spreading would still result from the dependence of $(v_g)_{l,m}$ on frequency. This spreading can be explained by the fact that a pulse with a spectral width $\Delta\omega$ will spread in a distance L by

$$\Delta\tau \approx 2L\left|\frac{d}{d\omega}\left(\frac{1}{v_g}\right)\right|\Delta\omega = \frac{2L}{v_g^2}\left|\frac{dv_g}{d\omega}\right|\Delta\omega \tag{2.9-20}$$

If the pulse is derived, say, by gating, from a coherent continuous source with a negligible spectral width, the pulse spectral width is related to the pulse duration τ by $\Delta\omega \sim 2/\tau$ and (2.9-20) becomes

$$\Delta\tau \approx \frac{4L}{v_g^2 \tau}\left|\frac{dv_g}{d\omega}\right| \tag{2.9-21}$$

If the source bandwidth $\Delta\omega_s$ exceeds τ^{-1}, then we need to replace $\Delta\omega$ in (2.9-20) by $\Delta\omega_s$. A rigorous treatment of this subject is reserved for Chapter 3.

2.10 PROPAGATION IN MEDIA WITH A QUADRATIC GAIN PROFILE

In many laser media the gain is a strong function of position. This variation can be due to a variety of causes, among them: (1) the radial distribution of energetic electrons in the plasma region of gas lasers [19], (2) the variation of pumping intensity in solid state lasers, and (3) the dependence of the degree of gain saturation on the radial position in the beam.

We can account for an optical medium with quadratic gain (or loss) variation by taking the complex propagation constant $k(r)$ in (2.4-5) as

$$k(r) = k \pm i(\alpha_0 - \tfrac{1}{2}\alpha_2 r^2) \tag{2.10-1}$$

where the plus (minus) sign applies to the case of gain (loss). Assuming $k_2 r^2 \ll k$ in (2.4-5), we have $k_2 = i\alpha_2$. Using this value in (2.4-11) to obtain the steady-state ($(1/q)' = 0$) solution of the complex beam radius yields[6]

$$\frac{1}{q} = -i\sqrt{\frac{k_2}{k}} = -i\sqrt{\frac{i\alpha_2}{k}} \tag{2.10-2}$$

The steady-state beam radius and spot size are obtained from (2.6-5) and (2.10-2)

$$\omega^2 = 2\sqrt{\frac{\lambda}{\pi n \alpha_2}}$$

$$R = 2\sqrt{\frac{\pi n}{\lambda \alpha_2}} \tag{2.10-3}$$

We thus find that the steady-state solution corresponds to a beam with a constant spot size but with a finite radius of curvature.

The general (non-steady-state) behavior of the Gaussian beam in a quadratic gain medium is described by (2.6-4), where $k_2 = i\alpha_2$.

Experimental data showing a decrease of the beam spot size with increasing gain parameter α_2 in agreement with (2.10-3) are shown in Figure 2-9.

---

[6] "Steady state" here refers not to the intensity, which according to (2.10-1) is growing or decaying with z, but to the beam radius of curvature and spot size.

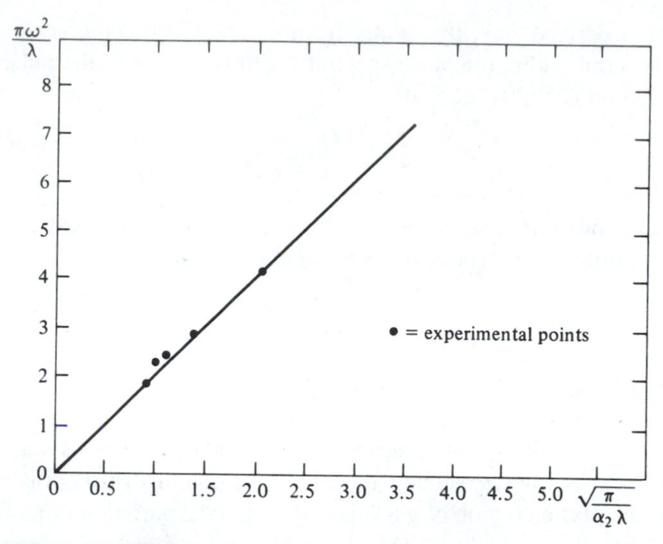

Figure 2-9 Theoretical curve showing the dependence of beam radius on quadratic gain constant α_2. Experimental points were obtained in a xenon 3.39-μm laser in which α_2 was varied by controlling the unsaturated laser gain. (After Reference [20].)

2.11 ELLIPTIC GAUSSIAN BEAMS

All the beam solutions considered up to this point have one feature in common. The field drops off as in (2.8-1), according to

$$E_{l,m} \propto \exp\left[-\frac{x^2+y^2}{\omega^2(z)}\right] \tag{2.11-1}$$

so that the locus in the x-y plane of the points where the field is down by a factor of e^{-1} from its value on the axis is a circle of radius $\omega(z)$. We will refer to such beams as *circular Gaussian beams.*

The wave equation (2.4-8) also admits solutions in which the variation in the x and y directions is characterized by

$$E_{l,m} \propto \exp\left[-\frac{x^2}{\omega_x^2(z)}-\frac{y^2}{\omega_y^2(z)}\right] \tag{2.11-2}$$

with $\omega_x \neq \omega_y$. Such beams, which we name *elliptic Gaussian,* result, for example, when a circular Gaussian beam passes through a cylindrical lens or when a laser beam emerges from an astigmatic resonator—that is, one whose mirrors possess different radii of curvature in the z-y and z-x planes.

We will not repeat the whole derivation for this case, but will indicate the main steps.

Instead of (2.4-9) we assume a solution

$$\psi = \exp\left\{-i\left[P(z) + \frac{k}{2q_x(z)}x^2 + \frac{k}{2q_y(z)}y^2\right]\right\} \tag{2.11-3}$$

that results, in a manner similar to (2.4-11), in[7]

$$\left(\frac{1}{q_x}\right)^2 + \left(\frac{1}{q_x}\right)' + \frac{k_{2x}}{k} = 0$$

$$\left(\frac{1}{q_y}\right)^2 + \left(\frac{1}{q_y}\right)' + \frac{k_{2y}}{k} = 0 \tag{2.11-4}$$

and

$$\frac{dP}{dz} = -\frac{i}{2}\left(\frac{1}{q_x} + \frac{1}{q_y}\right) \tag{2.11-5}$$

In the case of a homogeneous ($k_{2x} = k_{2y} = 0$) beam we obtain as in (2.5-4),

$$q_x(z) = z + C_x \tag{2.11-6}$$

where C_x is an arbitrary constant of integration. We find it useful to write C_x as

$$C_x = -z_x + q_{0x} \tag{2.11-7}$$

where z_x is real and q_{0x} is imaginary. The physical significance of these two constants will become clear in what follows. A similar result with $x \to y$ is obtained for $q_y(z)$. Using the solutions of $q_x(z)$ and $q_y(z)$ in (2.11-5) gives

$$P = -\frac{i}{2}\left[\ln\left(1 + \frac{z - z_x}{q_{0x}}\right) + \ln\left(1 + \frac{z - z_y}{q_{0y}}\right)\right]$$

Proceeding straightforwardly, as in the derivation connecting (2.5-6, . . . 14), results in

$$\begin{aligned} E(x, y, z) &= E_0 \frac{\sqrt{\omega_{0x}\omega_{0y}}}{\sqrt{\omega_x(z)\omega_y(z)}} \exp\left\{-i[kz - \eta(z)] - \frac{ikx^2}{2q_x(z)} - \frac{iky^2}{2q_y(z)}\right\} \\ &= E_0 \frac{\sqrt{\omega_{0x}\omega_{0y}}}{\sqrt{\omega_x(z)\omega_y(z)}} \exp\left\{-i[kz - \eta(z)] - x^2\left(\frac{1}{\omega_x^2(z)} + \frac{ik}{2R_x(z)}\right)\right. \\ &\quad \left. - y^2\left(\frac{1}{\omega_y^2(z)} + \frac{ik}{2R_y(z)}\right)\right\} \end{aligned} \tag{2.11-8}$$

---

[7]The parameters k_{2x} and k_{2y} are defined by

$$k^2(x, y) = k^2 - kk_{2x}x^2 - kk_{2y}y^2$$

which is a generalization of (2.4-5).

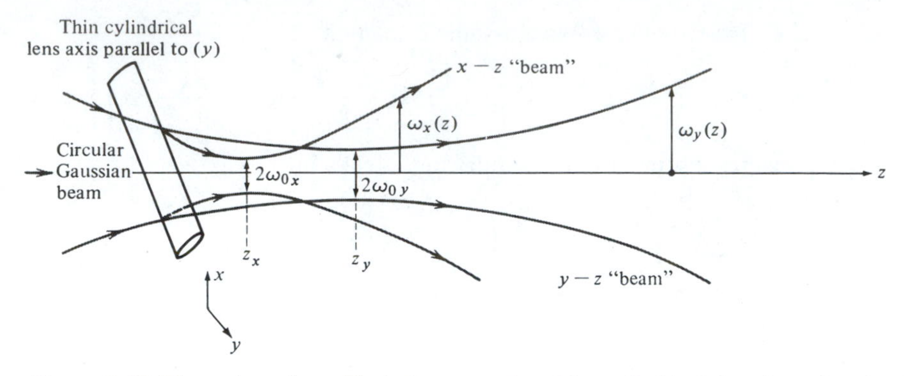

Figure 2-10 Illustration of an elliptic beam produced by cylindrical focusing of a circular Gaussian beam.

where

$$q_{0x} = i\,\frac{\pi\omega_{0x}^2 n}{\lambda}$$

$$\omega_x^2(z) = \omega_{0x}^2\left[1 + \left(\frac{\lambda(z - z_x)}{\pi\omega_{0x}^2 n}\right)^2\right] \tag{2.11-9}$$

$$R_x(z) = z\left[1 + \left(\frac{\pi\omega_{0x}^2 n}{\lambda(z - z_x)}\right)^2\right]$$

with similar expressions in which $x \rightarrow y$ for q_{0y}, ω_y, R_y.

The phase delay $\eta(z)$ in (2.11-8) is now given by

$$\eta(z) = \tfrac{1}{2}\tan^{-1}\left(\frac{\lambda(z - z_x)}{\pi\omega_{0x}^2 n}\right) + \tfrac{1}{2}\tan^{-1}\left(\frac{\lambda(z - z_y)}{\pi\omega_{0y}^2 n}\right) \tag{2.11-10}$$

It follows that *all* the results derived for the case of circular Gaussian beams apply, separately, to the *x-z* and to the *y-z* behavior of the elliptic Gaussian beam. For the purpose of analysis the elliptic beam can be considered as two independent "beams." The position of the waist is not necessarily the same for these two beams. It occurs at $z = z_x$ for the *x-z* beam and at $z = z_y$ for the *y-z* beam in the example of Figure 2-10, where z_x and z_y are arbitrary.

It also follows from the similarity between (2.11-4) and (2.4-11) that the ABCD transformation law (2.6-9) can be applied separately to $q_x(z)$ and $q_y(z)$ which, according to (2.11-8), are given by

$$\frac{1}{q_x(z)} = \frac{1}{R_x(z)} - i\,\frac{\lambda}{\pi n\omega_x^2(z)}$$

$$\frac{1}{q_y(z)} = \frac{1}{R_y(z)} - i\,\frac{\lambda}{\pi n\omega_y^2(z)} \tag{2.11-11}$$

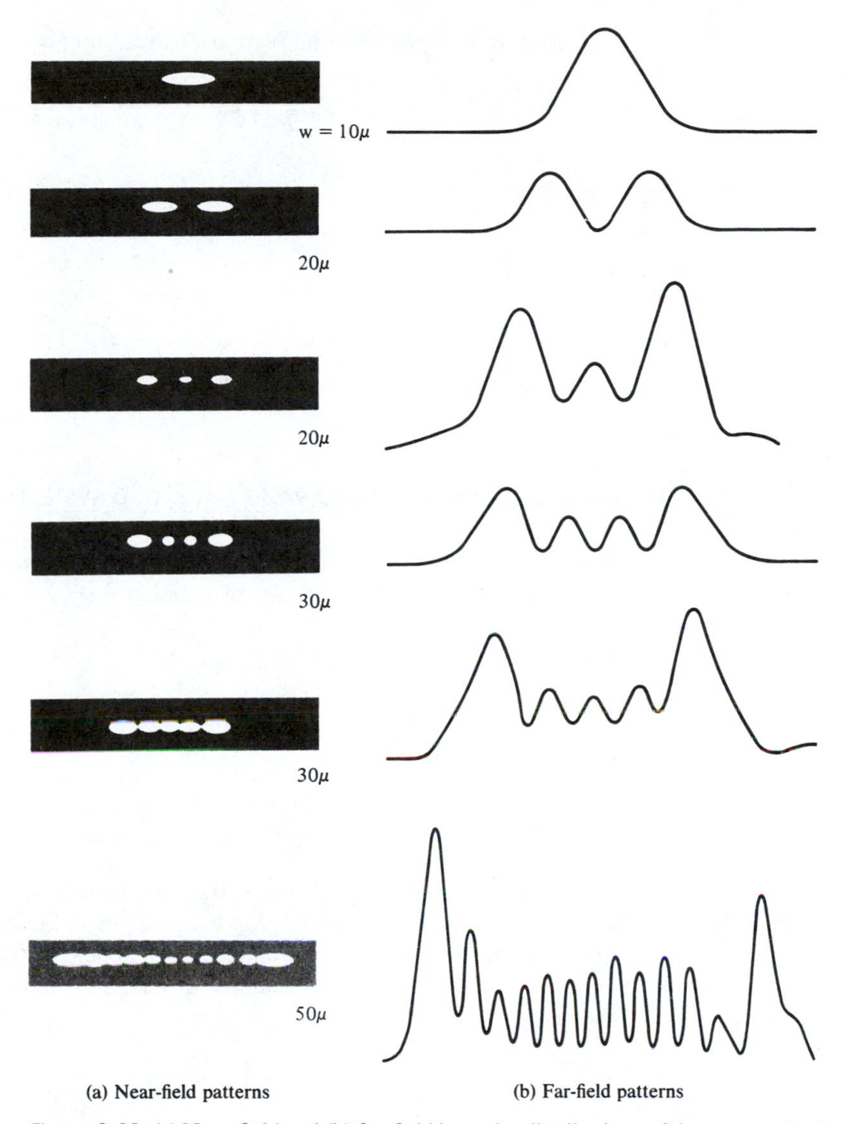

Figure 2-11 (a) Near-field and (b) far-field intensity distributions of the output of stripe contact GaAs-GaAlAs lasers. (After Reference [21].)

Elliptic Gaussian Beams in a Quadratic Lenslike Medium

Here we consider the *steady-state* elliptic beam propagating in a medium whose index of refraction is given by

$$n^2(\mathbf{r}) = n^2\left(1 - \frac{n_{2x}}{n}x^2 - \frac{n_{2y}}{n}y^2\right) \tag{2.11-12}$$

The derivation is identical to that presented in Section 2.9, resulting in

$$E_{l,m}(\mathbf{r}) = E_0 e^{-i\beta_{l,m}z} H_l\left(\sqrt{2}\,\frac{x}{\omega_x}\right) H_m\left(\sqrt{2}\,\frac{y}{\omega_y}\right) \exp\left(-\frac{x^2}{\omega_x^2} - \frac{y^2}{\omega_y^2}\right) \tag{2.11-13}$$

where

$$\omega_x = \left(\frac{\lambda}{\pi}\right)^{1/2}\left(\frac{1}{nn_{2x}}\right)^{1/4}$$

$$\omega_y = \left(\frac{\lambda}{\pi}\right)^{1/2}\left(\frac{1}{nn_{2y}}\right)^{1/4} \tag{2.11-14}$$

$$\beta_{l,m} = k\left\{1 - \frac{2}{k}\left[\sqrt{\frac{n_{2x}}{n}}\left(l + \frac{1}{2}\right) + \sqrt{\frac{n_{2y}}{n}}\left(m + \frac{1}{2}\right)\right]\right\}^{1/2} \tag{2.11-15}$$

2.12 DIFFRACTION INTEGRAL FOR A GENERALIZED PARAXIAL A, B, C, D SYSTEM (25)

In Sec. 1.6 we derived the, very important, Fresnel-Kirchhoff diffraction integral, Eq. (1.6-17), reproduced here with a slight change of notation.

$$f_1(x_1, y_1) = \frac{ik}{2\pi L}\exp(-ikL)\iint f_0(x_0, y_0)$$

$$\exp\left\{-\frac{ik}{2L}\left[(x_1 - x_0)^2 + (y_1 - y_0)^2\right]\right\} dx_0\, dy_0$$

$$k = \frac{2\pi}{\lambda} = \frac{2\pi}{\lambda_0}n \tag{2.12-1}$$

which relates the field f_1 at $z = L$ to the field f_0 at $z = 0$ in the case of a homogeneous and isotropic medium. This result can be extended to the case where the intervening medium consists of a cascade of paraxial axisymmetric components which can be described by an overall A, B, C, D matrix. An example of such a system is sketched in Figure 2-12. The result is

$$f_1(x_1, y_1) = \frac{ik_0}{2\pi B}\exp(-ik_0L)\iint f_0(x_0, y_0)$$

$$\exp\left\{-i\,\frac{k_0}{2B}\left[A(x_0^2 + y_0^2) - 2x_0x_1 - 2y_0y_1 + D(x_1^2 + y_1^2)\right]\right\} dx_0\, dy_0 \tag{2.12-2}$$

$$k_0 = \frac{2\pi}{\lambda_0} = \frac{\omega}{C}$$

$$L = \sum_i n_iL_i = \text{optical path length along the axis}$$

The proof of this important result is not given here. The interested reader can consult the original papers of Baues [26] and Collins [27], as well as Reference [28].

To demonstrate the power of this result, we will use it to derive the imaging condition of a generalized cascade of optical elements described by an overall

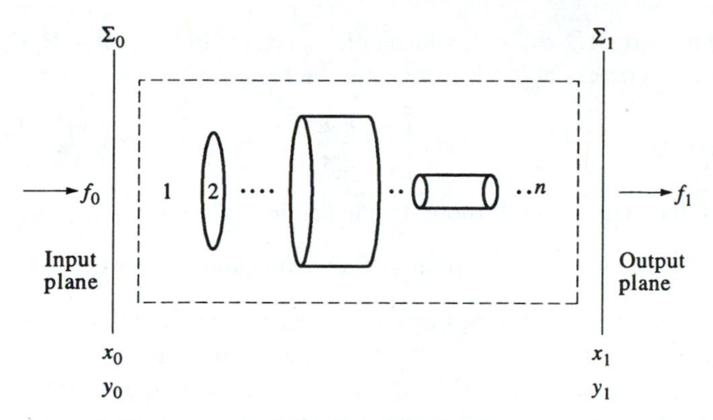

Figure 2-12 A cascade of lenslike (paraxial) elements with an overall *A*, *B*, *C*, *D* matrix.

(*A*, *B*, *C*, *D*) matrix. By rearranging and manipulating the exponent of Equation (2.12-2), we can rewrite it as

$$f_1(x_1, y_1) = \frac{ik_0}{2\pi B} \exp\left[-ik_0L - \frac{ik_0}{2B}\left(D - \frac{1}{A}\right)(x_1^2 + y_1^2)\right]$$

$$\iint f_0(x_0, y_0) \exp\left\{-\frac{ik_0}{2B}\left[A\left(x_0 - \frac{x_1}{A}\right)^2 + A\left(y_0 - \frac{y_1}{A}\right)^2\right]\right\} dx_0\, dy_0 \quad \textbf{(2.12-3)}$$

It is a straightforward matter to show [25] that

$$\lim_{B\to 0}\sqrt{\frac{i}{2\pi B}} \exp\left(-i\,\frac{x^2}{2B}\right) = \delta(x) \quad \textbf{(2.12-4)}$$

where $\delta(x)$ is the Dirac delta function. Using Equation (2.12-4) in (2.12-3), we obtain

$$\lim_{B\to 0} f_1(x_1, y_1) = \frac{\exp(-ik_0L)}{A}\, f_0\left(\frac{x_1}{A}, \frac{y_1}{A}\right) \exp\left[-i\,\frac{k_0(DA - 1)}{2AB}\,(x_1^2 + y_1^2)\right] \quad \textbf{(2.12-5)}$$

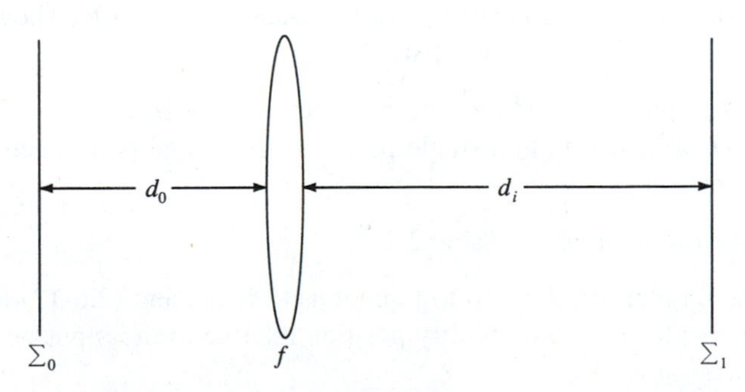

Figure 2-13 A single lens imaging system.

Using the property $AD\text{-}BC = 1$, which holds when the individual A, B, C, D matrices of the cascade possess unity determinants, we rewrite (2.12-5) as

$$\underset{B\to 0}{f_1(x_1, y_1)} = \frac{\exp(-ik_0L)}{A} \exp\left[-i\frac{k_0C}{2A}(x_1^2 + y_1^2)\right] \times f_0\left(\frac{x_1}{A}, \frac{y_1}{A}\right) \quad \text{(2.12-6)}$$

This is the main result. It shows that in the limit of $B \to 0$, $f_1(x_1, y_1)$ is magnified by A to $f_0\left(\frac{x_1}{A}, \frac{y_1}{A}\right)$ as well as multiplied by a quadratic phase factor (in x_1 and y_1). We thus associate $B = 0$ with the generalized imaging condition and A with the magnification. We leave it as a (simple) exercise to prove that in the case of imaging by a single thin lens, as shown in Figure 2.13, the imaging condition $B \to 0$ leads to the familiar, geometrical optics, imaging relation

$$\frac{1}{f} = \frac{1}{d_0} + \frac{1}{d_i} \quad \text{(2.12-7)}$$

while the magnification is $A = -\frac{d_i}{d_0}$.

Problems

2.1 Derive Equations (2.1-19) through (2.1-21).

2.2 Show that the eigenvalues λ of the equation

$$\begin{vmatrix} A & B \\ C & D \end{vmatrix} \begin{vmatrix} r_s \\ r'_s \end{vmatrix} = \lambda \begin{vmatrix} r_s \\ r'_s \end{vmatrix}$$

are $\lambda = e^{\pm i\theta}$ with $\exp(\pm i\theta)$ given by Equation (2.1-13). Note that, according to Equation (2.1-5), the foregoing matrix equation can also be written as

$$\begin{vmatrix} r_{s+1} \\ r'_{s+1} \end{vmatrix} = \lambda \begin{vmatrix} r_s \\ r'_s \end{vmatrix}$$

2.3 Derive Equation (2.3-6).

2.4 Make a plausibility argument to justify Equation (2.3-1) by showing that it holds for a plane wave incident on a lens.

2.5 Show that a lenslike medium occupying the region $0 \le z \le l$ will image a point on the axis at $z < 0$ onto a single point. (If the image point occurs at $z < l$, the image is virtual.)

2.6 Derive the ray matrices of Table 2-1.

2.7 Solve the problem leading up to Equations (2.6-11) and (2.6-12) for the case where the lens is placed in an arbitrary position relative to the input beam (that is, not at its waist).

2.8

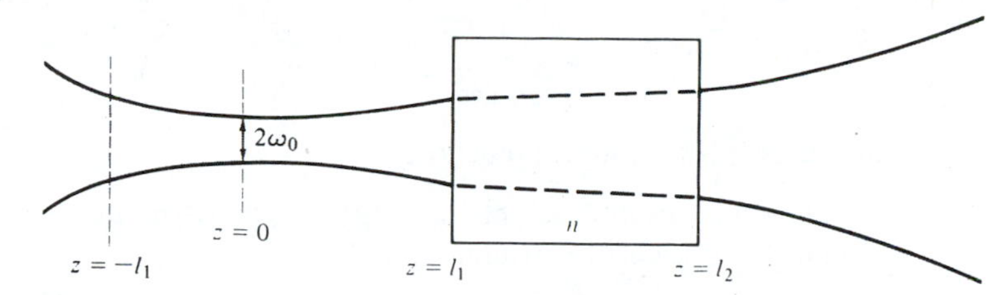

a. Assume a Gaussian beam incident normally on a solid prism with an index of refraction n as shown. What is the far-field diffraction angle of the output beam?

b. Assume that the prism is moved to the left until its input face is at $z = -l_1$. What is the new beam waist and what is its location? (Assume that the crystal is long enough that the beam waist is inside the crystal.)

2.9 A Gaussian beam with a wavelength λ is incident on a lens placed at $z = l$ as shown. Calculate the lens focal length, f, so that the output beam has a waist at the front surface of the sample crystal. Show that (given l and L) two solutions exist. Sketch the beam behavior for each of these solutions.

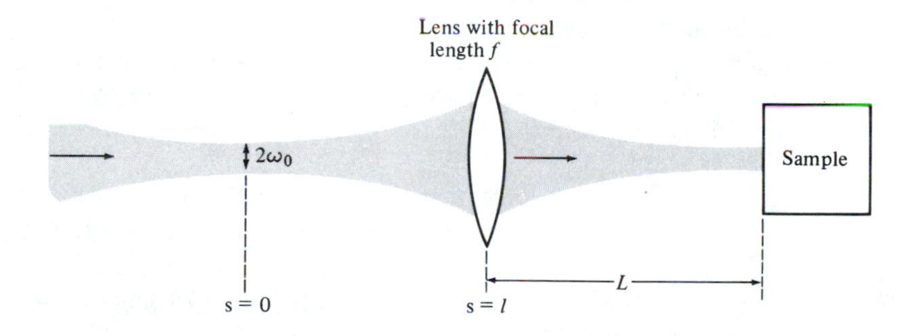

2.10 Complete all the missing steps in the derivation of Section 2.11.

2.11 Find the beam spot size and the maximum number of pulses per second that can be carried by an optical beam ($\lambda = 1\ \mu$m) propagating in a quadratic index glass fiber with $n = 1.5$, $n_2 = 5 \times 10^2$ cm^{-2}. (a) in the case of a single mode excitation $l = m = 0$; (b) in the case where all the modes with $l, m < 5$ are excited. Using dispersion data of any typical commercial glass and taking $n_2 = 5 \times 10^3$ cm^{-2}, $l_{max} = m_{max} = 30$, compare the relative contributions of modal and glass dispersion to pulse broadening.

2.12 Given a thick lens with radii of curvature R_1 and R_2 on its entrance and exit surfaces, an index of refraction n, and a thickness d,

a. Obtain the *ABCD* matrix of the lens.

b. What is its focal distance for light incident from the left?

2.13 Show that

$$\lim_{B \to 0} \sqrt{\frac{i}{2\pi B}} \exp\left(-i \frac{x^2}{2B}\right) = \delta(x)$$

where $\delta(x)$ is the Dirac δ function.

2.14 Show that in the case of imaging by a thin lens the generalized imaging condition $B = 0$ leads to Equation (2.12-7).

References

1. Pierce, J. R., *Theory and Design of Electron Beams*, 2d ed. Princeton, N.J.: Van Nostrand, 1954, Chapter 11.
2. Ramo, S., J. R. Whinnery, and T. Van Duzer, *Fields and Waves in Communication Electronics.* New York: Wiley, 1965, p. 576.
3. Yariv, A., *Quantum Electronics*, 2d ed. New York: Wiley, 1975.
4. Siegman, A. E., *An Introduction to Lasers and Masers.* New York: McGraw-Hill, 1968.
5. Kogelnik, H., and T. Li, ''Laser beams and resonators,'' *Proc. IEEE* 54:1312, 1966.
6. Herriot, D., H. Kogelnik, and R. Kompfner, ''Off-axis paths in spherical mirror interferometers,'' *Appl. Opt.* 3:523, 1964.
7. Kogelnik, H., ''On the propagation of Gaussian beams of light through lenslike media including those with a loss and gain variation,'' *Appl. Opt.* 4:1562, 1965.
8. Born, M., and E. Wolf, *Principles of Optics*, 3d ed. New York: Pergamon, 1965, p. 121.
9. Gordon, J. P., R. C. C. Leite, R. S. Moore, S. P. S. Porto, and J. R. Whinnery, ''Long-transient effects in lasers with inserted liquid samples,'' *J. Appl. Phys.* 36:3, 1965.
10. Dabby, F. W., and J. R. Whinnery, ''Thermal self-focusing of laser beams in lead glasses,'' *Appl. Phys. Lett.* 13:284, 1968.
11. Yariv, A., ''Three dimensional pictorial transmission in optical fibers,'' *Appl. Phys. Lett.* 2:88, 1976.
12. Marcuse, D., *Light Transmission Optics.* Princeton, N.J.: Van Nostrand, 1972.
13. Yariv, A., *Quantum Electronics*, 2d ed. New York: Wiley, 1975, Section 2.2.
14. Kogelnik, H., and W. Rigrod, *Proc. IRE* 50:230, 1962.
15. Kawakami, S., and J. Nishizawa, ''An optical waveguide with the optimum distribution of the refractive index with reference to waveform distortion,'' *IEEE Trans. Microwave Theory and Technique*, MTT-16, 10:814, 1968.
16. Miller, S. E., E. A. J. Marcatili, and T. Li, ''Research toward optical fiber transmission systems,'' *Proc. IEEE* 61:1703, 1973.
17. Cohen, L. G., and H. M. Presby, ''Shuttle pulse measurement of pulse spreading in a low loss graded index fiber,'' *Appl. Opt.* 14:1361, 1975.
18. Bloom, D. M., L. F. Mollenauer, Chinlon Lin, D. W. Taylor, and A. M. DelGaudio, ''Direct demonstration of distortionless picosecond-pulse propagation in kilometer-length optical fibers,'' *Opt. Lett.* 4:297, 1979.

19. Bennett, W. R., "Inversion Mechanisms in Gas Lasers," *Appl. Opt.*, Suppl. 2 (*Chemical Lasers*):3, 1965.
20. Casperson, L., and A. Yariv, "The Gaussian mode in optical resonators with a radial gain profile," *Appl. Phys. Lett.* 12:355, 1968.
21. Zachos, T. H., "Gaussian beams from GaAs junction lasers," *Appl. Phys. Lett.* 12:318, 1969.
22. Zachos, T. H., and J. E. Ripper, "Resonant modes of GaAs junction lasers," *IEEE J. of Quantum Electron. QE*-5:29, 1969.
23. H. Yonezu et al., "A GaAs–Al_xGa_{1-x} as double heterostructure planar stripe laser," *Jpn. J. Appl. Phys.* 12:1585, 1973.
24. Suematsu, Y., "Long Wavelength Optical Fiber Communication," *Proc. IEEE* 71:692, 1983.
25. Yariv, A., "Imaging of coherent fields through lenslike systems," *Opt. Lett.* 19:1607 (1994).
26. Baues, P., "Huygens' principle in inhomogeneous isotropic media," *Optoelectronics* 1:37 (1969).
27. Collins, S. A., "Lens-system diffraction integral written in terms of matrix optics," *J. Opt. Soc. Am.* 60:1168 (1970).
28. Siegman, A. E., "Lasers," University Science Books, Mill Valley, CA, 1986, pp. 779–782.

3 Propagation of Optical Beams in Fibers

3.0 INTRODUCTION

The silica glass fiber has become the most important transmission medium for long-distance, high-data-rate optical communication. It has caused what can be called, with very little exaggeration, a revolution in the art and practice of communication. This success is due mostly to the prediction [1] and realization [2] of low-loss propagation of confined optical modes in such fibers once the concentration of the absorbing impurities has been reduced to insignificance.

The technology of optical communication in fibers can be stated, in simple terms, as that of feeding optical pulses at a maximal rate into one end of a fiber and retrieving them at the other end. The length of the fiber may vary from a few meters, in the case of computer interconnect applications, to thousands of kilometers, in the case of transoceanic submarine cables. The main goal of a communication system is to receive the pulses at the output end with minimal loss of energy, minimal spread, and minimal contamination by noise. Nature, naturally, throws up many obstacles in the path of anyone trying to achieve these modest goals. Some of nature's tricks include: diffraction; group velocity dispersion, which causes pulse spreading; and a variety of nonlinear scattering mechanisms. Much of this book is devoted to the understanding of these phenomena and, using this knowledge. to devising strategies for optimal transmission. In this chapter, we will study the subject of optical guided modes in fibers. We will also study the problem of pulse spreading due to group velocity dispersion and various strategies of combatting it. The topics of detection, amplification, and noise will be taken up in Chapters 10 and 11.

3.1 WAVE EQUATIONS IN CYLINDRICAL COORDINATES

In Chapter 2 we have shown that optical waveguides with a quadratic index profile (see Equation 2.9-1a) can support guided nondiffracting modes. The effect of diffraction spreading is counterbalanced by the lensing effort of the index profile of the guide. Commercial silica-based optical fibers use a step index profile with a "high" index core and a "low" index cladding. These fibers form the backbone of most modern communication systems, and the study of their modes of propagation is the subject at hand.

Since the refractive index profiles $n(r)$ of most fibers are cylindrically symmetric, it is convenient to use the cylindrical coordinate system. The field components are E_r, E_ϕ, E_z, H_ϕ, H_r, and H_z. The wave equation (2.4-3) assumes its simple form only for the Cartesian components of the field vectors. Since the unit vectors $\mathbf{a}_r$ and $\mathbf{a}_\phi$ are not constant vectors, the wave equations involving the transverse components are very complicated. The wave equation for the z component of the field vectors, however, remains simple,

$$(\nabla^2 + k^2)\begin{Bmatrix} E_z \\ H_z \end{Bmatrix} = 0 \tag{3.1-1}$$

where $k^2 = \omega^2 n^2/c^2$ and ∇^2 is the Laplacian operator given by

$$\nabla^2 = \frac{\partial^2}{\partial r^2} + \frac{1}{r}\frac{\partial}{\partial r} + \frac{1}{r^2}\frac{\partial^2}{\partial \phi^2} + \frac{\partial^2}{\partial z^2}$$

The problems of wave propagation in a cylindrical structure are usually approached by solving for E_z and H_z first and then expressing E_r, E_ϕ, H_r, and H_ϕ in terms of E_z and H_z.

Since we are concerned with the propagation along the waveguide, we assume

$$\begin{bmatrix} \mathbf{E}(\mathbf{r}, t) \\ \mathbf{H}(\mathbf{r}, t) \end{bmatrix} = \begin{bmatrix} \mathbf{E}(r, \phi) \\ \mathbf{H}(r, \phi) \end{bmatrix} \exp[i(\omega t - \beta z)] \tag{3.1-2}$$

i.e., every component of the field vector assumes the same z- and t-dependence of $\exp[i(\omega t - \beta z)]$. Maxwell's curl equations are now written in terms of the cylindrical components and are given

$$i\omega\varepsilon E_r = i\beta H_\phi + \frac{1}{r}\frac{\partial}{\partial \phi} H_z \tag{3.1-3a}$$

$$i\omega\varepsilon E_\phi = -i\beta H_r - \frac{\partial}{\partial r} H_z \tag{3.1-3b}$$

$$i\omega\mu E_z = -\frac{1}{r}\frac{\partial}{\partial \phi} H_r + \frac{1}{r}\frac{\partial}{\partial r}(rH_\phi) \tag{3.1-3c}$$

and

$$-i\omega\mu H_r = i\beta E_\phi + \frac{1}{r}\frac{\partial}{\partial\phi}E_z \tag{3.1-4a}$$

$$-i\omega\mu H_\phi = -i\beta E_r - \frac{\partial}{\partial r}E_z \tag{3.1-4b}$$

$$-i\omega\mu H_z = -\frac{1}{r}\frac{\partial}{\partial\phi}E_r + \frac{1}{r}\frac{\partial}{\partial r}(rE_\phi) \tag{3.1-4c}$$

Using (3.1-3a), (3.1-3b), (3.1-4a), and (3.1-4b), we can solve for E_r, E_ϕ, H_r, and H_ϕ in terms of E_z and H_z. The results are

$$E_r = \frac{-i\beta}{\omega^2\mu\varepsilon - \beta^2}\left(\frac{\partial}{\partial r}E_z + \frac{\omega\mu}{\beta}\frac{\partial}{r\partial\phi}H_z\right)$$

$$E_\phi = \frac{-i\beta}{\omega^2\mu\varepsilon - \beta^2}\left(\frac{\partial}{r\partial\phi}E_z - \frac{\omega\mu}{\beta}\frac{\partial}{\partial r}H_z\right) \tag{3.1-5}$$

$$H_r = \frac{-i\beta}{\omega^2\mu\varepsilon - \beta^2}\left(\frac{\partial}{\partial r}H_z - \frac{\omega\varepsilon}{\beta}\frac{\partial}{r\partial\phi}E_z\right)$$

$$H_\phi = \frac{-i\beta}{\omega^2\mu\varepsilon - \beta^2}\left(\frac{\partial}{r\partial\phi}H_z + \frac{\omega\varepsilon}{\beta}\frac{\partial}{\partial r}E_z\right) \tag{3.1-6}$$

These relations show that it is sufficient to determine E_z and H_z in order to specify uniquely the wave solution. The remaining components can be calculated from (3.1-5) and (3.1-6).

With the assumed z-dependence of (3.1-2), the wave equation (3.1-1) becomes

$$\left[\frac{\partial^2}{\partial r^2} + \frac{1}{r}\frac{\partial}{\partial r} + \frac{1}{r^2}\frac{\partial^2}{\partial\phi^2} + (k^2 - \beta^2)\right]\begin{bmatrix}E_z\\ H_z\end{bmatrix} = 0 \tag{3.1-7}$$

This equation is separable, and the solution takes the form

$$\begin{bmatrix}E_z\\ H_z\end{bmatrix} = \psi(r)\exp(\pm il\phi) \tag{3.1-8}$$

where $l = 0, 1, 2, 3, \ldots$, so that E_z and H_z are single-valued functions of ϕ. Then (3.1-7) becomes

$$\frac{\partial^2\psi}{\partial r^2} + \frac{1}{r}\frac{\partial\psi}{\partial r} + \left(k^2 - \beta^2 - \frac{l^2}{r^2}\right)\psi = 0 \tag{3.1-9}$$

where $\psi = E_z, H_z$.

Equation (3.1-9) is the Bessel differential equation, and the solutions are called Bessel functions of order l. If $k^2 - \beta^2 > 0$, the general solution of (3.1-9) is

$$\psi(r) = c_1 J_l(hr) + c_2 Y_l(hr) \tag{3.1-10}$$

where $h^2 = k^2 - \beta^2$, c_1 and c_2 are constants, and J_l, Y_l are Bessel functions of the first and second kind, respectively, of order l. If $k^2 - \beta^2 < 0$, the general solution of (3.1-9) is

$$\psi(r) = c_1 I_l(qr) + c_2 K_l(qr) \tag{3.1-11}$$

where $q^2 = \beta^2 - k^2$, c_1 and c_2 are constants, and I_l, K_l are the modified Bessel functions of the first and second kind, respectively, of order l.

To proceed with our solution, we need the asymptotic forms of these functions for small and large arguments. Only leading terms will be given for simplicity.

For $x \ll 1$:

$$\begin{aligned}
J_l(x) &\to \frac{1}{l!}\left(\frac{x}{2}\right)^l \\
Y_0(x) &\to \frac{2}{\pi}\left(\ln\frac{x}{2} + 0.5772\ldots\right) \\
Y_l(x) &\to -\frac{(l-1)!}{\pi}\left(\frac{2}{x}\right)^l \qquad l = 1, 2, 3, \ldots \\
I_l(x) &\to \frac{1}{l!}\left(\frac{x}{2}\right)^l \\
K_0(x) &\to -\left(\ln\frac{x}{2} + 0.5772\ldots\right) \\
K_l(x) &\to \frac{(l-1)!}{2}\left(\frac{2}{x}\right)^l \qquad l = 1, 2, 3, \ldots
\end{aligned} \tag{3.1-12}$$

For $x \gg 1, l$:

$$\begin{aligned}
J_l(x) &\to \left(\frac{2}{\pi x}\right)^{1/2} \cos\left(x - \frac{l\pi}{2} - \frac{\pi}{4}\right) \\
Y_l(x) &\to \left(\frac{2}{\pi x}\right)^{1/2} \sin\left(x - \frac{l\pi}{2} - \frac{\pi}{4}\right) \\
I_l(x) &\to \left(\frac{1}{2\pi x}\right)^{1/2} e^x \\
K_l(x) &\to \left(\frac{\pi}{2x}\right)^{1/2} e^{-x}
\end{aligned} \tag{3.1-13}$$

In these formulas l is assumed to be a nonnegative integer. The transition from the small x behavior to the large x asymptotic form occurs in the region of $x \sim l$.

3.2 THE STEP-INDEX CIRCULAR WAVEGUIDE

The geometry of the step-index circular waveguide is shown in Figure 3-1. It consists of a core of refractive index n_1 and radius a, and a cladding of refractive index n_2 and radius b. The radius b of the cladding is usually chosen to be large enough so that the field of confined modes is virtually zero at $r = b$. In the calculation below we will put $b = \infty$; this is a legitimate assumption in most waveguides, as far as confined modes are concerned.

The radial dependence of the fields E_z and H_z is given by (3.1-10) or (3.1-11), depending on the sign of $k^2 - \beta^2$. For confined propagation, β must be larger than $n_2\omega/c$ (i.e., $\beta > n_2k_0 = n_2\omega/c$). This ensures that the wave is evanescent in the cladding region, $r > a$. The solution is thus given by (3.1-11) with $c_1 = 0$. This is evident from the asymptotic behavior for large r given by (3.1-13). The evanescent decay of the field also ensures that the power flow is along the direction of the z axis, i.e., no radial power flow exists. Thus the fields of a confined mode in the cladding ($r > a$) are given by

$$\begin{aligned} E_z(\mathbf{r}, t) &= CK_l(qr) \exp[i(\omega t + l\phi - \beta z)] \\ H_z(\mathbf{r}, t) &= DK_l(qr) \exp[i(\omega t + l\phi - \beta z)] \end{aligned} \qquad r > a \tag{3.2-1}$$

where C and D are two arbitrary constants, and q is given by

$$\begin{aligned} q^2 &= \beta^2 - n_2^2k_0^2 \\ k_0 &= \frac{\omega}{c} \end{aligned} \tag{3.2-2}$$

For the fields in the core, $r < a$, we must consider the behavior of the fields as $r \to 0$. According to (3.1-12), Y_l and K_l are divergent as $r \to 0$. Since the fields must remain finite at $r = 0$, the proper choice for the fields in the core ($r < a$) is (3.1-10) with $c_2 = 0$. This becomes evident only when matching, at the interface $r = a$, the tangential components of the field vectors **E** and **H** in the core with the cladding field components derived from (3.2-1); we are unable to accomplish this if the radial dependence of the core fields is given by I_l. Thus the propagation constant β must be less than n_1k_0, and the core fields are given by

$$\begin{aligned} E_z(\mathbf{r}, t) &= AJ_l(hr) \exp[i(\omega t + l\phi - \beta z)] \\ H_z(\mathbf{r}, t) &= BJ_l(hr) \exp[i(\omega t + l\phi - \beta z)] \end{aligned} \qquad r < a \tag{3.2-3}$$

where A and B are two arbitrary constants, and h is given by

$$h^2 = n_1^2k_0^2 - \beta^2 \tag{3.2-4}$$

In the field expressions (3.2-1) and (3.2-3), we have taken a "+" sign in front of $l\phi$ in the exponents. A negative sign would yield a set of independent solutions, but with the same radial dependence. Physically, l plays a role similar to the quantum number describing the z component of the orbital angular momentum of an electron in a cylindrically symmetric potential field. Thus, if the positive sign in front of $l\phi$ corresponds to a clockwise "circulation" of photons about the z axis, the negative

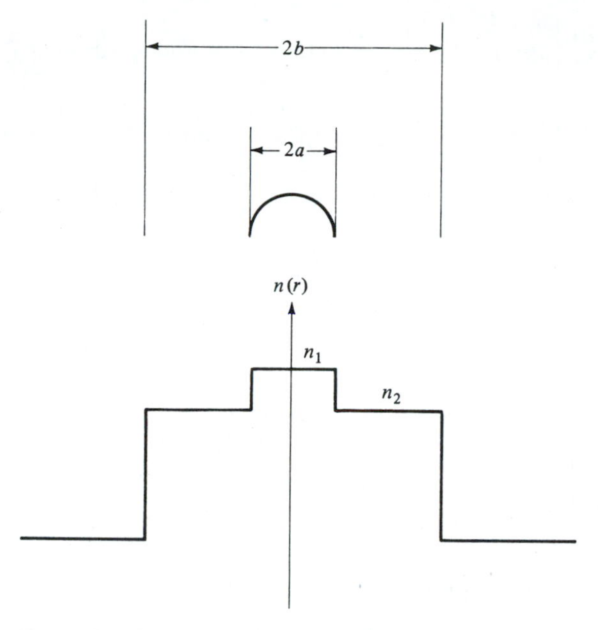

Figure 3-1 Structure and index profile of a step-index circular waveguide.

sign would correspond to a counterclockwise "circulation" of photons around the axis. Since the fiber itself does not possess any preferred sense of rotation, these two states are degenerate.

Equations (3.2-1) and (3.2-3) together require that $h^2 > 0$ and $q^2 > 0$, which translates to

$$n_1 k_0 > \beta > n_2 k_0 \tag{3.2-5}$$

which can be regarded as a necessary condition for confined modes to exist. This is identical to the condition discussed in Section 13.1 for the slab dielectric waveguide and can be expected on intuitive grounds from our discussions of total internal reflection at a dielectric interface.

Using (3.2-1) and (3.2-3) in conjunction with (3.1-5) and (3.1-6), we can calculate all the field components in both the cladding and the core regions. The result is

Core ($r < a$):

$$E_r = \frac{-i\beta}{h^2}\left[AhJ_l'(hr) + \frac{i\omega\mu l}{\beta r}BJ_l(hr)\right]\exp[i(\omega t + l\phi - \beta z)]$$

$$E_\phi = \frac{-i\beta}{h^2}\left[\frac{il}{r}AJ_l(hr) - \frac{\omega\mu}{\beta}BhJ_l'(hr)\right]\exp[i(\omega t + l\phi - \beta z)]$$

$$E_z = AJ_l(hr)\exp\left[i(\omega t + l\phi - \beta z)\right] \tag{3.2-6}$$

$$H_r = \frac{-i\beta}{h^2}\left[BhJ_l'(hr) - \frac{i\omega\varepsilon_1 l}{\beta r}AJ_l(hr)\right]\exp[i(\omega t + l\phi - \beta z)]$$

$$H_\phi = \frac{-i\beta}{h^2}\left[\frac{il}{r}BJ_l(hr) + \frac{\omega\varepsilon_1}{\beta}AhJ_l'(hr)\right]\exp[i(\omega t + l\phi - \beta z)]$$

$$H_z = BJ_l(hr)\exp[i(\omega t + l\phi - \beta z)] \qquad (3.2\text{-}7)$$

where

$$J_l'(hr) = dJ_l(hr)/d(hr), \quad \varepsilon_1 = \varepsilon_0 n_1^2$$

Cladding ($r > a$):

$$E_r = \frac{i\beta}{q^2}\left[CqK_l'(qr) + \frac{i\omega\mu l}{\beta r}DK_l(qr)\right]\exp[i(\omega t + l\phi - \beta z)]$$

$$E_\phi = \frac{i\beta}{q^2}\left[\frac{il}{r}CK_l(qr) - \frac{\omega\mu}{\beta}DqK_l'(qr)\right]\exp[i(\omega t + l\phi - \beta z)]$$

$$E_z = CK_l(qr)\exp[i(\omega t + l\phi - \beta z)] \qquad (3.2\text{-}8)$$

$$H_r = \frac{i\beta}{q^2}\left[DqK_l'(qr) - \frac{i\omega\varepsilon_2 l}{\beta r}CK_l(qr)\right]\exp[i(\omega t + l\phi - \beta z)]$$

$$H_\phi = \frac{i\beta}{q^2}\left[\frac{il}{r}DK_l(qr) + \frac{\omega\varepsilon_2}{\beta}CqK_l'(qr)\right]\exp[i(\omega t + l\phi - \beta z)]$$

$$H_z = DK_l(qr)\exp[i(\omega t + l\phi - \beta z)] \qquad (3.2\text{-}9)$$

where $K_l'(qr) = dK_l(qr)/d(qr)$, $\varepsilon_2 = \varepsilon_0 n_2^2$. These fields must satisfy the boundary conditions that E_ϕ, E_z, H_ϕ, and H_z be continuous at $r = a$. This leads to

$$AJ_l(ha) - CK_l(qa) = 0$$

$$A\left[\frac{il}{h^2 a}J_l(ha)\right] + B\left[-\frac{\omega\mu}{h\beta}J_l'(ha)\right] + C\left[\frac{il}{q^2 a}K_l(qa)\right] + D\left[-\frac{\omega\mu}{q\beta}K_l'(qa)\right] = 0$$

$$BJ_l(ha) - DK_l(qa) = 0$$

$$A\left[\frac{\omega\epsilon_1}{h\beta}J_l'(ha)\right] + B\left[\frac{il}{h^2 a}J_l(ha)\right] + C\left[\frac{\omega\varepsilon_2}{q\beta}K_l'(qa)\right] + D\left[\frac{il}{q^2 a}K_l(qa)\right] = 0 \qquad (3.2\text{-}10)$$

where the primes on J_l and K_l again refer to differentiation with respect to their arguments ha and qa, respectively. Equations (3.2-10) yield a nontrivial solution for

A, B, C, and D, provided the determinant of their coefficients vanishes. This requirement yields the following mode condition that determines the propagation constant

$$\left(\frac{J_l'(ha)}{haJ_l(ha)} + \frac{K_l'(qa)}{qaK_l(qa)}\right)\left(\frac{n_1^2 J_l'(ha)}{haJ_l(ha)} + \frac{n_2^2 K_l'(qa)}{qaK_l(qa)}\right) = l^2\left[\left(\frac{1}{qa}\right)^2 + \left(\frac{1}{ha}\right)^2\right]^2\left(\frac{\beta}{k_0}\right)^2 \tag{3.2-11}$$

Equation (3.2-11), together with (3.2-4) and (3.2-2), is a transcendental function of β for each l. The function $J_l'(x)/xJ_l(x)$ in (3.2-11) is a rapidly varying oscillatory function of $x = ha$. Therefore, (3.2-11) may be considered roughly as a quadratic equation in $J_l'(ha)/haJ_l(ha)$. For a given l and a given frequency ω, only a finite number of eigenvalues β can be found that satisfy (3.2-11) and (3.2-5). Once the eigenvalues have been found, we employ (3.2-10) to solve for the ratios B/A, C/A, and D/A that determine the six field components of the mode corresponding to each propagation constant β. These ratios are, from (3.2-10),

$$\frac{C}{A} = \frac{J_l(ha)}{K_l(qa)}$$

$$\frac{B}{A} = \frac{i\beta l}{\omega\mu}\left(\frac{1}{q^2a^2} + \frac{1}{h^2a^2}\right)\left(\frac{J_l'(ha)}{haJ_l(ha)} + \frac{K_l'(qa)}{aqK_l(qa)}\right)^{-1}$$

$$\frac{D}{A} = \frac{J_l(ha)}{K_l(qa)}\frac{B}{A} \tag{3.2-12}$$

The quantity B/A is of particular interest because it is a measure of the relative amount of E_z and H_z in a mode (i.e., $B/A = H_z/E_z$). Note that E_z and H_z are out of phase by $\pi/2$.

Mode Characteristics and Cutoff Conditions

In the treatment of slab waveguide modes in Section 13.2, we show that the solutions are easily separated into two classes, the TE and TM modes. In the circular waveguide, the solutions also separate into two classes. However, these are not in general TE or TM, each having in general nonvanishing E_z, H_z, E_ϕ, H_ϕ, E_r, and H_r components. The two classes in solutions can be obtained by noting that (3.2-11) is quadratic in $J_l'(ha)/haJ_l(ha)$, and when we solve for this quantity, we obtain two different equations corresponding to the two roots of the quadratic equation. The eigenvalues resulting from these two equations yield the two classes of solutions that are designated conventionally as the EH and HE modes.

By solving equation (3.2-11) for $J_l'(ha)/haJ_l(ha)$, we obtain

$$\frac{J_l'(ha)}{haJ_l(ha)} = -\left(\frac{n_1^2 + n_2^2}{2n_1^2}\right)\frac{K_l'}{qaK_l} \pm \left[\left(\frac{n_1^2 - n_2^2}{2n_1^2}\right)^2\left(\frac{K_l'}{qaK_l}\right)^2 + \frac{l^2}{n_1^2}\left(\frac{\beta}{k_0}\right)^2\left(\frac{1}{q^2a^2} + \frac{1}{h^2a^2}\right)^2\right]^{1/2} \tag{3.2-13}$$

where the arguments of K'_l and K_l are qa. We now use the Bessel function relations

$$J'_l(x) = -J_{l+1}(x) + \frac{l}{x} J_l(x)$$

$$J'_l(x) = J_{l-1}(x) - \frac{l}{x} J_l(x) \tag{3.2-14}$$

and (3.2-13) becomes

EH modes:

$$\frac{J_{l+1}(ha)}{haJ_l(ha)} = \frac{n_1^2 + n_2^2}{2n_1^2} \frac{K'_l(qa)}{qaK_l(qa)} + \left(\frac{l}{(ha)^2} - R\right) \tag{3.2-15a}$$

HE modes:

$$\frac{J_{l-1}(ha)}{haJ_l(ha)} = -\left(\frac{n_1^2 + n_2^2}{2n_1^2}\right) \frac{K'_l(qa)}{qaK_l(qa)} + \left(\frac{l}{(ha)^2} - R\right) \tag{3.2-15b}$$

where

$$R = \left[\left(\frac{n_1^2 - n_2^2}{2n_1^2}\right)^2 \left(\frac{K'_l(qa)}{qaK_l(qa)}\right)^2 + \left(\frac{l\beta}{n_1 k_0}\right)^2 \left(\frac{1}{q^2a^2} + \frac{1}{h^2a^2}\right)^2\right]^{1/2} \tag{3.2-16}$$

Equation (3.2-15) can be solved graphically by plotting both sides as functions of ha, letting $(qa)^2 = (n_1^2 - n_2^2)k_0^2 - (ha)^2$ on the right-hand side.

We consider first the special case when $l = 0$. At $l = 0$ we have $\partial/\partial\phi = 0$, and all the field components of the modes are radially symmetric. There are two families of solutions that correspond to (3.2-15b) and (3.2-15a) above. In the first case, the mode condition (3.2-15b) becomes

$$\frac{J_1(ha)}{haJ_0(ha)} = -\frac{K_1(qa)}{qaK_0(qa)} \quad \text{(TE)} \tag{3.2-17a}$$

where we used $K'_0(x) = -K_1(x)$. Under condition (3.2-17a), the constants A and C vanish according to (3.2-10) or (3.2-12). By substituting $A = C = 0$ and $l = 0$ in equations (3.2-6) through (3.2-9), we find that the only nonvanishing field components are H_r, H_z, and E_ϕ. These solutions are thus referred to as TE modes. If the eignevalues are β_m, $m = 1, 2, 3, \ldots$, the TE modes are designated as TE_{0m}, $m = 1, 2, 3, \ldots$, where the first subscript is $l = 0$.

In the second case, the mode condition (3.2-15a) at $l = 0$ becomes

$$\frac{J_1(ha)}{haJ_0(ha)} = -\frac{n_2^2 K_1(qa)}{qan_1^2 K_0(qa)} \quad \text{(TM)} \tag{3.2-17b}$$

where we used $K'_0(x) = -K_1(x)$ and $J_{-1}(x) = -J_1(x)$. In this case the constants B and D vanish according to (3.2-10) or (3.2-12). By substituting $B = D = 0$ and $l = 0$ in equations (3.2-6) through (3.2-9), we find that the only nonvanishing field components are E_r, E_z, and H_ϕ. These solutions are thus referred to as TM modes and are designated as TM_{0m}.

Now consider the graphical solution of (3.2-17a) and (3.2-17b). Confined modes require that q be real to achieve the exponential decay of the field in the cladding. Thus we need only consider ha in the range $0 \le ha \le V \equiv k_0 a(n_1^2 - n_2^2)^{1/2}$. The right-hand sides of (3.2-17) are always negative. Starting from $-K_1(V)/VK_0(V)$ for TE modes at $ha = 0$, the right side of (3.2-17a) is a monotonically decreasing function of ha and becomes asymptotical, according to (3.1-12)

$$-\frac{K_1(qa)}{qaK_0(qa)} \xrightarrow[ha \to V]{} \frac{2}{(V^2 - h^2a^2)\ln(V^2 - h^2a^2)} \tag{3.2-18}$$

which diverges to $-\infty$ at $ha = V$. The right side of (3.2-17b) for TM modes behaves identically except for a factor of n_2^2/n_1^2. On the left sides of (3.2-17a) and (3.2-17b), $J_1(ha)/haJ_0(ha)$ starts from 1/2 at $ha = 0$ and increases monotonically until it diverges to ∞ at $ha = 2.405$, which is the first zero of $J_0(ha)$. Beyond $ha = 2.405$, $J_1(ha)/haJ_0(ha)$ varies from $-\infty$ to $+\infty$ between the zeros of $J_0(ha)$. For large values of ha, $J_1(ha)/haJ_0(ha)$ is a function resembling $-(ha)^{-1}\tan(ha - \pi/4)$, according to Figure 3-2 which shows the two curves describing the right and left sides of (3.2-17a), respectively. The normalized frequency $V = k_0 a(n_1^2 - n_2^2)^{1/2}$ is assumed to be high enough so that two modes, marked by the circles at the intersection of the two curves, exist. The vertical asymptotes are given by the roots of $J_0(ha) = 0$. If the

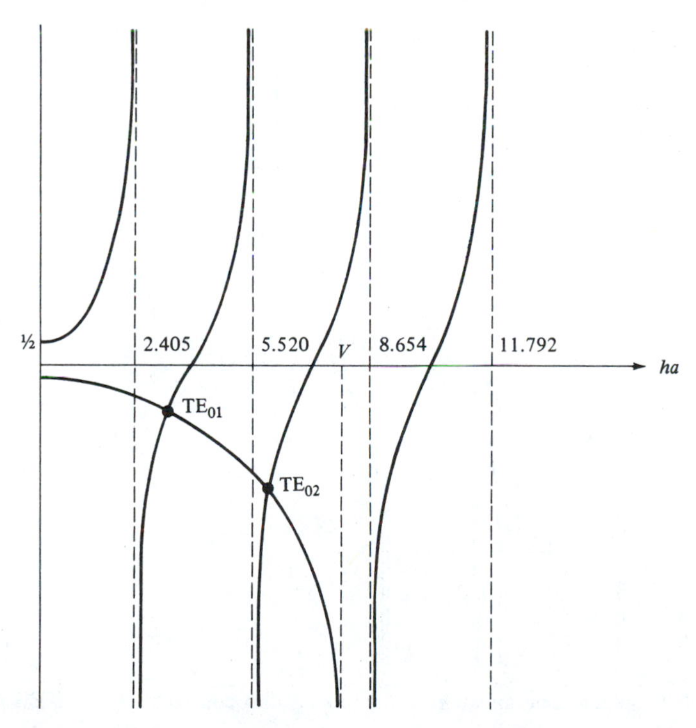

Figure 3-2 Graphical determination of the propagation constants of TE modes ($l = 0$) for a step-index waveguide.

maximum value of ha, $(ha)_{\max} = V$, is smaller than the first root of $J_0(x)$, 2.405, there can be no intersection of the two curves for real β. If V is between the first and the second zero of $J_0(x)$, there will be exactly one intersection of the two curves. Thus the cutoff value (a/λ) for TE_{0m} (or TM_{0m}) waves is given by

$$\left(\frac{a}{\lambda}\right)_{0m} = \frac{x_{0m}}{2\pi(n_1^2 - n_2^2)^{1/2}} \tag{3.2-19}$$

where x_{0m} is the mth zero of $J_0(x)$. The first three zeros are

$$x_{01} = 2.405 \qquad x_{02} = 5.520 \qquad x_{03} = 8.654$$

For higher zeros, the asymptotic formula

$$x_{0m} \simeq (m - \tfrac{1}{4})\pi$$

gives adequate accuracy (to at least three figures).

When $l \neq 0$ in equations (3.2-15), the modes are no longer TE or TM but become the EH or HE modes of the waveguide. These can still be solved graphically in a manner similar to that outlined for the $l = 0$ case. For $l = 1$, the two curves representing the two sides of the EH mode condition (3.2-15a) are shown in Figure 3-3. The normalized frequency $V = k_0 a(n_1^2 - n_2^2)^{1/2}$ is assumed to be 8, so that there are

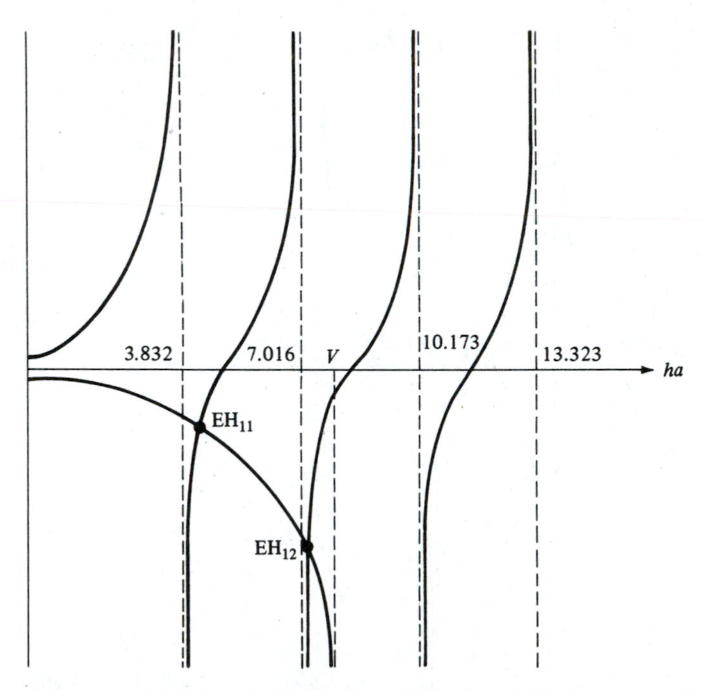

Figure 3-3 Graphical determination of the propagation constants of $l = 1$ EH modes for a step-index fiber.

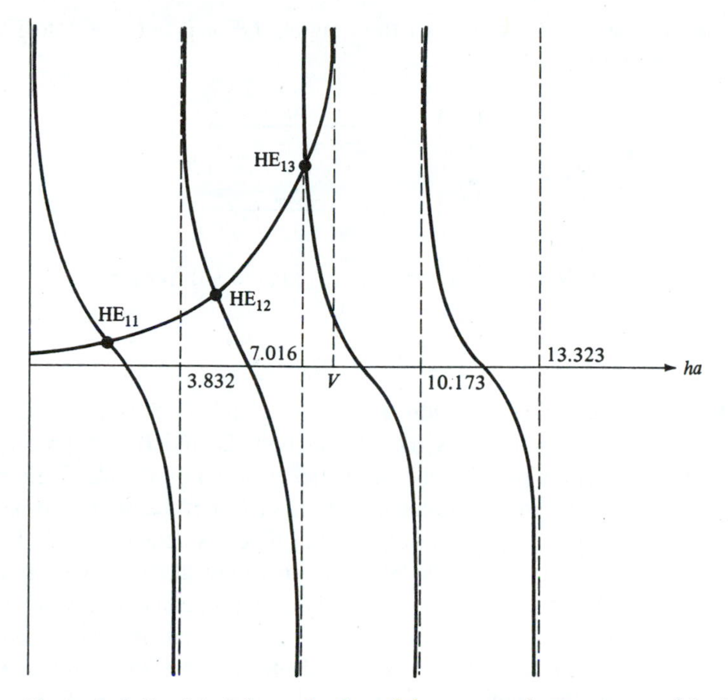

Figure 3-4 Graphical determination of the propagation constants of the $l = 1$ HE modes for a step-index dielectric waveguide.

two intersections. These are the EH_{11} and EH_{12} modes. The vertical asymptotes are given by the roots of $J_1(x) = 0$. Figure 3-4 shows those of the HE modes. At the same value of $V = 8$ there are three intersections that correspond to HE_{11}, HE_{12}, and HE_{13} modes, respectively. The vertical asymptotes are also given by the roots of $J_1(x) = 0$. Note that, as shown in Figure 3-4, the intersection for HE_{11} mode always exists regardless of the value of V. This means the HE_{11} mode does not have a cutoff. All other HE_{1m}, EH_{1m} modes have cutoff values of a/λ given by

$$\left(\frac{a}{\lambda}\right)_{1m} = \frac{x_{1m'}}{2\pi(n_1^2 - n_2^2)^{1/2}} \tag{3.2-20}$$

where $m' = m$ for EH_{1m} modes and $m' = m - 1$ for HE_{1m} modes; x_{1m} is the mth zero of $J_1(x)$, excluding the one at $x = 0$. The first three zeros are

$$x_{11} = 3.832 \qquad x_{12} = 7.016 \qquad x_{13} = 10.173$$

For higher zeros, the asymptotic formula

$$x_{1m} \simeq m\pi + \frac{\pi}{4}$$

gives adequate accuracy (to at least three figures). For $l > 1$, the cutoff values for a/λ are given by [3]

$$\left(\frac{a}{\lambda}\right)_{lm}^{\mathrm{EH}} = \frac{x_{lm}}{2\pi(n_1^2 - n_2^2)^{1/2}} \tag{3.2-21}$$

$$\left(\frac{a}{\lambda}\right)_{lm}^{\mathrm{HE}} = \frac{z_{lm}}{2\pi(n_1^2 - n_2^2)^{1/2}} \tag{3.2-22}$$

where x_{lm} is the mth zero of $J_l(x) = 0$, and z_{lm} is the mth root of

$$zJ_l(z) = (l - 1)\left(1 + \frac{n_1^2}{n_2^2}\right) J_{l-1}(z) \qquad l > 1 \tag{3.2-23}$$

If we substitute the propagation constant β for $l > 1$ into (3.2-12), we find that B/A is neither zero nor infinite. This means that both E_z and H_z are present in these modes. The designation of these hybrid modes is based on the relative contribution of E_z and H_z to a transverse component (e.g., E_r or E_ϕ) of the field at some reference point. If E_z makes the larger contribution, the mode is considered E-like and designated EH_{lm}, and so on. The mode HE_{11} can propagate at any wavelength, as noted earlier, since $(a/\lambda)_{11}^{\mathrm{HE}} = 0$. The next modes that can propagate, according to (3.2-19) are the TE_{01} and TM_{01} modes. Since x_{lm} or z_{lm} forms an increasing sequence for fixed l and increasing m, or for fixed m and increasing l, the number of allowed modes increases as the square of a/λ (see Problem 3.1).

For many applications, the important characteristic of a mode is the propagation constant β as a function of the frequency ω (or normalized frequency V). This information is often presented as the mode index of the confined mode

$$n = \frac{\beta}{k_0} \tag{3.2-24}$$

as a function of $V = k_0 a(n_1^2 - n_2^2)^{1/2}$; here $k_0 = \omega/c$. Since the phase velocity of a mode is ω/β, n is the ratio of the speed of light in vacuum to the mode phase velocity (n is also called the effective mode index). Figure 3-5 shows n for a number of the low-order modes of the step-index circular waveguide [4]. We note that at cutoff, each mode has a value of $(\beta/k_0) = n_2$. We can easily understand this by recalling that as the mode approaches cutoff, the fields extend well into the cladding layer. Thus, near cutoff the modes are poorly confined and most of the energy propagates in medium 2 and thus $n = n_2$. By similar reasoning, for frequencies far above cutoff, the mode is tightly confined to the core, and n approaches n_1.

As discussed earlier, for $V < 2.405$, only the fundamental HE_{11} mode can propagate. This is an important result, since for many applications single mode propagation is required. These applications include interferometry that calls for well-defined stationary phase fronts and optical communications by transmission of very short optical pulses. In the latter case the excitation of many modes would lead to pulse broadening, since the different modes possess different group velocities. This limits the number of pulses, i.e., bits, that can be packed into a given time slot and still be separable on the receiving end.

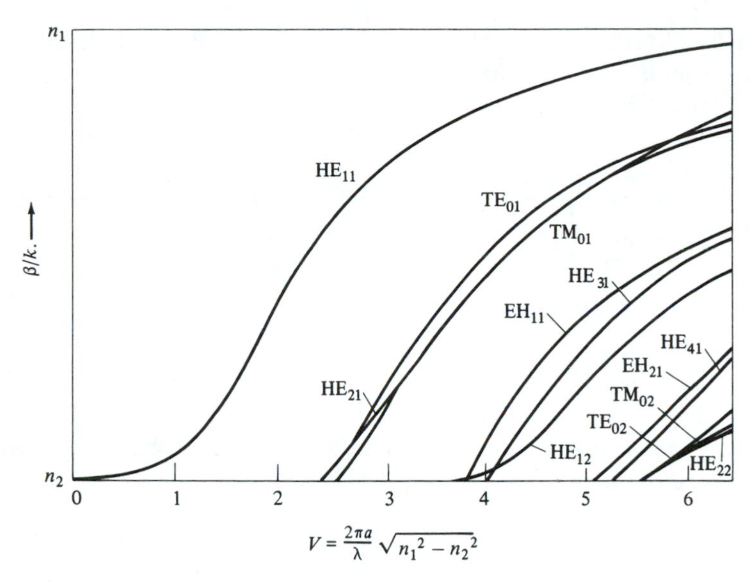

Figure 3-5 Normalized propagation constant as a function of V parameter for a few of the lowest-order modes of a step-index waveguide [4].

3.3 LINEARLY POLARIZED MODES

The mode condition (3.2-11) and the field components (3.2-6) through (3.2-9) and (3.2-12) are exact solutions of the wave equation (2.4-3) for the step-index dielectric waveguide. These exact expressions are very complicated especially for those hybrid modes (EH_{lm}, HE_{lm}) that have all six nonzero field components. A good approximation of the field components and mode condition can be obtained in most fibers whose core refractive index is only slightly higher than that of the cladding medium. Assuming that

$$n_1 - n_2 \ll 1 \tag{3.3-1}$$

the continuity condition on the tangential components of $\mathbf{H}$ at the interface between n_1 and n_2 becomes identical to that of the tangential components of the field vector $\mathbf{E}$. This leads to a tremendous simplification in matching the field components at the core-cladding interface. Thus we may use the Cartesian components of the field vectors without introducing much complexity in solving the wave equation.

This simplified solution of the linearly polarized modes for the round fiber using the assumption (3.3-1) is due to Gloge [5]. In the limit (3.3-1), all the transverse wave numbers (h, q) are much smaller compared to the propagation constant β, i.e.,

$$q, h \ll \beta \tag{3.3-2}$$

We now start by solving the wave equation for the transverse Cartesian field components E_x, E_y, H_x, and H_y. These field components also satisfy the wave equations (3.1-7) and (3.1-9). For a step-index dielectric waveguide, the general solutions are given by (3.1-10) and (3.1-11). We now look for solutions where either the x or y component of the electric field vanishes. Since E_ϕ can be expressed in terms of E_x and E_y as

$$E_\phi = -E_x \sin \phi + E_y \cos \phi \tag{3.3-3}$$

it is apparent that E_ϕ component is simply proportional to either E_x or E_y. Thus the continuity of E_ϕ becomes equivalent to the continuity of E_x or E_y in these new solutions. Take the **E** field of a y-polarized solution of the form

$$E_x = 0 \tag{3.3-4}$$

$$E_y = \begin{cases} AJ_l(hr)e^{il\phi} \exp[i(\omega t - \beta z)] & r < a \\ BK_l(qr)e^{il\phi} \exp[i(\omega t - \beta z)] & r > a \end{cases} \tag{3.3-5}$$

where A and B are constants. We assume that $E_z \ll E_y$. The magnetic field components are then given, according to (2.4-1) and (3.1-2), by

$$H_x = \frac{-i}{\omega\mu}\frac{\partial}{\partial z}E_y = \frac{-\beta}{\omega\mu}E_y$$

$$H_y \simeq 0$$

$$H_z = \frac{i}{\omega\mu}\frac{\partial}{\partial x}E_y \tag{3.3-6}$$

The longitudinal component of the electric field vector **E** is related to H_x, according to the Maxwell equation $\nabla \times \mathbf{H} = \varepsilon\, \partial\mathbf{E}/\partial t$

$$E_z = \frac{i}{\omega\varepsilon}\frac{\partial}{\partial y}H_x = \frac{-i\beta}{\omega^2\mu\varepsilon}\frac{\partial}{\partial y}E_y \tag{3.3-7}$$

where we used (3.3-6) in arriving at the last equality. We note that the field components E_x and H_y are zero in this solution. The other four field components can be expressed in terms of E_y. In order to calculate H_z and E_z, we need to carry out the differentiation with respect to x and y, respectively, according to (3.3-6) and (3.3-7). Since E_y is of the form (3.3-5), we need the relations

$$\frac{\partial}{\partial x} = \frac{\partial r}{\partial x}\frac{\partial}{\partial r} + \frac{\partial \phi}{\partial x}\frac{\partial}{\partial \phi} \tag{3.3-8}$$

and

$$\frac{\partial}{\partial y} = \frac{\partial r}{\partial y}\frac{\partial}{\partial r} + \frac{\partial \phi}{\partial y}\frac{\partial}{\partial \phi} \tag{3.3-9}$$

By using the definition of r and ϕ

$$r = (x^2 + y^2)^{1/2} \tag{3.3-10}$$

$$\phi = \tan^{-1}\left(\frac{y}{x}\right) \tag{3.3-11}$$

we obtain

$$\frac{\partial r}{\partial x} = \frac{x}{r} = \cos\phi \tag{3.3-12}$$

$$\frac{\partial r}{\partial y} = \frac{y}{r} = \sin\phi \tag{3.3-13}$$

$$\frac{\partial \phi}{\partial x} = -\frac{y}{r^2} = -\frac{1}{r}\sin\phi \tag{3.3-14}$$

and

$$\frac{\partial \phi}{\partial y} = \frac{x}{r^2} = \frac{1}{r}\cos\phi \tag{3.3-15}$$

We now substitute (3.3-5) for E_y in (3.3-6) and (3.3-7) and carry out the differentiation, using equations (3.3-8) through (3.3-15). After some laborious algebra and using the following functional relations of the Bessel function,

$$\begin{aligned} J_l'(x) &= \tfrac{1}{2}[J_{l-1}(x) - J_{l+1}(x)] \\ K_l'(x) &= -\tfrac{1}{2}[K_{l-1}(x) + K_{l+1}(x)] \end{aligned} \tag{3.3-16}$$

$$\begin{aligned} \frac{l}{x} J_l(x) &= \tfrac{1}{2}[J_{l-1}(x) + J_{l+1}(x)] \\ \frac{l}{x} K_l(x) &= -\tfrac{1}{2}[K_{l-1}(x) - K_{l+1}(x)] \end{aligned} \tag{3.3-17}$$

we obtain the following expressions for the field components.

Core ($r < a$):

$$\begin{aligned} E_x &= 0 \\ E_y &= AJ_l(hr)e^{il\phi}\exp[i(\omega t - \beta z)] \\ E_z &= \frac{h}{\beta}\frac{A}{2}[J_{l+1}(hr)e^{i(l+1)\phi} + J_{l-1}(hr)e^{i(l-1)\phi}]\exp[i(\omega t - \beta z)] \\ H_x &= -\frac{\beta}{\omega\mu}AJ_l(hr)e^{il\phi}\exp[i(\omega t - \beta z)] \\ H_y &\simeq 0 \\ H_z &= -\frac{ih}{\omega\mu}\frac{A}{2}[J_{l+1}(hr)e^{i(l+1)\phi} - J_{l-1}(hr)e^{i(l-1)\phi}]\exp[i(\omega t - \beta z)] \end{aligned} \tag{3.3-18}$$

Cladding ($r > a$):

$$\begin{aligned} E_x &= 0 \\ E_y &= BK_l(qr)e^{il\phi}\exp[i(\omega t - \beta z)] \end{aligned}$$

$$E_z = \frac{q}{\beta}\frac{B}{2}[K_{l+1}(qr)e^{i(l+1)\phi} - K_{l-1}(qr)e^{i(l-1)\phi}]\exp[i(\omega t - \beta z)]$$

$$H_x = -\frac{\beta}{\omega\mu}BK_l(qr)e^{il\phi}\exp[i(\omega t - \beta z)]$$

$$H_y \simeq 0$$

$$H_z = -\frac{iq}{\omega\mu}\frac{B}{2}[K_{l+1}(qr)e^{i(l+1)\phi} + K_{l-1}(qr)e^{i(l-1)\phi}]\exp[i(\omega t - \beta z)] \qquad \textbf{(3.3-19)}$$

In arriving at (3.3-18) and (3.3-19), we have also used $\beta = n_1k_0 \simeq n_2k_0$, since $n_2k_0 < \beta < n_1k_0$ and $n_2 \to n_1$. Note that E_y and H_x are the dominant field components because in the limit (3.3-1) $h, q \ll \beta$. In other words, the field is essentially transverse. The constant B is given by

$$B = \frac{AJ_l(ha)}{K_l(qa)} \qquad \textbf{(3.3-20)}$$

to ensure the continuity of $E_y(E_\phi \propto E_y)$ at the core boundary $r = a$. The constant A is then determined by the normalization condition.

The field solution (3.3-18) and (3.3-19) is a y-polarized wave ($E_x = 0$). For a complete field description, we also need the mode with the orthogonal polarization (i.e., an x-polarized wave). The field components E_x and E_y of this orthogonal mode are taken of the form

$$E_x = \begin{cases} AJ_l(hr)e^{il\phi}\exp[i(\omega t - \beta z)] & r < a \\ BK_l(qr)e^{il\phi}\exp[i(\omega t - \beta z)] & r > a \end{cases} \qquad \textbf{(3.3-21)}$$

$$E_y = 0 \qquad \textbf{(3.3-22)}$$

and the other field components are, according to the Maxwell equations,

$$E_z = \frac{-i}{\omega\varepsilon}\frac{\partial}{\partial x}H_y = \frac{-i\beta}{\omega^2\mu\varepsilon}\frac{\partial}{\partial x}E_x$$

$$H_x \simeq 0$$

$$H_y = \frac{i}{\omega\mu}\frac{\partial}{\partial z}E_x = \frac{\beta}{\omega\mu}E_x$$

$$H_z = \frac{-i}{\omega\mu}\frac{\partial}{\partial y}E_x \qquad \textbf{(3.3-23)}$$

where we have assumed that $E_z \ll E_x$. We note that $E_y = 0$ and $H_x \simeq 0$ in this solution. By substituting (3.3-21) for E_x in (3.3-23) and carrying out the differentiation, using equations (3.3-8) through (3.3-15), we obtain, again after some laborious algebra and using the relations (3.3-16) and (3.3-17), the following expressions for the field amplitudes:

Core ($r < a$):

$$E_x = AJ_l(hr)e^{il\phi} \exp[i(\omega t - \beta z)]$$

$$E_y = 0$$

$$E_z = i\frac{h}{\beta}\frac{A}{2}[J_{l+1}(hr)e^{i(l+1)\phi} - J_{l-1}(hr)e^{i(l-1)\phi}]\exp[i(\omega t - \beta z)]$$

$$H_x \simeq 0$$

$$H_y = \frac{\beta}{\omega\mu} AJ_l(hr)e^{il\phi} \exp[i(\omega t - \beta z)]$$

$$H_z = \frac{h}{\omega\mu}\frac{A}{2}[J_{l+1}(hr)e^{i(l+1)\phi} + J_{l-1}(hr)e^{i(l-1)\phi}]\exp[i(\omega t - \beta z)] \quad \textbf{(3.3-24)}$$

Cladding ($r > a$):

$$E_x = BK_l(qr)e^{il\phi} \exp[i(\omega t - \beta z)]$$

$$E_y = 0$$

$$E_z = i\frac{q}{\beta}\frac{B}{2}[K_{l+1}(qr)e^{i(l+1)\phi} + K_{l-1}(qr)e^{i(l-1)\phi}]\exp[i(\omega t - \beta z)]$$

$$H_x \simeq 0$$

$$H_y = \frac{\beta}{\omega\mu} BK_l(qr)e^{il\phi} \exp[i(\omega t - \beta z)]$$

$$H_z = \frac{q}{\omega\mu}\frac{B}{2}[K_{l+1}(qr)e^{i(l+1)\phi} - K_{l-1}(qr)e^{i(l-1)\phi}]\exp[i(\omega t - \beta z)] \quad \textbf{(3.3-25)}$$

In arriving at (3.3-24) and (3.3-25), we again made the assumption that $\beta \simeq n_1 k_0 \simeq n_2 k_0$ because of (3.3-1). We note that E_x and H_y are the dominant field components in this solution. Therefore, the mode is again nearly transverse and linearly polarized along the x direction. The constant B is again given by (3.3-20) to ensure the continuity of $E_x(E_\phi \propto E_x)$ at the core boundary $r = a$.

We have obtained the field expressions for two types of guided modes whose transverse fields are polarized orthogonally to each other. These field expressions are approximate solutions of Maxwell's equations, provided the tangential components of the field vectors are continuous at the dielectric interface $r = a$. The continuity of E_ϕ at $r = a$ leads to (3.3-20). The H_ϕ components are proportional to the E_ϕ components, according to the field expressions (3.3-18), (3.3-19), (3.3-24), and (3.3-25) in this approximation. Therefore the continuity of E_ϕ results in the continuity of H_ϕ.

We now consider the continuity of E_z at $r = a$. Since the continuity condition must hold for all azimuth angles ϕ, we must equate the coefficients of $\exp[i(l+1)\phi]$ and $\exp[i(l-1)\phi]$ separately. Using the field expressions (3.3-18) and (3.3-19) and (3.3-20), we obtain the following mode conditions:

$$h\frac{J_{l+1}(ha)}{J_l(ha)} = q\frac{K_{l+1}(qa)}{K_l(qa)} \tag{3.3-26}$$

and

$$h\frac{J_{l-1}(ha)}{J_l(ha)} = -q\frac{K_{l-1}(qa)}{K_l(qa)} \tag{3.3-27}$$

The same equations result from the continuity of H_z. In addition, if we use the field expressions (3.3-24) and (3.3-25) for the x-polarized mode, we will arrive at the same mode conditions (3.3-26) and (3.3-27). This means that these two transversely orthogonal modes are degenerate in the propagation constant β. The mode condition (3.3-27) is mathematically equivalent to (3.3-26) if we use the recurrence relation of the Bessel functions (3.3-17).

The mode condition (3.3-26) obtained in this approximation is much simpler than the exact expression (3.2-11). The exact mode condition (3.2-11) has twice as many solutions as the simple one (3.3-26) because (3.2-11) is quadratic in J_l'/J_l. This indicates that each solution of (3.3-26) is really twofold degenerate. In fact the propagation constants of the exact $\text{HE}_{l+1,m}$ and $\text{EH}_{l-1,m}$ modes are nearly degenerate [6]. They become exactly the same in the limit $n_1 \to n_2$. This can also be seen from the expressions of the field components E_z and H_z in (3.3-18), (3.3-19), (3.3-24), and (3.3-25). Comparison of the linearly polarized mode expressions with the exact modes (3.2-3) shows that the linearly polarized modes are actually a superposition of $\text{HE}_{l+1,m}$ and $\text{EH}_{l-1,m}$ modes [6]. Two independent linear superpositions lead to the x-polarized and y-polarized modes. The total number of modes is the same in both theories. The eigenvalues obtained from (3.3-26) are labeled as β_{lm} with $l = 0, 1, 2, 3, \ldots, m = 1, 2, 3, \ldots$, where the subscript m indicates the mth root of the transcendental equation (3.3-26). The modes are designated LP_{lm}. The lowest-order mode HE_{11} now has the propagation constant labeled β_{01} and the mode is designated as LP_{01}.

The mode conditions for those linearly polarized waves (3.3-26) or (3.3-27) can also be solved graphically. Figure 3-6 shows the normalized propagation constant as a function of the normalized frequency V. The mode cutoff corresponds to the condition $q = 0$ which, according to (3.3-27), leads to the condition

$$J_{l-1}(V) = 0 \tag{3.3-28}$$

where

$$V = k_0 a(n_1^2 - n_2^2)^{1/2} = 2\pi\frac{a}{\lambda}(n_1^2 - n_2^2)^{1/2} \tag{3.3-29}$$

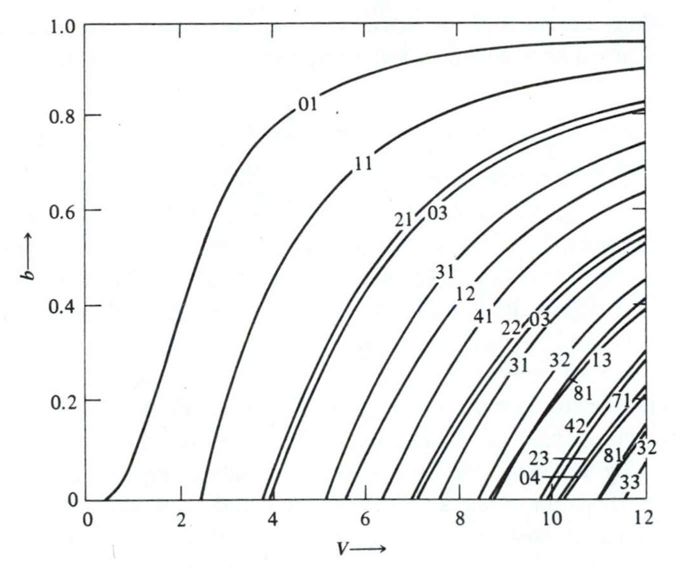

Figure 3-6 Normalized propagation constant b as function of normalized frequency V for the guided modes of the optical fiber, $b = (\beta/k_c - n_2)/(n_1 - n_2)$. (After Reference [5].)

It follows that the lowest-order mode, characterized by $l = 0$, has a cutoff given by the lowest root of the equation

$$J_{-1}(V) = -J_1(V) = 0 \tag{3.3-30}$$

Hence $V = 0$. In other words, the lowest-order mode does not have a cutoff. This is the HE_{11} mode and is now labeled LP_{01}. The next mode of the type $l = 0$, cuts off when $J_1(V)$ next equals zero, that is, when $V \simeq 3.832$. This mode is labeled LP_{02}. The cutoff values of V for some low-order LP_{lm} modes are given in Table 3-1.

Table 3-1 Cutoff Values of V for Some Low-Order LP Modes

V	$m = 1$	$m = 2$	$m = 3$	$m = 4$
$l = 0$	0	3.832	7.016	10.173
$l = 1$	2.405	5.520	8.654	11.792
$l = 2$	3.832	7.016	10.173	13.323
$l = 3$	5.136	8.417	11.620	14.796
$l = 4$	6.379	9.760	13.017	16.224

All these values are zeros of the Bessel function. For high-order modes, the cutoff value of V is given approximately according to (3.3-28) and (3.1-13)

$$V(\mathrm{LP}_{lm}) \simeq m\pi + \left(l - \frac{3}{2}\right)\frac{\pi}{2} \tag{3.3-31}$$

Figure 3-7 shows the regions in which a given mode is the highest one allowed for a given l value group, labeled in LP mode designation. Also shown in the figure are the associated HE, EH, TE, and TM mode notations that are the exact modes. Figure 3-8 shows the field distribution of the LP_{11} modes [6]. The LP_{01} mode has radially symmetric field distribution $J_0(hr)$ in the core.

One of the most important advantages of using the linearly polarized mode is that the modes are almost transversely polarized and are dominated by one transverse electric field component (E_x or E_y) and one transverse magnetic field component (H_y or H_x). The **E** vector can be chosen to be along any arbitrary radial direction with the **H** vector along a perpendicular radial direction. Once this mode is chosen, there exists another independent mode with E and H orthogonal to the first pair.

Power Flow and Power Density

We now derive expressions for the Poynting vector and the power flow in the core and cladding. The time-averaged Poynting vector along the waveguide is, acccording to (1.3-18)

$$S_z = \tfrac{1}{2}\mathrm{Re}[E_x H_y^* - E_y H_x^*) \tag{3.3-32}$$

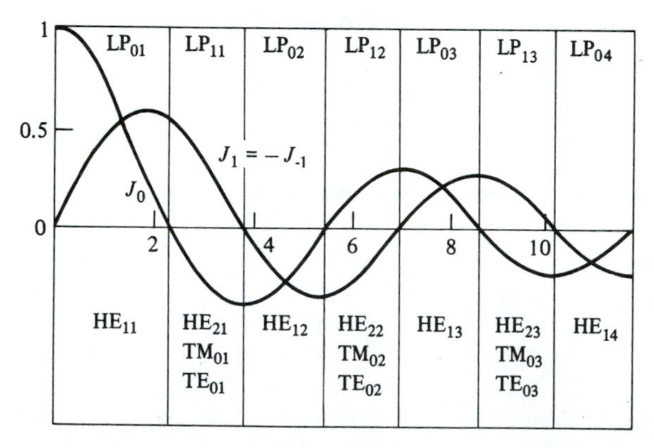

Figure 3-7 The regions of the parameter V for modes of order $l = 0, 1$.

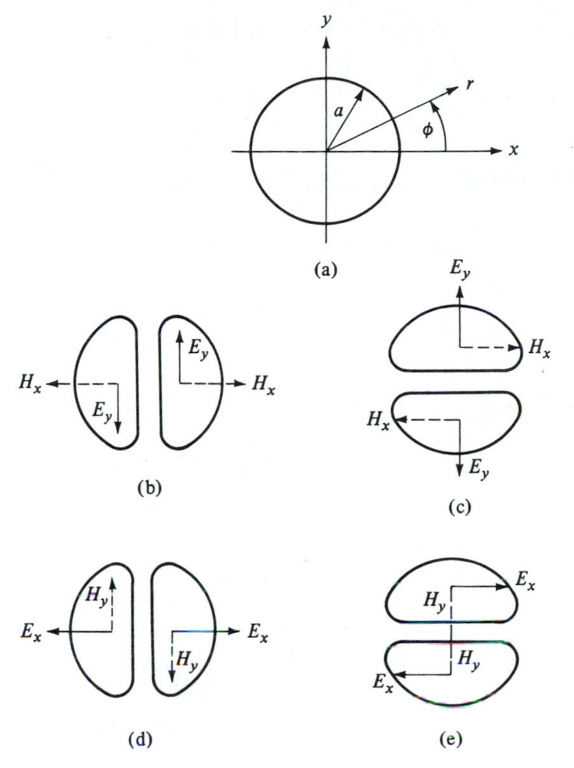

Figure 3-8 Sketch of the fiber cross section and the four possible distributions of LP_{11}.

Substituting the field components from (3.3-18) and (3.3-19) or (3.3-24) and (3.3-25) into (3.3-32), we obtain

$$S_z = \begin{cases} \dfrac{\beta}{2\omega\mu} |A|^2 J_l^2(hr) & r < a \\ \dfrac{\beta}{2\omega\mu} |B|^2 K_l^2(hr) & r > a \end{cases} \tag{3.3-33}$$

Note that the intensity distribution is cylindrically symmetric (i.e., no ϕ dependence). The amount of power that is contained in the core and the cladding is given by, respectively,

$$P_{\text{core}} = \int_0^{2\pi} \int_0^a S_z r \, dr \, d\phi \tag{3.3-34}$$

$$P_{\text{clad}} = \int_0^{2\pi} \int_a^{\infty} S_z r \, dr \, d\phi \tag{3.3-35}$$

Using the following integrals of Bessel functions [7]

$$\int_0^a rJ_l^2(hr)\,dr = \frac{a^2}{2}\,[J_l^2(ha) - J_{l-1}(ha)J_{l+1}(ha)] \tag{3.3-36a}$$

$$\int_a^\infty rK_l^2(qr)\,dr = \frac{a^2}{2}\,[-K_l^2(qa) + K_{l-1}(qa)K_{l+1}(qa)] \tag{3.3-36b}$$

the powers P_{core} and P_{clad} can be written, respectively, as

$$P_{\text{core}} = \frac{\beta}{2\omega\mu}\,\pi a^2|A|^2[J_l^2(ha) - J_{l-1}(ha)J_{l+1}(ha)] \tag{3.3-37}$$

$$P_{\text{clad}} = \frac{\beta}{2\omega\mu}\,\pi a^2|B|^2[-K_l^2(qa) + K_{l-1}(qa)K_{l+1}(qa)] \tag{3.3-38}$$

By using (3.3-20) for B and the mode conditions (3.3-26) and (3.3-27), the power P_{clad} can be written

$$P_{\text{clad}} = \frac{\beta}{2\omega\mu}\,\pi a^2|A|^2\left[-J_l^2(ha) - \left(\frac{h}{q}\right)^2 J_{l-1}(ha)J_{l+1}(ha)\right] \tag{3.3-39}$$

For those ha values that are allowed by the mode condition (3.3-26) or (3.3-27), $J_{l-1}(ha)J_{l+1}(ha)$ is always negative, so that P_{clad} is always positive. The negativeness of $J_{l-1}(ha)J_{l+1}(ha)$ can be seen from (3.3-26) and (3.3-27), since the $K_l(qa)$'s are always positive. According to (3.3-37) and (3.3-39), the total power flow is thus given by

$$P = \frac{\beta}{2\omega\mu}\,\pi a^2|A|^2\left(1 + \frac{h^2}{q^2}\right)[-J_{l-1}(ha)J_{l+1}(ha)] \tag{3.3-40}$$

The ratio of cladding power to the total power, $\Gamma_2 = (P_{\text{clad}}/P)$, which measures the fraction of mode power flowing in the cladding layer, is given, according to (3.3-39) and (3.3-40), by

$$\Gamma_2 = \frac{P_{\text{clad}}}{P} = \frac{1}{V^2}\left[(ha)^2 + \frac{(qa)^2 J_l^2(ha)}{J_{l-1}(ha)J_{l+1}(ha)}\right] \tag{3.3-41}$$

where we used $(ha)^2 + (qa)^2 = k_0^2a^2(n_1^2 - n_2^2) = V^2$. Figure 3-9 shows the ratio P_{clad}/P for several modes as a function of the normalized frequency V [5]. Note that the fundamental mode LP_{01} is best confined. Generally speaking, P_{clad}/P increases when the mode subscript lm increases.

3.4 OPTICAL PULSE PROPAGATION AND PULSE SPREADING IN FIBERS

Most of the traffic carried by optical fibers is in the form of digital pulses, each representing one bit of information. It thus follows that the narrower the pulses, the more of them can be crowded into a given (transmission) time slot, and thus more data (bits) can be transmitted during that time. As a matter of fact, modern com-

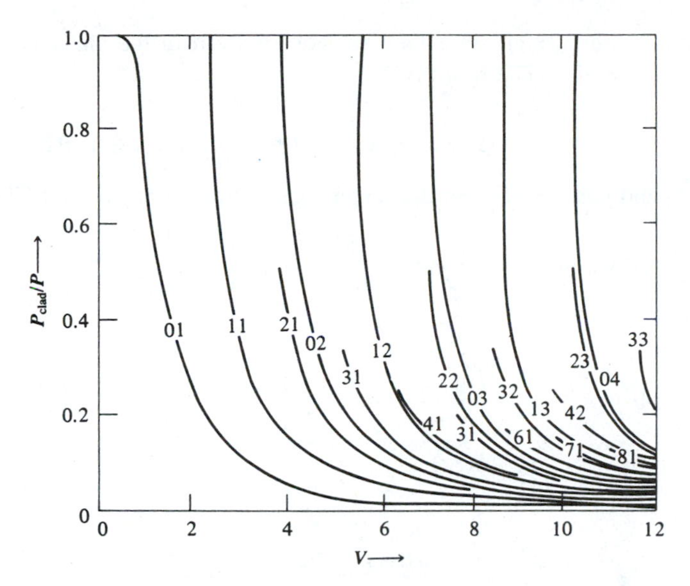

Figure 3-9 Fractional power contained in the cladding as a function of the frequency parameter V. (After Reference [5].)

munication systems are designed today with pulse widths as narrow as 3×10^{-11} s and with data rates exceeding 10^{10} bits/s. The trend to narrower pulses and higher rates continues unabated. What limits our ability to reduce the pulse width even further is the basic phenomenon of pulse lengthening due to the dependence of its group velocity on the frequency. This phenomenon is termed *group velocity dispersion*. To be specific, we assumed a single mode, usually the lowest-order fundamental mode, of the fiber excited at $z = 0$. We will take the temporal envelope as a Gaussian

$$E(x,y,0,t) = u_0(x,y)\ \mathrm{Re}[\exp(-\alpha t^2 + i\omega_0 t)] \tag{3.4-1}$$

where $u_0(x,y)$ specifies the transverse modal profile of the mode and is obtained from a solution such as that of Sections 3.1, 3.2. This solution also results in the dispersion relation for the propagation constant $\beta(\omega)$ of the mode. Since in practice the envelope function varies slowly compared to the optical oscillation, $\omega_0 \gg \alpha^{1/2}$, it is convenient to expand the pulse envelope in a Fourier transform integral

$$E(x,y,z,t) = \mathrm{Re}[u_0(x,y)\exp(i\omega_0 t)\int \tilde{f}(\Omega)\exp(i\Omega t)d\Omega] \tag{3.4-2}$$

$$\tilde{f}(\Omega) = \mathrm{FT}[\exp(-\alpha t^2)] = \left[\frac{\exp(-\Omega^2/2\alpha)}{4\pi\alpha}\right]^{1/2} \tag{3.4-3}$$

We may view (3.4-2) as an assembly of harmonic fields, each with its unique frequency $(\omega_0 + \Omega)$ and amplitude $\tilde{f}(\Omega)d\Omega$. To obtain the field at some other plane z, we need to multiply each frequency component $\tilde{f}(\Omega)d\Omega\exp[i(\omega_0 + \Omega)t]$ in (3.4-2) by its propagation delay factor, $\exp[-i\beta(\omega_0 + \Omega)z]$. We will also omit the "real part'' symbol and $u_0(x,y)$, since they are not involved in the analysis and can be restored when needed. The result is

$$E(z,t) = \int \tilde{f}(\Omega)\exp\{i[(\omega_0 + \Omega)t - \beta(\omega_0 + \Omega)z]\}d\Omega$$

We can expand $\beta(\omega_0 + \Omega)$ near the center (optical) frequency ω_0 in a Taylor series

$$\beta(\omega_0 + \Omega) = \beta(\omega_0) + \left.\frac{d\beta}{d\omega}\right|_{\omega_0} \Omega + \frac{1}{2}\left.\frac{d^2\beta}{d\omega^2}\right|_{\omega_0} \Omega + \ldots$$

and obtain

$$E(z,t) = \exp[i(\omega_0 t - \beta_0 z] \int_{-\infty}^{\infty} d\Omega \tilde{f}(\Omega)\exp\left\{i\left[\Omega t - \frac{\Omega z}{v_g} - \frac{1}{2}\frac{d}{d\omega}\left(\frac{1}{v_g}\right)\Omega^2 z\right]\right\}$$

$$\equiv \exp[i(\omega_0 t - \beta_0 z)]\mathscr{E}(z,t) \tag{3.4-4}$$

where

$$\beta_0 \equiv \beta(\omega_0), \left.\frac{d\beta}{d\omega}\right|_{\omega_0} = \frac{1}{v_g} = \frac{1}{\text{group velocity}} \tag{3.4-4a}$$

The field envelope is given by the integral (3.4-4)

$$\begin{aligned}\mathscr{E}(z,t) &= \int_{-\infty}^{\infty} d\Omega \tilde{f}(\Omega)\exp\left\{i\Omega\left[\left(t - \frac{z}{v_g}\right) - \frac{1}{2}\frac{d}{d\omega}\left(\frac{1}{v_g}\right)\Omega z\right]\right\} \\ &= \int_{-\infty}^{\infty} d\Omega \tilde{f}(\Omega)\exp\left\{i\Omega\left[\left(t - \frac{z}{v_g}\right) - a\Omega z\right]\right\}\end{aligned} \tag{3.4-5}$$

and it propagates, in the case $a = 0$, at the group velocity, v_g. The pulse spreading is caused by the group velocity dispersion characterized by the parameter

$$a \equiv \frac{1}{2}\left.\frac{d^2\beta}{d\omega^2}\right|_{\omega=\omega_0} = \frac{1}{2}\frac{d}{d\omega}\left(\frac{1}{v_g}\right) = -\frac{1}{2v_g^2}\frac{dv_g}{d\omega} \tag{3.4-5a}$$

After substituting for $\tilde{f}(\Omega)$ from (3.4-3), equation (3.4-5) becomes

$$\mathscr{E}(z,t) = \sqrt{\frac{1}{4\pi\alpha}} \int_{-\infty}^{\infty} \exp\left\{-\left[\Omega^2\left(\frac{1}{4\alpha} + iaz\right) + i\left(t - \frac{z}{v_g}\right)\Omega\right]\right\} d\Omega$$

Carrying out the integration yields

$$\mathscr{E}(z,t) = \frac{1}{\sqrt{1 + i4a\alpha z}} \exp\left(-\frac{(t - z/v_g)^2}{1/\alpha + 16a^2z^2\alpha}\right)\exp\left(i\frac{4az(t - z/v_g)^2}{1/\alpha^2 + 16a^2z^2}\right) \tag{3.4-6}$$

The pulse duration τ at z can be taken as the separation between the two times when the pulse envelope squared is smaller by a factor of 1/2 than its peak value, that is,

$$\tau(z) = \sqrt{2 \ln 2}\sqrt{\frac{1}{\alpha} + 16a^2z^2\alpha} \tag{3.4-7}$$

The initial pulse width is

$$\tau_0 = \left(\frac{2 \ln 2}{\alpha}\right)^{1/2} \tag{3.4-8}$$

The pulse width after propagating a distance L can thus be expressed as

$$\tau(L) = \tau_0 \sqrt{1 + \left(\frac{8aL \ln 2}{\tau_0^2}\right)^2} \tag{3.4-9}$$

At large distances such that $|aL| \gg \tau_0^2$ we obtain

$$\tau(L) \sim \frac{(8 \ln 2)aL}{\tau_0} \tag{3.4-10}$$

If we use the definition (Eq. 3.4-5a), of the factor a, the last expression becomes

$$\tau(L) = \frac{4 \ln 2}{v_g^2} \left|\frac{dv_g}{d\omega}\right| \frac{L}{\tau_0} \tag{3.4-11}$$

The group velocity dispersion is often characterized by $D \equiv L^{-1}(dT/d\lambda)$, where T is the pulse transmission time through length L of the fiber. This definition is related to the second-order derivative of β with respect to ω as

$$D = -\frac{2\pi c}{\lambda^2}\left(\frac{d^2\beta}{d\omega^2}\right) \tag{3.4-12}$$

and is related to the parameter a used above by

$$D = -\frac{4\pi c}{\lambda^2} a \tag{3.4-13}$$

With this new definition, the pulse-width expression (3.4-9) can be written as

$$\tau(L) = \tau_0 \sqrt{1 + \left(\frac{2 \ln 2}{\pi c} \frac{DL\lambda^2}{\tau_0^2}\right)^2} \tag{3.4-14}$$

If DL is in units of picoseconds per nanometer, λ is in units of micrometers, and τ is in units of picoseconds, the pulse width can be written as

$$\tau(L) = \tau_0 \sqrt{1 + \left(\frac{1.47DL\lambda^2}{\tau_0^2}\right)^2} \tag{3.4-15}$$

The group velocity dispersion, i.e., the dependence of v_g on ω, which according to (3.4-11) leads to pulse broadening, is due to two mechanisms:

a. Material dispersion. The ω indices of refraction $n_1(\omega)$ and $n_2(\omega)$ of the core and cladding materials depend on ω.

b. The confinement of the mode in a waveguide causes its propagation constant β and thus v_g to depend on ω. This is referred to as waveguide dispersion.

The propagation constant of a guided mode, say, LP_{lm}, is obtained as a solution of the mode condition (3.3-26) or (3.3-27). It is often expressed in terms of the mode index defined as

$$\beta_{lm} = n_{lm}k_0 = n_{lm}(n_1, n_2, \omega)\frac{\omega}{c} \tag{3.4-16}$$

where the mode index n_{lm} (often called effective index of mode lm) is considered as a function of n_1, n_2, and ω (see Figure 3-6). The velocity with which the mode energy in a light pulse travels down a waveguide is called the *group velocity* and is characterized by the expression

$$(v_g)_{lm} = \frac{d\omega}{d\beta_{lm}} = \left(\frac{d\beta_{lm}}{d\omega}\right)^{-1} \tag{3.4-17}$$

At a given frequency, different modes will thus have different group velocities. This is the modal dispersion discussed in Section 2.9. Pulse broadening and distortion in multimode waveguides where the energy is carried simultaneously by many modes is due mostly to modal dispersion, i.e., the lm-dependence of v_g.

In single-mode waveguides (e.g., LP_{01} mode $l = 0$, $m = 1$), modal dispersion is not operative, and the pulse broadening is caused by the group velocity dispersion alone. Dropping the subscript $lm = 01$, the group velocity in a single-mode step-index fiber can be written, using (3.4-16), as

$$\frac{1}{v_g} = \frac{d\beta}{d\omega} = \frac{\omega}{c}\left(\frac{\partial n}{\partial n_1}\frac{\partial n_1}{\partial \omega} + \frac{\partial n}{\partial n_2}\frac{\partial n_2}{\partial \omega} + \frac{\partial n}{\partial \omega}\right) + \frac{n}{c} \tag{3.4-18}$$

where n_1 is the refractive index of the core, n_2 is the refractive index of the cladding, and n is the mode index. The first two terms in the parentheses are the contribution from material dispersion, whereas the third term is a result of the waveguide dispersion. From the uniform dielectric perturbation theory, the change in the eigenvalue β^2 results from a uniform dielectric perturbation δn_1^2, and δn_2^2 in the core and cladding, respectively, is given by

$$\delta\beta^2 = \left(\frac{\omega}{c}\right)^2 (\Gamma_1 \delta n_1^2 + \Gamma_2 \delta n_2^2) \tag{3.4-19}$$

where Γ_1 and Γ_2 are the fraction of power flowing in the core and cladding, respectively. Using $\beta^2 = n^2(\omega/c)^2$, we obtain from (3.4-19)

$$\frac{\partial n}{\partial n_1} = \Gamma_1\left(\frac{n_1}{n}\right)$$

$$\frac{\partial n}{\partial n_2} = \Gamma_2\left(\frac{n_2}{n}\right) \tag{3.4-20}$$

The group velocity can thus be expressed as

$$\frac{1}{v_g} = \frac{d\beta}{d\omega} = \frac{\omega}{c}\left[\Gamma_1\left(\frac{n_1}{n}\right)\left(\frac{\partial n_1}{\partial \omega}\right) + \Gamma_2\left(\frac{n_2}{n}\right)\left(\frac{\partial n_2}{\partial \omega}\right) + \left(\frac{\partial n}{\partial \omega}\right)_w\right] + \frac{n}{c} \tag{3.4-21}$$

where we put a subscript w to indicate that $(\partial n/\partial\omega)_w$ is a waveguide dispersion. In a weakly guiding fiber $n_1 \simeq n_2$, we may assume that

$$\frac{\partial n_1}{\partial\omega} \simeq \frac{\partial n_2}{\partial\omega} \equiv \left(\frac{\partial n}{\partial\omega}\right)_m \tag{3.4-22}$$

where the subscript m indicates material dispersion. The group velocity (3.4-21) can thus be written

$$\frac{1}{v_g} = \frac{d\beta}{d\omega} = \frac{\omega}{c}\left[\left(\frac{\partial n}{\partial\omega}\right)_m + \left(\frac{\partial n}{\partial\omega}\right)_w\right] + \frac{n}{c} \tag{3.4-23}$$

Using $\omega = 2\pi c/\lambda$, (3.4-23) can be written in terms of λ as

$$\frac{1}{v_g} = \frac{d\beta}{d\omega} = -\frac{\lambda}{c}\left[\left(\frac{\partial n}{\partial\lambda}\right)_m + \left(\frac{\partial n}{\partial\lambda}\right)_w\right] + \frac{n}{c} \tag{3.4-24}$$

The group velocity dispersion D is thus given, according to (3.4-12) and (3.4-23), by

$$D = -\frac{\lambda}{c}\left[\left(\frac{\partial^2 n}{\partial\lambda^2}\right)_m + \left(\frac{\partial^2 n}{\partial\lambda^2}\right)_w\right] \tag{3.4-25}$$

Note that both material dispersion and waveguide dispersion contribute to the group velocity dispersion. The second-order derivatives $(\partial^2 n/\partial\lambda^2)_{m,w}$ vanish at the point of inflection on the curve $n(\lambda)$, i.e., point where $(\partial n/\partial\lambda)_{m,w}$ is minimum or maximum. For GeO_2-doped silica, $(\partial^2 n/\partial\lambda^2)_m$ passes through zero near $\lambda = 1.3\ \mu m$ [8, 9]. The waveguide dispersion $(\partial^2 n/\partial\lambda^2)_w$ vanishes at a wavelength that depends on core diameter a as well as n_1 and n_2. It is possible to tailor the zero-dispersion wavelength in single-mode fibers by balancing the (negative) material dispersion against the (positive) waveguide dispersion [10]. Thus, by choosing a core diameter a between 4 and 5 μm and relative refractive index difference of $(n_1 - n_2)/n_1 > 0.004$, the wavelength of minimum group velocity dispersion can be shifted to the 1.5- to 1.6-μm region where the loss is lowest [11–16]. Figure 3-10(a) shows the waveguide and material (chromatic) contributions to the group velocity dispersion of the "conventional" 1.3 μm fibers. Figure 3-10(b) plots the dispersion curves for dispersion-shifted fibers.

Figure 3-11 shows an input optical pulse and the output of the pulse after propagation in 2.5-km long fiber.

1. **Modal Dispersion.** Even if the material index of refraction of a fiber (or any optical guide) were independent of the frequency ω, the mere (transverse) confinement of the light to finite dimensions would cause its group (and phase) velocity to depend on ω. An example is Equation (2.9-17) for propagation in a quadratic index profile fiber. The physical reason for this dependence is that with increasing frequency the mode is more tightly confined, i.e., more of its power is contained in the core region, which results in an increase of the group velocity, since the core possesses a larger index of refraction than the cladding medium.

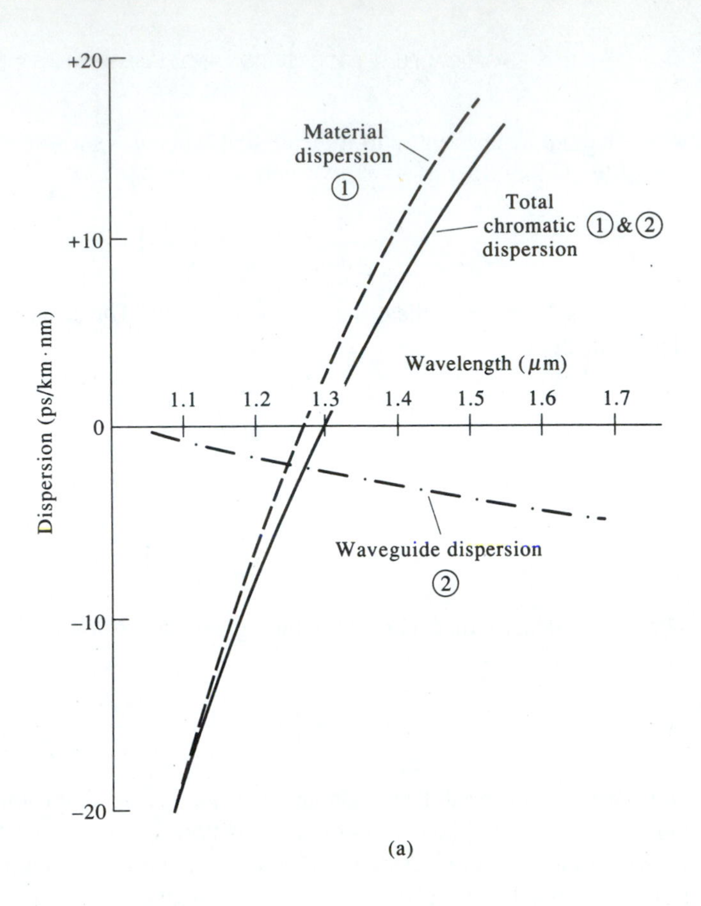

(a)

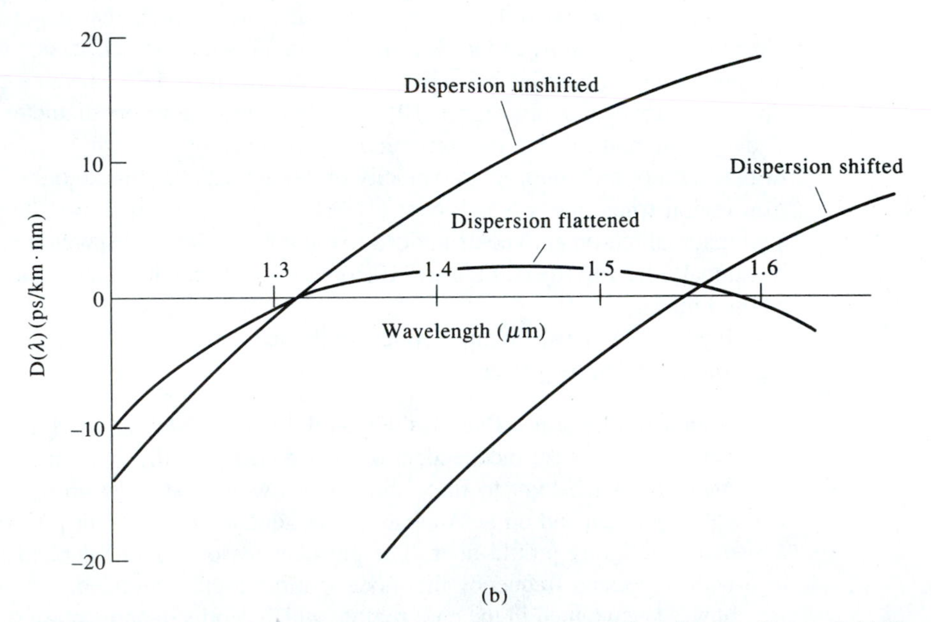

(b)

Figure 3-10 Group velocity disperion of (a) dispersion-unshifted 1.3 μm fiber and (b) dispersion- flattened and dispersion-shifted fibers. (After Reference [1].)

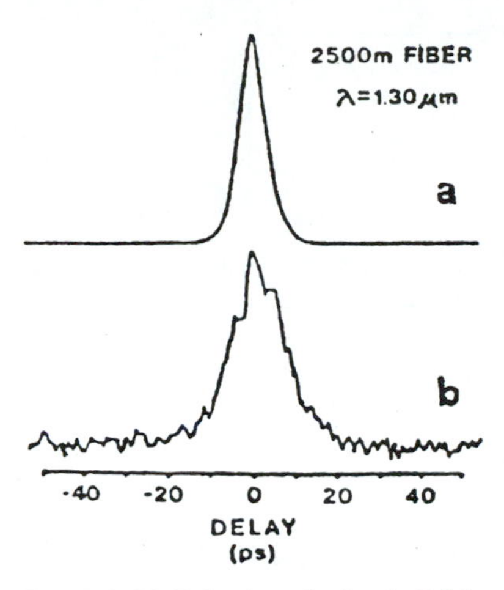

Figure 3-11 Pulse broadening in 2.5-km-long fiber resulting from chromatic (group velocity) dispersion [11].

2. **Material Dispersion** The group velocity, v_g, depends implicitly on ω because of the dependence of the indices of refraction of the waveguiding structure on ω. This will be discussed in some detail in what follows, with emphasis on the commercially dominant step index fiber.

Frequency Chirp

In addition to pulse broadening, the effect of pulse propagation in a dispersive fiber ($d^2\beta/d\omega^2 \neq 0$) is to modify the optical frequency. In the case of a Gaussian pulse after propagating a distance z, the result, according to Equation (3.4-6), is

$$E(z,t) = \frac{1}{\sqrt{1 + i4a\alpha z}} \exp\left[-\frac{(t - z/v_g)^2}{1/\alpha + 16a^2z^2\alpha}\right] \exp\left[i(\omega_0 t - \beta_0 z) + i\,\frac{4az(t - z/v_g)^2}{1/\alpha^2 + 16a^2z^2}\right],$$

$$a = \frac{1}{2}\frac{d^2\beta}{d\omega^2} = -\frac{1}{2v_g^2}\frac{dv_g}{d\omega} \tag{3.4-26}$$

The total optical phase is

$$\phi(z,t) = \omega_0 t - \beta_0 z + \frac{4az(t - zv_g)^2}{1/\alpha^2 + 16a^2z^2} \tag{3.4-27}$$

The optical frequency $\omega(z,t)$, fundamentally, is the number of (optical) oscillations per second as measured by a stationary observer at z. It is thus given by

$$\omega(z,t) = \frac{\partial}{\partial t}\phi(z,t) = \omega_0 + 8\,\frac{az}{(1/\alpha^2 + 16a^2z^2)}\,(t - z/v_g) \tag{3.4-28}$$

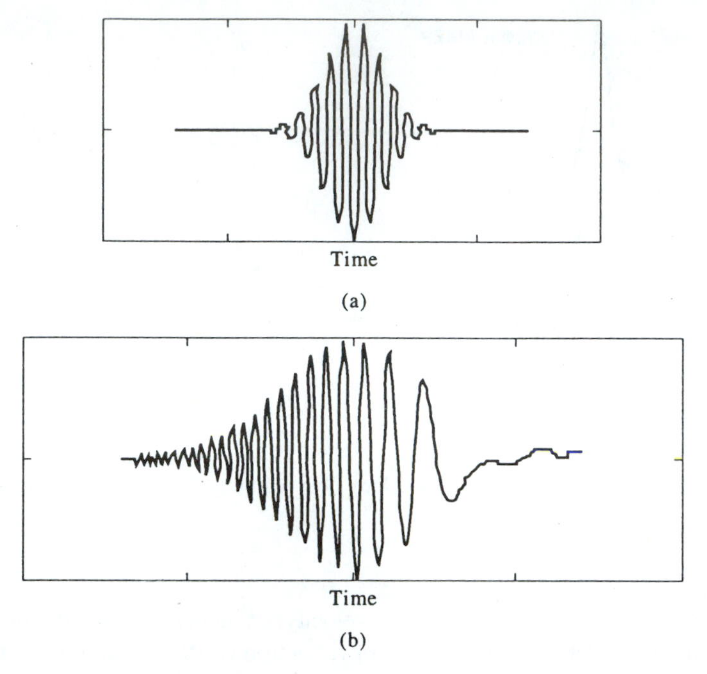

Figure 3-12 (a) An unchirped optical pulse with a Gaussian envelope. (b) The same pulse after propagation in a dispersive fiber ($a < 0$). The pulse is broader and is chirped. The vertical scale here is magnified, compared to (a), for clarity.

The frequency is not a constant, but is ''chirped.'' The linear chirp is a consequence of the group velocity dispersion causing different ''groups'' of frequencies to travel at different velocities and thus spread themselves along the pulse.

A chirped optical pulse resulting from propagation in a fiber with $\frac{d^2\beta}{d\omega^2} < 0$, $\left(\frac{dv_g}{d\omega} > 0\right)$ is shown in Figure 3-12. The blue (low-frequency) portion of the pulse spectrum travels faster and arrives at z before the ''red'' portion, so that the frequency decreases (linearly) with time.

3.5 COMPENSATION FOR GROUP VELOCITY DISPERSION

The broadening of optical pulses due to group velocity dispersion, discussed in Section 3.4, is a limiting factor in high bit rate communication. This is due to the fact that the broadening of the optical pulses makes the process of reconstructing the ''ones'' and ''zeros,'' of the data are in a digital format, on detection less certain, leading to errors. A number of different techniques are used to renarrow the optical pulses after propagation in a fiber, and some of the more important ones will be considered in what follows.

We start with Equation (3.4-5) for the pulse envelope after propagating a distance z in a fiber. The result, after a slight reshuffling of the exponent, is

$$\mathscr{E}(z,t) = \int \left[\tilde{f}(\Omega) \exp\left(-iaz\Omega^2 - i\frac{z\Omega}{v_g}\right)\right] \exp(i\Omega t)d\Omega \qquad (3.5\text{-}1)$$

where $\tilde{f}(\Omega)$ is the Fourier transform of the input envelope.

Recalling the Fourier integral theorem

$$g(t) = \int_{-\infty}^{\infty} \tilde{g}(\Omega)\exp(i\Omega t)d\Omega \qquad (3.5\text{-}2)$$

where $g(t)$ is an arbitrary function, we identify the expression within the square brackets of Equation (3.5-1) as the Fourier transform of the envelope $\mathscr{E}(z,t)$.

$$\text{FT}\ \{\mathscr{E}(z,t)\} = \tilde{f}(\Omega) \exp\left(-iaz\Omega^2 - i\frac{z}{v_g}\Omega\right)$$

or

$$\text{FT}\left\{\mathscr{E}\left(z,t + \frac{z}{v_g}\right)\right\} = \tilde{f}(\Omega)\exp(-iaz\Omega^2) \qquad (3.5\text{-}3)$$

Since our main interest here is in pulse shape and not in delay, we will, in the following, replace $(t + z/v_g)$ by t. We will consequently take the *transfer function* of a fiber of length L as

$$\text{Fiber transfer function} = \exp[-iaL\Omega^2] \qquad (3.5\text{-}4)$$

The delay can be reintroduced into the final result by replacing t by $(t - z/v_g)$. To obtain $\mathscr{E}(z,t)$, we need to take the inverse Fourier transform of the right side of Equation (3.5-3). This is accomplished most easily by using the convolution theorem (1.6-2) to express $\mathscr{E}(z,t)$ as

$$\mathscr{E}(z,t) = \frac{1}{\sqrt{i4\pi az}} \int_{-\infty}^{\infty} f(t')\exp\left[+\frac{i}{4az}(t - t')^2\right] dt' \qquad (3.5\text{-}4a)$$

We thus identify the function

$$\tau(t) = \frac{1}{\sqrt{i4\pi az}} \exp\left(+\frac{i}{4az}t^2\right) \qquad (3.5\text{-}4b)$$

as the envelope impulse response of a fiber of length z and dispersion parameter $a \equiv \beta''/2$.

To illustrate the power of our formalism, we will put it to work in addressing the problem of the spreading of optical pulses with distance in their propagation in an optical fiber. We will find it advantageous to operate in the Ω domain and use (3.5-4) as our point of departure. We will consider two different schemes for recovering the pulses at the receiving end.

Compensation for Pulse Broadening by Fibers with Opposite Dispersion

The first method is illustrated by Figure 3-13. An optical pulse with an input envelope $f_1(t)$ broadens in propagation through a fiber emerging as $f_2(t)$. It then enters a second fiber exiting it as $f_3(t)$. The envelope Fourier transform at the various stages is obtained by using (3.5-4).

1. $\tilde{f}_1(\Omega)$

2. $\tilde{f}_2(\Omega) = \tilde{f}_1(\Omega)\exp(-ia_1L_1\Omega^2)$ **(3.5-5)**

3. $\tilde{f}_3(\Omega) = \tilde{f}_2(\Omega)\exp(-ia_2L_2\Omega^2)$
$= \tilde{f}_1(\Omega)\exp[-i(a_1L_1 + a_2L_2)\Omega^2]$

The second fiber is chosen so that its group velocity dispersion has an opposite sign to that of fiber 1, so that the condition $a_1L_1 = -a_2L_2$ is satisfied. When this happens, $\tilde{f}_3(\Omega) = \tilde{f}_1(\Omega)$ and the *output pulse is identical in form and width to that of the input*. In physical terms, frequencies that travel faster in the first fiber are slowed down in the second, and vice versa. The powerful and concise use of the transfer function approach should be noted. It becomes even more important in the method of compensation, which will be discussed next.

Compensation for Pulse Broadening by Phase Conjugation (27, 28, 29)

Another method for pulse narrowing following group velocity dispersion in a fiber involves phase conjugation. The topic of phase conjugation is the concern of Chapter 17. For the purpose of this discussion, we need merely accept the operational characteristics of a phase conjugator, as illustrated in Figure 3-14. In the case of an optical pulse, it performs complex conjugation on the (complex) individual frequency amplitudes making up the pulse accompanied by *spectral inversion* about the center optical frequency ω_0. The central frequency ω_0 about which the spectral inversion takes place is that of a strong optical field $E_p \exp(i\omega_0 t)$, which is used to

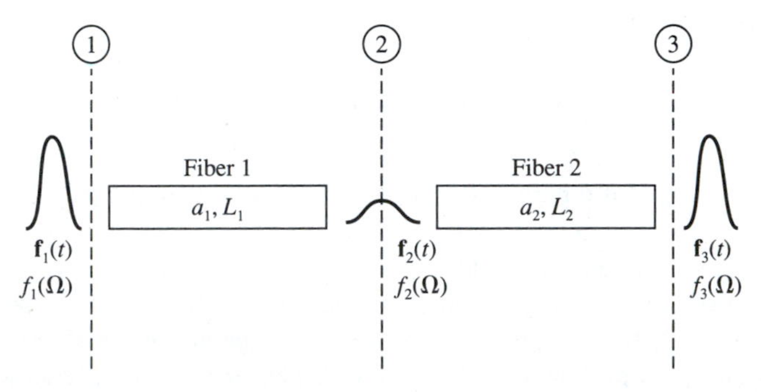

Figure 3-13 An optical pulse envelope evolving in propagation through two fibers.

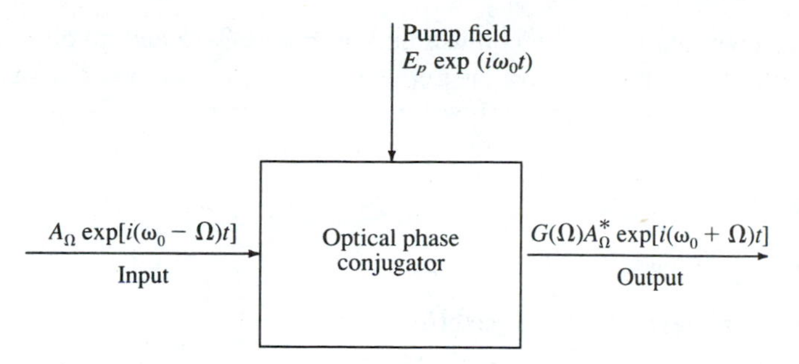

Figure 3-14 The ideal optical phase conjugator flips the frequency of each (monochromatic) single-frequency input while replacing the amplitude by its phase conjugate. A constant, G, ideally independent of Ω, accounts for possible gain or loss. The optical property of the material employed by the conjugator is its third-order nonlinear optical susceptibility $\chi^{(3)}$. (See Chapter 17.)

"pump" the conjugation. In terms of our formalism, if an incoming optical field has the form

$$\text{Input to conjugator} = f(t)\exp(i\omega_0 t) = \int \tilde{f}(\Omega)\exp[i(\omega_0 - \Omega)t]d\Omega$$

then the output field is given by

$$\text{Output from conjugator} \propto \int \tilde{f}^*(\Omega)\exp[i(\omega_0 + \Omega)t]d\Omega \qquad (3.5\text{-}6)$$

The conjugator thus replaces $\tilde{f}$ by its complex conjugate and *inverts* all the *optical* frequencies about ω_0. This is illustrated in Figure 3-14.

The student may be curious as to how one may perform optical spectral inversion. The process, which is treated in detail in Section 17.2, involves essentially a third-order optical multiplication

$$E_{\text{out}} \propto E^3_{\text{in}}$$

If the input field, E_{in} at some point in the conjugator is the sum of three fields entering the conjugator from different directions

$$E_{\text{in}} = E_{\text{signal}} \exp[i(\omega_0 - \Omega)t] + E_1 \exp(i\omega_0 t) + E_p \exp(i\omega_0 t) + \text{c.c.}$$

then simple substitution shows that there exists an output field component

$$E_{\text{out}} \propto E_1 E_2 E^*_{\text{signal}} \exp\{i[\omega_0 + \omega_0 - (\omega_0 - \Omega)]t\} + \text{c.c.}$$
$$= E_1 E_p E^*_{\text{signal}} \exp[i(\omega_0 + \Omega)t] + \text{c.c.}$$

We note that the input wave at $(\omega_0 - \Omega)$ gives rise to a proportional and complex conjugate output at $(\omega_0 + \Omega)$. In a multifrequency input, the spectrum is "reflected" about ω_0.

The compensation scheme is illustrated by Figure 3-15 and involves two fibers and an interposing phase conjugator. Let us follow the evolution of an optical pulse envelope through the system. Its Fourier transform at each of the four numbered planes is

$$
\begin{aligned}
&1.\quad \tilde{f}(\Omega) \\
&2.\quad \tilde{f}_2(\Omega) = \tilde{f}_1(\Omega)\exp(-ia_1L_1\Omega^2) \\
&3.\quad \tilde{f}_3(\Omega) = \tilde{f}_2^*(-\Omega) = \tilde{f}_1^*(-\Omega)\exp(ia_1L_1\Omega^2) \\
&4.\quad \tilde{f}_4(\Omega) = \tilde{f}_3(\Omega)\exp(-ia_2L_2\Omega^2) \\
&\qquad\quad = \tilde{f}_1^*(-\Omega)\exp[i(a_1L_1 - a_2L_2)\Omega^2]
\end{aligned}
\tag{3.5-6}
$$

It follows that if we choose the two fibers such that

$$a_1L_1 = a_2L_2 \tag{3.5-7}$$

then

$$\tilde{f}_4(\Omega) = \tilde{f}_1^*(-\Omega) \tag{3.5-8}$$

returning to the time domain at the output of the system

$$f_4(t) = \int \tilde{f}_4(\Omega)e^{i\Omega t}d\Omega = \int f_1^*(-\Omega)e^{i\Omega t}d\Omega = f_1^*(t) \tag{3.5-9}$$

Had we included the delay factor as in (3.5-3), then instead of (3.5-9) we would have obtained $f_4(t) = f_1^*(t - L_1/v_{g1} - L_2/v_{g2})$; $v_{gi} = d\omega/d\beta_i$ is the group velocity in fiber i. The (squared magnitude) output envelope is thus identical to that of the input. The student may appreciate the power of the transfer function approach used above by comparing the length of our derivation with that of the original 1978 proposal [27].

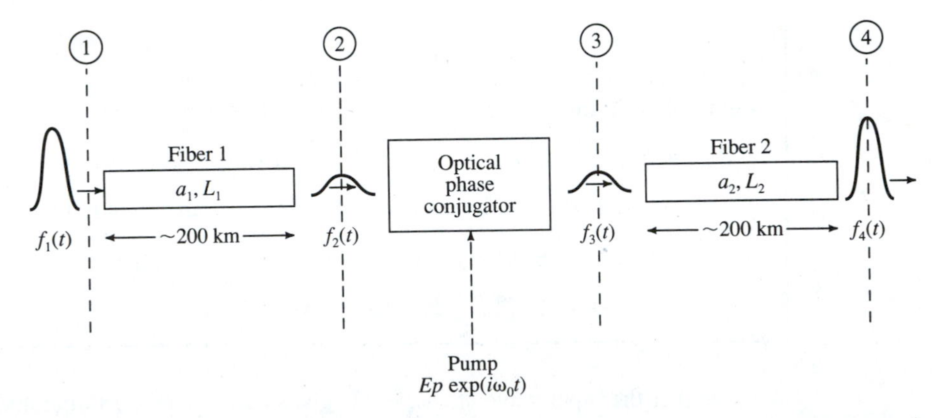

Figure 3-15 Compensation of pulse spreading in fiber links by optical phase conjugation [27, 28, 29].

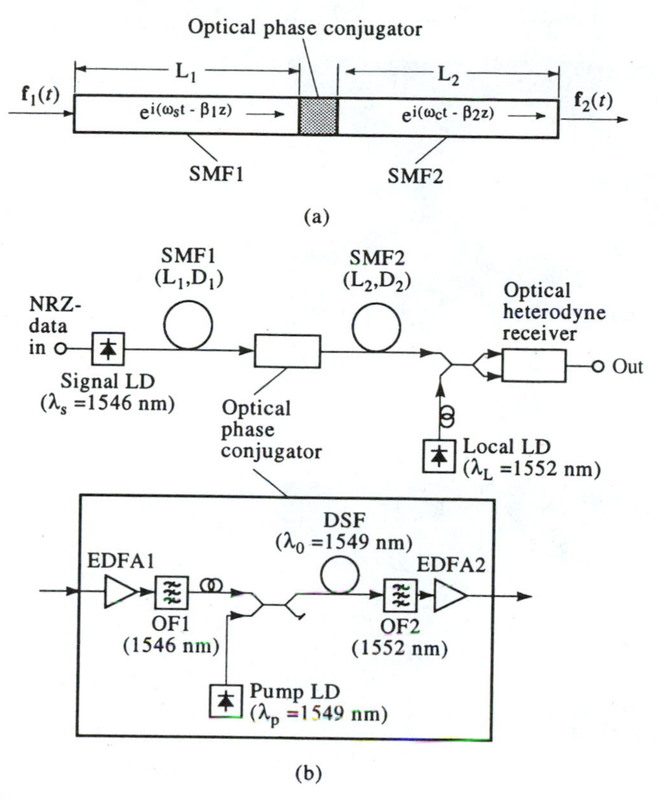

Figure 3-16 (a) Compensation for chromatic dispersion in a transmission line using optical phase conjugation. (b) The experimental setup. The enlarged portion shows the configuration of an optical phase conjugator. OF1 and OF2 are optical filters. (After Reference [28].)

Figure 3-16 shows an experimental setup used to compensate for dispersive pulse spread by phase conjugation [28]. The ''eye'' diagrams of Figure 3-17 demonstrate in (b) the pulse broadening and in (c) its renarrowing as evidenced by the closing in (b) and the opening in (c) of the ''eye'' to a value comparable to the initial pattern in (a).

> The ''eye'' diagram is a method used widely by the transmission and system engineer to depict the degradation of digital binary pulses due to spreading and noise. The stream of received pulses is displayed on a storage oscilloscope whose horizontal scan is triggered in synchronism with the bit rate. The storage oscilloscope thus records multiple sequences of rising, falling, and zero pulses adding up to an ''eye'' diagram.

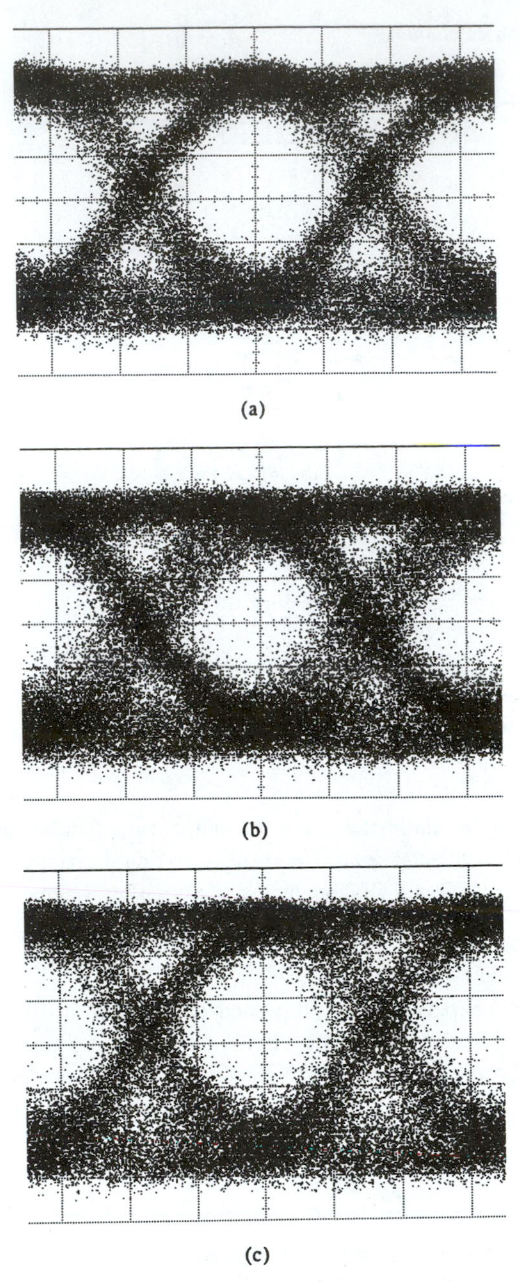

Figure 3-17 (a) An "eye" diagram of the pulse train at input to fiber (b) after 195-km transmission. (c) After 95-km transmission followed by optical phase conjugation and 101-km transmission. [28]

3.6 ANALOGY OF SPATIAL DIFFRACTION AND TEMPORAL DISPERSION

In Chapter 1, we derived the Fresnel-Kirchhoff integral [see Equation (1.6-17)]

$$u_2(x,y,z) = \frac{i}{\lambda z}\exp(-ikz)\iint u_1(x',y')\exp\left\{-\frac{ik}{2z}[(x-x')^2+(y-y')^2]\right\}dx'dy'$$

to describe the propagation of a monochromatic optical beam with a transverse profile $u_1(x,y)$ from an initial plane (taken without loss of generality as $z = 0$) to plane z. This last equation is *identical* in form to that of Equation (3.5-4a) describing the temporal evolution of an optical pulse in a dispersive ($\beta'' \equiv d^2\beta/d\omega^2 \neq 0$) fiber. This observation enables us to transfer directly many of the tools and concepts dealing with diffraction and refraction to the problem of temporal propagation.

The analogy is summarized in Figure 3-18, where, for simplicity, we consider diffraction in two dimensions (x,z). The comparison of the above equation to (3.5-4a) shows that we can apply results from spatial diffraction to temporal dispersion, and vice versa, by using the following transformations:

Diffraction		Dispersion
K	$\leftrightarrow$	Ω
x	$\leftrightarrow$	t
z	$\leftrightarrow$	z
k^{-1}	$\leftrightarrow$	$-\beta''$
$\dfrac{k}{f}$	$\leftrightarrow$	$\dfrac{\omega_0}{f_t}$

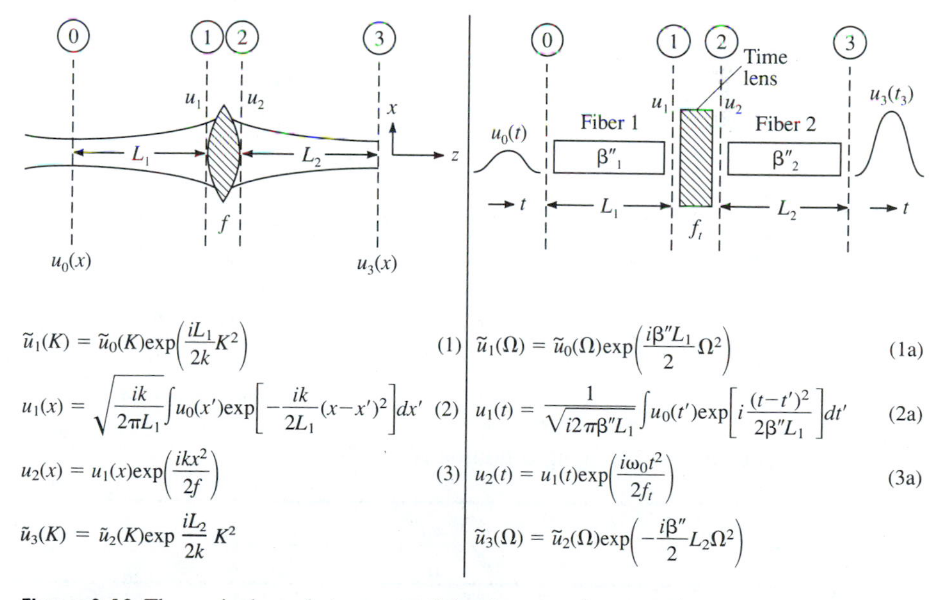

Figure 3-18 The equivalence between spatial and temporal propagation.

Before proceeding, we need to specify what we mean by a "time lens." We will define a time lens as an element that multiplies an incoming optical pulse, $u_{\text{in}}(t)\exp(i\omega_0 t)$, by a quadratic phase factor, $\exp\left(i\,\dfrac{\omega_0 t^2}{2f_t}\right)$. In terms of the pulse envelope $u_{\text{in}}(t)$, the time lens yields an output envelope

$$u_{\text{out}}(t) = u_{\text{in}}(t)\exp(i\omega_0 t^2/2f_t) \tag{3.6-1}$$

This is in exact analogy to the spatial factor $\exp[ik(x^2 + y^2)/2f]$ by which a conventional lens multiplies an incident optical beam. The constant f_t has the dimensions of time and is the focal time of the time lens. We will describe methods for realizing time lenses further below.

As an example of the use of this equivalence, we consider the case of temporal imaging. Referring to Figure 3-18, an "input" pulse envelope a distance L_1 to the left of the time lens is given by $u_0(t)$. Our task is to find the distance L_2 to the right of the lens, where the pulse regains its initial shape (this is the image plane), and find the magnification of the envelope.

Applying Equations (1a), (2a), and (3a) of Figure 3-18, we can relate $u_3(t)$ to $u_0(t)$.

$$u_1(t_1) = \frac{1}{\sqrt{i2\pi D_1}}\int u_0(t)\exp\left[i\,\frac{(t-t_1)^2}{2D_1}\right]dt \tag{3.6-2}$$

$$(D_1 \equiv \beta_1'' L_1)$$

$$u_2(t_1) = u_1(t_1)\exp\left(\frac{i\omega_0 t_1^2}{2f_t}\right) \tag{3.6-3}$$

$$u_3(t_3) = \frac{1}{\sqrt{i2\pi D_2}}\int u_2(t_1)\exp\left[i\,\frac{(t_3-t_1)^2}{2D_2}\right]dt_1 \tag{3.6-4}$$

$$(D_2 \equiv \beta_2'' L_2)$$

$$\begin{aligned} u_3(t_3) &= \frac{-i}{2\pi\sqrt{D_1 D_2}}\iint u_0(t)\exp\left\{i\left[\frac{\omega_0 t_1^2}{2f_t} + \frac{(t_3-t_1)^2}{2D_2} + \frac{(t-t_1)^2}{2D_1}\right]\right\}dt_1\,dt \\ &= \frac{-i}{2\pi\sqrt{D_1 D_2}}\iint u_0(t)\exp\left\{i\left[\frac{t_1^2}{2}\left(\frac{\omega_0}{f_t} + \frac{1}{D_1} + \frac{1}{D_2}\right) - t_1\left(\frac{t_3}{D_2} + \frac{t}{D_1}\right) + \frac{t^2}{2D_1} + \frac{t_3^2}{2D_2}\right]\right\}dt_1\,dt \end{aligned} \tag{3.6-5}$$

The temporal imaging condition is

$$\boxed{\frac{\omega_0}{f_t} + \frac{1}{D_1} + \frac{1}{D_2} = 0} \tag{3.6-6}$$

With this condition fulfilled, the integration in (3.6-5) can be carried out by using the relation

$$\int_{-\infty}^{\infty} \exp\left[-it_1\left(\frac{t_3}{D_3} + \frac{t}{D_1}\right)\right]dt_1 = 2\pi D_1 \delta\left(t + \frac{D_1}{D_2} t_3\right) \tag{3.6-7}$$

resulting in

$$u_3(t) = -i\sqrt{\frac{D_1}{D_2}} \exp\left[i\left(\frac{1}{2D_2} + \frac{D_1}{2D_2^2}\right)t^2\right] u_0\left(-\frac{D_1}{D_2} t\right) \tag{3.6-8}$$

If we define the temporal magnification factor as

$$M \equiv \frac{D_2}{D_1} = \frac{\beta_2'' L_2}{\beta_1'' L_1} \tag{3.6-9}$$

and make use of (3.6-6), we can rewrite (3.6-8) as

$$u_3(t) = -i\sqrt{\frac{1}{M}} \exp\left(i\frac{\omega_0 t^2}{2Mf_t}\right) u_0\left(-\frac{t}{M}\right) \tag{3.6-10}$$

At the image plane ③, we thus recover an inverted replica $u_0(-t/M)$ of the input pulse $u_0(t)$. Note that if $M < 1$, the output pulse is narrower than the input pulse. An accompanying frequency chirp term, $\exp[i(\omega_0 t^2/2Mf_t)]$ is present. A similar factor exists in the case of spatial imaging Equation (2.12-6).

A related discussion of time lenses and their application to pulse narrowing is given in Section 6.8. A description of a nonlinear optical technique for realizing a time lens is given in the boxed material in Section 6.8. The use of electrooptic phase modulation to achieve time lensing is discussed in Reference [24].

3.7 ATTENUATION IN SILICA FIBERS

Probably the single most important factor responsible for the emergence of the silica glass optical fiber as the premium information transmission medium is the low optical propagation losses in such fibers. Figure 3-19 shows the measured losses as a function of wavelength of a high-quality, germania-doped single-mode fiber. The loss peak around 1.4 μm is due to residual OH contamination of the glass. A low value of loss ~0.2 dB/km obtains near $\lambda = 1.55$ μm. Consequently, this region of the spectrum is now favored for long-distance optical communication. Recent experiments have taken advantage of the small pulse spreading near the zero group velocity dispersion wavelength and the low losses to demonstrate high-data-rate transmission (data rate exceeding 400 Mb/s) over a propagation path exceeding 100 km [20, 21] at $\lambda \sim 1.55$ μm.

For a more detailed discussion of propagation effects in optical fibers, the student can consult Reference [22].

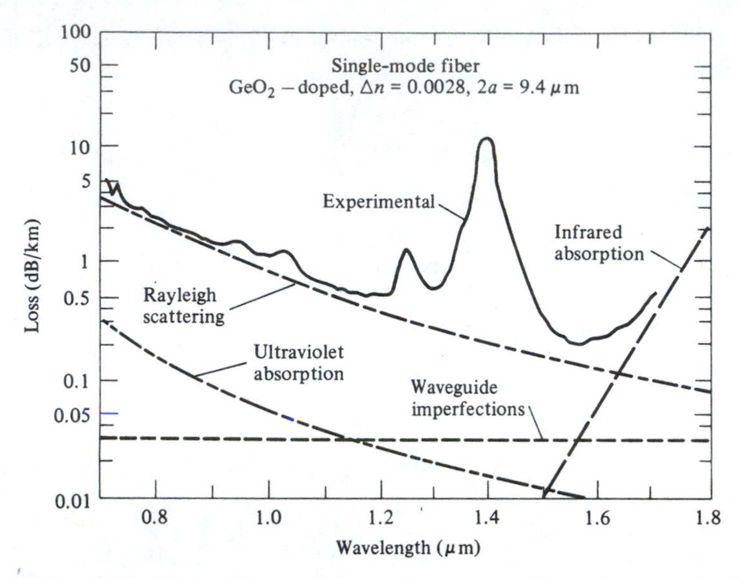

Figure 3-19 Observed loss spectrum of a germanosilicate single-mode fiber. Estimated loss spectra for various intrinsic materials effects and waveguide imperfections are also shown. (From Reference [20].)

Problems

3.1 The number of confined modes that can be supported by a circular dielectric waveguide depends on the refractive-index profile and the wavelength.

a. Using the cutoff value for the LP_{lm} mode, show that the mode subscripts (l, m) for a step-index fiber must satisfy the condition

$$m\pi + \left(l - \frac{3}{2}\right)\frac{\pi}{2} \leq V$$

where $V = k_0 a(n_1^2 - n_2^2)^{1/2}$. Show that each LP_{lm} mode is fourfold degenerate.

b. By counting the allowed mode subscripts (l, m), show that the total number of confined modes that can be supported by a step-index fiber is

$$N \simeq \frac{4}{\pi^2} V^2 \simeq \frac{1}{2} V^2$$

c. Using (3.4-25) show that for a truncated quadratic-index fiber

$$N \simeq \tfrac{1}{4}V^2$$

Note that the total number of modes in a truncated quadratic-index fiber is one-half of that of a step-index fiber.

d. Estimate the number of confined modes in a multimode step-index fiber with $a = 50$ μm, $n_1 = 1.52$, $n_2 = 1.50$ at a carrier wavelength of $\lambda = 1$ μm.

e. In a general, truncated graded-index fiber with a core radius a and a cladding index n_2, it is convenient to define an effective V number such that

$$V_{\text{eff}}^2 = 2k_0^2 \int_0^a [n^2(r) - n_2^2] r \, dr$$

and the number of confined modes is approximately given by

$$N \simeq \tfrac{1}{2} V_{\text{eff}}^2$$

Show that this approximation agrees with (b) and (c) for step-index and quadratic-index fibers, respectively.

f. Show that, according to (e), the number of confined modes in a power-law (power g) graded-index fiber with an index profile is given by

$$N = \frac{1}{2(1 + 2/g)} V^2 = \frac{k_0^2 a^2}{2(1 + 2/g)} (n_1^2 - n_2^2)$$

Show that this expression again agrees with the results obtained in (b) for step-index fibers ($g = \infty$) and in (c) for quadratic-index fibers ($g = 2$).

3.2 The numerical aperture (NA) is a measure of the light-gathering capability of a fiber. It is defined as the sine of the maximum external angle of the entrance ray (measured with respect to the axis of fiber) that is trapped in the core by total internal reflection.

a. Show that

$$\text{NA} = n_1 \sin \theta_1 = (n_1^2 - n_2^2)^{1/2}$$

b. Show that the solid acceptance angle in air is

$$\Omega = \pi(n_1^2 - n_2^2) = \pi(\text{NA})^2$$

c. Show that the solid angle (in air) for a single electromagnetic radiation mode leaving or entering the core aperture is

$$\Omega_{\text{mode}} = \frac{\lambda^2}{\pi a^2}$$

d. The total number of modes the fiber can support, couple to, and radiate into air is therefore

$$N = 2 \frac{\Omega}{\Omega_{\text{mode}}}$$

where 2 accounts for the two independent polarizations in air. Show that this estimate agrees with Problem 3.1.

e. Find the numerical aperture of a multimode fiber with $n_1 = 1.52$ and $n_2 = 1.50$.

3.3 A single-mode step-index fiber must have a V number less than 2.405; i.e.,

$$V = k_0 a (n_1^2 - n_2^2)^{1/2} < 2.405$$

a. Show that the expression derived in Problem 3.1(b) ($N \simeq 4V^2/\pi^2$) still applies, provided we realize that a single-mode fiber supports two independently polarized HE_{11} modes (or LP_{01} modes).
b. With $a = 5\ \mu m$, $n_2 = 1.50$, and $\lambda = 1\ \mu m$, find the maximum core index for a single-mode fiber. (*Answer*: $n_1 = 1.50195$.)
c. With $n_1 = 1.501$, $n_2 = 1.500$, and $\lambda = 1\ \mu m$, find the maximum core radius for a single-mode fiber. (*Answer*: $a = 7\ \mu m$.)
d. Show that the confinement factor for a single-mode fiber is

$$\Gamma_1 = \frac{P_{\text{core}}}{P} = \frac{(qa)^2}{V^2}\left(1 + \frac{J_0^2(ha)}{J_1(ha)}\right)$$

where ha satisfies the mode condition (3.3-26)

$$ha\,\frac{J_1(ha)}{J_0(ha)} = qa\,\frac{K_1(qa)}{K_0(qa)}$$

e. Show that, by using the table of Bessel functions, $ha = 1.647$ is an approximate solution to the mode condition for $V = 2.405$. Evaluate the confinement factor Γ_1 for the LP_{01} mode of this single-mode fiber (*Answer*: $\Gamma_1 = 83\%$). Note that this is the maximum confinement factor for a single-mode fiber. Compare this value with the curves in Figure 3-9.

3.4 Mode condition

a. Derive the mode condition for step-index fibers (3.2-11).
b. Derive the expressions for the constants B, C, D, in terms of A (3.2-12).
c. Derive the mode condition for *TE* and *TM* modes (3.2-17).
d. Show that $E_z = E_r = 0$ for *TE* modes and $H_z = H_r = 0$ for *TM* modes.
e. Show that in the limit $n_1 - n_2 \ll n_1$, *TE* and *TM* modes become identical.

3.5

a. Derive (3.3-6) and (3.3-7).
b. Derive (3.3-18) and (3.3-19).
c. Derive (3.3-23).
d. Derive (3.3-24) and (3.3-25).

3.6 Show that in a quadratic index fiber in which n_2 is independent of ω, the spreading of a Gaussian pulse of width τ can be described by

$$\Delta\tau \cong \frac{2L}{c}\left|\underbrace{\frac{nn_2}{ck^3}(\ell + m + 1)^2}_{\text{Modal dispersion}} - \underbrace{\frac{dn}{d\omega}}_{\text{Material dispersion}}\right|\Delta\omega$$

where ℓ and m are the transverse mode indices.

3.7 In the analysis of dispersive spread and chirp in Section 3.4, show that the total frequency excursion during the pulse duration satisfies the relation

$$|\omega_{1/2} - \omega_{-1/2}| \approx FWHM \text{ of } F(\Omega)$$

independent of z, where $\omega_{1/2}$ and $\omega_{-1/2}$ are the frequencies at the two times when the pulse envelope is at half its maximum value. [It would be all right to take the propagation distance z sufficiently large so that $z > (4a\alpha)^{-1}$.]

References

1. Kao, C. K., and T. W. Davies, ''Spectroscopic studies of ultra low loss optical glasses,'' *J. Sci. Instrum.* 1(no. 2):1063, 1968.
2. Kapron, F. P., D. B. Keck, and R. D. Maurer, ''Radiation losses in glass optical waveguides,'' *Appl. Phys. Lett.* 17:423, 1970.
3. Snitzer, E., ''Cylindrical dielectric waveguide modes,'' *J. Opt. Soc. Am.* 51:491, 1961.
4. Keck, D. B., *Fundamentals in Optical Fiber Communications* (M. K. Barnoski, ed.). New York: Academic Press, 1976, Chapter 1.
5. Gloge, D., ''Weakly guiding fibers,'' *Appl. Opt.* 10:2252, 1971.
6. See, for example, D. Marcuse, *Theory of Dielectric Optical Waveguides.* New York: Academic Press, 1974.
7. See, for example, I. S. Gradshteyn and I. M. Ryzhik, *Table of Integrals, Series, and Products.* New York: Academic Press, 1965, p. 634, Eq. (5.54-2).
8. Payne, D. N., and W. A. Gambling, ''Zero material dispersion in optical fibers,'' *Electron. Lett.* 11:176, 1975.
9. Cohen, L. G., and C. Lin, ''Pulse delay measurements in the zero material dispersion wavelength region for optical fibers,'' *Appl. Opt.* 12:3136, 1977.
10. Li, T., ''Structures, parameters, and transmission properties of optical fibers,'' *Proc. IEEE* 68:1175, 1980.
11. Cohen, L. G., W. L. Mammel, and H. M. Presby, ''Correlation between numerical predications and measurements of single-mode fiber dispersion characteristics,'' *Appl. Opt.* 19:2007, 1980.
12. Tsuchiya, H., and N. Imoto, ''Dispersion-free single mode fibers in 1.5 μm wavelength region,'' *Electron. Lett.* 15:476, 1979.
13. Cohen, L. G., C. Lin, and W. G. French, ''Tailoring zero chromatic dispersion into the 1.5–1.6 μm low-loss spectral region of single-mode fibers,'' *Electron. Lett.* 15:334, 1979.
14. White, K. I., and B. P. Nelson, ''Zero total dispersion in step-index monomode fibres at 1.30 and 1.55 m,'' *Electron. Lett.* 15:396, 1979.
15. Gambling, W. A., H. Matsumara, and C. M. Ragdale, ''Zero total dispersion in graded-index single-mode fibers,'' *Electron. Lett.* 15:474, 1979.
16. Bloom, D. M., L. F. Mollenauer, Chinlon Lin, and A. M. Del Gaudio, ''Demonstration of pulse propagation in km-length fibers,'' *Opt. Lett.* 4:297, 1979.
17. Morse, P. M., and H. Feshbach, *Methods of Theoretical Physics.* New York: McGraw-Hill, 1953.

18. Mathews, J., and R. L. Walker, *Mathematical Methods of Physics.* New York: Benjamin, 1965.
19. Landau, L. D., and E. M. Lifshitz, *Quantum Mechanics.* London: Pergamon, 1958.
20. Miya, T., Y. Terunuma, T. Hosaka, and T. Miyashita, ''Ultimate low-loss single-mode fiber at 1.55 μm,'' *Electron. Lett.* 15:106, 1979.
21. Suematsu, Y., ''Long wavelength optical fiber communication,'' *Proc. IEEE* 71:692, 1983.
22. Miller, S., and I. P. Kaminow, ed., *Optical Fiber Telecommunication II.* San Diego: Academic, 1988, Chaps. 2 and 3.
23. Figure 3-10(a) is courtesy of S. R. Nagel, AT&T Bell Laboratories. Figure 3-10(b) is from D. L. Frazen, ''Single mode fiber measurements,'' *Proceedings of the Tutorial Sessions Conference on Optical Fiber Communications.* Washington, D.C., Opt. Soc. Am., 1988, p. 101.
24. Kolner, B. H., and M. Nazarathy, ''Temporal imaging with a time lens,'' *Opt. Lett.* 14:630, 1989.
25. Lohman, A. W., D. Mendlovic, ''Temporal filtering with time lenses,'' *Appl. Opt.* 31:6212, 1992.
26. Papoulis, A. ''Pulse compression, fiber communication, and diffraction: a unified approach, *J. Opt. Soc. Am.* 11:3, 1994.
27. Yariv, A., D. Fekete, and D. M. Pepper, ''Compensation for channel dispersion by nonlinear optical phase conjugation,'' *Opt. Lett.* 4:52, 1979.
28. Watanabe, S., T. Naito, and T. Chikama, ''Compensation of chromatic dispersion in a single mode fiber by optical phase conjugation,'' *IEEE Photonics Technology Letters*, 5:92, 1993.
29. Kolner, B. H., and M. Nazarathy, ''Temporal imaging with a time lense,'' *Opt. Lett.* 14:630, 1989.

Optical Resonators

4.0 INTRODUCTION

Optical resonators, like their low-frequency, radio-frequency, and microwave counterparts, are used primarily in order to build up large field intensities with moderate power inputs. They consist in most cases of two, or more, curved mirrors that serve to "trap," by repeated reflections and refocusing, an optical beam that thus becomes the mode of the resonator. A universal measure of this property is the quality factor Q of the resonator. Q is defined by the relation

$$Q = \omega \times \frac{\text{field energy stored by resonator}}{\text{power dissipated by resonator}} \tag{4.0-1}$$

As an example, consider the case of a simple resonator formed by bouncing a plane TEM wave between two perfectly conducting planes of separation l so that the field inside is

$$e(z, t) = E \sin \omega t \sin kz \tag{4.0-2}$$

According to (1.3-22), the average electric energy stored in the resonator is

$$\mathscr{E}_{\text{electric}} = \frac{A\varepsilon}{2T} \int_0^l \int_0^T e^2(z, t)\, dz\, dt \tag{4.0-3}$$

where A is the cross-sectional area, ε is the dielectric constant, and $T = 2\pi/\omega$ is the period. Using (4.0-2) we obtain

$$\mathscr{E}_{\text{electric}} = \tfrac{1}{8}\varepsilon E^2 V \tag{4.0-4}$$

where $V = lA$ is the resonator volume. Since the average magnetic energy stored in a resonator is equal to the electric energy [1], the total stored energy is

$$\mathscr{E} = \tfrac{1}{4}\varepsilon E^2 V \tag{4.0-5}$$

Thus, recognizing that in steady state the input power is equal to the dissipated power, and designating the power input to the resonator by P, we obtain from (4.0-1)

$$Q = \frac{\omega \varepsilon E^2 V}{4P}$$

The peak field is given by

$$E = \sqrt{\frac{4QP}{\omega \varepsilon V}} \tag{4.0-6}$$

Mode Density in Optical Resonators

The main challenge in the optical frequency regime is to build resonators that possess a very small number, ideally only one, high Q modes in a given spectral region. The reason is that for a resonator to fulfill this condition, its dimensions need to be of the order of the wavelength.

Example: One-Dimensional Resonator

We consider the simple transverse electromagnetic (TEM) two-mirror resonator with a field distribution as given by Equation (4.0-2). The resonant frequencies are determined by requiring that the field vanish at $z = 0$ and at the location $z = L$ of the second reflector. This happens when

$$\sin k_m L = m\pi$$

$$m = 1, 2, \ldots$$

Using $k_m = \frac{\omega_m}{c} n$, where n is the index of refraction, we obtain $\omega_m = m(\pi c/nL)$ for the resonance frequencies corresponding to a frequency separation between adjacent modes of $\Delta\omega = \pi c/nL$. If we, arbitrarily, choose the criterion of sufficient mode spacing as $\Delta\omega = \omega$, we obtain $L = \lambda/2n$, i.e., the *linear dimension needs to be comparable to the wavelength* (in the medium).

Mode control in the optical regime would thus seem to require that we construct resonators with volume $\sim\lambda^3(\sim 10^{-12}\,\text{cm}^3$ at $\lambda = 1\ \mu\text{m})$. This is not easily achievable. An alternative is to build large $(L \gg \lambda)$ resonators but to use a geometry that endows only a small fraction of these modes with low losses (a high Q). In our two-mirror example, any mode that does not travel normally to the mirror will "walk off" after a few bounces and thus will possess a low Q factor. We will show later that when the resonator contains an amplifying (inverted population) medium, oscillation will occur preferentially at high Q modes, so that the strategy of modal discrimination by controlling Q is sensible. We shall also find that further modal discrimination is due to the fact that the atomic medium is capable of amplifying radiation only within a limited frequency region so that modes outside this region, even if possessing high Q, do not oscillate.

One question asked often is the following: Given a large $(L \gg \lambda)$ optical resonator, how many of its modes will have their resonant frequencies in a given frequency interval, say, between ν and $\nu + \Delta\nu$? To answer this problem, consider a large, perfectly reflecting box resonator with sides, a,b,c along the x,y,z directions. Without going into modal details, it is sufficient for our purpose to take the amplitude field solution in the form

$$E(x,y,z) \propto \sin k_x x \sin k_y y \sin k_z z \tag{4.0-7}$$

(Resonators of different shapes will differ in detail, but for large, $L \gg \lambda$, resonators, the results are similar.)

$$k_x^2 + k_y^2 + k_z^2 = \left(\frac{\omega}{c}n\right)^2 \tag{4.0-8}$$

For the field to vanish at the boundaries, we thus need to satisfy

$$k_x = \frac{r\pi}{a},\ k_y = \frac{s\pi}{b},\ k_z = \frac{t\pi}{c} \qquad r,s,t \text{ any integers} \tag{4.0-8a}$$

With each such mode, we may thus associate a propagation vector $\mathbf{k} = \hat{x}k_x + \hat{y}k_y + \hat{z}k_z$. The triplet r,s,t defines a mode. Since replacing any integer with its negative does not, according to Equation (4.0-7), generate an independent mode, we will restrict, without loss of generality, r,s,t to positive integers. It is convenient to describe the modal distribution in $\mathbf{k}$ space, as in Figure 4-1. Since each (positive) triplet r,s,t generates an independent mode, we can associate with each mode an elemental volume in $\mathbf{k}$ space.

$$V_{\text{mode}} = \frac{\pi^3}{abc} = \frac{\pi^3}{V} \tag{4.0-9}$$

where V is the physical volume of the resonator. We recall that the length of the vector $\mathbf{k}$ satisfies Equation (4.0-8), rewritten here as

$$k(r,s,t) = \frac{2\pi\nu(r,s,t)}{c}n \tag{4.0-10}$$

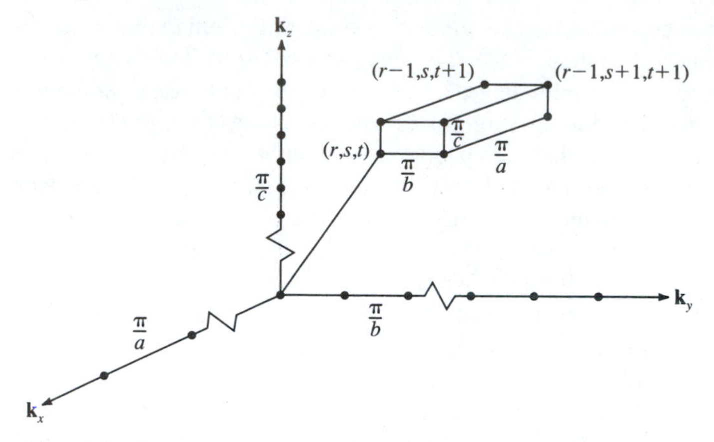

Figure 4-1 **k** space description of modes. Every positive triplet of integers r,s,t defines a unique mode. We can thus associate a primitive volume π^3/abc in **k** space with each mode.

To find the total number of modes with **k** values between 0 and k, we divide the corresponding volume in **k** space by the volume per mode:

$$N(k) = \frac{\left(\frac{1}{8}\right)\frac{4\pi}{3}k^3}{\frac{\pi^3}{V}} = \frac{k^3 V}{6\pi^2}$$

(The factor 1/8 is due to the restriction of $r,s,t > 0$.)

We next use (4.0-10) to obtain the number of modes with resonant frequencies between 0 and ν:

$$N(\nu) = \frac{4\pi\nu^3 n^3 V}{3c^3}$$

The mode density, that is, the number of modes per unit ν near ν in a resonator with volume $V(\gg\lambda^3)$, is thus

$$p(\nu) = \frac{dN(\nu)}{d\nu} = \frac{8\pi\nu^2 n^3 V}{c^3} \tag{4.0-11}$$

where we multiplied the final result by 2 to account for the two independent orthogonally polarized modes that are associated with each r,s,t triplet.

The number of modes that fall within the interval $d\nu$ centered on ν is thus

$$N \simeq \frac{8\pi n^3 \nu^2 V}{c^3}\, d\nu \tag{4.0-12}$$

where V is the volume of the resonator. For the case of $V = 1\ \text{cm}^3$, $\nu = 3 \times 10^{14}$ Hz and $d\nu = 3 \times 10^{10}$, as an example, (4.0-12) yields $N \sim 2 \times 10^9$ modes. If the resonator were closed, all these modes would have similar values of Q. This situation is to be avoided in the case of lasers, since it will cause the atoms to emit power (thus causing oscillation) into a large number of modes, which may differ in their frequencies as well as in their spatial characteristics.

This objection is overcome to a large extent by the use of open resonators, which consist essentially of a pair of opposing flat or curved reflectors. In such resonators the energy of the vast majority of the modes does not travel at right angles to the mirrors and will thus be lost in essentially a single traversal. These modes will consequently possess a very low Q. If the mirrors are curved, the few surviving modes will, as shown below, have their energy localized near the axis; thus the diffraction losses caused by the open sides can be made small compared with other loss mechanisms such as mirror transmission. (This point is considered in detail in Section 4.9. The subject of losses is also considered in Section 4.7.)

4.1 FABRY–PEROT ETALON

The Fabry–Perot etalon, or interferometer, named after its inventors [3], can be considered as the archetype of the optical resonator. It consists of a plane-parallel plate of thickness l and index n that is immersed in a medium of index n'.[1] Let a plane wave be incident on the etalon at an angle θ' to the normal, as shown in Figure 4-2(a). We can treat the problem of the transmission (and reflection) of the plane wave through the etalon by considering the infinite number of partial waves produced by reflections at the two end surfaces. The phase delay between two partial waves—which is attributable to one additional round trip—is given, according to Figure 4-2(a), by

$$\delta = \frac{4\pi n l \cos\theta}{\lambda} \tag{4.1-1}$$

where λ is the vacuum wavelength of the incident wave and θ is the internal angle of incidence. If the complex amplitude of the incident wave is taken as A_i, then the partial reflections, B_1, B_2, and so forth, are given by

$$B_1 = rA_i \qquad B_2 = tt'r'\,A_i e^{i\delta} \qquad B_3 = tt'r'^3\,A_i e^{2i\delta} \qquad \cdots$$

where r is the reflection coefficient (ratio of reflected to incident amplitude), t is the transmission coefficient for waves incident from n' toward n, and r' and t' are the corresponding quantities for waves traveling from n toward n'. The complex amplitude of the (total) reflected wave is $A_r = B_1 + B_2 + B_3 + \cdots$, or

$$A_r = \{r + tt'r'e^{i\delta}(1 + r'^2e^{i\delta} + r'^4e^{2i\delta} + \cdots)\}\,A_i \tag{4.1-2}$$

For the transmitted wave,

$$A_1 = tt'\,A_i \qquad A_2 = tt'r'^2e^{i\delta}\,A_i \qquad A_3 = tt'r'^4e^{2i\delta}\,A_i$$

[1] In practice, one often uses etalons made by spacing two partially reflecting mirrors a distance l apart so that $n = n' = 1$. Another common form of etalon is produced by grinding two plane-parallel (or curved) faces on a transparent solid and then evaporating a metallic or dielectric layer (or layers) on the surfaces.

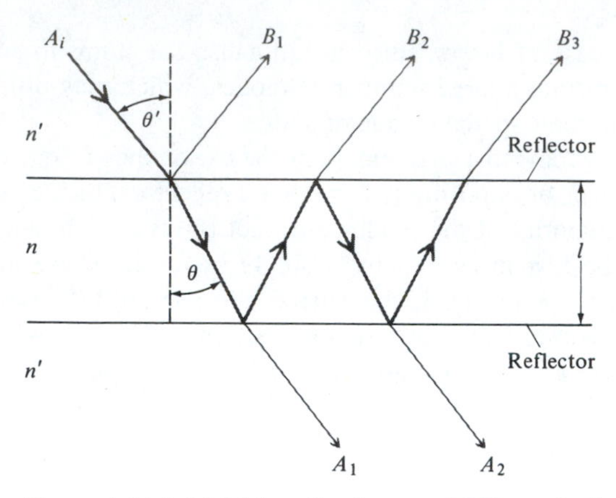

Figure 4-2(a) Multiple reflections model for analyzing the Fabry–Perot etalon.

where a phase factor $\exp\left(i\dfrac{\delta}{2}\right)$, which corresponds to a single traversal of the plate and is common to all the terms, has been left out. Adding up the A terms, we obtain

$$A_t = A_i\, tt'(1 + r'^2 e^{i\delta} + r'^4 e^{2i\delta} + \cdots) \tag{4.1-3}$$

for the complex amplitude of the total transmitted wave. We notice that the terms within the parentheses in (4.1-2) and (4.1-3) form an infinite geometric progression; adding them, we get

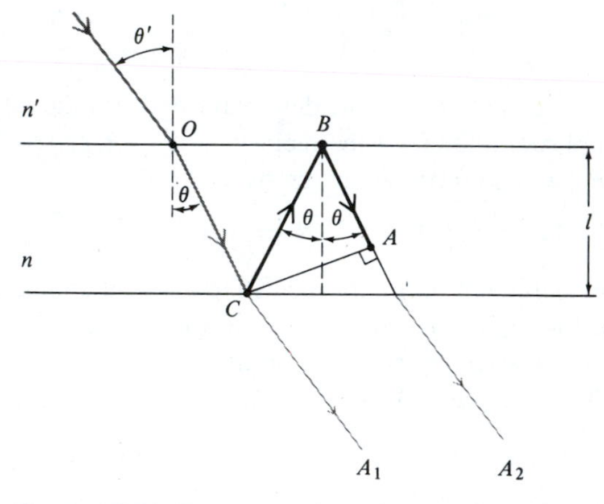

Figure 4-2(b) Two successive reflections, A_1 and A_2. Their path difference is given by

$$\delta L = AB + BC = l\,\frac{\cos 2\theta}{\cos\theta} + \frac{l}{\cos\theta} = 2l\cos\theta$$

$$\rightarrow \delta = \frac{2\pi(\delta L)n}{\lambda} = \frac{4\pi nl\cos\theta}{\lambda}$$

$$A_r = \frac{(1 - e^{i\delta})\sqrt{R}}{1 - Re^{i\delta}} A_i \tag{4.1-4}$$

and

$$A_t = \frac{T}{1 - Re^{i\delta}} A_i \tag{4.1-5}$$

where we used the fact that $r' = -r$, the conservation-of-energy relation that applies to lossless mirrors

$$r^2 + tt' = 1$$

as well as the definitions

$$R \equiv r^2 = r'^2 \qquad T \equiv tt'$$

R and T are, respectively, the fraction of the intensity reflected and transmitted at each interface and will be referred to in the following discussion as the mirrors' reflectance and transmittance.

If the incident intensity (watts per square meter) is taken as $A_i A_i^*$, we obtain from (4.1-4) the following expression for the fraction of the incident intensity that is reflected:

$$\frac{I_r}{I_i} = \frac{A_r A_r^*}{A_i A_i^*} = \frac{4R \sin^2(\delta/2)}{(1 - R)^2 + 4R \sin^2 (\delta/2)} \tag{4.1-6}$$

Moreover, from (4.1-5),

$$\frac{I_t}{I_i} = \frac{A_t A_t^*}{A_i A_i^*} = \frac{(1 - R)^2}{(1 - R)^2 + 4R \sin^2 (\delta/2)} \tag{4.1-7}$$

for the transmitted fraction. Our basic model contains no loss mechanisms, so conservation of energy requires that $I_t + I_r$ be equal to I_i, as is indeed the case.

Let us consider the transmission characteristics of a Fabry–Perot etalon. According to (4.1-7) the transmission is unity whenever

$$\delta = \frac{4\pi nl \cos \theta}{\lambda} = 2m\pi \qquad m = \text{any integer} \tag{4.1-8}$$

The condition (4.1-8) for maximum transmission can be written as

$$\nu_m = m \frac{c}{2nl \cos \theta} \qquad m = \text{any integer} \tag{4.1-9}$$

where $c = \nu\lambda$ is the velocity of light in vacuum and ν is the optical frequency. For a fixed l and θ, (4.1-9) defines the unity transmission (resonance) frequencies of the etalon. These are separated by the so-called *free spectral range*

$$\Delta\nu \equiv \nu_{m+1} - \nu_m = \frac{c}{2nl \cos \theta} \tag{4.1-10}$$

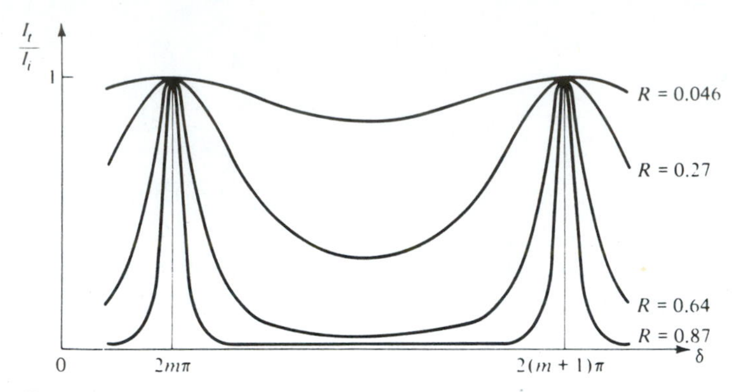

Figure 4-3 Transmission characteristics (theoretical) of a Fabry–Perot etalon. (After Reference [4].)

Theoretical transmission plots of a Fabry–Perot etalon are shown in Figure 4-3. The maximum transmission is unity, as stated previously. The minimum transmission, on the other hand, approaches zero as R approaches unity.

If we allow for the existence of losses in the etalon medium, we find that the peak transmission is less than unity. Taking the fractional intensity loss per pass as $(1 - A)$, we find that the maximum transmission drops from unity to

$$\left(\frac{I_t}{I_i}\right)_{\max} = \frac{(1 - R)^2 A}{(1 - RA)^2} \tag{4.1-11}$$

The proof of (4.1-11) is left as an exercise (Problem 4-2).

An experimental transmission plot of a Fabry–Perot etalon is shown in Figure 4-4.

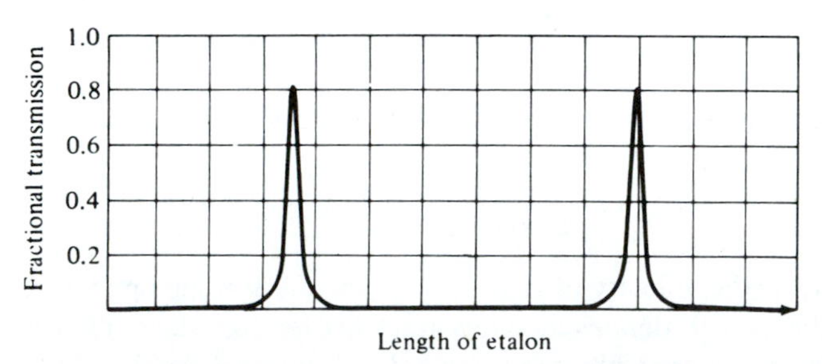

Figure 4-4 Experimental transmission characteristics of a Fabry–Perot etalon at 6328 Å as a function of the etalon optical length with $R = 0.9$ and $A = 0.98$. The two peaks shown correspond to a change in the optical length $\Delta(nl) = \lambda/2$. (After Reference [5].)

4.2 FABRY–PEROT ETALONS AS OPTICAL SPECTRUM ANALYZERS

According to (4.1-8), the maximum transmission of a Fabry–Perot etalon occurs when

$$\frac{2nl\cos\theta}{\lambda} = m \tag{4.2-1}$$

Taking, for simplicity, the case of normal incidence ($\theta = 0°$), we obtain the following expression for the change $d\nu$ in the resonance frequency of a given transmission peak due to a length variation dl

$$\frac{d\nu}{\Delta\nu} = -\frac{dl}{(\lambda/2n)} \tag{4.2-2}$$

where $\Delta\nu$ is the intermode frequency separation as given by (4.1-10). According to (4.2-2), we can tune the peak transmission frequency of the etalon by $\Delta\nu$ by changing its length by half a wavelength. This property is utilized in operating the etalon as a scanning interferometer. The optical signal to be analyzed passes through the etalon as its length is being swept. If the width of the transmission peaks is small compared to that of the spectral detail in the incident optical beam signal, the output of the etalon will constitute a replica of the spectral profile of the signal. In this application it is important that the spectral width of the signal beam be smaller than the intermode spacing of the etalon ($c/2nl$) so that the ambiguity due to simultaneous transmission through more than one transmission peak is avoided. For the same reason the total length scan is limited to $dl < \lambda/2n$. Figure 4-5 demonstrates the operation of a scanning Fabry–Perot etalon; Figure 4-6 shows intensity versus frequency data obtained by analyzing the output of a multimode He–Ne laser oscillating near 6328 Å. The peaks shown correspond to longitudinal laser modes, which will be discussed in Section 4-5.

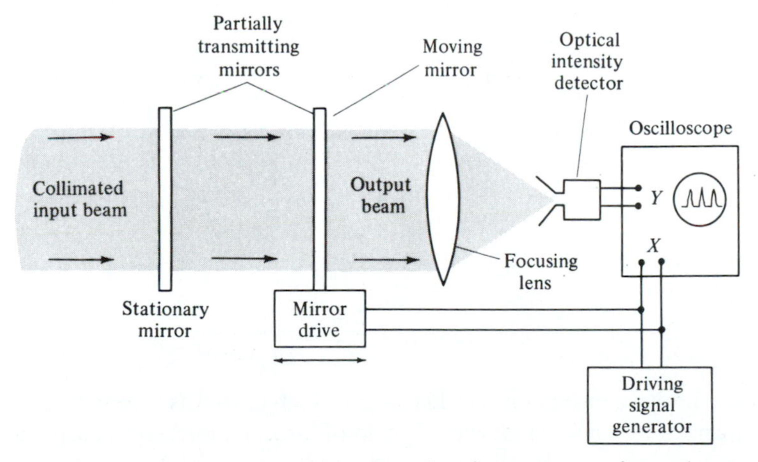

Figure 4-5 Typical scanning Fabry–Perot interferometer experimental arrangement.

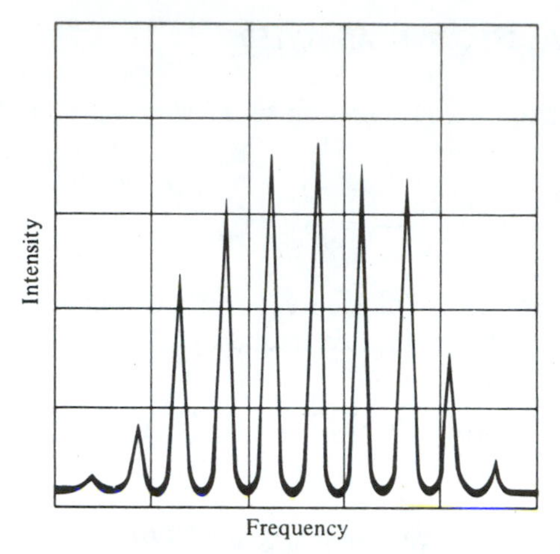

Figure 4-6 Intensity versus frequency analysis of the output of a He–Ne 6328 Å laser obtained with a scanning Fabry–Perot etalon. The horizontal scale is 250 MHz per division.

It is clear from the foregoing that when operating as a spectrum analyzer the etalon resolution—that is, its ability to distinguish details in the spectrum—is limited by the finite width of its transmission peaks. If we take, somewhat arbitrarily,[2] the limiting resolution of the etalon as the separation $\Delta\nu_{1/2}$ between the two frequencies at which the transmission is down to half of its peak value, from (4.1-7) we obtain

$$\sin^2\left(\frac{\delta_{1/2} - 2m\pi}{2}\right) = \frac{(1 - R)^2}{4R} \tag{4.2-3}$$

where $\delta_{1/2}$ is the value of δ corresponding to the two half-power points—that is, the value of δ at which the denominator of (4.1-7) is equal to $2(1 - R)^2$. If we assume $(\delta_{1/2} - 2m\pi) \ll \pi$, so that the width of the high-transmission regions in Figure 4-3 is small compared to the separation between the peaks, we obtain

$$\Delta\nu_{1/2} = \frac{c}{2\pi nl \cos\theta}(\delta_{1/2} - 2m\pi) \simeq \frac{c}{2\pi nl \cos\theta}\frac{1 - R}{\sqrt{R}} \tag{4.2-4}$$

or using (4.1-10) and defining the etalon *finesse* as

$$F \equiv \frac{\pi\sqrt{R}}{1 - R} \tag{4.2-5}$$

we obtain

$$\Delta\nu_{1/2} = \frac{\Delta\nu}{F} = \frac{c}{2nl \cos\theta F} \tag{4.2-6}$$

for the limiting resolution. The finesse F (which is used as a measure of the resolution of Fabry–Perot etalon) is, according to (4.2-6), the ratio of the separation between

[2]For a more complete discussion concerning the definition of resolution, see Reference [4].

peaks to the width of a transmission bandpass. This ratio can be read directly from the transmission characteristics such as those of Figure 4-4, for which we obtain $F \simeq 26$.

Numerical Example: Design of a Fabry–Perot Etalon

Consider the problem of designing a scanning Fabry–Perot etalon to be used in studying the mode structure of a He–Ne laser with the following characteristics: $l_{\text{laser}} = 100$ cm and the region of oscillation $= \Delta\nu_{\text{gain}} \simeq 1.5 \times 10^9$ Hz.

The free spectral range of the etalon (that is, its intermode spacing) must exceed the spectral region of interest, so from (4.1-10) we obtain

$$\frac{c}{2nl_{\text{etal}}} \geq 1.5 \times 10^9 \text{ Hz} \qquad \text{or} \qquad 2nl_{\text{etal}} \leq 20 \text{ cm} \tag{4.2-7}$$

The separation between longitudinal modes of the laser oscillation is $c/2nl_{\text{laser}} = 1.5 \times 10^8$ Hz (here we assume $n = 1$). We choose the resolution of the etalon to be a tenth of this value, so spectral details as narrow as 1.5×10^7 Hz can be resolved. According to (4.2-6), this resolution can be achieved if

$$\Delta\nu_{1/2} = \frac{c}{2nl_{\text{etal}}F} \leq 1.5 \times 10^7 \text{ Hz} \qquad \text{or} \qquad 2nl_{\text{etal}}F \geq 2 \times 10^3 \text{ cm} \tag{4.2-8}$$

To satisfy condition (4.2-7), we choose $2nl_{\text{etal}} = 20$ cm; thus (4.2-8) is satisfied when

$$F \geq 100 \tag{4.2-9}$$

A finesse of 100 requires, according to (4.2-5), a mirror reflectivity of approximately 97 percent.

As a practical note we may add that the finesse, as defined by the first equality in (4.2-6), depends not only on R but also on the mirror flatness and the beam angular spread. These points are taken up in Problems 4-3 and 4-4.

Another important mode of optical spectrum analysis performed with Fabry–Perot etalons involves the fact that a noncollimated monochromatic beam incident on the etalon will emerge simultaneously, according to (4.1-8), along many directions θ,[3] which corresponds to the various orders m. If the output is then focused by a lens, each such direction θ will give rise to a circle in the focal plane on the lens, and, therefore, each frequency component present in the beam leads to a family of circles. This mode of spectrum analysis is especially useful under transient conditions where scanning etalons cannot be employed. Further discussion of this topic is included in Problem 4-6.

[3]Each direction θ corresponds in three dimensions to the surface of a cone with a half-apex angle θ.

4.3 OPTICAL RESONATORS WITH SPHERICAL MIRRORS

In this section we study the properties of optical resonators formed by two opposing spherical mirrors, see References [6] and [7]. We will show that the field solutions inside the resonators are those of the propagating Gaussian beams, which were considered in Chapter 2. It is, consequently, useful to start by reviewing the properties of these beams.

The field distribution corresponding to the (l, m) transverse mode is given, according to (2.8-1), by

$$E_{l,m}(\mathbf{r}) = E_0 \frac{\omega_0}{\omega(z)} H_l\left(\sqrt{2}\frac{x}{\omega(z)}\right) H_m\left(\sqrt{2}\frac{y}{\omega(z)}\right) \times \exp\left[-\frac{x^2+y^2}{\omega^2(z)} - ik\frac{x^2+y^2}{2R(z)} - ikz + i(l+m+1)\eta\right] \quad \textbf{(4.3-1)}$$

where the spot size $\omega(z)$ is

$$\omega(z) = \omega_0\left[1 + \left(\frac{z}{z_0}\right)^2\right]^{1/2} \qquad z_0 = \frac{\pi\omega_0^2 n}{\lambda} \quad \textbf{(4.3-2)}$$

and where ω_0, the minimum spot size, is a parameter characterizing the beam. The radius of curvature of the wavefront is

$$R(z) = z\left[1 + \left(\frac{\pi\omega_0^2 n}{\lambda z}\right)^2\right] = \frac{1}{z}[z^2 + z_0^2] \quad \textbf{(4.3-3)}$$

and the phase factor η is as follows:

$$\eta = \tan^{-1}\left(\frac{\lambda z}{\pi\omega_0^2 n}\right) \quad \textbf{(4.3-4)}$$

The sign of $R(z)$ is taken as positive when the center of curvature is to the left of the wavefront, and vice versa. According to (4.3-1) and (4.3-2), the loci of the points at which the beam intensity (watts per square meter) is a given fraction of its intensity on the axis are the hyperboloids

$$x^2 + y^2 = \text{const.} \times \omega^2(z) \quad \textbf{(4.3-5)}$$

The hyperbolas generated by the intersection of these surfaces with planes that include the z axis are shown in Figure 4-7. These hyperbolas are normal to the phase fronts and thus correspond to the local direction of energy flow. The hyperboloid $x^2 + y^2 = \omega^2(z)$ is, according to (4.3-1), the locus of the points where the exponential factor in the field amplitude is down to e^{-1} from its value on the axis. The quantity $\omega(z)$ is thus defined as the *mode spot size* at the plane z.

Given a beam of the type described by (4.3-1), we can form an optical resonator merely by inserting at points z_1 and z_2 two reflectors with radii of curvature that match those of the propagating beam spherical phase fronts at these points. Since the surfaces are normal to the direction of energy propagation as shown in Figure 4-7, the reflected beam retraces itself; thus, if the phase shift between the mirrors is some multiple of 2π radians, a *self-reproducing stable field* configuration results.

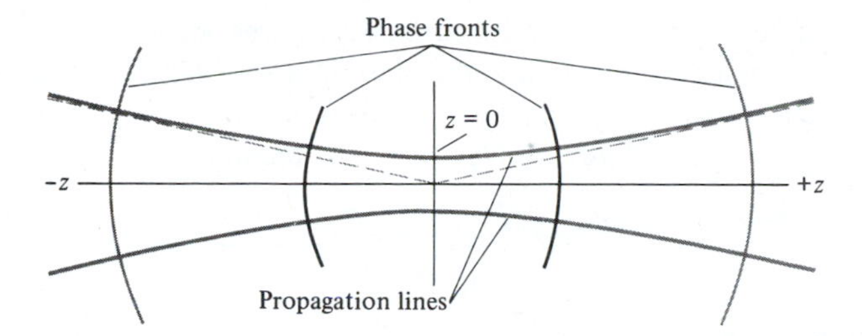

Figure 4-7 Hyperbolic curves corresponding to the local directions of propagation. The nearly spherical phase fronts represent possible positions for reflectors. Any two reflectors form a resonator with a transverse field distribution given by (4.3-1).

Alternatively, given two mirrors with spherical radii of curvature R_1 and R_2 and some distance of separation l, we can, under certain conditions to be derived later, adjust the position $z = 0$ and the parameter ω_0 so that the mirrors coincide with two spherical wavefronts of the propagating beam defined by the position of the waist ($z = 0$) and ω_0. If, in addition, the mirrors can be made large enough to intercept the majority (99 percent, say) of the incident beam energy in the fundamental ($l = m = 0$) transverse mode, we may expect this mode to have a larger Q than higher-order transverse modes, which, according to Figure 2-7, have fields extending farther from the axis and consequently lose a larger fraction of their energy by "spilling" over the mirror edges (diffraction losses).

Optical Resonator Algebra

As mentioned in the preceding paragraphs, we can form an optical resonator by using two reflectors, one at z_1 and the other at z_2, chosen so that their radii of curvature are the same as those of the beam wavefronts at the two locations. The propagating beam mode (4.3-1) is then reflected back and forth between the reflectors without a change in the transverse profile. The requisite radii of curvature, determined by (4.3-3), are

$$R_1 = z_1 + \frac{z_0^2}{z_1}$$

$$R_2 = z_2 + \frac{z_0^2}{z_2}$$

from which we get

$$z_1 = \frac{R_1}{2} \pm \frac{1}{2}\sqrt{R_1^2 - 4z_0^2}$$

$$z_2 = \frac{R_2}{2} \pm \frac{1}{2}\sqrt{R_2^2 - 4z_0^2} \qquad \textbf{(4.3-6)}$$

For a given minimum spot size $\omega_0 = (\lambda z_0/\pi n)^{1/2}$, we can use (4.3-6) to find the positions z_1 and z_2 at which to place mirrors with curvatures R_1 and R_2, respectively.

In practice, we often start with given mirror curvatures R_1 and R_2 and a mirror separation l. The problem is then to find the minimum spot size ω_0, its location with respect to the reflectors, and the mirror spot sizes ω_1 and ω_2. Taking the mirror spacing as $l = z_2 - z_1$, we can solve (4.3-6) for z_0^2, obtaining

$$z_0^2 = \frac{l(-R_1 - l)(R_2 - l)(R_2 - R_1 - l)}{(R_2 - R_1 - 2l)^2} \tag{4.3-7}$$

where z_2 is to the right of z_1 (so that $l = z_2 - z_1 > 0$) and the mirror curvature is taken as positive when the center of curvature is to the left of the mirror.

The minimum spot size $\omega_0 = (\lambda z_0/\pi n)^{1/2}$ and its position is next determined from (4.3-6). The mirror spot sizes are then calculated by the use of (4.3-2).

The Symmetrical Mirror Resonator

The special case of a resonator with symmetrically (about $z = 0$) placed mirrors merits a few comments. The planar phase front at which the minimum spot size occurs is, by symmetry, at $z = 0$. Putting $R_2 = -R_1 = R$ in (4.3-7) gives

$$z_0^2 = \frac{(2R - l)l}{4} \tag{4.3-8}$$

and

$$\omega_0 = \left(\frac{\lambda z_0}{\pi n}\right)^{1/2} = \left(\frac{\lambda}{\pi n}\right)^{1/2} \left(\frac{l}{2}\right)^{1/4} \left(R - \frac{l}{2}\right)^{1/4} \tag{4.3-9}$$

which, when substituted in (4.3-2) with $z = l/2$, yields the following expression for the spot size at the mirrors:

$$\omega_{1,2} = \left(\frac{\lambda l}{2\pi n}\right)^{1/2} \left[\frac{2R^2}{l(R - l/2)}\right]^{1/4} \tag{4.3-10}$$

A comparison with (4.3-9) shows that, for $R \gg l$, $\omega_{1,2} \simeq \omega_0$ and the beam spread inside the resonator is small.

The value of R (for a given l) for which the mirror spot size is a minimum, is readily found from (4.3-10) to be $R = l$. When this condition is fulfilled we have what is called a symmetrical *confocal resonator,* since the two foci, occurring at a distance of $R/2$ from the mirrors, coincide. From (4.3-9) we obtain

$$(\omega_0)_{\text{conf}} = \left(\frac{\lambda l}{2\pi n}\right)^{1/2} \tag{4.3-11}$$

whereas from (4.3-10) we get

$$(\omega_{1,2})_{\text{conf}} = (\omega_0)_{\text{conf}}\sqrt{2} \tag{4.3-12}$$

so the beam spot size increases by $\sqrt{2}$ between the center and the mirrors.

Numerical Example: Design of a Symmetrical Resonator

Consider the problem of designing a symmetrical resonator for $\lambda = 10^{-4}$ cm with a mirror separation $l = 2m$. If we were to choose the confocal geometry with $R = l = 2m$, the minimum spot size (at the resonator center) would be, from (4.3-11) and for $n = 1$

$$(\omega_0)_{\text{conf}} = \left(\frac{\lambda l}{2\pi n}\right)^{1/2} = 0.0564 \text{ cm}$$

whereas, using (4.3-12), the spot size at the mirrors would have the value

$$(\omega_{1,2})_{\text{conf}} = \omega_0\sqrt{2} \simeq 0.0798 \text{ cm}$$

Assume next that a mirror spot size $\omega_{1,2} = 0.3$ cm is desired. Using this value in (4.3-10) and assuming $R \gg l$, we get

$$\frac{\omega_{1,2}}{(\lambda l/2\pi n)^{1/2}} = \frac{0.3}{0.056} = \left(\frac{2R}{l}\right)^{1/4}$$

whence

$$R \simeq 400l \simeq 800 \text{ meters}$$

so that the assumption $R \gg l$ is valid. The minimum beam spot size ω_0 is found, through (4.3-2) and (4.3-8), to be

$$\omega_0 = 0.994\omega_{1/2} \simeq 0.3 \text{ cm}$$

Thus, to increase the mirror spot size from its minimum (confocal) value of 0.0798 cm to 0.3 cm, we must use exceedingly plane mirrors ($R = 800$ meters). This also shows that even small mirror curvatures (that is, large R) give rise to ''narrow'' beams.

The numerical example we have worked out applies equally well to the case in which a plane mirror is placed at $z = 0$. The beam pattern is equal to that existing in the corresponding half of the symmetric resonator in the example, so the spot size on the planar reflector is ω_0.

4.4 MODE STABILITY CRITERIA

The ability of an optical resonator to support low (diffraction) loss[4] modes depends on the mirrors' separation l and their radii of curvature R_1 and R_2. To illustrate this point, consider first the symmetric resonator with $R_2 = R_1 = R$.

[4]By diffraction loss we refer to the fact that due to the beam spread [see (2.5-18)], a fraction of the Gaussian beam energy ''misses'' the mirror and is not reflected and is thus lost.

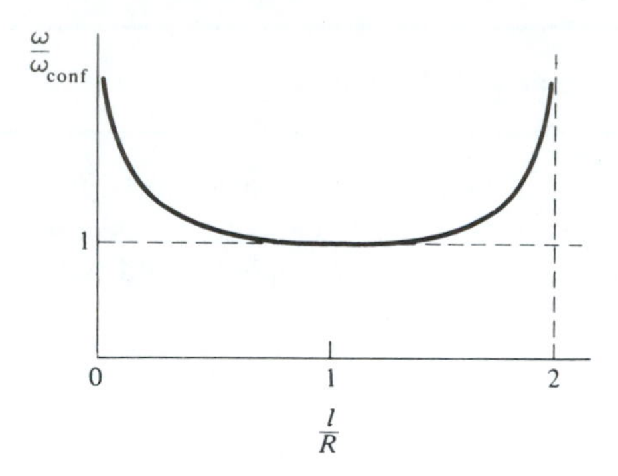

Figure 4-8 Ratio of beam spot size at the mirrors of a symmetrical resonator to its confocal ($l/R = 1$) value.

The ratio of the mirror spot size at a given l/R to its minimum confocal ($l/R = 1$) value, given by the ratio of (4.3-10) to (4.3-12), is

$$\frac{\omega_{1,2}}{(\omega_{1,2})_{\mathrm{conf}}} = \left[\frac{1}{(l/R)[2 - (l/R)]}\right]^{1/4} \tag{4.4-1}$$

This ratio is plotted in Figure 4-8. For $l/R = 0$ (plane-parallel mirrors) and for $l/R = 2$ (two concentric mirrors), the spot size becomes infinite. It is clear that the diffraction losses from these cases are very high, since most of the beam energy ''spills over'' the reflector edges. Since, according to Table 2-1, the reflection of a Gaussian beam from a mirror with a radius of curvature R is formally equivalent to its transmission through a lens with a focal length $f = R/2$, the problem of the existence of stable confined optical modes in a resonator is formally the same as that of the existence of stable solutions for the propagation of a Gaussian beam in a biperiodic lens sequence, as shown in Figure 4-9. This problem was considered in Section 2.1 and led to the stability condition (2.1-16).

If, in (2.1-16), we replace f_1 by $R_1/2$ and f_2 by $R_2/2$, we obtain the stability condition for optical resonators[5]

$$0 \le \left(1 - \frac{l}{R_1}\right)\left(1 - \frac{l}{R_2}\right) \le 1 \tag{4.4-2}$$

A convenient representation of the stability condition (4.4-2) is by means of the diagram [7] shown in Figure 4-10. From this diagram, for example, it can be seen that the symmetric concentric ($R_1 = R_2 = l/2$), confocal ($R_1 = R_2 = l$), and the plane-parallel ($R_1 = R_2 = \infty$) resonators are all on the verge of instability and thus may become extremely lossy by small deviations of the parameters in the direction of instability.

---

[5]This causes the sign convention of R_1 and R_2 to be different from that used in the preceding sections. The sign of R is the same as that of the focal length of the equivalent lens. This makes R_1 (or R_2) positive when the center of curvature of mirror 1 (or 2) is in the direction of mirror 2 (or 1), and negative otherwise.

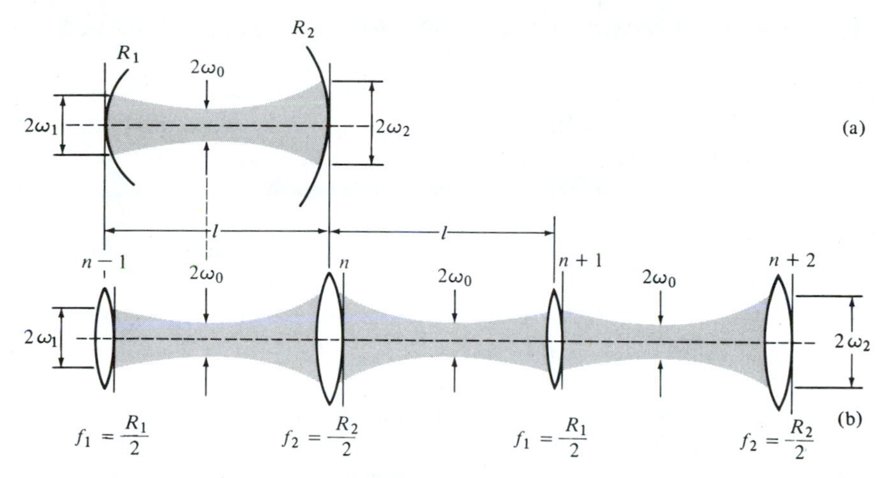

Figure 4-9 (a) Asymmetric resonator ($R_1 \neq R_2$) with mirror curvatures R_1 and R_2. (b) Biperiodic lens system (lens waveguide) equivalent to resonator shown in (a).

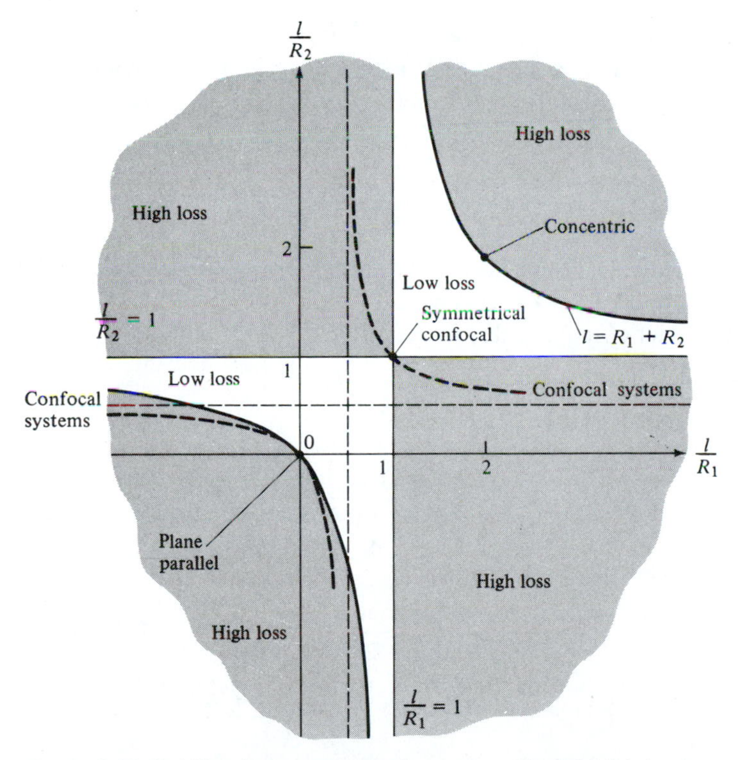

Figure 4-10 Stability diagram of optical resonator. Shaded (high-loss) areas are those in which the stability condition $0 \leq (1 - l/R_1)(1 - l/R_2) \leq 1$ is violated and the clear (low-loss) areas are those in which it is fulfilled. The sign convention for R_1 and R_2 is discussed in footnote 5. (After Reference [7].)

4.5 MODES IN A GENERALIZED RESONATOR—THE SELF-CONSISTENT METHOD

Up to this point we have treated resonators consisting of two opposing spherical mirrors. We may, sometimes, wish to consider the properties of more complex resonators made up of an arbitrary number of lenslike elements such as those shown in Table 2-1. A simple case of such a resonator may involve placing a lens between two spherical reflectors or constructing an off-axis three-reflector resonator. Yet another case is that of a traveling wave resonator in which the beam propagates in one sense only.

In either of these cases we need to find if low-loss (that is, "stable") modes exist in the complex resonator, and if so, to solve for the spot size $\omega(z)$ and the radius of curvature $R(z)$ everywhere.

We apply the self-consistency condition and require that a stable eigenmode of the resonator is one that *reproduces itself after one round trip.* We choose an *arbitrary* reference plane in the resonator, denote the steady-state complex beam parameter at this plane as q_s, and, using the ABCD law (2.6-6), require that

$$q_s = \frac{Aq_s + B}{Cq_s + D} \tag{4.5-1}$$

where A, B, C, D are the "ray" matrix elements for one complete round trip—starting and ending at the chosen reference plane.

Solving (4.5-1) for $1/q_s$ gives

$$\frac{1}{q_s} = \frac{(D - A) \pm \sqrt{(D - A)^2 + 4BC}}{2B} \tag{4.5-2}$$

since the individual elements in the resonator are described by unimodular matrices, that is, $A_iD_i - B_iC_i = 1$ (see Table 2-1), it follows that the matrix A, B, C, D, which is the product of individual matrices, satisfies

$$AD - BC = 1$$

and (4.5-2) can, consequently, be written as

$$\frac{1}{q_s} = \frac{D - A}{2B} \pm i\,\frac{\sqrt{1 - [(D + A)/2]^2}}{B} = \frac{D - A}{2B} + \frac{i \sin \theta}{B} \tag{4.5-3}$$

where

$$\cos \theta = \frac{D + A}{2}$$

$$\theta = \pm \left| \cos^{-1} \left(\frac{D + A}{2} \right) \right| \tag{4.5-4}$$

According to (2.5-11) the condition for a confined Gaussian beam is that the square of the beam spot size ω^2 be a finite positive number. Recalling that q is related to the spot size ω and the radius of curvature R as

$$\frac{1}{q} = \frac{1}{R} - i\frac{\lambda}{\pi\omega^2 n}$$

we find by comparing the last expression to (4.5-3) that the condition for a confined beam is satisfied by choosing θ in (4.5-4), so that $\sin\theta/\beta < 0$ provided

$$\left|\frac{D + A}{2}\right| < 1 \tag{4.5-5}$$

and the steady-state beam parameter is

$$\frac{1}{q_s} = \frac{D - A}{2B} - i\frac{\sqrt{1 - [(D + A)/2]^2}}{|B|} = \frac{D - A}{2B} + \frac{i\sin\theta}{|B|} \qquad \theta < 0 \tag{4.5-6}$$

Equation (4.5-5) can thus be viewed as the generalization of the stability condition (4.4-2) to the case of an arbitrary resonator. When applied to a resonator composed of two spherical reflectors, it reduces to (4.4-2).

The radius of curvature R and the spot size ω at the reference plane are obtained from (4.5-6) by using (2.6-5)

$$R = \frac{2B}{D - A}$$

$$\omega = \left(\frac{\lambda}{\pi n}\right)^{1/2} \frac{(|B|)^{1/2}}{[1 - [(D + A)/2]^2]^{1/4}} \tag{4.5-7}$$

The complex beam parameter q, and hence ω and R, at any other plane can be obtained by applying the ABCD law (2.6-6) to q_s.

Stability of the Resonator Modes

The treatment just concluded dealt with the existence of steady-state (self-reproducing) resonator modes. Having found that such modes do exist, we need to inquire whether the modes are stable. This can be done by perturbing the steady-state solution $1/q_s$ as given by (4.5-6) and following the evolution of the perturbation with propagation [8].

We start with (2.6-6), which relates the beam parameter q_{out} to the beam parameter q_{in}, after one round trip

$$q_{\text{out}} = \frac{Aq_{\text{in}} + B}{Cq_{\text{in}} + D}$$

where A, B, C, D are the ray matrix elements for one complete round trip inside the optical resonator. Rewriting the last expression as

$$q_{\text{out}}^{-1} = \frac{C + Dq_{\text{in}}^{-1}}{A + Bq_{\text{in}}^{-1}} \tag{4.5-8}$$

we obtain by differentiation

$$\frac{dq_{\text{out}}^{-1}}{dq_{\text{in}}^{-1}} = \frac{D - [(C + Dq_{\text{in}}^{-1})/(A + Bq_{\text{in}}^{-1})]B}{A + Bq_{\text{in}}^{-1}}$$

$$= \frac{D - Bq_{\text{out}}^{-1}}{A + Bq_{\text{in}}^{-1}} \tag{4.5-9}$$

At steady state $q_{\text{out}} = q_{\text{in}} \equiv q_s$

$$\left.\frac{dq_{\text{out}}^{-1}}{dq_{\text{in}}^{-1}}\right|_{q_{\text{in}}=q_s} = \frac{D - Bq_s^{-1}}{A + Bq_s^{-1}} \tag{4.5-10}$$

Using (4.5-6) we obtain

$$D - Bq_s^{-1} = \frac{D + A}{2} - i \sin \theta = e^{-i\theta}$$

$$A + Bq_s^{-1} = \frac{D + A}{2} + i \sin \theta = e^{i\theta}$$

so that

$$\left.\frac{dq_{\text{out}}^{-1}}{dq_{\text{in}}^{-1}}\right|_{q_{\text{in}}=q_s} = e^{-2i\theta} \tag{4.5-11}$$

Because confined modes require, according to (4.5-4) and (4.5-5), that θ be real, it follows from (4.5-11) that a small perturbation Δq_{in}^{-1} of the beam parameter q^{-1} from the steady-state value q_s^{-1} does not decay, since the perturbation after one round trip ($\Delta q_{\text{out}}^{-1}$) satisfies

$$|\Delta q_{\text{out}}^{-1}| = |\Delta q_{\text{in}}^{-1}| \tag{4.5-12}$$

We thus find that the theory predicts that mode perturbations in Gaussian mode resonators do not decay. This does not agree with experience, which shows that the mode characteristics of laser oscillators are highly stable, thus implying a strong perturbational decay, that is, $|\Delta q_{\text{out}}^{-1}| < |\Delta q_{\text{in}}^{-1}|$. The discrepancy is resolved if we include in the analysis leading to (4.5-11) the fact that the resonator mirrors are of finite extent.

4.6 RESONANCE FREQUENCIES OF OPTICAL RESONATORS

Up to this point we have considered only the dependence of the spatial mode characteristics on the resonator mirrors (their radii of curvature and separation). Another important consideration is that of determining the resonance frequency of a given spatial mode.

The frequencies are determined by the condition that the complete round-trip phase delay of a resonant mode be some multiple of 2π. This requirement is equivalent to that in microwave waveguide resonators where the resonator length must be equal to an integral number of half-guide wavelengths [1]. This requirement makes it possible for a stable standing wave pattern to establish itself along the axis with a transverse field distribution equal to that of the propagating mode.

If we consider a spherical mirror resonator with mirrors at z_2 and z_1, the resonance condition for the l, m mode can be written as[6]

$$\theta_{l,m}(z_2) - \theta_{l,m}(z_1) = q\pi \tag{4.6-1}$$

where q is some integer and $\theta_{l,m}(z)$, the phase shift, is given according to (2.8-2) by

$$\theta_{l,m}(z) = kz - (l + m + 1)\tan^{-1}\frac{z}{z_0}$$

$$(z_0 = \pi\omega_0^2 n/\lambda) \tag{4.6-2}$$

The resonance condition (4.6-1) is thus

$$k_q d - (l + m + 1)\left(\tan^{-1}\frac{z_2}{z_0} - \tan^{-1}\frac{z_1}{z_0}\right) = q\pi$$

where $d = z_2 - z_1$ is the resonator length. It follows that

$$k_{q+1} - k_q = \frac{\pi}{d}$$

or, using $k = 2\pi\nu n/c$,

$$\nu_{q+1} - \nu_q = \frac{c}{2nd} \tag{4.6-3}$$

for the intermode frequency spacing.

Let us consider, next, the effect of varying the transverse mode indices l and m in a mode with a fixed q. We notice from (4.6-3) that the resonant frequencies depend on the sum $(l + m)$ and not on l and m separately, so for a given q all the modes with the same value of $l + m$ are degenerate (that is, they have the same resonance frequencies). Considering (4.6-3) at two different values of $l + m$ gives

$$k_1 d - (l + m + 1)_1\left(\tan^{-1}\frac{z_2}{z_0} - \tan^{-1}\frac{z_1}{z_0}\right) = q\pi$$

$$k_2 d - (l + m + 1)_2\left(\tan^{-1}\frac{z_2}{z_0} - \tan^{-1}\frac{z_1}{z_0}\right) = q\pi \tag{4.6-4}$$

and, by subtraction,

$$(k_1 - k_2)d = [(l + m + 1)_1 - (l + m + 1)_2]\left(\tan^{-1}\frac{z_2}{z_0} - \tan^{-1}\frac{z_1}{z_0}\right) \tag{4.6-5}$$

---

[6]In obtaining (4.6-1) we did not allow for the phase shift upon reflection. This correction does not affect any of the results of this section, since these shifts cancel out in the subtraction of Equation (4.6-3).

and

$$\Delta\nu = \frac{c}{2\pi nd}\,\Delta(l + m)\left(\tan^{-1}\frac{z_2}{z_0} - \tan^{-1}\frac{z_1}{z_0}\right) \qquad \textbf{(4.6-6)}$$

for the change $\Delta\nu$ in the resonance frequency caused by a change $\Delta(l + m)$ in the sum $(l + m)$. As an example, in the case of a confocal resonator ($R = d$) we have, according to (4.3-6), $z_2 = -z_1 = z_0$; therefore, $\tan^{-1}(z_2/z_0) = -\tan^{-1}(z_1/z_0) = \pi/4$, and (4.6-6) becomes

$$\Delta\nu_{\text{conf}} = \frac{1}{2}\,[\Delta(l + m)]\,\frac{c}{2nd} \qquad \textbf{(4.6-7)}$$

Comparing (4.6-7) to (4.6-4), we find that in the confocal resonator the resonance frequencies of the transverse modes, resulting from changing l and m, either coincide or fall halfway between those resulting from a change of the longitudinal mode index q. This situation is depicted in Figure 4-11.

To see what happens to the transverse resonance frequencies (that is, those due to a variation of l and m) in a confocal resonator, we may consider the nearly planar resonator in which $|z_1|$ and z_2 are small compared to z_0 (that is, $d \ll |R_1|$ and R_2). In this case, (4.6-6) becomes

$$\Delta\nu \simeq \frac{c}{2\pi n z_0}\,\Delta(l + m) \qquad \textbf{(4.6-8)}$$

The mode grouping for this case is illustrated in Figure 4-12.

In the general case where $|z_1|$ and z_2 are comparable to z_0, the approximation used to derive Equation (4.6-8) does not hold. In this case, it is possible to show using a lengthy, but straightforward, algebra that

$$\tan^{-1}\frac{z_2}{z_0} - \tan^{-1}\frac{z_1}{z_0} = \cos^{-1}\left[\pm\sqrt{\left(1 - \frac{d}{R_1}\right)\left(1 - \frac{d}{R_2}\right)}\right] \qquad \textbf{(4.6-9)}$$

The plus (+) sign is used when both $[1 - (d/R)]$ factors are positive while the minus (−) sign applies when both are negative. (The other options correspond to unstable resonators.) We can then solve either of Equations (4.6-4) for

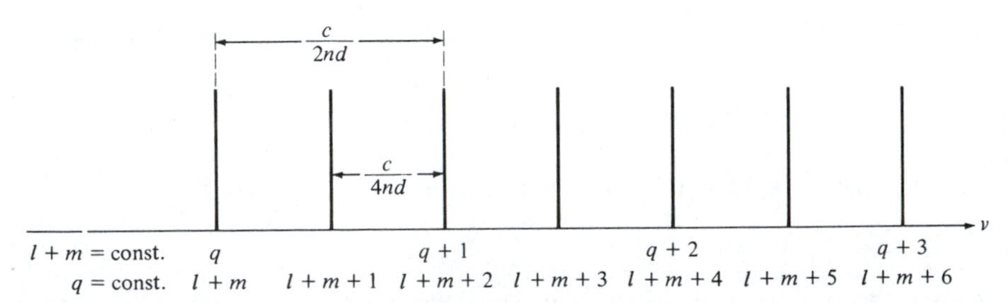

Figure 4-11 Position of resonance frequencies of a confocal ($d = R$) optical resonator as a function of the mode indices l, m, and q.

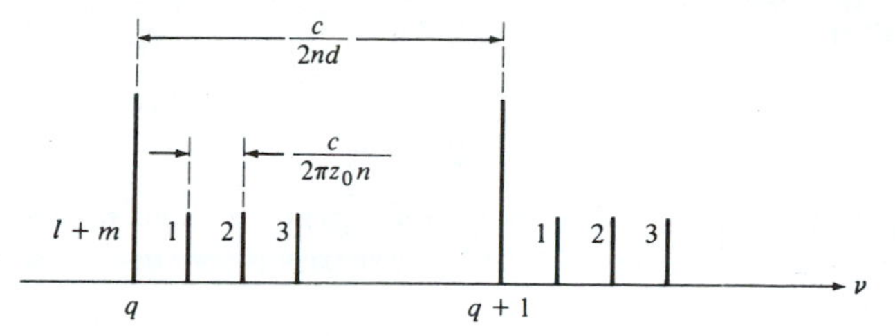

Figure 4-12 Resonant frequencies of a near-planar ($R \gg d$) optical resonator as a function of the mode indices l, m, and q.

$$\nu_{q,\ell,m} = \frac{c}{2nd}\left[q + (\ell + m + 1)\frac{\cos^{-1}\left[\pm\sqrt{\left(1 - \frac{d}{R_2}\right)\left(1 - \frac{d}{R_2}\right)}\right]}{\pi}\right] \quad \textbf{(4.6-10)}$$

the resonant frequency of mode q, ℓ, m.

The situation depicted in Figure 4-12 is highly objectionable if the resonator is to be used as a scanning interferometer. The reason is that in reconstructing the spectral profile of the unknown signal, an ambiguity is caused by the simultaneous transmission of more than one frequency. This ambiguity is resolved by using a confocal etalon whose mode spacing is as shown in Figure 4-11 and by choosing d to be small enough that the intermode spacing $c/4nd$ exceeds the width of the spectral region that is scanned.

4.7 LOSSES IN OPTICAL RESONATORS

An understanding of the mechanisms by which electromagnetic energy is dissipated in optical resonators and the ability to control them are of major importance in understanding and operating a variety of optical devices. For historical reasons as well as for reasons of convenience, these losses are often characterized by a number of different parameters. This book uses, in different places, the concepts of loss per pass, photon lifetime, and quality factor Q to describe losses in resonators. Let us see how these quantities are related to each other.

The decay lifetime (photon lifetime) t_c of a cavity mode is defined by means of the equation

$$\frac{d\mathscr{E}}{dt} = -\frac{\mathscr{E}}{t_c} \quad \textbf{(4.7-1)}$$

where $\mathscr{E}$ is the energy stored in the mode so that in a passive resonator $\mathscr{E}(t) = \mathscr{E}(0)$ $\exp(-t/t_c) = \mathscr{E}(0)\exp(-\omega t/Q)$. If the fractional (intensity) loss per pass is L and the length of the resonator is l, then the fractional loss per unit time is cL/nl; therefore

$$\frac{d\mathscr{E}}{dt} = -\frac{cL}{nl}\mathscr{E}$$

and, from (4.7-1),

$$t_c = \frac{nl}{cL} \tag{4.7-2}$$

for the case of a resonator with mirrors' reflectivities R_1 and R_2 and an average distributed loss constant α, the average loss per pass is for small losses $L = \alpha l - \ln\sqrt{R_1R_2}$ so that

$$t_c = \frac{n}{c[\alpha - (1/l)\ln\sqrt{R_1R_2}]} \approx \frac{nl}{c[\alpha l + (1 - \sqrt{R_1R_2})]} \tag{4.7-3}$$

where the approximate equality applies when $R_1R_2 \approx 1$.

The quality factor of the resonator is defined universally as

$$Q = \frac{\omega\mathscr{E}}{P} = -\frac{\omega\mathscr{E}}{d\mathscr{E}/dt} \tag{4.7-4}$$

where $\mathscr{E}$ is the stored energy, ω is the resonant frequency, and $P = -d\mathscr{E}/dt$ is the power dissipated. By comparing (4.7-4) and (4.7-1) we obtain

$$Q = \omega t_c \tag{4.7-5}$$

The Q factor is related to the full width $\Delta\nu_{1/2}$ (at the half-power points) of the resonator's Lorentzian response curve as ([4] and Section 5.1).

$$\Delta\nu_{1/2} = \frac{\nu}{Q} = \frac{1}{2\pi t_c} \tag{4.7-6}$$

so that, according to (4.7-3)

$$\Delta\nu_{1/2} = \frac{c[\alpha - (1/l)\ln\sqrt{R_1R_2}]}{2\pi n} \tag{4.7-7}$$

The most common loss mechanisms in optical resonators are the following.

1. *Loss resulting from nonperfect reflection.* Reflection loss is unavoidable, since without some transmission no power output is possible. In addition, no mirror is ideal; and even when mirrors are made to yield the highest possible reflectivities, some residual absorption and scattering reduce the reflectivity to somewhat less than 100 percent.
2. *Absorption and scattering in the laser medium.* Transitions from some of the atomic levels, which are populated in the process of pumping, to higher-lying levels constitute a loss mechanism in optical resonators when they are used as laser oscillators. Scattering from inhomogeneities and imperfections is especially serious in solid-state laser media.
3. *Diffraction losses.* From (4.3-1) or from Figure 2-7, we find that the energy of propagating-beam modes extends to considerable distances from the axis. When a resonator is formed by "trapping" a propagating beam between two reflectors, it is clear that for finite-dimension reflectors some of the beam energy

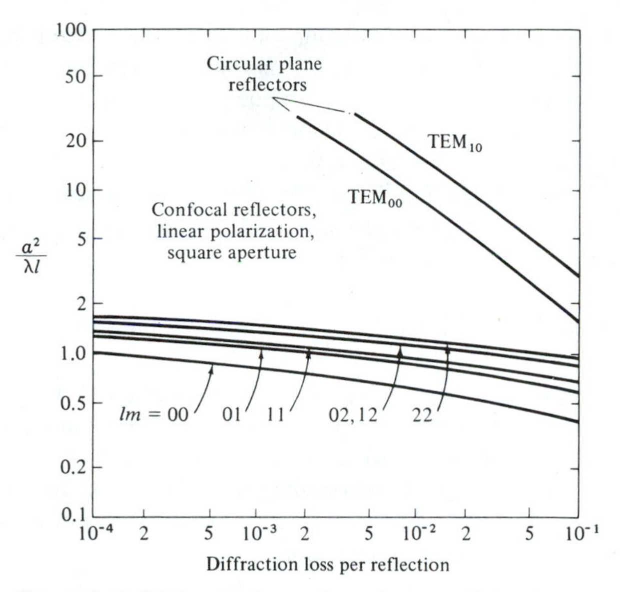

Figure 4-13 Diffraction losses for a plane-parallel and several low-order confocal resonators; a is the mirror radius and l is their spacing. The pairs of numbers under the arrows refer to the transverse-mode indices l, m. (After Reference [6].)

will not be intercepted by the mirrors and will therefore be lost. For a given set of mirrors this loss will be greater, the higher the transverse mode indices l, m, since in this case the energy extends farther. This fact is used to prevent the oscillation of higher-order modes by inserting apertures into the laser resonator whose opening is large enough to allow most of the fundamental $(0, 0, q)$ mode energy through, but small enough to increase substantially the losses of the higher-order modes. Figure 4-13 shows the diffraction losses of a number of low-order confocal resonators. Of special interest is the dramatic decrease of the diffraction losses that results from the use of spherical reflectors instead of the plane-parallel ones.

4.8 OPTICAL RESONATORS—DIFFRACTION THEORY APPROACH

The approach used in this chapter to develop the formalism of optical resonators is to start with the electromagnetic propagating beam modes whose properties were derived in Chapter 2 and then "construct" the resonator by placing two spherical reflectors whose radii of curvature are identical to those of the beam wavefronts at the mirrors' positions. This approach is elegant in that it treats both beams and resonators with the same formalism. It also leads to closed form expressions for many of the important resonator quantities such as beam spot sizes, resonant frequency, and others.

There exists a second method of treating optical resonators that is based on the diffraction theory of electromagnetic waves. This approach was the one used by Fox and Li [11] to prove the existence of optical modes. It also lends itself in a natural manner to numerical calculations of resonator properties and is used a great deal in the practical design of high-power laser systems.

According to scalar diffraction theory, the complex amplitude $U_2(\mathbf{P}_2)$ at some point $\mathbf{P}_2$ of a monochromatic electromagnetic field (frequency $\omega/2\pi$) is related to that on some aperture Σ by the integral [9]

$$U_2(\mathbf{P}_2) = \frac{i}{\lambda} \iint_{\Sigma} U_1(\mathbf{P}_1) \frac{e^{-ikr_{21}}}{r_{21}} \cos(\mathbf{n}, \mathbf{r}_{21})\, dx_1\, dy_1 \qquad \textbf{(4.8-1)}$$

where $k = \omega n/c = 2\pi/\lambda$ (so that $\lambda = \lambda_{\text{vac}}/n$ is the wavelength in the medium) $\mathbf{n}$ is the unit vector normal to Σ, and the other symbols are as shown in Figure 4-14.

Equation (4.8-1) can be simplified in situations where the angle $(\mathbf{n}, \mathbf{r}_{21})$ is very small. This is true when the distance z from the plane of observation to the aperture plane far exceeds the transverse dimensions (x_1, y_1) and (x_2, y_2) of interest. Under these conditions we can use the approximation

$$\cos(\mathbf{n}, \mathbf{r}_{21}) \simeq 1$$

and replace r_{21} in the denominator of (4.8-1) by z. The result is

$$U_2(\mathbf{P}_2) = \frac{i}{\lambda z} \iint_{\Sigma} U_1(\mathbf{P}_1) e^{-ikr_{21}}\, dx_1\, dy_1 \qquad \textbf{(4.8-2)}$$

Note that we have not replaced r_{21} by z in the exponent, since there it is multiplied by ik whose magnitude at $\lambda = 1\ \mu\text{m}$, for example, is $\sim 10^7\ \text{m}^{-1}$. Under these

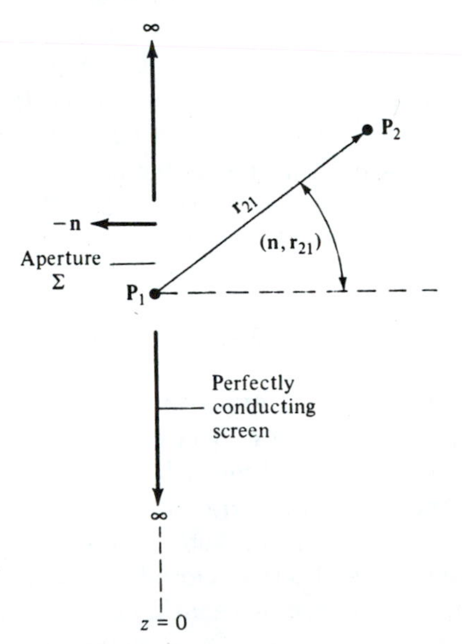

Figure 4-14 The geometry used in Equation (4.8-1). $\mathbf{n}$ is the outward normal unit vector in the aperture plane.

conditions the difference between kr_{21} and kz could be comparable to or larger than 2π.

We can use Equation (4.8-1) to derive formally an integral equation for the modes of an optical resonator. This method is equivalent to the propagating beam method of Section 4.3. It is, however, more amenable to numerical solutions. To be specific, we will consider the modes of the generic resonator shown in Figure 4-15.

In deriving the modes of the resonator by the diffraction integral method, we assume that the mode corresponds to a wave that bounces continuously back and forth between reflectors 1 and 2. It is thus analogous to a transmission of the same wave through an infinite biperiodic sequence of equivalent lenses ($f_{1,2} = R_{1,2}$) that are separated by L and set in opaque absorbing screens as shown in Figure 4-15(b). We shall seek to find whether such a structure possesses field solutions that repeat themselves within a complex multiplicative constant after each round trip. These will correspond to the *eigenmodes* of the equivalent resonator.

For $\mathbf{r}_{21}$ making a small angle with the z axis we obtain

$$r_{21} \simeq \sqrt{d^2 + (x_2 - x_1)^2 + (y_2 - y_1)^2} - \frac{x_2^2 + y_2^2}{2R_2} - \frac{x_1^2 + y_1^2}{2R_1}$$

$$\simeq d\left(1 + \frac{(x_2 - x_1)^2}{2d^2} + \frac{(y_2 - y_1)^2}{2d^2}\right) - \frac{x_2^2 + y_2^2}{2R_2} - \frac{x_1^2 + y_1^2}{2R_1}$$

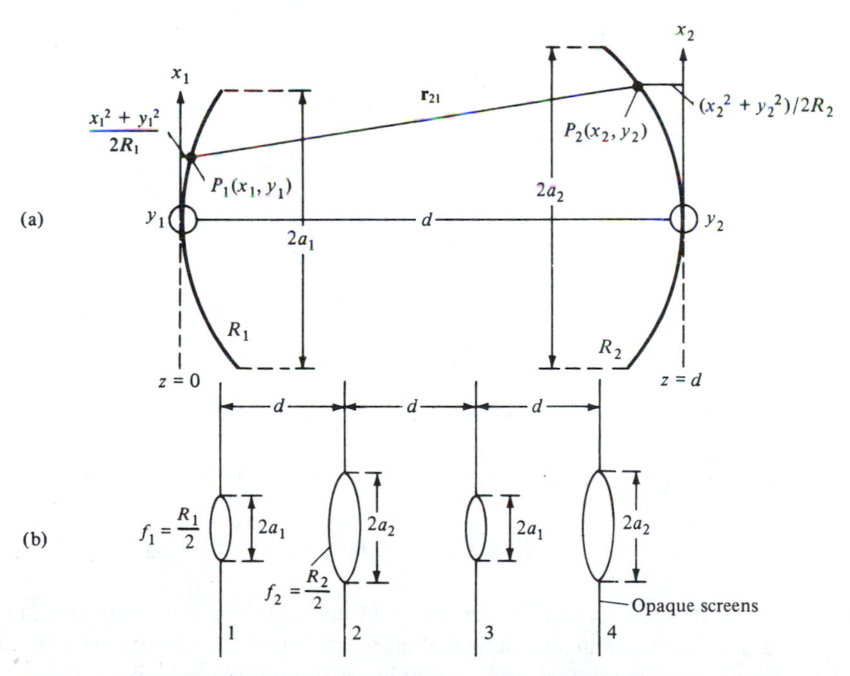

Figure 4-15 (a) The two spherical reflectors separated by d have radii of curvature R_1 and R_2. (b) The biperiodic sequence of lenses that is equivalent to the resonator (a). The focal distances of the lenses are the same as those of the corresponding reflectors, i.e., $f_{1,2} = R_{1,2}/2$. The spacing d of the lenses is the same as that of the reflectors.

where the factors involving R_1 and R_2 account for the additional path differences due to the sphericity of the mirrors as shown in Figure 4-15(a).

The diffraction integral (4.8-2) becomes

$$U_2(x_2, y_2) = \frac{ie^{-ikd}}{\lambda d} \iint_{\Sigma_1} U_1(x_1, y_1) \exp\left[-ik\left(\frac{(x_2 - x_1)^2 + (y_2 - y_1)^2}{2d} - \frac{x_1^2 + y_1^2}{2R_1} - \frac{x_2^2 + y_2^2}{2R_2}\right)\right] dx_1\, dy_1 \tag{4.8-3}$$

$$= \iint_{\Sigma_1} h(x_2, y_2; x_1, y_1) U_1(x_1, y_1)\, dx_1\, dy_1 \tag{4.8-4}$$

where the kernel h is defined implicitly by (4.8-4) as

$$h(x_2, y_2; x_1, y_1) = \frac{ie^{-ikd}}{\lambda d} \exp\left[-ik\left(\frac{(x_2 - x_1)^2 + (y_2 - y_1)^2}{2d} - \frac{x_1^2 + y_1^2}{2R_1} - \frac{x_2^2 + y_2^2}{2R_2}\right)\right] \tag{4.8-5}$$

The integration in (4.8-4) is limited to the area Σ_1 of the aperture, thus assuming tacitly that the field outside it is negligibly small.

In the following we will simplify our notation by writing the address (x, y) of a point $\mathbf{x}$ so that Equation (4.8-4) can be written as

$$U_2(\mathbf{x}_2) = \int_{\Sigma_1} h(\mathbf{x}_2, \mathbf{x}_1) U_1(\mathbf{x}_1) d^2\mathbf{x}_1 \tag{4.8-6}$$

In a similar fashion we can express the field on mirror 1 due to $U_2(\mathbf{x}_2)$ as

$$U_3(\mathbf{x}) = \int_{\Sigma_2} h(\mathbf{x}, \mathbf{x}_2) U_2(\mathbf{x}_2) d^2\mathbf{x}_2 \tag{4.8-7}$$

Replacing U_2 by the right side of (4.8-6) and interchanging the order of integration leads to

$$U_3(\mathbf{x}) = \int_{\Sigma_1} K(\mathbf{x}, \mathbf{x}_1) U_1(\mathbf{x}_1) d^2\mathbf{x}_1 \tag{4.8-8}$$

where $K(\mathbf{x}, \mathbf{x}_1)$, the *round trip* kernel $1 \rightarrow 2 \rightarrow 1$, is given by

$$K(\mathbf{x}, \mathbf{x}_1) = \int_{\Sigma_2} h(\mathbf{x}, \mathbf{x}_2) h(\mathbf{x}_2, \mathbf{x}_1) d^2\mathbf{x}_2 \tag{4.8-9}$$

A necessary condition for the field calculated by our technique to correspond to that of a resonator mode is that after one round trip the field on each mirror, say 1, returns to its original value to within some multiplicative constant γ. Using (4.8-7) we express this condition as

$$U_3(\mathbf{x}) = \gamma U_1(\mathbf{x})$$

or, using (4.8-8) and letting $U_1(\mathbf{x}) \rightarrow U(\mathbf{x})$,

$$\int_{\Sigma_1} K(\mathbf{x}, \mathbf{x}_1)U(\mathbf{x}_1)d^2\mathbf{x}_1 = \gamma U(\mathbf{x}) \tag{4.8-10}$$

Such an integral equation can be shown to possess a discrete set of solutions (eigenfunctions), which we will denote by U_{mn}. The function U_{mn} can be shown [10] to approach in the limit of large Fresnel numbers ($a_{1,2}^2/d\lambda \gg 1$) the Hermite–Gaussian solutions obtained in Chapter 2 by the propagating beam mode method. The solution yields also the associated complex eigenvalue γ_{mn}. γ_{mn} corresponds physically to the factor by which the amplitude changes in one round trip. It we write

$$\gamma_{mn} = |\gamma_{mn}|e^{-i\phi_{mn}}$$

then $\alpha_{mn} = 1 - |\gamma_{mn}|^2$ is the loss in mode power per round trip, while ϕ_{mn} is the phase shift per round trip.

The oscillation frequency is determined by the requirement that the phase delay ϕ_{mn} per one round trip be some integer, say q, of 2π. Since the round trip kernel K contains, according to (4.8-5) and (4.8-9), the phase delay factor $\exp(-2ikd)$, it is convenient to define

$$\phi_{mn} = \psi_{mn} + 2kd$$

and write the oscillation frequency condition as

$$\psi_{mn} + 2k_{\mathbf{mn}}d = q2\pi$$

and using $k_{mn} = 2\pi f_{mn}/c$, obtain

$$f_{mn\,q} = \frac{c}{2d}\left(q - \frac{\psi_{mn}}{2\pi}\right) \tag{4.8-11}$$

as the oscillation frequency of mode (m, n, q). The integer q corresponds to the number of maxima of the standing wave interference pattern between the two reflectors.

Equivalent Resonator Systems

Equation (4.8-3) can be rewritten as

$$U_2(x_2, y_2) = \frac{ie^{-ikd}}{\lambda d}\iint_{\Sigma_1} U_1(x_1,y_1) \exp\left\{-\frac{ik}{2d}[g_2(x_2^2 + y_2^2) + g_1(x_1^2 + y_1^2) - 2x_1x_2 - 2y_1y_2]\right\} dx_1\, dy_1 \tag{4.8-12}$$

$$g_{1,2} = 1 - \frac{d}{R_{1,2}}$$

A number of equivalence properties can be deduced directly from the above integral relation:

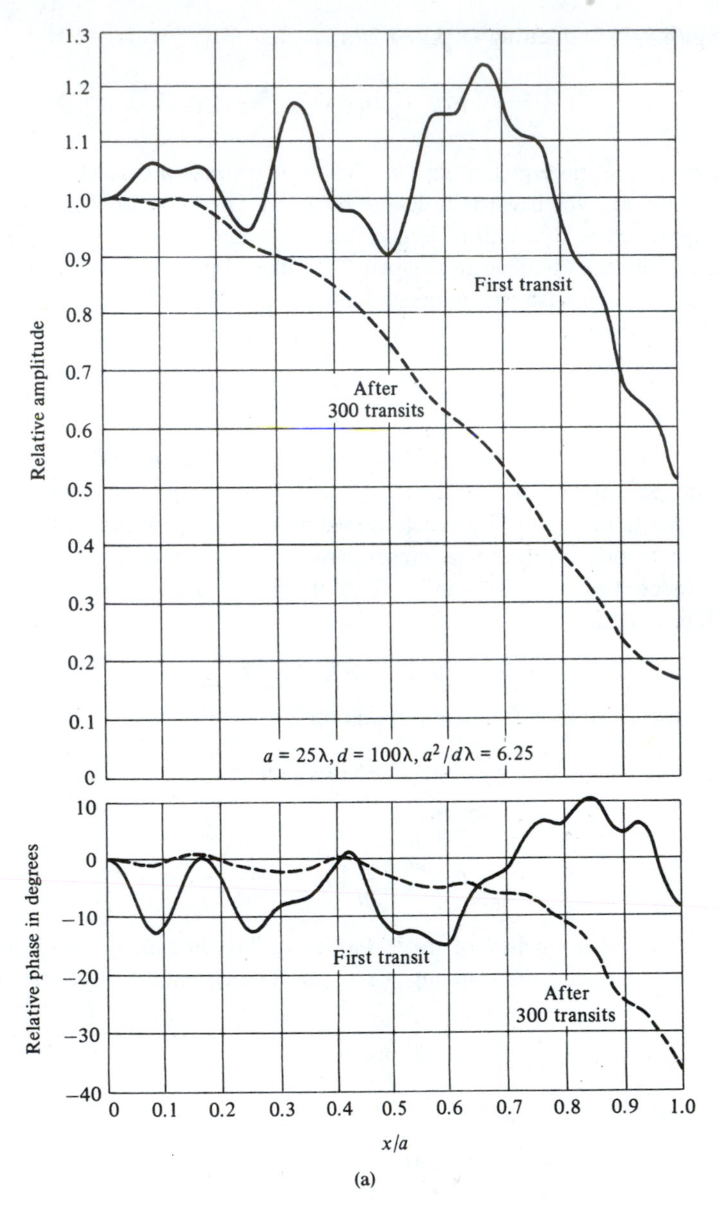

Figure 4-16 (a) Relative amplitude and phase distributions of field for finite-width strip mirrors. (The initially launched wave has a uniform distribution.) The mirror height is $2a$. (b) Relative steady-state amplitude and phase distributions of field intensity of the lowest-order even-symmetry mode for infinite strip mirrors. (After Reference [11].)

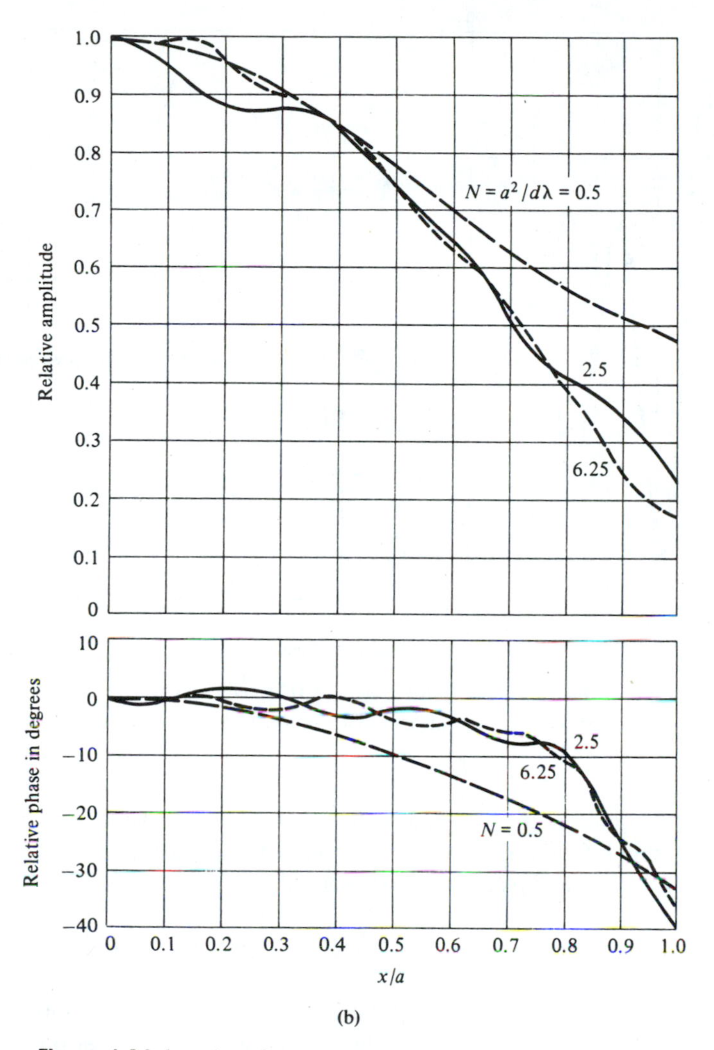

Figure 4-16 *(continued)*

1. Since the subscripts 1 and 2 can be interchanged, the reflectors of a resonator can be interchanged without affecting the field distribution at the mirror.
2. The diffraction loss and the intensity pattern of a mode remain invariant if both g_1 and g_2 are reversed in the sign. The field eigenfunctions $U_{m,n}$ and the eigenvalues γ_{mn} are merely replaced by their complex conjugates. One example for such equivalent systems is that of a planar parallel ($g_1 = g_2 = 1$) and concentric ($g_1 = g_2 = -1$) resonators.

Mode Solution by Numerical Iteration

The Kirchhoff–Fresnel integral formulation of the propagation of electromagnetic field between two planes, Equation (4.8-3), can be used to calculate numerically the

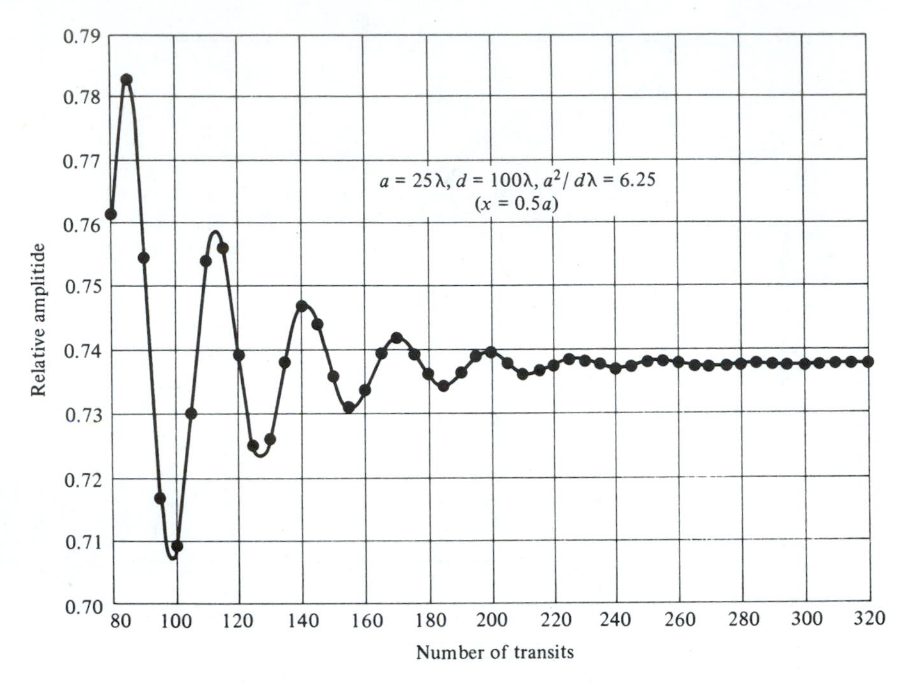

Figure 4-17 Fluctuation of field amplitude at $x = 0.5a$ as a function of number of transits. (The initially launched wave has a uniform distribution.) (After Reference [11].)

mode field distribution, the losses, and the round-trip phase shift of optical resonators. This method, first employed by Fox and Li [15], has played a key role in the understanding of optical resonator modes and in the practical design of such resonators.

The basic approach is intuitively simple. We assume some arbitrary (both in phase and amplitude) field distribution over one of the mirrors, say 1, of the resonator of Figure 4-15(a). We will call this field distribution $U^{(0)}$. Using the integral relation (4.8-8), we can calculate the resulting field $U(1)$ on mirror 1 after one round trip. This procedure can be repeated n times, resulting in

$$U^{(n)}(\mathbf{x}) = \int_{\Sigma_1} K(\mathbf{x}, \mathbf{x}')U^{(n-1)}\, d^2\mathbf{x}' \tag{4.8-13}$$

K being the round-trip propagation kernel as given by (4.8-9). We would expect that if the resonator possesses a stable fundamental mode, then after a large number of iterations the field $U^{(n)}$ will converge toward a steady state where the only change per iteration of the field is due to the multiplicative constant, i.e.,

$$\frac{U^{(n)}(\mathbf{x})}{U^{(n-1)}(\mathbf{x})} = \gamma = |\gamma|\, e^{-i\phi}$$

The mode intensity loss per round trip is thus $1 - |\gamma|^2$, while ϕ is the phase shift.

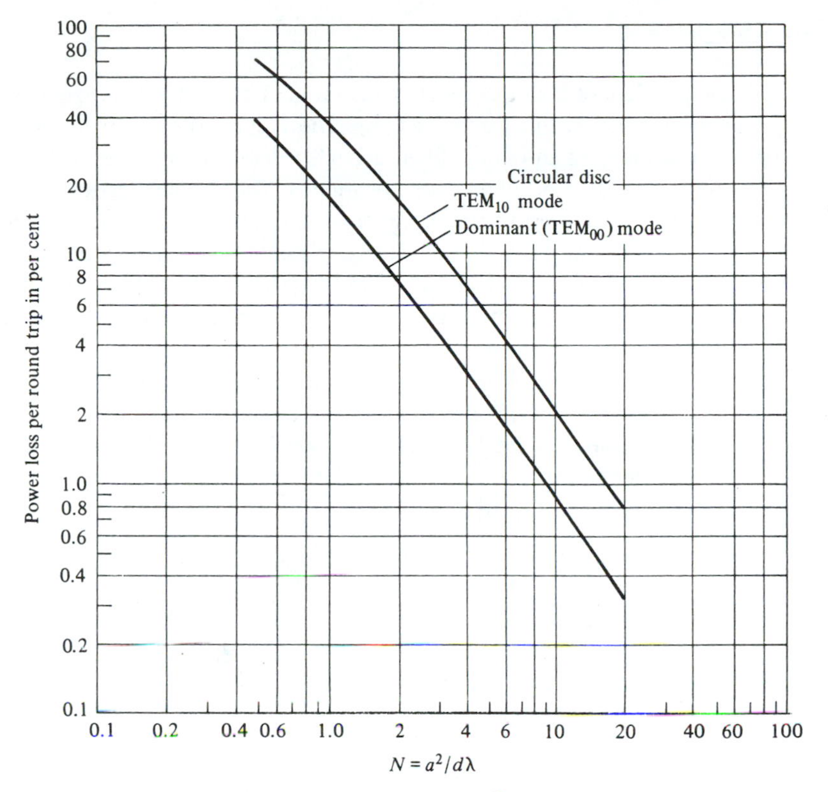

Figure 4-18 Power loss per transit versus $N = a^2/d\lambda$ for circular reflectors. (After Reference [11].)

The physical intuitive basis for the convergence of this procedure to the unique resonator mode can be, perhaps, better appreciated by reference to Figure 4-15. The arbitrary initial field distribution $U^{(0)}$ is incident on aperture 1 of the biperiodic lens sequence. This excites a large number of the propagating eigenmodes U_{mn} of this structure [i.e., of the equivalent resonator of Figure 4-15(b)]. Since the diffraction loss per period $1 - |\gamma_{mn}|^2$ (loss due to energy spreading by diffraction beyond the finite apertures of the resonators) of the high-order modes is larger than that of the fundamental mode, we expect that after a sufficiently long distance corresponding to many bounces in the resonator, the high-order modes will have decayed to a degree where the remaining field distribution is predominantly that of the lowest loss eigenmode. This is analogous to spatial filtering in optical engineering. It also follows from this argument of spatial filtering that if the starting distribution $U^{(0)}(\mathbf{x})$ possesses odd symmetry in one plane, the resulting ''steady-state'' solution will correspond to the lowest-order odd-symmetry mode TE_{10}.

In Figures 4-16 to 4-18, we reproduce some of the numerical results of Fox and Li and their original calculation which illustrate how the procedure described above yields the steady-state field configuration and the corresponding eigenvalues.

4.9 MODE COUPLING

A basic problem of both theoretical and practical interest is how to couple efficiently an incident beam with a complex amplitude $E_{in}(x,y,z)$ to a given mode m,n of an optical resonator or an optical fiber, and also derive a measure of the residual, undesirable, excitation of other resonator modes. The problem arises frequently when the resonator is used as a scanning Fabry–Perot etalon to obtain the spectrum of an optical beam. In this case, the transmitted (resonant) frequencies are functions of the resonator mode (q,l,m), which is excited by the incident beam. An unambiguous spectral determination requires that just one such mode, usually the fundamental, be excited.

Referring to Figure (4-19), we designate the incoming incident field at the "input" plane z_1 as $E_{in}(x,y)$ and the eigen (resonant) modes of the resonator as $E_{qlm}(x,y,z)$, where q is the longitudinal mode integer while l,m are the transverse integers, as defined in Section 4.6. We define $E_{in}(x,y,z_1) \equiv E_{in}(x,y)$ and recall the basic resonator mode Equation (4.3-1).

$$E_{lm}(x,y,z) \equiv E_{lm}(x,y) \propto H_1\left(\sqrt{2}\,\frac{x}{\omega_1}\right) H_m\left(\sqrt{2}\,\frac{y}{\omega_1}\right) \exp\left[-\frac{x^2+y^2}{\omega_1^2} - ik\frac{x^2+y^2}{2R_1} - i\theta\right] \tag{4.9-1}$$

where $\omega_1 = \omega(z = z_1)$ and

$$\theta = kz_1 - (1 + m + 1)1 \tan^{-1}\frac{z_1}{z_0} \tag{4.9-2}$$

The set $E_{mn}(x,y,z)$ of resonator modes as given by (4.3-1) constitutes a complete orthonormal set in which to expand an arbitrary function of x and y. They satisfy

$$\iint E_{lm}(x,y,z)\, E^*_{l'm'}(x,y,z)dx\,dy = 0 \quad \text{unless } l' = l, \text{ and } m' = m \tag{4.9-3}$$

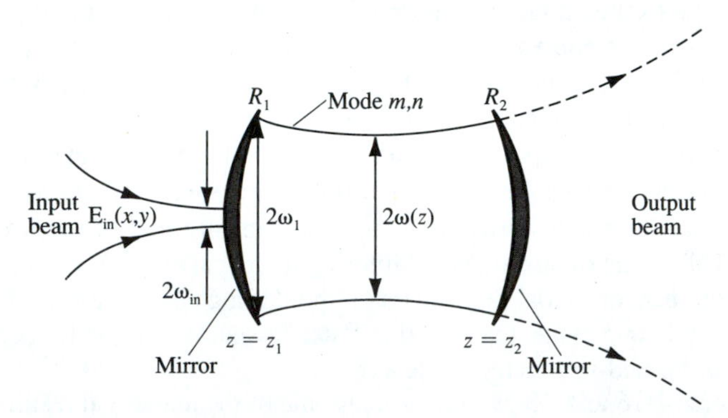

Figure 4-19 Excitation of a transverse mode m,n of a resonator (or a fiber) by an incident monochromatic beam.

We can choose the norm of these functions arbitrarily. We will do so by requiring that when the electric field is given by

$$E(\text{of mode } lm) = a_{lm} E_{lm}(x,y,z)$$

that $|a_{lm}|^2$ is equal to the total power (in watts) carried by that mode. Using Equation (1.3-19), the normalization condition becomes

$$\iint |E_{lm}(x,y,z)|^2 \, dx \, dy = 2\eta \qquad \eta = \sqrt{\frac{\mu}{\varepsilon}} \tag{4.9-4}$$

We are now ready to address the coupling problem. We take advantage of the completeness of the set E_{lm} and expand the incident field

$$E_{\text{in}}(x,y) = \sum_{l',m'} a_{l'm'} E_{l'm'}(x,y,z_1)$$

where z_1 is the position of the mirror. Multiplying both sides by E^*_{lm}, integrating over the entire x,y plane at z_1, and making use of Equations (4.9-3, 4) yields

$$a_{lm} = \frac{1}{2\eta} \iint E^*_{\text{in}}(x,y) \, E^*_{lm}(x, y) \, dx \, dy \tag{4.9-5}$$

The total power in the input beam is

$$\begin{aligned} P_{\text{in}} &= \frac{1}{2\eta} \iint |E_{\text{in}}(x,y)|^2 \, dx \, dy \\ &= \frac{1}{2\eta} \iint \sum_{lm} a_{mn} E_{lm}(x,y) \sum_{l'm'} a^*_{l'm'} E^*_{l'm'}(x,y) \, dx \, dy \\ &= \sum_{lm} |a_{lm}|^2 \end{aligned}$$

where we made use of Equations (4.9-3) and (4.9-4).

The coupling efficiency of the incident field into a given spatial mode, say, mn, is defined as

$$\begin{aligned} \eta_{mn} &= \frac{\text{Power coupled into mode } mn}{\text{Total incident power}} = \frac{|a_{lm}|^2}{\sum_{l'm'} |a_{l'm'}|^2} \\ &= \frac{\left| \iint E_{\text{in}}(x,y) E^*_{lm}(x,y) \, dx \, dy \right|^2}{\sum_{l'm'} \left| \iint E_{\text{in}}(x,y) E^*_{l'm'}(x,y) \, dx \, dy \right|^2} \end{aligned} \tag{4.9-6}$$

It follows from (4.9-6) that if the input beam at the input plane z_1 has the *same* spatial dependence as that of the mode to be excited, that is, when

$$E_{\text{in}}(x,y) \propto E_{lm}(x,y)$$

then $\eta_{lm} = 1$, and all other $\eta_{l'm'}$ are zero, and all the incident power goes into exciting the single l,m mode. In practice, mode matching requires that at the input resonator mirror, the incident beam possesses the same transverse mode numbers m and n as the mode to be excited, as well as possessing the same spot size and radius of curvature. An additional condition is that the mode m,n thus excited fulfill the longitudinal Fabry–Perot resonance condition (4.6-10) of the resonator. Otherwise most of the incident beam power will be reflected. In the situation depicted in Figure 4-18, where $\omega_{\text{in}} \ll \omega_1$, a large number of modes will be excited. It will be left as an exercise to show that the number of modes that are excited "substantially" is $\sim(\omega_1/\omega_{\text{in}})^2$.

By "mode matching" we can avoid the excitation of higher-order modes when we use a spherical mirror scanning Fabry–Perot etalon, as an example, to analyze the spectrum of an incident fundamental Gaussian beam. The transmission peaks of these modes, illustrated in Figure 4-12, introduce a deleterious ambiguity as to the frequency transmitted at a given mirror spacing and invariably degrade the etalon performance.

Problems

4.1 Plot I_t/I_i vs. δ of a Fabry–Perot etalon with $R = 0.9$.

4.2 Show that if a Fabry–Perot etalon has a fractional intensity loss per pass of $(1 - A)$, its peak transmission is given as $(1 - R)^2A/(1 - RA)^2$.

4.3 Starting with the definition (4.2-6)

$$F \equiv \frac{\nu_{m+1} - \nu_m}{\Delta\nu_{1/2}}$$

for the Finesse of a Fabry–Perot etalon and using semiquantitative arguments, show why in the case where the root-mean-square surface deviation from perfect flatness is approximately λ/N, the finesse cannot exceed $F \simeq N/2$. [*Hint:* Consider the spreading of the transmission peak due to a small number of etalons of nearly equal length transmitting in parallel.]

4.4 Show that the angular spread of a beam that is incident normally on a plane-parallel Fabry–Perot etalon must not exceed

$$\theta_{1/2} = \sqrt{\frac{2n\lambda}{lF}}$$

if its peak transmission is not to deviate substantially from unity.

4.5 Complete the derivation of Equations (4.1-4), (4.1-5), (4.1-6), and (4.1-7).

4.6 Consider a diverging monochromatic beam that is incident on a plane-parallel Fabry–Perot etalon.

a. Obtain an expression for the various angles along which the output energy is propagating. [*Hint:* These correspond to the different values of θ in (4.1-8) that result from changing m.]
b. Let the output beam in (a) be incident on a lens with a focal length f. Show that the energy distribution in the focal plane consists of a series of circles, each corresponding to a different value of m. Obtain an expression for the radii of the circles.
c. Consider the effect in (*b*) of having simultaneously two frequencies ν_1 and ν_2 present in the input beam. Derive an expression for the separation of the respective circles in the focal plane. Show that the smallest separation $\nu_1 - \nu_2$ that can be resolved by this technique is given by $(\Delta\nu)_{\min} \sim c/2nlF$.

4.7 Calculate the fraction of the power of a fundamental ($l = m = 0$) Gaussian beam that passes through an aperture with a radius equal to the beam spot size.

4.8 Show that in the case of a conventional two-reflector resonator the stability condition [Equation (4.5-5)] reduces to Equation (4.4-2).

4.9 Consider a spherical mirror with a radius of curvature R whose reflectivity varies as

$$\rho(r) = \rho_0 \exp(-r^2/a^2)$$

where r is the radial distance from the center.

Show that the (A, B, C, D) matrix of this mirror is given by

$$\begin{vmatrix} A & B \\ C & D \end{vmatrix} = \begin{vmatrix} 1 & 0 \\ -\dfrac{2}{2R} - i\dfrac{\lambda}{\pi a^2} & 1 \end{vmatrix}$$

4.10 Given an optical resonator

$$(\longleftarrow l = 30 \text{ cm} \longrightarrow)$$
$$R_1 = -20 \text{ cm} \qquad R_2 = 15 \text{ cm}$$

a. Calculate the position of the waist of the mode at $\lambda = 1\ \mu$m.
b. Calculate the diameter of the waist.

4.11 Consider a Fabry–Perot etalon with a front mirror with intensity reflectivity R_f and perfectly reflecting back mirror ($R_b = 1$).

a. Show that $|E_r/E_i|^2 = 1$ at all frequencies. (The notation is that of Sec. 4.1)
b. Derive an approximate analytic expression for the phase delay (phase of E_r/E_i) vs. δ near the zero crossing points. (Clue: $\dfrac{D\phi}{d\delta} \approx \dfrac{4\sqrt{R_f}}{1 - R_f}$ $(R_f \geq 0.6)$. The above etalon is named after its inventor the "Gires-Tournois" etalon.

4.12 Show by simple arguments that in the situation depicted in Fig. 4-18 the number of Hermite-Gaussian resonator modes which are excited substantially (say within an order of magnitude of the maximal value) by a fundamental (incident) Gaussian beam is $\sim(\omega_1/\omega_{\text{in}})^2$. Ignore longitudinal resonance and consider only transverse (x,y) modes. Clue: Consider the integral in the numerator of expression (4.9-6) for η_{mn}. Reason why the main contribution to the integral comes from a circle of radius ω_{in} or ω_1/ℓ (ℓ,m are the Hermite polynomial integers) whichever is smaller.

4.13 Obtain an expression for the coupling efficiency of an incident Gaussian fundamental beam into an optical resonator whose mirrors have a radius of curvature R and a spot radius at the mirrors of ω. The incident beam has a radius R_b and spot size radius ω_b at the mirror.

References

1. Ramo, S., J. R. Whinnery, and T. Van Duzer, *Fields and Waves in Communication Electronics.* New York: Wiley, 1965.
2. Yariv, A., *Quantum Electronics,* 3rd ed. New York: Wiley, 1989, p. 99.
3. Fabry, C., and A. Perot, ''Théorie et applications d'une nouvelle methode de spectroscopie interférentielle,'' *Ann. Chim. Phys.* 16:115, 1899.
4. Born, M., and E. Wolf, *Principles of Optics,* 3rd ed. New York: Pergamon, 1965, Chap. 7.
5. Peterson, D. G., and A. Yariv, ''Interferometry and laser control with Fabry-Perot etalons,'' *Appl. Opt.* 5:985, 1966.
6. Boyd, G. D., and J. P. Gordon, ''Confocal multimode resonator for millimeter through optical wavelength masers,'' *Bell System Tech. J.* 40:489, 1961.
7. Boyd, G. D., and H. Kogelnik, ''Generalized confocal resonator theory,'' *Bell System Tech. J.* 41:1347, 1962.
8. Casperson, L., ''Gaussian light beams in inhomogeneous media,'' *Appl. Opt.* 12:2434, 1973.
9. Born, M., and E. Wolf, *Principles of Optics.* New York: Macmillan, 1967.
10. Kogelnik, H., and T. Li, ''Laser beams and resonators,'' *Proc. IEEE* 54:1312–1329, 1966.
11. Fox, A. G., and T. Li, ''Resonant modes in a maser interferometer,'' *Bell Syst. Tech. J.* 40:453–488, 1961.

Interaction of Radiation and Atomic Systems

5.0 INTRODUCTION

In this chapter we consider what happens to an electromagnetic wave propagating in an atomic medium. We are chiefly concerned with the possibility of growth (or attenuation) of the radiation resulting from its interaction with atoms. We also consider the changes in the velocity of propagation of light due to such interaction. The concepts derived in this chapter will be used in the next one in treating the laser oscillator.

5.1 SPONTANEOUS TRANSITIONS BETWEEN ATOMIC LEVELS—HOMOGENEOUS AND INHOMOGENEOUS BROADENING

One of the basic results of the theory of quantum mechanics is that each physical system can be found, upon measurement, in only one of a predetermined set of energy states—the so-called *eigenstates* of the system. With each of these states we associate an energy that corresponds to the total energy of the system when occupying the state. Some of the simpler systems, which are treated in any basic text on quantum mechanics, include the free electron, the hydrogen atom, and the harmonic oscillator. Examples of more complicated systems include the hydrogen molecule and the semiconducting crystal. With each state, the state i of the hydrogen atom say, we associate an eigenfunction [1]

$$\psi_i(\mathbf{r},t) = u_i(\mathbf{r})e^{-iE_i t/\hbar} \tag{5.1-1}$$

where $|u_i(\mathbf{r})|^2\, dx\, dy\, dz$ gives the probability of finding the electron, once it is known to be in the state i, within the volume element $dx\, dy\, dz$, which is centered on the point **r**. E_i is the state energy described above and $\hbar \equiv h/2\pi$ where $h = 6.626 \times 10^{-34}$ joule-second is Planck's constant.

One of the main tasks of quantum mechanics is the determination of the eigenfunctions $u_i(\mathbf{r})$ and the corresponding energies E_i of various physical systems. In this book, however, we will accept the existence of these states, their energy levels, as well as a number of other related results whose justification is provided by the experimentally proved formalism of quantum mechanics. Some of these results are discussed in the following.

The Concept of Spontaneous Emission

In Figure 5-1 we show a system of energy levels that are associated with a given physical system—an atom, say. Let us concentrate on two of these levels—1 and 2, for example. If the atom is known to be in state 2 at $t = 0$, there is a finite probability per unit time that it will undergo a transition to state 1, emitting in the process a photon of energy $h\nu = E_2 - E_1$. This process, occurring as it does without the inducement of a radiation field, is referred to as *spontaneous emission.*

Another, equivalent, way of thinking about spontaneous transitions, which corresponds more closely to experimental situations, is the following: Consider a large number N_2 of identical atoms that are known to be in state 2 at $t = 0$. The average number of these atoms undergoing spontaneous transition to state 1 per unit time is

$$-\frac{dN_2}{dt} = A_{21}N_2 \equiv \frac{N_2}{(t_{\text{spont}})_{21}} \tag{5.1-2}$$

where A_{21} is the spontaneous transition rate and $(t_{\text{spont}})_{21} \equiv A_{21}^{-1}$ is called the spontaneous lifetime associated with the transition $2 \rightarrow 1$. It follows from quantum mechanical considerations that spontaneous transitions take place from a given state only to states lying lower in energy, so no spontaneous transitions take place from 1 to 2. The rate A_{21} can be calculated using the eigenfunctions of states 2 and 1. In

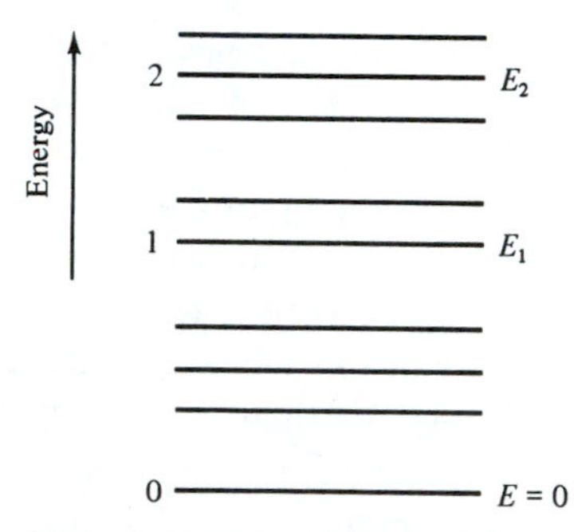

Figure 5-1 Some of the energy levels of an atomic system. Level 0, the ground state, is the lowest energy state. Levels 1 and 2 represent two excited states.

this book we *accept* the existence of spontaneous emission A_{21} and regard A_{21} as a parameter characterizing the transition $2 \rightarrow 1$ of the given physical system.[1]

Lineshape Function—Homogeneous and Inhomogeneous Broadening

If one performs a spectral analysis of the radiation emitted by spontaneous $2 \rightarrow 1$ transitions, one finds that the radiation is not strictly monochromatic (that is, of one frequency) but occupies a finite frequency bandwidth. The function describing the distribution of emitted intensity versus the frequency ν is referred to as the lineshape function $g(\nu)$ (of the transition $2 \rightarrow 1$) and its arbitrary scale factor is usually chosen so that the function is normalized according to

$$\int_{-\infty}^{+\infty} g(\nu)\, d\nu = 1 \tag{5.1-3}$$

We can consequently view $g(\nu)\, d\nu$ as the *a priori* probability that a given spontaneous emission from level 2 to level 1 will result in a photon whose frequency is between ν and $\nu + d\nu$.

Another method of determining $g(\nu)$ is to apply an electromagnetic field to the sample containing the atoms and then plot the amount of energy absorbed by $1 \rightarrow 2$ transitions as a function of the frequency. This function, when normalized according to (5.1-3), is again $g(\nu)$.

The fact that both the emission and the absorption are described by the same lineshape function $g(\nu)$ can be verified experimentally, and follows from basic quantum mechanical considerations. The proof is beyond the scope of this book, but we can perhaps make a plausibility argument using the following example. Consider a parallel *RLC* circuit that is excited into oscillation by connecting it to a signal source of frequency $\nu_0 = 1/2\pi\sqrt{LC}$. The excitation is then discontinued and the transient decay of the oscillation is observed. It is a straightforward problem to show that the intensity spectrum of the decaying oscillation, which is analogous to spontaneous emission since the total energy is decreasing, is the same as a plot of the absorption power vs. frequency of the same circuit, this last process being equivalent to induced absorption in the atomic system.

It will be left as an exercise to show that in the case of the *RLC* circuit the spectrum characterizing the decay or absorption is proportional to

---

[1]The quantum mechanical derivation gives [1]

$$A_{21} = \frac{2e^2\omega^3(x_{12}^2 + y_{12}^2 + z_{12}^2)n^3}{3hc^3\varepsilon}$$

for a class of transitions known as electric dipole transitions. The parameter ε is the dielectric constant at ω and

$$x_{12} = \int_{\substack{\text{all}\\ \text{space}}} u_1^*(\mathbf{r})xu_2(\mathbf{r})d^3\mathbf{r}$$

where x, y, and z are the coordinates of the electron. n is the index of refraction.

$$f(\nu) = \frac{1}{(\nu - \nu_0)^2 + (\nu_0/2Q)^2} \tag{5.1-4}$$

where $Q = 2\pi\nu_0 CR$ is the quality factor of the circuit.

The formal equivalence between an atomic transition and an oscillator goes even further than this *RLC* circuit example indicates. Later in this chapter we will use it extensively to describe the interaction between an atomic system and an electromagnetic field.

Homogeneous and Inhomogeneous Broadening (2)

One of the possible causes for the frequency spread of spontaneous emission is the finite lifetime τ of the emitting state. If we consider the emission from the excited state as that corresponding to a damped oscillator and choose the decay time of the oscillator as τ, we can take the radiated field as

$$\begin{aligned} e(t) &= E_0 e^{-t/\tau} \cos \omega_0 t \\ &= \frac{E_0}{2} [e^{i(\omega_0 + i\sigma/2)t} + e^{-i(\omega_0 - i\sigma/2)t}] \end{aligned} \tag{5.1-5}$$

where $\sigma/2 = \tau^{-1}$ is the field decay rate (the intensity decay rate is σ). The Fourier transform of $e(t)$ is

$$\begin{aligned} E(\omega) &= \int_0^{+\infty} e(t) e^{-i\omega t}\, dt \\ &= \frac{E_0}{2}\left[\frac{i}{(\omega_0 - \omega + i\sigma/2)} - \frac{i}{(\omega_0 + \omega - i\sigma/2)}\right] \end{aligned} \tag{5.1-6}$$

where the lower limit of integration is taken as $t = 0$ (instead of $t = -\infty$) to correspond with the start of our observation period. The spectral density of the spontaneous emission is proportional to $|E(\omega)|^2$. If we limit our attention to the vicinity of the resonant frequency $\omega \simeq \omega_0$, we obtain

$$|E(\omega)|^2 \propto \frac{1}{(\omega - \omega_0)^2 + (\sigma/2)^2} \tag{5.1-7}$$

which is of the same form as (5.1-4).

Curves with the functional dependence of (5.1-7) are called Lorentzian. They occur often in physics and engineering, since, as shown, they characterize the response of damped resonant systems.

The separation $\Delta\nu$ between the two frequencies at which the Lorentzian is down to half its peak value is referred to as the linewidth and is given by

$$\Delta\nu = \frac{\sigma}{2\pi} = \frac{1}{\pi\tau} \tag{5.1-8}$$

In the case of atomic transitions between an upper level (u) and a lower level (l), the coherent interaction of an atom in either state (u or l) with the field can be interrupted by the finite lifetime of the state (τ_u, τ_l) or by an elastic collision that erases any phase memory (τ_{cu}, τ_{cl}). We thus generalize (5.1-8) to read

$$\Delta\nu = \frac{1}{\pi}(\tau_u^{-1} + \tau_l^{-1} + \tau_{cu}^{-1} + \tau_{cl}^{-1})$$

Rewriting (5.1-7) in terms of $\Delta\nu$ and, at the same time, normalizing it according to (5.1-3), we obtain the normalized Lorentzian lineshape function

$$g(\nu) = \frac{\Delta\nu}{2\pi[(\nu - \nu_0)^2 + (\Delta\nu/2)^2]} \tag{5.1-9}$$

The type of broadening (that is, the finite width of the emitted spectrum) described above is called *homogeneous broadening.* It is characterized by the fact that the spread of the response over a band $\sim\Delta\nu$ is characteristic of *each* atom in the sample. The function $g(\nu)$ thus describes the response of any of the atoms, which are indistinguishable.

As mentioned above, homogeneous broadening is due most often to the finite interaction lifetime of the emitting or absorbing atoms. Some of the most common mechanisms are:

1. The spontaneous lifetime of the excited state.
2. Collision of an atom embedded in a crystal with a phonon. This may involve the emission or absorption of acoustic energy. Such a collision does not terminate the lifetime of the atom in its absorbing or emitting state. It does, however, interrupt the relative phase between the atomic oscillation (see Section 5.4) and that of the field, thus causing a broadening of the response according to (5.1-6) where τ now represents the mean uninterrupted interaction time.
3. Pressure broadening of atoms in a gas. At sufficiently high atomic densities, the collisions between atoms become frequent enough that lifetime termination and phase interruption as in the preceding mechanism dominate the broadening mechanism.

There are, however, many physical situations in which the individual atoms are distinguishable, each having a slightly different transition frequency ν_0. If one observes, in this case, the spectrum of the spontaneous emission, its spectral distribution will reflect the spread in the individual transition frequencies and not the broadening due to the finite lifetime of the excited state. Two typical situations give rise to this type of broadening, referred to as *inhomogeneous* broadening.

First of all, the energy levels, hence the transition frequencies, of ions present as impurities in a host crystal depend on the immediate crystalline surroundings. The ever present random strain, as well as other types of crystal imperfections, cause the crystal surroundings to vary from one ion to the next, thus effecting a spread in the transition frequencies.

Second, the transition frequency ν of a gaseous atom (or molecule) is Doppler-shifted due to the finite velocity of the atom according to

$$\nu = \nu_0 + \frac{v_x}{c}\nu_0 \tag{5.1-10}$$

where v_x is the component of the velocity along the direction connecting the observer with the moving atom, c is the velocity of light in the medium, and ν_0 is the frequency corresponding to a stationary atom. The Maxwell velocity distribution function of a gas with atomic mass M that is at equilibrium at temperature T is [3]

$$f(v_x, v_y, v_z) = \left(\frac{M}{2\pi kT}\right)^{3/2} \exp\left[-\frac{M}{2kT}(v_x^2 + v_y^2 + v_z^2)\right] \tag{5.1-11}$$

$k = 1.38 \times 10^{-23}$ J/°K is the Boltzmann constant, and $f(v_x, v_y, v_z)\, dv_x\, dv_y\, dv_z$ is thus the fraction of all the atoms whose x component of velocity is contained in the interval v_x to $v_x + dv_x$ while, simultaneously, their y and z components lie between v_y and $v_y + dv_y$, v_z and $v_z + dv_z$, respectively. Alternatively, we may view $f(v_x, v_y, v_z)\, dv_x\, dv_y\, dv_z$ as the *a priori* probability that the velocity vector $\mathbf{v}$ of any given atom terminates within the differential volume $dv_x\, dv_y\, dv_z$ centered on $\mathbf{v}$ in velocity space so that

$$\iiint_{-\infty}^{\infty} f(v_x, v_y, v_z)\, dv_x\, dv_y\, dv_z = 1 \tag{5.1-12}$$

According to (5.1-10) the probability $g(\nu)\, d\nu$ that the transition frequency is between ν and $\nu + d\nu$ is equal to the probability that v_x will be found between $v_x = (\nu - \nu_0)(c/\nu_0)$ and $(\nu + d\nu - \nu_0)(c/\nu_0)$ irrespective of the values of v_y and v_z [since if $v_x = (\nu - \nu_0)(c/\nu_0)$, the Doppler-shifted frequency will be equal to ν regardless of v_y and v_z]. This probability is thus obtained by substituting $v_x = (\nu - \nu_0)c/\nu_0$ in $f(v_x, v_y, v_z)\, dv_x\, dv_y\, dv_z$, and then integrating over all values of v_y and v_z. The result is

$$g(\nu)\, d\nu = \left(\frac{M}{2\pi kT}\right)^{3/2} \int_{-\infty}^{\infty} \int_{-\infty}^{\infty} e^{-(M/2kT)(v_4^{2y} + v_z^2)}\, dv_y\, dv_z$$
$$\times\, e^{-(M/2kT)(c^2/\nu_0^2)(\nu - \nu_0)^2} \left(\frac{c}{\nu_0}\right) d\nu \tag{5.1-13}$$

Using the definite integral

$$\int_{-\infty}^{\infty} e^{-(M/2kT)v_z^2}\, dv_z = \left(\frac{2\pi kT}{M}\right)^{1/2}$$

we obtain, from (5.1-13),

$$g(\nu) = \frac{c}{\nu_0}\left(\frac{M}{2\pi kT}\right)^{1/2} e^{-(M/2kT)(c^2/\nu_0^2)(\nu - \nu_0)^2} \tag{5.1-14}$$

for the *normalized Doppler-broadened lineshape.* The functional dependence of $g(\nu)$ in (5.1-14) is referred to as Gaussian. The width of $g(\nu)$ in this case is taken as the frequency separation between the points where $g(\nu)$ is down to half its peak value. It is obtained from (5.1-14) as

$$\Delta\nu_D = 2\nu_0 \sqrt{\frac{2kT}{Mc^2} \ln 2} \tag{5.1-15}$$

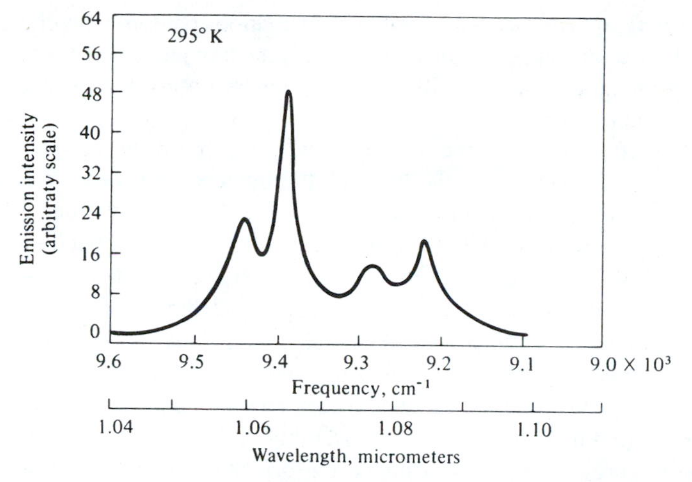

Figure 5-2 Emission spectrum of Nd^{3+}: $CaWO_4$ in the vicinity of the 1.06-μm laser transition. The main peak corresponds to the laser transition. (After Reference [4].)

where the subscript D stands for *Doppler.* We can reexpress $g(\nu)$ in terms of $\Delta\nu_D$, obtaining

$$g(\nu) = \frac{2(\ln 2)^{1/2}}{\pi^{1/2}\,\Delta\nu_D}\, e^{-[4(\ln 2)(\nu - \nu_0)^2/\Delta\nu_D^2]} \tag{5.1-16}$$

In Figure 5-2 we show, as an example of a lineshape function, the spontaneous emission spectrum of Nd^{3+} when present as an impurity ion in a $CaWO_4$ lattice. The spectrum consists of a number of transitions, which are partially overlapping.

Numerical Example: The Doppler Linewidth of Ne

Consider the 6328 Å transition in Ne, which is used in the popular He–Ne lasers. Using the atomic mass 20 for neon in (5.1-15) and taking $T = 300°$K, we obtain

$$\Delta\nu_D \simeq 1.5 \times 10^9 \text{ Hz}$$

for the Doppler linewidth. The 10.6 μm transition in the CO_2 laser has, according to (5.1-15), a linewidth $\Delta\nu_D \approx 6 \times 10^7$ Hz.

5.2 INDUCED TRANSITIONS

In the presence of an electromagnetic field of frequency $\nu \sim (E_2 - E_1)/h$, an atom whose energy levels are shown in Figure 5-1 can undergo a transition from state 1 to 2, *absorbing* in the process a quantum of excitation (photon) with energy $h\nu$ from the field. If the atom happens to occupy state 2 at the moment when it is first subjected to the electromagnetic field, it will make a downward transition to state 1, *emitting* a photon of energy $h\nu$.

What distinguishes the process of induced transition from the spontaneous one described in the last section is the fact that the induced rate for $2 \rightarrow 1$ and $1 \rightarrow 2$ transitions is *equal,* whereas the spontaneous $1 \rightarrow 2$ (that is, the one in which the atomic energy increases) transition rate is zero. Another fundamental difference—one that, again, follows from quantum mechanical considerations—is that the induced rate is *proportional* to the *intensity* of the electromagnetic field, whereas the spontaneous rate is independent of it. The relationship between the induced transition rate and the (inducing) field intensity is of fundamental importance in treating the interaction of atomic systems with electromagnetic fields. Its derivation follows.

Consider first the interaction of an assembly of identical atoms with a radiation field whose energy density is distributed uniformly in frequency in the vicinity of the transition frequency. Let the energy density per unit frequency be $\rho(\nu)$. We assume that the induced transition rates per atom from $2 \rightarrow 1$ and $1 \rightarrow 2$ are both proportional to $\rho(\nu)$ and take them as

$$(W'_{21})_{\text{induced}} = B_{21}\rho(\nu)$$
$$(W'_{12})_{\text{induced}} = B_{12}\rho(\nu) \qquad \textbf{(5.2-1)}$$

where B_{21} and B_{12} are constants to be determined. The total downward $(2 \rightarrow 1)$ transition rate is the sum of the induced and spontaneous contributions

$$W'_{21} = B_{21}\rho(\nu) + A_{21} \qquad \textbf{(5.2-2)}$$

The spontaneous rate A_{21} was discussed in Section 5.1. The total upward $(1 \rightarrow 2)$ transition rate is

$$W'_{12} = (W'_{12})_{\text{induced}} = B_{12}\rho(\nu) \qquad \textbf{(5.2-3)}$$

Our first task is to obtain an expression for B_{12} and B_{21}. Since the magnitude of the coefficients B_{21} and B_{12} depends on the atoms and not on the radiation field, we consider, without loss of generality, the case where the atoms are in thermal equilibrium with a blackbody (thermal) radiation field at temperature T. In this case the radiation density is given by [5]

$$\rho(\nu) = \frac{8\pi n^3 h\nu^3}{c^3} \frac{1}{e^{h\nu/kT} - 1} \qquad \textbf{(5.2-4)}$$

Since at thermal equilibrium the average populations of levels 2 and 1 are constant with time, it follows that the number of $2 \rightarrow 1$ transitions in a given time interval is equal to the number of $1 \rightarrow 2$ transitions; that is,

$$N_2 W'_{21} = N_1 W'_{12} \qquad \textbf{(5.2-5)}$$

where N_1 and N_2 are the population densities of levels 1 and 2, respectively. Using (5.2-2) and (5.2-3) in (5.2-5), we obtain

$$N_2[B_{21}\rho(\nu) + A_{21}] = N_1 B_{12}\rho(\nu)$$

and, substituting for $\rho(\nu)$ from (5.2-4),

$$N_2\left[B_{21}\frac{8\pi n^3 h\nu^3}{c^3(e^{h\nu/kT}-1)} + A_{21}\right] = N_1\left[B_{12}\frac{8\pi n^3 h\nu^3}{c^3(e^{h\nu/kT}-1)}\right] \tag{5.2-6}$$

Since the atoms are in thermal equilibrium, the ratio N_2/N_1 is given by the Boltzmann factor [5] as

$$\frac{N_2}{N_1} = e^{-h\nu/kT} \tag{5.2-7}$$

Equating (N_2/N_1) as given by (5.2-6) to (5.2-7) gives

$$\frac{8\pi n^3 h\nu^3}{c^3(e^{h\nu/kT}-1)} = \frac{A_{21}}{B_{12}e^{h\nu/kT} - B_{21}} \tag{5.2-8}$$

The last equality can be satisfied only when

$$B_{12} = B_{21} \tag{5.2-9}$$

and simultaneously

$$\frac{A_{21}}{B_{21}} = \frac{8\pi n^3 h\nu^3}{c^3} \tag{5.2-10}$$

The last two equations were first given by Einstein [6]. We can, using (5.2-10), rewrite the induced transition rate (5.2-1) as

$$W_i' = \frac{A_{21}c^3}{8\pi n^3 h\nu^3}\rho(\nu) = \frac{c^3}{8\pi n^3 h\nu^3 t_{\text{spont}}}\rho(\nu) \tag{5.2-11}$$

where, because of (5.2-9) the distinction between $2 \rightarrow 1$ and $1 \rightarrow 2$ induced transition rates is superfluous.

Equation (5.2-11) gives the transition rate per atom due to a field with a uniform (white) spectrum with energy density per unit frequency $\rho(\nu)$. In quantum electronics our main concern is in the transition rates that are induced by a monochromatic (that is, single-frequency) field of frequency ν. Let us denote this transition rate as $W_i(\nu)$. We have established in Section 5.1 that the strength of interaction is proportional to the lineshape function $g(\nu)$, so $W_i(\nu) \propto g(\nu)$. Furthermore, we would expect $W_i(\nu)$ to go over into W_i' as given by (5.2-11) if the spectral width of the radiation field is gradually increased from zero to a point at which it becomes large compared to the transition linewidth. These two requirements are satisfied if we take $W_i(\nu)$ as

$$W_i(\nu) = \frac{c^3\rho_\nu}{8\pi n^3 h\nu^3 t_{\text{spont}}}g(\nu) \tag{5.2-12}$$

where ρ_ν is the energy density (joules per cubic meter) of the electromagnetic field inducing the transitions. To show that $W_i(\nu)$ as given by (5.2-12) indeed goes over smoothly into (5.2-11) as the spectrum of the field broadens, we may consider the broad spectrum field as made up of a large number of closely spaced monochromatic components at ν_k with random phases, and then by adding the individual transition rates obtained from (5.2-12),

$$W_i' = \sum_{\nu_k} W_i(\nu_k) = \frac{c^3}{8\pi n^3 h t_{\text{spont}}} \sum_k \frac{\rho_{\nu_k}}{\nu_k^3} g(\nu_k) \tag{5.2-13}$$

where ρ_{ν_k} is the energy density of the field component oscillating at ν_k. We can replace the summation of (5.2-13) by an integral if we replace ρ_{ν_k} by $\rho(\nu)\,d\nu$ where $\rho(\nu)$ is the energy density per unit frequency; thus, (5.2-13) becomes

$$W_i' = \frac{c^3}{8\pi n^3 h t_{\text{spont}}} \int_{-\infty}^{+\infty} \frac{\rho(\nu) g(\nu) d\nu}{\nu^3} \tag{5.2-14}$$

In situations where $\rho(\nu)$ is sufficiently broad compared with $g(\nu)$, and thus the variation of $\rho(\nu)/\nu^3$ over the region of interest [where $g(\nu)$ is appreciable] can be neglected, we can pull $\rho(\nu)/\nu^3$ outside the integral sign, obtaining

$$W_i' = \frac{c^3}{8\pi n^3 h \nu^3 t_{\text{spont}}} \rho(\nu)$$

where we used the normalization condition

$$\int_{-\infty}^{+\infty} g(\nu)\, d\nu = 1$$

This agrees with (5.2-11).

Returning to our central result, Equation (5.2-12), we can rewrite it in terms of the intensity $I_\nu = c\rho_\nu/n$ (watts per square meter) of the optical wave as

$$W_i(\nu) = \frac{A_{21} c^2 I_\nu}{8\pi n^2 h \nu^3} g(\nu) = \frac{\lambda^2 I_\nu}{8\pi n^2 h \nu t_{\text{spont}}} g(\nu) \tag{5.2-15}$$

where c is the velocity of propagation of light in vacuum, λ is the vacuum wavelength, and $t_{\text{spont}} \equiv 1/A_{21}$.

5.3 ABSORPTION AND AMPLIFICATION

Consider the case of a monochromatic plane wave of frequency ν and intensity I_ν propagating through an atomic medium with N_2 atoms per unit volume in level 2 and N_1 in level 1. According to (5.2-15) there will occur $N_2 W_i$ induced transitions per unit time per unit volume from level 2 to level 1 and $N_1 W_i$ transitions from 1 to 2. The net power generated within a unit volume is thus

$$\frac{P}{\text{Volume}} = (N_2 - N_1) W_i h\nu$$

This radiation is added coherently (that is, with a definite phase relationship) to that of the traveling wave so that it is equal, in the absence of any dissipation mechanisms, to the increase in the intensity per unit length, or, using (5.2-15),

$$\frac{dI_\nu}{dz} = (N_2 - N_1)\frac{c^2 g(\nu)}{8\pi n^2 \nu^2 t_{\text{spont}}} I_\nu \tag{5.3-1}$$

The solution of (5.3-1) is

$$I_\nu(z) = I_\nu(0)e^{\gamma(\nu)z} \tag{5.3-2}$$

where

$$\gamma(\nu) = (N_2 - N_1)\frac{c^2}{8\pi n^2 \nu^2 t_{\text{spont}}} g(\nu) \tag{5.3-3}$$

that is, the intensity grows exponentially when the population is inverted ($N_2 > N_1$) or is attenuated when $N_2 < N_1$. The first case corresponds to laser-type amplification, whereas the second case is the one encountered in atomic systems at thermal equilibrium. The two situations are depicted in Figure 5-3. We recall that at thermal equilibrium

$$\frac{N_2}{N_1} = e^{-h\nu/kT} \tag{5.3-4}$$

so that systems at thermal equilibrium are always absorbing. The inversion condition $N_2 > N_1$ can still be represented by (5.3-4), provided we take T as negative. As a matter of fact, the condition $N_2 > N_1$ is often referred to as one of "negative temperature"—the "temperature" in this case serving as an indicator of the population ratio, in accordance with (5.3-4).

The absorption, or amplification, of electromagnetic radiation by an atomic transition can be described not only by means of the exponential gain constant $\gamma(\nu)$

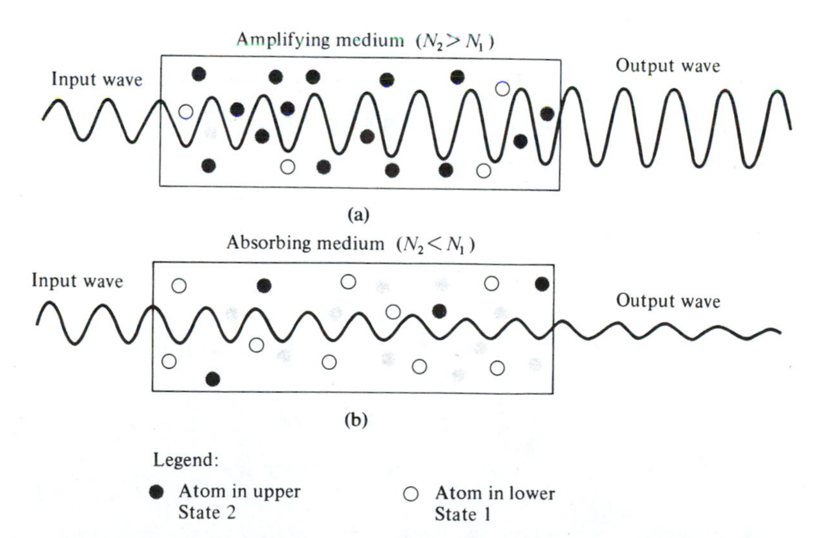

Figure 5-3 Amplification of a traveling electromagnetic wave in (a) an inverted population ($N_2 > N_1$), and (b) attenuation in an absorbing ($N_2 < N_1$) medium.

but also, alternatively, in terms of the imaginary part of the electric susceptibility $\chi_e''(\nu)$ of the propagation medium. According to (1.2-19) the density of absorbed power is

$$\frac{\overline{\text{Power}}}{\text{Volume}} = \frac{\omega\varepsilon_0\chi''(\nu)}{2}|E|^2 \tag{5.3-5}$$

where, since we are concerned here only with electric susceptibilities, we replace $\chi_e(\nu)$ by the symbol $\chi(\nu)$. This last result must agree with a derivation using the concept of the induced transition rate $W_i(\nu)$ according to which

$$\frac{\overline{\text{Power}}}{\text{Volume}} = (N_1 - N_2)W_i(\nu)h\nu \tag{5.3-6}$$

Equating (5.3-5) to (5.3-6), substituting (5.2-15) for $W_i(\nu)$, and using the relation $I_\nu = (c/n)\varepsilon|E|^2/2$ [see (1.3-26)], we obtain

$$\chi''(\nu) = \frac{(N_1 - N_2)\lambda^3}{16\pi^2 n t_{\text{spont}}} g(\nu) \tag{5.3-7}$$

where $n^2 \equiv \varepsilon/\varepsilon_0$ and λ is the wavelength in vacuum. In the case of a Lorentzian lineshape function $g(\nu)$, we use (5.1-9) to rewrite the last result as

$$\chi''(\nu) = \frac{(N_1 - N_2)\lambda^3}{8\pi^3 n t_{\text{spont}}\Delta\nu} \frac{1}{1 + [4(\nu - \nu_0)^2]/(\Delta\nu)^2} \tag{5.3-8}$$

This is a key result and it will be used on numerous occasions.

Numerical Example: The Exponential Gain Constant in a Ruby Laser

Let us estimate the exponential gain constant at line center of a ruby (Al_2O_3 doped with Cr^{3+} ions) crystal having the following characteristics:

$$N_2 - N_1 = 5 \times 10^{17}/\text{cm}^3$$

$$\Delta\nu \simeq \frac{1}{g(\nu_0)} = 2 \times 10^{11} \text{ Hz at } 300^\circ \text{ K}$$

$$t_{\text{spont}} = 3 \times 10^{-3} \text{ second}$$

$$\nu = 4.326 \times 10^{14} \text{ Hz}$$

$$\frac{c}{n} \text{ (in ruby)} \simeq 1.69 \times 10^{10} \text{ cm/sec}$$

Using these values in (5.3-3) gives

$$\gamma(\nu) \simeq 5 \times 10^{-2} \text{ cm}^{-1}$$

Thus, the intensity of a wave with a frequency corresponding to the center of the transition is amplified by approximately 5 percent per cm in its passage through a ruby rod with the foregoing characteristics.

5.4 DERIVATION OF $\chi'(\nu)$

In Section 5.3, we derived an expression for the imaginary part $-\chi''(\nu)$ of the optical susceptibility $\chi(\nu) = \chi'(\nu) - i\chi''(\nu)$. To complete the picture, we also need an expression for $\chi'(\nu)$. A first principles quantum mechanical derivation would lead to an expression for the complex $x(\nu)$, thus yielding both $\chi'(\nu)$ and $\chi''(\nu)$. Since such a derivation prerequires concepts not covered in this book, we will employ instead a powerful analytical tool—the Kramers–Kroning relations [12]—to obtain $\chi'(\nu)$ from $\chi''(\nu)$. These relations were derived initially in the context of the interaction of light and matter. They apply, however, to the response function of any time-invariant linear passive system and relate the real and imaginary part of its frequency response function. One condition that must be obeyed by a dissipative system is that if shocked by an impulse $e(t) = \delta(t)$, its response $p(t)$ will decay with time. This requires (the proof will be assigned as a problem) that its response function, which is the Fourier transform of the impulse response, *$\chi(\nu)$ possess no poles in the lower half ν plane.* (If we had adopted a convention for harmonic time dependence $\exp(-i\omega t)$ instead of $\exp(i\omega t)$, then $\chi(\nu)$ could possess no poles in the upper half plane). In this case, a simple contour integration over the lower half ν plane of the function $\chi(\nu)/(\nu - \nu')$ leads to

$$\chi'(\nu) = \frac{1}{\pi} P \int_{-\infty}^{\infty} \frac{\chi''(\nu')}{\nu' - \nu} d\nu'$$

$$\chi''(\nu) = -\frac{1}{\pi} P \int_{-\infty}^{\infty} \frac{\chi'(\nu')}{\nu' - \nu} d\nu' \tag{5.4-4}$$

where P stands for the Cauchy principal value of the integral that follows. The derivation of these very important relations is given in Appendix A. The great practical importance of the *KK* relations is due to the fact that in practice one uses them most often to obtain $\chi'(\nu)$ from the measured spectral dependence of the (exponential) absorption coefficient $\alpha(\nu)$, which is proportional to $\chi''(\nu)$. We will now apply the first of Equations (5.4-4) to derive $\chi'(\nu)$ from $\chi''(\nu)$. We start with (5.3-8) in which, to simplify notation, we replace $\Delta\nu \rightarrow 2/\tau$.

$$\chi''(\nu)_{\nu\approx\nu 0} = \frac{(N_1 - N_2)\lambda^3 \tau}{16\pi^3 n t_{\text{spont}}} \frac{1}{1 + (\nu - \nu_0)^2\tau^2} \tag{5.4-5}$$

Substituting (5.4-5) in the first of the *KK* relation Equation (5.4-4) gives

$$\chi'(\nu) = +\frac{(N_1 - N_2)\lambda^3\tau}{16\pi^4 n t_{\text{spont}}} P\int_{-\infty}^{\infty} \frac{d\nu'}{[1 + (\nu' - \upsilon_0)^2\tau^2](\nu' - \nu)}$$

$$= +\frac{(N_1 - N_2)\lambda^3\tau}{16\pi^4 n t_{\text{spont}}} P\int_{-\infty}^{\infty} \frac{\left(\frac{1}{\tau}\right)^2 d\nu'}{\left[\nu' - \left(\nu_0 + i\frac{1}{\tau}\right)\right]\left[\nu' - \left(\nu_0 - i\frac{1}{\tau}\right)\right](\nu' - \nu)} \tag{5.4-6}$$

To evaluate (5.4-6), we integrate over the contour shown in Figure 5-4. The two poles of $\chi'(\nu')$ at $\nu_0 \pm i(1/\tau)$ are shown. By applying Cauchy's residue theorem, we obtain

$$\oint_c f(\nu')d\nu' = \int_{-R}^{\nu-\varepsilon} f(\nu')d\nu' + \int_{\nu+\varepsilon}^{R} f(\nu')d\nu' + \int_{c_1} f(\nu')d\nu' + \int_{c_2} f(\nu')d\nu' = -2\pi i \text{ (residue)} \tag{5.4-7}$$

where $f(\nu')$ is the integrand of (5.4-6). The negative sign in front of the residue is due to the clockwise sense of the integration. We take the limit of $R \to \infty$, $\varepsilon \to 0$. In this limit, the integral over c_2 is $o(R^{-2}) \to 0$. The sum of the first two terms on the right side of (5.4-7) is, by definition, the principal value of the integral. The integral over c_1 yields

$$\int_{c_1} f(\nu')d\nu' = \pi i[f(\nu')(\nu' - \nu)]_{\nu'=\nu}$$

We can thus rewrite (5.4-7) as

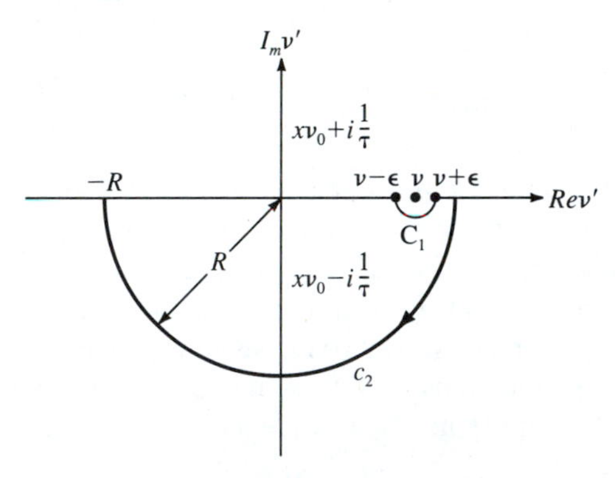

Figure 5-4 The contour c in the complex frequency (ν') plane used to derive $x'(\nu)$. The semicircle c has a radius ε.

$$P\int_{-\infty}^{\infty}\frac{\left(\frac{1}{\tau}\right)^2}{\left[\nu'-\left(\nu_0+i\frac{1}{\tau}\right)\right]\left[\nu'-\left(\nu_0-i\frac{1}{\tau}\right)\right](\nu'-\nu)}\,d\nu' =$$

$$-\pi i\frac{\left(\frac{1}{\tau}\right)^2}{\left[\nu-\left(\nu_0+i\frac{1}{\tau}\right)\right]\left[\nu-\left(\nu_0-i\frac{1}{\tau}\right)\right]}-2\pi i\frac{\left(\frac{1}{\tau}\right)^2}{\left(-2i\frac{1}{\tau}\right)\left(\nu_0-\nu-i\frac{1}{\tau}\right)} \tag{5.4-8}$$

The first term on the right side of (5.4-8) is the result of integrating over c_1. (For details see an identical integration in Appendix A.) The second term is $-2\pi i$ (residue at $\nu'=\nu_0-i/\tau$). Adding the two terms on the right side of (5.4-8), using the result in (5.4-6), and returning to our initial notation $\Delta\nu = 2/\tau$ leads, after some straightforward algebra, to

$$\chi'(\nu)=\frac{(N_1-N_2)\lambda^3}{8\pi^3 n t_{\text{spont}}\Delta\nu}\,\frac{2(\nu_0-\nu)/\Delta\nu}{1+[4(\nu-\nu_0)^2/(\Delta\nu)^2]} \tag{5.4-9}$$

which is the sought result.

We can combine Equations (5.3-8) and (5.4-9) to obtain

$$\chi(\nu)=\chi'(\nu)-i\chi''(\nu)$$

$$\nu\approx\nu_0=-\frac{(N_1-N_2)}{16\pi^3 n t_{\text{spont}}\Delta\nu}\left[\frac{\Delta\nu}{\nu-(\nu_0+i\Delta\nu/2)}\right] \tag{5.4-10}$$

The total complex susceptibility $\chi(\nu)$ has but a single pole, in the upper half ν plane, while $\chi'(\nu)$ and $\chi''(\nu)$ have, each, a pole in the upper and lower half plane (these are indicated in Figure 5-4). This, as claimed above, is a necessary condition for the response function of a linear passive system.

The susceptibility $\chi(\nu)$ Equation (5.4-10) is defined by the relation

$$P(\nu)=\varepsilon_0\chi(\nu)E(\nu) \tag{5.4-11}$$

which relates the induced (complex) polarization and the inducing field amplitude. It is tempting to use this relation in a Fourier integral of the form

$$p(t)=\int_{-\infty}^{\infty}\tilde{p}(\nu)\exp(i2\pi\nu t)\,d\nu$$

$$=\int_{-\infty}^{\infty}\varepsilon_0\chi(\nu)\tilde{e}(\nu)\exp(i2\pi\nu t)\,d\nu \tag{5.4-12}$$

where $\tilde{e}(\nu)$ is the Fourier integral transform of the electric field $e(t)$. Since $e(t)$ and $p(t)$ are real, it follows that $\tilde{e}(-\nu) = \tilde{e}^*(\nu)$ and that

$$\chi(-\nu) = \chi^*(\nu) \tag{5.4-13}$$

Our derived expression (5.4-10) does not satisfy condition (5.4-13). The reason is that in deriving $\chi''(\nu)$ in (5.4-5), we assumed $\nu \approx \nu_0$, thus restricting its validity to *positive* frequencies near the resonance $+\nu_0$. We can take advantage of the fact that in practice our resonances are very narrow, i.e., ν_0 >>> [typically $(\nu_0/\Delta\nu) \sim 10^2 - 10^4$] and simply extend $\chi(\nu)$ to negative frequencies in a manner that satisfies Equation (5.4-13). The result is

$$\underset{-\infty<\nu<\infty}{\chi(\nu)} = \frac{(N_1 - N_2)\lambda^3}{8\pi n^3 t_{\text{spont}} \Delta\nu} \left[\frac{(\Delta\nu/2)}{\nu - (-\nu_0 + i\Delta\nu/2)} - \frac{(\Delta\nu/2)}{\nu - (\nu_0 + i\Delta\nu/2)} \right] \tag{5.4-14}$$

$$= \frac{(N_1 - N_2)\lambda^3}{8\pi n^3 t_{\text{spont}} \Delta\nu} \left[\frac{\dfrac{2(\nu_0 + \nu)}{\Delta\nu} + i}{1 + \dfrac{4(\nu + \nu_0)^2}{(\Delta\nu)^2}} + \frac{\dfrac{2(\nu_0 - \nu)}{\Delta\nu} + i}{1 + \dfrac{4(\nu_0 + \nu)^2}{(\Delta\nu)^2}} \right] \tag{5.4-15}$$

This function has poles at $\nu_{1,2} = \pm\,\nu_0 + i\Delta\nu/2$, which are both in the upper half of the ν plane as required by causality, or equivalently, by the requirement that transients of $p(t)$ decay rather than grow with time. It can be used in (5.4-12). A plot of $\chi'(\nu)$ and $\chi''(\nu)$ is shown in Figure 5-5.

5.5 THE SIGNIFICANCE OF $\chi(\nu)$

According to (1.2-3) the electric displacement vector is defined by

$$\mathbf{D} = \varepsilon_0\,\mathbf{E} + \mathbf{P} + \mathbf{P}_{\text{transition}} = \varepsilon\mathbf{E} + \varepsilon_0\chi\mathbf{E}$$

where the complex notation is used and the polarization is separated into a resonant component $\mathbf{P}_{\text{transition}}$ due to the specific atomic transition and a nonresonant component $\mathbf{P}$ that accounts for all the other contributions to the polarization. We can rewrite the last equation as

$$\mathbf{D} = \varepsilon\left[1 + \frac{\varepsilon_0}{\varepsilon}\chi(\omega)\right]\mathbf{E} = \varepsilon'(\omega)\mathbf{E} \tag{5.5-1}$$

so that the complex dielectric constant becomes

$$\varepsilon'(\omega) = \varepsilon\left[1 + \frac{\varepsilon_0}{\varepsilon}\chi(\omega)\right] \tag{5.5-2}$$

We have thus accounted for the effect of the atomic transition by modifying ε according to (5.5-2). Having derived $\chi(\omega)$, using detailed atomic information, we can ignore its physical origin and proceed to treat the wave propagation in the medium with ε' given by (5.5-2), using Maxwell's equations.

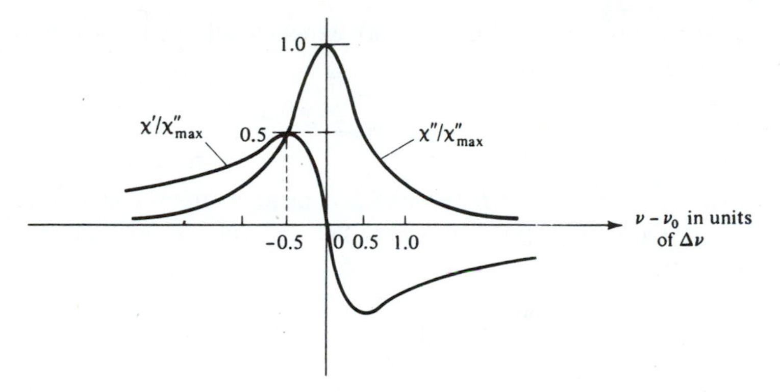

Figure 5-5 A plot of the real (χ') and (negative) imaginary (χ'') parts of the electronic susceptibility.

As an example of this point of view we consider the propagation of a plane electromagnetic wave in a medium with a dielectric constant $\varepsilon'(\omega)$. According to (1.3-17), the wave has the form of

$$e(z, t) = \text{Re}\,[Ee^{i(\omega t - k'z)}] \tag{5.5-3}$$

where, using (1.3-13) and (5.5-2) and assuming $(\varepsilon_0/\varepsilon)|\chi| \ll 1$, we obtain

$$k' = \omega\sqrt{\mu\varepsilon'} \simeq k\left[1 + \frac{\varepsilon_0}{2\varepsilon}\chi\right]$$

where $k = \omega\sqrt{\mu\varepsilon}$.

Expressing $\chi(\omega)$ in terms of its real and imaginary components as in (5.4-9) leads to

$$k' = k\left[1 + \frac{\chi'(\omega)}{2n^2}\right] - i\frac{k\chi''(\omega)}{2n^2} \tag{5.5-4}$$

where $n = (\varepsilon/\varepsilon_0)^{1/2}$ is the index of refraction in the medium[2] far away from resonance. Substituting (5.5-4) back into (5.5-3), we find that the atomic transition results in a wave propagating according to

$$e(z, t) = \text{Re}\,[Ee^{i\omega t - i(k+\Delta k)z}e^{(\gamma/2)z}] \tag{5.5-5}$$

The result of the atomic polarization is thus to change the phase delay per unit length from k to $k + \Delta k$, where

$$\Delta k = \frac{k\chi'(\omega)}{2n^2} \tag{5.5-6}$$

[2]Since the velocity of light is $c = (\mu\varepsilon)^{-1/2}$, n is the ratio of the velocity of light in vacuum to that in the medium at frequencies sufficiently removed from resonance that the effect of the specific atomic transition can be ignored.

as well as to cause the amplitude to vary exponentially with distance according to $e^{(\gamma/2)z}$, where

$$\gamma(\omega) = -\frac{k\chi''(\omega)}{n^2} \tag{5.5-7}$$

It is quite instructive to rederive (5.5-7) using a different approach. According to (1.2-13), the average power absorbed per unit volume from an electromagnetic field with an x component only is

$$\overline{\frac{\text{Power}}{\text{Volume}}} = \overline{e_x(t)\frac{dp_x(t)}{dt}} = \tfrac{1}{2}\text{Re}\,[E(i\omega P)^*] \tag{5.5-8}$$

where E and P are the complex electric field and polarization in the x direction, respectively, and horizontal bars denote time-averaging. Using (1.2-16) and (1.2-18) in (5.5-8), we obtain

$$\overline{\frac{\text{Power}}{\text{Volume}}} = \frac{\omega\varepsilon_0}{2}\chi''|E|^2 \tag{5.5-9}$$

The absorption of energy at a rate given by (5.5-9) must lead to a variation of the wave intensity I, according to

$$I(z) = I_0 e^{\gamma(\omega)z} \tag{5.5-10}$$

where

$$\gamma(\omega) = I^{-1}\frac{dI}{dz} \tag{5.5-11}$$

Conservation of energy thus requires that

$$\frac{dI}{dz} = -(\text{power absorbed per unit volume}) = -\frac{\omega\varepsilon_0}{2}\chi''|E|^2$$

Using the last result in (5.5-11), as well as relation (1.3-26),

$$I = \frac{c\varepsilon}{2n}|E|^2$$

where $c/n = \omega/k$ is the velocity of light in the medium, gives

$$\gamma(\omega) = -\frac{k\chi''(\omega)}{n^2}$$

in agreement with (5.5-7).

5.6 GAIN SATURATION IN HOMOGENEOUS LASER MEDIA

In Section 5.3 we derived an expression (5.3-3) for the exponential gain constant due to a population inversion. It is given by

$$\gamma(\nu) = (N_2 - N_1)\frac{c^2}{8\pi n^2 \nu^2 t_{\text{spont}}} g(\nu) \tag{5.6-1}$$

where N_2 and N_1 are the population densities of the two atomic levels involved in the induced transition. There is nothing in (5.6-1) to indicate what causes the inversion $(N_2 - N_1)$, and this quantity can be considered as a parameter of the system. In practice the inversion is caused by a "pumping" agent, hereafter referred to as the pump, that can take various forms such as the electric current in injection lasers, the flashlamp light in pulsed ruby lasers, or the energetic electrons in plasma-discharge gas lasers.

Consider next the situation prevailing at some point *inside* a laser medium in the presence of an optical wave. The pump establishes a population inversion, which in the absence of any optical field has a value ΔN^0. The presence of the optical field induces $2 \rightarrow 1$ and $1 \rightarrow 2$ transitions. Since $N_2 > N_1$ and the induced rates for $2 \rightarrow 1$ and $1 \rightarrow 2$ transitions are equal, it follows that more atoms are induced to undergo a transition from level 2 to level 1 than in the opposite direction and that, consequently, the new equilibrium population inversion is smaller than ΔN^0.

The reduction in the population inversion and hence of the gain constant brought about by the presence of an electromagnetic field is called gain saturation. Its understanding is of fundamental importance in quantum electronics. As an example, which will be treated in the next chapter, we may point out that gain saturation is the mechanism that reduces the gain inside laser oscillators to a point where it just balances the losses so that steady oscillation can result.

In Figure 5-6 we show the ground state 0 as well as the two laser levels 2 and 1 of a four-level laser system. The density of atoms pumped per unit time into level 2 is taken as R_2, and that pumped into 1 is R_1. Pumping into 1 is, of course, undesirable since it leads to a reduction of the inversion. In many practical situations it

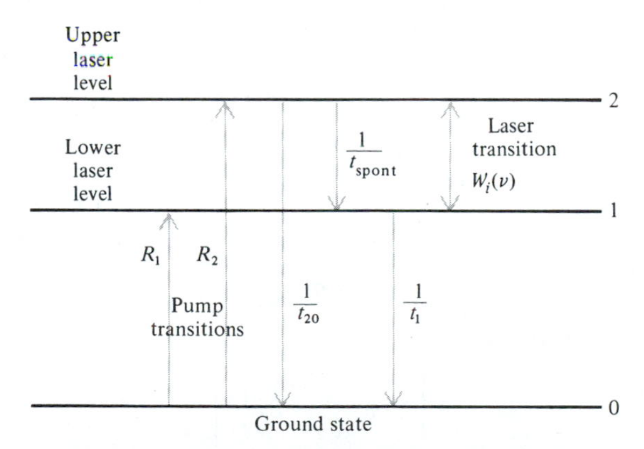

Figure 5-6 Energy levels and transition rates of a four-level laser system. (The fourth level, which is involved in the original excitation by the pump, is not shown and the pumping is shown as proceeding directly into levels 1 and 2.) The total lifetime of level 2 is t_2, where $1/t_2 = 1/t_{\text{spont}} + 1/t_{20}$.

cannot be avoided. The actual "decay" lifetime of atoms in level 2 at the absence of any radiation field is taken as t_2. This decay rate has a contribution t_{spont}^{-1} that is due to spontaneous (photon emitting) $2 \rightarrow 1$ transitions as well as to additional nonradiative relaxation from 2 to 1. The lifetime of atoms in level 1 is t_1. The induced rate for $2 \rightarrow 1$ and $1 \rightarrow 2$ transitions due to a radiation field at frequency ν is denoted by $W_i(\nu)$ and, according to (5.2-15), is given by

$$W_i(\nu) = \frac{\lambda^2 g(\nu)}{8\pi n^2 h\nu t_{spont}} I_\nu \tag{5.6-2}$$

where $g(\nu)$ is the normalized lineshape of the transition and I_ν is the intensity (watts per square meter) of the optical field.

The equations describing the populations of levels 2 and 1 in the combined presence of a radiation field at ν and a pump are:

$$\frac{dN_2}{dt} = R_2 - \frac{N_2}{t_2} - (N_2 - N_1)W_i(\nu) \tag{5.6-3}$$

$$\frac{dN_1}{dt} = R_1 - \frac{N_1}{t_1} + \frac{N_2}{t_{spont}} + (N_2 - N_1)W_i(\nu) \tag{5.6-4}$$

N_2 and N_1 are the population densities (m^{-3}) of levels 2 and 1 respectively. R_2 and R_1 are the pumping rates ($m^{-3} - s^{-1}$) into these levels. N_2/t_2 is the change per unit time in the population of 2 due to decay out of level 2 to all levels. This includes spontaneous transitions to 1 but *not* induced transitions. The rate for the latter is $N_2W_i(\nu)$ so that the net change in N_2 due to induced transitions is given by the last term of (5.6-3). At steady state the populations are constant with time, so putting $d/dt = 0$ in the two preceding equations, we can solve for N_1, N_2, and obtain[3]

$$N_2 - N_1 = \frac{R_2 t_2 - (R_1 + \delta R_2)t_1}{1 + [t_2 + (1 - \delta)t_1]W_i(\nu)} \tag{5.6-5}$$

where $\delta = t_2/t_{spont}$. If the optical field is absent, $W_i(\nu) = 0$, and the inversion density is given by

$$\Delta N^0 = R_2 t_2 - (R_1 + \delta R_2)t_1 \tag{5.6-6}$$

we can use (5.6-6) to rewrite (5.6-5) as

$$N_2 - N_1 = \frac{\Delta N^0}{1 + \phi t_{spont} W_i(\nu)} \tag{5.6-7}$$

where the parameter ϕ is defined by

$$\phi = \delta\left[1 + (1 - \delta)\frac{t_1}{t_2}\right]$$

[3]Levels 1 and 2 are assumed to be high enough (in energy) that the role of thermal processes in populating them can be neglected.

We note that in efficient laser systems $t_2 \simeq t_{\text{spont}}$, so $\delta \simeq 1$, and that $t_1 \ll t_2$, so $\phi \simeq 1$. Substituting (5.6-2) for $W_i(\nu)$, the last equation becomes

$$N_2 - N_1 = \frac{\Delta N^0}{1 + [\phi\lambda^2 g(\nu)/8\pi n^2 h\nu]I_\nu} = \frac{\Delta N^0}{1 + I_\nu/I_s(\nu)} \tag{5.6-8}$$

where $I_s(\nu)$, the saturation intensity, is given by

$$I_s(\nu) = \frac{8\pi n^2 h\nu}{\phi\lambda^2 g(\nu)} = \frac{8\pi n^2 h\nu}{(t_2/t_{\text{spont}})\lambda^2 g(\nu)} = \frac{8\pi n^2 h\nu\Delta\nu}{(t_2/t_{\text{spont}})\lambda^2} \tag{5.6-9}$$

and corresponds to the intensity level (watts per square meter) that causes the inversion to drop to one half of its nonsaturated value (ΔN^0). By using (5.6-8) in the gain expression (5.6-1), we obtain our final result

$$\begin{aligned}\gamma(\nu) &= \frac{1}{1 + I_\nu/I_s(\nu)}\left(\frac{\Delta N^0\lambda^2}{8\pi n^2 t_{\text{spont}}}\right) g(\nu) \\ &= \frac{\gamma_0(\nu)}{1 + I_\nu/I_s(\nu)}\end{aligned} \tag{5.6-10}$$

which shows the dependence of the gain constant on the optical intensity.

In closing we recall that (5.6-10) applies to a homogeneous laser system. This is due to the fact that in the rate equations (5.6-3) and (5.6-4) we considered all the atoms as equivalent and, consequently, experiencing the same transition rates. This assumption is no longer valid in inhomogeneous laser systems. This case is treated in the next section.

5.7 GAIN SATURATION IN INHOMOGENEOUS LASER MEDIA

In Section 5.6 we considered the reduction in optical gain—that is, saturation—due to the optical field in a homogeneous laser medium. In this section we treat the problem of gain saturation in inhomogeneous systems.

According to the discussion of Section 5.1, in an inhomogeneous atomic system the individual atoms are distinguishable, with each atom having a unique transition frequency $(E_2 - E_1)/h$. We can thus imagine the inhomogeneous medium as made up of classes of atoms each designated by a continuous variable ξ.[4] Furthermore, we define a function $p(\xi)$ so that the *a priori* probability that an atom has its ξ parameter between ξ and $\xi + d\xi$ is $p(\xi)\, d\xi$. It follows that

$$\int_{-\infty}^{\infty} p(\xi)\, d\xi = 1 \tag{5.7-1}$$

since any atom has a unit probability of having its ξ value between $-\infty$ and ∞.

[4]The variable ξ can, as an example, correspond to the center frequency of the lineshape function $g^{\xi}(\nu)$ of atoms in group ξ.

The atoms within a given class ξ are considered as homogeneously broadened, having a lineshape function $g^{\xi}(\nu)$ that is normalized so that

$$\int_{-\infty}^{\infty} g^{\xi}(\nu)\, d\nu = 1 \tag{5.7-2}$$

In Section 5.1 we defined the transition lineshape $g(\nu)$ by taking $g(\nu)\, d\nu$ to represent the *a priori* probability that a spontaneous emission will result in a photon whose frequency is between ν and $\nu + d\nu$. Using this definition we obtain

$$g(\nu)\, d\nu = \left[\int_{-\infty}^{\infty} p(\xi) g^{\xi}(\nu)\, d\xi\right] d\nu \tag{5.7-3}$$

which is a statement of the fact that the probability of emitting a photon of frequency between ν and $\nu + d\nu$ is equal to the probability $g^{\xi}(\nu)\, d\nu$ of this occurrence, given that the atom belongs to class ξ, summed up over all the classes.

Next we proceed to find the contribution to the inversion that is due to a single class ξ. The rate equations are

$$\frac{dN_2^{\xi}}{dt} = R_2 p(\xi) - \frac{N_2^{\xi}}{t_2} - [N_2^{\xi} - N_1^{\xi}] W_i^{\xi}(\nu)$$

$$\frac{dN_1^{\xi}}{dt} = R_1 p(\xi) - \frac{N_1^{\xi}}{t_1} + \frac{N_2^{\xi}}{t_{\text{spont}}} + [N_2^{\xi} - N_1^{\xi}] W_i^{\xi}(\nu) \tag{5.7-4}$$

and are similar to (5.6-3) and (5.6-4), except that N_2^{ξ} and N_1^{ξ} refer to the upper and lower level densities of atoms in class ξ only. The pumping rate (atoms/m^3-sec) into levels 2 and 1 is taken to be proportional to the probability of finding an atom in class ξ and is given by $R_2 p(\xi)$ and $R_1 p(\xi)$, respectively. The total pumping rate into level 2 is, as in Section 5.6, R_2 since

$$\int_{-\infty}^{\infty} R_2 p(\xi)\, d\xi = R_2 \int_{-\infty}^{\infty} p(\xi)\, d\xi = R_2$$

where we made use of (5.7-1). The induced transition rate $W_i^{\xi}(\nu)$ is given, according to (5.2-15), by

$$W_i^{\xi}(\nu) = \frac{\lambda^2}{8\pi n^2 h\nu t_{\text{spont}}} g^{\xi}(\nu) I_{\nu} \tag{5.7-5}$$

which is of a form identical to (5.6-2) except that $g^{\xi}(\nu)$ refers to the lineshape function of atoms in class ξ. The steady-state $d/dt = 0$ solution of (5.7-4) yields

$$N_2^{\xi} - N_1^{\xi} = \frac{\Delta N^0 p(\xi)}{1 + \phi t_{\text{spont}} W_i^{\xi}(\nu)} \tag{5.7-6}$$

where ΔN^0 and ϕ have the same significance as in Section 5.6. The total power emitted by induced transitions per unit volume by atoms in class ξ is thus

$$\frac{P^{\xi}(\nu)}{V} = (N_2^{\xi} - N_1^{\xi})h\nu W_i^{\xi}(\nu) = \frac{\Delta N^0 p(\xi)h\nu}{1/W_i^{\xi}(\nu) + \phi t_{\text{spont}}} \tag{5.7-7}$$

where the spontaneous lifetime is assumed the same for all the groups ξ.

Summing (5.7-7) over all the classes, we obtain an expression for the total power at ν per unit volume emitted by the atoms

$$\frac{P(\nu)}{V} = \frac{\Delta N^0 h\nu}{t_{\text{spont}}} \int_{-\infty}^{\infty} \frac{p(\xi)\, d\xi}{1/(W_i^{\xi}(\nu)t_{\text{spont}}) + \phi} \tag{5.7-8}$$

which, by the use of (5.7-5), can be rewritten as

$$\frac{P(\nu)}{V} = \frac{\Delta N^0 h\nu}{t_{\text{spont}}} \int_{-\infty}^{\infty} \frac{p(\xi)\, d\xi}{8\pi n^2 h\nu/(\lambda^2 I_\nu g^{\xi}(\nu)) + \phi} \tag{5.7-9}$$

The stimulated emission of power causes the intensity of the traveling optical wave to increase with distance z according to $I_\nu = I_\nu(0) \exp[\gamma(\nu)z]$, where

$$\gamma(\nu) = \frac{dI_\nu/dz}{I_\nu} = \frac{P(\nu)/V}{I_\nu}$$
$$= \frac{\Delta N^0 \lambda^2}{8\pi n^2 t_{\text{spont}}} \int_{-\infty}^{\infty} \frac{p(\nu_\xi) d\nu_\xi}{[1/g^{\xi}(\nu)] + (\phi\lambda^2 I_\nu/8\pi n^2 h\nu)} \tag{5.7-10}$$

where we replaced $p(\xi)\, d\xi$ by $p(\nu_\xi)\, d\nu_\xi$. This is our basic result.

As a first check on (5.7-10), we shall consider the case in which $I_\nu \ll 8\pi n^2 h\nu/\phi\lambda^2 g^{\xi}(\nu)$ and therefore the effects of saturation can be ignored. Using (5.7-3) in (5.7-10), we obtain

$$\gamma(\nu) = \frac{\Delta N^0 \lambda^2}{8\pi n^2 t_{\text{spont}}} g(\nu)$$

which is the same as (5.3-3). This shows that in the absence of saturation the expressions for the gain of a homogeneous and an inhomogeneous atomic system are identical.

Our main interest in this treatment is in deriving the saturated gain constant for an inhomogeneously broadened atomic transition. If we assume that in each class ξ all the atoms are identical (homogeneous broadening), we can use (5.1-9) for the lineshape function $g^{\xi}(\nu)$, and therefore,

$$g^{\xi}(\nu) = \frac{\Delta\nu}{2\pi[(\Delta\nu/2)^2 + (\nu - \nu_\xi)^2]} \tag{5.7-11}$$

where $\Delta\nu$ is called the homogeneous linewidth of the inhomogeneous line. Atoms with transition frequencies that are clustered within $\Delta\nu$ from each other can be considered as indistinguishable. The term "homogeneous packet" is often used to describe them. Using (5.7-11) in (5.7-10) leads to

$$\gamma(\nu) = \frac{\Delta N^0 \lambda^2 \Delta\nu}{16\pi^2 n^2 t_{\text{spont}}} \int_{-\infty}^{\infty} \frac{p(\nu_\xi) d\nu_\xi}{(\nu - \nu_\xi)^2 + (\Delta\nu/2)^2 + (\phi\lambda^2 I_\nu \Delta\nu / 16\pi^2 n^2 h\nu)} \tag{5.7-12}$$

In the extreme inhomogeneous cases, the width of $p(\nu_\xi)$ is by definition very much larger than the remainder of the integrand in (5.7-12) and thus it is essentially a constant over the region in which the integrand is appreciable. In this case we can pull $p(\nu_\xi)_{\nu_\xi=\nu} = p(\nu)$ outside the integral sign in (5.7-12), obtaining

$$\gamma(\nu) = \frac{\Delta N^0 \lambda^2 \Delta\nu}{16\pi^2 n^2 t_{\text{spont}}} p(\nu) \times \int_{-\infty}^{\infty} \frac{d\nu_\xi}{(\nu - \nu_\xi)^2 + (\Delta\nu/2)^2 + \phi\lambda^2 \Delta\nu I_\nu / 16\pi^2 n^2 h\nu} \tag{5.7-13}$$

Using the definite integral

$$\int_{-\infty}^{\infty} \frac{dx}{x^2 + a^2} = \frac{\pi}{a}$$

to evaluate (5.7-13), we obtain

$$\gamma(\nu) = \frac{\Delta N^0 \lambda^2 p(\nu)}{8\pi n^2 t_{\text{spont}}} \frac{1}{\sqrt{1 + \phi\lambda^2 I_\nu / 4\pi^2 n^2 h\nu\, \Delta\nu}} \tag{5.7-14}$$

$$= \gamma_0(\nu) \frac{1}{\sqrt{1 + I_\nu / I_s}} \tag{5.7-15}$$

where $I_s = 4\pi^2 n^2 h\nu\, \Delta\nu / \phi\lambda^2$ is the saturation intensity. A comparison of (5.7-15) with (5.6-10) shows that, because of the square root, the saturation—that is, decrease in gain—sets in more slowly as the intensity I_ν is increased in the case of inhomogeneous broadening.

Problems

5.1 Consider a parallel *RLC* circuit that is connected to a signal generator so that the voltage across it is

$$v(t) = V_0 \cos 2\pi\nu t$$

At $t = 0$ the circuit is disconnected from the signal generator.

a. What is the voltage $v(t)$ for $t > 0$?
b. Find the Fourier transform $V(\omega)$ of $v(t)$. Show that in the high-Q case (where $Q = 2\pi\nu_0 RC$) and for frequencies $\nu \simeq \nu_0 \equiv 1/2\pi\sqrt{LC}$,

$$|V(\nu)|^2 \propto \frac{1}{(\nu - \nu_0)^2 + (\nu_0/2Q)^2}$$

c. Obtain the expression for the amount of average power $P(\nu)$ absorbed by the *RLC* circuit from a signal generator with an output current

$$i(t) = I_0 \cos 2\pi\nu t$$

Show that the expression for $P(\nu)$ is proportional to that of $|V(\nu)|^2$ obtained in (b).

5.2 Calculate the maximum absorption coefficient for the R_1 transition in pink ruby with a Cr^{3+} concentration of 2×10^{19} cm^{-3}. Assume that $t_{spont} = 3 \times 10^{-3}$ second and $\Delta\nu = 11$ cm^{-1}. Compare the result to the absorption data of Figure 7-4.

5.3 Show that if $\chi(\nu)$ possesses a pole in the upper half plane, a step excitation $e(t)$ would lead to a response $p(t)$ that grows exponentially in time. *Hint:* Use the Fourier integral relation

$$p(t) = \int_{-\infty}^{\infty} \tilde{p}(\nu)\exp(i2\pi\nu t)\, d\nu$$

$$\tilde{p}(\nu) = \varepsilon_0\chi(\nu)\tilde{e}(\nu)$$

5.4 Show that for $p(t)$ to be a real function

$$\chi(-\nu) = \chi^*(\nu) \tag{1}$$

or equivalently

$$\begin{aligned} \chi'(\nu) &= \chi'(-\nu) \\ \chi''(\nu) &= -\chi''(-\nu) \end{aligned} \tag{2}$$

5.5 The equation of motion of a one-dimensional electron oscillator due to an electric field $e(t)$ is

$$\frac{d^2x}{dt^2} + \sigma\frac{dx}{dt} + \frac{k}{m}x(t) = -\frac{e}{m}e(t)$$

where $x(t)$ is the excursion from the equilibrium position. σ, k, and m are, respectively, the damping constant, the restoring (''spring constant'') coefficient, and electron mass. The charge is $-e$.

If $e(t) = \text{Re}[E\exp(i2\pi\nu t)]$ and $x(t) = \text{Re}[x(\nu)\exp(i2\pi\nu t)]$

a. Solve for the complex medium polarization, $P(\nu) = -Nex(\nu)$, where N is the density (m^{-3}) of electrons.
b. Obtain an expression for the complex susceptibility $\chi(\nu)$ defined by $\varepsilon_0\chi(\nu) = P(\nu)/E(\nu)$
   The result should be

$$\chi(\nu) = \frac{Ne^2/m\varepsilon_0}{4\pi^2(\nu_0^2 - \nu^2) + i2\pi\nu\sigma}$$

where $\nu_0 \equiv 1/2\pi\sqrt{k/m}$.

References

1. See, for example, A. Yariv, *Quantum Electronics,* 3d ed. New York: Wiley, 1989.
2. Portis, A. M., ''Electronic structure of F centers: Saturation of the electron spin resonance,'' *Phys. Rev.* 91:1071, 1953.

3. See, for example, R. Kubo, *Statistical Mechanics.* Amsterdam: North Holland, 1964, p. 31.
4. Johnson, L. F., ''Optically pumped pulsed crystal lasers other than ruby.'' In *Lasers,* vol. 1, A. K. Levine, ed. New York: Marcel Dekker, Inc., 1966, p. 137.
5. Kittel, C., *Elementary Statistical Physics.* New York: Wiley, 1958, p. 197.
6. Einstein, A., ''Zur Quantentheorie der Strahlung,'' *Phys. Z.* 18:121–128, March 1917.
7. See, for example, R. H. Pantell and H. E. Puthoff, *Fundamentals of Quantum Electronics,* New York: Wiley, 1969, p. 31.
8. Ditchburn, R. W., *Light.* New York: Interscience, 1963, Chap. 15.
9. Gordon, J. P., unpublished memorandum, Bell Telephone Laboratories.
10. *Lasers and Light—Readings from Scientific American.* San Francisco: Freeman, 1969.
11. Mitchell, A. C. G., and M. W. Zemansky, *Resonance Radiation and Excited Atoms.* New York: Cambridge, 1961.
12. Kroning, R. L., *J. Opt. Soc. Am.* 12:547, 1926.
13. Kramers, H. A., *Atti. Cong. Intern. Fis.* 2:545, 1927.

6 Theory of Laser Oscillation and Its Control in the Continuous and Pulsed Regimes

6.0 INTRODUCTION

In Chapter 5 we found that an atomic medium with an inverted population ($N_2 > N_1$) is capable of amplifying an electromagnetic wave if the latter's frequency falls within the transition lineshape. Consider next the case in which the laser medium is placed inside an optical resonator. As the electromagnetic wave bounces back and forth between the two reflectors, it passes through the laser medium and is amplified. If the amplification exceeds the losses caused by imperfect reflection in the mirrors and scattering in the laser medium, the field energy stored in the resonator will increase with time. This causes the amplification constant to decrease as a result of gain saturation (see (5.6-10) and the discussion surrounding it.) The oscillation level will keep increasing until the saturated gain per pass just equals the losses. At this point the net gain per pass is unity and no further increase in the radiation intensity is possible—that is, steady-state oscillation obtains.

In this chapter we will derive the start-oscillation inversion needed to sustain laser oscillation, beginning with the theory of the Fabry–Perot etalon. We will also obtain an expression for the oscillation frequency of the laser oscillator and show how it is affected by the dispersion of the atomic medium. We will conclude by considering the problem of optimum output coupling and laser pulses.

6.1 FABRY–PEROT LASER

A two-mirror laser oscillator is basically a Fabry–Perot etalon, as studied in detail in Chapter 4, in which the space between the two mirrors contains an amplifying medium with an inverted atomic population. We can account for the inverted population by using (5.5-4). Taking the propagation constant of the medium as

$$k'(\omega) = k + k\frac{\chi'(\omega)}{2n^2} - ik\frac{\chi''(\omega)}{2n^2} - i\frac{\alpha}{2} \qquad (6.1\text{-}1)$$

where $k - i\alpha/2$ is the propagation constant of the medium at frequencies well removed from that of the laser transition, $\chi(\omega) = \chi'(\omega) - i\chi''(\omega)$ is the complex dielectric susceptibility due to the laser transition and is given by (5.3-8) and (5.4-9). Since α accounts for the distributed passive losses of the medium,[1] the intensity loss factor per pass is $\exp(-\alpha l)$.

Figure 6-1 shows a plane wave of (complex) amplitude E_i that is incident on the left mirror of a Fabry–Perot etalon containing a laser medium. The ratio of transmitted to incident fields at the left mirror is taken as t_1 and that at the right mirror as t_2. The ratios of reflected to incident fields inside the laser medium at the left and right boundaries are r_1 and r_2, respectively.

The propagation factor corresponding to a single transit is $\exp(-ik'l)$ where k' is given by (6.1-1) and l is the length of the etalon.

Adding the partial waves at the output to get the total outgoing wave E_t we obtain

$$E_t = t_1 t_2 E_i e^{-ik'l}[1 + r_1 r_2 e^{-i2k'l} + r_1^2 r_2^2 e^{-i4k'l} + \cdots]$$

which is a geometric progression with a sum

$$\begin{aligned} E_t &= E_i\left[\frac{t_1 t_2 e^{-ik'l}}{1 - r_1 r_2 e^{-i2k'l}}\right] \\ &= E_i\left[\frac{t_1 t_2 e^{-i(k+\Delta k)l} e^{(\gamma-\alpha)l/2}}{1 - r_1 r_2 e^{-2i(k+\Delta k)l} e^{(\gamma-\alpha)l}}\right] \end{aligned} \qquad (6.1\text{-}2)$$

where we used (5.3-3), (6.1-1), and the relation $k' = k + \Delta k + i(\gamma - \alpha)/2$ with

$$\Delta k = k\frac{\chi'(\omega)}{2n^2}$$

$$\gamma = -k\frac{\chi''(\omega)}{n^2} \qquad (6.1\text{-}3)$$

$$= (N_2 - N_1)\frac{\lambda^2}{8\pi n^2 t_{\text{spont}}} g(\nu) \qquad (6.1\text{-}4)$$

---

[1] In addition to and in the presence of the gain attributable to the inverted laser transition, the medium may possess a residual attenuation due to a variety of mechanisms, such as scattering at imperfections, absorption by excited atomic levels, and others. The attenuation resulting from all of these mechanisms is lumped into the distributed loss constant α.

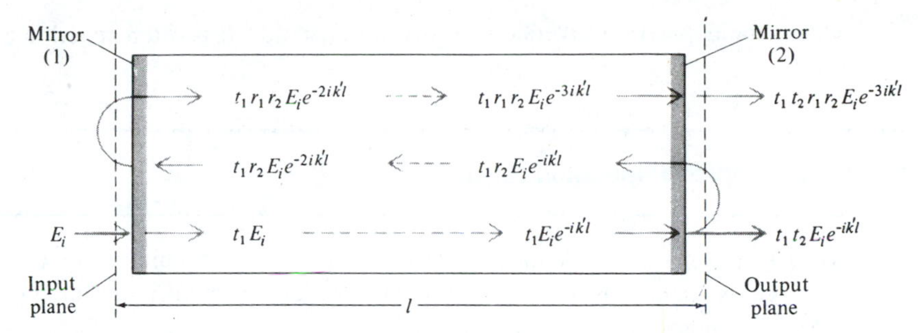

Figure 6-1 Model used to analyze a laser oscillator. A laser medium (that is, one with an inverted atomic population) with a complex propagation constant $k'(\omega)$ is placed between two reflecting mirrors.

If the atomic transition is inverted ($N_2 > N_1$), then $\gamma > 0$ and the denominator of (6.1-2) can become very small. The transmitted wave E_t can thus become larger than the incident wave E_i. The Fabry–Perot etalon (with the laser medium) in this case acts as an amplifier with a power gain $|E_t/E_i|^2$. We recall that in the case of the passive Fabry–Perot etalon (that is, one containing no laser medium), whose transmission is given by (4.1-7), $|E_t| \leq |E_i|$ and thus no power gain is possible. In the case considered here, however, the inverted population constitutes an energy source, so the transmitted wave can exceed the incident one.

If the denominator of (6.1-2) becomes zero, which happens when

$$r_1 r_2 e^{-2i[k+\Delta k(\omega)]l} e^{[\gamma(\omega)-\alpha]l} = 1 \tag{6.1-5}$$

then the ratio E_t/E_i becomes infinite. This corresponds to a finite transmitted wave E_t with a *zero* incident wave ($E_i = 0$)—that is, to *oscillation*. Physically, condition (6.1-5) represents the case in which a wave making a complete round trip inside the resonator returns to the starting plane with the *same amplitude* and, except for some integral multiple of 2π, with the *same phase*. Separating the oscillation condition (6.1-5) into the amplitude and phase requirements gives

$$r_1 r_2 e^{[\gamma_t(\omega)-\alpha]l} = 1 \tag{6.1-6}$$

for the threshold gain constant $\gamma_t(\omega)$ and

$$2[k + \Delta k(\omega)]l = 2\pi m \qquad m = 1, 2, 3, \ldots \tag{6.1-7}$$

for the phase condition. The amplitude condition (6.1-6) can be written as

$$\gamma_t(\omega) = \alpha - \frac{1}{l} \ln r_1 r_2 \tag{6.1-8}$$

which, using (6.1-4), becomes

$$N_t \equiv (N_2 - N_1)_t = \frac{8\pi n^2 t_{\text{spont}}}{g(\nu)\lambda^2}\left(\alpha - \frac{1}{l} \ln r_1 r_2\right) \tag{6.1-9}$$

This is the population inversion density at threshold.[2] It is often stated in a different form.[3]

Numerical Example: Population Inversion

To get an order of magnitude estimate of the critical population inversion $(N_2 - N_1)_t$ we use data typical of a 6328 Å He–Ne laser (which is discussed in Section 7.5). The appropriate constants are

$$\lambda = 6.328 \times 10^{-5} \text{ cm}$$

$$t_{\text{spont}} = 10^{-7} \text{ sec}$$

$$l = 12 \text{ cm}$$

$$\frac{1}{g(\nu_0)} \simeq \Delta\nu \simeq 10^9 \text{ Hz}$$

(The last figure is the Doppler-broadened width of the laser transition.)

The cavity decay time t_c is calculated from (6.1-10) assuming $\alpha = 0$ and $R_1 = R_2 = 0.98$. Since $R_1 = R_2 \simeq 1$, we can use the approximation $-\ln x = 1 - x$, $x \simeq 1$, to write

$$t_c \simeq \frac{nl}{c(1 - R)} = 2 \times 10^{-8} \text{ second}$$

Using the foregoing data in (6.1-11), we obtain

$$N_t \simeq 10^9 \text{ cm}^{-3}$$

[2]It was derived originally by Schawlow and Townes in their classic paper on the feasibility of lasers; see Reference [1].

[3]Consider the case in which the mirror losses and the distributed losses are all small, and therefore $r_1^2 \approx 1$, $r_2^2 \approx 1$ and $\exp(-\alpha l) \approx 1$. A wave starting with a unit intensity will return after one round trip with an intensity $R_1 R_2 \exp(-2\alpha l)$, where $R_1 \equiv r_1^2$ and $R_2 \equiv r_2^2$ are the mirrors' reflectivities. The fractional intensity loss per round trip is thus $1 - R_1 R_2 \exp(-2\alpha l)$. Since this loss occurs in a time $2ln/c$, it corresponds to an exponential decay time constant t_c (of the intensity) given by

$$\frac{1}{t_c} = \frac{(1 - R_1 R_2 e^{-2\alpha l})c}{2ln}$$

Therefore, the energy $\mathscr{E}$ stored in the passive resonator decays as $d\mathscr{E}/dt = -\mathscr{E}/t_c$. Since $R_1 R_2 e^{-2\alpha l} \approx 1$, we can use the relation $1 - x \approx -\ln x$, $x \approx 1$, to write $1/t_c$ as

$$\frac{1}{t_c} \simeq \frac{c}{n}\left[\alpha - \frac{1}{l}\ln r_1 r_2\right] \tag{6.1-10}$$

and the threshold condition (6.1-9) becomes

$$N_t \equiv (N_2 - N_1)_t = \frac{8\pi n^3 \nu^2 t_{\text{spont}}}{c^3 t_c g(\nu)} \tag{6.1-11}$$

where $N \equiv N_2 - N_1$ and the subscript t signifies threshold.

Figure 6-2 consists of a plot (a) of the transmission factor $|E_t/E_i|^2$ and (b) and reflection factor $|E_r/E_i|^2$ of a Fabry–Perot etalon as a function of the phase delay per round trip. Each curve is for a different value of the distributed gain constant γ. We note that when $e^{\gamma l} > 1$; i.e., when the net gain per pass exceeds unity, the transmission exceeds unity and the etalon functions as an amplifier.

It is especially interesting to note the narrowing of the peaks as the oscillation condition $e^{\gamma l} = 1/R = 0.9^{-1}$ is approached. The spectral distribution of the output of a laser oscillator can be viewed as made up of one of these peaks with the effective input being that of the spontaneous emission. This point of view is explored further in Section 10.6.

6.2 OSCILLATION FREQUENCY

The phase part of the start oscillation condition as given by (6.1-7) is satisfied at an infinite set of frequencies, which correspond to the different values of the integer m. If, in addition, the gain condition (6.1-6) is satisfied at one or more of these frequencies, the laser will oscillate at this frequency.

To solve for the oscillation frequency we use (6.1-3) to rewrite (6.1-7) as

$$kl\left[1 + \frac{\chi'(\nu)}{2n^2}\right] = m\pi \tag{6.2-1}$$

Introducing

$$\nu_m = \frac{mc}{2ln} \tag{6.2-2}$$

so that it corresponds to the mth resonance frequency of the passive $[N_2 - N_1 = 0]$ resonator and, using relations,

$$\chi'(\nu) = \frac{2(\nu_0 - \nu)}{\Delta\nu}\chi''(\nu)$$

$$\gamma(\nu) = -\frac{k\chi''(\nu)}{n^2}$$

we obtain from (6.2-1)

$$\nu\left[1 - \left(\frac{\nu_0 - \nu}{\Delta\nu}\right)\frac{\gamma(\nu)}{k}\right] = \nu_m \tag{6.2-3}$$

where ν_0 is the center frequency of the atomic lineshape function. Let us assume that the laser length is adjusted so that one of its resonance frequencies ν_m is very near ν_0. We anticipate that the oscillation frequency ν will also be close to ν_m and take advantage of the fact that when $\nu \simeq \nu_0$, the gain constant $\gamma(\nu)$ is a slowly varying function of ν; see Figure 5-5 for $\chi''(\nu)$, which is proportional to $\gamma(\nu)$. We can consequently replace $\gamma(\nu)$ in (6.2-3) by $\gamma(\nu_m)$, and $(\nu_0 - \nu)$ by $(\nu_0 - \nu_m)$ obtaining

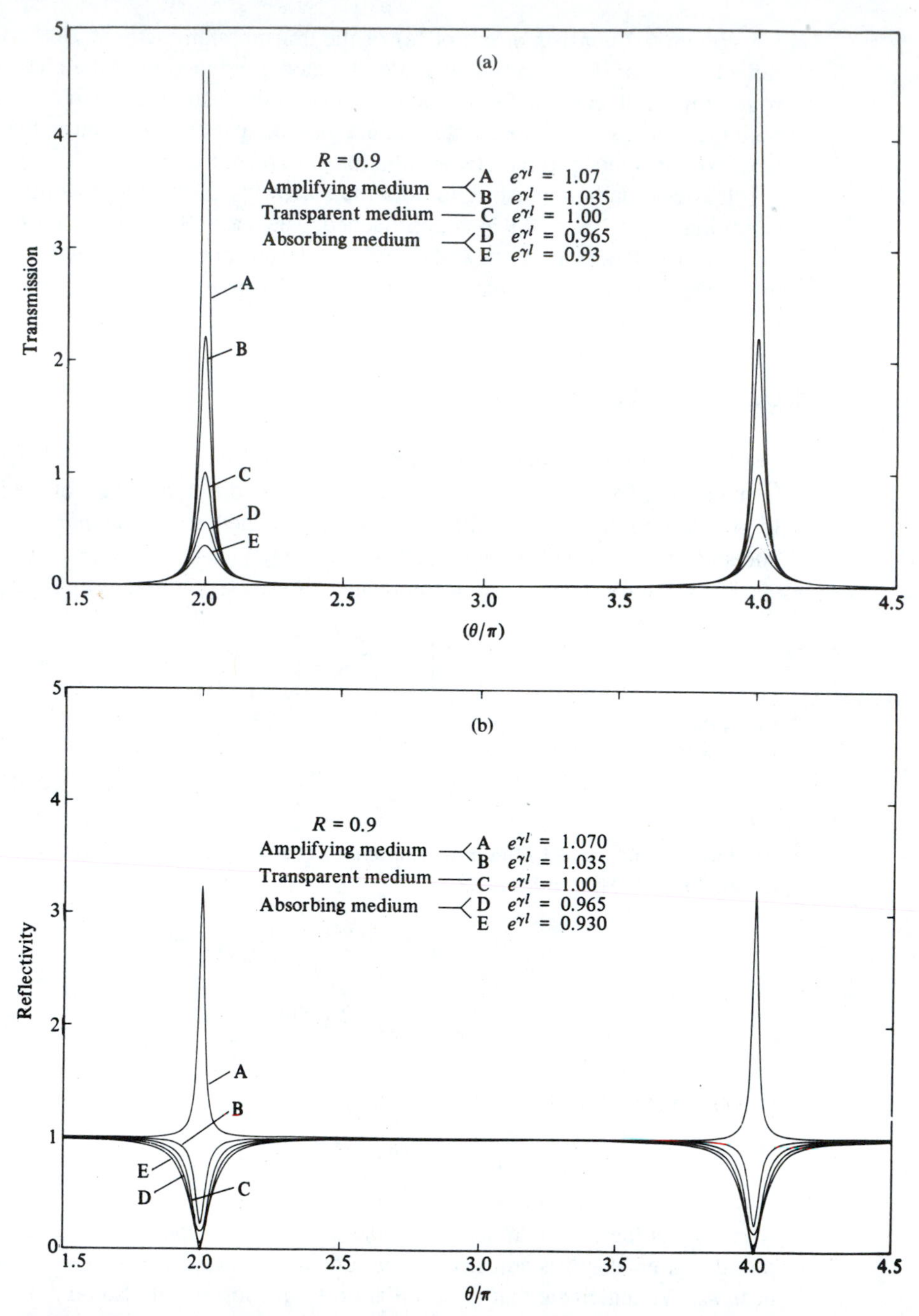

Figure 6-2 (a) The transmission $|E_t/E_i|^2$ versus phase shift per round trip $\theta = 2(kl - m\pi)$ of a Fabry–Perot etalon filled with an atomic medium. The different curves correspond to different values of gain (or loss) of the medium. Curve $C(e^{\gamma l} = 1)$ corresponds to transparency. (b) The reflection $|E_r/E_i|^2$ of a Fabry–Perot etalon.

$$\nu = \nu_m - (\nu_m - \nu_0)\frac{\gamma(\nu_m)c}{2\pi n\,\Delta\nu} \tag{6.2-4}$$

as the solution for the oscillation frequency ν.

We can recast (6.2-4) in a slightly different, and easier to use, form by starting with the gain threshold condition (6.1-6). Taking, for simplicity $r_1 = r_2 = \sqrt{R}$ and assuming that $R \simeq 1$ and $\alpha = 0$, we can write (6.1-8) as[4]

$$\gamma_t(\nu) \simeq \frac{1 - R}{l}$$

We also take advantage of the relation

$$\Delta\nu_{1/2} \simeq \frac{c(1 - R)}{2\pi nl}$$

which relates the passive resonator linewidth $\Delta\nu_{1/2}$ to R (this relation follows from (4.7-7) for $\alpha = 0$ and $R \approx 1$) and write (6.2-4) as

$$\nu = \nu_m - (\nu_m - \nu_0)\frac{\Delta\nu_{1/2}}{\Delta\nu} \tag{6.2-5}$$

A study of (6.2-5) shows that if the passive cavity resonance ν_m coincides with the atomic line center—that is, $\nu_m = \nu_0$—oscillation takes place at $\nu = \nu_0$. If $\nu_m \neq \nu_0$, oscillation takes place near ν_m but is shifted slightly toward ν_0. This phenomenon is referred to as *frequency pulling* and is demonstrated by Figure 6-3.

---

[4]This result can be obtained by putting $R = 1 - \Delta$, where $\Delta \ll 1$. Equation (6.1-6) becomes $1 + \gamma_t l \approx 1 + \Delta \Rightarrow \gamma_t \approx \Delta/l = (1 - R)/l$.

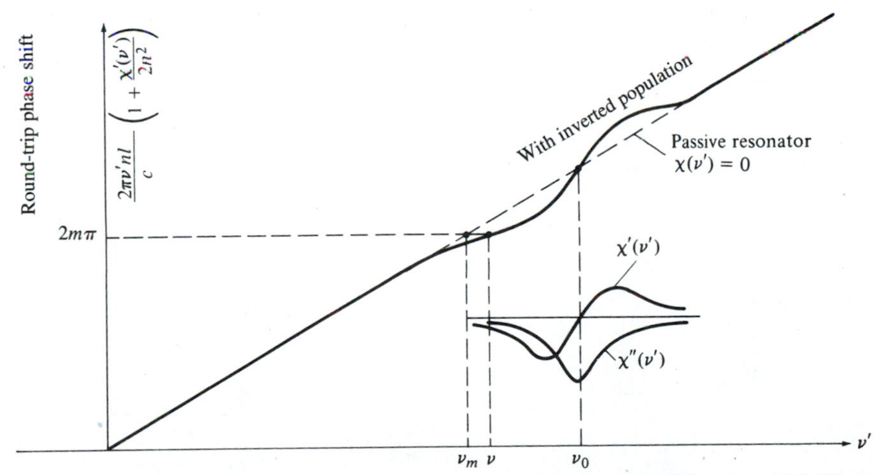

Figure 6-3 A graphical illustration of the laser frequency condition [Equation (6.2-1)] showing how the atomic dispersion $\chi'(\nu)$ "pulls" the laser oscillation frequency, ν, from the passive resonator value, ν_m, toward that of the atomic resonance at ν_0.

6.3 THREE- AND FOUR-LEVEL LASERS

Lasers are commonly classified into the so-called "three-level" or "four-level" lasers. An idealized model of a four-level laser is shown in Figure 6-4. The feature characterizing this laser is that the separation E_1 of the terminal laser level from the ground state is large enough that at the temperature T at which the laser is operated, $E_1 \gg kT$. This guarantees that the thermal equilibrium population of level 1 can be neglected. If, in addition, the lifetime t_1 of atoms in level 1 is short compared to t_2, we can neglect N_1 compared to N_2 and the threshold condition (6.1-11) is satisfied when

$$N_2 \simeq N_t \tag{6.3-1}$$

Therefore, laser oscillation begins when the upper laser level acquires, by pumping, a population density equal to the threshold value N_t.

A three-level laser is one in which the lower laser level is either the ground state or a level whose separation E_1 from the ground state is small compared to kT, so that at thermal equilibrium a substantial fraction of the total population occupies this level. An idealized three-level laser system is shown in Figure 6-5.

At a pumping level that is strong enough to create a population $N_2 = N_1 = N_0/2$ in the upper laser level,[5] the optical gain γ is zero, since $\gamma \propto N_2 - N_1 = 0$. To satisfy the oscillation condition the pumping rate has to be further increased until

$$N_2 = \frac{N_0}{2} + \frac{N_t}{2}$$

and

$$N_1 = \frac{N_0}{2} - \frac{N_t}{2} \tag{6.3-2}$$

[5]Here we assume that because of the very fast transition rate ω_{32} out of level 3, the population of this level is negligible and $N_1 + N_2 = N_0$, where N_0 is the density of the active atoms.

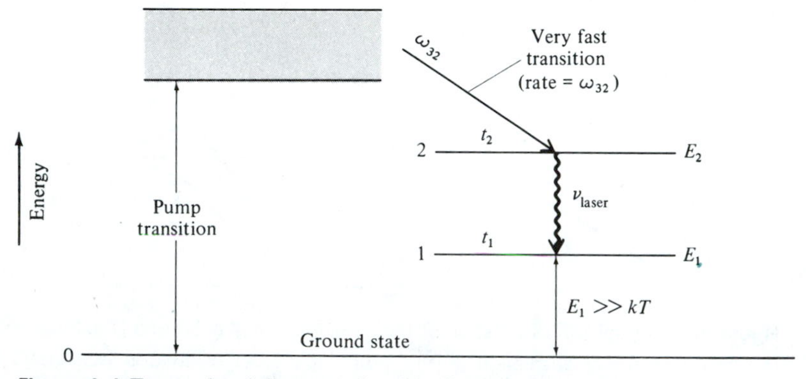

Figure 6-4 Energy-level diagram of an idealized four-level laser.

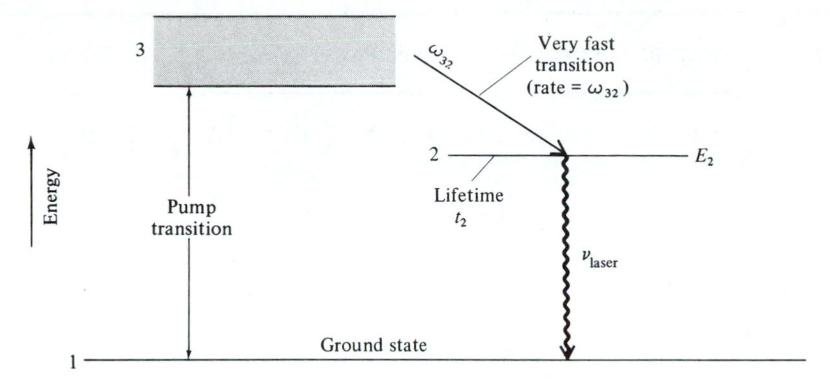

Figure 6-5 Energy-level diagram of an idealized three-level laser.

so $N_2 - N_1 = N_t$. Since in most laser systems $N_0 \gg N_t$, we find by comparing (6.3-1) to (6.3-2) that the pump rate at threshold in a three-level laser must exceed that of a four-level laser—all other factors being equal—by

$$\frac{(N_2)_{\text{3-level}}}{(N_2)_{\text{4-level}}} \sim \frac{N_0}{2N_t}$$

In the numerical example given in the next chapter we will find that in the case of the ruby laser this factor is ~ 100.

The need to maintain about $N_0/2$ atoms in the upper level of a three-level laser calls for a *minimum* expenditure of power of

$$(P_s)_{\text{3-level}} = \frac{N_0 h\nu V}{2t_2} \tag{6.3-3}$$

and of

$$(P_s)_{\text{4-level}} = \frac{N_t h\nu V}{t_2} \tag{6.3-4}$$

in a four-level laser. V is the volume. The last two expressions are derived by multiplying the decay rate (atoms per second) from the upper level at threshold, which is $N_0 V/2t_2$ and $N_t V/t_2$ in the two cases, by the energy $h\nu$ per transition. If the decay rate per atom t_2^{-1} (seconds^{-1}) from the upper level is due to spontaneous emission only, we can replace t_2 by t_{spont}. P_s is then equal to the power emitted through fluorescence by atoms within the (mode) volume V at threshold. We will refer to it as the *critical fluorescence* power. In the case of the four-level laser we use (6.1-11) for N_t and obtain

$$(P_s)_{\text{4-level}} = \frac{N_t h\nu V}{t_2} = \frac{8\pi n^3 h\Delta\nu V}{\lambda^3 t_c}\frac{t_{\text{spont}}}{t_2} \tag{6.3-5}$$

where $\Delta\nu \cong 1/g(\nu_0)$ is the width of the laser transition lineshape.

Numerical Example: Critical Fluorescence Power of an Nd^{3+}: Glass Laser

The critical fluorescence power of an Nd^{3+}:glass laser is calculated using the following data:

$$l = 10 \text{ cm}$$

$$V = 10 \text{ cm}^3$$

$$\lambda = 1.06 \times 10^{-6} \text{ meter}$$

$$R = (\text{mirror reflectivity}) = 0.95$$

$$n \simeq 1.5$$

$$t_c \simeq \frac{nl}{(1 - R)c} = 10^{-8} \text{ second}$$

$$\Delta\nu = 3 \times 10^{12} \text{ Hz}$$

The Nd^{3+}:glass is a four-level laser system (see Figure 7-11), since level 1 is about 2,000 cm^{-1} above the ground state so that at room temperature $E_1 \approx 10kT$. We can thus use (6.3-5), obtaining $N_t = 8.5 \times 10^{15}$ cm^{-3} and

$$P_s \simeq 150 \text{ watts}$$

6.4 POWER IN LASER OSCILLATORS

In Section 6.1 we derived an expression for the threshold population inversion N_t at which the laser gain becomes equal to the losses. We would expect that as the pumping intensity is increased beyond the point at which $N_2 - N_1 = N_t$ the laser will break into oscillation and emit power. In this section we obtain the expression relating the laser power output to the pumping intensity. We also treat the problem of optimum coupling—that is, of the mirror transmission that results in the maximum power output.

Rate Equations

Consider an ideal four-level laser such as the one shown in Figure 6-4. We take $E_1 \gg kT$ so that the thermal population of the lower laser level 1 can be neglected. We assume that the critical inversion density N_t is very small compared to the ground-state population, so during oscillation the latter is hardly affected. We can consequently characterize the pumping intensity by R_2 and R_1, the density of atoms pumped per second into levels 2 and 1, respectively. Process R_1, which populates the lower level 1, causes a reduction of the gain and is thus detrimental to the laser operation. In many laser systems, such as discharge gas lasers, considerable pumping into the lower laser level is unavoidable, and therefore a realistic analysis of such systems must take R_1 into consideration.

The rate equations that describe the populations of levels 1 and 2 become

$$\frac{dN_2}{dt} = -N_2\omega_{21} - W_i(N_2 - N_1) + R_2 \tag{6.4-1}$$

$$\frac{dN_1}{dt} = -N_1\omega_{10} + N_2\omega_{21} + W_i(N_2 - N_1) + R_1 \tag{6.4-2}$$

ω_{ij} is the decay rate per atom from level i to j; thus the density of atoms per second undergoing decay from i to j is $N_i\omega_{ij}$. If the decay rate is due entirely to spontaneous transitions, then ω_{ij} is equal to the Einstein A_{ij} coefficient introduced in Section 5.1. W_i is the probability per unit time that an atom in level 2 will undergo an *induced* (stimulated) transition to level 1 (or vice versa). W_i, given by (5.2-15), is proportional to the energy density of the radiation field inside the cavity.

Implied in the foregoing rate equations is the fact that we are dealing with a homogeneously broadened system. In an inhomogeneously broadened atomic transition, atoms with different transition frequencies $(E_2 - E_1)/h$ experience different induced transition rates and a single parameter W_i is not sufficient to characterize them.

In a steady-state situation we have $\dot{N}_1 = \dot{N}_2 = 0$. In this case we can solve (6.4-1) and (6.4-2) for N_1 and N_2, obtaining

$$N_2 - N_1 = \frac{R_2[1 - (\omega_{21}/\omega_{10})(1 + R_1/R_2)]}{W_i + \omega_{21}} \tag{6.4-3}$$

A necessary condition for population inversion in our model is thus $\omega_{21} < \omega_{10}$, which is equivalent to requiring that the lifetime of the upper laser level ω_{21}^{-1} exceed that of the lower one. The effectiveness of the pumping is, according to (6.4-3), reduced by the finite pumping rate R_1 and lifetime ω_{10}^{-1} of level 1 to an effective value

$$R = R_2\left[1 - \frac{\omega_{21}}{\omega_{10}}\left(1 + \frac{R_1}{R_2}\right)\right] \tag{6.4-4}$$

so (6.4-3) can be written as

$$N_2 - N_1 = \frac{R}{W_i + \omega_{21}} \tag{6.4-5}$$

Below the oscillation threshold the induced transition rate W_i is zero (since the oscillation energy density is zero) and $N_2 - N_1$ is, according to (6.4-5), proportional to the pumping rate R. This state of affairs continues until $R = N_t\omega_{21}$, at which point $N_2 - N_1$ reaches the threshold value [see (6.1-11)]

$$N_t = \frac{8\pi n^3\nu^2 t_{\text{spont}}}{c^3 t_c g(\nu_0)} = \frac{8\pi n^3\nu^2 t_{\text{spont}}\Delta\nu}{c^3 t_c} \tag{6.4-6}$$

This is the point at which the gain at ν_0 due to the inversion is large enough to make up *exactly* for the cavity losses (the criterion that was used to derive N_t). Further increase of $N_2 - N_1$ with pumping is impossible in a *steady-state situation*, since it would result in a rate of induced (energy) emission that exceeds the losses so that the field energy stored in the resonator will increase with time in violation of the steady-state assumption.

This argument suggests that, under steady-state conditions, $N_2 - N_1$ must remain equal to N_t regardless of the amount by which the threshold pumping rate is exceeded. An examination of (6.4-5) shows that this is possible, provided W_i is allowed to increase once R exceeds its threshold value $\omega_{21}N_t$, so that the equality

$$N_t = \frac{R}{W_i + \omega_{21}} \tag{6.4-7}$$

is satisfied. Since, according to (5.2-15), W_i is proportional to the energy density in the resonator, (6.4-7) relates the electromagnetic energy stored in the resonator to the pumping rate R. To derive this relationship we first solve (6.4-7) for W_i, obtaining

$$W_i = \frac{R}{N_t} - \omega_{21} \qquad R \geqslant N_t\omega_{21} \tag{6.4-8}$$

The total power generated by stimulated emission is

$$P_e = (N_tV)W_ih\nu \tag{6.4-9}$$

where V is the volume of the oscillating mode. Using (6.4-8) in (6.4-9) gives

$$\frac{P_e}{Vh\nu} = N_t\omega_{21}\left(\frac{R}{N_t\omega_{21}} - 1\right) \qquad R \geqslant N_t\omega_{21} \tag{6.4-10}$$

This expression may be recast in a slightly different form, which we will find useful later on. We use expression (6.4-6) for N_t and, recalling that in our idealized model $\omega_{21}^{-1} = t_{\text{spont}}$, obtain

$$\frac{P_e}{Vh\nu} = N_t\omega_{21}\left(\frac{R}{p/t_c} - 1\right) \qquad R \geqslant \frac{p}{t_c} \tag{6.4-11}$$

where

$$p = \frac{8\pi n^3\nu^2}{c^3 g(\nu_0)} = \frac{8\pi n^3\nu^2\Delta\nu}{c^3} \tag{6.4-12}$$

According to (4.0-12), p corresponds to the density (meters^{-3}) of radiation modes whose resonance frequencies fall within the atomic transition linewidth $\Delta\nu$—that is, the density of radiation modes that are capable of interacting with the transition.

Returning to the expression for the power output of a laser oscillator (6.4-11), we find that the term $R/(p/t_c)$ is the factor by which the pumping rate R exceeds its threshold value p/t_c. In addition, in an ideal laser system, $\omega_{21} = t_{\text{spont}}^{-1}$, so we can identify $N_t\omega_{21}h\nu V$ with the power P_s going into spontaneous emission at threshold, which is defined by (6.3-5). We can consequently rewrite (6.4-11) as

$$P_e = P_s\left(\frac{R}{R_t} - 1\right) \tag{6.4-13}$$

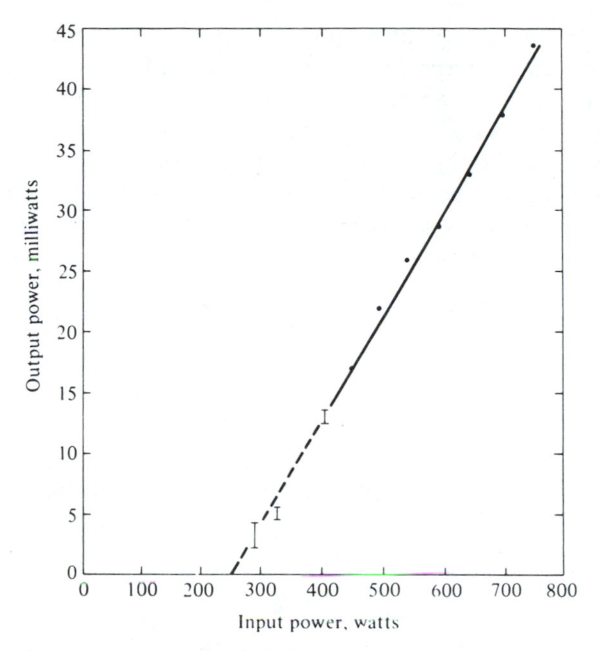

Figure 6-6 Plot of output power versus electric power input to a xenon lamp in a CW $CaF_2:U^{3+}$ laser. Mirror transmittance at 2.61 μm is 0.2 percent, $T = 77$ K. (After Reference [2].)

The main attraction of (6.4-13) is in the fact that, in addition to providing an extremely simple expression for the power emitted by the laser atoms, it shows that for each increment of pumping, measured relative to the threshold value, the power increases by P_s. An experimental plot showing the linear relation predicted by (6.4-13) is shown in Figure 6-6.

In the numerical example of Section 6.3, which was based on an Nd^{3+}: glass laser, we obtained $P_s = 150$ watts. We may expect on this basis that the power from this laser for, say $(R/R_t) \simeq 2$ (that is, twice above threshold) will be of the order of 150 watts.

6.5 OPTIMUM OUTPUT COUPLING IN LASER OSCILLATORS

The total loss encountered by the oscillating laser mode can conveniently be attributed to two different sources: (a) the inevitable residual loss due to absorption and scattering in the laser material and in the mirrors, as well as diffraction losses in the finite diameter reflectors; (b) the (useful) loss due to coupling of output power through the partially transmissive reflector. It is obvious that loss (a) should be made as small as possible since it raises the oscillation threshold without contributing to the output power. The problem of the coupling loss (b), however, is more subtle. At zero coupling (that is, both mirrors have zero transmission) the threshold will be at its minimum value and the power P_e emitted by the atoms will be maximum. But since none of this power is available as output, this is not a useful state of affairs. If, on the other hand, we keep increasing the coupling loss, the increasing threshold pumping will at some point exceed the actual pumping level. When this happens, oscillation will cease and the power output will again be zero. Between these two extremes there exists an optimum value of coupling (that is, mirror transmission) at which the power output is a maximum.

The expression for the population inversion was shown in (6.4-5) to have the form

$$N_2 - N_1 = \frac{R/\omega_{21}}{1 + W_i/\omega_{21}} \tag{6.5-1}$$

Since the exponential gain constant $\gamma(\nu)$ is, according to (5.3-3), proportional to $N_2 - N_1$, we can use (6.5-1) to write it as

$$\gamma = \frac{\gamma_0}{1 + W_i/\omega_{21}} \tag{6.5-2}$$

where γ_0 is the unsaturated ($W_i = 0$) gain constant (that is, the gain exercised by a very weak field, so that $W_i \ll \omega_{21}$). We can use (6.4-9) to express W_i in (6.5-2) in terms of the total emitted power P_e and then, in the resulting expression, replace $N_t V h\nu\omega_{21}$ by P_s. The result is

$$\gamma = \frac{\gamma_0}{1 + P_e/P_s} \tag{6.5-3}$$

where P_s, the saturation power, is given by (6.3-4). The oscillation condition (6.1-6) can be written as

$$e^{\gamma_t l}(1 - L) = 1 \tag{6.5-4}$$

where $L = 1 - r_1 r_2 \exp(-\alpha l)$ is the fraction of the intensity lost per pass. In the case of small losses ($L \ll 1$), (6.5-4) can be written as

$$\gamma_t l = L \tag{6.5-5}$$

According to the discussion in the introduction to this chapter, once the oscillation threshold is exceeded, the actual gain γ exercised by the laser oscillation is clamped at the threshold value γ_t regardless of the pumping. We can thus replace γ by γ_t in (6.5-3) and, solving for P_e, obtain

$$P_e = P_s\left(\frac{g_0}{L} - 1\right) \tag{6.5-6}$$

where $g_0 = \gamma_0 l$ (that is, the unsaturated gain per pass in nepers). P_e, we recall, is the *total* power given off by the atoms due to stimulated emission. The total loss per pass L can be expressed as the sum of the residual (unavoidable) loss L_i and the useful mirror transmission[6] T, so

$$L = L_i + T \tag{6.5-7}$$

The fraction of the total power P_e that is coupled out of the laser as useful output is thus $T/(T + L_i)$. Therefore, using (6.5-6) we can write the (useful) power output as

$$P_o = P_s\left(\frac{g_0}{L_i + T} - 1\right)\frac{T}{T + L_i} \tag{6.5-8}$$

Replacing P_s in (6.5-8) by the right side of (6.3-5), and recalling from (4.7-2) that for small losses

$$t_c = \frac{nl}{(L_i + T)c} = \frac{nl}{Lc} \tag{6.5-9}$$

Equation (6.5-8) becomes

$$P_o = \frac{8\pi n^2 h\nu\Delta\nu A}{\lambda^2(t_2/t_{\text{spont}})} T\left(\frac{g_0}{L_i + T} - 1\right) = I_s AT\left(\frac{g_0}{L_i + T} - 1\right) \tag{6.5-10}$$

where $A = V/l$ is the cross-sectional area of the mode (assumed constant) and I_s is the saturation intensity as given in (5.6-9). Maximizing P_0 with respect to T by setting $\partial P_0/\partial T = 0$ yields

$$T_{\text{opt}} = -L_i + \sqrt{g_0 L_i} \tag{6.5-11}$$

as the condition for the mirror transmission that yields the maximum power output.

The expression for the power output at optimum coupling is obtained by substituting (6.5-11) for T in (6.5-10). The result, using (5.6-9), is

$$\begin{aligned}(P_o)_{\text{opt}} &= \frac{8\pi n^2 h\nu\Delta\nu A}{(t_2/t_{\text{spont}})\lambda^2}(\sqrt{g_0} - \sqrt{L_i})^2 = I_s A(\sqrt{g_0} - \sqrt{L_i})^2 \\ &\equiv S(\sqrt{g_0} - \sqrt{L_i})^2\end{aligned} \tag{6.5-12}$$

where the parameter $S = I_s A$ is defined by (6.5-12) and is independent of the excitation level (pumping) or losses.

Theoretical plots of (6.5-10) with L_i as a parameter are shown in Figure 6-7. Also shown are experimental data points obtained in a He–Ne 6328-Å laser. Note that the value of g_0 is given by the intercept of the $L_i = 0$ curve and is equal to 12 percent. The existence of an optimum coupling resulting in a maximum power output for each L_i is evident.

It is instructive to consider what happens to the energy $\mathscr{E}$ stored in the laser resonator as the coupling T is varied. A little thinking will convince us that $\mathscr{E}$ is

[6]For the sake of simplicity we can imagine one mirror as being perfectly reflecting, whereas the second (output) mirror has a transmittance T.

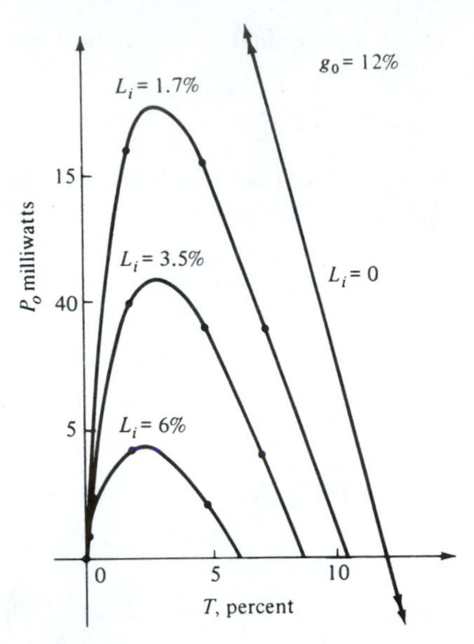

Figure 6-7 Useful power output (P_o) versus mirror transmission T for various values of internal loss L_i in an He–Ne 6328 Å laser. (After Laures, *Phys. Lett.* 10:61, 1964.)

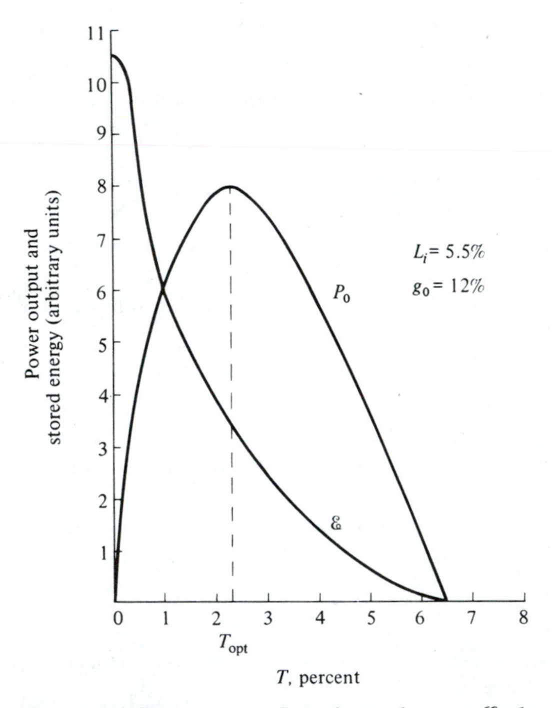

Figure 6-8 Power output P_o and stored energy ℰ plotted against mirror transmission T.

proportional to P_o/T.[7] A plot of P_o (taken from Figure 6-7) and $\mathscr{E} \propto P_o/T$ as a function of the coupling T is shown in Figure 6-8. As we may expect, $\mathscr{E}$ is a monotonically decreasing function of T.

6.6 MULTIMODE LASER OSCILLATION AND MODE LOCKING

In this section we contemplate the effect of homogeneous or inhomogeneous broadening (in the sense described in Section 5.1) on the laser oscillation.

We start by reminding ourselves of some basic results pertinent to this discussion:

1. The actual gain constant prevailing inside a laser oscillator *at the oscillation frequency* ν is clamped, at steady state, at a value that is equal to the losses

$$\gamma_t(\nu) = \alpha - \frac{1}{l} \ln r_1 r_2 \qquad \textbf{(6.1-8)}$$

where l is the length of the gain medium as well as the distance between the mirrors which are taken here to be the same.

2. The gain constant of a distributed laser medium is given, according to (5.3-3), by

$$\gamma(\nu) = (N_2 - N_1) \frac{c^2}{8\pi n^2 \nu^2 t_{\text{spont}}} g(\nu)$$

3. The optical resonator can support oscillations, provided sufficient gain is present to overcome losses, at frequencies[8] ν_q separated according to (4.6-3) by

$$\nu_{q+1} - \nu_q = \frac{c}{2nl}$$

Now consider what happens to the gain constant $\gamma(\nu)$ inside a laser oscillator as the pumping is increased from some value below threshold. Operationally, we can imagine an extremely weak wave of frequency ν launched into the laser medium and then measuring the gain constant $\gamma(\nu)$ as "seen" by this signal as ν is varied.

We treat first the case of a homogeneous laser. Below threshold the inversion $N_2 - N_1$ is proportional to the pumping rate and $\gamma(\nu)$, which is given by (5.3-3), is proportional to $g(\nu)$. This situation is illustrated by curve A in Figure 6-9(a). The spectrum (4.6-3) of the passive resonances is shown in Figure 6-9(b). As the pumping rate is increased, the point is reached at which the gain per pass at the center resonance frequency ν_0 is equal to the average loss per pass. This is shown in curve B. At this point, oscillation at ν_0 starts. An increase in the pumping cannot increase the inversion since this will cause $\gamma(\nu_0)$ to increase beyond its clamped value as given by Equation (6.1-8). Since the spectral lineshape function $g(\nu)$ describes the response

[7]The internal one-way power P_i incident on the mirrors is related, by definition, to P_o by $P_o = P_i T$. The total energy $\mathscr{E}$ is proportional to P_i.

[8]The high-order transverse modes discussed in Section 4.5 are ignored here.

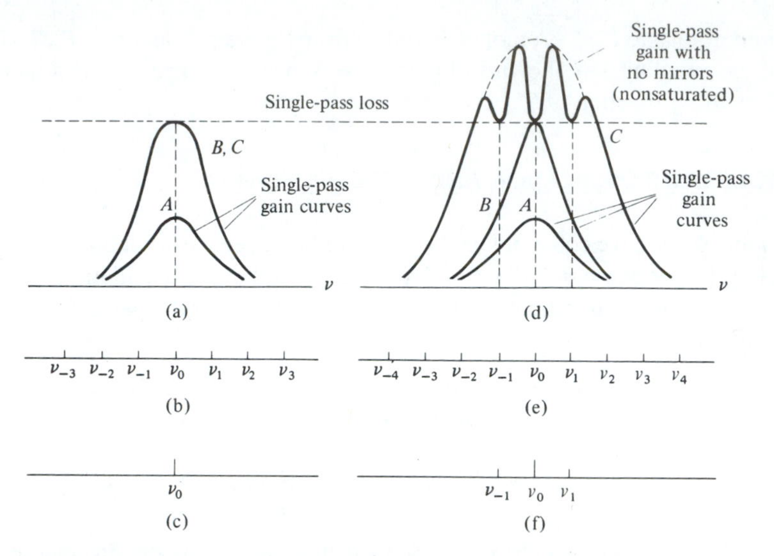

Figure 6-9 (a) Single-pass gain curves for a homogeneous atomic system (*A*—below threshold; *B*—at threshold; *C*—well above threshold). (b) Mode spectrum of optical resonator. (c) Oscillation spectrum (only one mode oscillates). (d) Single-pass gain curves for an inhomogeneous atomic system (*A*—below threshold; *B*—at threshold; *C*—well above threshold). (e) Mode spectrum of optical resonator. (f) Oscillation spectrum for pumping level *C*, showing three oscillating modes.

of each individual atom, all the atoms being identical, it follows that the gain profile $\gamma(\nu)$ above threshold as in curve C is identical to that at threshold curve B.[9] The gain at other frequencies—such as ν_{-1}, ν_1, ν_{-2}, ν_2, and so forth—remains below the threshold value so that the ideal homogeneously broadened laser can oscillate only at a single frequency.

In the extreme inhomogeneous case, the individual atoms can be considered as being all different from one another and as acting independently. The lineshape function $g(\nu)$ reflects the distribution of the transition frequencies of the individual atoms. The gain profile $\gamma(\nu)$ below threshold is proportional to $g(\nu)$, and its behavior is similar to that of the homogeneous case. Once threshold is reached as in curve B, the gain at ν_0 remains clamped at the threshold value. There is no reason, however, why the gain at other frequencies should not increase with further pumping. This gain is due to atoms that do not communicate with those contributing to the gain at ν_0. Further pumping will thus lead to oscillation at additional longitudinal-mode frequencies as shown in curve C. Since the gain at each oscillating frequency is clamped, the gain profile curve acquires depressions at the oscillation frequencies. This phenomenon is referred to as "hole burning" [7].

[9]Further increase in pumping, and the resulting increase in optical intensity, will eventually cause a broadening of $\gamma(\nu)$ due to the shortening of the lifetime by induced emission.

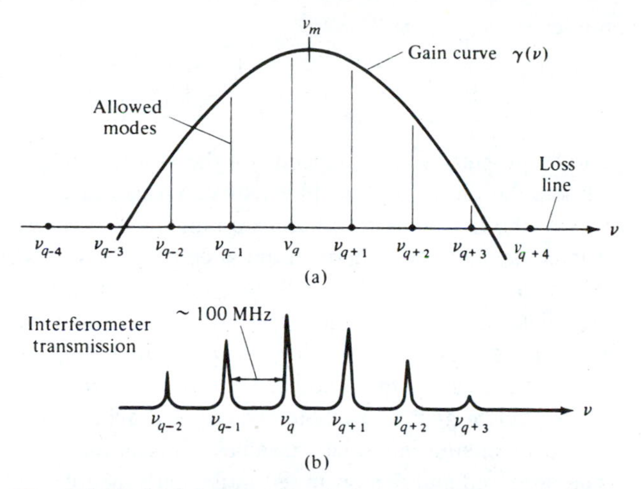

Figure 6-10 (a) Inhomogeneously broadened Doppler gain curve of the 6328 Å Ne transition and position of allowed longitudinal-mode frequencies. (b) Intensity versus frequency profile of an oscillating He–Ne laser. Six modes have sufficient gain to oscillate (After Reference [8].)

A plot of the output frequency spectrum showing the multimode oscillation of a He–Ne 0.6328-μm laser is shown in Figure 6-10.

Mode Locking

We have argued above that in an inhomogeneously broadened laser, oscillation can take place at a number of frequencies, which are separated by (assuming $n = 1$)

$$\omega_q - \omega_{q-1} = \frac{\pi c}{l} \equiv \omega$$

Now consider the total optical electric field resulting from such multimode oscillation at some arbitrary point, say next to one of the mirrors, in the optical resonator. It can be taken, using complex notation, as

$$e(t) = \sum_n E_n e^{i[(\omega_0 + n\omega)t + \phi_n]} \tag{6.6-1}$$

where the summation is extended over the oscillating modes and ω_0 is chosen, arbitrarily, as a reference frequency. ϕ_n is the phase of the nth mode. One property of (6.6-1) is that $e(t)$ is periodic in $T \equiv 2\pi/\omega = 2l/c$, which is the round-trip transit time inside the resonator

$$\begin{aligned} e(t+T) &= \sum_n E_n \exp\left\{i\left[(\omega_0 + n\omega)\left(t + \frac{2\pi}{\omega}\right) + \phi_n\right]\right\} \\ &= \sum_n E_n \exp\{i[(\omega_0 + n\omega)t + \phi_n]\} \exp\left\{i\left[2\pi\left(\frac{\omega_0}{\omega} + n\right)\right]\right\} \\ &= e(t) \end{aligned} \tag{6.6-2}$$

Since ω_0/ω is an integer ($\omega_0 = m\pi c/l$),

$$\exp\left[2\pi i\left(\frac{\omega_0}{\omega} + n\right)\right] = 1$$

Note that the periodic property of $e(t)$ depends on the fact that the phases ϕ_n are fixed. In typical lasers the phases ϕ_n are likely to vary randomly with time. This causes the intensity of the laser output to fluctuate randomly[10] and greatly reduces its usefulness for many applications where temporal coherence is important.

Two ways in which the laser can be made coherent are: First, make it possible for the laser to oscillate at a single frequency only so that mode interference is eliminated. This can be achieved in a variety of ways, including shortening the resonator length l, thus increasing the mode spacing ($\omega = \pi c/l$) to a point where only one mode has sufficient gain to oscillate. The second approach is to force the modes' phases ϕ_n to maintain their relative values. This is the so-called ''mode locking'' technique proposed and demonstrated in the early history of the laser [9, 10, 11]. This mode locking causes the oscillation intensity to consist of a periodic train with a period of $T = 2l/c = 2\pi/\omega$.

One of the most useful forms of mode locking results when the phases ϕ_n are made equal to zero. To simplify the analysis of this case, assume that there are N oscillating modes with equal amplitudes. Taking $E_n = 1$ and $\phi_n = 0$ in (6.6-1) gives

$$e(t) = \sum_{-(N-1)/2}^{(N-1)/2} e^{i(\omega_o + n\omega)t} \tag{6.6-3}$$

$$= e^{i\omega_o t}\frac{\sin(N\omega t/2)}{\sin(\omega t/2)} \tag{6.6-4}$$

The average laser power output is proportional to $e(t)e^*(t)$ and is given by[11]

$$P(t) \propto \frac{\sin^2(N\omega t/2)}{\sin^2(\omega t/2)} \tag{6.6-5}$$

Some of the analytic properties of $P(t)$ are immediately apparent:

1. The power is emitted in a form of a train of pulses with a period $T = 2\pi/\omega = 2l/c$, i.e., the round-trip delay time.
2. The peak power, $P(sT)$ (for $s = 1, 2, 3, \ldots$), is equal to N times the average power, where N is the number of modes locked together.
3. The peak field amplitude is equal to N times the amplitude of a single mode.
4. The individual pulse width, defined as the time from the peak to the first zero is $\tau_0 = T/N$. The number of oscillating modes can be estimated by $N \simeq \Delta\omega/\omega$—that is, the ratio of the transition lineshape width $\Delta\omega$ to the frequency

[10] It should be noted that this fluctuation takes place because of random interference between modes and not because of intensity fluctuations of individual modes.

[11] The averaging is performed over a time that is long compared with the optical period $2\pi/\omega_0$ but short compared with the modulation period $2\pi/\omega$.

spacing ω between modes. Using this relation, as well as $T = 2\pi/\omega$ in $\tau_0 = T/N$, we obtain

$$\tau_0 \sim \frac{2\pi}{\Delta\omega} = \frac{1}{\Delta\nu} \tag{6.6-6}$$

Thus the length of the mode-locked pulses is approximately the inverse of the gain linewidth.

A theoretical plot of $\sqrt{P(t)}$ as given by (6.6-5) for the case of five modes ($N = 5$) is shown in Figure 6-11. The ordinate may also be considered as being proportional to the instantaneous field amplitude.

The foregoing discussion was limited to the consideration of mode locking as a function of time. It is clear, however, that since the solution of Maxwell's equations in the cavity involves traveling waves (a standing wave can be considered as the sum of two waves traveling in opposite directions), mode locking causes the oscillation energy of the laser to be condensed into a packet that travels back and forth between the mirrors with the velocity of light c. The pulsation period $T = 2l/c$ corresponds simply to the time interval between two successive arrivals of the pulse at the mirror. The spatial length of the pulse L_p must correspond to its time duration multiplied by its velocity c. Using $\tau_0 = T/N$ we obtain

$$L_p \sim c\tau_0 = \frac{cT}{N} = \frac{2\pi c}{\omega N} = \frac{2l}{N} \tag{6.6-7}$$

We can verify the last result by taking the basic resonator mode as a standing wave $\sin k_n z \sin \omega_n t$; the total optical field is then

$$e(z, t) = \sum_{n=-(N-1)/2}^{(N-1)/2} \sin\left[\frac{(m+n)\pi}{l} z\right] \sin\left[(m+n)\frac{\pi c}{l} t\right] \tag{6.6-8}$$

where, using (4.6-3), $\omega_n = (m + n)(\pi c/l)$, $k_n = \omega_n/c$, and m is the integer (equal to the number of half wavelengths $m = (l/\lambda/2)$) corresponding to the central mode. We can rewrite (6.6-8) as

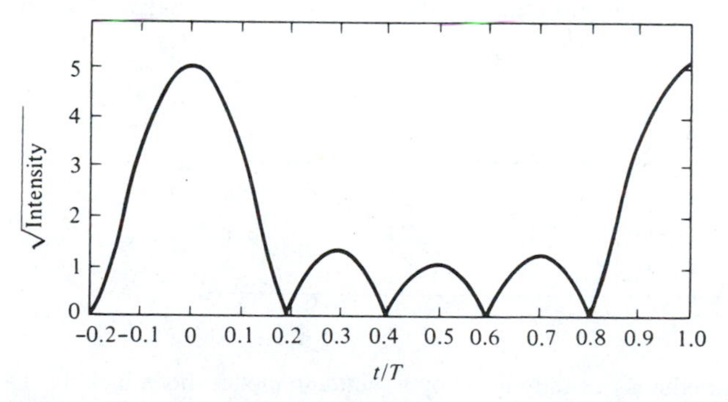

Figure 6-11 Theoretical plot of optical field amplitude $\sqrt{P(t)} \propto |\sin(N\omega t/2)/\sin(\omega t/2)|$ resulting from phase locking of five ($N = 5$) equal-amplitude modes separated from each other by a frequency interval $\omega = 2\pi/T$.

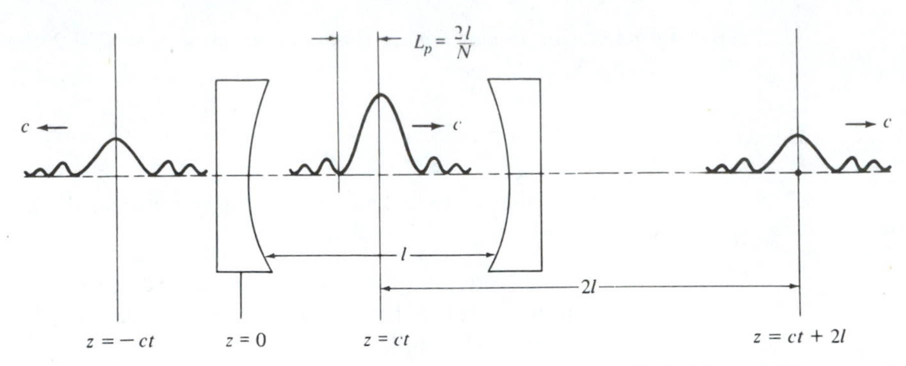

Figure 6-12 Traveling pulse of energy resulting from the mode locking of N laser modes; based on Equation (6.6-9).

$$e(z, t) = \frac{1}{2} \sum_{n=-(N-1)/2}^{(N-1)/2} \left\{ \cos\left[(m+n)\frac{\pi}{l}(z-ct)\right] - \cos\left[(m+n)\frac{\pi}{l}(z+ct)\right]\right\} \quad (6.6\text{-}9)$$

which can be shown to have the spatial and temporal properties described previously. Figure 6-12 shows a spatial plot of (6.6-9) at time t.

Methods of Mode Locking

In the preceding discussion we considered the consequences of fixing the phases of the longitudinal modes of a laser—mode locking. Mode locking can be achieved by modulating the losses (or gain) of the laser at a radian frequency $\omega = \pi c/l$, which is equal to the intermode frequency spacing. The theoretical proof of mode locking by loss modulation (References [2, 9, and 10]) is rather formal, but a good plausibility argument can be made as follows: As a form of loss modulation consider a thin shutter inserted inside the laser resonator. Let the shutter be closed (high optical loss) most of the time except for brief periodic openings for a duration of τ_{open} every $T = 2l/c$ seconds. This situation is illustrated by Figure 6-13. A single laser mode

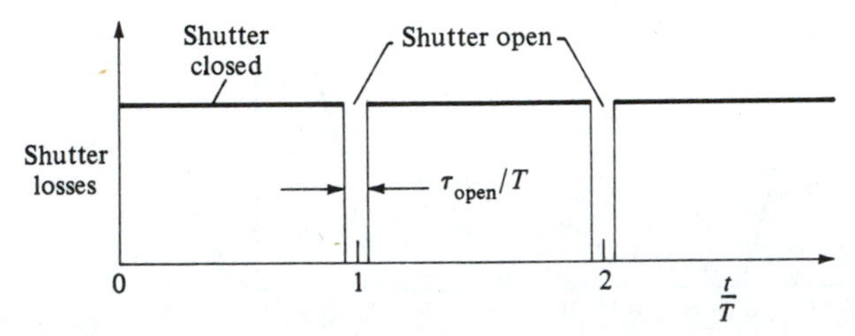

Figure 6-13 Periodic losses introduced by a shutter to induce mode locking. The presence of these losses favors the choice of mode phases that results in a pulse passing through the shutter during open intervals—that is, mode locking.

will not oscillate in this case because of the high losses (we assume that τ_{open} is too short to allow the oscillation to build up during each opening). The same applies to multimode oscillation with arbitrary phases. There is one exception, however. If the phases were "locked" as in (6.6-3), the energy distribution inside the resonator would correspond to that shown in Figure 6-12 and would consist of a narrow ($L_p \simeq 2l/N$) traveling pulse. If this pulse should arrive at the shutter's position when it is open and if the pulse (temporal) length τ_0 is short compared to the opening time τ_{open}, the mode-locked pulse will be "unaware" of the shutter's existence and, consequently, *will not be attenuated by it*. We may thus reach the conclusion that loss modulation causes mode locking through some kind of "survival of the fittest" mechanism. In reality the periodic shutter chops off any intensity tails acquired by the mode-locked pulses due to a "wandering" of the phases from their ideal (ϕ_n = 0) values. This has the effect of continuously restoring the phases.

An experimental setup used to mode-lock a He–Ne laser is shown in Figure 6-14; the periodic loss [11] is introduced by Bragg diffraction (see Sections 12.2 and 12.3) of a portion of the laser intensity from a standing acoustic wave. The standing-wave nature of the acoustic oscillation causes the strain to have a form

$$S(z, t) = S_0 \cos \omega_a t \cos k_a z \tag{6.6-10}$$

where the acoustic velocity is $v_a = \omega_a/k_a$. Since the change in the index of refraction is to first order, proportional to the strain $S(z, t)$, we can interpret (6.6-10) as a phase

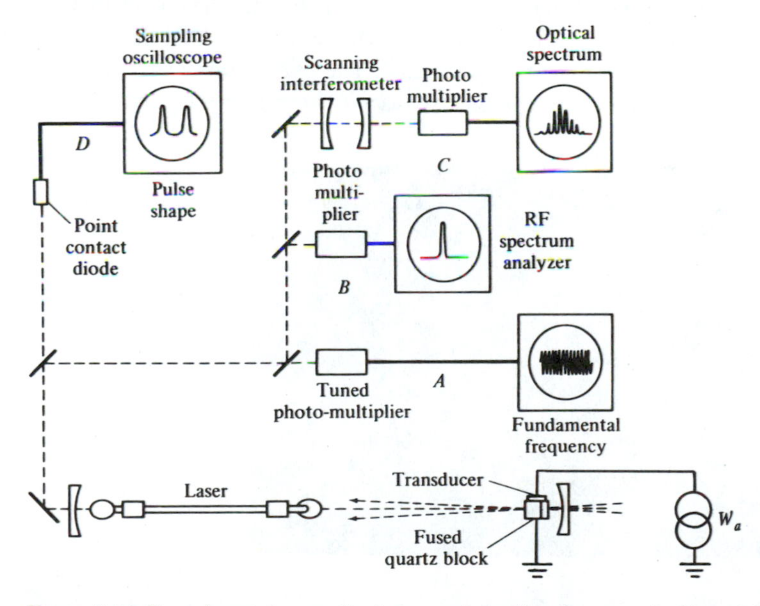

Figure 6-14 Experimental setup for laser mode locking by acoustic (Bragg) loss modulation. The loss is due to Bragg diffraction of the main laser beam by a standing acoustic wave. Parts *A*, *B*, *C*, and *D* of the experimental setup are designed to display the fundamental component of the intensity modulation, the power spectrum of the intensity modulation, the power spectrum of the optical field $e(t)$, and the optical intensity, respectively. (After Reference [12].)

diffraction grating (see Sections 12.2, 3) with a spatial period $2\pi/k_a$, which is equal to the acoustic wavelength. The diffraction loss of the incident laser beam due to the grating reaches its peak twice in each acoustic period when $S(z, t)$ has its maximum and minimum values. The loss modulation frequency is thus $2\omega_a$, and mode locking occurs when $2\omega_a = \omega$, where ω is the (radian) frequency separation between two longitudinal laser modes.

Figure 6-15 shows the pulses resulting from mode locking a Rhodamine 6G dye laser.

Mode locking occurs spontaneously in some lasers if the optical path contains a saturable absorber (an absorber whose opacity decreases with increasing optical intensity). This method is used to induce mode locking in the high-power pulsed solid-state lasers [13, 15] and in continuous dye lasers. This is due to the fact that such a dye will absorb less power from a mode-locked train of pulses than from a random phase oscillation of many modes [2], since the first form of oscillation leads to the highest possible peak intensities, for a given average power from the laser, and is consequently attenuated less severely. From arguments identical with those advanced in connection with the periodic shutter (see discussion following [6.6-9]), it follows that the presence of a saturable absorber in the laser cavity will ''force'' the laser, by a ''survival of the fittest'' mechanism, to lock its modes' phases as in (6.6-9).

Some of the shortest mode-locked pulses to date were obtained from dye lasers employing Rhodamine 6G as the gain medium. The mode locking is caused by synchronous gain modulation that is due to the fact that the pumping (blue-green) argon gas laser is itself mode-locked. The pump pulses are synchronized exactly to the pulse repetition rate of the dye laser. (This requires that both lasers have precisely

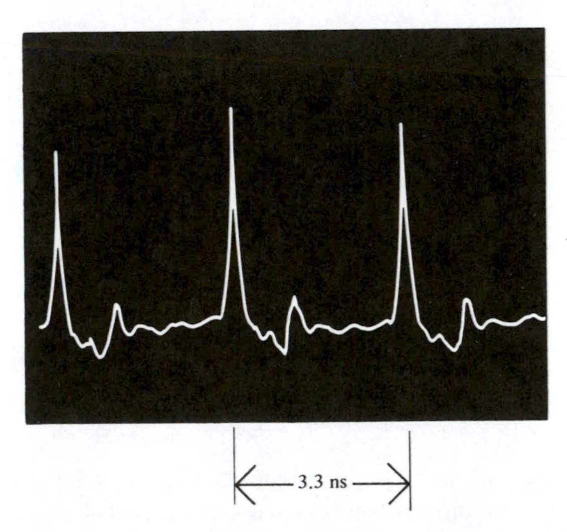

Figure 6-15 Power output as a function of time of a mode-locked dye laser, using Rhodamine 6G. The oscillation is at $\lambda = 0.61\ \mu$m. The pulse width is detector limited (After Reference [12].)

the same optical length.) When this is done, the dye laser gain medium will be pumped once in each round-trip period so that the pumping pulse and the mode-locked pulse overlap spatially and temporally in the dye cell.

Additional sharpening of the mode-locked pulses can result from the inclusion of a saturable (dye) absorber in the resonator.

A sketch of a synchronously mode-locked dye laser configuration is shown in Figure 6-16.

Additional amplification of the output pulses of the dye laser by a sequence of three to four dye laser amplifier cells (consisting of Rhodamine 6G pumped by the pulsed second harmonic of Q-switched Nd^{3+}:YAG lasers) has yielded subpicosecond pulses with peak power exceeding 10^9 watts.

The shortest pulses obtained to date are $\sim 30 \times 10^{-15}$ s [28]. These pulses have been narrowed down further to $\sim 6 \times 10^{-15}$ s by the use of nonlinear optical techniques that are described in Section 6.8.

Ultrashort mode-locked pulses are now used in an ever-widening circle of applications involving the measurement and study of short-lived molecular and electronic phenomena. The use of ultrashort optical pulses has led to an improvement of the temporal resolution of such experiments by more than three orders of magnitude. For a description of many of these applications as well as of the many methods used to measure the pulse duration, the student should consult References [31–33].

Mode locking in semiconductor lasers is of particular interest owing to the very large gain bandwidth in these media. These lasers offer potential operation in the 10–20 femtosecond range although present results are far of this goal [43, 44]. Of

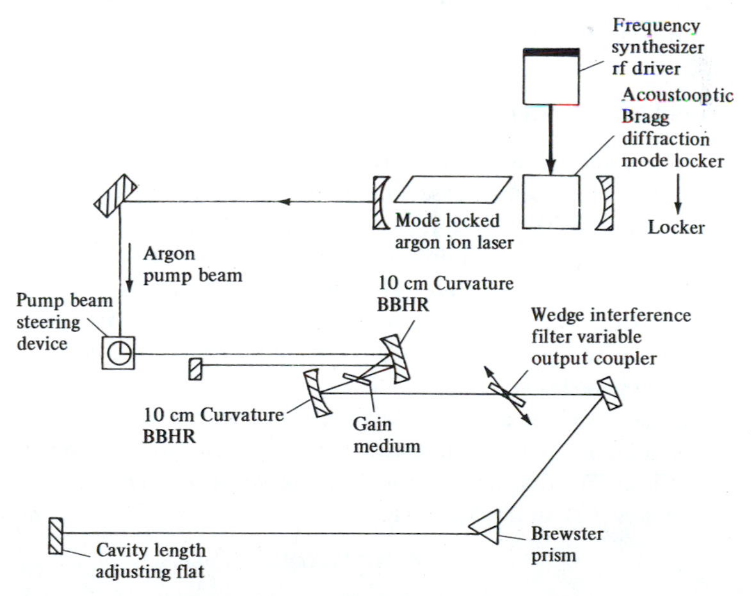

Figure 6-16 Synchronously mode-locked dye laser configuration. (After Reference [29].)

Table 6-1 Some Laser Systems, Their Gain Linewidth $\Delta\nu$, and the Length of Their Pulses in the Mode-Locked Operations

Laser Medium	$\Delta\nu$, Hz	$(\Delta\nu)^{-1}$, Seconds	Observed Pulse Duration τ_0, Seconds
He–Ne (0.6328 μm) CW	$\sim 1.5 \times 10^9$	6.66×10^{-10}	$\sim 6 \times 10^{-10}$
Nd:YAG (1.06 μm) CW	$\sim 1.2 \times 10^{10}$	8.34×10^{-11}	$\sim 7.6 \times 10^{-11}$
Ruby (0.6934 μm) pulsed	6×10^{10}	1.66×10^{-11}	$\sim 1.2 \times 10^{-11}$
Nd^{3+}:glass pulsed	3×10^{12}	3.33×10^{-13}	$\sim 3 \times 10^{-13}$
Rhodamine 6G (dye laser)(0.6 μm)	10^{13}	10^{-13}	3×10^{-14}
Diode lasers	10^{13}	10^{-13}	4×10^{-13}

special interest is the possibility of controlling the gain and loss by means of multiple electrodes [36].

Table 6-1 lists some of the lasers commonly used in mode locking, along with the achievable pulse durations.

Theory of Mode Locking (9, 10)

We have argued above that the introduction of a time-periodic loss, or gain, into a laser oscillator will cause phase locking between the otherwise independent, longitudinal modes of the resonator. A necessary condition is that the frequency of this modulation be equal to a multiple integer of the intermode frequency separation (free spectral range) of the resonator. This follows from the gating argument leading to Figure 6-14 or, equivalently, from the requirement that optical sidebands at $\omega_0 \pm \omega_m$, which arise from loss modulation of optical mode ω_0, coincide in frequency with those of modes $\omega_{\pm 1}$.

The theoretical treatment of this problem consists of solving Maxwell's equations for oscillation of an optical resonator with gain (due to inversion) and a time-periodic loss. The loss and gain will be accounted for by using a time-dependent conductivity, $\sigma(t)$, in the equations

$$\nabla \times \mathbf{H} = \sigma(t)\mathbf{E} + i\omega\varepsilon\mathbf{E} \tag{6.6-11}$$

$$\nabla \times \mathbf{E} = -i\omega\mu\mathbf{H} \tag{6.6-12}$$

By taking the curl of (6.6-12), using $\nabla \times \nabla \times \mathbf{E} = \nabla(\nabla \cdot \mathbf{E}) - \nabla^2\mathbf{E}$, $\nabla \cdot \mathbf{E} = 0$, and replacing $\nabla \times \mathbf{H}$ by the right side of (6.6-11), we obtain

$$\nabla^2 E + \omega^2 \mu\varepsilon\left(1 - i\,\frac{\sigma(t)}{\omega\varepsilon}\right) E = 0 \tag{6.6-13}$$

for any of the Cartesian components of the resonator field **E**.

We first obtain the resonant mode solution of (6.6-13) with $\sigma(t) = 0$. The simplest such solution is that of modes in a plane, two (infinite) mirror, resonator of length ℓ. These modes are given by (6.6-8). A linear combination of these modes is

$$E(z, t) = \sum_q a_q \sin\left(\frac{q\pi}{\ell} z\right) \sin\left(\frac{q\pi c}{\ell} t\right) \tag{6.6-14}$$

Since $c^2 = (\mu\varepsilon)^{-1}$, each term in (6.6-14) satisfies Equation (6.6-12) with $\sigma = 0$, and so does the total field $E(z, t)$ for any *arbitrary choice* of the mode amplitudes a_q.

If we now turn on the periodic loss modulation $\sigma(t)$, the field solution (6.6-14) no longer satisfies the oscillator wave equation (6.6-13). However, we can still use the complete and orthonormal basis function set $\sin\left(\frac{q\pi}{\ell} z\right)$ to expand $E(z, t)$ at any instant t in the segment $0 < z < \ell$.

$$E(z, t) = \sum_q a_q(t) \sin\left(\frac{q\pi c}{\ell} t\right) \sin\left(\frac{q\pi}{\ell} z\right) \tag{6.6-15}$$

Here we view $a_q(t) \sin\left(\frac{q\pi c}{\ell} t\right)$ merely as the Fourier series expansion coefficient of the field at instant t, i.e.,

$$E(z, t) = \sum_q b_q(t) \sin\left(\frac{q\pi}{\ell} z\right)$$

$$b_q(t) = a_q(t) \sin\left(\frac{q\pi c}{\ell} t\right) \tag{6.6-16}$$

The major and fundamental difference is that in comparison to free oscillation, (6.6-14), the coefficients $a_q(t)$ are no longer *arbitrary* and are related to each other. They must be obtained by solving (6.6-13).

In the case of mode locking, we represent the periodic loss modulation by taking

$$\sigma(t) = \sigma_0 + \sigma_1(z) \cos \omega_m t \tag{6.6-17}$$

We use (6.6-17) in (6.6-13), in which the field E is taken in the form of (6.6-15). A detailed derivation is presented, for example, in References [10, 37]. The key result is that when the modulation frequency ω_m is equal to the intermode frequency separation, $\omega_m = \frac{\pi c}{\ell}$, we obtain a solution wherein all the modes have the *same* phase, i.e., $a_q = a_0$, a_0 any constant. This corresponds to the ultra-short pulse form of mode locking discussed at the beginning of this section.

6.7 MODE LOCKING IN HOMOGENEOUSLY BROADENED LASER SYSTEMS

The analysis of mode locking in inhomogeneous laser systems in Section 6.6 assumed that the role of internal modulation was that of locking together the phases of modes that, in the absence of modulation, oscillate with random phases. In the case of homogeneous broadening, only one mode can normally oscillate. Experiments, however, reveal that mode locking leads to short pulses in a manner quite similar to that described in Section 6.6 and therefore must involve multimode oscillation. One way to reconcile the two points of view and the experiments is to realize that *in the presence of internal modulation,* power is transferred continuously from the high gain mode to those of lower gain (that is, those which would not normally oscillate). This power can be viewed simply as that of the sidebands at $(\omega_0 \pm n\omega)$ of the mode at ω_0 created by a modulation at ω. Armed with this understanding we see that the physical phenomenon is not only one of mode locking but one of mode generation. The net result, however, is that of a large number of oscillating modes with equal frequency spacing and fixed phases, as in the inhomogeneous case, leading to ultrashort pulses.

The analytical solution to this case [24, 25, 26] follows an approach used originally to analyze short pulses in traveling wave microwave oscillators [40].

Referring to Figure 6-17, we consider an optical resonator with mirror reflectivites R_1 and R_2 that contains, in addition to the gain medium, a periodically modulated loss cell. The method of solution is to follow one pulse through a complete round trip through the resonator and to require that the pulse reproduce itself. The temporal pulse shape at each stage is assumed to be Gaussian.

Before proceeding, we need to characterize the effect of the gain medium and the loss cell on a traveling Gaussian pulse.

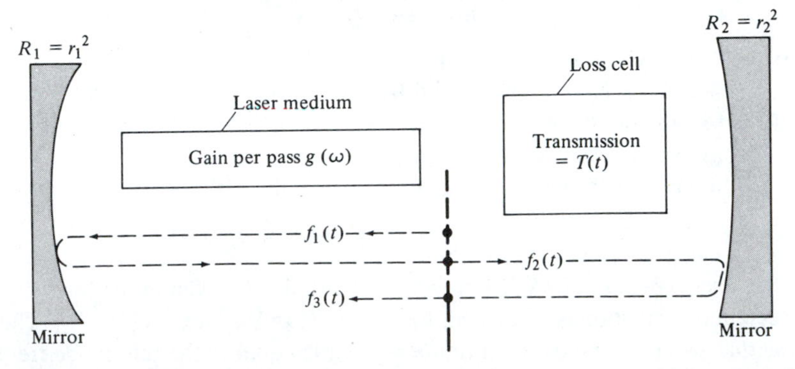

Figure 6-17 The experimental arrangement assumed in the theoretical analysis of mode locking in homogeneously broadened lasers.

TRANSFER FUNCTION OF THE GAIN MEDIUM

Assume that an optical pulse with a field $E_{in}(t)$ is incident on an amplifying optical medium of length l. Taking the Fourier transform of $E_{in}(t)$ as $E_{in}(\omega)$, the amplifier can be characterized by a transfer function $g(\omega)$ where

$$E_{out}(\omega) = E_{in}(\omega)g(\omega) \tag{6.7-1}$$

is the Fourier transform of the output field. Equation (6.7-1) is a linear relationship and applies only in the limit of negligible saturation.

Using (6.1-1), (5.3-8), and (5.4-9) we have

$$\begin{aligned} g(\omega) &= \exp\left\{-ikl\left[1 + \frac{1}{2n^2}(\chi' - i\chi'')\right]\right\} \\ &= \exp\left\{-ikl - \left(\frac{kl}{2n^2}\right)\frac{\Delta N\lambda^3 T_2}{8\pi^2 n t_{spont}}\left[\frac{1}{1 + i(\omega - \omega_0)T_2}\right]\right\} \\ &\simeq \exp\left\{-ikl + \frac{\gamma_{max} l}{2}[1 - i(\omega - \omega_0)T_2 - (\omega - \omega_0)^2 T_2^2]\right\} \end{aligned}$$

where $k = \omega n/c$ and l is the length of the amplifying medium and we define $T_2 = (\pi\,\Delta v)^{-1}$. The approximation is good for $(\omega - \omega_0)T_2 \ll 1$. We recall that $\Delta N < 0$ for gain. Since the pulse is making two passes through the cell, we take

$$\frac{E_{out}(\omega)}{E_{in}(\omega)} = [g(\omega)]^2 = \exp\{-i2kl + \gamma_{max} l[1 - i(\omega - \omega_0)T_2 - (\omega - \omega_0)^2 T_2^2]\}$$

The imaginary terms in the exponent correspond to a time delay (due to the finite group velocity of the pulse) of

$$\tau_d = \frac{2l}{c} + l\gamma_{max} T_2$$

We are considering here only the effect on the pulse shape so that, ignoring the imaginary term,[12] we obtain

$$[g(\omega)^2] = e^{\gamma_{max} l[1-(\omega-\omega_0)^2 T_2^2]} \tag{6.7-2}$$

TRANSFER FUNCTION OF THE LOSS CELL

Here we need to express the effect of the loss cell on the pulse in the time domain.

Assume that the single pass amplitude transmission factor $T(t)$ of the loss cell is given by

$$E_{out}(t) = E_{in}(t)T(t) = E_{in}(t)\exp[-2\delta_l^2 \sin^2(\pi\Delta v_{axial}\, t)]$$

[12] The finite propagation delay affects the round-trip pulse propagation time that must be equal to the period of the loss modulation.

where Δv_{axial}, the longitudinal mode spacing, is given by

$$\Delta v_{\text{axial}} = \frac{c}{2l_c}$$

where l_c is the effective optical length of the resonator. The transmission peaks are thus separated by $2l_c/c$ sec so that a mode-locked pulse can pass through the cell on successive trips with minimum loss. Since the pulses pass through the cell centered on the point of maximum transmission, we approximate the expression for $E_{\text{out}}(t)$ by

$$E_{\text{out}}(t) = E_{\text{in}}(t)T(t) = E_{\text{in}}(t)\exp[-2\delta_l^2(\pi\Delta v_{\text{axial}}t)^2] \tag{6.7-3}$$

We can view the form of (6.7-3) as the prescribed transmission function of the cell. The form, however, is suggested by physical considerations. In the case of an electrooptic shutter with a retardation (see Section 9.3) $\Gamma(t) = \Gamma_m \sin \omega_m t$, the transmission factor is $T(t) = \cos^2(\Gamma(t)/2)$. Near the transmission peaks $\Gamma(t) \ll 1$ and $T(t)$ is given by

$$T(t) \simeq \exp[-\tfrac{1}{4}(\Gamma_m^2\omega_m^2 t^2)] = \exp[-2\delta_l^2(\pi\Delta v_{\text{axial}}t)^2]$$

where $\omega_m = \pi\Delta v_{\text{axial}}$ and $\Gamma_m = 2\sqrt{2}\delta_l$.

We now return to the main analysis. The starting pulse $f_1(t)$ in Figure 6-17 is taken as

$$f_1(t) = Ae^{-\alpha_1 t^2}\, e^{i(\omega_0 t + \beta_1 t^2)} \tag{6.7-4}$$

corresponding to a "chirped" frequency

$$\omega(t) = \omega_0 + 2\beta_1 t \tag{6.7-5}$$

Its Fourier transform is

$$\begin{aligned} F_1(\omega) &= \frac{1}{2\pi}\int_{-\infty}^{\infty} f_1(t)e^{-i\omega t}\, dt \\ &= \frac{A}{2}\sqrt{\frac{1}{\pi(\alpha_1 - i\beta_1)}}\exp\left(\frac{-(\omega - \omega_0)^2}{4(\alpha_1 - i\beta_1)}\right) \end{aligned} \tag{6.7-6}$$

A double pass through the amplifier and one mirror reflection (r_1) are accounted for by multiplying $F_1(\omega)$ by the transfer factor $[g(\omega)]^2 r_1$

$$\begin{aligned} F_2(\omega) &= F_1(\omega)[g(\omega)]^2 r_1 \\ &= \frac{r_1 A}{2} e^{g_0}\sqrt{\frac{1}{\pi(\alpha_1 - i\beta_1)}}\exp\left\{[-(\omega - \omega_0)^2]\left(\frac{1}{4(\alpha_1 - i\beta_1)} + g_0 T_2^2\right)\right\} \end{aligned} \tag{6.7-7}$$

where $g_0 \equiv \gamma_{\max} l$ and $[g(\omega)]^2$ is given by (6.7-2). Transforming back to the time domain

$$f_2(t) = \int_{-\infty}^{\infty} F_2(\omega)e^{i\omega t}\, d\omega$$
$$= \frac{r_1 A e^{g_0}}{2\pi} \sqrt{\frac{\pi}{\alpha_1 - i\beta_1}}\, e^{-\omega_0^2 Q} \sqrt{\frac{\pi}{Q}} \exp[-(2i\omega_0 Q - t)^2/4Q] \quad \textbf{(6.7-8)}$$

where

$$Q \equiv \frac{1}{4(\alpha_1 - i\beta_1)} + g_0 T_2^2 \quad \textbf{(6.7-9)}$$

A reflection from mirror 2 and a passage through the loss cell lead according to (6.7-3) to

$$f_3(t) = r_2 f_2(t) e^{-2\delta_l^2 \pi^2 (\Delta\nu_{\text{axial}})^2 t^2}$$
$$= \frac{r_1 r_2 A e^{g_0}}{2} \sqrt{\frac{1}{(\alpha_1 - i\beta_1)Q}}\, e^{i\omega_0 t}\, e^{-[2\delta_l^2(\pi\Delta\nu_{\text{axial}})^2 + (1/4Q)]t^2} \quad \textbf{(6.7-10)}$$

For self-consistency we require that $f_3(t)$ be a replica of $f_1(t)$. We thus equate the exponent of (6.7-10) to that of (6.7-4)

$$\alpha_1 = 2\delta_l^2(\pi\,\Delta\nu_{\text{axial}})^2 + \text{Re}\left(\frac{1}{4Q}\right)$$
$$\beta_1 = -\text{Im}\left(\frac{1}{4Q}\right) \quad \textbf{(6.7-11)}$$

Using (6.7-9), the second equation of (6.7-11) gives

$$\beta_1 = \frac{\beta_1}{(1 + 4g_0 T_2^2 \alpha_1)^2 + (4g_0 T_2^2 \beta_1)^2}$$

so that a self-consistent solution requires that

$$\beta_1 = 0$$

that is, no chirp. With $\beta_1 = 0$ the first of (6.7-11), becomes

$$2\delta_l^2(\pi\Delta\nu_{\text{axial}})^2 + \frac{\alpha_1}{(1 + 4g_0 T_2^2 \alpha_1)} = \alpha_1 \quad \textbf{(6.7-12)}$$

that, assuming

$$4g_0 T_2^2 \alpha_1 \ll 1 \quad \textbf{(6.7-13)}$$

results in

$$\alpha_1 = \left(\frac{\delta_l^2}{2g_0}\right)^{1/2} \frac{\pi\Delta\nu_{\text{axial}}}{T_2}$$

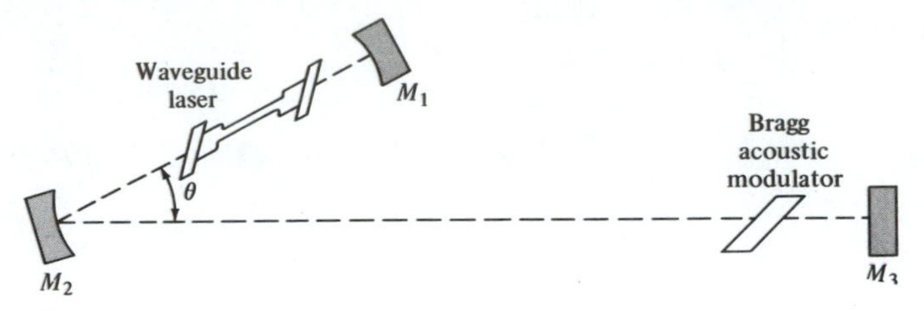

Figure 6-18 A schematic drawing of the mode-locking experiment in a high pressure CO_2 laser. (After Reference [39].)

The pulse width at the half intensity points is from (6.7-4)

$$\tau_p = (2 \ln 2)^{1/2} \alpha_1^{-1/2}$$

so that the self-consistent pulse has a width

$$\tau_p = \frac{(2 \ln 2)^{1/2}}{\pi} \left(\frac{2g_0}{\delta_l^2}\right)^{1/4} \left(\frac{1}{\Delta v_{\text{axial}} \, \Delta v}\right)^{1/2} \tag{6.7-14}$$

where $\Delta v \equiv (\pi T_2)^{-1}$. The condition (6.7-13) can now be interpreted as requiring that $\tau_p \gg 2\sqrt{g_0}T_2$, which is true in most cases.

An experimental setup demonstrating mode locking in a pressure broadened CO_2 laser is sketched in Figure 6-18. The inverse square root dependence of τ_p on Δv is displayed by the data of Figure 6-19, while the dependence on the modulation parameter δ_l is shown in Figure 6-20.

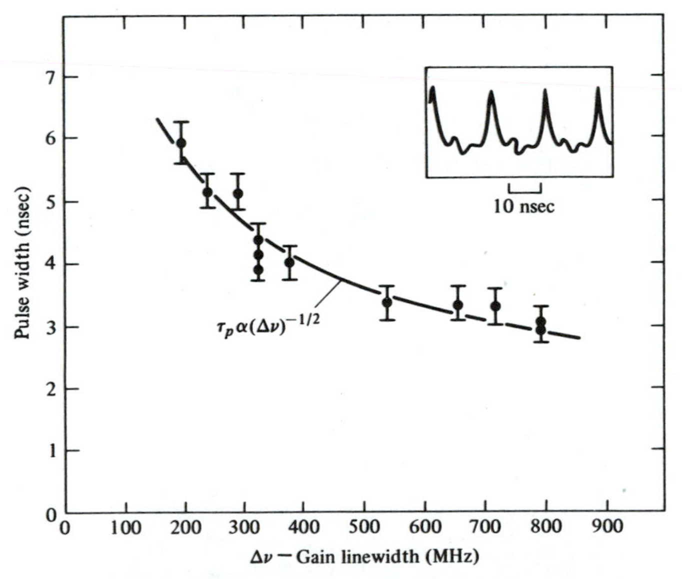

Figure 6-19 The dependence of the pulse width on the gain linewidth, Δv, that is controlled by varying the pressure ($\Delta v = 8 \times 10^8$ at 150 torr). (After Reference [39].)

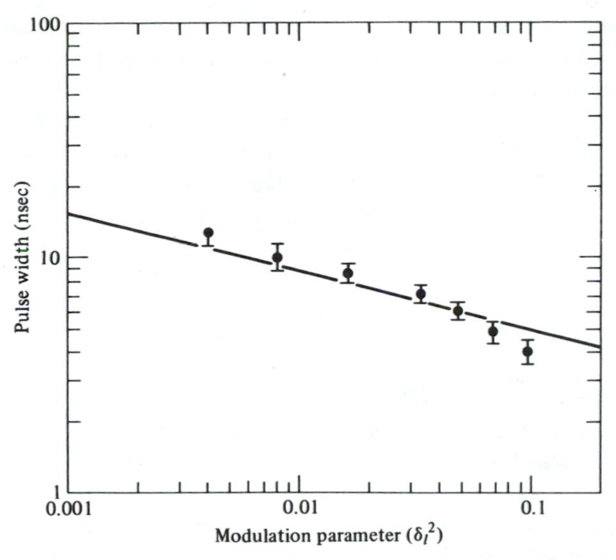

Figure 6-20 The mode-locked pulse width as a function of the modulation parameter, δ_l^2. (After Reference [39].)

MODE LOCKING BY PHASE MODULATION

Mode locking can be induced by internal phase, rather than loss, modulation. This is usually done by using an electrooptic crystal inside the resonator oriented in the basic manner of Figure 9-7 such that the passing wave undergoes a phase delay proportional to the instantaneous electric field across the crystal. The frequency of the modulating signal is equal, as in the loss modulation case, to the inverse of the round-trip group delay time, that is, to the longitudinal intermode frequency separation.

We employ an analysis similar to that of the homogeneous case except that the transfer function through the modulation cell is taken, instead of (6.7-3), as

$$E_{\text{out}}(t) = E_i(t) \exp(-i2\delta_\phi \cos 2\pi\Delta\upsilon_{\text{axial}}\, t) \tag{6.7-15}$$

For pulses passing near the extrema of the phase excursion, we can approximate the last equation as

$$E_{\text{out}}(t) = E_i(t) \exp(\mp i2\delta_\phi \pm i\delta_\phi 4\pi^2\, \Delta\upsilon_{\text{axial}}^2\, t^2) \tag{6.7-16}$$

An analysis identical to that leading to (6.7-14) yields

$$\tau_p = \frac{(2\ \ln 2)^{1/2}}{\pi} \left(\frac{2g_0}{\delta_\phi}\right)^{1/4} \left(\frac{1}{\Delta\upsilon_{\text{axial}}\ \Delta\upsilon}\right)^{1/2} \tag{6.7-17}$$

In this case self-consistency leads to a chirped pulse with

$$\beta = \pm\alpha = \pm\pi^2\Delta\upsilon_{\text{axial}}\ \Delta\upsilon \sqrt{\frac{\delta_\phi}{2g_0}} \tag{6.7-18}$$

The upper and lower signs in (6.7-16) and (6.7-18) correspond to two possible pulse solutions, one passing through the cell near the maximum of the phase excursion and the other near its minimum.

We note that (6.7-17) is similar to the loss modulation result (6.7-14) except that δ_ϕ appears instead of δ_l^2. The difference can be traced into a difference between (6.7-3) and (6.7-15). The choice of notation in both cases is such that δ *corresponds* to the *retardation* (phase delay per pass) induced by the electrooptic crystal.

6.8 PULSE LENGTH MEASUREMENT AND NARROWING OF CHIRPED PULSES

The problem of measuring the duration of mode-locked ultrashort pulses is of great practical and theoretical interest. Since the fastest conventional optical detectors possess response times of $\sim 2 \times 10^{-11}$ s, it is impossible to use these optical detectors to measure directly the short ($\tau < 10^{-11}$ s) mode-locked pulses. A number of techniques invented for this purpose all take advantage of some nonlinear process to obtain a spatial autocorrelation trace of the optical intensity pulse. The measurement of a pulse of a duration, say, of $\tau_0 = 10^{-12}$ s is thus replaced with measuring the spatial extent of an autocorrelation trace of length $c\tau_0 = 0.3$ mm, which is a relatively simple task.

In what follows we will describe one such method, the one most widely used, that is based on the phenomenon of optical second harmonic generation. The process of second harmonic generation is developed in detail in Chapter 8. It will suffice for the purpose of the present discussion to state that when an optical pulse

$$e_1(t) = \text{Re}[\mathscr{E}_1(t)e^{i\omega t}] \tag{6.8-1}$$

is incident on a nonlinear optical crystal it generates an output optical pulse $e_2(t)$ at twice the frequency with

$$e_2(t) = \text{Re}[\mathscr{E}_2(t)e^{2i\omega t}] \propto \text{Re}[\mathscr{E}_1^2(t)e^{2i\omega t}] \tag{6.8-2}$$

A sketch of a second harmonic system for measuring the pulse length is shown in Figure 6-21. The laser emits a continuous stream of mode-locked pulses. Each individual pulse $\mathscr{E}(t)e^{i\omega t}$ is divided by a beam splitter into two equal intensity pulses. One of these pulses is advanced (or delayed) by τ seconds relative to the other. The two pulses recombine again in a nonlinear optical crystal. The second harmonic (2ω) pulse generated by the crystal is incident on a "slow" detector whose output current is integrated over a time long compared to the optical pulse duration.

The total optical field incident on the nonlinear crystal is the sum of the direct and retarded fields

$$\begin{aligned} e_{\text{tot}}(t) &= \text{Re}\{[\mathscr{E}_1(t) + \mathscr{E}_1(t-\tau)e^{-i\omega\tau}]e^{i\omega t}\} \\ &= \text{Re}[\mathscr{E}(t)e^{i\omega t}] \end{aligned}$$

$$\mathscr{E}(t) = \mathscr{E}_1(t) + \mathscr{E}_1(t-\tau)e^{-i\omega\tau} \tag{6.8-3}$$

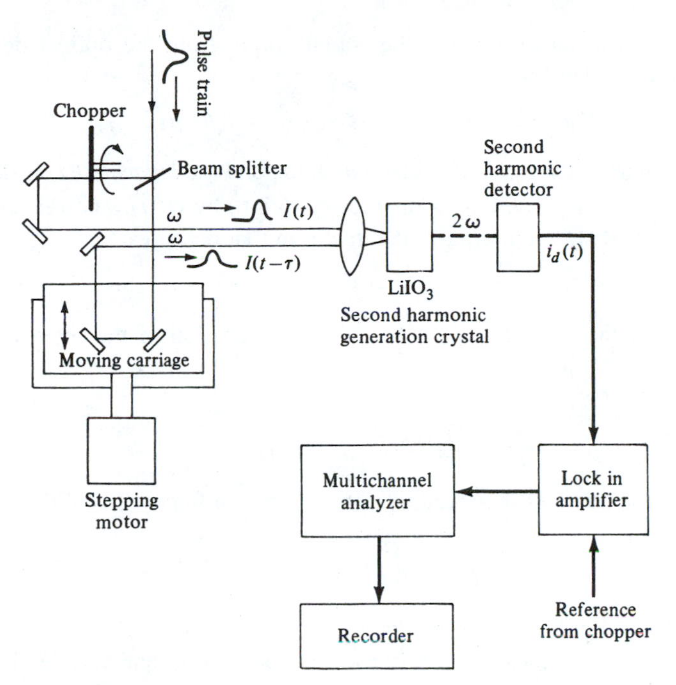

Figure 6-21 The second harmonic generation autocorrelation setup for measuring the width of mode-locked ultrashort pulses.

According to (6.8-2), the second harmonic field radiated by the crystal has a complex amplitude that is proportional to the square of the complex amplitude $\mathscr{E}(t)$ of the incident fundamental field.

$$\begin{aligned}\mathscr{E}_2(t) &\propto [\mathscr{E}_1(t) + \mathscr{E}_1(t-\tau)e^{-i\omega\tau}]^2 \\ &= \mathscr{E}_1^2(t) + \mathscr{E}_1^2(t-\tau)e^{-2i\omega\tau} + 2\mathscr{E}_1(t)\mathscr{E}_1(t-\tau)e^{-i\omega\tau}\end{aligned} \quad \textbf{(6.8-4)}$$

The second harmonic field, $e_2(t) = \text{Re}[\mathscr{E}_2(t) \exp(i2\omega t)]$ is incident next on the optical detector (photomultiplier, diode, etc.) whose output current i_d (see Section 11.1) is proportional to the incident intensity. Using (6.8-4) we can obtain

$$\begin{aligned}i_d(t) \propto \mathscr{E}_2(t)\mathscr{E}_2^*(t) &= [\mathscr{E}_1(t)\mathscr{E}_1^*(t)]^2 + [\mathscr{E}_1(t-\tau)\mathscr{E}_1^*(t-\tau)]^2 \\ &\quad + 4\mathscr{E}_1(t)\mathscr{E}_1^*(t)\mathscr{E}_1(t-\tau)\mathscr{E}_1^*(t-\tau) + s(\tau)\end{aligned} \quad \textbf{(6.8-5)}$$

where $s(\tau)$ is composed of terms with cos $\omega\tau$ and cos $2\omega\tau$ dependence. Since these terms fluctuate with a delay period $\Delta\tau \sim 10^{-15}$ s, a small unintentional or deliberate integration (smearing) over the delay τ averages them out to near zero. The term $s(\tau)$ is consequently left out.

Since the temporal (t) variation of the first three terms in (6.8-5) is on the scale of picoseconds (or less), the much slower optical detector inevitably integrates the current $i_d(t)$, with the result that the actual output from the optical detector is a function of the delay (τ) only

$$i_d(\tau) \propto \langle I^2(t)\rangle + \langle I^2(t-\tau)\rangle + 4\langle I(t)I(t-\tau)\rangle \tag{6.8-6}$$

where the angle brackets signify time-averaging and the *intensity* $I(t)$ is defined[13] as $I(t) = \mathscr{E}_1(t)\mathscr{E}_1^*(t)$. By dividing both sides of (6.8-6) by $\langle I^2(t)\rangle$ and recognizing that $\langle I^2(t)\rangle = \langle I^2(t-\tau)\rangle$, the normalized detector output becomes

$$i_d(\tau) = 1 + 2G^{(2)}(\tau) \tag{6.8-7}$$

where $G^{(2)}(\tau)$, the second-order autocorrelation function of the intensity pulse, is defined by

$$G^{(2)}(\tau) \equiv \frac{\langle I(t)I(t-\tau)\rangle}{\langle I^2(t)\rangle} \tag{6.8-8}$$

In the case of a well-behaved ultrashort coherent light pulse of duration τ_0, we have

$$i_d(0) = 3 \qquad i_d(\tau \gg \tau_0) = 1$$

since $G^{(2)}(0) = 1$ and $G^{(2)}(\tau \gg \tau_0) = 0$.

A plot of $i_d(\tau)$ versus τ will consist of a peak of (normalized) height of 3 atop a background of unity height. The central peak will have a width $\sim\tau_0$.

It is important in practice to be able to distinguish between the case just discussed and that of incoherent light (such as light due to a laser oscillating in a large number of independent modes). In this case we have $i_d(0) = 3$ (since even incoherent light is correlated with itself at zero delay). For $\tau > 0$, or more precisely for τ longer than the coherence time of the light, we have

$$G^{(2)}(\tau) = \frac{\langle I(t)I(t-\tau)\rangle}{\langle I^2(t)\rangle} = \frac{\langle I(t)\rangle^2}{\langle I^2(t)\rangle} \tag{6.8-9}$$

since $I(t)$ and $I(t-\tau)$ are completely uncorrelated. For truly incoherent light of the type we are considering here, the time-averaging indicated in (6.6-19) can be replaced by ensemble averaging so that

$$\langle I^2(t)\rangle = \int_0^\infty p(I)I^2\,dI \tag{6.8-10}$$

where $p(I)$ is the intensity distribution function so that $p(I)dI$ is the probability that a measurement of I will result in a value between I and $I + dI$. For incoherent light[14]

$$p(I) = \frac{1}{\langle I\rangle} e^{-I/\langle I\rangle}$$

---

[13]A proportionality constant involved in this definition is left out since it cancels out in the subsequent division of Equation (6.8-9).

[14]This follows directly from the fact that the optical field distribution $p(e)$ for the field of an incoherent beam function is Gaussian.

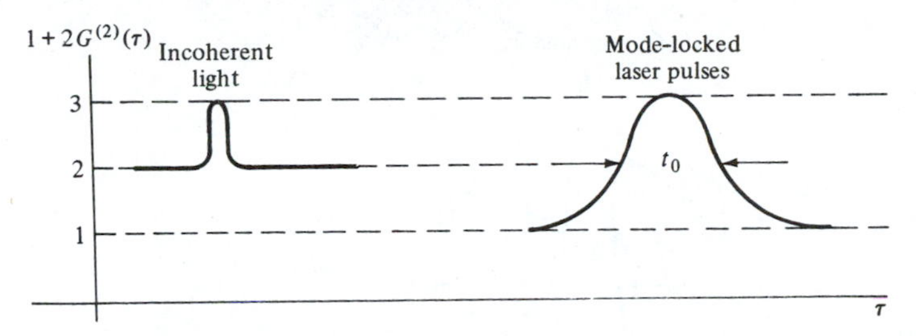

Figure 6-22 The second harmonic (averaged) integrated intensity due to two optical pulses as a function of time delay between them for the case (left) of an incoherent source and (right) coherent mode-locked optical pulses.

which when used in (6.8-10) gives

$$\langle I^2(t)\rangle = 2\langle I\rangle^2$$

and, returning to (6.8-7),

$$i_d(\tau > 0) \propto 1 + 2\frac{\langle I(t)\rangle^2}{\langle I^2(t)\rangle} = 2$$

A plot of $i_d(\tau)$ versus τ in the case of incoherent light should thus consist of a very narrow peak of height 3 on a background of height 2. The general features of the coherent mode-locked pulses and the incoherent light is depicted in Figure 6-22.

The determination of the original pulse width from the width of the second harmonic correlation trace is somewhat ambiguous. We can show by performing the integration indicated by (6.8-8) that the width (at half-maximum) t_0 of $G^{(2)}(\tau)$ and τ_0 of $I(t)$ are related as in the case of the ''popular'' waveforms tabulated in the left column of Table 6-2.

We conclude this section by showing in Figure 6-23 the autocorrelation trace of one of the shortest optical pulses ($\tau_0 \sim 30 \times 10^{-15}$ s) produced to date. It is interesting to note that within such a pulse the light rises and falls a mere 15(!) times.

Table 6-2 Some Simple Pulse Widths

$I(t)$	t_0/τ_0
$1(0 \le t \le t_0)$, zero otherwise	1
$\exp\left\{-\dfrac{(4\ln 2)t^2}{\tau_0^2}\right\}$	$\sqrt{2}$
$\text{sech}^2\left(\dfrac{1.76t}{\tau_0^2}\right)$	1.55
$\exp\left(-\dfrac{(\ln 2)t}{\tau_0}\right) \quad (t \ge 0)$	2

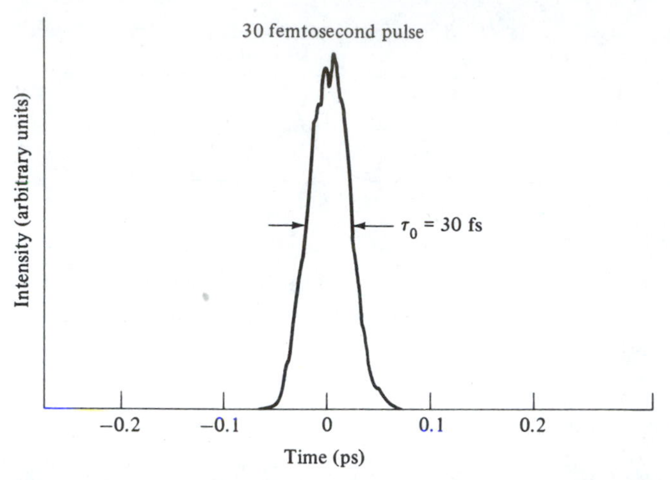

Figure 6-23 The autocorrelation trace of a mode-locked dye laser ($\lambda \sim 6100$ Å) pulse.

With just one order of magnitude improvement we should thus be able to isolate a single optical cycle. There may, however, be some compelling reasons (and the student is encouraged to think of some) why this development may not take place anytime soon.

Pulse Narrowing by Chirping and Compression

Mode locking of a laser oscillator was shown in Section 6.6 to lead to a mode of oscillation consisting of a continuous train of ultra-short pulses. The width of each pulse is $\tau_p \sim \dfrac{1}{(\Delta\omega)_{\text{gain}}}$ where $(\Delta\omega)_{\text{gain}}$ is the spectral width over which the gain is capable of sustaining oscillation. It is, however, possible to further reduce the pulse width beyond the limit $(\Delta\omega)_{\text{gain}}^{-1}$ by employing nonlinear optical techniques. To lend some plausibility to the new ideas that we will soon introduce, we remind ourselves of the exact formal analogy that exists between the diffractive propagation of an optical beam in free homogeneous space and that of an optical pulse propagation in a dispersive channel (fiber). This analogy was discussed in Section 3.6.

The propagation of an optical pulse, $f_1(t)\exp(i\omega_0 t)$, through a length L of a dispersive fiber was found in Equation (3.5-4) to be described by

$$\tilde{f}_2(\Omega) = \tilde{f}_1(\Omega)\exp\left[-i\,\frac{\beta''}{2}\,L\Omega^2\right] \tag{6.8-11}$$

where $\tilde{f}_1(\Omega)$ and $\tilde{f}_2(\Omega)$ are the Fourier transforms of the input and output pulse envelopes $f_1(t)$ and $f_2(t)$, respectively, and $\beta'' \equiv d^2\beta/d\omega^2$.

The propagation of a confined optical beam with a transverse profile $u_1(x)$ was found in Section 1.6 to obey

$$\tilde{u}_2(K) = \tilde{u}_1(K)\exp\left[i\,\frac{LK^2}{2k}\right] \tag{6.8-12}$$

The exact formal analogy of (6.8-11) and (6.8-12) suggests that it should be possible to construct a temporal analog of the well-familiar spatial optical demagnification by a lens as illustrated in Figure 6-24. Such a system can be used to narrow optical pulses. All that is needed is a time lens, a device that multiplies an incoming optical field by $\exp(iat^2)$, which is the exact analog of the factor $\exp(ikx^2/2f)$ by which the spatial lens multiplies an incoming beam.

Our temporal pulse narrower is illustrated in Figure 6-24(a). The input pulse envelope, before narrowing, is taken as a Gaussian

$$f_1(t) = \exp(-t^2/\tau_p^2) \tag{6.8-13}$$

with a (FWHM) intensity width $\Delta\tau = \sqrt{2\ell n 2}t_p$. Passage through the time lens multiplies $f_1(t)$ by a phase factor, $\exp(iAt^2/\tau_p^2)$

$$f_2(t) = \exp[-(1 - iA)(t/\tau_p)^2] \tag{6.8-14}$$

The total field at plane 2 is

$$E_2(t) = f_2(t)\exp(i\omega_0 t) = \exp[-(t/\tau_p)^2] \exp i\left[\omega_0 t + A\left(\frac{t}{\tau_p}\right)^2\right] \tag{6.8-15}$$

If we write $E_2(t) = \exp[-(t/\tau_p)^2 + i\phi(t)]$, the instantaneous frequency is $\omega(t) = d\phi/dt$

$$\omega(t) = \frac{d}{dt}\left[\omega_0 t + A\left(\frac{t}{\tau_p}\right)^2\right] = \omega_0 + 2A\frac{t}{\tau_p^2} \tag{6.8-16}$$

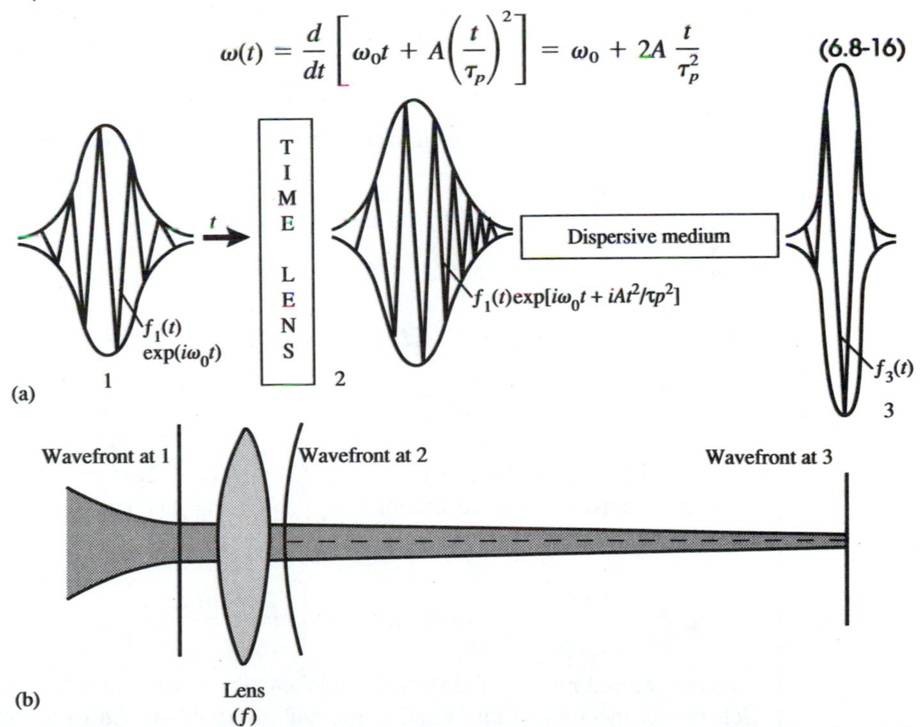

Figure 6-24 The equivalence between temporal focusing (pulse narrowing) in (a) and spatial focusing by a conventional lens in (b). Notice that a chirped optical pulse corresponds to a Gaussian spatial beam with a curve wavefront.

corresponding to a linear "chirp." The pulse width, however, is unaffected by the lens. The lens, however, modifies the spectrum of the pulse. To find the spectrum we take the Fourier transform of $f_2(t)$

$$\tilde{f}_2(\Omega) = \sqrt{\frac{\tau_p^2}{4\pi(1 - iA)}} \exp\left[-\frac{1 + iA}{4(1 + A^2)} (\Omega\tau_p)^2\right] \qquad (6.8\text{-}17)$$

The effect of the chirp is thus to increase the spectral width (FWHM) by the factor $\sqrt{1 + A^2}$.

Time Lenses

The fact that the time lens has increased the spectral width while preserving the total pulse energy implies redistribution of power among the spectral components. This, by definition, cannot be accomplished by a linear passive device. We will, indeed, find that our practical candidates for time lenses all employ nonlinear techniques. One such method employs self-phase modulation in a medium with a large Kerr coefficient, i.e., a medium where the index of refraction depends on the light intensity, $n = n_0 + n_2 I$ (Kerr effect). $I =$ intensity (watts/m^2), n_2 a constant of the material. Consider, for example, a Gaussian intensity pulse,

$$I = I_0 \exp(-2\alpha t^2)$$

entering a nonlinear medium characterized by n_2 so that the index of refraction is $n = n_0 + n_2 I(t)$

$$E_{\text{in}} = E_0 \exp(-\alpha t^2)$$

$$I_{\text{in}} = I_0 \exp(-2\alpha t^2)$$

The pulse will emerge at $z = L$ with a phase delay factor, $\exp\left[i \frac{\omega}{c} (n_0 + n_2 I(t) L\right]$.

If we expand the Gaussian pulse as $I(t) = I_0(1 - 2\alpha t^2 + \ldots)$ and keep only the first two terms, we find that the delay factor becomes

$$E_{\text{out}} = E_{\text{in}} \times \text{delay factor}$$

$$= \exp(-\alpha t^2) \exp\left[i \frac{2\omega n_2 \alpha I_0 L}{c} t^2\right]$$

and has, according to (6.8-15), the requisite chirp to serve as a time lens. We left out an inconsequential fixed phase and group delay. An exact treatment of this topic will be given in Chapter 17.

The pulse $f_2(t)$ is thus no longer transform-limited since its time-bandwidth product

$$\Delta\nu\Delta\tau = \frac{2\ell n2}{\pi}(1 + A^2)^{1/2} = 0.4413(1 + A^2)^{1/2} \tag{6.8-19}$$

exceeds the minimum value for a Gaussian pulse by $(1 + A^2)^{1/2}$. This excess bandwidth should make it possible, in principle, to reduce the pulse width, employing passive means, by the factor $(1 + A^2)^{1/2}$, which will result, again, in a compressed transform-limited pulse.

In the spatial example of Figure 6-24(a), the narrowing of the spatially "chirped" pulse is accomplished by propagating it through the appropriate distance L in space. The temporal equivalent is to propagate the chirped pulse through a dispersive device with transfer characteristics of the form of (6.8-11). This results in

$$\begin{aligned}\tilde{f}_3(\Omega) &= \tilde{f}_2(\Omega)\exp\left[-i\frac{\beta''}{2}L\Omega^2\right] \\ &= \sqrt{\frac{\tau_p^2}{4\pi(1 - iA)}}\exp\left[-\frac{\Omega^2\tau_p^2}{4(1 + A^2)} - i\left(\frac{A\tau_p^2}{4(1 + A^2)} + \frac{\beta''L}{2}\right)\Omega^2\right]\end{aligned} \tag{6.8-20}$$

where L is the length of the dispersive channel. If the condition

$$\beta''L = -\frac{A\tau_p^2}{2(1 + A^2)} \tag{6.8-21}$$

is satisfied, the output is given by

$$\tilde{f}_3(\Omega) = \sqrt{\frac{\tau_p^2}{4\pi(1 - iA)}}\exp\left[-\frac{(\Omega\tau_p)^2}{4(1 + A^2)}\right] \tag{6.8-22}$$

which corresponds in the time domain to

$$f_3(t) = F^{-1}\{\tilde{f}_3(\Omega)\} = \sqrt{1 + iA}\exp\left(-(1 + A^2)\frac{t^2}{\tau_p^2}\right) \tag{6.8-23}$$

The width of the output is thus compressed to

$$(\Delta\tau)_{\text{comp}} = \frac{\Delta\tau}{\sqrt{1 + A^2}} \tag{6.8-24}$$

For $A \gg 1$, the compression ratio is $\sim A$. This compression ratio is thus exactly the factor by which the spectral width is increased by the chirping. A simple integration shows that

$$\int_{-\infty}^{\infty} |f_1(t)|^2\,dt = \int_{-\infty}^{\infty} |f_2(t)|^2\,dt = \int_{-\infty}^{\infty} |f_3(t)|^2\,dt \tag{6.8-24}$$

so that the pulse energy is conserved.

The Grating Pair Compressor

In the above example, a fiber with group velocity dispersion ($\beta'' \neq 0$) was employed in order to compress the chirped pulse. The essential feature of the fiber was its transfer function, given by Equation (6.8-20) as

$$\text{Transfer function of fiber (length } L) = \exp\left[-\frac{i\beta''}{2} L\Omega^2\right]$$

Any other device with a transfer function of the form $\exp(ib\Omega^2)$, where b is some real constant, can thus serve as compressor provided b can be adjusted in magnitude as well as sign. A commonly used pulse-narrowing configuration is the dual grating telescope compressor [41, 42, 43] illustrating in Figure (6-25). It is based on the fact that different Fourier components (Ω) of the incident optical beam are diffracted by the gratings along different directions, thus following paths of different length between the input and output, accumulating in the process a differential phase $b\Omega^2$.

Figure 6-26 shows intensity profiles, obtained by second harmonic autocorrelation, of a chirped and a compressed pulse in the experiment depicted by Figure 6-25. The input pulse is compressed from an initial width of 5.2 ps to 0.32 ps. This corresponds, according to (6.8-24), to a compression ratio $A \sim 16$. The $\Delta\nu(\Delta\tau)_{\text{comp}}$ product is $\sim$0.45, close to the theoretical limit of 0.44, which indicates that the pulse is as narrow as allowed by its spectral content, i.e., it is transform-limited.

A semiconductor laser excited by current pulses or pulsating due to mode locking is naturally chirped. This is due to carrier density, hence, index of refraction transients in the active region. This makes the output pulses of such lasers natural candidates for compression. Chirping in semiconductor laser is treated in Chapter 15.

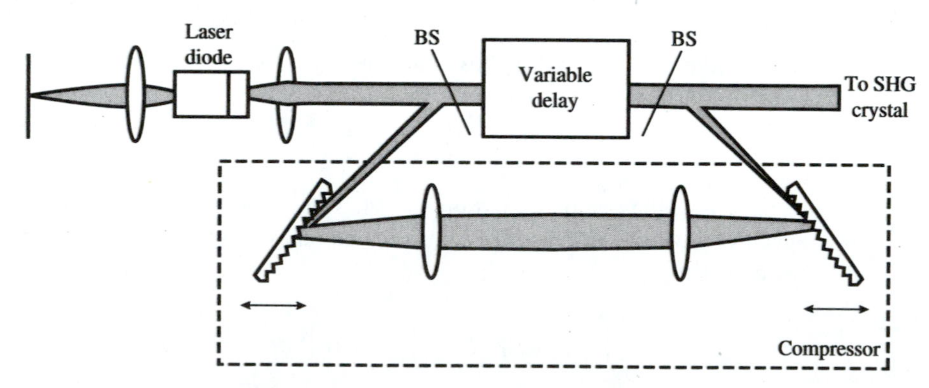

Figure 6-25 A dual-grating telescope pulse compressor (after Reference [43]).

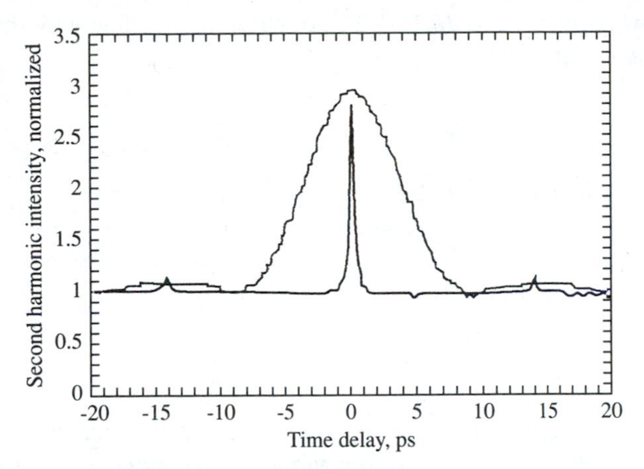

Figure 6-26 Intensity autocorrelation of uncompressed and compressed pulses. The compression ratio is $\sim A = 16$. (After Reference [43]).

6.9 GIANT PULSE (Q-switched) LASERS (19)

The technique "Q-switching" is used to obtain intense and short bursts of oscillation from lasers; see References [16–18]. The quality factor Q of the optical resonator is degraded (lowered) by some means during the pumping so that the gain (that is, inversion $N_2 - N_1$) can build up to a very high value without oscillation. (The spoiling of the Q raises the threshold inversion to a value higher than that obtained by pumping.) When the inversion reaches its peak, the Q is restored abruptly to its (ordinary) high value. The gain (per pass) in the laser medium is now well above threshold. This causes an extremely rapid buildup of the oscillation and a simultaneous exhaustion of the inversion by stimulated $2 \rightarrow 1$ transitions. This process converts most of the energy that was stored by atoms pumped into the upper laser level into photons, which are now inside the optical resonator. These proceed to bounce back and forth between the reflectors with a fraction $(1 - R)$ "escaping" from the resonator each time. This causes a decay of the pulse with a characteristic time constant (the "photon lifetime") given in (4.7-3) as

$$t_c \simeq \frac{nl}{c(1 - R)}$$

Both experiment and theory indicate that the total evolution of a giant laser pulse as described above is typically completed in $\sim 2 \times 10^{-8}$ second. We will consequently neglect the effect of population relaxation and pumping that take place during the pulse. We will also assume that the switching of the Q from the low to the high value is accomplished instantaneously.

The laser is characterized by the following variables: ϕ; the total number of photons in the optical resonator, $n \equiv (N_2 - N_1)V$; the total inversion; and t_c, the decay time constant for photons in the *passive* resonator. The exponential gain constant γ is proportional to n. The radiation intensity I thus grows with distance as $I(z)$

$= I_0 \exp(\gamma z)$ and $dI/dz = \gamma I$. An observer traveling with the wave velocity will see it grow at a rate

$$\frac{dI}{dt} = \frac{dI}{dz}\frac{dz}{dt} = \gamma\left(\frac{c}{n}\right) I$$

and thus the temporal exponential growth constant is $\gamma(c/n)$. If the laser rod is of length L while the resonator length is l, then only a fraction L/l of the photons is undergoing amplification at any one time and the average growth constant is $\gamma c(L/nl)$. We can thus write

$$\frac{d\phi}{dt} = \phi\left(\frac{\gamma c L}{nl} - \frac{1}{t_c}\right) \tag{6.9-1}$$

where $-\phi/t_c$ is the decrease in the number of resonator photons per unit time due to incidental resonator losses and to the output coupling. Defining a dimensionless time by $\tau = t/t_c$ we obtain, upon multiplying (6.9-1) by t_c,

$$\frac{d\phi}{d\tau} = \phi\left[\left(\frac{\gamma}{nl/cLt_c}\right) - 1\right] = \phi\left[\frac{\gamma}{\gamma_t} - 1\right]$$

where $\gamma_t = (nl/cLt_c)$ is the minimum value of the gain constant at which oscillation (that is, $d\phi/d\tau = 0$) can be sustained. Since, according to (5.3-3) γ is proportional to the inversion n, the last equation can also be written as

$$\frac{d\phi}{d\tau} = \phi\left[\frac{n}{n_t} - 1\right] \tag{6.9-2}$$

where $n_t = N_t V$ is the total inversion at threshold as given by (6.1-9).

The term $\phi(n/n_t)$ in (6.9-2) gives the number of photons generated by induced emission per unit of normalized time. Since each generated photon results from a single transition, it corresponds to a decrease of $\Delta n = -2$ in the total inversion. We can thus write directly

$$\frac{dn}{d\tau} = -2\phi\frac{n}{n_t} \tag{6.9-3}$$

The coupled pair of equations, (6.9-2) and (6.9-3), describes the evolution of ϕ and n. It can be solved easily by numerical techniques. Before we proceed to give the results of such calculation, we will consider some of the consequences that can be deduced analytically.

Dividing (6.9-2) by (6.9-3) results in

$$\frac{d\phi}{dn} = \frac{n_t}{2n} - \frac{1}{2}$$

and, by integration,

$$\phi - \phi_i = \frac{1}{2}\left[n_t \ln\frac{n}{n_i} - (n - n_i)\right]$$

Assuming that ϕ_i, the initial number of photons in the cavity, is negligible, we obtain

$$\phi = \frac{1}{2}\left[n_t \ln \frac{n}{n_i} - (n - n_i)\right] \tag{6.9-4}$$

for the relation between the number of photons ϕ and the inversion n at any moment. At $t \gg t_c$ the photon density ϕ will be zero so that setting $\phi = 0$ in (6.7-4) results in the following expression for the final inversion n_f:

$$\frac{n_f}{n_i} = \exp\left[\frac{n_f - n_i}{n_t}\right] \tag{6.9-5}$$

This equation is of the form $(x/a) = \exp(x - a)$, where $x = n_f/n_t$ and $a = n_i/n_t$, so that it can be solved graphically (or numerically) for n_f/n_i as a function of n_i/n_t.[15] The result is shown in Figure 6-27. We notice that the fraction of the energy originally stored in the inversion that is converted into laser oscillation energy is $(n_i - n_f)/n_i$ and that it tends to unity as n_i/n_t increases.

[15]This can be done by assuming a value of a and finding the corresponding x at which the plots of x/a and $\exp(x - a)$ intersect.

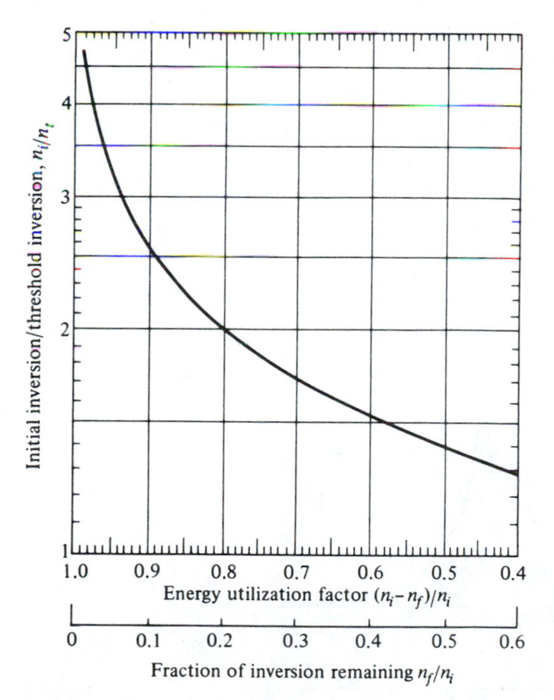

Figure 6-27 Energy utilization factor $(n_i - n_f)/n_i$ and inversion remaining after the giant pulse. (After Reference [19].)

The instantaneous power output of the laser is given by $P = \phi h\nu/t_c$, or, using (6.9-4), by

$$P = \frac{h\nu}{2t_c}\left[n_t \ln \frac{n}{n_i} - (n - n_i)\right] \quad \textbf{(6.9-6)}$$

Of special interest to us is the peak power output. Setting $\partial P/\partial n = 0$ we find that maximum power occurs when $n = n_t$. Putting $n = n_t$ in (6.9-6) gives

$$P_p = \frac{h\nu}{2t_c}\left[n_t \ln \frac{n_t}{n_i} - (n_t - n_i)\right] \quad \textbf{(6.9-7)}$$

for the peak power. If the initial inversion is well in excess of the (high-Q) threshold value (that is, $n_i \gg n_t$), we obtain from (6.9-7)

$$(P_p)_{n_i \gg n_t} \simeq \frac{n_i h\nu}{2t_c} \quad \textbf{(6.9-8)}$$

Since the power P at any moment is related to the number of photons ϕ by $P = \phi h\nu/t_c$, it follows from (6.9-8) that the maximum number of stored photons inside the resonator is $n_i/2$. This can be explained by the fact that if $n_i \gg n_t$, the buildup of the pulse to its peak value occurs in a time short compared to t_c so that at the peak of the pulse, when $n = n_t$, most of the photons that were generated by stimulated emission are still present in the resonator. Moreover, since $n_i \gg n_t$, the number of these photons $(n_i - n_t)/2$ is very nearly $n_i/2$.

A typical numerical solution of (6.9-2) and (6.9-3) is given in Figure 6-28.

To initiate the pulse we need, according to (6.9-2) and (6.9-3), to have $\phi_i \neq 0$. Otherwise the solution is trivial ($\phi = 0$, $n = n_i$). The appropriate value of ϕ_i is

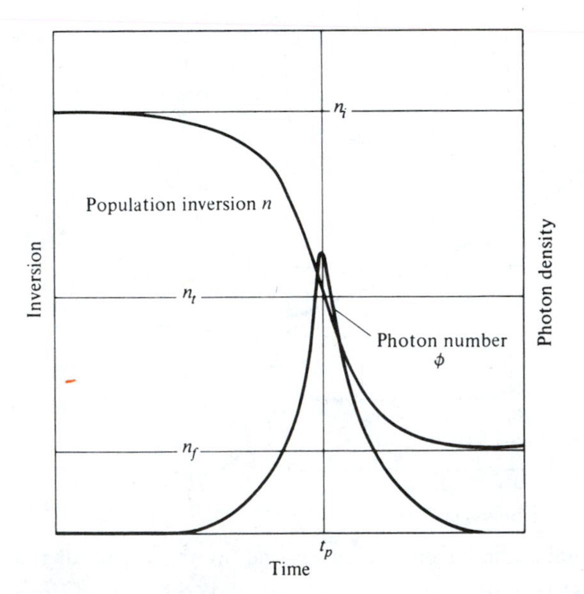

Figure 6-28 Inversion and photon density during a giant pulse. (After Reference [19].)

usually estimated on the basis of the number of spontaneously emitted photons within the acceptance solid angle of the laser mode at $t = 0$. We also notice, as discussed above, that the photon density, hence the power, reaches a peak when $n = n_t$. The energy stored in the cavity ($\propto \phi$) at this point is maximum, so stimulated transitions from the upper to the lower laser levels continue to reduce the inversion to a final value $n_f < n_t$.

Numerical solutions of (6.9-2) and (6.9-3) corresponding to different initial inversions n_i/n_t are shown in Figure 6-29. We notice that for $n_i \gg n_t$ the rise time becomes short compared to t_c but the fall time approaches a value nearly equal to

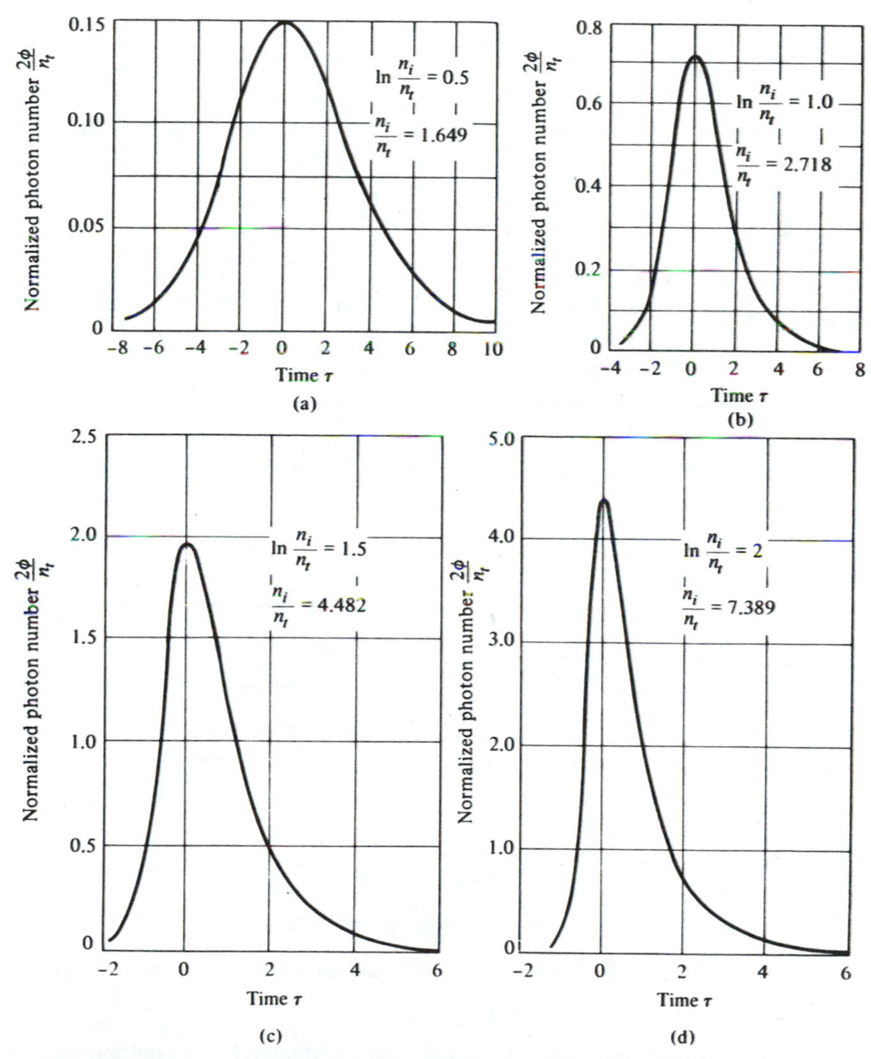

Figure 6-29 Photon number vs. time in central region of giant pulse. Time is measured in units of photon lifetime. (After Reference [19].)

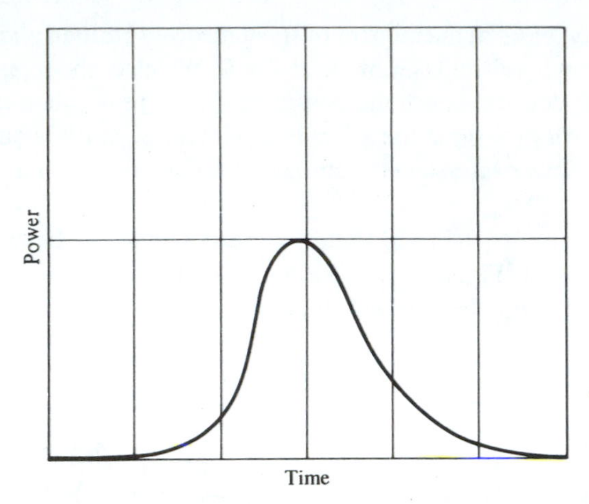

Figure 6-30 An oscilloscope trace of the intensity of a giant pulse. Time scale is 20 ns per division.

t_c. The reason is that the process of stimulated emission is essentially over at the peak of the pulse ($\tau = 0$) and the observed output is due to the free decay of the photons in the resonator.

In Figure 6-30 we show an actual oscilloscope trace of a giant pulse. Giant laser pulses are used extensively in applications that require high peak powers and short duration. These applications include experiments in nonlinear optics, ranging, material machining and drilling, initiation of chemical reactions, and plasma diagnostics.

Numerical Example: Giant Pulse Ruby Laser

Consider the case of pink ruby with a chromium ion density of $N = 1.58 \times 10^{19}$ cm^{-3}. Its absorption coefficient is taken from Figure 7-4, where it corresponds to that of the R_1 line at 6943 Å, and is $\alpha \simeq 0.2$ cm^{-1} (at 300 K). Other assumed characteristics are:

$$l = \text{length of ruby rod} = 10 \text{ cm}$$

$$A = \text{cross-sectional area of mode} = 1 \text{ cm}^2$$

$$(1 - R) = \text{fractional intensity loss per pass} = 20 \text{ percent}$$

$$n = 1.78$$

Since, according to (5.3-3), the exponential loss coefficient is proportional to $N_1 - N_2$, we have

$$\alpha(\text{cm}^{-1}) = 0.2 \frac{N_1 - N_2}{1.58 \times 10^{19}} \tag{6.9-9}$$

Thus, at room temperature, where $N_2 \ll N_1$ when $N_1 - N_2 \cong 1.58 \times 10^{19}$ cm^{-3}, and (6.9-9) yields $\alpha = 0.2$ cm^{-1} as observed. The expression for the gain coefficient follows directly from (6.9-9):

$$\gamma(\text{cm}^{-1}) = 0.2 \frac{N_2 - N_1}{1.58 \times 10^{19}} = 0.2 \frac{n}{1.58 \times 10^{19}\, V} \tag{6.9-10}$$

where n is the total inversion and $V = AL$ is the crystal volume in cm^3.

Threshold is achieved when the net gain per pass is unity. This happens when

$$e^{\gamma_t l} R = 1 \quad \text{or} \quad \gamma_t = -\frac{1}{l} \ln R \tag{6.9-11}$$

where the subscript t indicates the threshold condition.

Using (6.9-10) in the threshold condition (6.9-11) plus the appropriate data from above gives

$$n_t = 1.8 \times 10^{19} \tag{6.9-12}$$

Assuming that the initial inversion is $n_i = 5n_t = 9 \times 10^{19}$, we find from (6.9-8) that the peak power is approximately

$$P_p = \frac{n_t h\nu}{2t_c} = 5.1 \times 10^9 \text{ watts} \tag{6.9-13}$$

where $t_c = nl/c(1 - R) \simeq 2.5 \times 10^{-9}$ s.

The total pulse energy is

$$\mathscr{E} \sim \frac{n_i h\nu}{2} \sim 13 \text{ joules}$$

while the pulse duration (see Figure 6-29) $\simeq 3t_c \simeq 7.5 \times 10^{-9}$ s.

Methods of Q-Switching

Some of the schemes used in Q-switching are:

1. Mounting one of the two end reflectors on a rotating shaft so that the optical losses are extremely high except for the brief interval in each rotation cycle in which the mirrors are nearly parallel.
2. The inclusion of a saturable absorber (bleachable dye) in the optical resonator, see References [13–15]. The absorber whose opacity decreases (saturates) with increasing optical intensity prevents rapid inversion depletion due to buildup of oscillation by presenting a high loss to the early stages of oscillation during which the slowly increasing intensity is not high enough to saturate the absorp-

tion. As the intensity increases the loss decreases, and the effect is similar, but not as abrupt, as that of a sudden increase of Q.

3. The use of an electrooptic crystal (or liquid Kerr cell) as a voltage-controlled gate inside the optical resonator. It provides a more precise control over the losses (Q) than schemes 1 and 2. Its operation is illustrated by Figure 6-31 and is discussed in some detail in the following. The control of the phase delay in

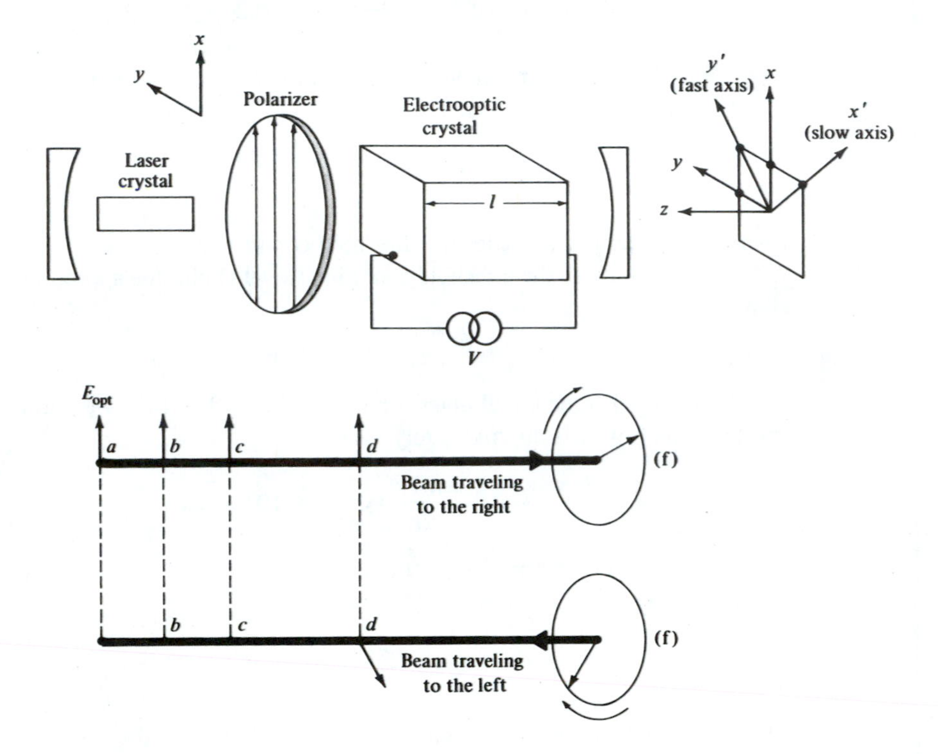

For beam traveling to right:

At point d,

$$E_{x'} = \frac{E}{\sqrt{2}} \cos \omega t, \qquad E_{y'} = \frac{E}{\sqrt{2}} \cos \omega t$$

The optical field is linearly polarized with its electric field vector parallel to x

At point f,

$$E_{x'} = \frac{E}{\sqrt{2}} \cos \left(\omega t + kl + \frac{\pi}{2}\right), \qquad E_{y'} = \frac{E}{\sqrt{2}} \cos (\omega t + kl)$$

Circularly polarized

For beam traveling to left:

At point f,

$$E_{x'} = -\frac{E}{\sqrt{2}} \cos \left(\omega t + kl + \frac{\pi}{2}\right), \qquad E_{y'} = -\frac{E}{\sqrt{2}} \cos (\omega t + kl)$$

Circularly polarized

At point d,

$$E_{x'} = -\frac{E}{\sqrt{2}} \cos (\omega t + 2kl + \pi), \qquad E_{y'} = -\frac{E}{\sqrt{2}} \cos (\omega t + 2kl)$$

Linearly polarized along y

Figure 6-31 Electrooptic crystal used as voltage-controlled gate in Q-switching a laser.

the electrooptic crystal by the applied voltage is discussed in detail in Chapter 9.

During the pumping of the laser by the light from a flashlamp, a voltage is applied to the electrooptic crystal of such magnitude as to introduce a $\pi/2$ relative phase shift (retardation) between the two mutually orthogonal components (x' and y') that make up the linearly polarized (x) laser field. On exiting from the electrooptic crystal at point f, the light traveling to the right is circularly polarized. After reflection from the right mirror, the light passes once more through the crystal. The additional retardation of $\pi/2$ adds to the earlier one to give a total retardation of π, thus causing the emerging beam at d to be linearly polarized along y and consequently to be blocked by the polarizer.

It follows that with the voltage on, the losses are high, so oscillation is prevented. The Q-switching is timed to coincide with the point at which the inversion reaches its peak and is achieved by a removal of the voltage applied to the electrooptic crystal. This reduces the retardation to zero so that state of polarization of the wave passing through the crystal is unaffected and the Q regains its high value associated with the ordinary losses of the system.

6.10 HOLE-BURNING AND THE LAMB DIP IN DOPPLER-BROADENED GAS LASERS

In this section we concern ourselves with some of the consequences of Doppler broadening in low-pressure gas lasers.

Consider an atom with a transition frequency $\nu_0 = (E_2 - E_1)/h$ where 2 and 1 refer to the upper and lower laser levels, respectively. Let the component of the velocity of the atom parallel to the wave propagation direction be v. This component, thus, has the value

$$v = \frac{\mathbf{v}_{\text{atom}} \cdot \mathbf{k}}{k} \tag{6.10-1}$$

where the electromagnetic wave is described by

$$\mathbf{E} = \mathbf{E}e^{i(2\pi\nu t - \mathbf{k}\cdot\mathbf{r})} \tag{6.10-2}$$

An atom moving with a constant velocity $\mathbf{v}$, so that $\mathbf{r} = \mathbf{v}t + \mathbf{r}_0$, will experience a field

$$\begin{aligned} \mathbf{E}_{\text{atom}} &= \mathbf{E}e^{i[2\pi\nu t - \mathbf{k}\cdot(\mathbf{r}_0 + \mathbf{v}t)]} \\ &= \mathbf{E}e^{i[(2\pi\nu - \mathbf{v}\cdot\mathbf{k})t - \mathbf{k}\cdot\mathbf{r}_0]} \end{aligned} \tag{6.10-3}$$

and will thus "see" a Doppler-shifted frequency

$$\nu_D = \nu - \frac{\mathbf{v} \cdot \mathbf{k}}{2\pi} = \nu - \frac{v}{c}\,\nu \tag{6.10-4}$$

where in the second equality we took $n = 1$ so that $k = 2\pi\nu/c$ and used (6.10-1).

The condition for the maximum strength of interaction (that is, emission or absorption) between the moving atom and the wave is that the apparent (Doppler) frequency ν_D "seen" by the atom be equal to the atomic resonant frequency ν_0

$$\nu_0 = \nu - \frac{v}{c}\,\nu \tag{6.10-5}$$

or reversing the argument, a wave of frequency ν moving through an ensemble of atoms will "seek out" and interact most strongly with those atoms whose velocity component v satisfies

$$\nu = \frac{\nu_0}{1 - \dfrac{v}{c}} \approx \nu_0\left(1 + \frac{v}{c}\right) \tag{6.10-6}$$

where the approximation is valid for $v \ll c$.

Now consider a gas laser oscillating at a single frequency ν where, for the sake of definiteness, we take $\nu > \nu_0$. The standing wave electromagnetic field at ν inside the laser resonator consists of two waves traveling in opposite directions. Consider, first, the wave traveling in the positive x direction (the resonator axis is taken parallel to the axis). Since $\nu > \nu_0$ the wave interacts, according to Equation (6.10-6) with atoms having $v > 0$, that is, atoms with

$$v_x = +\frac{c}{\nu}(\nu - \nu_0) \tag{6.10-7}$$

The wave traveling in the opposite direction $(-x)$ must also interact with atoms moving in the same direction so that the Doppler shifted frequency is reduced from ν to ν_0. These are atoms with

$$v_x = -\frac{c}{\nu}(\nu - \nu_0) \tag{6.10-8}$$

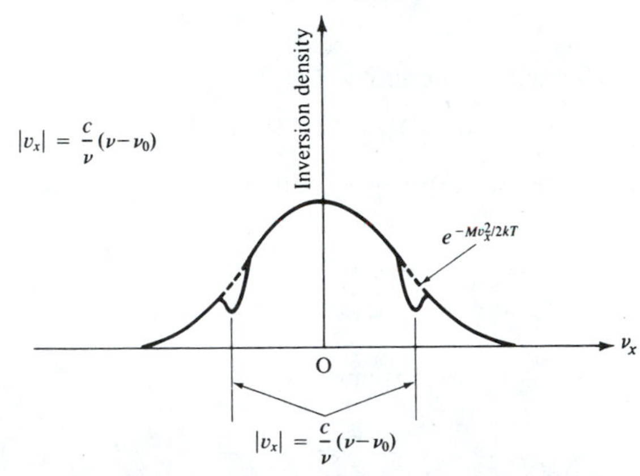

Figure 6-32 The distribution of inverted atoms as a function of v_x. The dashed curve that is proportional to $\exp(-Mv_x^2/2kT)$ corresponds to the case of zero field intensity. The solid curve corresponds to a standing wave field at $\nu = \nu_0/(1 - v_x/c)$ or one at $\nu = \nu_0/(1 + v_x/c)$.

We conclude that due to the standing wave nature of the field inside a conventional two-mirror laser oscillator, a given frequency of oscillation interacts with two velocity classes of atoms.

Consider, next, a four-level gas laser oscillating at a frequency $\nu > \nu_0$. At negligibly low levels of oscillation and at low gas pressure, the velocity distribution function of atoms in the upper laser level is given, according to (5.1-11), by

$$f(v_x) \propto e^{-Mv_x^2/2kT} \tag{6.10-9}$$

where $f(v_x)\, dv_x$ is proportional to the number of atoms (in the upper laser level) with x component of velocity between v_x and $v_x + dv_x$. As the oscillation level is increased, say by reducing the laser losses, we expect the number of atoms in the upper laser level, with x velocities near $v_x = \pm(c/\nu)(\nu - \nu_0)$, to decrease from their equilibrium value as given by (6.10-9). This is due to the fact that these atoms undergo stimulated downward transitions from level 2 to 1, thus reducing the number of atoms in level 2. The velocity distribution function under conditions of oscillation consequently has two depressions as shown schematically in Figure 6-32.

If the oscillation frequency ν is equal to ν_0, only a single "hole" exists in the velocity distribution function of the inverted atoms. This "hole" is centered on $v_x = 0$. We may, thus, expect the power output of a laser oscillating at $\nu = \nu_0$ to be

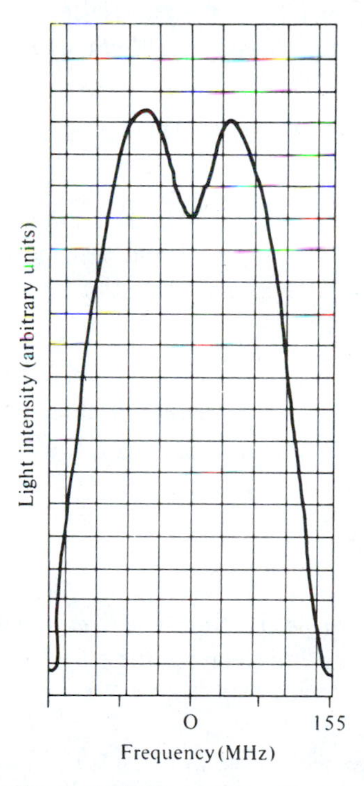

Figure 6-33 The power output as a function of the frequency of a single-mode 1.15 μm He–Ne laser using the ^{20}Ne isotope. (After Reference [21].)

less than that of a laser in which ν is tuned slightly to one side or the other of ν_0 (this tuning can be achieved by moving one of the laser mirrors). This power dip first predicted by Lamb [20] is indeed observed in gas lasers [21]. An experimental plot of the power versus frequency in a He–Ne 1.15-μm laser is shown in Figure 6-26. The phenomenon is referred to as the "Lamb dip" and is used in frequency stabilization schemes of gas lasers [22].

Problems

6.1 Show that the effect of frequency pulling by the atomic medium is to reduce the intermode frequency separation from $c/2l$ to

$$\frac{c}{2l}\left(1 - \frac{\gamma c}{2\pi\Delta\nu}\right)$$

where the symbols are defined in Section 6.2. Calculate the reduction for the case of a laser with $\Delta\nu = 10^9$ Hz, $\gamma = 4 \times 10^{-2}$ meter^{-1}, and $l = 100$ cm.

6.2 Derive Equation (6.4-3).

6.3 Derive the optimum coupling condition (Equation 6.5-11).

6.4 Calculate the critical fluorescence power P_s of the He–Ne laser operating at 6328 Å. Assume $V = 2$ cm^3, $L = 1$ percent per pass, $l = 30$ cm, and $\Delta\nu = 1.5 \times 10^9$ Hz.

6.5 Calculate the critical inversion density N_t of the He–Ne laser described in Problem 6.4.

6.6 Derive an expression for the finesse of a Fabry–Perot etalon containing an inverted population medium. Assume that $r_1^2 = r_2^2 \simeq 1$ and that the inversion is insufficient to result in oscillation. Compare the finesse to that of a passive Fabry–Perot etalon.

6.7 Derive an expression for the maximum gain–bandwidth product of a Fabry–Perot regenerative amplifier, i.e., with an amplifying medium. Define the bandwidth as the frequency region in which the intensity gain $(E_tE_t^*)/(E_iE_i^*)$ exceeds half its peak value. Assume that $\nu_0 = \nu_m$.

6.8

a. Derive Equation (6.6-4).
b. Show that if in (6.6-3) the phases are taken as $\phi_n = n\phi$, where ϕ is some constant, instead of $\phi_n = 0$, the result is merely one of delaying the pulses by $-\phi/\omega$.

6.9

a. Describe qualitatively what one may expect to see in parts A, B, C, and D of the mode-locking experiment sketched in Figure 6-14. (The reader may find it useful to read first the section on photomultipliers in Chapter 11.)

b. What is the effect of mode locking on the intensity of the beat signal (at $\omega = \pi c/l$) displayed by the RF spectrum analyzer in *B*? Assume *N* equal amplitude modes spaced by ω whose phases before mode locking are random. (*Answer:* Mode locking increases the beat signal power by *N*.)

c. Show that a standing wave at $\nu_0 + \delta\nu_0$ is the center frequency of the Doppler-broadened lineshape function) in a gas laser will burn the same two holes in the velocity distribution function (see Figure 6-32) as a field at $\nu_0 - \delta$.

d. Can two traveling waves, one at $\nu_0 + \delta$ the other at $\nu_0 - \delta$, interact with the same class of atoms? If the answer is yes, under what conditions?

6.10 Design a frequency stabilization scheme for gas lasers based on the Lamb dip (see Figure 6-33). [*Hint:* You may invent a new scheme, but, failing that, consider what happens to the phase of the modulation in the power output when the cavity length is modulated sinusoidally near the bottom of the Lamb dip. Can you derive an error correction signal from this phase that will control the cavity length?]

6.11 Verify the relations of Table 6-2.

6.12 A helium-neon laser ($\lambda = 0.63\ \mu$m) operating in the fundamental transverse mode has mirrors separated by $l = 30$ cm. The Doppler width is $\Delta\nu_D = 1.5$ GHz, and the effective refractive index is $n = 1$. The output mirror is flat, and the other mirror is spherical with a radius of curvature 16 m.

a. What is the frequency difference between longitudinal modes in the resonator?

b. Show that the resonator is stable.

c. What would the Doppler width become if the temperature of the laser medium were doubled?

d. What is the spot size at the flat mirror?

e. If the output is taken from the flat mirror, what is the spot size 16 km away?

f. Given that the internal cavity loss is $L_i = 10^{-1}$, the small signal gain coefficient is $\gamma_0 = 10^{-3}$ cm^{-1} and the reflection coefficient of the spherical mirror is 1.0, what is the reflection coefficient (*R*) of the output mirror that will give the maximum output power?

g. A thin lens of focal length *f* is placed against the output mirror. Find the radius of curvature of the Gaussian beam at a distance $d = f$ from the lens.

References

1. Schawlow, A. L., and C. H. Townes, "Infrared and optical masers," *Phys. Rev.* 112:1940, 1958.
2. Yariv, A., *Quantum Electronics*, 2d ed. New York: Wiley, 1975.
3. Smith, W. V., and P. P. Sorokin, *The Laser*. New York: McGraw-Hill, 1966.
4. Lengyel, B. A., *Introduction to Laser Physics*. New York: Wiley, 1966.
5. Birnbaum, G., *Optical Masers*. New York: Academic Press, 1964.
6. *Lasers and Light—Readings from Scientific American*. San Francisco: Freeman, 1969.
7. Bennett, W. R., Jr., "Gaseous optical masers," *Appl. Opt. Suppl. 1, Optical Masers*, 1962, p. 24.

8. Fork, R. L., D. R. Herriott, and H. Kogelnik, ''A scanning spherical mirror interferometer for spectral analysis of laser radiation,'' *Appl. Opt.* 3:1471, 1964.
9. DiDomenico, M., Jr., ''Small signal analysis of internal modulation of Lasers,'' *J. Appl. Phys.* 35:2870, 1964.
10. Yariv, A., ''Internal modulation in multimode laser oscillators,'' *J. Appl. Phys.* 36:388, 1965.
11. Hargrove, L. E., R. L. Fork, and M. A. Pollack, ''Locking of He–Ne laser modes induced by synchronous intracavity modulation,'' *Appl. Phys. Lett.* 5:4, 1964.
12. DiDomenico, M., Jr., J. E. Geusic, H. M. Marcos, and R. G. Smith, ''Generation of ultrashort optical pulses by mode locking the Nd^{3+} : YAG laser,'' *Appl. Phys. Lett.* 8:180, 1966.
13. Mocker, H., and R. J. Collins, ''Mode competition and self-locking effects in a *Q*-switched ruby laser,'' *Appl. Phys. Lett.* 7:270, 1965.
14. De Maria, A. J., ''Picosecond laser pulses,'' *Proc. IEEE* 57:3, 1969.
15. DeMaria, A. J., ''Mode locking,'' *Electronics*, Sept. 16, 1968, p. 112.
16. Hellwarth, R. W., ''Control of fluorescent pulsations.'' In *Advances in Quantum Electronics*, J. R. Singer, ed. New York: Columbia University Press, 1961, p. 334.
17. McClung, F. J., and R. W. Hellwarth, *J. Appl. Phys.* 33:828, 1962.
18. Hellwarth, R. W., ''*Q* modulation of lasers.'' In *Lasers*, vol. 1, A. K. Levine, ed. New York: Marcel Dekker, Inc., 1966, p. 253.
19. Wagner, W. G., and B. A. Lengyel, ''Evolution of the giant pulse in a laser,'' *J. Appl. Phys.* 34:2042, 1963.
20. Lamb, W. E., Jr., ''Theory of an optical maser,'' *Phys. Rev.* 134:A1429, 1964.
21. Szöke, A., and A. Javan, ''Isotope shift and saturation behavior of the 1.15μ transition of neon,'' *Phys. Rev. Lett.* 10:512, 1963.
22. Bloom, A., *Gas Lasers*. New York: Wiley, 1963, p. 93.
23. Collins, R. J., D. F. Nelson, A. L. Schawlow, W. Bond, C. G. B. Garrett, and W. Kaiser, *Phys. Rev. Lett.* 5:303, 1960.
24. Siegman, A. E., and D. J. Kuizenga, ''Simple analytic expressions for AM and FM mode locked pulses in homogeneous lasers,'' *Appl. Phys. Lett.* 14:181, 1969.
25. Kuizenga, D. J., and A. E. Siegman, ''FM and AM mode locking of the homogeneous laser: Part I, Theory; Part II, Experiment,'' *J. Quant. Elec.* QE-6:694, 1970.
26. Siegman, A. E., *Lasers*, University Science Books, 1986.
27. Cutler, C. C., ''The regenerative pulse generator,'' *Proc. IRE* 43:140, 1955.
28. Valdmanis, J. A., R. L. Ford, and J. P. Gordon, *Opt. Lett.* 10:131, 1985.
29. Koch, Thomas L., ''Gigawatt picosecond dye lasers and ultrafast processes in semiconductor lasers,'' Ph.D. thesis, California Institute of Technology, 1982.
30. Fork, R. L., C. H. Brito Cruz, P. C. Becker, and C. V. Shank, *Opt. Lett.* 12:483, 1987.
31. Shapiro, S. L. (ed.), *Ultrashort Light Pulses* (Topics in *Appl. Phys.*, vol. 18). Berlin–New York: Springer-Verlag, 1977.

32. Hochstrasser, R. M., W. Kaiser, and C. V. Shank, eds., *Picosecond Phenomena II* (Series in Chemical Physics, vol. 14). Berlin–New York: Springer-Verlag, 1980.
33. Eisenthal, K. B., R. M. Hochstrasser, W. Kaiser, and A. Laubereau, eds., *Picosecond Phenomena III* (Series in Chemical Physics, vol. 23). Berlin–New York: Springer-Verlag, 1982.
34. Suematsu, Y., ''Long wavelength optical fiber communication,'' *Proc. IEEE* 71:692, 1983.
35. Lau, K. Y., N. Bar-Chaim, I. Ury, Ch. Harder, and A. Yariv, ''Direct amplitude modulation of short cavity GaAs lasers up to X-band frequencies,'' *Appl. Phys. Lett.* 43:1, 1983.
36. Sanders, S., L. Eng, J. Paslaski, and A. Yariv, ''108 GHz passive mode locking of a multiple quantum well semiconductor laser with an intracavity absorber,'' *Appl. Phys. Lett.* 56:310, 1990.
37. Yariv, A. *Quantum Electronics*, New York: Wiley, 3d ed., 1989, p. 550.
38. DiDomenico, Jr., ''Small signal analysis of internal coupling modulation of lasers,'' *J. Appl. Phys.* 35, 1954.
39. Smith, P. J., T. J. Bridges, and E. J. Burkhardt, ''Mode locked high pressure CO_2 laser,'' *Appl. Phys. Lett.* 21:470, 1972.
40. E. B. Tracy, *IEEE J. Quantum Electron.*, 5:454, 1969.
41. Martinez, O. E., J. P. Gordon, and R. L. Fork, *J. Opt. Soc. Am. A.*, 1:1003, 1984.
42. Martinez, O. E., *IEEE J. Quantum Electron.*, 23:59, 1987.
43. Schrans, Thomas, ''Part I: Longitudinal static and dynamic effects in semiconductor lasers; Part II: Spectral characteristics of passively mode-locked quantum well lasers,'' Ph.D. thesis, California Institute of Technology, 1994.

Some Specific Laser Systems

7.0 INTRODUCTION

The pumping of the atoms into the upper laser level is accomplished in a variety of ways, depending on the type of laser. In this chapter we will review some of the more common laser systems and in the process describe their pumping mechanisms. The laser systems described include: ruby, Nd^{3+}: YAG, Nd^{3+}: glass, He–Ne, CO_2, Ar^+, excimer, and organic-dye lasers. The semiconductor current-pumped laser, because of its unique technological importance, receives special billing in Chapters 15 and 16.

7.1 PUMPING AND LASER EFFICIENCY

Figure 7-1 shows the pumping–oscillation cycle of some (hypothetical) representative laser. The pumping agent elevates the atoms into some excited state 3 from which they relax into the upper laser level 2. The stimulated laser transition takes place between levels 2 and 1 and results in the emission of a photon of frequency ν_{21}.

It is evident from this figure that the minimum energy input per output photon is $h\nu_{30}$, so the power efficiency of the laser cannot exceed

$$\eta_{\text{atomic}} = \frac{\nu_{21}}{\nu_{30}} \tag{7.1-1}$$

to which quantity we will refer as the ''atomic quantum efficiency.'' The overall laser efficiency depends on the fraction of the total pump power that is effective in transferring atoms into level 3 and on the pumping quantum efficiency defined as

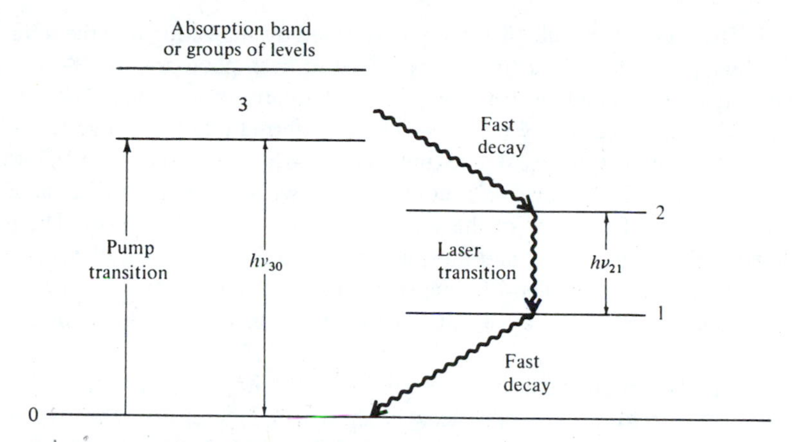

Figure 7-1 Pumping–oscillation cycle of a typical laser.

the fraction of the atoms that once in 3, make a transition to 2. The product of the last two factors, which constitutes an upper limit on the efficiency of optically pumped lasers, ranges from about 1 percent for solid-state lasers such as Nd^{3+} : YAG, to about 30 percent in the CO_2 laser, and to near unity in the GaAs junction laser. We shall discuss these factors when we get down to some specific laser systems. We may note, however, that according to (7.1-1), in an efficient laser system ν_{21} and ν_{30} must be of the same order of magnitude, so the laser transition should involve low-lying levels.

7.2 RUBY LASER

The first material in which laser action was demonstrated [1] and still one of the most useful laser materials is ruby, whose output is at $\lambda_0 = 0.6943\ \mu m$. The active laser particles are Cr^{3+} ions present as impurities in Al_2O_3 crystal. Typical Cr^{3+} concentrations are ~0.05 percent by weight. The pertinent energy level diagram is shown in Figure 7-2.

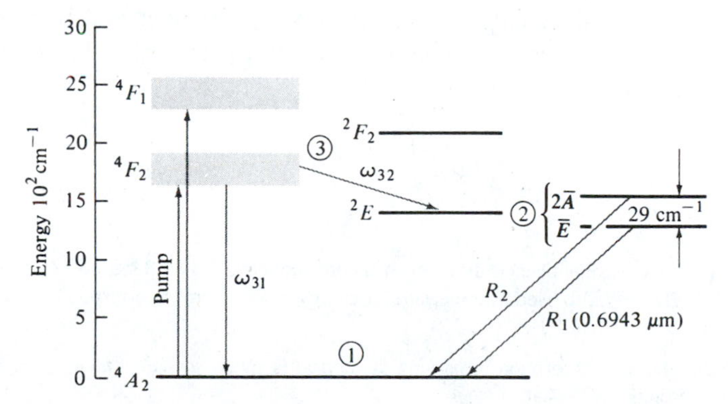

Figure 7-2 Energy levels pertinent to the operation of a ruby laser. (After Reference [2].)

The pumping of ruby is usually performed by subjecting it to the light of intense flashlamps (quite similar to the types used in flash photography). A portion of this light that corresponds in frequency to the two absorption bands $^4F_2$ and $^4F_1$ is absorbed, thereby causing Cr^{3+} ions to be transferred into these levels. The ions proceed to decay, within an average time of $\omega_{32}^{-1} \simeq 5 \times 10^{-8}$ seconds [2], into the upper —laser level ^2E. The level ^2E is composed of two separate levels $2\overline{A}$ and $\overline{E}$ separated by 29 cm^{-1}.[1] The lower of these two, $\overline{E}$, is the upper laser level. The lower laser level is the ground state, and thus, according to the discussion of Section 6.3, ruby is a three-level laser. The lifetime of atoms in the upper laser level $\overline{E}$ is $t_2 \simeq 3 \times 10^{-3}$ second. Each decay results in the (spontaneous) emission of a photon, so $t_2 \simeq t_{spont}$.

An absorption spectrum of a typical ruby with two orientations of the optical field relative to the c (optic) axis is shown in Figure 7-3. The two main peaks correspond to absorption into the useful $^4F_1$ and $^4F_2$ bands, which are responsible for the characteristic (ruby) color.

The ordinate is labeled in terms of the absorption coefficient and in terms of the transition cross section σ, which may be defined as the absorption coefficient per unit inversion per unit volume and has consequently the dimension of area. According to this definition, $\alpha(\nu)$ is given by

$$\alpha(\nu) = (N_1 - N_2)\sigma(\nu) \tag{7.2-1}$$

A more detailed plot of the absorption near the laser emission wavelength is shown in Figure 7-4. The width $\Delta\nu$ of the laser transition as a function of temperature is shown in Figure 7-5. At room temperature, $\Delta\nu = 11 \text{ cm}^{-1}$.

We can use ruby to illustrate some of the considerations involved in optical pumping of solid-state lasers. Figure 7-6 shows a typical setup of an optically pumped laser, such as ruby. The helical flashlamp surrounds the ruby rod. The flash excitation is provided by the discharge of the charge stored in a capacitor bank across the lamp.

The typical flash output consists of a pulse of light of duration $t_{flash} \simeq 5 \times 10^{-4}$ seconds. Let us, for the sake of simplicity, assume that the flash pulse is rectangular in time and of duration t_{flash} and that it results in an optical flux at the crystal surface having $s(\nu)$ watts per unit area per unit frequency at the frequency ν. If the absorption coefficient of the crystal is $\alpha(\nu)$, then the amount of energy absorbed by the crystal per unit volume is[2]

$$t_{flash} \int_0^\infty s(\nu)\alpha(\nu)\, d\nu$$

[1]The unit 1 cm^{-1} (one wavenumber) is the frequency corresponding to $\lambda_0 = 1$ cm, so 1 cm^{-1} is equivalent to $\nu = 3 \times 10^{10}$ Hz. It is also used as a measure of energy where 1 cm^{-1} corresponds to the energy $h\nu$ of a photon with $\nu = 3 \times 10^{10}$ Hz.

[2]We assume that the total absorption in passing the crystal is small, so $s(\nu)$ is taken to be independent of the distance through the crystal.

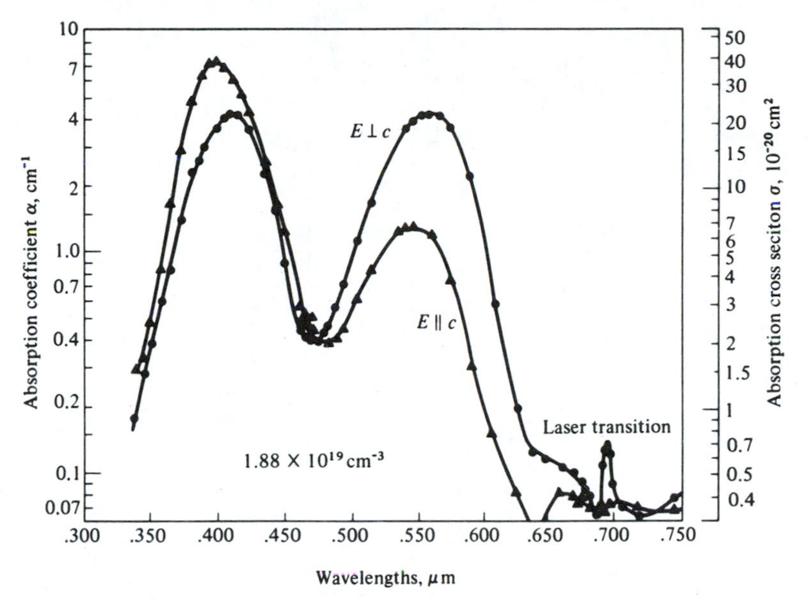

Figure 7-3 Absorption coefficient and absorption cross section as functions of wavelength for $E \parallel c$ and $E \perp c$. The 300 K data were derived from transmittance measurements on pink ruby with an average Cr ion concentration of 1.88×10^{19} cm^{-3}. (After Reference [3].)

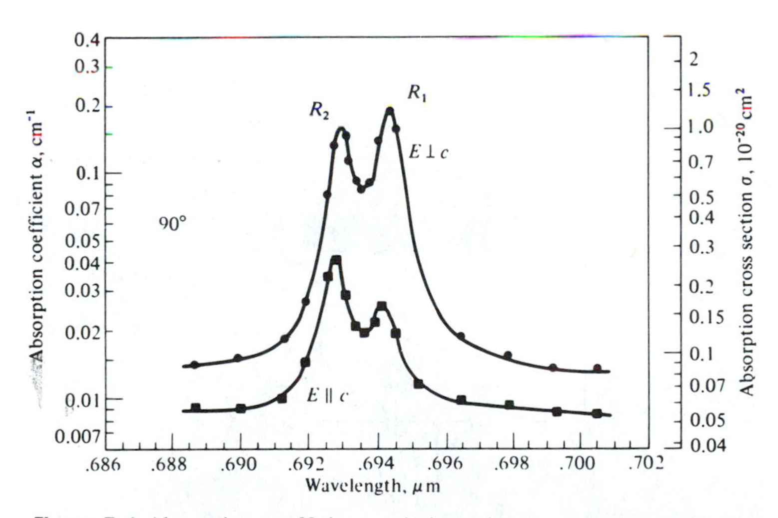

Figure 7-4 Absorption coefficient and absorption cross section as functions of wavelength for $E \parallel c$ and $E \perp c$. Sample was a pink ruby laser rod having a 90° c-axis orientation with respect to the rod axis and a Cr concentration of 1.58×10^{19} cm^{-3}. (After Reference [3].)

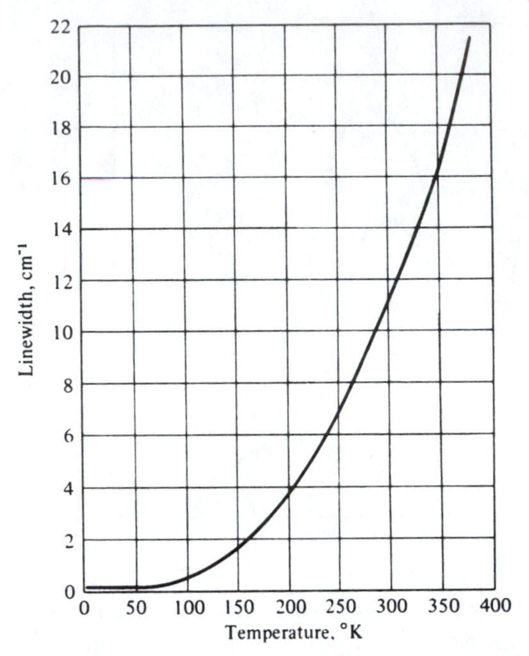

Figure 7-5 Linewidth of the R_1 line of ruby as a function of temperature. (After Reference [4].)

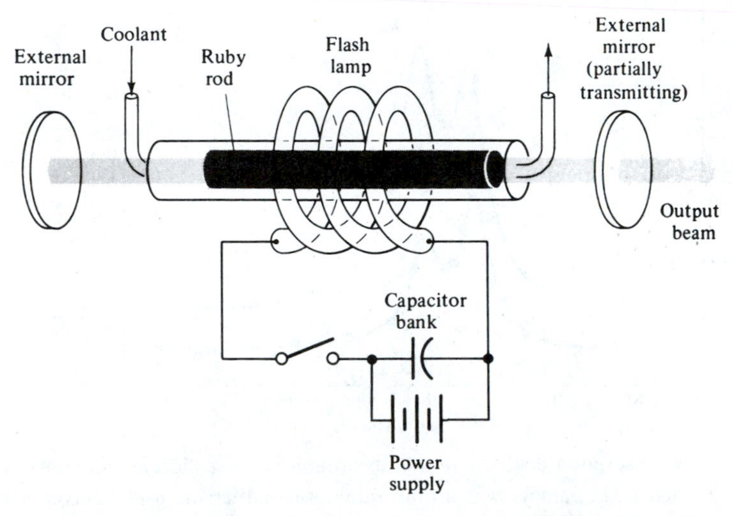

Figure 7-6 Typical setup of a pulsed ruby laser using flashlamp pumping and external mirrors.

If the absorption quantum efficiency (the probability that the absorption of a pump photon at ν results in transferring one atom into the upper laser level) is $\eta(\nu)$, the number of atoms pumped into level 2 per unit volume is

$$N_2 = t_{\text{flash}} \int_0^\infty \frac{s(\nu)\alpha(\nu)\eta(\nu)}{h\nu}\, d\nu \tag{7.2-2}$$

Since the lifetime $t_2 \simeq 3 \times 10^{-3}$ second of atoms in level 2 is considerably longer than the flash duration ($\sim 5 \times 10^{-4}$ s) we may neglect the spontaneous decay out of level 2 during the time of the flash pulse, so N_2 represents the population of level 2 after the flash.

Numerical Example: Flash Pumping of a Pulsed Ruby Laser

Consider the case of a ruby laser with the following parameters:

$$N_0 = 2 \times 10^{19} \text{ atoms/cm}^3 \ (\text{Cr}^{3+})$$

$$t_2 \simeq t_{\text{spont}} \simeq 3 \times 10^{-3} \text{ s}$$

$$t_{\text{flash}} = 5 \times 10^{-4} \text{ s}$$

If the useful absorption is limited to relatively narrow spectral regions, we may approximate (7.2-2) by

$$N_2 = \frac{t_{\text{flash}}\, \overline{s(\nu)}\, \overline{\alpha(\nu)}\, \overline{\eta(\nu)}\, \overline{\Delta\nu}}{h\bar{\nu}} \tag{7.2-3}$$

where the bars represent average values over the useful absorption region whose width is $\overline{\Delta\nu}$.

From Figure 7-3 we deduce an average absorption coefficient of $\overline{\alpha(\nu)} \simeq 2 \text{ cm}^{-1}$ over the two central peaks. Since ruby is a three-level laser, the upper level population is, according to (6.3-2), $N_2 \simeq N_0/2 = 10^{19} \text{ cm}^{-3}$. Using $\bar{\nu} \simeq 5 \times 10^{14}$ Hz, $\overline{\eta(\nu)} \simeq 1$, (7.2-3) yields

$$\bar{s}\, \overline{\Delta\nu}\, t_{\text{flash}} \simeq 1.5 \text{ J/cm}^2$$

for the pump energy in the useful absorption region that must fall on each square centimeter of crystal surface in order to obtain threshold inversion. To calculate the total lamp energy that is incident on the crystal we need to know the spectral characteristics of the lamp output. Typical data of this sort are shown in Figure 7-7. The mercury-discharge lamp is seen to contain considerable output in the useful absorption regions (near 4000 Å and 5500 Å) of ruby. If we estimate the useful fraction of the lamp output at 10 percent, the fraction of the lamp light actually incident on the crystal as 20 percent, and the conversion of electrical-to-optical energy as 50

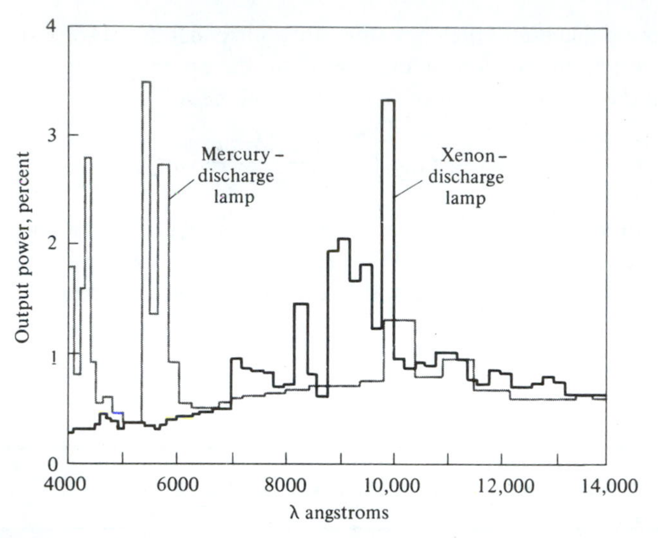

Figure 7-7 Spectral output characteristics of two commercial high-pressure lamps. Output is plotted as a fraction of electrical input to lamp over certain wavelength intervals (mostly 200 Å) between 0.4 and 1.4 μm. (After Reference [5].)

percent, we find the threshold electric energy input to the flashlamp per square centimeter of laser surface is

$$\frac{1.5}{0.1 \times 0.2 \times 0.5} = 150 \text{ J/cm}^2$$

These are, admittedly, extremely crude calculations. They are included not only to illustrate the order of magnitude numbers involved in laser pumping, but also as an example of the quick and rough estimates needed to discriminate between feasible ideas and "pie-in-the-sky" schemes.

7.3 Nd^{3+}: YAG LASER

One of the most important laser systems is that using trivalent neodymium ions (Nd^{3+}), which are present as impurities in yttrium aluminum garnet (YAG = $Y_3Al_5O_{12}$); see References [6, 7]. The laser emission occurs at $\lambda_0 = 1.0641$ μm at room temperature. The relevant energy levels are shown in Figure 7-8. The lower laser level is at $E_2 \simeq 2111$ cm^{-1} from the ground state so that at room temperature its population is down by a factor of $\exp(-E_2/kT) \simeq e^{-10}$ from that of the ground state and can be neglected. The Nd^{3+} : YAG thus fits our definition (see Section 6.3) of a four-level laser.

The spontaneous emission spectrum of the laser transition is shown in Figure 7-9. The width of the gain linewidth at room temperature is $\Delta\nu \simeq 6$ cm^{-1}. The

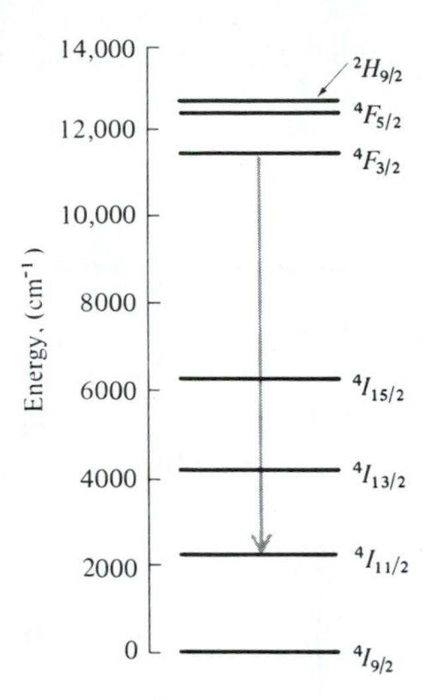

Figure 7-8 Energy-level diagram of Nd^{3+} in YAG. (After Reference [6].)

spontaneous lifetime for the laser transition has been measured [7] as $t_{spont} = 5.5 \times 10^{-4}$ s. The room-temperature cross section at the center of the laser transition is $\sigma = 9 \times 10^{-19}$ cm^2. If we compare this number to $\sigma = 1.22 \times 10^{-20}$ cm^2 in ruby (see Figure 7-4), we expect that at a given inversion the optical gain constant γ in Nd^{3+}:YAG is approximately 75 times that of ruby. This causes the oscillation threshold to be very low and explains the easy continuous (CW) operation of this laser compared to ruby.

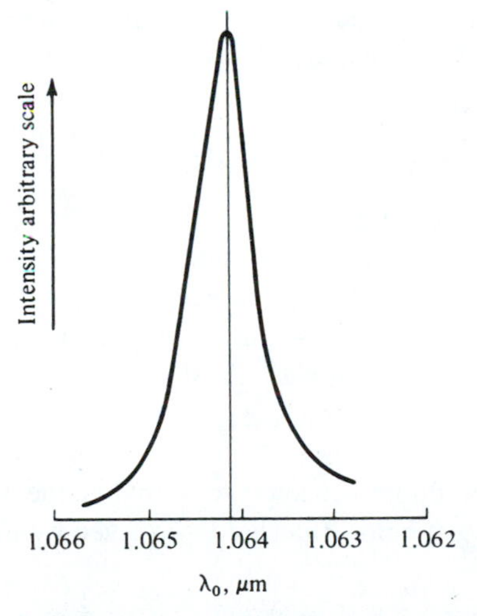

Figure 7-9 Spontaneous-emission spectrum of Nd^{3+} in YAG near the laser transition at $\lambda_0 = 1.064$ μm. (After Reference [7].)

The absorption responsible for populating the upper laser level takes place in a number of bands between 13,000 and 25,000 cm^{-1}.

Numerical Example: Threshold of an Nd^{3+}:YAG Laser

Pulsed threshold First we estimate the energy needed to excite a typical Nd^{3+}:YAG laser on a pulse basis so that we can compare it with that of ruby. We use the following data:

$$\left.\begin{aligned} l &= 20 \text{ cm (length optical resonator)} \\ L &= 4 \text{ percent } (= \text{loss per pass}) \\ n &= 1.5 \end{aligned}\right\} t_c = \frac{nl}{Lc} = 2.5 \times 10^{-8} \text{ s}$$

$$\Delta\nu = 6 \text{ cm}^{-1} (= 6 \times 3 \times 10^{10} \text{ Hz})$$

$$t_{\text{spont}} = 5.5 \times 10^{-4} \text{ s}$$

$$\lambda = 1.06 \ \mu\text{m}$$

Using the foregoing data in (6.1-11) gives

$$N_t = \frac{8\pi n^3 t_{\text{spont}} \Delta\nu}{c t_c \lambda^2} \simeq 1.0 \times 10^{15} \text{ cm}^{-3}$$

Assuming that 5 percent of the exciting light energy falls within the useful absorption bands, that 5 percent of this light is actually absorbed by the crystal, that the average ratio of laser frequency to the pump frequency is 0.5, and that the lamp efficiency (optical output/electrical input) is 0.5, we obtain

$$\mathscr{E}_{\text{lamp}} = \frac{N_t h\nu_{\text{laser}}}{5 \times 10^{-2} \times 5 \times 10^{-2} \times 0.5 \times 0.5} \simeq 0.3 \text{ J/cm}^3$$

for the energy input to the lamp at threshold.

It is interesting to compare this last number to the figure of 150 joules per square centimeter of surface area obtained in the ruby example of Section 7.2. For reasonable dimension crystals (say, length = 5 cm, r = 2 mm) we obtain $\mathscr{E}_{\text{lamp}} = 0.19$ J. We expect the ruby threshold to exceed that of Nd^{3+}:YAG by three orders of magnitude, which is indeed the case.

Continuous operation The critical fluorescence power—that is, the actual power given off by spontaneous emission just below threshold—is given by (6.3-4) as

$$\left(\frac{P_s}{V}\right) = \frac{N_t h\nu}{t_{\text{spont}}} \simeq 0.34 \text{ W/cm}^3$$

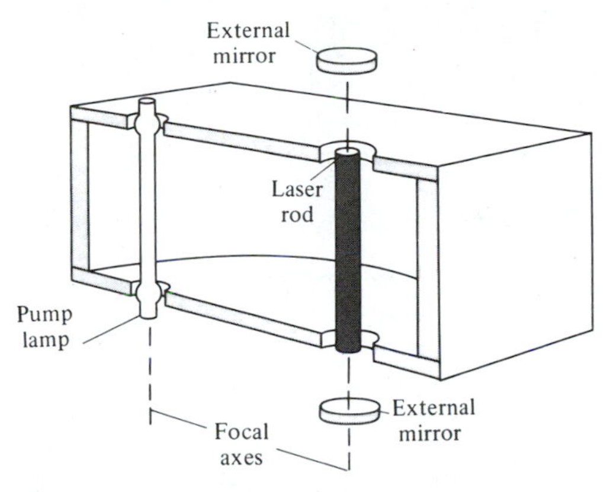

Figure 7-10 Typical continuous solid-state laser arrangement employing an elliptic cylinder housing for concentrating lamp light onto laser.

Taking the crystal diameter as 0.25 cm and its length as 3 cm and using the same efficiency factors assumed in the first part of this example, we can estimate the power input to the lamp at threshold as

$$P_{(\text{lamp})} = \frac{0.34 \times (\pi/4) \times (0.25)^2 \times 3}{5 \times 10^{-2} \times 5 \times 10^{-2} \times 0.5 \times 0.5} \simeq 81 \text{ watts}$$

which is in reasonable agreement with experimental values [6].

A typical arrangement used in continuous solid state lasers is shown in Figure 7-10. The highly polished elliptic cylinder is used to concentrate the light from the lamp, which is placed along one focal axis, onto the laser rod, which occupies the other axis. This configuration guarantees that most of the light emitted by the lamp passes through the laser rod. The reflecting mirrors are placed outside the cylinder.

7.4 NEODYMIUM-GLASS LASER

One of the most useful laser systems is that which results when the Nd^{3+} ion is present as an impurity atom in glass [8].

The energy levels involved in the laser transition in a typical glass are shown in Figure 7-11. The laser emission wavelength is at $\lambda = 1.059$ μm and the lower level is approximately 1950 cm^{-1} above the ground state. As in the case of Nd^{3+}:YAG described in Section 7.3, we have here a four-level laser, since the thermal population of the lower laser level is negligible. The fluorescent emission near $\lambda_0 = 1.06$ μm is shown in Figure 7-12. The fluorescent linewidth can be measured off directly and ranges, for the glasses shown, around 300 cm^{-1}. This width is approximately a factor of 50 larger than that of Nd^{3+} in YAG. This is due to the amorphous structure of glass, which causes different Nd^{3+} ions to "see"

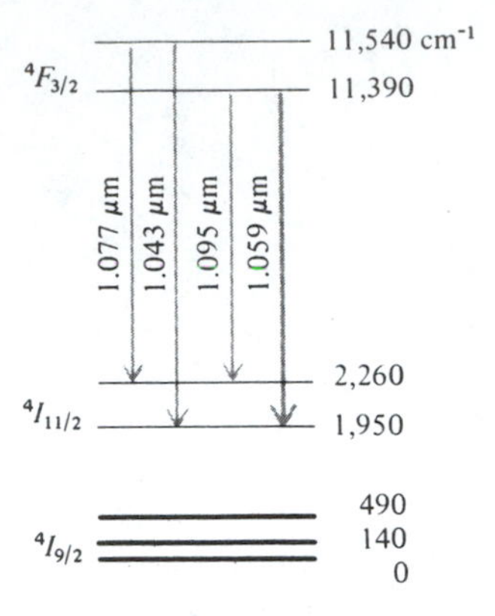

Figure 7-11 Energy-level diagram for the ground state and the states involved in laser emission at 1.059 μm for Nd^{3+} in a rubidium potassium barium silicate glass. (After Reference [8].)

slightly different surroundings. This causes their energy splittings to vary slightly. Different ions consequently radiate at slightly different frequencies, causing a broadening of the spontaneous emission spectrum. The absorption bands responsible for pumping the laser level are shown in Figure 7-13. The probability that the absorption of a photon in any of these bands will result in pumping an atom to the upper laser level (that is, the absorption quantum efficiency) has been estimated [8] at about 0.4.

The lifetime t_2 of the upper laser level depends on the host glass and on the Nd^{3+} concentration. This variation in two glass series is shown in Figure 7-14.

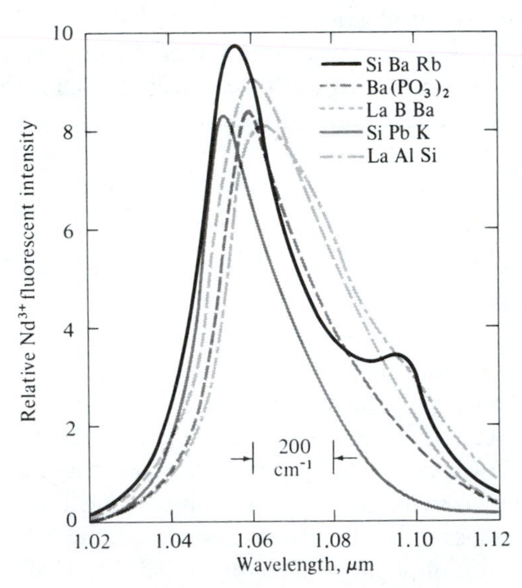

Figure 7-12 Fluorescent emission of the 1.06-μm line of Nd^{3+} at 300 K in various glass bases. (After Reference [8].)

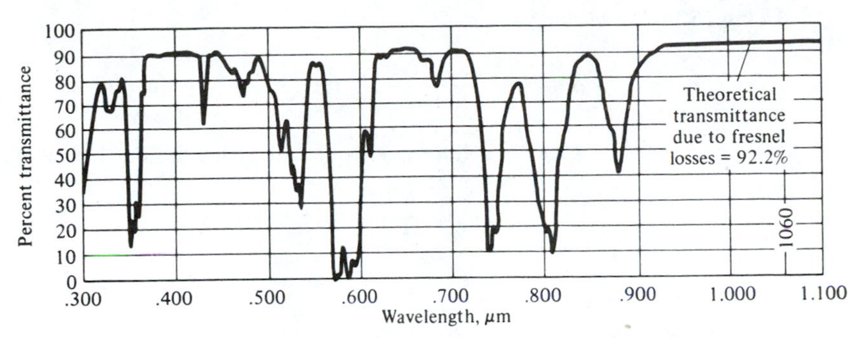

Figure 7-13 Nd^{3+} absorption spectrum for a sample of glass 6.4 mm thick with the composition 66 wt.% SiO_2, 5 wt.% Nd_2O_3, 16 wt.% Na_2O, 5 wt.% BaO, 2 wt.% Al_2O_3, and 1 wt.% Sb_2O_3. (After Reference [8].)

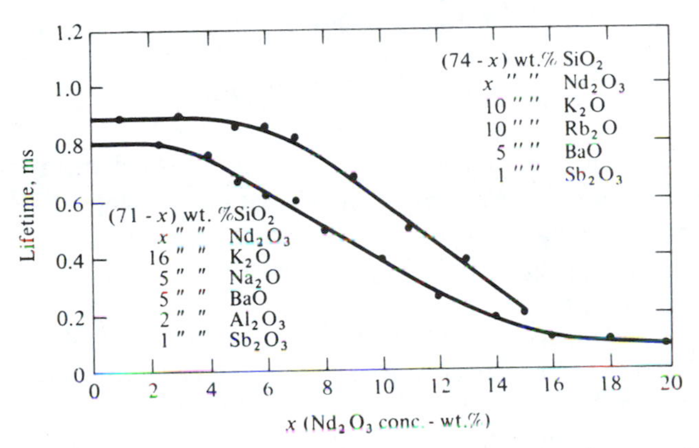

Figure 7-14 Lifetime as a function of concentration for two glass series. (After Reference [8].)

Numerical Example: Thresholds for CW and Pulsed Operation of Nd^{3+}: Glass Lasers

Let us estimate first the threshold for continuous (CW) laser action in a Nd^{3+}: glass laser using the following data:

$$\Delta\nu = 200 \text{ cm}^{-1} \qquad \text{(see Figure 7-12)}$$

$$n = 1.5$$

$$t_{\text{spont}} \simeq t_2 = 3 \times 10^{-4} \text{ s}$$

$$\left.\begin{aligned} l &= \text{length of resonator} = 20 \text{ cm} \\ L &= \text{loss per pass} = 2 \text{ percent} \end{aligned}\right\} t_c \simeq \frac{nl}{Lc} = 5 \times 10^{-8} \text{ s}$$

Using (6.1-11) we obtain

$$N_t = \frac{8\pi t_{\text{spont}} n^3 \Delta\nu}{c t_c \lambda^2} = 9.05 \times 10^{15} \text{ atoms/cm}^3$$

for the critical inversion. The fluorescence power at threshold P_s is thus [see (6.3-5)]

$$P_s = \frac{N_t h\nu V}{t_{\text{spont}}} = 5.65 \text{ watts}$$

in a crystal volume $V = 1 \text{ cm}^3$.

We assume (a) that only 10 percent of the pump light lies within the useful absorption bands, (b) that because of the optical coupling inefficiency and the relative transparency of the crystal only 10 percent of the energy leaving the lamp within the absorption bands is actually absorbed, (c) that the absorption quantum efficiency is 40 percent, and (d) that the average pumping frequency is twice that of the emitted radiation. The lamp output at threshold is thus

$$\frac{2 \times 5.65}{0.1 \times 0.1 \times 0.4} = 2825 \text{ watts}$$

If the efficiency of the lamp in converting electrical to optical energy is about 50 percent, we find that continuous operation of the laser requires about 5 kW of power. This number is to be contrasted with a threshold of approximately 100 watts for the Nd:YAG laser, which helps explain why Nd:glass lasers are not operated continuously.

If we consider the pulsed operation of a Nd:glass laser by flash excitation, we have to estimate the minimum energy needed to pump the laser at threshold. Let us assume here that the losses (attributable mostly to the output mirror transmittance) are $L = 20$ percent.[3] A recalculation of N_t gives

$$N_t = 9.05 \times 10^{16} \text{ atoms/cm}^3$$

The minimum energy needed to pump N_t atoms into level 2 is then

$$\frac{\mathscr{E}_{\min}}{V} = N_t(h\nu) = 1.7 \times 10^{-2} \text{ J/cm}^3$$

Assuming a crystal volume $V = 10 \text{ cm}^3$ and the same efficiency factors used in the CW example above, we find that the input energy to the flashlamp at threshold $\simeq 2 \times 1.7 \times 10^{-2} \times 10/(0.1 \times 0.1 \times 0.4) = 85$ J. Typical Nd^{3+}:glass lasers with characteristics similar to those used in this example are found to require an input of about 150–300 joules at threshold.

[3]Because of the higher pumping rate available with flash pumping, optimum coupling (see Section 6.5) calls for larger mirror transmittances compared to the CW case.

7.5 He-Ne LASER

The first CW laser, as well as the first gas laser, was one in which a transition between the $2S$ and the $2p$ levels in atomic Ne resulted in the emission of 1.15 μm radiation [9]. Since then transitions in Ne were used to obtain laser oscillation at $\lambda_0 = 0.6328$ μm [10] and at $\lambda_0 = 3.39$ μm. The operation of this laser can be explained with the aid of Figure 7-15. A dc (or rf) discharge is established in the gas mixture containing typically, 1.0 mm Hg of He and 0.1 mm Hg of Ne. The energetic electrons in the discharge excite helium atoms into a variety of excited states. In the normal cascade of these excited atoms down to the ground state, many collect in the long-lived metastable states 2^3S and 2^1S whose lifetimes are 10^{-4} second and 5×10^{-6} second, respectively. Since these long-lived (metastable) levels nearly coincide in energy with the $2S$ and $3S$ levels of Ne, they can excite Ne atoms into these two excited states. This excitation takes place when an excited He atom collides with a Ne atom in the ground state and exchanges energy with it. The small difference in energy ($\sim$400 cm^{-1} in the case of the $2S$ level) is taken up by the kinetic energy of the atoms after the collision. This is the main pumping mechanism in the He–Ne system.

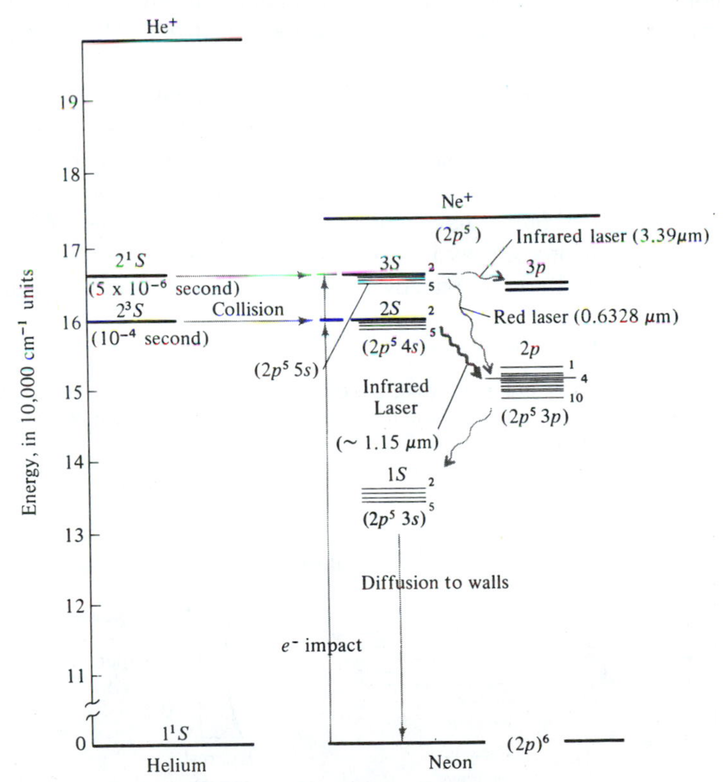

Figure 7-15 He–Ne energy levels. The dominant excitation paths for the red and infrared laser transitions are shown. (After Reference [11].)

1. *The 0.6328 μm oscillation.* The upper level is one of the Ne 3*S* levels, whereas the terminal level belongs to the 2*p* group. The terminal (2*p*) level decays radiatively with a time constant of about 10^{-8} second into the long-lived 1*S* state. This time is much shorter than the 10^{-7} second lifetime of the upper laser level 3*S*. The condition $t_1 < t_2$ for population inversion in the 3*S*-2*p* transition (see Section 6.4) is thus fulfilled.

   Another important point involves the level 1*S*. Because of its long life it tends to collect atoms reaching it by radiative decay from the lower laser level 2*p*. Atoms in 1*S* collide with discharge electrons and are excited back into the lower laser level 2*p*. This reduces the inversion. Atoms in the 1*S* states relax back to the ground state mostly in collisions with the wall of the discharge tube. For this reason the gain in the 0.6328 μm transition is found to increase with decreasing tube diameter.
2. *The 1.15 μm oscillation.* The upper laser level 2*S* is pumped by resonant (that is, energy-conserving) collisions with the metastable 2^3S He level. It uses the same lower level as the 0.6328 μm transition and, consequently, also depends on wall collisions to depopulate the 1*S* Ne level.
3. *The 3.39 μm oscillation.* This involves a 3*S*–3*p* transition and thus uses the same upper level as the 0.6328 μm oscillation. It is remarkable for the fact that it provides a small-signal optical gain of about 50 dB/m.[4] This large gain reflects partly the inverse dependence of γ on ν^2 [see Equation (5.3-3)] as well as the short lifetime of the 3*p* level, which allows the buildup of a large inversion.

   Because of the high gain in this transition, oscillation would normally occur at 3.39 μm rather than at 0.6328 μm. The reason is that the threshold condition will be reached first at 3.39 μm and, once that happens, the gain "clamping" will prevent any further buildup of the population of 3*S*. The 0.6328 μm lasers overcome this problem by introducing into the optical path elements, such as glass or quartz Brewster windows, that absorb strongly at 3.39 μm but not at

[4]This is not the actual gain that exists inside the laser resonator, but the one-pass gain exercised by a very small input wave propagating through the discharge. In the laser the gain per pass is reduced by saturation until it equals the loss per pass.

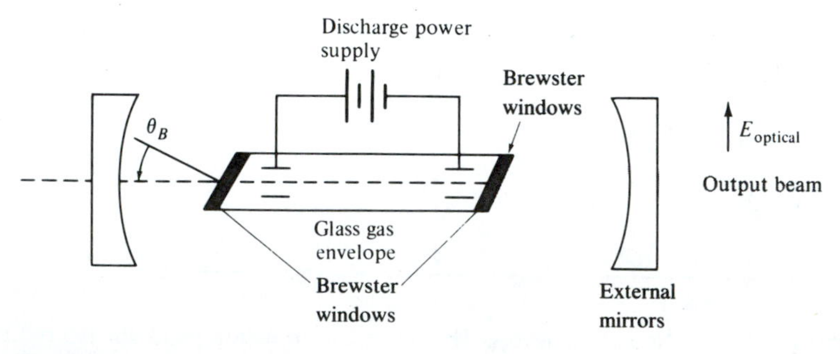

Figure 7-16 Typical gas laser.

0.6328 μm. This raises the threshold pumping level for the 3.39 μm oscillation above that of the 0.6328 μm oscillation.

A typical gas laser setup is illustrated by Figure 7-16. The gas envelope windows are tilted at Brewster's angle θ_B, so radiation with the electric field vector in the plane of the paper suffers no reflection losses at the windows. This causes the output radiation to be polarized in the sense shown, since the orthogonal polarization (the E vector out of the plane of the paper) undergoes reflection losses at the windows and, consequently, has a higher threshold.

7.6 CARBON DIOXIDE LASER

The lasers described so far in this chapter depend on electronic transitions between states in which the electronic orbitals (that is, charge distributions around the atomic nucleus) are different. As an example, consider the red (0.6328 μm) transition in Ne shown in Figure 7-15. It involves levels $2p^5 5s$ and $2p^5 3p$ so that in making a transition from the upper to the lower laser level one of the six outer electrons changes from a hydrogen-like state $5s$ (that is, $n = 5$, $l = 0$) to one in which $n = 3$ and $l = 1$.

The CO_2 laser [12] is representative of the so-called molecular lasers in which the energy levels of concern involve the internal vibration of the molecules—that is, the relative motion of the constituent atoms. The atomic electrons remain in their lowest energetic states and their degree of excitation is not affected.

As an illustration, consider the simple case of the nitrogen molecule. The molecular vibration involves the relative motion of the two atoms with respect to each other. This vibration takes place at a characteristic frequency of $\nu_0 = 2326 \text{ cm}^{-1}$, which depends on the molecular mass as well as the elastic restoring force between the atoms [13]. According to basic quantum mechanics, the degrees of vibrational excitation are discrete (that is, quantized) and the energy of the molecule can take on the values $h\nu_0(v + \frac{1}{2})$, where $v = 0, 1, 2, 3, \ldots$. The energy-level diagram of N_2 (in its lowest electronic state) would then ideally consist of an equally spaced set of levels with a spacing of $h\nu_0$. The ground state ($v = 0$) and the first excited state ($v = 1$) are shown on the right side of Figure 7-17.

The CO_2 molecule presents a more complicated case. Since it consists of three atoms, it can execute three basic internal vibrations, the so-called normal modes of vibration. These are shown in Figure 7-18. In (a) the molecule is at rest. In (b) the atoms vibrate along the internuclear axis in a symmetric manner. In (c) the molecules vibrate symmetrically along an axis perpendicular to the internuclear axis—the bending mode. In (d) the atoms vibrate asymmetrically along the internuclear axis. The mode is referred to as the asymmetric stretching mode. In the first approximation one can assume that the three normal modes are independent of each other, so the state of the CO_2 molecule can be described by a set of three integers (v_1, v_2, v_3) that correspond respectively to the degree of excitation of the three modes described. The total energy of the molecule is thus

$$E(v_1, v_2, v_3) = h\nu_1(v_1 + \tfrac{1}{2}) + h\nu_2(v_2 + \tfrac{1}{2}) + h\nu_3(v_3 + \tfrac{1}{2}) \tag{7.6-1}$$

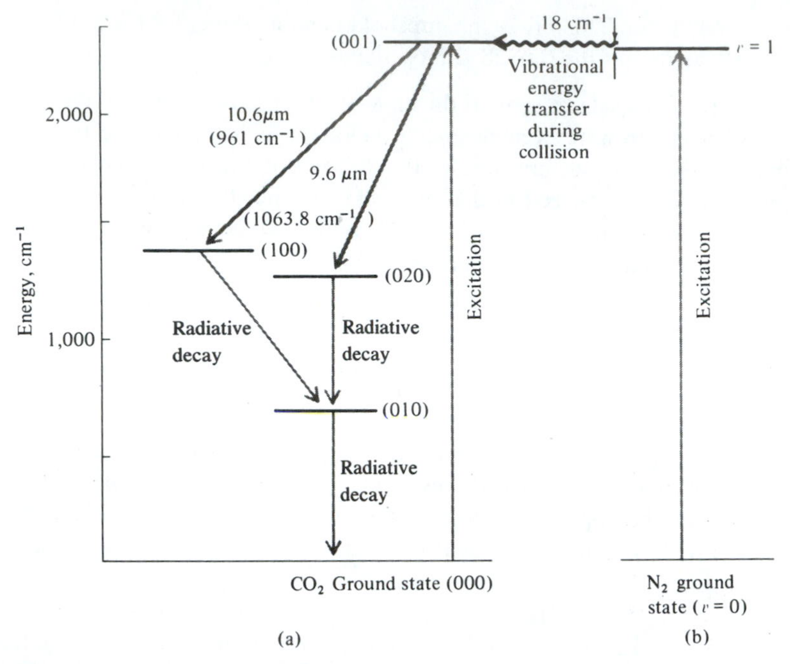

Figure 7-17 (a) Some of the low-lying vibrational levels of the carbon dioxide (CO_2) molecule, including the upper and lower levels for the 10.6 μm and 9.6 μm laser transitions. (b) Ground state ($v = 0$) and first excited state ($v = 1$) of the nitrogen molecule, which plays an important role in the selective excitation of the (001) CO_2 level.

where ν_1, ν_2, ν_3 are the frequencies of the symmetric stretch, bending, and asymmetric stretch modes, respectively.

Some of the low vibrational levels of CO_2 are shown in Figure 7-17. The upper laser level (001) is thus one in which only the asymmetric stretch mode, Figure 7-18(d), is excited and contains a single quantum $h\nu_3$ of energy.

The laser transition at 10.6 μm takes place between the (001) and (100) levels of CO_2. The excitation is provided usually in a plasma discharge that, in addition to CO_2, typically contains N_2 and He. The CO_2 laser possesses a high overall working efficiency of about 30 percent. This efficiency results primarily from three factors: (a) The laser levels are all near the ground state, and the atomic quantum efficiency ν_{21}/ν_{30}, which was discussed in Section 7.1, is about 45 percent; (b) a large fraction of the CO_2 molecules excited by electron impact cascade down the energy ladder from their original level of excitation and tend to collect in the long-lived (001) level; (c) a very large fraction of the N_2 molecules that are excited by the discharge tend to collect in the $v = 1$ level. Collisions with ground-state CO_2 molecules result in transferring their excitation to the latter, thereby exciting them to the (001) state as shown in Figure 7-17. The slight deficiency in energy (about 18 cm^{-1}) is made up by a decrease of the total kinetic energy of the molecules following the collision. This collision can be represented by

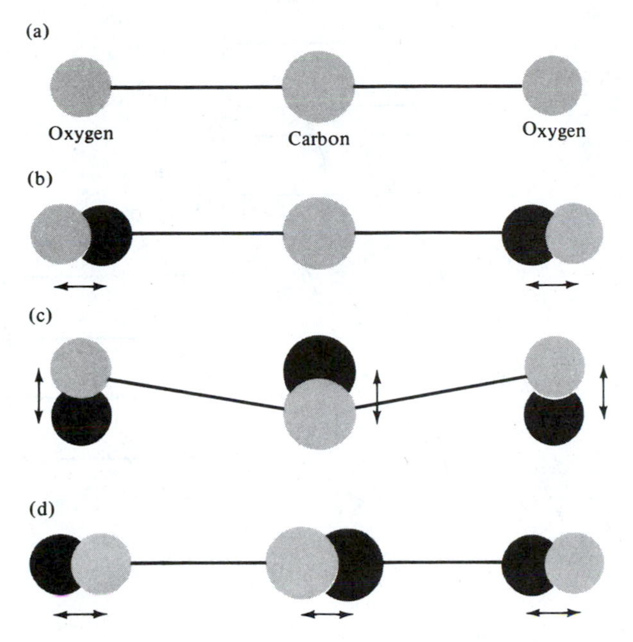

Figure 7-18 (a) Unexcited CO_2 molecule. (b), (c), and (d) The three normal modes of vibration of the CO_2 molecule. (After Reference [14].)

$$(v = 1) + (000) + \text{K.E.} = (v = 0) + (001) \tag{7.6-2}$$

and has a sufficiently high cross section that at the pressures and temperatures involved in the operation of a CO_2 laser most of the N_2 molecules in the $v = 1$ lose their excitation energy by this process.[5]

Carbon dioxide lasers are not only efficient but can emit large amounts of power. Laboratory-size lasers with discharge envelopes of a few feet in length can yield an output of a few kilowatts. This is due not only to the very *selective* excitation of the low-lying upper laser level, but also to the fact that once a molecule is stimulated to emit a photon it returns quickly to the ground state, where it can be used again. This is accomplished mostly through collisions with other molecules—such as that of He, which is added to the gas mixture.

7.7 Ar$^+$ LASER

Transitions between highly excited states of the singly ionized argon atom can be used to obtain oscillation at a number of visible (or near visible) wavelengths be-

[5]The cross section σ was defined in Section 7.2. In the present context it follows directly from the definition that the number of collisions of the type described by (7.6-2) per unit volume per unit time is equal to $N(v = 1)N(000)\sigma\bar{v}$ where $N(v = 1)$ and $N(000)$ are the densities of molecules in the states $v = 1$ of N_2 and (000) of CO_2, respectively. $\bar{v}$ is the (mean) relative velocity of the colliding molecules.

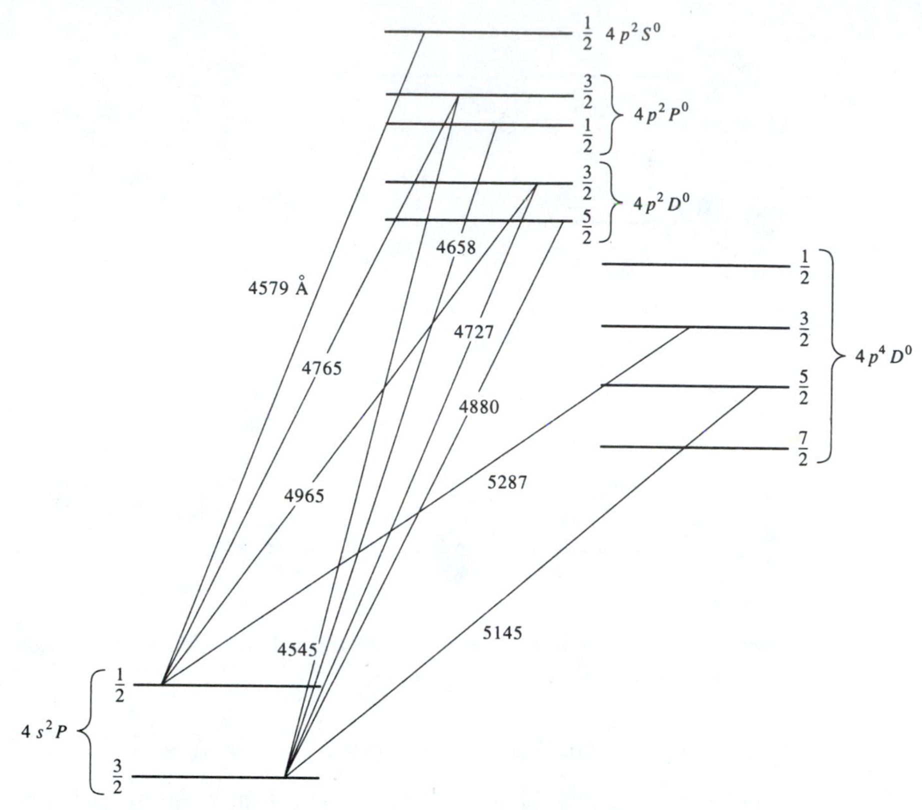

Figure 7-19 Energy levels of the $4p \rightarrow 4s$ Ar II laser transitions. (After Reference [15] with a correction supplied by the author.)

tween 0.35 and 0.52 μm; see References [15, 16]. The Ar^+ laser is consequently one of the most important lasers in use today. The pertinent energy level scheme is shown in Figure 7-19. The most prominent transition is the one at 4880 Å.

The Ar^+ laser can be operated in a pure Ar discharge that contains no other gases. The excitation mechanism involves collisions with energetic (~4–5 eV) electrons. Since the mean electron energy is small compared to the energy of the upper laser level (~20 eV above the ground state of the ion), it is clear that pumping is achieved by multiple collisions of Ar^+ ground state ions with electrons followed by a number of cascading paths. The details of the collision and cascading processes are not clearly understood.

7.8 EXCIMER LASERS

The term *excimer* was introduced originally [17, 20] to describe a homopolar dimer such as Xe_2 or Hg_2 that is bound (i.e., the atoms are attracted to each other thus forming a stable molecule) in an excited state, but which dissociates in its ground state. The term *exciplex* was used to describe heteropolar cases such as XeF, where the constituent atoms are different.

The distinction between the two terms has been lost to a large extent, and bowing to popular usage we will refer to exciplex molecules as excimers.

The interest in excimer lasers principally of heavy noble gases (Xe, Kr, Ar) and the halogens (F, Cl, Br, I) is due to the relatively efficient production of their excited state by electron beam collisions and the fact that their emission wavelengths lie in the ultraviolet and vacuum ultraviolet ($0.2 < \lambda < 0.4\ \mu$m) region of the spectrum, a region not covered well by other types of lasers. In what follows, we will limit our discussion to the noble gas halide lasers [20] which have, to date, yielded the best laser performance.

When the ionization energy (energy to remove an electron from the outermost shell) of an atom A is less than the sum of the electron affinity (energy released during electron attachment) of X, plus the electrostatic attraction energy between A^+ and X^-, the process of forming the ionic molecule (A^+X^-) via the process (A + X) $\rightarrow$ (A^+X^-) is exothermic (energy is released) and is favored. In the case of KrF, for example, this is when Kr is in an excited state (Kr*), since the ionization energy of Kr* is less than that of Kr. The ionization energies of Ar and Kr, as an example, are 15.68 and 13.93 eV, respectively, while in the excited state it is ~5 eV. The electron affinity of Cl is ~3.75 eV, while the repulsive energy is ~1 eV. The Coulomb attraction is ~8 eV. It follows that the process of forming KrCl (starting with Kr and Cl in their ground state) is endothermic and requires the investment of ~4 eV per molecule, while if we start with excited Kr (Kr*) the process is exothermic and releases some 6 eV per molecule. To summarize:

$$A + X \longrightarrow (A^+X^-) - 4\ \text{eV}$$

$$A^* + X \longrightarrow (A^+X^-)^* + 6\ \text{eV}$$

Typical generalized potential curves for an excimer molecule AX in its excited state (AX)* and ground state AX are shown in Figure 7-20.

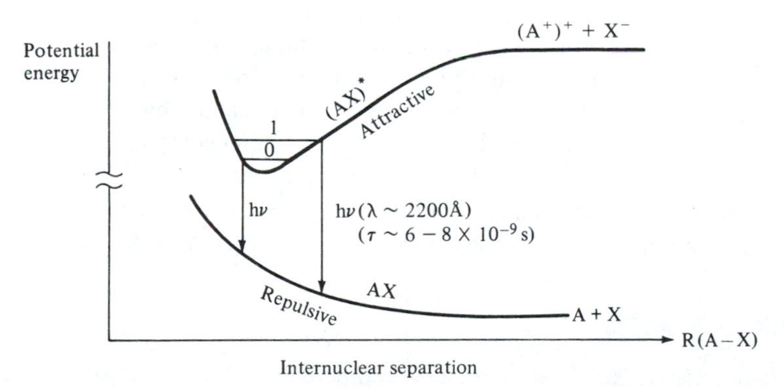

Figure 7-20 Potential energy curves for the ground state and an excited state of a typical nobel gas (A)–halide (X) molecule. The excited state is bound (i.e., possesses a minimum), while the ground state is repulsive. The two lowest vibrational states of the excimer state (AX)* are shown.

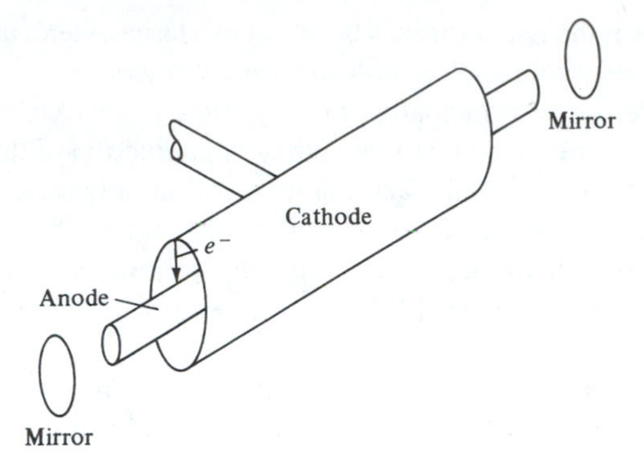

Figure 7-21 An excimer laser with a coaxial e-beam exciting geometry. (After Reference [20].)

The lifetime of the upper bound state in KrF is ~6–10 ns, while that of the lower (repulsive) state is $\sim 10^{-13}$ s. This results in broad spontaneous emission with typical widths of 200–400 cm^{-1}. The emission wavelength of a laser that uses the excimer as the gain medium can be tuned over most of the region spanned by the spontaneous emission.

The noble gas lasers offer the possibility of generating and amplifying the pulses (2–8 ns) to very high (~50 kJ) energies. The lasers are relatively efficient when pumped by energetic electron beams (e-beam). High-current (1–10 kA) high-voltage (0.25–2 MeV) beams are utilized. A typical e-beam excited KrF excimer laser configuration is shown in Figure 7-21. The basic kinetics of the $Ar/Kr/F_2$ gas mixture used in KrF has been elucidated by Rokni et al. [18]. The main effect of the pumping e-beam is to form Kr_2^+ and F^-. This is followed by the ionic recombination reaction [19]

$$Kr_2^+ + F^- + Ar \longrightarrow (KrF)^* + Kr + Ar$$

The excited molecules KrF* thus form the inverted population gain medium.

Due to the shortness of the excited state lifetime ($\sim 6-8 \times 10^{-9}$ s), the excimer laser is used mostly for amplifying short pulses. In a gas mixture of 93.5% Ar, 6% Kr, and 0.3% F_2 with a total pressure of 1 atm excited by a high-energy electron beam, as much as 30 percent of the pumping beam energy results in the production of KrF* excited-state molecules. Total amplified output of ~100 kJ and overall wall plug to optical output amplifier efficiencies of ~8 percent appear feasible [19]. This is the main interest for the current high level of activity in this laser system.

7.9 ORGANIC-DYE LASERS

Many organic dyes (that is, organic compounds that absorb strongly in certain visible-wavelength regions) also exhibit efficient luminescence, which often spans a large wavelength region in the visible portion of the spectrum. This last property makes it possible to obtain an appreciable tuning range from dye lasers; see References [21–30].

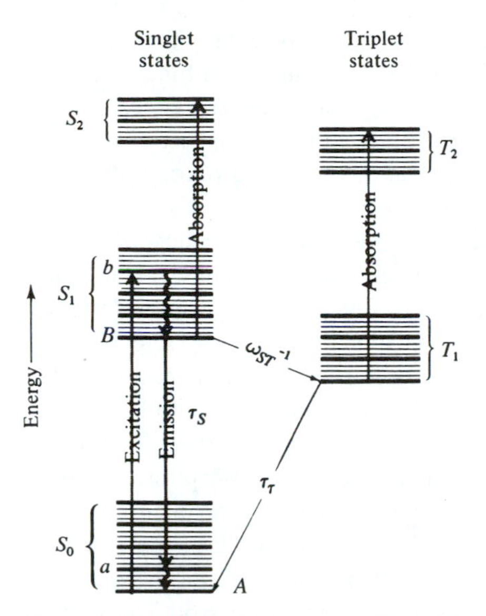

Figure 7-22 Schematic representation of the energy levels of an organic dye molecule. The heavy horizontal lines represent vibrational states and the lighter lines represent the rotational fine structure. Excitation and laser emission are represented by the transitions $A \rightarrow b$ and $B \rightarrow a$, respectively.

A schematic representation of an organic dye molecule (such as rhodamine 6G, for example) is shown in Figure 7-22.

State S_0 is the ground state. S_1, S_2, T_1, and T_2 are excited electronic states—that is, states in which one ground-state electron is elevated to an excited orbit. Typical energy separation, such as S_1–S_0 is about 20,000 cm^{-1}. In a singlet (S) state, the magnetic spin of the excited electron is antiparallel to the spin of the remaining molecule. In a triplet (T) state, the spins are parallel. Singlet → triplet, or triplet → singlet transitions thus involve a spin flip and are far less likely than transitions between two singlet or between two triplet states.

Transitions between two singlet states or between two triplet states, which are spin-allowed (that is, they do not involve a spin flip), give rise to intense absorption and fluorescence. The characteristic color of organic dyes is due to the $S_0 \rightarrow S_1$ absorption.

The singlet and triplet states, in turn, are split further into vibrational levels shown as heavy horizontal lines in Figure 7-22. These correspond to the quantized vibrational states of the organic molecule, as discussed in detail in Section 7.6. Typical energy separation between two adjacent vibrational levels within a given singlet or triplet state is about 1500 cm^{-1}. The fine splitting shown corresponds to rotational levels whose spacing is about 15 cm^{-1}.[6]

[6]A transition between two adjacent rotational levels involves a change in the total angular momentum of the molecule about some axis.

In the process of pumping the laser, the molecule is first excited, by absorbing a pump photon, into a rotational–vibrational state b within S_1. This is followed by a very fast decay to the bottom of the S_1 group, with the excess energy taken up by the vibrational and rotational energy of the molecules. Most of the excited molecules will then decay spontaneously to state a, emitting a photon of energy $\nu = (E_B - E_a)/h$. The lifetime for this process is τ_S.

There is, however, a small probability, approximately $\omega_{ST}\tau_S$, that an excited molecule will decay instead to the triplet state T_1, where ω_{ST} is the rate per molecule for undergoing an $S_1 \rightarrow T_1$ transition. Since this is a spin-forbidden transition, its rate is usually much smaller than the spontaneous decay rate τ_S^{-1}, so that $\omega_{ST}\tau_S \ll 1$. The lifetime τ_T for decay of T_1 to the ground state is relatively long (since this too is a spin-forbidden transition) and may vary from 10^{-7} to 10^{-3} second, depending on the experimental conditions [24]. Owing to its relatively long lifetime, the triplet state T_1 acts as a trap for excited molecules. The absorption of molecules due to a $T_1 \rightarrow T_2$ transition is spin-allowed and is therefore very strong. If the wavelength region of this absorption coincides with that of the laser emission [at $\nu \simeq (E_B - E_a)/h$], an accumulation of molecules in T_1 increases the laser losses and at some critical value quenches the laser oscillation. For this reason, many organic-dye lasers operate only on a pulsed basis. In these cases fast-rise-time pump pulses—often derived from another laser [22]—cause a buildup of the S_1 population with oscillation taking place until an appreciable buildup of the T_1 population occurs.

Another basic property of molecules is that the peak of the absorption spectrum usually occurs at shorter wavelengths than the peak of the corresponding emission spectrum. This is illustrated in Figure 7-23, which shows the absorption and emission spectra of rhodamine 6G, which when dissolved in H_2O is used as a CW laser medium [26]. Laser oscillation occurring near the peak of the emission curve is thus absorbed weakly. But for this fortunate circumstance, laser action involving electronic transitions in molecules would not be possible.

Typical excitation and oscillation waveforms of a dye laser are shown in Figure 7-24. The possibility of quenching the laser action by triplet state absorption is evident.

A list of some common laser dyes is given in Table 7-1.

The broad fluorescence spectrum of the organic dyes suggests a broad tunability range for lasers using them as the active material. The spectrum in Figure 7-23, as an example, corresponds to a width of $\Delta\nu \simeq 1000\ \text{cm}^{-1}$. One elegant solution for realizing this tuning range [25] consists of replacing one of the laser mirrors with a diffraction grating, as shown in Figure 7-25. A diffraction grating has the property that (for a given order) an incident beam will be reflected back *exactly* along the direction of incidence, provided

$$2d \cos \theta = m\lambda \qquad m = 1, 2, \ldots \tag{7.9-1}$$

where d is the ruling distance, θ is the angle between the propagation direction and its projection on the grating surface, λ is the optical wavelength in the medium next to the grating, and m is the order of diffraction. This type of operation of a grating

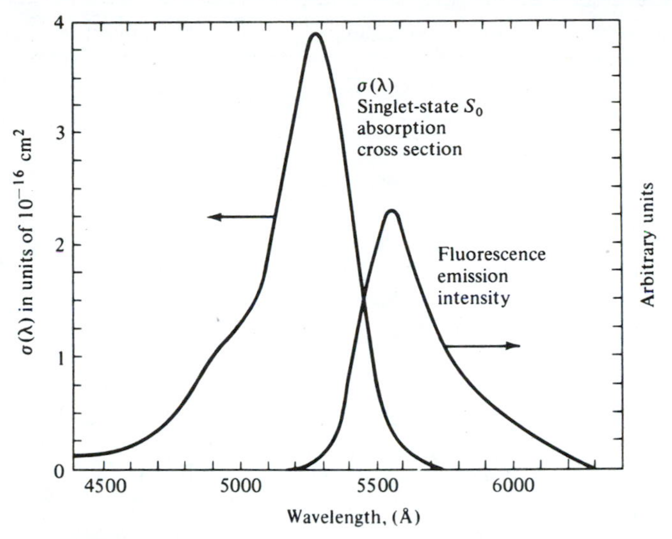

Figure 7-23 Singlet-state absorption and fluorescence spectra of rhodamine 6G obtained from measurements with a 10^{-4} molar ethanol solution of the dye. (After Reference [24].)

is usually referred to as the Littrow arrangement. When a grating is used as one of the laser mirrors, it is clear that the oscillation wavelength will be that which satisfies (7.9-1), since other wavelengths are not reflected along the axis of the optical resonator and will consequently "see" a very lossy (low-Q) resonator. The tuning (wavelength selection) is thus achieved by a rotation of the grating. It follows also that any other means of introducing a controlled, wavelength-dependent loss into the optical resonator can be used for tuning the output.

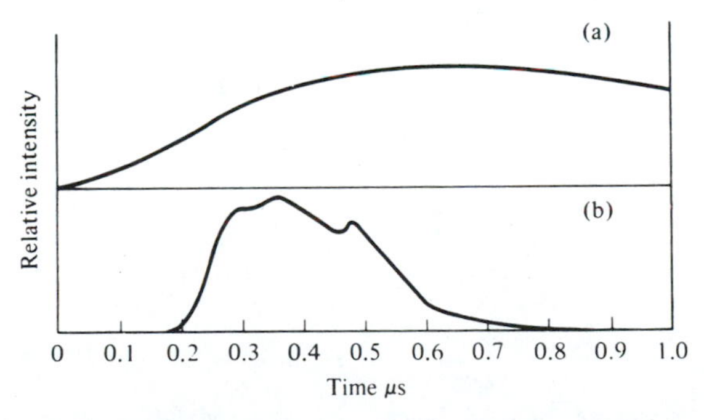

Figure 7-24 (a) Flashlamp pulse produced by a linear xenon flashlamp in a low-inductance circuit. (b) Laser pulse from a 10^{-3} molar solution of rhodamine 6G in methanol. (After Reference [23].)

Table 7-1 Molecular Structure, Laser Wavelength, and Solvents for Some Laser Dyes (After Reference [24].)

Dye	Structure	Solvent	Wavelength
Acridine red	$(H_3C)NH$, O, $\overset{+}{N}H(CH_3)$ Cl^-, H	EtOH	Red 600–630 nm
Puronin B	$(C_2H_5)_2N$, O, $\overset{+}{N}H(C_2H_5)_2$ Cl^-, H	MeOH H_2O	Yellow
Rhodamine 6G	C_2H_5HN, O, $\overset{+}{N}HC_2H_5$ Cl^-, H_3C, CH_3, $COOC_2H_5$	EtOH MeOH H_2O DMSO Polymethyl-methacrylate	Yellow 570–610 nm
Rhodamine B	$(C_2H_5)_2N$, O, $\overset{+}{N}(C_2H_5)_2$ Cl^-, COOH	EtOH MeOH Polymethyl-methacrylate	Red 605–635 nm
Na-fluorescein	NaO, O, O, COONa	EtOH H_2O	Green 530–560 nm
2,7-Dichloro-fluorescein	HO, O, O, Cl, Cl, COOH	EtOH	Green 530–560 nm
7-Hydroxycoumarin	O, O, OH	H_2O (pH ~ 9)	Blue 450–470 nm
4-Methylumbelli-ferone	O, O, OH, CH_3	H_2O (pH ~ 9)	Blue 450–470 nm
Esculin	O, O, OH, O, H OH H H, HC–C–C–C–C–CH_2OH, OH H OH, O	H_2O (pH ~ 9)	Blue 450–470 nm
7-Diethylamino-4-Methylcoumarin	O, O, N, C_2H_5, C_2H_5, CH_5	EtOH	Blue
Acetamidopyrene-trisulfonate	NaO_3S, SO_3Na, NaO_3S, N, $COCH_3$, H	MeOH H_2O	Green-yellow
Pyrylium salt	BF_4, H_3CO, O, +, OCH_3	MeOH	Green

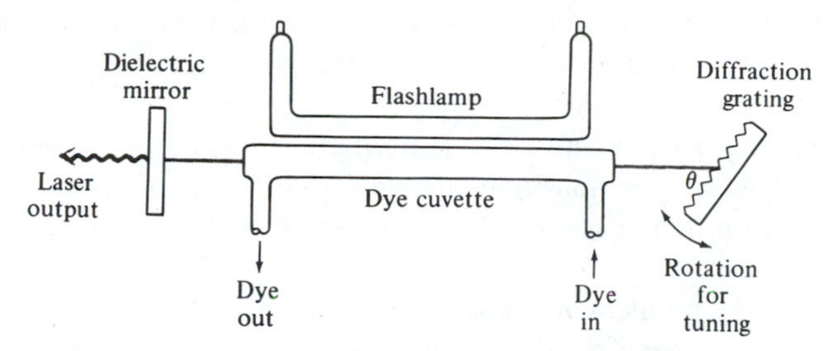

Figure 7-25 A typical pulsed dye laser experimental setup employing a linear flashlamp and a wavelength-selecting diffraction-grating reflector.

7.10 HIGH-PRESSURE OPERATION OF GAS LASERS

Consider a laser medium with an inversion density of ΔN atoms/m^3 at some transition with energy spacing near $h\nu_0$. If this medium is to be used as an amplifier of a pulsed signal at ν_0, then the maximum energy that can be extracted by the signal, through stimulated emission, is $\sim\Delta N h\nu_0$ joules per unit volume of the laser medium. It would follow straightforwardly that to increase the energy gain (= energy out/energy in) of the amplifier we need to increase the inversion density ΔN which, according to (6.4-5), can be done by stronger pumping.

Unfortunately, a mere increase in the pumping strength will increase, according to (5.6-10), the unsaturated gain $\gamma_0(\nu_0)$ of the medium, which will lead at some point to parasitic oscillation off spurious reflections or to energy depletion by amplification of the spontaneous emission [27].

One way around this problem in gas lasers is to increase the density (and pressure) of the amplifying medium. The increase in molecular density causes a proportionate decrease in molecular collision time τ which, according to (5.1-8), causes the transition linewidth $\Delta\nu$ to increase. At a given inversion, this would cause, according to (5.6-10), a reduction in the gain [recall here that $g(\nu_0) = (\Delta\nu)^{-1}$]. Alternatively, if the maximum tolerable gain is $\gamma_{\max}$, the reduction in gain due to increased pressure makes it possible to increase the inversion ΔN (by increased pumping) relative to its low-pressure value, until the maximum allowable gain $\gamma_{\max}$ is achieved. This, as discussed above, leads to increased stored energy density that can be "milked" by the signal pulse.

Let us look, somewhat more formally, at the problem of operating a continuous gas laser oscillator at increased pressures. Much of the work in this field was done on CO_2 lasers so that the following discussion will refer to this particular system, although the considerations are quite general.

The transition linewidth of the mixture of CO_2 and other gases used in CO_2 lasers can be written according to (5.1-8) as

$$\Delta\nu = \Delta\nu_D + \sum_i \frac{1}{\pi\tau_i} \tag{7.10-1}$$

where $\Delta\nu_D$ is the Doppler linewidth (5.1-15) and τ_i is the mean collision lifetime of a CO_2 molecule with a molecule of the ith molecular species (N_2, He, and so on) present in the mixture.

For a large range of pressures, τ_i^{-1} is proportional to the pressure [28] so that once $\Sigma_i(\pi\tau_i)^{-1} > \Delta\nu_D$, the transition linewidth $\Delta\nu$ is essentially proportional to pressure. This region is referred to as the *pressure-broadened regime* and is illustrated by Figure 7-26.

Consider now the problem of maintaining the laser oscillation in a high-pressure discharge. First, to achieve a given gain (that is equal to the resonator loss) we need, according to (5.6-10), to increase the inversion density by an amount proportional to the pressure P in order to compensate for the increase of $\Delta\nu$.[7] Second, since the lifetime t_2 in the upper laser level varies as P^{-1}, the pumping power per molecule increases, according to (6.3-4), as P. The result is that the pumping power, for a given gain, increases as P^2. It follows that the output power, along with the excitation power, increases with P^2. This conclusion follows more formally, from (6.5-10), for the power output

$$P_0 = \frac{8\pi n^2 h\nu\Delta\nu A}{\lambda^2(t_2/t_{\text{spont}})} T\left(\frac{g_0}{L_i + T} - 1\right)$$

since $\Delta\nu \propto P$ and $t_2 \propto 1/P$.

The increase of power with pressure is seen in Figure 7-27. The roll-off near $P = 150$ torr reflects the reduction in gain at the higher pressures. A more fundamental measure of the pressure effects is the variation of the saturation intensity (5.6-9)

$$I_s = \frac{8\pi n^2 \Delta\nu h\nu}{(t_2/t_{\text{spont}})\lambda^2}$$

that, for the reasons given above, should increase as P^2. Experimental data of I_s versus P is shown in Figure 7-28.

[7]Recall here that in (5.6-10) $g(\nu_0) = (\Delta\nu)^{-1}$.

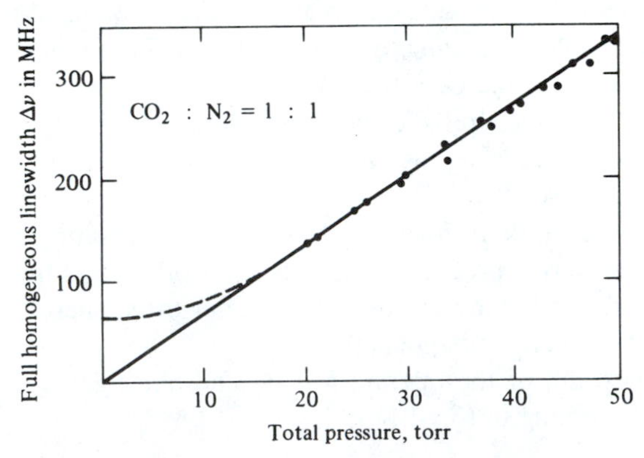

Figure 7-26 The 10.6 μm transition linewidth versus pressure for a gas mixture with equal partial pressures of CO_2 and N_2 at 300 K.

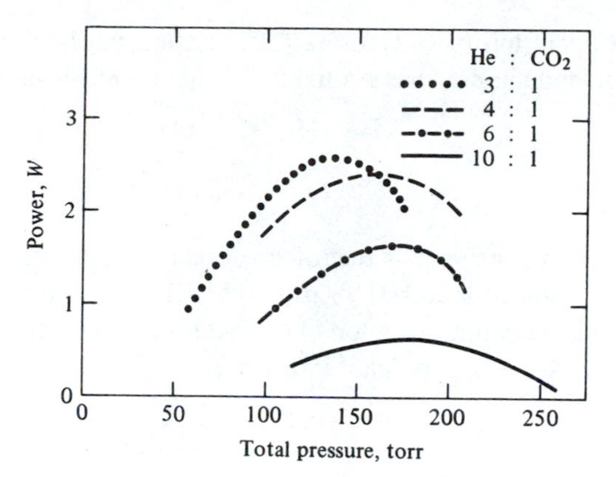

Figure 7-27 Output power versus total pressure under optimum pumping for He:CO_2 mixtures. (After Reference [29].)

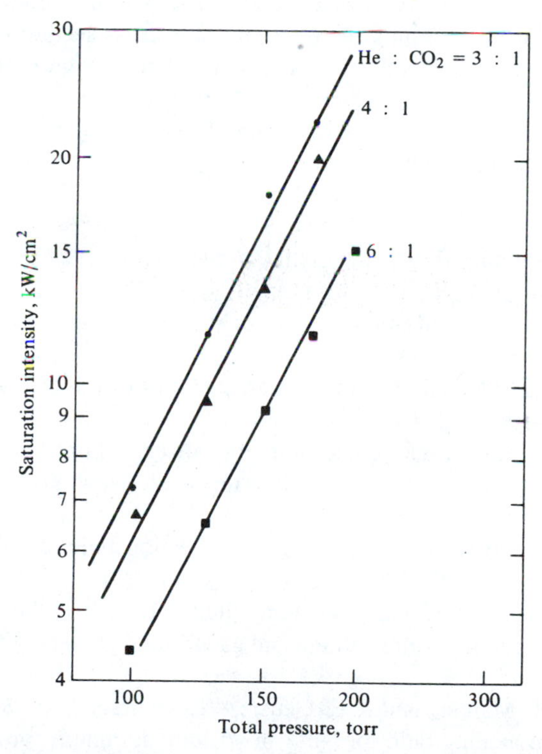

Figure 7-28 Measured saturation intensity versus pressure in a CO_2 laser. (After Reference [29].)

High-energy pulsed operation of CO_2 lasers [30] at atmospheric pressure has been responsible for large and simple lasers suitable for many industrial uses.

7.11 THE *Er*-SILICA LASER

One of the most important laser systems is that of *Er*-doped silica fibers [31–36] at $\lambda = 1.55\ \mu m$. Such fibers pumped at $\lambda = 0.98\ \mu m$ or $\lambda = 1.48\ \mu m$ are used as in-line optical amplifiers in optical communication fiber systems and have major system implications [33]. They are discussed in detail in Section 11.11.

Problems

7-1 Derive the expression relating the absorption cross section at ν in a given $a \rightarrow b$ transition to the spontaneous $b \rightarrow a$ lifetime.

7-2 Derive condition (7.9-1) for the Littrow arrangement of a diffraction grating for which the reflection is parallel to the direction of incidence.

7-3

a. Estimate the exponential gain coefficient $\gamma(\nu_0)$ of a 10^{-4} molar solution of rhodamine 6G in ethanol by assuming the peak emission cross section to be comparable to the peak absorption cross section. Use the data of Figure 7-23.
b. Estimate the spontaneous lifetime for an $S_1 \rightarrow S_0$ transition.
c. Estimate the CW pump power threshold assuming 50 percent absorption of pump and 100 percent pumping quantum efficiency.

References

1. Maiman, T. H., "Stimulated optical radiation in ruby masers," *Nature* 187:493, 1960.
2. Maiman, T. H., "Optical and microwave-optical experiments in ruby," *Phys. Rev. Lett.* 4:564, 1960.
3. Cronemeyer, D. C., "Optical absorption characteristics of pink ruby," *J. Opt. Soc. Am.* 56:1703, 1966.
4. Schawlow, A. L., "Fine structure and properties of chromium fluorescence." In *Advances in Quantum Electronics*, J. R. Singer, ed. New York: Columbia University Press, p. 53, 1961.
5. Yariv, A., "Energy and power considerations in injection and optically pumped lasers," *Proc. IEEE* 51:1723, 1963.
6. Geusic, J. E., H. M. Marcos, and L. G. Van Uitert, "Laser oscillations in Nd-doped yttrium aluminum, yttrium gallium and gadolinium garnets," *Appl. Phys. Lett.* 4:182, 1964.
7. Kushida, T., H. M. Marcos, and J. E. Geusic, "Laser transition cross section and fluorescence branching ratio for Nd^{3+} in yttrium aluminum garnet," *Phys. Rev.* 167:1289, 1968.

8. Snitzer, E., and C. G. Young, ''Glass lasers.'' In *Lasers*, vol. 2, A. K. Levine, ed. New York: Marcel Dekker, Inc., p. 191, 1968.
9. Javan, A., W. R. Bennett, Jr., and D. R. Herriott, ''Population inversion and continuous optical maser oscillation in a gas discharge containing a He–Ne mixture,'' *Phys. Rev. Lett.* 6:106, 1961.
10. White, A. D., and J. D. Rigden, ''Simultaneous gas maser action in the visible and infrared,'' *Proc. IRE* 50:2366, 1962.
11. Bennett, W. R., ''Gaseous optical masers,'' *Appl. Opt., Suppl. 1, Optical Masers*, p. 24, 1962.
12. Patel, C. K. N., ''Interpretation of CO_2 optical maser experiments,'' *Phys. Rev. Lett.* 12:588, 1964; also, ''Continuous-wave laser action on vibrational rotational transitions of CO_2,'' *Phys. Rev.* 136:A1187, 1964.
13. Herzberg, G. H., *Spectra of Diatomic Molecules*. Princeton, N.J.: Van Nostrand, 1963.
14. Patel, C. K. N., ''High power CO_2 lasers,'' *Sci. Am.* 219:22, Aug. 1968.
15. Bridges, W. B., ''Laser oscillation in singly ionized argon in the visible spectrum,'' *Appl. Phys. Lett.* 4:128, 1964.
16. Gordon, E. I., E. F. Labuda, and W. B. Bridges, ''Continuous visible laser action in singly ionized argon, krypton and xenon,'' *Appl. Phys. Lett.* 4:178, 1964.
17. Stevens, B., and E. Hutton, *Nature* 186:1045, 1960.
18. Rokni, M., J. Jacob, and J. Mangano, *Phys. Rev.* A16:2216, 1977.
19. Holzrichter, J. F., D. Eimerl, E. V. George, J. B. Trenholme, W. W. Simmons, and J. T. Hunt, ''High Powered Lasers,'' *J. Fusion Energy* 2:5, 1982.
20. Hutchinson, M. H. R., ''Excimers and Excimer Lasers,'' *Appl. Phys.* (Springer-Verlag) 21:95, 1980.
21. Stockman, D. L., W. R. Mallory, and K. F. Tittel, ''Stimulated emission in aromatic organic compounds,'' *Proc. IEEE* 52:318, 1964.
22. Sorokin, P. P., and J. R. Lankard, ''Stimulated emission observed from an organic dye, chloroaluminum phtalocyanine,'' *IBM J. Res. Dev.* 10:162, 1966.
23. Schafer, F. P., W. Schmidt, and J. Volze, ''Organic dye solution laser,'' *Appl. Phys. Lett.* 9:306, 1966.
24. Snavely, B. B., ''Flashlamp-excited dye lasers,'' *Proc. IEEE* 57:1374, 1969.
25. Soffer, B. H., and B. B. McFarland, ''Continuously tunable, narrow band organic dye lasers,'' *Appl. Phys. Lett.* 10:266, 1967.
26. Peterson, O. G., S. A. Tuccio, and B. B. Snavely, ''CW operation of an organic dye laser,'' *Appl. Phys. Lett.* 17:266, 1970.
27. Yariv, A., *Quantum Electronics*, 3d ed. New York: Wiley, 1989.
28. Taylor, R. L., and S. Bitterman, ''Survey of vibrational and relaxation data for processes important in the CO_2–N_2 laser system,'' *Rev. Mod. Phys.* 41:26, 1969.
29. Abrams, R. L., and W. B. Bridges, ''Characteristics of sealed-off waveguide CO_2 lasers,'' *IEEE J. Quant. Elec.* QE-9:940, 1973.
30. Beaulieu, J. A., ''High peak power gas lasers,'' *Proc. IEEE* 59:667, 1971.
31. Simon J. C., ''Semiconductor laser amplifier for single mode optical fiber communications,'' *J. Opt. Commun.* 4:51, 1983.

32. Mears, R. J., L. Reekie, I. M. Jauncey, and D. N. Payne, "Low noise Erbium-doped fiber amplifier operating at 1.54 mm," *Electron. Lett.* 23:1026, 1987.

33. Hagimoto, K. et al., "A 212 km non-repeatered transmission experiment at 1.8 Gb/s using LD pumped Er^{3+}-doped fiber amplifiers in an Im/direct-detection repeater system." In *Proc. Opt. Fiber Conf., Houston, TX,* Postdeadline Paper PD15, 1989.

34. Olshansky, R., "Noise figure for Er-doped optical fibre amplifiers," *Elect. Lett.* 24:1363, 1988.

35. Payne, David N., "Tutorial session abstracts," *Optical Fiber Communication (OFC 1990) Conference*, San Francisco, 1990.

36. See, for example, Eisenstein, G., U. Koren, G. Raybon, T. L. Koch, M. Wiesenfeld, M. Wegener, R. S. Tucker, and B. I. Miller, "Large-signal and small-signal gain characteristics of 1.5 mm quantum well optical amplifiers," *Appl. Phys. Lett.* 56:201, 1990.

Second-Harmonic Generation and Parametric Oscillation

8.0 INTRODUCTION

In Chapter 1 we considered the propagation of electromagnetic radiation in linear media in which the polarization is proportional to the electric field that induces it. In this chapter we consider some of the consequences of the nonlinear dielectric properties of certain classes of crystals in which, in addition to the linear response, a field produces a polarization proportional to the square of the field.

The nonlinear response can give rise to exchange of energy between a number of electromagnetic fields of different frequencies. Two of the most important applications of this phenomenon are: (1) second-harmonic generation in which part of the energy of an optical wave of frequency ω propagating through a crystal is converted to that of a wave at 2ω, and (2) parametric oscillation in which a strong pump wave at ω_3 causes the simultaneous generation in a nonlinear crystal of radiation at ω_1 and ω_2, where $\omega_3 = \omega_1 + \omega_2$. These will be treated in detail in this chapter.

8.1 ON THE PHYSICAL ORIGIN OF NONLINEAR POLARIZATION

The optical polarization of dielectric crystals is due mostly to the outer, loosely bound valence electrons that are displaced by the optical field. Denoting the electron deviation from the equilibrium position by x and the density of electrons by N, the polarization p is given by

$$p(t) = -Nex(t)$$

In symmetric crystals the potential energy of an electron must reflect the crystal symmetry, so that, using a one-dimensional analog, it can be written as

$$V(x) = \frac{m}{2}\,\omega_0^2 x^2 + \frac{m}{4}\,Bx^4 + \cdots \tag{8.1-1}$$

where ω_0^2 and B are constants[1] and m is the electron mass. Because of the symmetry $V(x)$ contains only even powers of x, so $V(-x) = V(x)$. The restoring force on an electron is

$$F = -\frac{\partial V}{\partial x} = -m\omega_0^2 x - mBx^3 \tag{8.1-2}$$

and is zero at the equilibrium position $x = 0$.

The linear polarization of crystals in which the polarization is proportional to the electric field is accounted for by the first term in (8.1-1). To see this, consider a "low" frequency electric field $E(t)$—that is, a field whose Fourier components are at frequencies small compared to ω_0. The excursion $x(t)$ caused by this field is found by equating the total force on the electron to zero[2]

$$-eE(t) - m\omega_0^2 x(t) = 0$$

so that

$$x(t) = -\frac{e}{m\omega_0^2}\,E(t) \tag{8.1-3}$$

thus resulting in a polarization $p(t) = -Nex(t)$, which is instantaneously proportional to the field.

Now in an asymmetric crystal in which the condition $V(x) = V(-x)$ is no longer fulfilled, the potential function can contain odd powers of x and thus

$$V(x) = \frac{m\omega_0^2}{2}\,x^2 + \frac{m}{3}\,Dx^3 + \cdots \tag{8.1-4}$$

which corresponds to a restoring force on the electron

$$F = -\frac{\partial V(x)}{\partial x} = -(m\omega_0^2 x + mDx^2 + \cdots) \tag{8.1-5}$$

An examination of (8.1-5) reveals that a positive excursion ($x > 0$) results in a larger restoring force, assuming $D > 0$, than does the same excursion in the opposite direction. It follows immediately that if the electric force on the electron is positive ($E < 0$), the induced polarization is smaller than when the field direction is reversed. This situation is depicted in Figure 8-1.

---

[1]The constant ω_0 corresponds to the resonance frequency of the electronic oscillator.

[2]The "low" frequency assumption makes it possible to neglect the acceleration term $m\, d^2x/dt^2$ in the force equation.

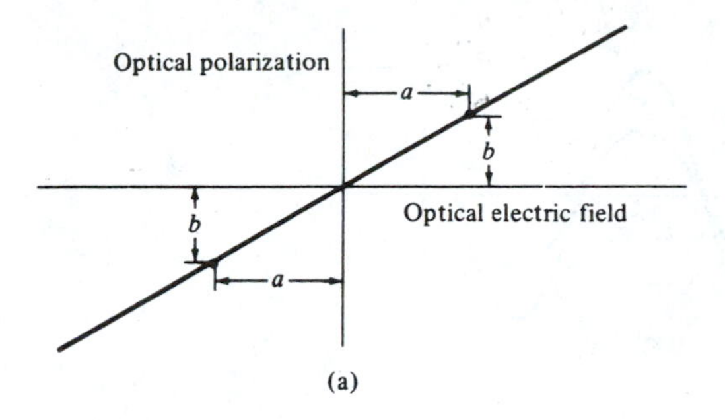

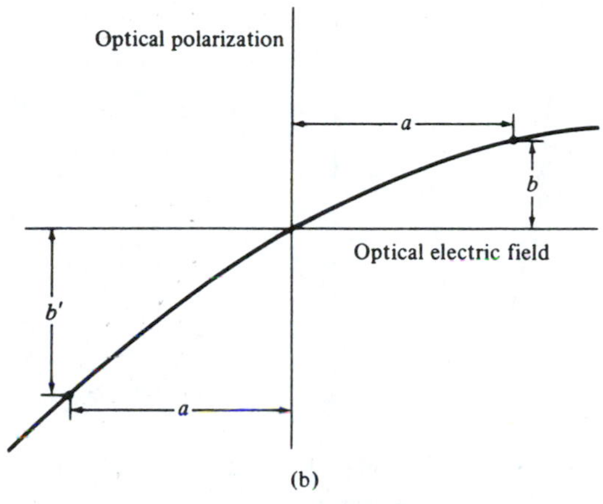

Figure 8-1 Relation between induced polarization and the electric field causing it; (a) in a linear dielectric and (b) in a crystal lacking inversion symmetry.

Next consider an alternating electric field at an (optical) frequency ω applied to the crystal. In a linear crystal the induced polarization will be proportional, at any moment, to the field, resulting in a polarization oscillating at ω as shown in Figure 8-2(a). In a nonlinear crystal we can use Figure 8-1(b) to obtain the induced polarization corresponding to a given field and then plot it (vertically) as in Figure 8-2(b). The result is a polarization wave in which the stiffer restoring force at $x > 0$ results in positive peaks (b), which are smaller than the negative ones (b'). A Fourier analysis of the nonlinear polarization wave in Figure 8-2(b) shows that it contains the second harmonic of ω as well as an average (dc) term. The average, fundamental, and second-harmonic components are plotted in Figure 8-3.

To relate the nonlinear polarization formally to the inducing field, we use Equation (8.1-5) for the restoring force and take the driving electric field as $E^{(\omega)} \cos \omega t$. The equation of motion of the electron $F = m\ddot{x}$ is then

$$\frac{d^2x(t)}{dt^2} + \sigma \frac{dx(t)}{dt} + \omega_0^2 x(t) + Dx^2(t) = -\frac{eE^{(\omega)}}{2m}(e^{i\omega t} + e^{-i\omega t}) \qquad \textbf{(8.1-6)}$$

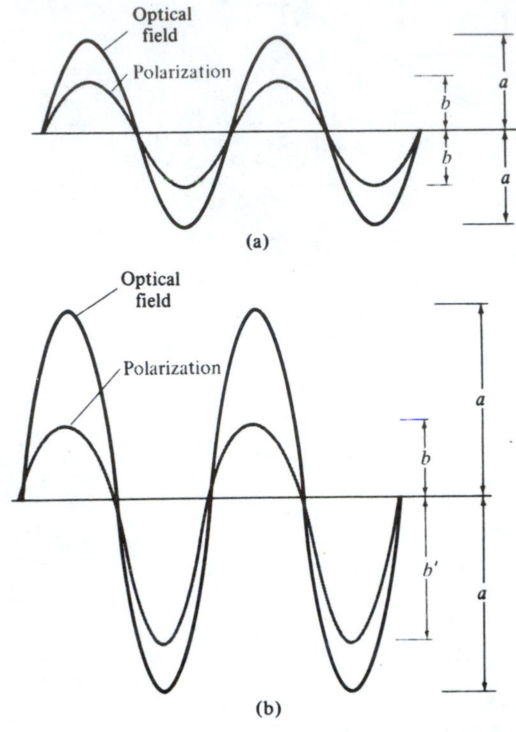

Figure 8-2 An applied sinusoidal electric field and the resulting polarization; (a) in a linear crystal and (b) in a crystal lacking inversion symmetry.

where we account for the losses by a frictional force $-m\sigma\dot{x}$. An inspection of (8.1-6) shows that the term Dx^2 gives rise to a component oscillating at 2ω, so we assume the solution for $x(t)$ in the form[3]

$$x(t) = \tfrac{1}{2}(q_1 e^{i\omega t} + q_2 e^{2i\omega t} + \text{c.c.}) \tag{8.1-7}$$

where c.c. stands for "complex conjugate."

Substituting the last expression into (8.1-6) gives

$$\begin{aligned}
&-\frac{\omega^2}{2}(q_1 e^{i\omega t} + \text{c.c.}) - 2\omega^2(q_2 e^{2i\omega t} + \text{c.c.}) + \frac{i\omega\sigma}{2}(q_1 e^{i\omega t} - \text{c.c.}) \\
&+ i\omega\sigma(q_2 e^{2i\omega t} - \text{c.c.}) + \frac{\omega_0^2}{2}(q_1 e^{i\omega t} + q_2 e^{2i\omega\sigma} + \text{c.c.}) \\
&+ \frac{D}{4}\left(q_1^2 e^{2i\omega t} + q_2^2 e^{4i\omega t} + \frac{q_1 q_1^*}{2} + 2q_1 q_2 e^{3i\omega t}\right. \\
&\left. + 2q_1 q_2^* e^{-i\omega t} + \frac{q_2 q_2^*}{2} + \text{c.c.}\right) = \frac{-eE^{(\omega)}}{2m}(e^{i\omega t} + \text{c.c.})
\end{aligned} \tag{8.1-8}$$

[3]Here we must use the real form of $x(t)$ instead of the complex one since, as discussed in Section 1.1, the differential equation involves x^2.

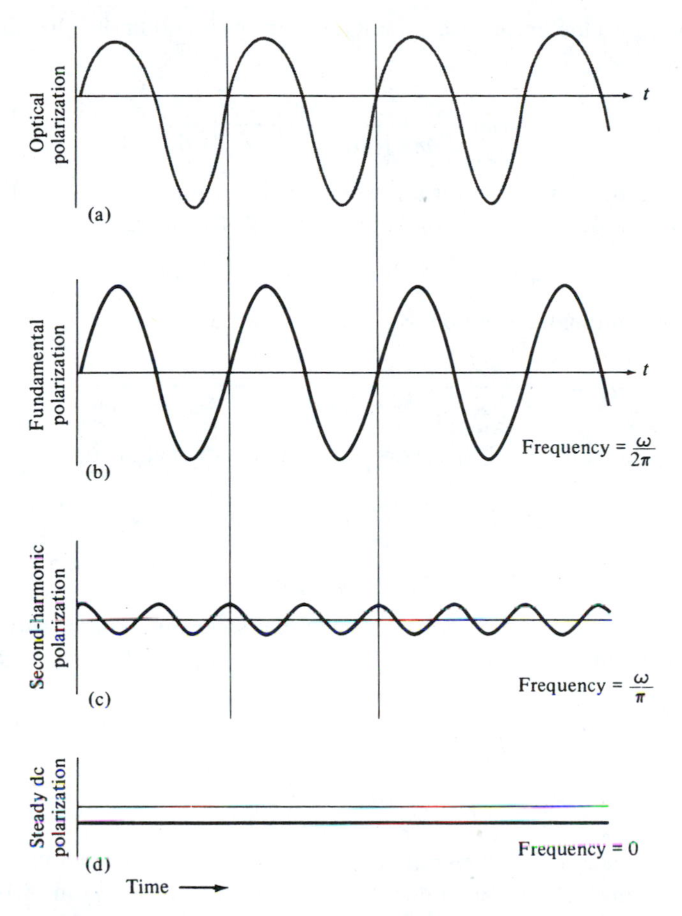

Figure 8-3 Analysis of the nonlinear polarization wave (a) of Figure 8.2 (b) shows that it contains components oscillating at (b) the same frequency (ω) as the wave inducing it, (c) twice that frequency (2ω), and (d) an average (dc) negative component.

If (8.1-8) is to be valid for all times t, the coefficients of $e^{\pm i\omega t}$ and $e^{\pm 2i\omega t}$ on both sides of the equation must be equal. Equating first the coefficients of $e^{i\omega t}$, assuming that $|Dq_2| \ll [(\omega_0^2 - \omega^2)^2 + \omega^2\sigma^2]^{1/2}$, gives

$$q_1 = -\frac{eE^{(\omega)}}{m}\frac{1}{(\omega_0^2 - \omega^2) + i\omega\sigma} \tag{8.1-9}$$

The polarization at ω is related to the electronic deviation at ω by

$$p^{(\omega)}(t) = -\frac{Ne}{2}(q_1 e^{i\omega t} + \text{c.c.})$$

$$\equiv \frac{\epsilon_0}{2}[\chi(\omega)E^{(\omega)}e^{i\omega t} + \text{c.c.}] \tag{8.1-10}$$

where $\chi(\omega)$ is thus the linear susceptibility. By using (8.1-9) in (8.1-10) and solving for $\chi(\omega)$, we obtain

$$\chi(\omega) = \frac{Ne^2}{m\epsilon_0[(\omega_0^2 - \omega^2) + i\omega\sigma]} \tag{8.1-11}$$

We now proceed to solve for the amplitude q_2 of the electronic motion at 2ω. Equating the coefficients of $e^{2i\omega t}$ on both sides of (8.1-8) leads to

$$q_2(-4\omega^2 + 2i\omega\sigma + \omega_0^2) = -\tfrac{1}{2}Dq_1^2$$

and, after substituting the solution (8.1-9) for q_1, we obtain

$$q_2 = \frac{-De^2(E^{(\omega)})^2}{2m^2[(\omega_0^2 - \omega^2) + i\omega\sigma]^2(\omega_0^2 - 4\omega^2 + 2i\omega\sigma)} \tag{8.1-12}$$

In a manner similar to (8.1-10), the nonlinear polarization at 2ω is

$$\begin{aligned} p^{(2\omega)}(t) &= -\frac{Ne}{2}(q_2 e^{2i\omega t} + \text{c.c.}) \\ &\equiv \tfrac{1}{2}\{d^{(2\omega)}[E^{(\omega)}]^2 e^{2i\omega t} + \text{c.c.}\} \end{aligned} \tag{8.1-13}$$

The second of equations (8.1-13) defines the *nonlinear optical coefficient* $d^{(2\omega)}$. If we denote the complex amplitude of the polarization as $P^{(2\omega)}$ we have, from (8.1-13),

$$p^{(2\omega)}(t) = \tfrac{1}{2}[P^{(2\omega)}e^{2i\omega t} + \text{c.c.}]$$

and

$$P^{(2\omega)} = d^{(2\omega)}E^{(\omega)}E^{(\omega)} \tag{8.1-14}$$

that is, $d^{(2\omega)}$ is the ratio of the (complex) amplitude of the polarization at 2ω to the square of the fundamental amplitude. Substituting (8.1-12) for q_2 in (8.1-13), then solving for $d^{(2\omega)}$, results in

$$d^{(2\omega)} = \frac{DNe^3}{2m^2[(\omega_0^2 - \omega^2) + i\omega\sigma]^2(\omega_0^2 - 4\omega^2 + 2i\omega\sigma)} \tag{8.1-15}$$

Using (8.1-11) we can rewrite (8.1-15) as

$$d^{(2\omega)} = \frac{mD[\chi^{(\omega)}]^2\chi^{(2\omega)}\epsilon_0^3}{2N^2e^3} \tag{8.1-16}$$

Equation (8.1-16) is important since it relates the nonlinear optical coefficient d to the linear optical susceptibilities χ and to the anharmonic coefficient D. Estimates based on this relation are quite successful in predicting the size of the coefficient d in a large variety of crystals; see References [1, 2].

Relation (8.1-14) is scalar. In actual crystals we must consider the symmetry so that the second harmonic polarization along, say, the x direction, is related to the electric field at ω by a third rank tensor d_{ijk}.

$$\begin{aligned} P_x^{(2\omega)} &= d_{xxx}^{(2\omega)}E_x^{(\omega)}E_x^{(\omega)} + d_{xyy}^{(2\omega)}E_y^{(\omega)}E_y^{(\omega)} + d_{xzz}^{(2\omega)}E_z^{(\omega)}E_z^{(\omega)} \\ &\quad + 2d_{xzy}^{(2\omega)}E_z^{(\omega)}E_y^{(\omega)} + 2d_{xzx}^{(2\omega)}E_z^{(\omega)}E_x^{(\omega)} + 2d_{xxy}^{(2\omega)}E_x^{(\omega)}E_y^{(\omega)} \end{aligned} \tag{8.1-17}$$

Similar relations give $P_y^{(2\omega)}$ and $P_z^{(2\omega)}$. Considerations of crystal symmetry reduce the number of nonvanishing $d_{ijk}^{(2\omega)}$ coefficients—or, in certain cases to be discussed in the following, cause them to vanish altogether. Table 8-1 lists the nonlinear coefficients of a number of crystals.

Crystals are usually divided into two main groups, depending on whether the

Table 8-1 The Nonlinear Optical Coefficients of a Number of Crystals*

Crystal	$d_{ijk}^{(2\omega)}$ in Units of $1/9 \times 10^{-22}$ MKS
$LiIO_3$	$d_{15} = 4.4$
$NH_4H_2PO_4$	$d_{36} = 0.45$
(ADP)	$d_{14} = 0.50 \pm 0.02$
KH_2PO_4	$d_{36} = 0.45 \pm 0.03$
(KDP)	$d_{14} = 0.35$
KD_2PO_4	$d_{36} = 0.42 \pm 0.02$
	$d_{14} = 0.42 \pm 0.02$
KH_2ASO_4	$d_{36} = 0.48 \pm 0.03$
	$d_{14} = 0.51 \pm 0.03$
Quartz	$d_{11} = 0.37 \pm 0.02$
$AlPO_4$	$d_{11} = 0.38 \pm 0.03$
ZnO	$d_{33} = 6.5 \pm 0.2$
	$d_{31} = 1.95 \pm 0.2$
	$d_{15} = 2.1 \pm 0.2$
CdS	$d_{33} = 28.6 \pm 2$
	$d_{31} = 30 \pm 10$
	$d_{36} = 33$
GaP	$d_{14} = 80 \pm 14$
GaAs	$d_{14} = 72$
$BaTiO_3$	$d_{33} = 6.4 \pm 0.5$
	$d_{31} = 18 \pm 2$
	$d_{15} = 17 \pm 2$
$LiNbO_3$	$d_{15} = 4.4$
	$d_{22} = 2.3 \pm 1.0$
Te	$d_{11} = 517$
Se	$d_{11} = 130 \pm 30$
$Ba_2NaNb_5O_{15}$	$d_{33} = 10.4 \pm 0.7$
	$d_{32} = 7.4 \pm 0.7$
Ag_3AsS_3	$d_{22} = 22.5$
(proustite)	$d_{36} = 13.5$
CdSe	$d_{31} = 22.5 \pm 3$
$CdGeAs_2$	$d_{36} = 363 \pm 70$
$AgGaSe_2$	$d_{36} = 27 \pm 3$
$AgSbS_3$	$d_{36} = 9.5$
ZnS	$d_{36} = 13$

*Some authors define the nonlinear coefficient d by $P = \epsilon_0 dE^2$ rather than by the relation $P = dE^2$ used here.

crystal structure remains unchanged upon inversion (that is, replacing the coordinate **r** by −**r**) or not. Crystals belonging to the first group are called centrosymmetric, whereas crystals of the second group are called noncentrosymmetric [3]. In Figure 8-4 we show the crystal structure of NaCl, a centrosymmetric crystal; an example of a crystal lacking inversion symmetry (noncentrosymmetric) is provided by crystals of the ZnS (zinc blende) class such as GaAs, CdTe, and others. The crystal structure of ZnS is shown in Figure 8-5. The lack of inversion symmetry is evident in the projection of the atomic positions given by Figure 8-6.

In crystals possessing an inversion symmetry, all the nonlinear optical coefficients $d_{ijk}^{(2\omega)}$ must be zero. This follows directly from the relation

$$P_i^{(2\omega)} = \sum_{j,k=x,y,z} d_{ijk}^{(2\omega)} E_j^{(\omega)} E_k^{(\omega)} \tag{8.1-18}$$

which is a compact notation for relation (8.1-17). Let us reverse the direction of the electric field so that in (8.1-18) $E_j^{(\omega)}$ becomes $-E_j^{(\omega)}$ and $E_k^{(\omega)}$ becomes $-E_k^{(\omega)}$. Since the crystal is centrosymmetric, the reversed field "sees" a crystal identical to the original one so that the polarization produced by it must bear the same relationship to the field as originally; that is, the new polarization is $-P_i^{(2\omega)}$. Since the new polarization and the electric field causing it are still related by (8.1-18), we have

$$-P_i^{(2\omega)} = \sum_{j,k} d_{ijk}^{(2\omega)}(-E_j^{(\omega)})(-E_k^{(\omega)}) \tag{8.1-19}$$

Equations (8.1-18) and (8.1-19) can hold simultaneously only if the coefficients $d_{ijk}^{(2\omega)}$ are all zero. We may thus summarize: *In crystals possessing an inversion symmetry there is no second-harmonic generation.*

In the actual practice and design of experiments involving second-harmonic generation or any second-order nonlinear optics in general, it is crucial to take into account the vectorial nature of the interaction and the tensorial aspect of the d_{ijk}

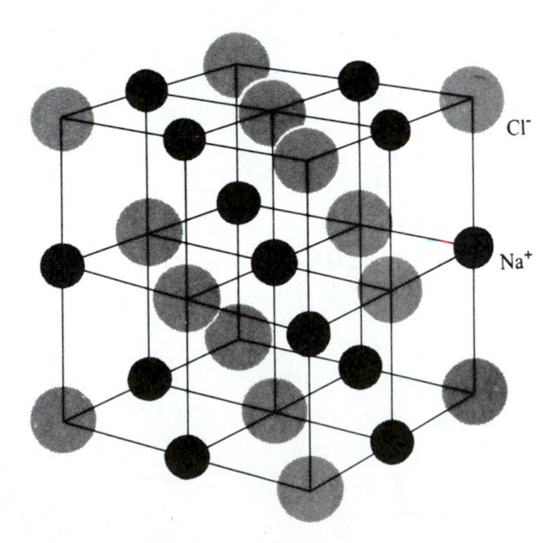

Figure 8-4 The crystal structure of NaCl. The crystal is centrosymmetric, since an inversion of any ion about the central Na^+ ion, as an example, leaves the crystal structure unchanged.

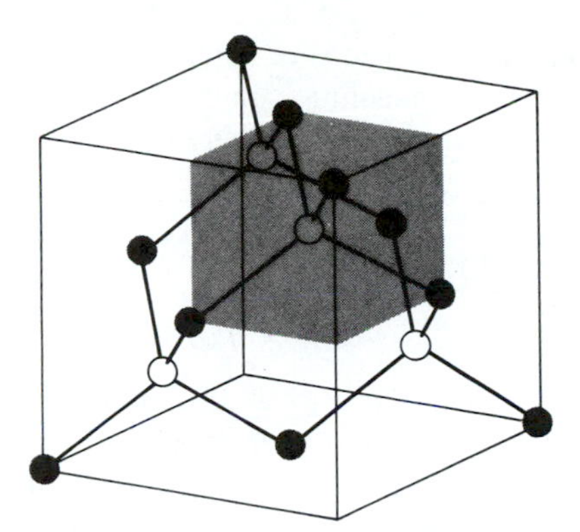

Figure 8-5 The crystal structure of cubic zinc sulfide.

coefficients. A tabulation of the symmetry properties of these coefficients is included in [12] as well as a detailed example of their use in the case of KH_2PO_4. Alternatively we can generate these ''symmetry tables'' by replacing rows by columns in Table 9-1, i.e., by applying the transformation rule $d_{ij} \leftrightarrow r_{ji}$ to generate the 3×6 d_{ij} matrices from the 6×3 r_{ji} matrices. In the following sections we will employ a simplified scalar approach that, although retaining most of the physical considerations, needs to be supplemented in practice by vectorial considerations.

As an example that illustrates this point, consider a second-harmonic generation experiment in KH_2PO_4 (KDP). The incident beam at ω propagates along the crystal

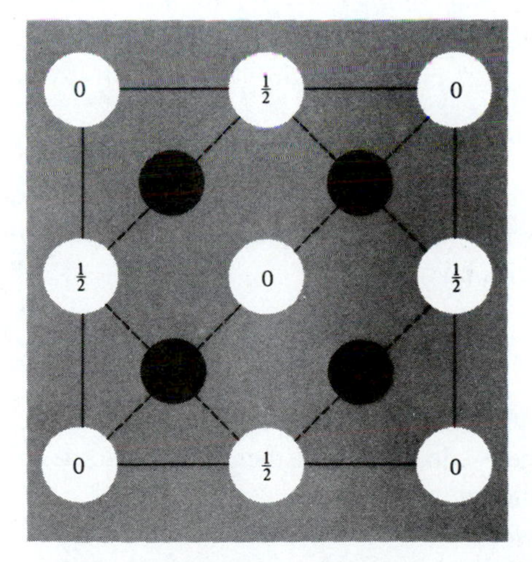

Figure 8-6 The atomic positions in the unit cell of ZnS projected on a cube face. The fractions denote height above base in units of a cube edge. The dark spheres correspond to zinc (or sulfur) atoms and are situated on a face-centered cubic (fcc) lattice, and the white spheres correspond to sulfur (or zinc) atoms and are situated on another fcc lattice displayed by ($\frac{1}{4}$, $\frac{1}{4}$, $\frac{1}{4}$) from the first one. Note the lack of inversion symmetry.

z (optic) axis and is polarized along the x or y axis or some intermediate direction between the two. Since second-harmonic generation in KDP (see Table 16.1 in Reference [12] for class $\bar{4}2$m crystals) is described uniquely by

$$\begin{aligned} P_x^{(2\omega)} &= d_{xyz} E_y^{(\omega)} E_z^{(\omega)} \qquad (d_{xyz} = d_{14}) \\ P_y^{(2\omega)} &= d_{yxz} E_x^{(\omega)} E_z^{(\omega)} \qquad (d_{yxz} = d_{25}) \\ P_z^{(2\omega)} &= d_{zxy} E_x^{(\omega)} E_y^{(\omega)} \qquad (d_{zxy} = d_{36}) \end{aligned} \tag{8.1-19a}$$

our choice of propagation direction is such that $E_z^{(\omega)} = 0$. This results in the production of only $P_z^{(2\omega)}$. This polarization cannot, however, radiate a wave at 2ω propagating along the z axis, since the field E (more exactly D) must be normal to the propagation direction. We thus must choose a propagation direction at some, hopefully large, angle with respect to the z axis. The choice of this direction is dictated by "phase-matching" considerations as discussed in Section 8.3.

It follows, by a direct extension of (8.1-19), that if the optical field at a point $\mathbf{r}$ consists of two beams

$$e(t) = \text{Re}[\mathbf{E}^{(\omega_1)} e^{i\omega_1 t} + \mathbf{E}^{(\omega_2)} e^{i\omega_2 t}] \tag{8.1-20}$$

there is induced in the material a polarization at the sum frequency $\omega_1 + \omega_2$

$$p_i^{(\omega_1+\omega_2)} = \text{Re}[d_{ijk}^{(\omega=\omega_1+\omega_2)} E_j^{(\omega_1)} E_k^{(\omega_2)} e^{i(\omega_1+\omega_2)t}] \tag{8.1-21}$$

as well as at the difference frequency $\omega_1 - \omega_2$

$$p_i^{(\omega_1-\omega_2)} = \text{Re}[d_{ijk}^{(\omega=\omega_1-\omega_2)} E_j^{(\omega_1)} (E_k^{(\omega_2)})^* e^{i(\omega_1-\omega_2)t}] \tag{8.1-22}$$

It follows that the complex amplitude of the induced polarization is related to those of the inducing fields according to

$$\begin{aligned} P_i^{(\omega_1+\omega_2)} &= d_{ijk}^{(\omega=\omega_1+\omega_2)} E_j^{(\omega_1)} E_k^{(\omega_2)} \\ P_i^{(\omega_1-\omega_2)} &= d_{ijk}^{(\omega=\omega_1-\omega_2)} E_j^{(\omega_1)} (E_k^{(\omega_2)})^* \end{aligned} \tag{8.1-23}$$

The material constants $d_{ijk}^{(\omega=\omega_1+\omega_2)}$ and $d_{ijk}^{(\omega=\omega_1-\omega_2)}$ are, in general, not equal to each other since the physical processes contributing to the nonlinear polarization are usually dependent on the frequencies involved.

8.2 FORMALISM OF WAVE PROPAGATION IN NONLINEAR MEDIA

In this section we derive the equations governing the propagation of electromagnetic waves in nonlinear media. These equations will then be used to describe second-harmonic generation and parametric oscillation.

The starting point is Maxwell's equations (1.2-1), (1.2-2):

$$\begin{aligned} \nabla \times \mathbf{h} &= \mathbf{i} + \frac{\partial \mathbf{d}}{\partial t} \\ \nabla \times \mathbf{e} &= -\mu \frac{\partial \mathbf{h}}{\partial t} \end{aligned} \tag{8.2-1}$$

and

$$\mathbf{d} = \epsilon_0 \mathbf{e} + \mathbf{p}$$
$$\mathbf{i} = \sigma \mathbf{e} \tag{8.2-2}$$

where σ is the conductivity. If we separate the total polarization $\mathbf{p}$ into its linear and nonlinear portions according to

$$\mathbf{p} = \epsilon_0 \chi_e \mathbf{e} + \mathbf{p}_{NL} \tag{8.2-3}$$

the first of equations (8.2-1) becomes

$$\nabla \times \mathbf{h} = \sigma \mathbf{e} + \epsilon \frac{\partial \mathbf{e}}{\partial t} + \frac{\partial}{\partial t} \mathbf{p}_{NL} \tag{8.2-4}$$

with $\epsilon \equiv \epsilon_0(1 + \chi_e)$. Taking the curl of both sides of the second of (8.2-1), using (8.2-4) and the vector identity

$$\nabla \times \nabla \times \mathbf{e} = \nabla\nabla\cdot\mathbf{e} - \nabla^2\mathbf{e}$$

and taking $\nabla\cdot\mathbf{e} = 0$, we get

$$\nabla^2\mathbf{e} = \mu\sigma \frac{\partial \mathbf{e}}{\partial t} + \mu\epsilon \frac{\partial^2 \mathbf{e}}{\partial t^2} + \mu \frac{\partial^2}{\partial t^2} \mathbf{p}_{NL} \tag{8.2-5}$$

Next we go over to a scalar notation and rewrite (8.2-5) as

$$\nabla^2 e = \mu\sigma \frac{\partial e}{\partial t} + \mu\epsilon \frac{\partial^2 e}{\partial t^2} + \mu \frac{\partial^2}{\partial t^2} p_{NL}(\mathbf{r}, t) \tag{8.2-6}$$

where we assumed, for simplicity, that $\mathbf{p}_{NL}$ is parallel to $\mathbf{e}$. Let us limit our consideration to a field made up of three plane waves propagating in the z direction with frequencies ω_1, ω_2, and ω_3 according to

$$e^{(\omega_1)}(z, t) = \tfrac{1}{2}[E_1(z)e^{i(\omega_1 t - k_1 z)} + \text{c.c.}]$$
$$e^{(\omega_2)}(z, t) = \tfrac{1}{2}[E_2(z)e^{i(\omega_2 t - k_2 z)} + \text{c.c.}]$$
$$e^{(\omega_3)}(z, t) = \tfrac{1}{2}[E_3(z)e^{i(\omega_3 t - k_3 z)} + \text{c.c.}] \tag{8.2-7}$$

Then the total instantaneous field is

$$e = e^{(\omega_1)}(z, t) + e^{(\omega_2)}(z, t) + e^{(\omega_3)}(z, t) \tag{8.2-8}$$

Next we substitute (8.2-8), using (8.2-7), into the wave equation (8.2-6) and separate the resulting equation into three equations, each containing only terms oscillating at one of the three frequencies. The nonlinear polarization $p_{NL}(\mathbf{r}, t)$ in (8.2-6) contains, according to (8.1-21) and (8.1-22), the terms

$$\text{Re}[d^{(\omega_1+\omega_2)} E_1 E_2 e^{i[(\omega_1+\omega_2)t - (k_1+k_2)z]}]$$

or

$$\text{Re}[d^{(\omega_3-\omega_2)} E_3 E_2^* e^{i[(\omega_3-\omega_2)t - (k_3-k_2)z]}]$$

These oscillate at the new frequencies $(\omega_1 + \omega_2)$ and $(\omega_3 - \omega_2)$ and, in general being nonsynchronous, will not be able to drive the oscillation at ω_1, ω_2, or ω_3. An exception to the last statement is the case when

$$\omega_3 = \omega_1 + \omega_2 \tag{8.2-9}$$

In this case the term

$$\mu d \frac{\partial^2}{\partial t^2} E_1 E_2 e^{i[(\omega_1+\omega_2)t-(k_1+k_2)z]}$$

oscillates at $\omega_1 + \omega_2 = \omega_3$ and can thus act as a source for the wave at ω_3. In physical terms, we have power flow from the fields at ω_1 and ω_2 into that at ω_3, or vice versa. Assuming that (8.2-9) holds, we return to (8.2-6) and, writing it for the oscillation at ω_1, obtain

$$\begin{aligned}\nabla^2 e^{(\omega_1)} &= \mu\sigma_1 \frac{\partial e^{(\omega_1)}}{\partial t} + \mu\epsilon_1 \frac{\partial^2 e^{(\omega_1)}}{\partial t^2} \\ &\quad + \mu d \frac{\partial^2}{\partial t^2}\left[\frac{E_3(z)E_2^*(z)}{2} e^{i[(\omega_3-\omega_2)t-(k_3-k_2)z]} + \text{c.c.}\right]\end{aligned} \tag{8.2-10}$$

Next we observe that, in view of (8.2-7),

$$\begin{aligned}\nabla^2 e^{(\omega_1)} &= \frac{1}{2}\frac{\partial^2}{\partial z^2}[E_1(z)e^{i(\omega_1 t-k_1 z)} + \text{c.c.}] \\ &= -\frac{1}{2}\left[k_1^2 E_1(z) + 2ik_1 \frac{dE_1(z)}{dz}\right] e^{i(\omega_1 t-k_1 z)} + \text{c.c.}\end{aligned}$$

where we assumed that

$$\left|k_1 \frac{dE_1(z)}{dz}\right| \gg \left|\frac{d^2E_1(z)}{dz^2}\right| \tag{8.2-11}$$

If we use (8.2-9) and (8.2-10), and take $\partial/\partial t = i\omega_1$, we obtain

$$\begin{aligned}-\frac{1}{2}&\left[k_1^2 E_1(z) + 2ik_1 \frac{dE_1(z)}{dz}\right] e^{i(\omega_1 t-k_1 z)} + \text{c.c.} \\ &= [i\omega_1\mu\sigma_1 - \omega_1^2\mu\epsilon_1]\left[\frac{E_1(z)}{2} e^{i(\omega_1 t-k_1 z)}\right] + \text{c.c.} \\ &\quad - \left[\frac{\omega_1^2\mu d}{2} E_3(z)E_2^*(z)e^{i[\omega_1 t-(k_3-k_2)z]} + \text{c.c.}\right]\end{aligned} \tag{8.2-12}$$

Recognizing that $k_1^2 = \omega_1^2\mu\epsilon_1$, we can rewrite (8.2-12) after multiplying all the terms by

$$\frac{i}{k_1}\exp(-i\omega_1 t + ik_1 z)$$

as

$$\frac{dE_1}{dz} = -\frac{\sigma_1}{2}\sqrt{\frac{\mu}{\epsilon_1}}\,E_1 - \frac{i\omega_1}{2}\sqrt{\frac{\mu}{\epsilon_1}}\,dE_3E_2^*e^{-i(k_3-k_2-k_1)z}$$

and, similarly,

$$\frac{dE_2^*}{dz} = -\frac{\sigma_2}{2}\sqrt{\frac{\mu}{\epsilon_2}}\,E_2^* + \frac{i\omega_2}{2}\sqrt{\frac{\mu}{\epsilon_2}}\,dE_1E_3^*e^{-i(k_1-k_3+k_2)z}$$

$$\frac{dE_3}{dz} = -\frac{\sigma_3}{2}\sqrt{\frac{\mu}{\epsilon_3}}\,E_3 - \frac{i\omega_3}{2}\sqrt{\frac{\mu}{\epsilon_3}}\,dE_1E_2e^{-i(k_1+k_2-k_3)z} \tag{8.2-13}$$

for the fields at ω_2 and ω_3. These are the basic equations describing nonlinear parametric interactions [4]. We notice that they are coupled to each other via the nonlinear constant d.

8.3 OPTICAL SECOND-HARMONIC GENERATION

The first experiment in nonlinear optics [5] consisted of generating the second harmonic ($\lambda = 0.3470$ μm) of a ruby laser beam ($\lambda = 0.694$ μm) that was focused on a quartz crystal. The experimental arrangement is depicted in Figure 8-7. The conversion efficiency of this first experiment ($\sim 10^{-8}$) was improved by methods to be described below to a point where about 30 percent conversion has been observed in a single pass through a few centimeters length of a nonlinear crystal. This technique is finding important applications in generating short-wave radiation from longer-wave lasers.

In the case of second-harmonic generation, two of the three fields that figure in (8.2-13) are of the same frequency. We may thus put $\omega_1 = \omega_2 = \omega$ and $E_1 = E_2 = E(\omega)$, for which case the first two equations are the complex conjugate of one another

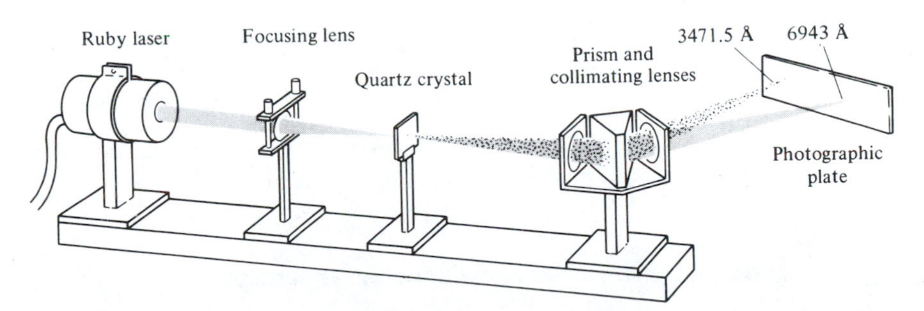

Figure 8-7 Arrangement used in first experimental demonstration of second-harmonic generation [5]. Ruby laser beam at $\lambda_0 = 0.694$ μm is focused on a quartz crystal, causing generation of a (weak) beam at $\lambda_0/2 = 0.347$ μm. The two beams are then separated by a prism and detected on a photographic plate.

and we need to consider only one of them. We take the input field at ω to correspond to E_1 in (8.2-13) and the second-harmonic field to E_3, and we put $\omega_3 = \omega_1 + \omega_2 = 2\omega$, neglecting the absorption, so $\sigma_{1,2,3} = 0$. The last equation becomes

$$\frac{dE^{(2\omega)}}{dz} = -i\omega \sqrt{\frac{\mu}{\epsilon}}\, d[E^{(\omega)}(z)]^2 e^{i(\Delta k)z} \tag{8.3-1}$$

where

$$\Delta k \equiv k_3 - 2k_1 = k^{(2\omega)} - 2k^{(\omega)} \tag{8.3-2}$$

To simplify the analysis further, we may assume that the depletion of the input wave at ω due to conversion of its power to 2ω is negligible. Under those conditions, which apply in the majority of the experimental situations, we can take $E^{(\omega)}(z) =$ constant in (8.3-1) and neglect its dependence on z. Assuming no input at 2ω—that is, $E^{(2\omega)}(0) = 0$—we obtain from (8.3-1) by integration the output field at the end of a crystal of length l:

$$E^{(2\omega)}(l) = -i\omega \sqrt{\frac{\mu}{\epsilon}}\, d[E^{(\omega)}]^2 \frac{e^{i\Delta kl} - 1}{i\Delta k}$$

The output intensity is proportional to

$$E^{(2\omega)}(l)E^{(2\omega)*}(l) = \left(\frac{\mu}{\epsilon_0}\right) \frac{\omega^2 d^2}{n^2} |E^{(\omega)}|^4 l^2 \frac{\sin^2 (\Delta kl/2)}{(\Delta kl/2)^2} \tag{8.3-3}$$

Here we used $\epsilon/\epsilon_0 = n^2$, where n is the index of refraction. If the input beam is confined to a cross section $A(m^2)$, then, according to (1.3-26), the power per unit area (intensity) is related to the field by

$$I \equiv \frac{P_{2\omega}}{A} = \frac{1}{2} \sqrt{\frac{\epsilon}{\mu}}\, |E^{(2\omega)}|^2 \tag{8.3-4}$$

and (8.3-3) can be written as

$$\eta_{\text{SHG}} \equiv \frac{P_{2\omega}}{P_\omega} = 2\left(\frac{\mu}{\epsilon_0}\right)^{3/2} \frac{\omega^2 d^2 l^2}{n^3} \frac{\sin^2(\Delta kl/2)}{(\Delta kl/2)^2} \frac{P_\omega}{A} \tag{8.3-5}$$

for the conversion efficiency from ω to 2ω. We notice that the conversion efficiency is proportional to the intensity P_ω/A of the fundamental beam.

Phase-Matching in Second-Harmonic Generation

According to (8.3-5), a prerequisite for efficient second-harmonic generation is that $\Delta k = 0$—or, using (8.3-2),

$$k^{(2\omega)} = 2k^{(\omega)} \tag{8.3-6}$$

If $\Delta k \neq 0$, the second-harmonic power generated at some plane, say z_1, having propagated to some other plane (z_2), is not in phase with the second-harmonic wave generated at z_2. This results in the interference described by the factor

$$\frac{\sin^2(\Delta kl/2)}{(\Delta kl/2)^2}$$

in (8.3-5). The main peak and the first zero of this spatial interference pattern are separated by the so-called "coherence length"

$$l_c = \frac{2\pi}{\Delta k} = \frac{2\pi}{k^{(2\omega)} - 2k^{(\omega)}} \tag{8.3-7}$$

The coherence length l_c is thus a *measure* of the *maximum crystal length that is useful in producing the second-harmonic power.* Under ordinary circumstances it may be no larger than 10^{-2} cm. This is because the index of refraction n^ω normally increases with ω so Δk is given by

$$\Delta k = k^{(2\omega)} - 2k^{(\omega)} = \frac{2\omega}{c}[n^{2\omega} - n^\omega] \tag{8.3-8}$$

where we used the relation $k^{(\omega)} = \omega n^\omega/c$. The coherence length is thus

$$l_c = \frac{\pi c}{\omega[n^{2\omega} - n^\omega]} = \frac{\lambda}{2[n^{2\omega} - n^\omega]} \tag{8.3-9}$$

where λ is the free-space wavelength of the fundamental beam. If we take a typical value of $\lambda = 1$ μm and $n^{2\omega} - n^\omega \simeq 10^{-2}$, we get $l_c \simeq 50$ μm. If l_c were to increase from 100 μm to 2 cm, as an example, according to (8.3-5) the second-harmonic power would go up by a factor of 4×10^4.

The technique that is used widely (see [6, 7]) to satisfy the *phase-matching* requirement $\Delta k = 0$ takes advantage of the natural birefringence of anisotropic crystals, which was discussed in Section 1.4. Using the relation $k^{(\omega)} = \omega\sqrt{\mu\epsilon_0}n^\omega$, (8.3-6) becomes

$$n^{2\omega} = n^\omega \tag{8.3-10}$$

so the indices of refraction at the fundamental and second-harmonic frequencies must be equal. In normally dispersive materials the index of the ordinary wave or the extraordinary wave along a given direction increases with ω, as can be seen from Table 8-2. This makes it impossible to satisfy (8.3-10) when both the ω and 2ω beams are of the same type—that is, when both are extraordinary or ordinary. We can, however, under certain circumstances, satisfy (8.3-10) by making the two waves be of different types. To illustrate the point, consider the dependence of the index of refraction of the extraordinary wave in a uniaxial crystal on the angle θ between the propagation direction and the crystal optic (z) axis. It is given by (1.4-12) as

$$\frac{1}{n_e^2(\theta)} = \frac{\cos^2\theta}{n_o^2} + \frac{\sin^2\theta}{n_e^2} \tag{8.3-11}$$

Table 8-2 Index of Refraction Dispersion Data of KH_2PO_4 (After Reference [8].

	Index	
Wavelength, μm	n_o (ordinary ray)	n_e (extraordinary ray)
0.2000	1.622630	1.563913
0.3000	1.545570	1.498153
0.4000	1.524481	1.480244
0.5000	1.514928	1.472486
0.6000	1.509274	1.468267
0.7000	1.505235	1.465601
0.8000	1.501924	1.463708
0.9000	1.498930	1.462234
1.0000	1.496044	1.460993
1.1000	1.493147	1.459884
1.2000	1.490169	1.458845
1.3000	1.487064	1.457838
1.4000	1.483803	1.456838
1.5000	1.480363	1.455829
1.6000	1.476729	1454797
1.7000	1.472890	1.453735
1.8000	1.468834	1.452636
1.9000	1.464555	1.451495
2.0000	1.460044	1.450308

If $n_e^{2\omega} < n_o^{\omega}$, there exists an angle θ_m at which $n_e^{2\omega}(\theta_m) = n_o^{\omega}$; so if the fundamental beam (at ω) is launched along θ_m as an ordinary ray, the second-harmonic beam will be generated along the *same direction* as an extraordinary ray. The situation is illustrated by Figure 8-8. The angle θ_m is determined by the intersection between the sphere (shown as a circle in the figure) corresponding to the index surface of the ordinary beam at ω, and the index surface of the extraordinary ray $n_e^{2\omega}(\theta)$. The angle θ_m, which defines a cone, for negative uniaxial crystals—that is, crystals in which $n_e^{\omega} < n_o^{\omega}$—is that satisfying $n_e^{2\omega}(\theta_m) = n_o^{\omega}$ or, using (8.3-11),

$$\frac{\cos^2\theta_m}{(n_o^{2\omega})^2} + \frac{\sin^2\theta_m}{(n_e^{2\omega})^2} = \frac{1}{(n_o^{\omega})^2} \tag{8.3-12}$$

and, solving for θ_m,

$$\sin^2\theta_m = \frac{(n_o^{\omega})^{-2} - (n_o^{2\omega})^{-2}}{(n_e^{2\omega})^{-2} - (n_o^{2\omega})^{-2}} \tag{8.3-13}$$

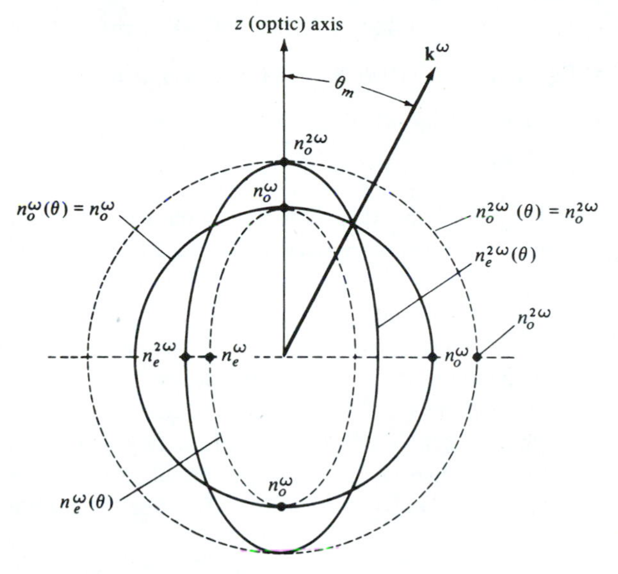

Figure 8-8 Normal (index) surfaces for the ordinary and extraordinary rays in a negative ($n_e < n_o$) uniaxial crystal. If $n_e^{2\omega} < n_o^{\omega}$, the condition $n_e^{2\omega}(\theta) = n_o^{\omega}$ is satisfied at $\theta = \theta_m$. The eccentricities shown are vastly exaggerated.

Numerical Example: Second-Harmonic Generation

Consider the problem of second-harmonic generation using the output of a pulsed ruby laser ($\lambda_0 = 0.6940\ \mu$m) in a KH_2PO_4 crystal (KDP) under the following conditions:

$$l = 1 \text{ cm}$$

$$P_\omega/A = 10^8 \text{ W/cm}^2$$

The appropriate d coefficient is, according to Table 8-1, $d = d_{312} \cos \theta_m \times 3 \times 10^{-24}$ MKS units. Using these data in (8.3-5) and assuming $\Delta k = 0$ gives a conversion efficiency of

$$\frac{P_{(\lambda_0=0.347\ \mu\text{m})}}{P_{(\lambda_0=0.694\ \mu\text{m})}} \simeq 0.21$$

The angle θ_m between the z axis and the direction of propagation for which $\Delta k = 0$ is given by (8.3-13). The appropriate indices are taken from Table 8-2, and are

$$n_e(\lambda = 0.694\ \mu\text{m}) = 1.466 \qquad n_e(\lambda = 0.347\ \mu\text{m}) = 1.490$$

$$n_o(\lambda = 0.694\ \mu\text{m}) = 1.506 \qquad n_o(\lambda = 0.347\ \mu\text{m}) = 1.534$$

Substituting the foregoing data into (8.3-13) gives

$$\theta_m = 52°$$

To obtain phase-matching along this direction, the fundamental beam in the crystal must be polarized as appropriate to an ordinary ray in accordance with the discussion following (8.3-11).

We conclude from this example that very large intensities are needed to obtain high-efficiency second-harmonic generation. This efficiency will, according to (8.3-5), increase as the square of the nonlinear optical coefficient d and will consequently improve as new materials are developed. Another approach is to take advantage of the dependence of η_{SHG} on P_ω/A and to place the nonlinear crystal inside the laser resonator where the energy flux P_ω/A can be made very large.[4] This approach has been used successfully [10] and it will be discussed in considerable detail further in this chapter.

Experimental Verification of Phase-Matching

According to (8.3-5), if the phase-matching condition $\Delta k = 0$ is violated, the output power is reduced by a factor

$$F = \frac{\sin^2(\Delta kl/2)}{(\Delta kl/2)^2} \tag{8.3-14}$$

from its (maximum) phase-matched value. The phase mismatch $\Delta kl/2$ is given, according to (8.3-8), by

$$\frac{\Delta kl}{2} = \frac{\omega l}{c}\left[n_e^{2\omega}(\theta) - n_o^{\omega}\right] \tag{8.3-15}$$

and is thus a function of θ. If we use (8.3-11) to expand $n_e^{2\omega}(\theta)$ as a Taylor series near $\theta \simeq \theta_m$, retain the first two terms only, and assume perfect phase-matching at $\theta = \theta_m$ so $n_e^{2\omega}(\theta_m) = n_o^{\omega}$, we obtain

$$\begin{aligned}\Delta k(\theta)l &= -\frac{2\omega l}{c}\sin(2\theta_m)\frac{(n_e^{2\omega})^{-2} - (n_o^{2\omega})^{-2}}{2(n_o^{\omega})^{-3}}(\theta - \theta_m)\\ &\equiv 2\beta(\theta - \theta_m)\end{aligned} \tag{8.3-16}$$

where β, as defined by (8.3-16), is a constant depending on $n_e^{2\omega}$, $n_o^{2\omega}$, n_o^{ω}, ω, and l. If we plot the output power at 2ω as a function of θ we would expect, according to (8.3-5) and (8.3-16), to find it varying as

[4] The one-way power flow inside the optical resonator P_i is related to the power output P_e as $P_i = P_e/(1 - R)$, where R is the reflectivity.

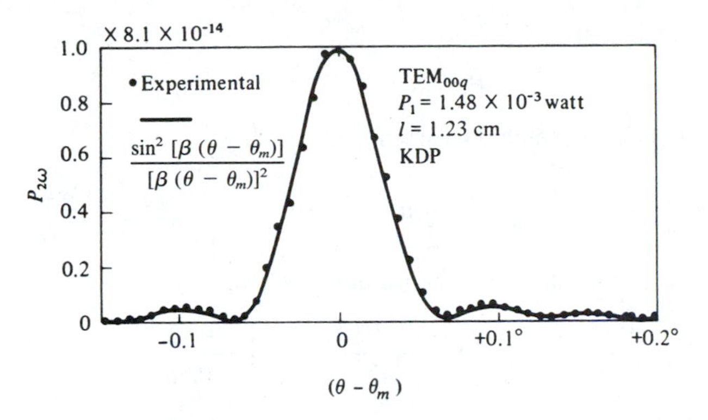

Figure 8-9 Variation of the second- harmonic power $P_{2\omega}$ with the angular departure $(\theta - \theta_m)$ from the phase-matching angle. (After Reference [11].)

$$P_{2\omega}(\theta) \propto \frac{\sin^2[\beta(\theta - \theta_m)]}{[\beta(\theta - \theta_m)]^2} \tag{8.3-17}$$

Figure 8-9 shows an experimental plot of $P_{2\omega}(\theta)$ as well as a plot of (8.3-17).

Another phase-matching technique involves the introduction of an artificial spatial periodicity $\Delta l = 2\pi/\Delta k$ into the beams' path. This method is discussed in Problem 8.10 and the references quoted therein.

Second-Harmonic Generation with Focused Gaussian Beams

The analysis of second-harmonic generation leading to (8.3-5) is based on a plane wave model. In practice one uses Gaussian beams that are focused so as to reach their minimum radius (waist) inside the crystal. A typical situation is depicted in Figure 8-10. The incident Gaussian beam is characterized by confocal parameter z_0, which according to (2.5-11) is the distance from the beam waist in which the beam "area" $\pi\omega^2$ is double that of the waist. We recall that $z_0 = \pi\omega_0^2 n/\lambda$, where ω_0 is the minimum beam radius (waist). If $z_0 \gg l$ (l is the crystal length), the beam area, hence the intensity, of the incident wave is nearly independent of z within the crystal, and we may apply the plane wave result (8.3-3) to write

$$|E^{(2\omega)}(r)|^2 = \frac{\mu}{\epsilon}\, \omega^2 d^2 |E^{(\omega)}(r)|^4 l^2 \frac{\sin^2(\Delta kl/2)}{(\Delta kl/2)^2} \tag{8.3-18}$$

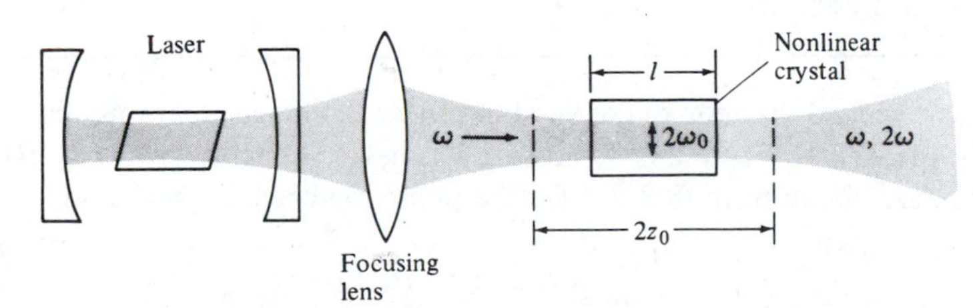

Figure 8-10 Second-harmonic generation with a focused Gaussian beam.

where $E^{(\omega)}(r)$ is taken as

$$E^{(\omega)}(r) \cong E_0 e^{-r^2/\omega_0^2} \tag{8.3-19}$$

as appropriate to a fundamental Gaussian beam. Using

$$P^{(\omega)} = \frac{1}{2}\sqrt{\frac{\epsilon}{\mu}} \int_{\text{cross section}} |E^{(\omega)}|^2 \, dx \, dy \cong \sqrt{\frac{\epsilon}{\mu}} E_0^2 \left(\frac{\pi\omega_0^2}{4}\right)$$

as well as (8.3-19), we obtain, by integrating (8.3-18),

$$\frac{P^{(2\omega)}}{P^{(\omega)}} = 2\left(\frac{\mu}{\epsilon_0}\right)^{3/2} \frac{\omega^2 d^2 l^2}{n^3} \left(\frac{P^{(\omega)}}{\pi\omega_0^2}\right) \frac{\sin^2(\Delta k l/2)}{(\Delta k l/2)^2} \tag{8.3-20}$$

where we used $(n^{\omega})^2 n^{2\omega} \equiv n^3$.

Equation (8.3-20) is identical to (8.3-5). We must recall, however, that it was derived for a Gaussian beam input with $z_0 \gg l$. According to (8.3-20) in a crystal of length l and with a given input $P^{(\omega)}$, the output power $P^{(2\omega)}$ can be increased by decreasing ω_0. This is indeed the case until $z_0(=\pi\omega_0^2 n/\lambda)$ becomes comparable to l. Further reduction of ω_0 (and z_0) will lead to a situation in which the beam begins to spread appreciably within the crystal, thus leading to a reduced intensity and a reduced second-harmonic generation. It is thus reasonable to focus the beam until $l = 2z_0$. At this point $\omega_0^2 = \lambda l/2\pi n$, which is referred to as confocal focusing, and (8.3-20) becomes

$$\eta \equiv \left.\frac{P^{(2\omega)}}{P^{(\omega)}}\right|_{\text{confocal focusing}} = \frac{2}{\pi c}\left(\frac{\mu}{\epsilon_0}\right)^{3/2} \frac{\omega^3 d^2 l}{n^2} P^{(\omega)} \frac{\sin^2(\Delta k l/2)}{(\Delta k l/2)^2} \tag{8.3-21}$$

A more exact analysis of second-harmonic generation with focused Gaussian beam shows that the maximum conversion efficiency is approximately 20 percent higher than the confocal result (8.3-21).

The main difference between (8.3-21) and the plane wave result (8.3-5) is that the conversion efficiency in this case increases as l instead of l^2. This reflects the fact that a longer crystal entails the use of a larger beam spot size ω_0 so as to keep $z_0 \approx l/2$, which reduces the intensity of the fundamental beam.

Example: Optimum Focusing

Consider second harmonic conversion under confocal focusing conditions, in KH_2PO_4 from $\lambda = 1$ μm to $\lambda = 0.5$ μm. Using $l = 1$ cm, $d_{\text{eff}} = 3.6 \times 10^{-24}$ MKS, $n = 1.5$, we obtain from (8.3-21) for the phase matched $\Delta k = 0$ case

$$\frac{P^{(2\omega)}}{P^{(\omega)}} = 4.4 \times 10^{-6} \, P^{(\omega)}$$

Second-Harmonic Generation with a Depleted Input

The expression (8.3-5) for the conversion efficiency in second harmonic generation was derived assuming negligible depletion of the fundamental beam at ω. It is, therefore, valid only for cases where the conversion efficiency is small, i.e., $\eta_{\text{SHG}} \ll 1$. A study of Equation (8.2-13) or the intuitive understanding of parametric processes which the student may have acquired by now shows that, assuming phase matching and a sufficiently long crystal, the conversion process $\omega \rightarrow 2\omega$ continues with distance and that it is not unreasonable to expect conversion efficiencies approaching unity. To consider this possibility, we return to Equation (8.2-13), but this time, anticipating pump depletion, the fundamental beams $E_1(z)$ and $E_2(z)$ are allowed to depend on z. We transform to a new set of field variables A_l defined by[5]

$$A_l \equiv \sqrt{\frac{n_l}{\omega_l}}\, E_l \qquad l = 1, 2, 3 \tag{8.3-22}$$

where $n_l^2 = \epsilon_l/\epsilon_0$, i.e., n_l is the index of refraction of wave l. See discussion following Equation (8.6-17) to better appreciate the transformation (8.3-22). The result is

$$\begin{aligned}
\frac{dA_1}{dz} &= -\frac{\alpha_1}{2} A_1 - \frac{i}{2} \kappa A_2^* A_3 e^{-i(\Delta k)z} \\
\frac{dA_2^*}{dz} &= -\frac{\alpha_2}{2} A_2^* + \frac{i}{2} \kappa A_1 A_3^* e^{i(\Delta k)z} \\
\frac{dA_3}{dz} &= -\frac{\alpha_3}{2} A_3 - \frac{i}{2} \kappa A_1 A_2 e^{i(\Delta k)z}
\end{aligned} \tag{8.3-23}$$

where

$$\begin{aligned}
\alpha_l &\equiv \sigma_l \sqrt{\frac{\mu}{\epsilon_l}} \\
\kappa &\equiv d \sqrt{\left(\frac{\mu}{\epsilon_0}\right) \frac{\omega_1 \omega_2 \omega_3}{n_1 n_2 n_3}} \\
\Delta k &\equiv k_3 - (k_1 + k_2)
\end{aligned} \tag{8.3-24}$$

In the case of second-harmonic generation, $A_1 = A_2$ and Equations (8.3-23) become

$$\begin{aligned}
\frac{dA_1}{dz} &= -i \frac{\kappa}{2} A_3 A_1^* \\
\frac{dA_3}{dz} &= -i \frac{\kappa}{2} A_1^2
\end{aligned} \tag{8.3-25}$$

[5] It follows from (8.3-22) that $|A_l|^2$ is proportional to the photon density at ω_l [see Equation (8.6-17)].

where we assumed transparent ($\alpha_l = 0$) media and phase matching ($\Delta k = 0$). It follows from (8.3-25) that if we choose, without loss of generality, $A_1(0)$ as a real number, then $A_1(z)$ is real and (8.3-25) can be rewritten in the form

$$\frac{dA_1}{dz} = -\frac{1}{2}\kappa A_3' A_1$$

$$\frac{dA_3'}{dz} = \frac{1}{2}\kappa A_1^2 \tag{8.3-26}$$

where $A_3 \equiv -iA_3'$. It follows from (8.3-26) that

$$\frac{d}{dz}(A_1^2 + A_3'^2) = 0$$

(i.e., for each photon "removed" from beam 1, one photon is added to beam 3; energy is conserved, since a photon is also removed simultaneously from beam 2).

Assuming no input at ω_3, we have $A_1^2 + A_3'^2 = A_1^2(0)$, and the second of (8.3-26) becomes

$$\frac{dA_3'}{dz} = \frac{1}{2}\kappa(A_1^2(0) - A_3'^2)$$

leading to a solution

$$A_3'(z) = A_1(0)\tanh[\tfrac{1}{2}\kappa A_1(0)z] \tag{8.3-27}$$

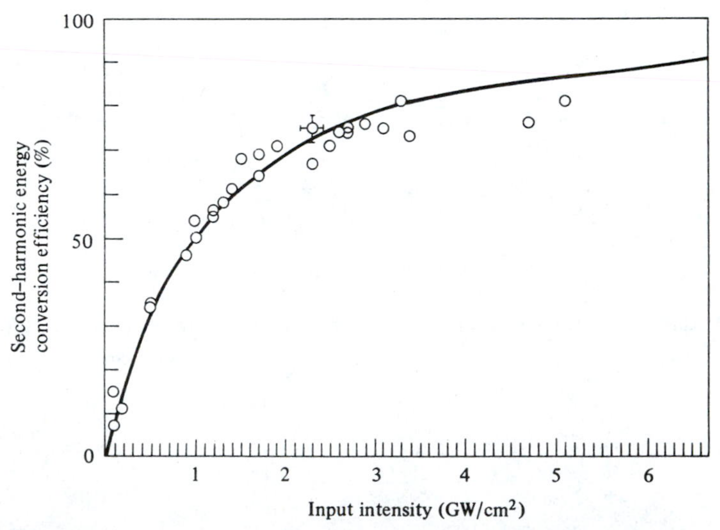

Figure 8-11 Frequency doubling energy conversion efficiency. Solid curve: theoretical prediction of Equation (8.3-28) (recall that $A_1(0) \propto \sqrt{I_{in}}$). The circles correspond to experimental points. (After supplementary Reference [30].)

We note that as $\kappa A_1(0)z \to \infty$, $A_3'(z) \to A_1(0)$ so that all the input photons at ω are converted into half (since $A_1 = A_2$) as many photons at 2ω and the power conversion efficiency approaches unity. In the general case

$$\eta_{\mathrm{SHG}} \equiv \frac{P^{(2\omega)}}{P^{(\omega)}} = \frac{|A_3(z)|^2}{|A_1(0)|^2} = \tanh^2[\tfrac{1}{2}\kappa A_1(0)z] \tag{8.3-28}$$

A plot of the theoretically predicted relation (8.3-28) as well as of experimental data obtained in converting from $\lambda = 1.06$ to $\lambda = 0.53$ μm is shown in Figure 8-11.

8.4 SECOND-HARMONIC GENERATION INSIDE THE LASER RESONATOR

According to the numerical example of Section 8.3 and Figure 8-11, we need to use large power densities at the fundamental frequency ω to obtain appreciable conversion from ω to 2ω in typical nonlinear optical crystals. These power densities are not usually available from continuous (CW) lasers. The situation is altered, however, if the nonlinear crystal is placed within the laser resonator. The intensity (one-way power per unit area in watts per square meter) inside the resonator exceeds its value outside a mirror by $(1 - R)^{-1}$, where R is the mirror reflectivity. If $R \simeq 1$, the enhancement is very large and since the second-harmonic conversion efficiency is, according to (8.3-5), proportional to the intensity, we may expect a far more efficient conversion inside the resonator. We will show below that under the proper conditions we can extract the *total available power* of the laser at 2ω instead of at ω and in that sense obtain 100 percent conversion efficiency. In order to appreciate the last statement, consider as an example the case of a (CW) laser in which the maximum power output, at a given pumping rate, is available when the output mirror has a (optimal) transmission of 5 percent.

The output mirror is next replaced with one having 100 percent reflection at ω and a nonlinear crystal is placed inside the laser resonator. If with the crystal inside the conversion efficiency from ω to 2ω in a *single pass* is 5 percent, the laser is loaded optimally as in the previous case except that the coupling is attributable to loss of power caused by second-harmonic generation instead of by the output mirror. It follows that the power generated at 2ω is the same as that coupled previously through the mirror and that the total available power of a laser can thus be converted to the second harmonic.

An experimental setup similar to the one used in the first internal second-harmonic generation experiment [10] is shown in Figure 8-12. The Nd^{3+}:YAG laser (see Chapter 7 for a description of this laser) emits a (fundamental) wave at $\lambda_0 = 1.06$ μm. The mirrors are, as nearly as possible, totally reflecting at $\lambda_0 = 1.06$ μm. A $Ba_2NaNb_5O_{15}$ crystal is used to generate the second harmonic at $\lambda_0 = 0.53$ μm. The latter is coupled through the mirror—which, ideally, transmits all the radiation at this wavelength.

In the mathematical treatment of internal second-harmonic generation that follows we use the results of the analysis of optimum power coupling in laser oscillators of Section 6.5.

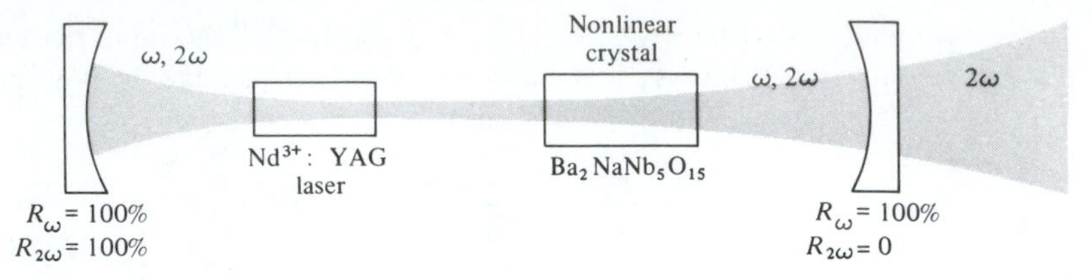

Figure 8-12 Typical setup for second-harmonic conversion inside a laser resonator. (After Reference [9].)

The mirror transmission T_{opt} that results in the maximum power output from a laser oscillator is given by (6.5-11) as

$$T_{\text{opt}} = \sqrt{g_0 L_i} - L_i \tag{8.4-1}$$

where L_i is the residual (that is, unavoidable) fractional intensity loss per pass and g_0 is the fractional unsaturated gain per pass.[6] The useful power output under optimum coupling is, according to (6.5-12),

$$P_o = I_s A(\sqrt{g_0} - \sqrt{L_i})^2 \tag{8.4-2}$$

where the saturation intensity of the laser transition $I_s A$ was given by (5.6-9) as[7]

$$I_s = \frac{8\pi n^2 h\nu \Delta\nu}{\lambda^2 (t_2/t_{\text{spont}})} \tag{8.4-3}$$

In the present problem the conversion from ω to 2ω can be considered, as far as the ω oscillation is concerned, just as another loss mechanism. We may think of it as due to a mirror with a transmission T' taken as equal to the conversion efficiency (from ω to 2ω) per pass, which, according to (8.3-5), is

$$T' \equiv \frac{P_{2\omega}}{P_\omega} = 2\left(\frac{\mu_0}{\epsilon_0}\right)^{3/2} \frac{\omega^2 d^2 l^2}{n^3} \left[\frac{\sin^2(\Delta k l/2)}{(\Delta k l/2)^2}\right] \frac{P_\omega}{A} \tag{8.4-4}$$

where d is the crystal nonlinear coefficient, l its length, A its cross-sectional area, Δk the wave-vector mismatch, and P_ω the one-way traveling power *inside* the laser. We can rewrite T' in the form

$$T' = \kappa P_\omega \tag{8.4-5}$$

where the value of the constant κ is evident from Equation (8.4-4). The equivalent mirror transmission T' is thus proportional to the power.

---

[6]We may recall here that the residual losses include all loss mechanisms except those representing useful power coupling. The unsaturated gain g_0 is that exercised by a very weak wave and represents the maximum available gain at a given pumping strength.

[7]I_s is, according to (5.6-8) [and putting $g(\nu)^{-1} = \Delta\nu$], the optical intensity (watts per square meter) that reduces the inversion, hence the gain, to one-half its zero intensity (unsaturated) value.

Using the last result in (8.4-1), we find immediately that at optimum conversion the product κP_ω must have the value

$$(\kappa P_\omega)_{\text{opt}} = \sqrt{g_0 L_i} - L_i \tag{8.4-6}$$

The total loss per pass seen by the fundamental beam is the sum of the conversion loss (κP_ω) and the residual losses, which, under optimum coupling, becomes

$$L_{\text{opt}} = L_i + (\kappa P_\omega)_{\text{opt}} = \sqrt{g_0 L_i} \tag{8.4-7}$$

Our next problem is to find the internal power P_ω at optimum coupling so that using (8.4-4) we may calculate the second-harmonic power. We start with the expression (6.5-6) for the total power P_e extracted from the laser atoms and replace the loss L by its optimum value (8.4-7) to obtain

$$\begin{aligned}(P_e)_{\text{opt}} &= P_s\left(\frac{g_0}{L_{\text{opt}}} - 1\right) = P_s\left(\sqrt{\frac{g_0}{L_i}} - 1\right) \\ &= \frac{8\pi n^3 h \Delta\nu V}{\lambda^3 (t_c)_{\text{opt}}}\left(\frac{t_{\text{spont}}}{t_2}\right)\left(\sqrt{\frac{g_0}{L_i}} - 1\right) = L_{\text{opt}} I_s A\left(\sqrt{\frac{g_0}{L_i}} - 1\right)\end{aligned} \tag{8.4-8}$$

where to get the last equality we used relation (4.7-2)

$$t_c = \frac{nl}{cL}$$

to relate the resonator decay time t_c to the loss per pass L. The fraction of the total power P_e emitted by the atoms that is available as useful output is T'/L. This power is also given by the product $P_\omega T'$ of the one-way internal power P_ω and the fraction T' of this power that is converted per pass. Equating these two forms gives

$$P_\omega = \frac{P_e}{L}$$

and using (8.4-8) we get

$$(P_\omega)_{\text{opt}} = I_s A\left(\sqrt{\frac{g_0}{L_i}} - 1\right) \tag{8.4-9}$$

for the one-way fundamental power inside the laser under optimum coupling conditions. The amount of second-harmonic power generated under optimum coupling is

$$(P_{2\omega})_{\text{opt}} = (\kappa P_\omega)_{\text{opt}} (P_\omega)_{\text{opt}}$$

which, through the use of (8.4-6) and (8.4-9), results in

$$(P_{2\omega})_{\text{opt}} = I_s A(\sqrt{g_0} - \sqrt{L_i})^2 \tag{8.4-10}$$

This is the same expression as the one previously obtained in (6.5-12) for the maximum available power output from a laser oscillator.

The nonlinear coupling constant κ was defined by (8.4-4) and (8.4-5) as

$$\kappa = 2\left(\frac{\mu_0}{\epsilon_0}\right)^{3/2} \frac{\omega^2 d^2 l^2}{n^3 A} \left(\frac{\sin^2(\Delta k l/2)}{(\Delta k l/2)^2}\right) \tag{8.4-11}$$

Its value under optimum coupling can be derived from (8.4-6) and (8.4-9) and is

$$\kappa_{\text{opt}} = \frac{(\kappa P_\omega)_{\text{opt}}}{(P_\omega)_{\text{opt}}} = \frac{L_i}{I_s A} \tag{8.4-12}$$

and is thus *independent of the pumping strength.*[8] It follows that once κ is adjusted to its optimum value $L_i/I_s A$, it remains optimal at any pumping level. This is quite different from the case of optimum coupling in ordinary lasers, in which optimum mirror transmission was found [see (6.5-11)] to depend on the pumping strength.

In closing we may note that apart from its dependence on the crystal length l, the nonlinear coefficient d, and the beam cross section A, κ depends also on the phase mismatch Δkl. Since Δk was shown in (8.3-15) to depend on the direction of propagation in the crystal, we can use the crystal orientation as a means of varying κ.

Numerical Example: Internal Second-Harmonic Generation

Consider the problem of designing an internal second harmonic generator of the type illustrated in Figure 8-12. The Nd^{3+}:YAG laser is assumed to have the following characteristics:

$\lambda_0 = 1.06\ \mu\text{m} = 1.06 \times 10^{-6}$ meter

$\Delta\nu = 1.35 \times 10^{11}$ Hz (width of the spectral gain profile)

Beam diameter (averaged over entire resonator length) = 2 mm

L_i = internal loss per pass = 2×10^{-2}

$n = 1.5$

The crystal used for second-harmonic generation is $Ba_2NaNb_5O_{15}$, whose second-harmonic coefficient (see Table 8-1) is $d \simeq 1.1 \times 10^{-22}$ MKS units.

Our problem is to calculate the length l of the nonlinear crystal that results in a full conversion of the optimally available fundamental power into the second harmonic at $\lambda = 0.53\ \mu\text{m}$. The crystal is assumed to be oriented at the phase-matching angle, so $\Delta k = k^{2\omega} - 2k^{\omega} = 0$.

[8]We recall here that the pumping strength in our analysis is represented by the unsaturated gain g_0.

The optimum coupling parameter is given by (8.4-12) as $\kappa_{\text{opt}} = L_i/I_sA$, where I_s is the saturation intensity defined by (8.4-3). Using the foregoing data in (8.4-3) gives

$$I_sA = 2 \text{ watts}$$

which, taking $L_i = 2 \times 10^{-2}$, yields

$$\kappa_{\text{opt}} = 10^{-2}$$

Next we use the definition (8.4-11)

$$\kappa = 2\left(\frac{\mu_0}{\epsilon_0}\right)^{3/2} \frac{\omega^2 d^2 l^2}{n^3 A}$$

where we put $\Delta k = 0$ and take the beam diameter at the crystal as 50 μm. (The crystal can be placed near a beam waist so the diameter is a minimum.) Equating the last expression to $\kappa_{\text{opt}} = 10^{-2}$ using the numerical data given above, and solving for the crystal length, results in

$$l_{\text{opt}} = 0.804 \text{ cm}$$

8.5 PHOTON MODEL OF SECOND-HARMONIC GENERATION

A very useful point of view and one that follows directly from the quantum mechanical analysis of nonlinear optical processes [12] is based on the photon model illustrated in Figure 8-13. According to this picture, the basic process of second-

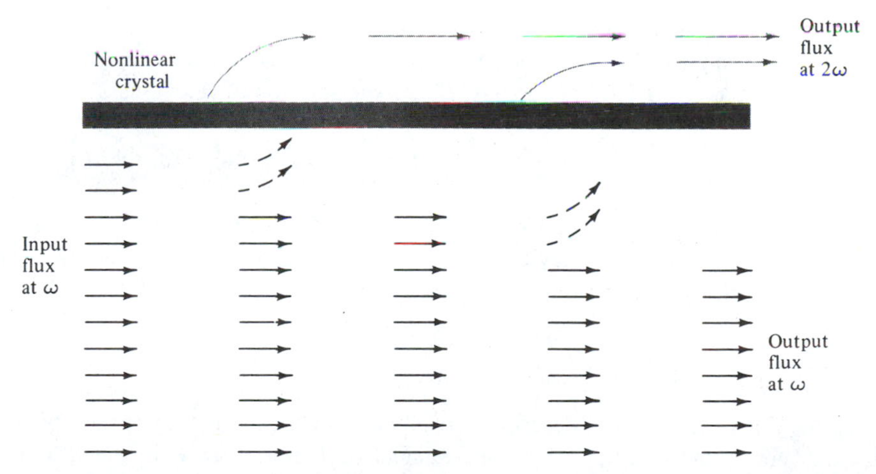

Figure 8-13 Schematic representation of the process of second-harmonic generation. Input photons (each arrow represents one photon) at ω are ''annihilated'' by the nonlinear crystal in pairs, with a new photon at 2ω being created for each annihilated pair. (Note that in reality both ω and 2ω occupy the same space inside the crystal.)

harmonic generation can be viewed as an annihilation of two photons at ω and a simultaneous creation of a photon at 2ω. Recalling that a photon has an energy $\hbar\omega$ and a momentum $\hbar\mathbf{k}$, it follows that if the fundamental conversion process is to conserve momentum as well as energy that

$$\mathbf{k}^{(2\omega)} = 2\mathbf{k}^{(\omega)} \tag{8.5-1}$$

which is a generalization to three dimensions of the condition $\Delta k = 0$ shown in Section 8.3 to lead to maximum second-harmonic generation.

8.6 PARAMETRIC AMPLIFICATION

Optical parametric amplification in its simplest form involves the transfer of power from a "pump" wave at ω_3 to waves at frequencies ω_1 and ω_2, where $\omega_3 = \omega_1 + \omega_2$. It is fundamentally similar to the case of second-harmonic generation treated in Section 8.3. The only difference is in the direction of power flow. In second-harmonic generation, power is fed from the low-frequency optical field at ω to the field at 2ω. In parametric amplification, power flow is from the high-frequency field (ω_3) to the low-frequency fields at ω_1 and ω_2. In the special case where $\omega_1 = \omega_2$, we have the exact reverse of second-harmonic generation. This is the case of the so-called degenerate parametric amplification.

Before we embark on a detailed analysis of the optical case it may be worthwhile to review some of the low-frequency beginnings of parametric oscillation.

Consider a classical nondriven oscillator whose equation of motion is given by

$$\frac{d^2v}{dt^2} + \kappa \frac{dv}{dt} + \omega_0^2 v = 0 \tag{8.6-1}$$

The variable v may correspond to the excursion of a mass M, which is connected to a spring with a constant $\omega_0^2 M$, or to the voltage across a parallel RLC circuit, in which case $\omega_0^2 = (LC)^{-1}$ and $\kappa = (RC)^{-1}$. The solution of (8.6-1) is

$$v(t) = v(0) \exp\left(-\frac{\kappa t}{2}\right) \exp\left(\pm i \sqrt{\omega_0^2 - \frac{\kappa^2}{4}}\, t\right) \tag{8.6-2}$$

that is, a damped sinusoid.

In 1883 Lord Rayleigh [13], investigating parasitic resonances in pipe organs, considered the consequences of the following equation

$$\frac{d^2v}{dt^2} + \kappa \frac{dv}{dt} + (\omega_0^2 + 2\alpha \sin \omega_p t)v = 0 \tag{8.6-3}$$

This equation may describe an oscillator in which an energy storage parameter (mass or spring constant in the mechanical oscillator, L or C in the RLC oscillator) is modulated at a frequency ω_p. As an example consider the case of the RLC circuit shown in Figure 8-14, in which the capacitance is modulated according to

$$C = C_0\left(1 - \frac{\Delta C}{C_0} \sin \omega_p t\right) \tag{8.6-4}$$

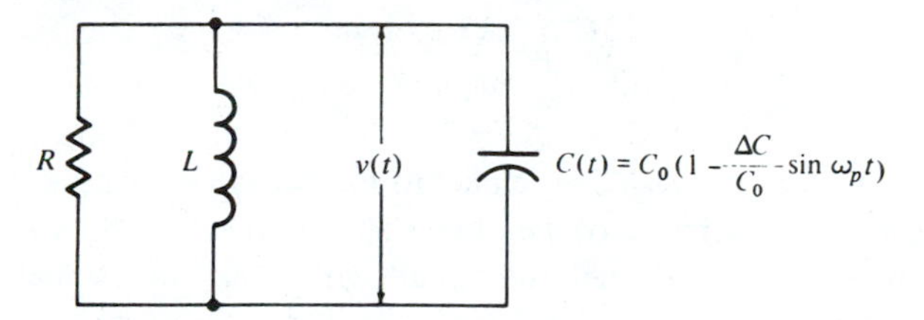

Figure 8-14 A degenerate parametric oscillator with a sinusoidally modulated capacitance.

The equation of the voltage across the *RLC* circuit is given by (8.6-1) with $\omega_0^2 = (LC)^{-1}$.

Using (8.6-4) and assuming $\Delta C \ll C_0$, (8.6-1) becomes

$$\frac{d^2v}{dt^2} + \kappa\frac{dv}{dt} + \frac{1}{LC_0}\left(1 + \frac{\Delta C}{C_0}\sin\omega_p t\right)v = 0 \tag{8.6-5}$$

which, if we make the identification

$$\omega_0^2 = \frac{1}{LC_0}, \qquad \alpha = \frac{\omega_0^2\Delta C}{2C_0} \tag{8.6-6}$$

is identical to (8.6-3).

The most important feature of the parametrically driven oscillator described by (8.6-3) is that it is capable of *sustained oscillation* at ω_0. To show this let us assume a solution

$$v = a\cos[\omega t + \phi] \tag{8.6-7}$$

Expanding $\sin\omega_p t$ in (8.6-3) in terms of exponentials, substituting (8.6-7) and neglecting nonsynchronous terms oscillating at $(\omega_p + \omega)$ leads to

$$(\omega_0^2 - \omega^2)e^{i(\omega t+\phi)} + i\omega\kappa e^{i(\omega t+\phi)} - i\alpha e^{i[(\omega_p-\omega)t-\phi]} = 0 \tag{8.6-8}$$

From (8.6-8) it follows that steady-state oscillation is possible if

$$\omega_p = 2\omega \qquad (\text{so that } \omega_p - \omega = \omega)$$

$$\omega = \omega_0 \qquad \phi = 0 \text{ or } \pi \qquad \alpha = \omega_0\kappa \tag{8.6-9}$$

or, in words:

The pump frequency ω_p is twice the oscillation frequency ω_0. The oscillation phase[9] is $\phi = 0$ or π and the strength of the pumping α must satisfy $\alpha = \omega_0\kappa$. The last condition is referred to as the "start-oscillation condition" or "threshold condition," since it gives the pumping strength (α) needed to overcome the losses (κ) at the oscillation threshold. In the case of the *RLC* circuit, whose capacitance is modulated according to (8.6-4), the threshold oscillation condition $\alpha = \omega_0\kappa$ can be written with the aid of (8.6-6) as

[9]The phase ϕ is of fundamental importance and it is defined relative to that of the pump oscillation as given by (8.6-4).

$$\frac{\Delta C}{2C_0} = \frac{\kappa}{\omega_0} = \frac{1}{Q} \tag{8.6-10}$$

where the quality factor $Q = \omega_0 RC$ is related to the decay rate κ by $\kappa = \omega_0/Q$.

In practice, if the capacitance of the circuit shown in Figure 8-14 is modulated so that condition (8.6-10) is satisfied, the circuit will break into spontaneous oscillation at a frequency $\omega_0 = \omega_p/2$. This constitutes a transfer of energy from ω_p to $\omega_p/2$.

The physical nature of this transfer may become clearer if we consider the time behavior of the voltage $v(t)$, the charge $q(t)$, and the capacitance $C(t)$ as illustrated in Figure 8-15.

$C(t)$ is a parallel-plate capacitor whose capacitance is periodically varied. Assume first that $C(t)$ is varied as in Figure 8-15(a) by pulling the capacitor plates apart and pushing them together again [$C \propto$ (plate separation)$^{-1}$]. At the same time the circuit is caused to oscillate so that the charge $q(t)$ on the capacitor plates varies as in Figure 8-15(b). Now, according to Figure 8-15(a), when the charge on the plates is a maximum, the plates are pulled apart slightly. The charge cannot change instantaneously, but since work must be done (against the Coulomb attraction of the opposite charges on the capacitor plates) to separate the plates, energy is fed into the capacitor and appears as a sudden increase in the voltage ($v = q/C$, $\mathscr{E} = \frac{1}{2}q^2/C$), as in Figure 8-15(c). One quarter of a period later, the charge and thus the field between the plates is zero and the plates can be returned to their original position with no energy expenditure. At the end of half a cycle, the charge has reversed sign and is again a maximum, so the plates are pulled apart once more. This process is then repeated many times, causing the total voltage to increase twice in each oscillation cycle. In this way, energy at *twice* the resonant frequency is pumped into the circuit where it appears as an increase in energy of the resonant frequency.

There are two noteworthy features to this degenerate oscillator. First, the frequency of the pump *must* be very nearly twice the resonant frequency of the oscillator for gain to occur, in agreement with the previous conclusions, see (8.6-9). In addition, the phase of the pump relative to the charge on the capacitor plates must be chosen properly. Consider the case where $C(t) = C_0 \pm \Delta C \sin 2\omega_0 t$, as in Figure 8-15(d). If we take the minus sign, which corresponds to the $\phi = 0°$ curve, then energy is continuously fed *into* the system as described above. If, however, the pumping phase is inverted (that is, the plus sign), then the capacitor plates are pushed together when the charge is a maximum, thus performing work, giving up energy, and decreasing the total voltage. Any initial oscillations that may be present will be damped out. The phase condition ($\phi = 0$) agrees with the second of (8.6-9).

To make a connection between the lumped-circuit parametric oscillator and the optical nonlinearity discussed in (8.1-14) we show that the (time) modulation of a capacitance at some frequency ω_p which was shown to give rise to oscillation at $\omega_p/2$ is formally equivalent to applying a field at ω_p to a nonlinear dielectric in which the polarization p and the electric field e are related by

$$p = \epsilon_0 \chi e + de^2 \tag{8.6-11}$$

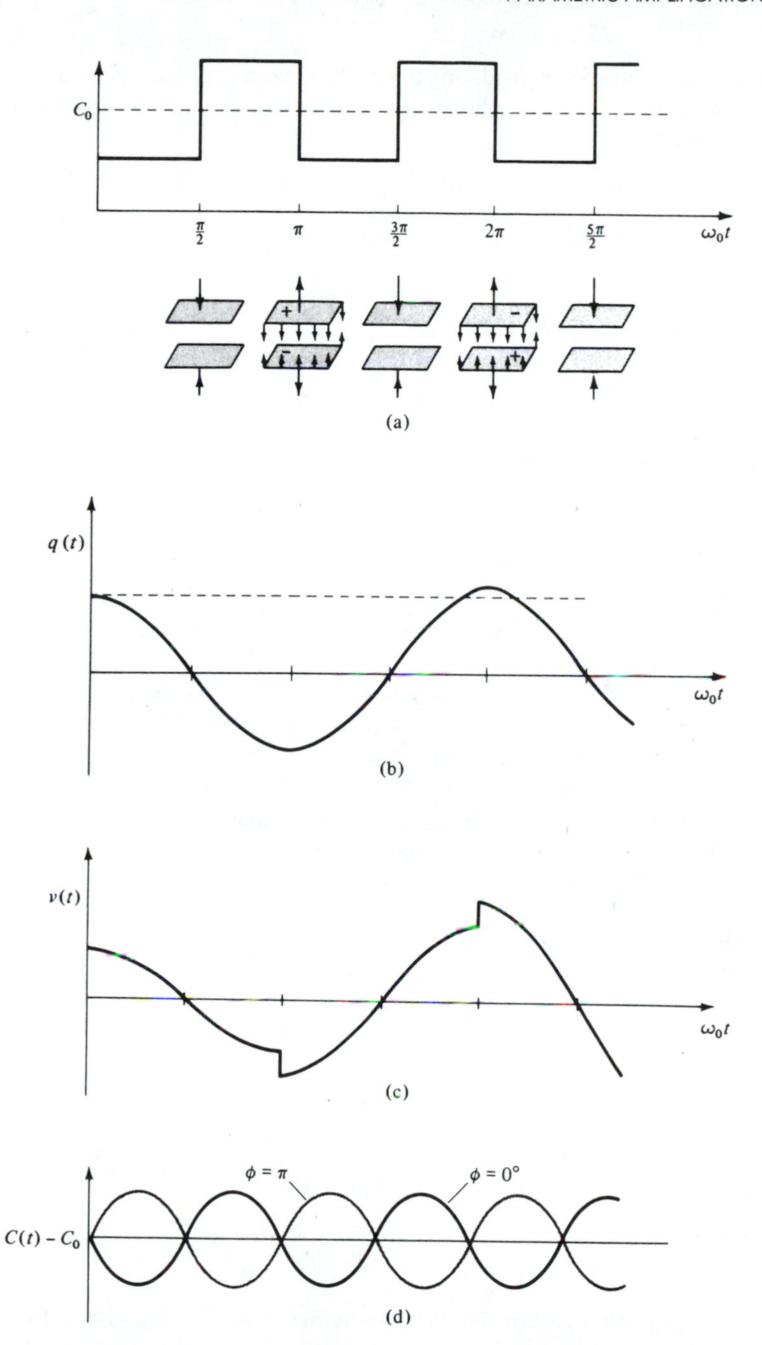

Figure 8-15 Physical model of a capacitively pumped parametric oscillator. (a) Square-wave capacitance variation at twice the circuit oscillation frequency. (Also shown is the motion of the capacitor plates, the charge, and the forces on the plates.) (b) The charge on one of the capacitor plates. (c) The voltage across the circuit. (d) Variation of the capacitance $C(t)$ at two phases relative to that of the charge.

This can be done by considering a parallel-plate capacitance of area A and separation s that is filled with a medium whose polarization is given by (8.6-11). Using the relations[10]

$$d(t) = \epsilon_0 e(t) + p(t) = \epsilon e(t) \tag{8.6-12}$$

the dielectric constant ϵ can be written as

$$\epsilon = \epsilon_0(1 + \chi) + de$$

and the capacitance $C = \epsilon A/s$ as

$$C = \frac{\epsilon_0(1 + \chi)A}{s} + \frac{Ad}{s} e \tag{8.6-13}$$

If the electric field is given by

$$e = -E_0 \sin \omega_p t$$

the capacitance becomes

$$C = \frac{\epsilon_0(1 + \chi)A}{s} - \frac{AdE_0}{s} \sin \omega_p t \tag{8.6-14}$$

which is of a form identical to (8.6-4). It follows that the two points of view used to describe parametric processes—the one represented by (8.6-4), in which an energy-storage parameter is modulated, and that in which the electric (or magnetic) response is nonlinear, as in (8.6-11)—are equivalent.

We return now to the basic nonlinear parametric equations (8.2-13) to analyze the case of optical parametric amplification. We find it convenient as in (8.3-22) to introduce a new field variable, defined by

$$A_l \equiv \sqrt{\frac{n_l}{\omega_l}} E_l \qquad l = 1, 2, 3 \tag{8.6-15}$$

so that the power flow per unit area at ω_l is given by

$$\frac{P_l}{A} = \frac{1}{2}\sqrt{\frac{\epsilon_0}{\mu}}\, n_l|E_l|^2 = \frac{1}{2}\sqrt{\frac{\epsilon_0}{\mu}}\, \omega_l|A_l|^2 \tag{8.6-16}$$

when n_l is the index of refraction at ω_l. The power flow P_l/A per unit area is related to the flux N_l (photons per square meter per second) by

$$\frac{P_l}{A} = N_l\hbar\omega_l = \frac{1}{2}\sqrt{\frac{\epsilon_0}{\mu}}\, |A_l|^2\omega_l \tag{8.6-17}$$

so that $|A_l|^2$ is proportional to the photon flux at ω_l. The equations of motion (8.2-13) for the A_l variables become

[10]The electric displacement $d(t)$ should not be confused with the nonlinear constant d in (8.6-11).

$$\frac{dA_1}{dz} = -\frac{1}{2}\alpha_1 A_1 - \frac{i}{2}\kappa A_2^* A_3 e^{-i(\Delta k)z}$$

$$\frac{dA_2^*}{dz} = -\frac{1}{2}\alpha_2 A_2^* + \frac{i}{2}\kappa A_1 A_3^* e^{i(\Delta k)z}$$

$$\frac{dA_3}{dz} = -\frac{1}{2}\alpha_3 A_3 - \frac{i}{2}\kappa A_1 A_2 e^{i(\Delta k)z} \quad (8.6\text{-}18)$$

where

$$\Delta k \equiv k_3 - (k_1 + k_2)$$

$$\kappa \equiv d\sqrt{\left(\frac{\mu}{\epsilon_0}\right)\frac{\omega_1\omega_2\omega_3}{n_1 n_2 n_3}}$$

$$\alpha_l \equiv \sigma_l\sqrt{\frac{\mu}{\epsilon_l}} \qquad l = 1, 2, 3 \quad (8.6\text{-}19)$$

The advantage of using the A_l instead of E_l is now apparent since, unlike (8.2-13), relations (8.6-18) involve a single coupling parameter κ.

We will now use (8.6-18) to solve for the field variables $A_1(z)$, $A_2(z)$, and $A_3(z)$ for the case in which three waves with amplitudes $A_1(0)$, $A_2(0)$, and $A_3(0)$ at frequencies ω_1, ω_2, and ω_3, respectively, are incident on a nonlinear crystal at $z = 0$. We take $\omega_3 = \omega_1 + \omega_2$, $\alpha_1 = \alpha_2 = \alpha_3 = 0$ (no losses), and $\Delta k = k_3 - k_1 - k_2 = 0$. In addition, we assume that $\omega_1|A_1(z)|^2$ and $\omega_2|A_2(z)|^2$ remain small compared to $\omega_3|A_3(0)|^2$ throughout the interaction region. This last condition, in view of (8.6-17), is equivalent to assuming that the power drained off the "pump" (at ω_3) by the "signal" (ω_1) and idler (ω_2) is negligible compared to the input power at ω_3. This enables us to view $A_3(z)$ as a constant. With the assumptions stated above, equations (8.6-18) become

$$\frac{dA_1}{dz} = -\frac{ig}{2}A_2^* \qquad \frac{dA_2^*}{dz} = \frac{ig^*}{2}A_1 \quad (8.6\text{-}20)$$

where

$$g \equiv \kappa A_3(0) = \sqrt{\left(\frac{\mu}{\epsilon_0}\right)\frac{\omega_1\omega_2}{n_1 n_2}}\, dE_3(0) \quad (8.6\text{-}21)$$

The solution of the coupled equations (8.6-20) subject to the initial conditions $A_1(z = 0) \equiv A_1(0)$, $A_2(z = 0) \equiv A_2(0)$, $A_3(0) = A_3^*(0)$, i.e., g is real, is

$$A_1(z) = A_1(0)\cosh\frac{g}{2}z - iA_2^*(0)\sinh\frac{g}{2}z$$

$$A_2^*(z) = A_2^*(0)\cosh\frac{g}{2}z + iA_1(0)\sinh\frac{g}{2}z \quad (8.6\text{-}22)$$

Equations (8.6-22) describe the growth of the signal and idler waves under phase-matching conditions. In the case of parametric amplification the input will consist of the pump (ω_3) wave and one of the other two fields, say, ω_1. In this case $A_2(0) = 0$, and using the relation $N_i \propto A_i A_i^*$ for the photon flux we obtain from (8.6-22)

$$N_1(z) \propto A_1^*(z)A_1(z) = |A_1(0)|^2 \cosh^2 \frac{gz}{2} \xrightarrow[gz \gg 1]{} \frac{|A_1(0)|^2}{4} e^{gz}$$

$$N_2(z) \propto A_2^*(z)A_2(z) = |A_1(0)|^2 \sinh^2 \frac{gz}{2} \xrightarrow[gz \gg 1]{} \frac{|A_1(0)|^2}{4} e^{gz} \qquad (8.6\text{-}23)$$

Thus, for $gz \gg 1$, the photon fluxes at ω_1 and ω_2 grow exponentially. If we limit our attention to the wave at ω_1, it undergoes an amplification by a factor

$$\frac{A_1^*(z)A_1(z)}{A_1^*(0)A_1(0)} \underset{gz \gg 1}{=} \tfrac{1}{4} e^{gz} \qquad (8.6\text{-}24)$$

Numerical Example: Parametric Amplification

The magnitude of the gain coefficient g available in a traveling-wave parametric interaction is estimated for the following case involving the use of a $LiNbO_3$ crystal.

$$d_{311} = 5 \times 10^{-23} \text{ MKS} \quad \text{(see Table 8-1)}$$

$$\nu_1 \cong \nu_2 = 3 \times 10^{14} \text{ Hz}$$

$$P_3/\text{Area} = \text{(pump power)} = 5 \times 10^6 \text{ W/cm}^2$$

$$n_1 \simeq n_3 = 2.2$$

Converting P_3 to $|E_3|^2$ with the use of (8.6-16) and then substituting in (8.6-21), yields

$$g = 0.7 \text{ cm}^{-1}$$

This shows that traveling-wave parametric amplification is not expected to lead to large values of gain except for extremely large pump-power densities. The main attraction of the parametric amplification just described is probably in giving rise to parametric oscillation, which will be described in Section 8.8.

8.7 PHASE-MATCHING IN PARAMETRIC AMPLIFICATION

In the preceding section the analysis of parametric amplification assumed that the phase-matching condition

$$k_3 = k_1 + k_2 \tag{8.7-1}$$

is satisfied. It is important to determine the consequences of violating this condition. We start with equations (8.6-18) taking the loss coefficients $\alpha_1 = \alpha_2 = 0$ and $g = g^*$:

$$\frac{dA_1}{dz} = -i\frac{g}{2}A_2^* e^{-i(\Delta k)z}$$

$$\frac{dA_2^*}{dz} = +i\frac{g}{2}A_1 e^{i(\Delta k)z} \tag{8.7-2}$$

The solution of (8.7-2) is facilitated by the substitution

$$A_1(z) = m_1 e^{[s-i(\Delta k/2)]z}$$

$$A_2^*(z) = m_2 e^{[s+i(\Delta k/2)]z} \tag{8.7-3}$$

where m_1 and m_2 are coefficients independent of z. The exponential growth constant s is to be determined. Substitution of (8.7-3) in (8.7-2) leads to

$$\left(s - i\frac{\Delta k}{2}\right)m_1 + i\frac{g}{2}m_2 = 0$$

$$-i\frac{g}{2}m_1 + \left(s + i\frac{\Delta k}{2}\right)m_2 = 0 \tag{8.7-4}$$

By equating the determinant of the coefficients of m_1 and m_2 in (8.7-4) to zero, we obtain the two solutions

$$s_{\pm} = \pm\tfrac{1}{2}\sqrt{g^2 - (\Delta k)^2} \equiv \pm b \tag{8.7-5}$$

The general solution of (8.7-2) is the sum of the two independent solutions

$$A_1(z) = m_1^+ e^{[s_+ - i(\Delta k/2)]z} + m_1^- e^{[s_- - i(\Delta k/2)]z}$$

$$A_2^*(z) = m_2^+ e^{[s_+ + i(\Delta k/2)]z} + m_2^- e^{[s_- + i(\Delta k/2)]z} \tag{8.7-6}$$

The coefficients m_1^+, m_1^-, m_2^+, m_2^- are next determined by requiring that at $z = 0$ the solution (8.7-6) agree with the input amplitudes $A_1(0)$ and $A_2^*(0)$. This leads straightforwardly to the result

$$A_1(z)e^{i(\Delta k/2)z} = A_1(0)\left[\cosh(bz) - \frac{i(\Delta k)}{2b}\sinh(bz)\right] - i\frac{g}{2b}A_2^*(0)\sinh(bz) \tag{8.7-7}$$

$$A_2^*(z)e^{-i(\Delta k/2)z} = A_2^*(0)\left[\cosh(bz) + \frac{i(\Delta k)}{2b}\sinh(bz)\right] + i\frac{g}{2b}A_1(0)\sinh(bz)$$

The last result reduces, as it should, to (8.6-22) if we put $\Delta k = 0$.

The most noteworthy feature of (8.7-5) and (8.7-7) is that the exponential gain coefficient b is a function of Δk and that unless

$$g \geqq \Delta k \tag{8.7-8}$$

no sustained growth of the signal (A_1) and idler (A_2) waves is possible, since in this case the sinh and cosh functions in (8.7-7) become

$$i\sin\{\tfrac{1}{2}[(\Delta k)^2 - g^2]^{1/2}z\}$$
$$\cos\{\tfrac{1}{2}[(\Delta k)^2 - g^2]^{1/2}z\}$$

respectively, and the energies at ω_1 and ω_2 oscillate as functions of the distance z.

The problem of phase-matching in parametric amplification is fundamentally the same as that in second-harmonic generation. Instead of satisfying the condition (8.3-6), $k^{2\omega} = 2k^{\omega}$, we have, according to (8.7-1), to satisfy the condition

$$k_3 = k_1 + k_2$$

This is done, as in second-harmonic generation, by using the dependence of the phase velocity of the extraordinary wave in uniaxial crystals on the direction of propagation. In a negative uniaxial crystal ($n_e < n_0$), we can, as an example, choose the signal and idler waves as ordinary while the pump at ω_3 is applied as an extraordinary wave. Using (8.3-11) and the relation $k^{\omega} = (\omega/c)n^{\omega}$, the phase-matching condition (8.7-1) is satisfied when all three waves propagate at an angle θ_m to the z (optic) axis where

$$n_e^{\omega_3}(\theta_m) = \left[\left(\frac{\cos\theta_m}{n_o^{\omega_3}}\right)^2 + \left(\frac{\sin\theta_m}{n_e^{\omega_3}}\right)^2\right]^{-1/2} = \frac{\omega_1}{\omega_3}n_o^{\omega_1} + \frac{\omega_2}{\omega_3}n_o^{\omega_2} \tag{8.7-9}$$

8.8 PARAMETRIC OSCILLATION[11]

In the two preceding sections we have demonstrated that a pump wave at ω_3 can cause a simultaneous amplification in a nonlinear medium of "signal" and "idler" waves at frequencies ω_1 and ω_2, respectively, where $\omega_3 = \omega_1 + \omega_2$. If the nonlinear crystal is placed within an optical resonator (as shown in Figure 8-16) that provides resonances for the signal or idler waves (or both), the parametric gain will, at some threshold pumping intensity, cause a simultaneous oscillation at the signal and idler frequencies. The threshold pumping corresponds to the point at which the parametric gain just balances the losses of the signal and idler waves. This is the physical basis of the optical parametric oscillator. Its practical importance derives from its ability to convert the power output of the pump laser to power at the signal and idler frequencies that, as will be shown below, can be tuned continuously over large ranges.

To analyze this situation we return to (8.6-18). We take $\Delta k = 0$ and neglect the depletion of the pump waves, so $A_3(z) = A_3(0)$. The result is

[11]See References [14–16].

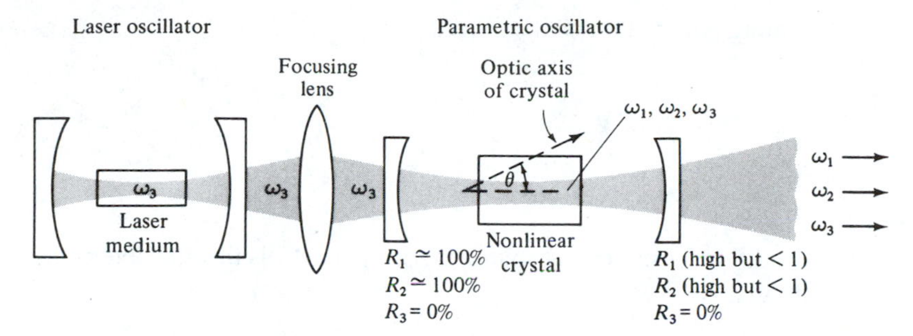

Figure 8-16 Schematic diagram of an optical parametric oscillator in which the laser output at ω_3 is used as the pump, giving rise to oscillations at ω_1 and ω_2 (where $\omega_3 = \omega_1 + \omega_2$) in an optical cavity that contains the nonlinear crystal and resonates at ω_1 and ω_2.

$$\frac{dA_1}{dz} = -\frac{1}{2}\alpha_1 A_1 - i\frac{g}{2}A_2^*$$

$$\frac{dA_2^*}{dz} = -\frac{1}{2}\alpha_2 A_2^* + i\frac{g}{2}A_1 \tag{8.8-1}$$

where, as in (8.6-21),

$$g \equiv \sqrt{\left(\frac{\mu}{\epsilon_0}\right)\frac{\omega_1\omega_2}{n_1 n_2}}\, dE_3(0)$$

$$\alpha_{1,2} \equiv \sigma_{1,2}\sqrt{\frac{\mu}{\epsilon_{1,2}}} \tag{8.8-2}$$

Equations (8.8-1) describe traveling-wave parametric interaction. We will use them to describe the interaction inside a resonator such as the one shown in Figure 8-16. This procedure seems plausible if we think of propagation inside an optical resonator as a folded optical path. The magnitude of the spatial distributed loss constants α_1 and α_2 must then be chosen so that they account for the actual losses in the resonator. The latter will include losses caused by the less than perfect reflection at the mirrors, as well as distributed loss in the nonlinear crystal and that due to diffraction.[12]

If the parametric gain is sufficiently high to overcome the losses, steady-state oscillation results. When this is the case,

$$\frac{dA_1}{dz} = \frac{dA_2^*}{dz} = 0 \tag{8.8-3}$$

and thus the power gained via the parametric interaction just balances the losses.

[12] The effective loss constant α_i is chosen so that $\exp(-\alpha_i l)$ is the total attenuation in intensity per resonator pass at ω_i, where l is the crystal length.

Putting $d/dz = 0$ in (8.8-1) gives

$$-\frac{\alpha_1}{2} A_1 - i\frac{g}{2} A_2^* = 0$$

$$i\frac{g}{2} A_1 - \frac{\alpha_2}{2} A_2^* = 0 \tag{8.8-4}$$

The condition for nontrivial solutions for A_1 and A_2^* is that the determinant at (8.8-4) vanish; that is,

$$\det \begin{vmatrix} -\frac{\alpha_1}{2} & -i\frac{g}{2} \\ i\frac{g}{2} & -\frac{\alpha_2}{2} \end{vmatrix} = 0$$

and, therefore,

$$g^2 = \alpha_1 \alpha_2 \tag{8.8-5}$$

This is the *threshold condition* for *parametric oscillation.*

If we choose to express the mode losses at ω_1 and ω_2 by the quality factors Q_1 and Q_2, respectively, we have[13]

$$\alpha_i = \frac{\omega_i n_i}{Q_i c} \tag{8.8-6}$$

By the use of (8.8-2), condition (8.8-5) can be written as

$$\frac{d(E_3)_t}{\sqrt{\epsilon_1 \epsilon_2}} = \frac{1}{\sqrt{Q_1 Q_2}} \tag{8.8-7}$$

where $(E_3)_t$ is the value of E_3 at threshold. This relation can be shown to be formally analogous to that obtained in (8.6-10) for the lumped-circuit parametric oscillator. According to (8.6-14), $\Delta C/C_0 = dE_0/\epsilon$; therefore, apart from a factor of two, if we put $Q_1 = Q_2$ and $\epsilon_1 = \epsilon_2$, (8.8-7) is the same as (8.6-10).

Another useful form of the threshold relation results from representing the quality factor Q in terms of the (effective) mirror reflectivities as in (4.7-3) and (4.7-5). If, furthermore, we express E_3 in terms of the power flow per unit area according to (8.6-16)

$$E_3^2 = 2\frac{P_3}{A}\sqrt{\frac{\mu}{\epsilon_0 n_3^2}}$$

[13]This relation follows from recognizing that the temporal decay rate $\sigma = \omega/Q$ is related to α by $\sigma = \alpha c/n$.

we can rewrite (8.8-7) as

$$\left(\frac{P_3}{A}\right)_t = \frac{1}{2}\left(\frac{\epsilon_0}{\mu}\right)^{3/2} \frac{n_1 n_2 n_3 (1 - R_1)(1 - R_2)}{\omega_1 \omega_2 l^2 d^2} \tag{8.8-8}$$

where l is the length of the nonlinear crystal, and $(1 - R_1)$ and $(1 - R_2)$ are the losses per pass at ω_1 and ω_2, respectively.

Numerical Example: Parametric Oscillation Threshold

Let us estimate the threshold pump requirement P_3/A (watts per square centimeter) of a parametric oscillator of the kind shown in Figure 8-16, which utilizes an $LiNbO_3$ crystal. We use the following set of parameters:

$(1 - R_1) = (1 - R_2) = 2 \times 10^{-2}$ (that is, loss per pass at ω_1 and $\omega_2 = 2$ percent)

$(\lambda)_1 = (\lambda)_2 = 1\ \mu\text{m}$

$l = 5$ cm (crystal length)

$n_1 = n_2 = n_3 = 1.5$

$d_{311}(LiNbO_3) = 5 \times 10^{-23}$ (MKS)

Substitution in (8.8-8) yields

$$\left(\frac{P_3}{A}\right)_t \cong 41\ \text{watts/cm}^2$$

This is a very modest intensity so that the example helps us appreciate the attractiveness of optical parametric oscillation as a means for generating coherent optical frequency at new optical frequencies.

8.9 FREQUENCY TUNING IN PARAMETRIC OSCILLATION

We have shown above that the pair of signals (ω_1) and idler frequencies that are caused to oscillate by parametric pumping at ω_3 satisfy the condition $k_3 = k_1 + k_2$. Using $k_i = \omega_i n_i/c$ we can write it as

$$\omega_3 n_3 = \omega_1 n_1 + \omega_2 n_2 \tag{8.9-1}$$

In a crystal the indices of refraction generally depend, as shown in Section 8.3, on the frequency, crystal orientation (if the wave is extraordinary), electric field (in electrooptic crystals), and on the temperature. If, as an example, we change the crystal orientation in the oscillator shown in Figure 8-16, the oscillation frequencies

ω_1 and ω_2 will change so as to compensate for the change in indices, and thus condition (8.9-1) will be satisfied at the new frequencies.

To be specific, we consider the case of a parametric oscillator pumped by an extraordinary beam at a fixed frequency ω_3. The signal (ω_1) and the idler (ω_2) are ordinary waves. At some crystal orientation θ_0 the oscillation takes place at frequencies ω_{10} and ω_{20}. Let the indices of refraction at ω_{10}, ω_{20}, and ω_3 under those conditions be n_{10}, n_{20}, and n_{30}, respectively. We want to find the change in ω_1 and ω_2 due to a small change $\Delta\theta$ in the crystal orientation.

From (8.9-1) we have, at $\theta = \theta_0$,

$$\omega_3 n_{30} = \omega_{10} n_{10} + \omega_{20} n_{20} \tag{8.9-2}$$

After the crystal orientation has been changed from θ_0 to $\theta_0 + \Delta\theta$, the following changes occur:

$$\begin{aligned} n_{30} &\to n_{30} + \Delta n_3 \\ n_{10} &\to n_{10} + \Delta n_1 \\ n_{20} &\to n_{20} + \Delta n_2 \\ \omega_{10} &\to \omega_{10} + \Delta\omega_1 \end{aligned}$$

Since $\omega_1 + \omega_2 = \omega_3 = \text{constant}$,

$$\omega_{20} \to \omega_{20} + \Delta\omega_2 = \omega_{20} - \Delta\omega_1$$

that is, $\Delta\omega_2 = -\Delta\omega_1$. Since (8.9-1) must be satisfied at $\theta = \theta_0 + \Delta\theta$, we have

$$\omega_3(n_{30} + \Delta n_3) = (\omega_{10} + \Delta\omega_1)(n_{10} + \Delta n_1) + (\omega_{20} - \Delta\omega_1)(n_{20} + \Delta n_2)$$

Neglecting the second-order terms $\Delta n_1 \Delta\omega_1$ and $\Delta n_2 \Delta\omega_1$ and using (8.9-2), we obtain

$$\Delta\omega_1 \Big|_{\substack{\omega_1 \simeq \omega_{10} \\ \omega_2 \simeq \omega_{20}}} = \frac{\omega_3 \Delta n_3 - \omega_{10}\Delta n_1 - \omega_{20}\Delta n_2}{n_{10} - n_{20}} \tag{8.9-3}$$

According to our starting hypotheses the pump is an extraordinary ray; therefore, according to (1.4-12), its index depends on the orientation θ, giving

$$\Delta n_3 = \frac{\partial n_3}{\partial \theta}\Big|_{\theta_0} \Delta\theta \tag{8.9-4}$$

The signal and idler are ordinary rays, so their indices depend on the frequencies but not on the direction. It follows that

$$\begin{aligned} \Delta n_1 &= \frac{\partial n_1}{\partial \omega_1}\Big|_{\omega_{10}} \Delta\omega_1 \\ \Delta n_2 &= \frac{\partial n_2}{\partial \omega_2}\Big|_{\omega_{20}} \Delta\omega_2 \end{aligned} \tag{8.9-5}$$

Using the last two equations in (8.9-3) results in

$$\frac{\partial \omega_1}{\partial \theta} = \frac{\omega_3(\partial n_3/\partial\theta)}{(n_{10} - n_{20}) + [\omega_{10}(\partial n_1/\partial\omega_1) - \omega_{20}(\partial n_2/\partial\omega_2)]} \tag{8.9-6}$$

for the rate of change of the oscillation frequency with respect to the crystal orientation. Using (1.4-12) and the relation $d(1/x^2) = -(2/x^3)\,dx$, we obtain

$$\frac{\partial n_3}{\partial \theta} = -\frac{n_3^3}{2}\sin(2\theta)\left[\left(\frac{1}{n_e^{\omega 3}}\right)^2 - \left(\frac{1}{n_o^{\omega 3}}\right)^2\right]$$

which, when substituted in (8.9-6), gives

$$\frac{\partial \omega_1}{\partial \theta} = \frac{-\dfrac{1}{2}\,\omega_3 n_{30}^3\left[\left(\dfrac{1}{n_e^{\omega_3}}\right)^2 - \left(\dfrac{1}{n_o^{\omega_3}}\right)^2\right]\sin(2\theta_0)}{(n_{10} - n_{20}) + \left(\omega_{10}\dfrac{\partial n_1}{\partial \omega_1} - \omega_{20}\dfrac{\partial n_2}{\partial \omega_2}\right)} \tag{8.9-7}$$

An experimental curve showing the dependence of the signal and idler frequencies on θ in $NH_4H_2PO_4$ (ADP) is shown in Figure 8-17. Also shown is a theoretical curve based on a quadratic approximation of (8.9-7), which was plotted using the dispersion (that is, n versus ω) data of ADP; see Reference [17].

Reasoning similar to that used to derive the angle-tuning expression (8.9-7) can be applied to determine the dependence of the oscillation frequency on temperature. Here we need to know the dependence of the various indices on temperature. This is discussed further in Problem 8-6. An experimental temperature-tuning curve is shown in Figure 8-18.

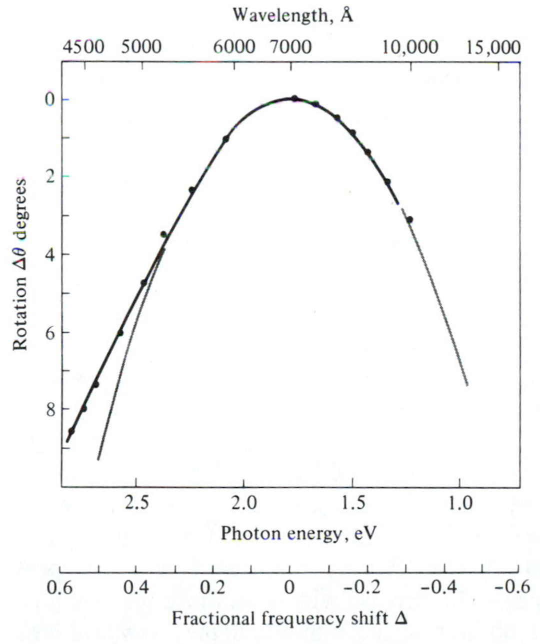

Figure 8-17 Dependence of the signal (ω_1) frequency on the angle between the pump propagation direction and the optic axis of the ADP crystal. The angle θ is measured with respect to the angle for which $\omega_1 = \omega_3/2$. $\Delta \equiv (\omega_1 - \omega_3/2)/(\omega_3/2)$. (After Reference [17].)

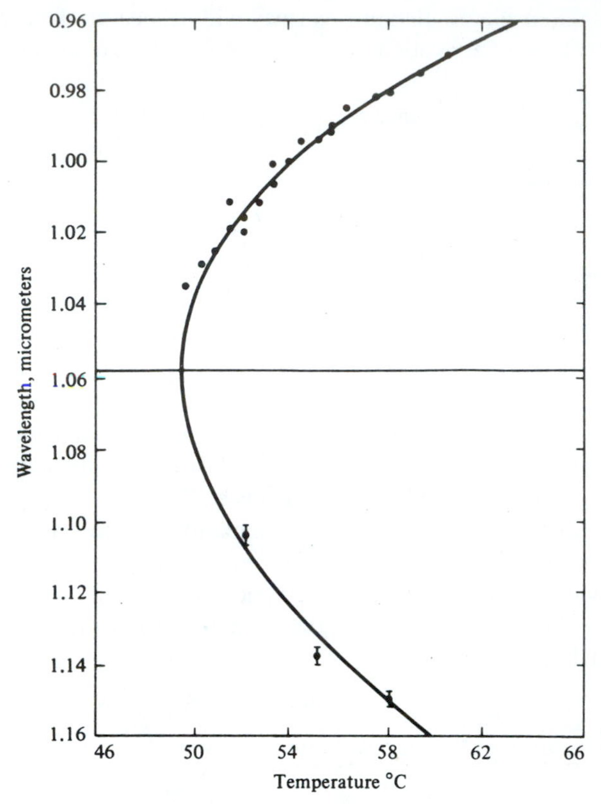

Figure 8-18 Signal and idler wavelength as a function of the temperature of the oscillator crystal. (After Reference [15].)

8.10 POWER OUTPUT AND PUMP SATURATION IN OPTICAL PARAMETRIC OSCILLATORS

In the treatment of the laser oscillator in Section 6.5 we showed that in the steady state the gain could not exceed the threshold value regardless of the intensity of the pump. A closely related phenomenon exists in the case of parametric oscillation. The pump field E_3 gives rise to amplification of the signal and idler waves. When E_3 reaches its critical (threshold) value given by (8.8-7), the gain just equals the losses and the device is on the threshold of oscillation. If the pump field E_3 is increased beyond its threshold value, the gain can no longer follow it and must be "clamped" at its threshold value. This follows from the fact that if the gain constant g exceeds its threshold value (8.8-5), a steady state is no longer possible and the signal and idler intensities will increase with time. Since the gain g is proportional to the pump field E_3, it follows that above threshold the *pump field inside* the optical resonator must saturate at its level just prior to oscillation. As power is conserved it follows that any additional pump power input must be diverted into power at the signal and idler fields. Since $\omega_3 = \omega_1 + \omega_2$, it follows that for each input pump photon above threshold we generate one photon at the signal (ω_1) and one at the idler (ω_2) frequencies, so [18]

$$\frac{P_1}{\omega_1} = \frac{P_2}{\omega_2} = \frac{(P_3)_t}{\omega_3}\left(\frac{P_3}{(P_3)_t} - 1\right) \tag{8.10-1}$$

The last argument shows that in principle the parametric oscillator can attain high efficiencies. This requires operation well above threshold, and thus $P_3/(P_3)_t \gg 1$. These considerations are borne out by actual experiments [19].

Figure 8-19 shows experimental confirmation of the phenomenon of pump saturation; see References [18, 21]. After a transient buildup the pump intensity inside the resonator settles down to its threshold value.

Figure 8-19(b) shows that the signal power is proportional to the excess (above threshold) pump input power. This is in agreement with Equation (8.10-1).

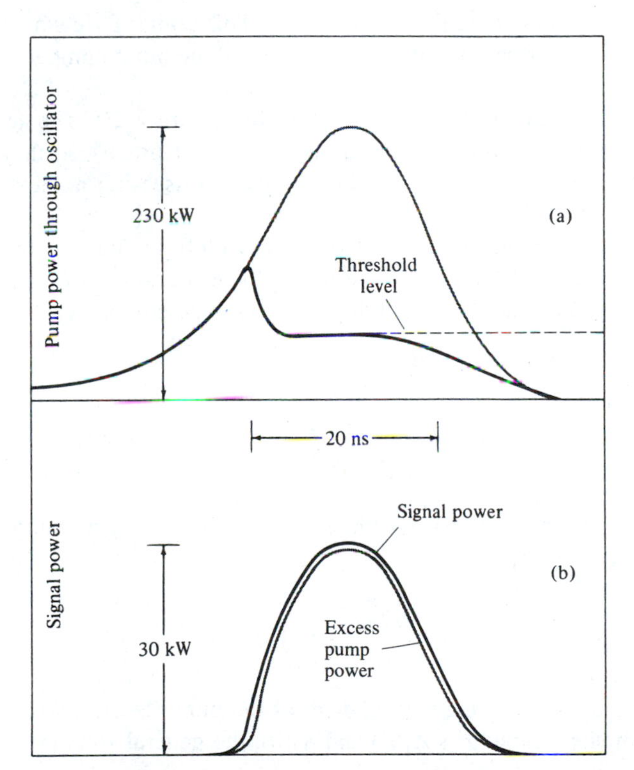

Figure 8-19 Power levels and pumping in a parametric oscillator. (a) Waveforms of P_3, the pump power passing through the oscillator. The gray waveform was obtained when the crystal was rotated so that oscillation did not occur; the solid waveform was obtained when oscillation took place. (b) Signal power and excess pump power. The gray waveform is the normalized difference between the waveforms in (a). (After Reference [19].)

8.11 FREQUENCY UP-CONVERSION

Parametric interactions in a crystal can be used to convert a signal from a "low" frequency ω_1 to a "high" frequency ω_3 by mixing it with a strong laser beam at ω_2, where

$$\omega_1 + \omega_2 = \omega_3 \tag{8.11-1}$$

Using the quantum mechanical photon picture described in Section 8.5, we can consider the basic process taking place in frequency up-conversion as one in which a signal (ω_1) photon and a pump (ω_2) photon are annihilated while, simultaneously, one photon at ω_3 is generated; see References [12, 22–25]. Since a photon energy is $\hbar\omega$, conservation of energy dictates that $\omega_3 = \omega_1 + \omega_2$ and, in a manner similar to (8.5-1), the conservation of momentum leads to the relationship

$$\mathbf{k}_3 = \mathbf{k}_1 + \mathbf{k}_2 \tag{8.11-2}$$

between the wave vectors at the three frequencies. This point of view also suggests that the number of output photons at ω_3 cannot exceed the input number of photons at ω_1.

The experimental situation is demonstrated by Figure 8-20. The ω_1 and ω_2 beams are combined in a partially transmissive mirror (or prism), so they traverse together (in near parallelism) the length l of a crystal possessing nonlinear optical characteristics.

The analysis of frequency up-conversion starts with Equation (8.6-18). Assuming negligible depletion of the pump wave A_2, no losses ($\alpha = 0$) at ω_1 and ω_3, and taking $\Delta k = 0$, we can write the first and third of these equations as

$$\frac{dA_1}{dz} = -i\frac{g}{2}A_3$$

$$\frac{dA_3}{dz} = -i\frac{g}{2}A_1 \tag{8.11-3}$$

where, using (8.6-15) and (8.6-19) and choosing without loss of generality the pump phase as zero so that $A_2(0) = A_2^*(0)$,

$$g \equiv \sqrt{\frac{\omega_1\omega_3}{n_1 n_3}\left(\frac{\mu}{\epsilon_0}\right)}\, dE_2 \tag{8.11-4}$$

where E_2 is the amplitude of the electric field of the pump laser. Taking the input waves with (complex) amplitudes $A_1(0)$ and $A_3(0)$, the general solution of (8.11-3) is

$$A_1(z) = A_1(0)\cos\left(\frac{g}{2}z\right) - iA_3(0)\sin\left(\frac{g}{2}z\right)$$

$$A_3(z) = A_3(0)\cos\left(\frac{g}{2}z\right) - iA_1(0)\sin\left(\frac{g}{2}z\right) \tag{8.11-5}$$

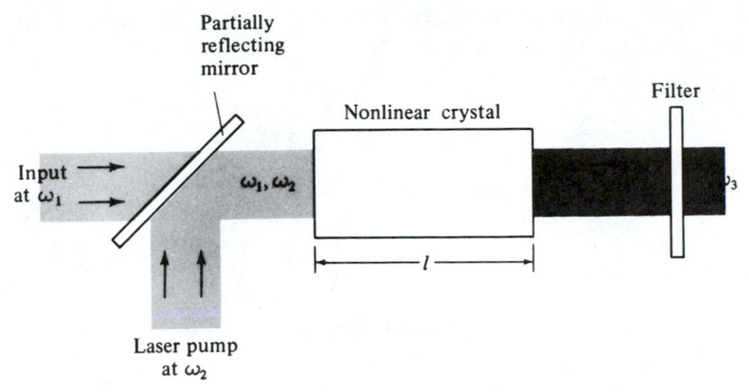

Figure 8-20 Parametric frequency up-conversion in which a signal at ω_1 and a strong laser beam at ω_2 combine in a nonlinear crystal to generate a beam at the sum frequency $\omega_3 = \omega_1 + \omega_2$.

In the case of a single (low) frequency input at ω_1, we have $A_3(0) = 0$. In this case,

$$|A_1(z)|^2 = |A_1(0)|^2 \cos^2\left(\frac{g}{2} z\right)$$

$$|A_3(z)|^2 = |A_1(0)|^2 \sin^2\left(\frac{g}{2} z\right) \tag{8.11-6}$$

therefore,

$$|A_1(z)|^2 + |A_3(z)|^2 = |A_1(0)|^2$$

In the discussion following (8.6-17) we pointed out that $|A_l(z)|^2$ is proportional to the photon flux (photons per square meter per second) at ω_l. Using this fact we may interpret (8.11-6) as stating that the photon flux at ω_1 plus that at ω_3 at any plane z is a constant equal to the input ($z = 0$) flux at ω_1. If we rewrite (8.11-6) in terms of powers, we obtain

$$P_1(z) = P_1(0) \cos^2\left(\frac{g}{2} z\right)$$

$$P_3(z) = \frac{\omega_3}{\omega_1} P_1(0) \sin^2\left(\frac{g}{2} z\right) \tag{8.11-7}$$

In a crystal of length l, the conversion efficiency is thus

$$\frac{P_3(l)}{P_1(0)} = \frac{\omega_3}{\omega_1} \sin^2\left(\frac{g}{2} l\right) \tag{8.11-8}$$

and can have a maximum value of ω_3/ω_1, corresponding to the case in which all the input (ω_1) photons are converted to ω_3 photons.

In most practical situations the conversion efficiency is small (see the following numerical example) so using $\sin x \simeq x$ for $x \ll 1$, we get

$$\frac{P_3(l)}{P_1(0)} \simeq \frac{\omega_3}{\omega_1}\left(\frac{g^2 l^2}{4}\right)$$

which, by the use of (8.11-4) and (8.6-16), can be written as

$$\frac{P_3(l)}{P_1(0)} \simeq \frac{\omega_3^2 l^2 d^2}{2n_1 n_2 n_3}\left(\frac{\mu}{\epsilon_0}\right)^{3/2}\left(\frac{P_2}{A}\right) \tag{8.11-9}$$

where A is the cross-sectional area of the interaction region.

Numerical Example: Frequency Up-Conversion

The main practical interest in parametric frequency up-conversion stems from the fact that it offers a means of detecting infrared radiation (a region where detectors are either inefficient, very slow, or require cooling to cryogenic temperatures) by converting the frequency into the visible or near-visible part of the spectrum. The radiation can then be detected by means of efficient and fast detectors such as photomultipliers or photodiodes; see References [23–26].

As an example of this application, consider the problem of up-converting a 10.6-μm signal, originating in a CO_2 laser to 0.96 μm by mixing it with the 1.06-μm output of an Nd^{3+}:YAG laser. The nonlinear crystal chosen for this application has to have low losses at 1.06 μm and 10.6 μm, as well as at 0.96 μm. In addition, its birefringence has to be such as to make phase matching possible. The crystal proustite (Ag_3AsS_3) listed in Table 8-1 meets these requirements [26].

Using the data

$$\frac{P_{1.06\mu\text{m}}}{A} = 10^4 \text{ W/cm}^2 = 10^8 \text{ W/m}^2$$

$$l = 10^{-2} \text{ meter}$$

$$n_1 \simeq n_2 \simeq n_3 = 2.6 \quad \text{(an average number based on the data of Reference [26])}$$

$$d_{\text{eff}} = 1.1 \times 10^{-22} \quad \text{(MKS) (taken conservatively as a little less than half the value given in Table 8.1 for } d_{22}\text{)}$$

we obtain, from (8.11-9),

$$\frac{P_{\lambda=0.96\mu\text{m}}(l = 1\text{ cm})}{P_{\lambda=10.6\mu\text{m}}(l = 0)} = 7.1 \times 10^{-4}$$

indicating a useful amount of conversion efficiency.

8.12 QUASI PHASE-MATCHING

An alternative technique for achieving phase-matching in crystals is referred to as *quasi phase-matching* [27], a reference to a crystal fashioned in such a way that the direction of one of its principal axes, say z, is reversed periodically. This, in a properly chosen crystal orientation and polarization directions of the participating optical fields, results in a periodic modulation of the nonlinear coefficient tensor element d_{ij} responsible for the interaction. The coupled wave equations (8.2-13) remain unchanged, except that d is replaced by $d(z)$, which, being periodic, can be expanded in a Fourier series

$$d(z) = d_{\text{bulk}}\left(\sum_{m=-\infty}^{\infty} a_m \exp\left(im \frac{2\pi}{\Lambda} z\right)\right) \tag{8.12-1}$$

where Λ is the period of $d(z)$. The effect on the first of Equations (8.2-13), as an example, is to transform it to

$$\frac{dE_1}{dz} = -\frac{\sigma_1}{2}\sqrt{\frac{\mu}{\epsilon_1}}E_1 - \frac{i\omega_1}{2}\sqrt{\frac{\mu}{\epsilon_1}}d_{\text{bulk}}E_3E_2^* \sum_{m=-\infty}^{\infty} a_m \exp\left[i\left(m\frac{2\pi}{\Lambda} - k_3 + k_2 + k_1\right)z\right] \tag{8.12-2}$$

Phase-matching obtains if for some integer m the condition

$$m\frac{2\pi}{\Lambda} = k_3 - k_2 - k_1 \tag{8.12-3}$$

is satisfied. Ignoring non-phase-matched terms in (8.12-2), (their contribution averages out to zero over distances that are large compared to the coherence length), we rewrite (8.12-2) as

$$\frac{dE}{dz} = -\frac{\sigma_1}{2}\sqrt{\frac{\mu}{\epsilon_1}}E_1 - i\frac{\omega_1}{2}\sqrt{\frac{\mu}{\epsilon_1}}d_{\text{bulk}}a_m \exp\left[i\left(m\frac{2\pi}{\Lambda} - k_3 + k_2 + k_1\right)z\right] \tag{8.12-4}$$

$$a_m = \frac{1}{\Lambda}\int_0^{\Lambda}\frac{d(z)}{d_{\text{bulk}}}\exp\left(-im\frac{2\pi}{\Lambda}z\right)dz \tag{8.12-5}$$

The simplest case of a spatially periodic $d(z)$ is one in which $d(z)$ switches from d_{bulk} to $-d_{\text{bulk}}$ every $\Lambda/2$. In this case

$$a_{\text{m}} = \frac{1 - \cos m\pi}{m\pi} \text{ for } m \neq 0 \tag{8.12-6}$$

so that, choosing $m = 1$, the effective nonlinear constant is

$$d_{\text{eff}} = a_m d_{\text{bulk}} = \frac{2}{\pi}d_{\text{bulk}} \tag{8.12-7}$$

It is clear from (8.12-4) that, in principle, quasi phase-matched configurations can give rise to the same conversion efficiency as in the ideal, $\Delta k = 0$, phase-matched case, except that we require a longer interaction path to achieve it. The length penalty factor is $d_{\text{bulk}}/d_{\text{eff}} = a_m^{-1}$.

To appreciate quasi phase-matching on an intuitive basis, we note that it involves reversing of the sign of the nonlinear interaction at

$$z = L_c, 2L_c, \ldots L_c \equiv \frac{\pi}{\Delta k} \tag{8.12-8}$$

These are the locations where the power flow would reverse direction in the non-phase-matched case. This keeps the power flowing in the same sense along the length of the crystal and leads to cumulative buildup of $E^{2\omega}(z)$. This can be best visualized using a phasor plot of the interaction. Taking the specific case of second harmonic generation as an example, we can divide the interaction path L into sufficiently short segments, each of length Δ, such that $(\Delta k)\Delta \ll \pi$ and obtain from (8.3-1)

$$\Delta E^{(2\omega)}(z) = -i\omega \sqrt{\frac{\mu}{\epsilon}}\, d(E^{(\omega)})^2 e^{i\Delta kz}\Delta \tag{8.12-9}$$

where $\Delta E^{(2\omega)}$ is the complex increment to the phasor $E^{(2\omega)}(z)$ due to the segment of length Δ centered on z. By adding the increments $\Delta E^{(2\omega)}$ vectorially, or rather phasorially, we obtain the phasor diagram shown in Figure 8-21.

In the non-phase-matched case (a), the generated second-harmonic field keeps growing, reaching a maximum, usually insignificantly small, at $z = \pi/\Delta k$. At longer distances, $E^{(2\omega)}$ begins to shrink, returning to zero at $z = 2\pi/\Delta k$. In the quasi phase-matched case (b), the sign of the interaction reverses every $\Delta z = \pi/\Delta k$. This is done by reversing the sign of $d(z)$. The resultant $E^{(2\omega)}(z)$ thus keeps growing monotonically, albeit at a (spatial) rate smaller than in the case of ideal bulk phase-matched (c).

Quasi Phase-Matching in Crystal Dielectric Waveguides

Quasi phase-matching has been practiced almost exclusively in configurations involving dielectric waveguiding. The main reason is that in a single mode waveguide, the small cross-sectional area ($\sim 10^{-8}$ cm^2) leads to very large intensities even at modest input power levels. This enables efficient conversion efficiencies with reasonable, say < 1 cm, crystal length. Secondly, the lack of diffraction makes it possible to maintain the high intensity throughout the crystal length. Most of the work, to date, involves optical waveguides fabricated in $LiNbO_3$ [35]. The spatial modulation of the nonlinear coefficient is achieved by reversing locally the direction of the c axis of the crystal by means of periodically applied electric fields [36, 37] or diffusion of T_i through openings in mask with a period $\Lambda = 2L_c$ [37]. This periodic reversal is caused by a periodic reversal of the direction of the crystal's permanent electric polarization by a periodically (spatially) reversing electric field or impurity diffusion.

The main attraction of quasi phase-matched second-harmonic generation in crystal waveguides is due to the real prospect of efficient conversion from the infra-

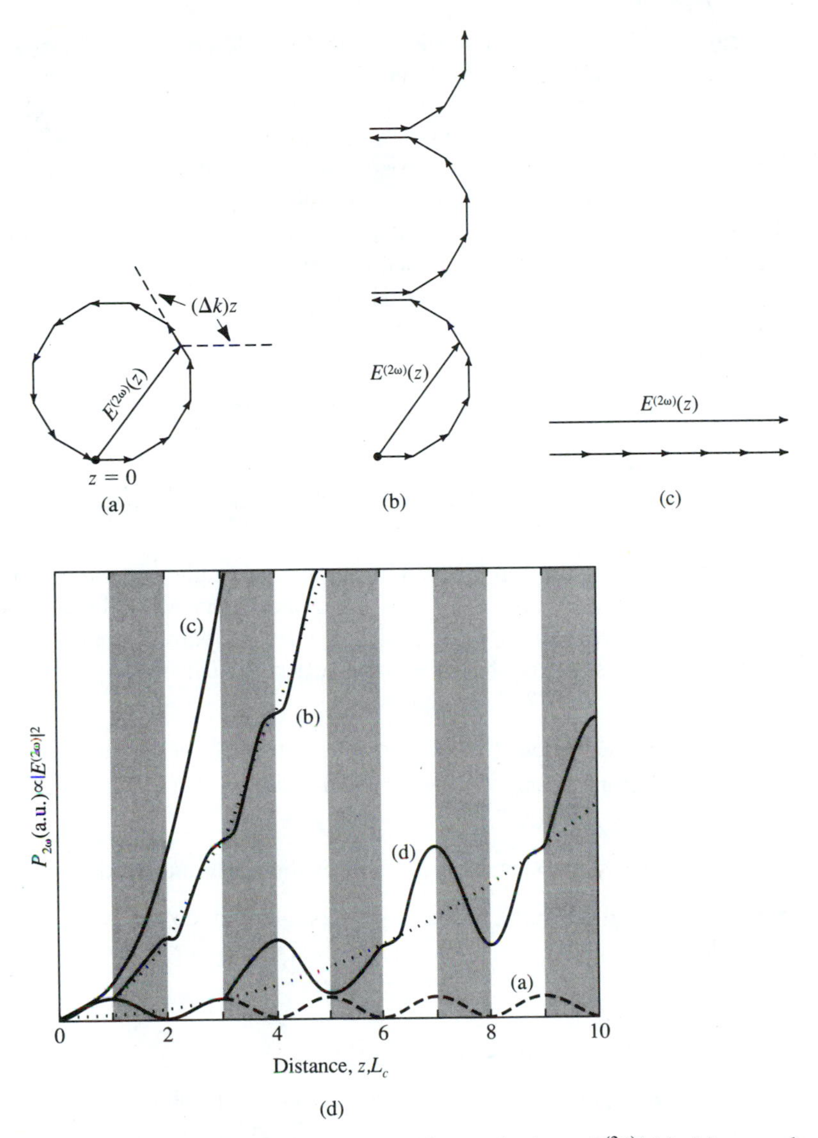

Figure 8-21 The evolution of the second-harmonic phasor $E^{(2\omega)}(z)$ in (a) a non-phase-matched case, (b) quasi phase-matched, and (c) bulk birefringent phase-matching ($\Delta k = 0$). (d) The second-harmonic power field $E^{(2\omega)}$ in a crystal for cases (a), (b), (c) above, as well as curve (d) for quasi phase-matched operation using the third Fourier coefficient $m = 3$. $L_c \equiv \dfrac{\pi}{\Delta k}$. Part (d) is reproduced from Reference [34].

red, say $\lambda \sim 1\ \mu$m, output of semiconductor diode lasers to the visible, $\lambda \sim 0.5\ \mu$m [38]. Especially attractive are configurations using the long-lived and efficient GaInAs lasers emitting near $\lambda = 1\ \mu$m. The ready availability of visible solid-state sources with power outputs in the 5 to 20-mw range will have a major impact on many applications. Chief among these applications is that of optical data storage as in compact disks. In areal storage applications, the (stored) bit density goes as λ^{-2} due to diffraction, so that frequency doubling enables the storage capacity of a given area disk to be quadrupled. In volumetric storage, such as in holographic data storage discussed in Chapter 14, the storage density goes as λ^{-3}.

Problems

8.1 Show that if θ_m is the phase-matching angle for an ordinary wave at ω and an extraordinary wave at 2ω, then

$$\Delta k(\theta)l|_{\theta=\theta_m} = -\frac{2\omega l}{c_0}\sin(2\theta_m)\frac{(n_e^{2\omega})^{-2}-(n_o^{2\omega})^{-2}}{2(n_o^{\omega})^{-3}}(\theta-\theta_m)$$

8.2 Derive the expression for the phase-matching angle of a parametric amplifier using KDP in which two of the waves are extraordinary while the third is ordinary. Which of the three waves (that is, signal, idler, or pump) would you choose as ordinary? Can this type of phase-matching be accomplished with $\omega_3 = 10{,}000\ \text{cm}^{-1}$, $\omega_1 = \omega_2 = 5000\ \text{cm}^{-1}$? If so, what is θ_m?

8.3 Show that Equations (8.6-22) are consistent with the fact that the increases in the photon flux at ω_1 and ω_2 are identical—that is, $A_1^*(z)A_1(z) - A_1^*(0)A_1(0) = A_2^*(z)A_2(z) - A_2^*(0)A_2(0)$.

8.4 Complete the missing steps in the derivation of Equation (8.7-7).

8.5 Show that the voltage $v(t)$ across an open-circuited parallel RLC circuit obeys

$$\frac{d^2v}{dt^2} + \frac{1}{RC}\frac{dv}{dt} + \frac{1}{LC}v = 0$$

and is thus of the form of Equation (8.6-1).

8.6 Consider a parametric oscillator setup such as that shown in Figure 8-16. The crystal orientation angle is θ, its temperature is T, and the signal and idler frequencies are ω_{10} and ω_{20}, respectively, with $\omega_{10} + \omega_{20} = \omega_3$. Show that a small temperature change ΔT causes the signal frequency to change by

$$\Delta\omega_1 = \Delta T \times \left\{\omega_3\left[\cos^2\theta\left(\frac{n_e^{\omega_3}(\theta)}{n_0^{\omega_3}}\right)^3\frac{\partial n_0^{\omega_3}}{\partial T} + \sin^2\theta\left(\frac{n_e^{\omega_3}(\theta)}{n_e^{\omega_3}}\right)^3\frac{\partial n_e^{\omega_3}}{\partial T}\right]\right.$$
$$\left. - \omega_{10}\frac{\partial n_o^{\omega_1}}{\partial T} - \omega_{20}\frac{\partial n_o^{\omega_2}}{\partial T}\right\} \times \frac{1}{n_{10} - n_{20}}$$

The pump is taken as an extraordinary ray, whereas the signal and idler are ordinary. [*Hint:* The starting point is Equation (8.9-3), which is valid regardless of the nature of the perturbation.]

8.7 Using the published dispersion data of proustite (Reference [26]), calculate the maximum angular deviation of the input beam at ν_1 (from parallelism with the pump beam at ν_2) that results in a reduction by a factor of 2 in the conversion efficiency. Take $\lambda_1 = 10.6\ \mu$m, $\lambda_2 = 1.06\ \mu$m, $\lambda_3 = 0.964\ \mu$m. [*Hint:* A proper choice must be made for the polarizations at ω_1, ω_2, and ω_3 so that phase-matching can be achieved along some angle.] The maximum angular deviation is that for which

$$\frac{\sin^2[\Delta k(\theta)l/2]}{[\Delta k(\theta)l/2]^2} = \frac{1}{2}$$

where, at the phase-matching angle θ_m, $\Delta k(\theta_m) = 0$. Approximate the dispersion data by a Taylor-series expansion about the nominal ($\Delta k = 0$) frequencies.

8.8 Using the dispersion data of Reference [26], discuss what happens to phase-matching in an up-conversion experiment due to a deviation of the input frequency from the nominal $(\Delta k = 0)\nu_{10}$ value. Derive an expression for the spectral width of the output in the case where the input spectral density (power per unit frequency) in the vicinity of ν_{10} is uniform. [*Hint:* Use a Taylor-series expansion of the dispersion data about the phase-matching ($\Delta k = 0$) frequencies to obtain an expression for $\Delta k(\nu_3)$.] Define the output spectral width as twice the frequency deviation at which the output is one-half its maximum ($\Delta k = 0$) value.

8.9 Explain using qualitative reasoning why (8.6-8) admits $\phi = 0$ and $\phi = \pi$ solutions.

8.10 Show that if the nonlinear optical constant is spatially periodic, i.e.,

$$d = \frac{d_0}{2}\,(e^{ikz} + e^{-ikz})$$

then a proper choice of the period $2\pi/k$ can lead to phase-matched operation. [*Hint:* Try Equation (8.2-13) and justify the neglect of terms with nonzero exponents compared with phase-matched terms (where the exponent is zero).] This method of phase-matching is referred to as quasi phase-matching [27–29].

8.11 Applying the transformation rule $d_{ij} \leftrightarrow r_{ji}$ from Section 8.1, generate the symmetry table d_{ij} of KH_2PO_4. Compare the result to that of Equation (8.1-19a).

References

1. Miller, R. C., ''Optical second harmonic generation in piezoelectric crystals,'' *Appl. Phys. Lett.* 5:17, 1964.
2. Garret, C. G. B., and F. N. H. Robinson, ''Miller's phenomenological rule for computing nonlinear susceptibilities,'' *IEEE J. Quant. Elec.* QE-2:328, 1966.
3. See, for example, J. F. Nye, *Physical Properties of Crystals.* New York: Oxford, 1957.
4. Armstrong, J. A., N. Bloembergen, J. Ducuing, and P. S. Pershan, ''Interactions between light waves in a nonlinear dielectric,'' *Phys. Rev.* 127:1918, 1962.
5. Franken, P. A., A. E. Hill, C. W. Peters, and G. Weinreich, ''Generation of optical harmonics,'' *Phys. Rev. Lett.* 7:118, 1961.

6. Maker, P. D., R. W. Terhune, M. Nisenoff, and C. M. Savage, ''Effects of dispersion and focusing on the production of optical harmonics,'' *Phys. Rev. Lett.* 8:21, 1962.
7. Giordmaine, J. A., ''Mixing of light beams in crystals,'' *Phys. Rev. Lett.* 8:19, 1962.
8. Zernike, F., Jr., ''Refractive indices of ammonium dihydrogen phosphate and potassium dihydrogen phosphate between 2000 Å and 1.5 μ,'' *J. Opt. Soc. Am.* 54:1215, 1964.
9. Thorsos, E. I., ed. *Laboratory for Laser Analytics Review*, vol. II, p. 7, 1979–1980.
10. Geusic, J. E., H. J. Levinstein, S. Singh, R. G. Smith, and L. G. Van Uitert, ''Continuous 0.53-μm solid-state source using $Ba_2NaNb_5O_{15}$,'' *IEEE J. Quant. Elec.* QE-4:352, 1968.
11. Ashkin, A., G. D. Boyd, and J. M. Dziedzic, ''Observation of continuous second harmonic generation with gas lasers,'' *Phys. Rev. Lett.* 11:14, 1963.
12. Yariv, A., *Quant. Elec.* 3rd ed. New York: Wiley, 1989.
13. Lord Rayleigh, ''On maintained vibrations,'' *Phil. Mag.*, vol. 15, ser. 5, pt. I, p. 229, 1883.
14. Parametric amplification was first demonstrated by C. C. Wang and G. W. Racette, ''Measurement of parametric gain accompanying optical difference frequency generation,'' *Appl. Phys. Lett.* 6:169, 1965.
15. The first demonstration of optical parametric oscillation is that of J. A. Giordmaine and R. C. Miller, ''Tunable optical parametric oscillation in $LiNbO_3$ at optical frequencies,'' *Phys. Rev. Lett.* 14:973, 1965.
16. Some of the early theoretical analyses of optical parametric oscillation are attributable to R. H. Kingston, ''Parametric amplification and oscillation at optical frequencies,'' *Proc. IRE* 50:472, 1962, and N. M. Kroll, ''Parametric amplification in spatially extended media and applications to the design of tunable oscillators at optical frequencies,'' *Phys. Rev.* 127:1207, 1962.
17. Magde, D., and H. Mahr, ''Study in ammonium dihydrogen phosphate of spontaneous parametric interaction tunable from 4400 to 16000 Å,'' *Phys. Rev. Lett.* 18:905, 1967.
18. Yariv, A., and W. H. Louisell, ''Theory of the optical parametric oscillator,'' *IEEE J. Quant. Elec.* QE-2:418, 1966.
19. Bjorkholm, J. E., ''Efficient optical parametric oscillation using doubly and singly resonant cavities,'' *Appl. Phys. Lett.* 13:53, 1968.
20. Kreuzer, L. B., ''High-efficiency optical parametric oscillation and power limiting in $LiNbO_3$,'' *Appl. Phys. Lett.* 13:57, 1968.
21. Siegman, A. E., ''Nonlinear optical effects: An optical power limiter,'' *Appl. Opt.* 1:739, 1962.
22. Louisell, W. H., A. Yariv, and A. E. Siegman, ''Quantum fluctuations and noise in parametric processes,'' *Phys. Rev.* 124:1646, 1961.
23. Johnson, F. M., and J. A. Durado, ''Frequency up-conversion,'' *Laser Focus* 3:31, 1967.

24. Midwinter, J. E., and J. Warner, ''Up-conversion of near infrared to visible radiation in lithium-meta-niobate,'' *J. Appl. Phys.* 38:519, 1967.

25. Warner, J., ''Photomultiplier detection of 10.6 μ radiation using optical up-conversion in proustite,'' *Appl. Phys. Lett.* 12:222, 1968.

26. Hulme, K. F., O. Jones, P. H. Davies, and M. V. Hobden, ''Synthetic proustite (Ag_3AsS_3): A new material for optical mixing,'' *Appl. Phys. Lett.* 10:133, 1967.

27. Somekh, S., and A. Yariv, ''Phase matching by periodic modulation of the nonlinear optical properties,'' *Opt. Commun.* 6(3):301, 1972.

28. Somekh, S., and A. Yariv, ''Phase matchable nonlinear optical interactions in periodic thin films,'' *Appl. Phys. Lett.* 21:140, 1972.

29. Bierlein, J. D., D. B. Laubacher, J. B. Brown, and C. J. van der Poel, ''Balanced phase matching in segmented KT_iOPO_4 waveguides,'' *Appl. Phys. Lett.* 56:1725, 1990.

30. Seka, W., S. D. Jacobs, J. E. Rizzo, R. Boni, and R. S. Craxton, ''Demonstration of high efficiency third harmonic conversion of high power Nd-glass laser radiation,'' *Optics Commun.* 34:469, 1980. Also Craxton, R. S., ''High efficiency frequency tripling schemes for high power Nd: glass lasers,'' *IEEE J. Quant. Elec.* QE-17:177, 1981. (Additional articles on doubling and frequency conversion are to be found in the same issue.)

31. Holzrichter, J. F., D. Eimerl, E. V. George, J. B. Trenholme, W. W. Simmons, and J. T. Hunt, *Physics of Laser Fusion*, vol. III. Lawrence Livermore National Laboratory, September 1982. For availability, see Supplementary Reference [32].

32. George, E. V., ed., *1981 Laser Program Annual Report—Lawrence Livermore Laboratory*, Chapter 7. (Available from the National Technical Information Service, U.S. Department of Commerce, Springfield, VA 22161.)

33. Armstrong, J. A., N. Bloembergen, J. Ducuing, and P. S. Pershan, ''Interactions between light waves in a nonlinear dielectric,'' *Phys. Rev.* 127:1918, 1962.

34. Bortz, M. L., ''Quasi phase matched optical frequency conversion in lithium niobate waveguides,'' Ph.D. thesis, Stanford University, 1994.

35. Jackel, J. L., C. E. Rice, and J. J. Vesseka, ''Proton exchange for high index waveguides in $LiNbO_3$,'' *Appl. Phys. Lett.* 41:607, 1982.

36. Fiesst, A., and P. Koidl, ''Current induced periodic ferroelectric domain reversal in $LiNbO_3$ for efficient second harmonic generation,'' *Appl. Phys. Lett.* 47:1125–1127, 1985.

37. Yamada, M., N. Nada, M. Saito, and K. Watanabe, ''First order quasi phase-matched $LiNbO_3$ waveguide periodically poled by applying an external field for efficient blue second harmonic generation,'' *Appl. Phys. Lett.* 62:435–438, 1993.

38. Ou, Z. Y., S. F. Pereira, E. S. Polzek, and H. J. Kimble, ''85% efficiency for CW doubling from 1.08 to 0.54 μm,'' *Opt. Lett.* 17:640, 1992.

Electrooptic Modulation of Laser Beams

9.0 INTRODUCTION

In Chapter 1 we treated the propagation of electromagnetic waves in anisotropic crystal media. It was shown how the properties of the propagating wave can be determined from the index ellipsoid surface.

In this chapter we consider the problem of propagation of optical radiation in crystals in the presence of an applied electric field. We find that in certain types of crystals it is possible to effect a change in the index of refraction that is proportional to the field. This is the linear electrooptic effect. It affords a convenient and widely used means of controlling the intensity or phase of the propagating radiation. This modulation is used in an ever expanding number of applications including: the impression of information onto optical beams, Q-switching of lasers (Section 6.9) for generation of giant optical pulses, mode locking, and optical beam deflection. Some of these applications will be discussed further in this chapter. Modulation and deflection of laser beams by acoustic beams are considered in Chapter 12.

9.1 ELECTROOPTIC EFFECT

In Chapter 1 we found that, given a direction in a crystal, in general two possible linearly polarized modes exist: the so-called rays of propagation. Each mode possesses a unique direction of polarization (that is, direction of **D**) and a corresponding index of refraction (that is, a velocity of propagation). The mutually orthogonal polarization directions and the indices of the two rays are found most easily by using the index ellipsoid

$$\frac{x^2}{n_x^2} + \frac{y^2}{n_y^2} + \frac{z^2}{n_z^2} = 1 \tag{9.1-1}$$

where the directions x, y, and z are the principal dielectric axes—that is, the directions in the crystal along which **D** and **E** are parallel. The existence of two rays (one "ordinary"; the other "extraordinary") with different indices of refraction is called *birefringence.*

The linear electrooptic effect is the change in the indices of the ordinary and extraordinary rays that is caused by and is proportional to an applied electric field. This effect exists only in crystals that do not possess inversion symmetry.[1] This statement can be justified as follows: Assume that in a crystal possessing an inversion symmetry, the application of an electric field E along some direction causes a change $\Delta n_1 = sE$ in the index, where s is a constant characterizing the linear electrooptic effect. If the direction of the field is reversed, the change in the index is given by $\Delta n_2 = s(-E)$, but because of the inversion symmetry the two directions are physically equivalent, so $\Delta n_1 = \Delta n_2$. This requires that $s = -s$, which is possible only for $s = 0$, so no linear electrooptic effect can exist. The division of all crystal classes into those that do and those that do not possess an inversion symmetry is an elementary consideration in crystallography and this information is widely tabulated [1].

Since the propagation characteristics in crystals are fully described by means of the index ellipsoid (9.1-1), the effect of an electric field on the propagation is expressed most conveniently by giving the changes in the constants $1/n_x^2$, $1/n_y^2$, $1/n_z^2$ of the index ellipsoid.

Following convention [1–2], we take the equation of the index ellipsoid in the presence of an electric field as

$$\left(\frac{1}{n^2}\right)_1 x^2 + \left(\frac{1}{n^2}\right)_2 y^2 + \left(\frac{1}{n^2}\right)_3 z^2 + 2\left(\frac{1}{n^2}\right)_4 yz + 2\left(\frac{1}{n^2}\right)_5 xz + 2\left(\frac{1}{n^2}\right)_6 xy = 1 \tag{9.1-2}$$

If we choose x, y, and z to be parallel to the principal dielectric axes of the crystal, then with zero applied field, Equation (9.1-2) must reduce to (9.1-1); therefore,

---

[1]If a crystal contains points (one in each unit cell) such that inversion (replacing each atom at **r** by one at $-\mathbf{r}$, with **r** being the position vector relative to the point) about any one of these points leaves the crystal structure invariant, the crystal is said to possess inversion symmetry.

$$\left(\frac{1}{n^2}\right)_1\bigg|_{E=0} = \frac{1}{n_x^2} \qquad \left(\frac{1}{n^2}\right)_2\bigg|_{E=0} = \frac{1}{n_y^2}$$

$$\left(\frac{1}{n^2}\right)_3\bigg|_{E=0} = \frac{1}{n_z^2} \qquad \left(\frac{1}{n^2}\right)_4\bigg|_{E=0} = \left(\frac{1}{n^2}\right)_5\bigg|_{E=0} = \left(\frac{1}{n^2}\right)_6\bigg|_{E=0} = 0$$

The linear change in the coefficients

$$\left(\frac{1}{n^2}\right)_i \qquad i = 1, \ldots, 6$$

due to an arbitrary dc electric field $\mathbf{E}(E_x, E_y, E_z)$ is defined by

$$\Delta\left(\frac{1}{n^2}\right)_i = \sum_{j=1}^{3} r_{ij}E_j \tag{9.1-3}$$

where in the summation over j we use the convention $1 = x$, $2 = y$, $3 = z$. Equation (9.1-3) can be expressed in a matrix form as

$$\begin{vmatrix} \Delta\left(\frac{1}{n^2}\right)_1 \\ \Delta\left(\frac{1}{n^2}\right)_2 \\ \Delta\left(\frac{1}{n^3}\right)_3 \\ \Delta\left(\frac{1}{n^2}\right)_4 \\ \Delta\left(\frac{1}{n^2}\right)_5 \\ \Delta\left(\frac{1}{n^2}\right)_6 \end{vmatrix} = \begin{vmatrix} r_{11} & r_{12} & r_{13} \\ r_{21} & r_{22} & r_{23} \\ r_{31} & r_{32} & r_{33} \\ r_{41} & r_{42} & r_{43} \\ r_{51} & r_{52} & r_{53} \\ r_{61} & r_{62} & r_{63} \end{vmatrix} \begin{vmatrix} E_1 \\ E_2 \\ E_3 \end{vmatrix} \tag{9.1-4}$$

where, using the rules for matrix multiplication, we have, for example,

$$\Delta\left(\frac{1}{n^2}\right)_6 = r_{61}E_1 + r_{62}E_2 + r_{63}E_3$$

The 6×3 matrix with elements r_{ij} is called the electrooptic tensor. We have shown above that in crystals possessing an inversion symmetry (centrosymmetric), $r_{ij} = 0$. The form, but not the magnitude, of the tensor r_{ij} can be derived from symmetry considerations [1], which dictate which of the 18 r_{ij} coefficients are zero, as well as the relationships that exist between the remaining coefficients. In Table 9-1 we give the form of the electrooptic tensor for all the noncentrosymmetric crystal classes. The electrooptic coefficients of some crystals are given in Table 9-2.

Table 9-1 The Form of the Electrooptic Tensor for all Crystal Symmetry Classes

Symbols:

- • zero element
- ● nonzero element
- ●—● equal nonzero elements
- ●—○ equal nonzero elements, but opposite in sign

The symbol at the upper left corner of each tensor is the conventional symmetry group designation.

Centrosymmetric—All elements zero

Triclinic

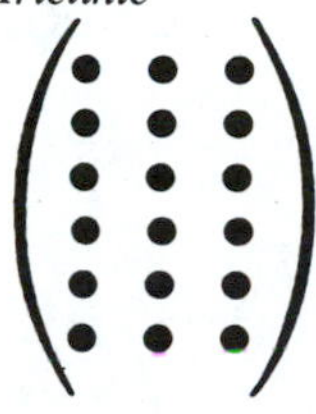

Monoclinic

2 (parallel to x_2)

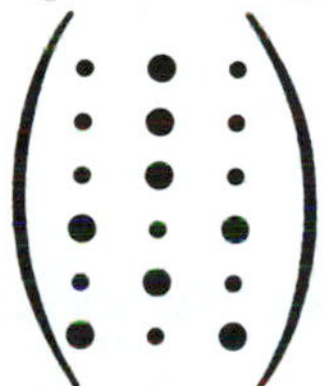

(parallel to x_3)

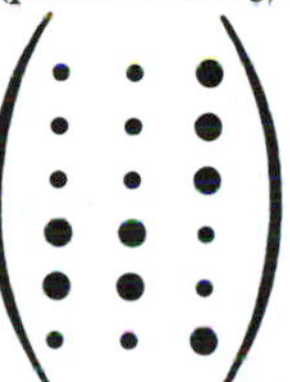

m (perpendicular to x_2)

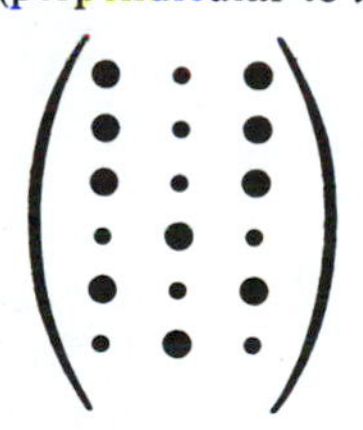

(perpendicular to x_3)

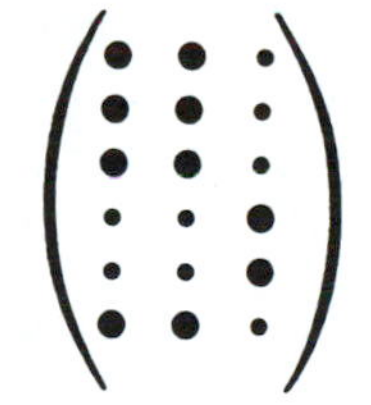

Orthorhombic

222

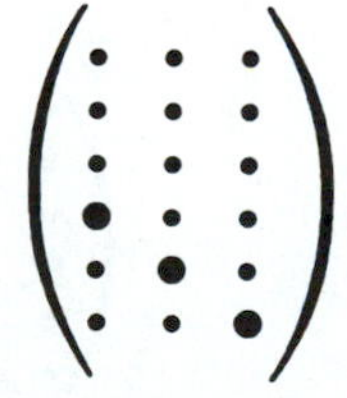

*mm*2

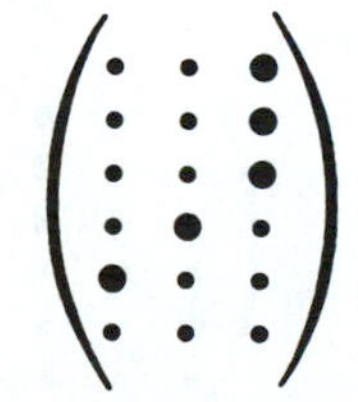

Table 9-1 *(continued)*

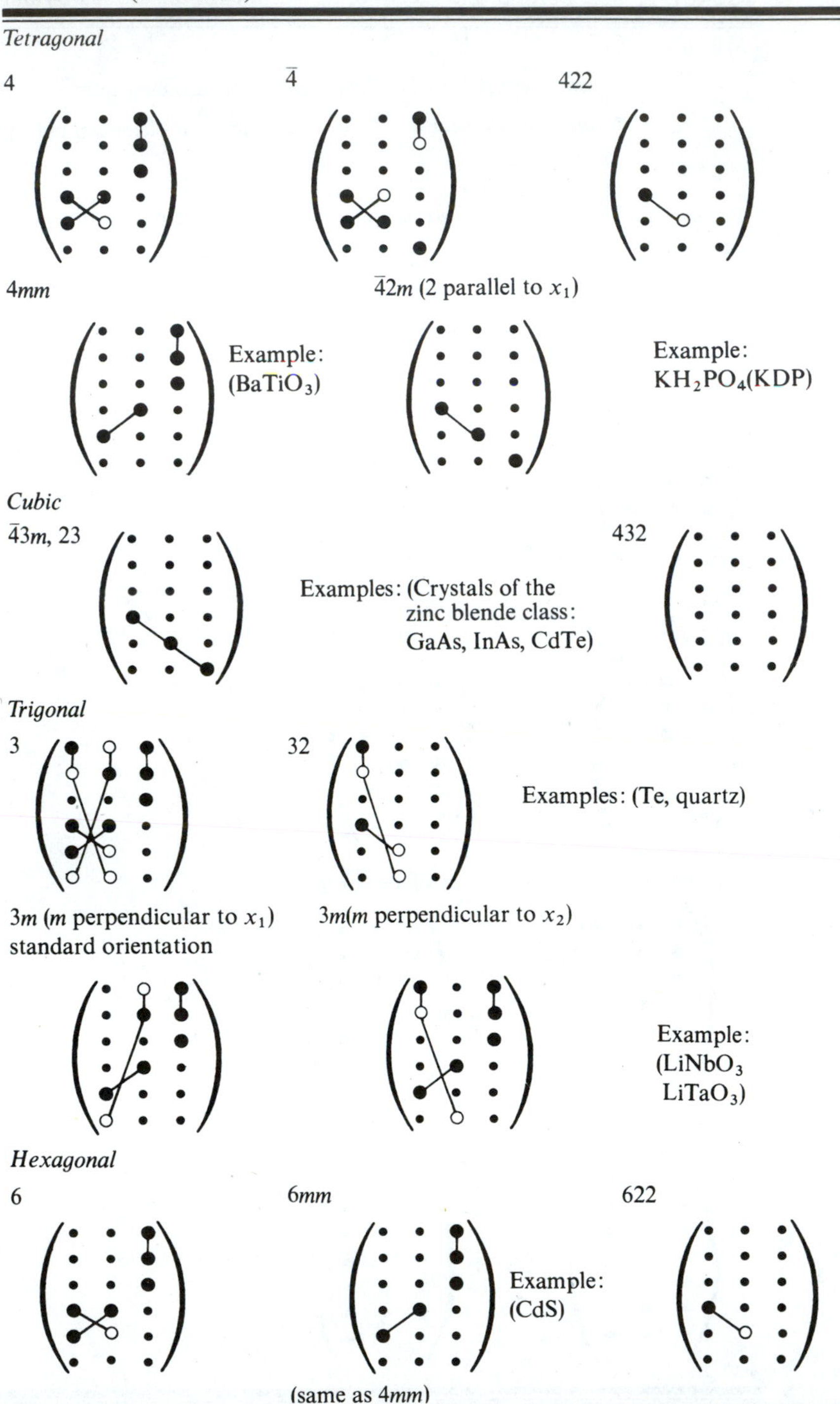

Table 9-1 *(continued)*

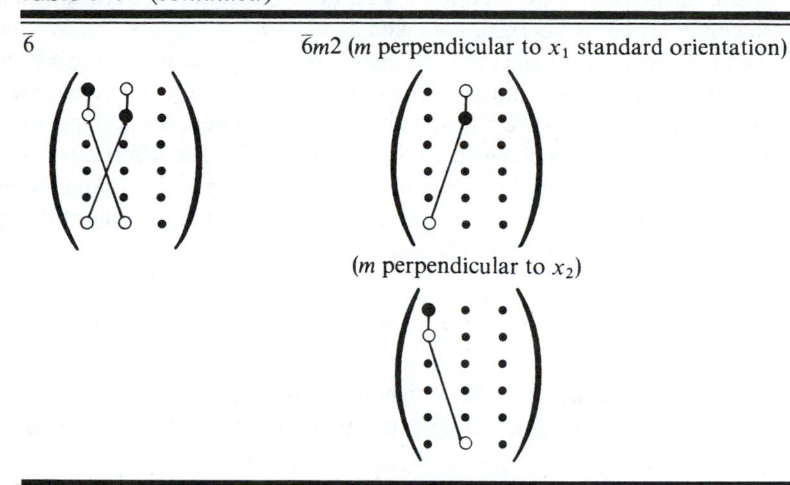

Example: The Electrooptic Effect in KH_2PO_4

Consider the specific example of a crystal of potassium dihydrogen phosphate (KH_2PO_4), also known as KDP. The crystal has a fourfold axis of symmetry,[2] which by strict convention is taken as the z (optic) axis, as well as two mutually orthogonal twofold axes of symmetry that lie in the plane normal to z. These are designated as the x and y axes. The symmetry group of this crystal is $\bar{4}2m$.[3] Using Table 9-1, we take the electrooptic tensor in the form of

$$r_{ij} = \begin{vmatrix} 0 & 0 & 0 \\ 0 & 0 & 0 \\ 0 & 0 & 0 \\ r_{41} & 0 & 0 \\ 0 & r_{41} & 0 \\ 0 & 0 & r_{63} \end{vmatrix} \qquad (9.1\text{-}5)$$

so the only nonvanishing elements are $r_{41} = r_{52}$ and r_{63}. Using (9.1-2), (9.1-4), and (9.1-5), we obtain the equation of the index ellipsoid in the presence of a field $\mathbf{E}(E_x, E_y, E_z)$ as

$$\frac{x^2}{n_o^2} + \frac{y^2}{n_o^2} + \frac{z^2}{n_e^2} + 2r_{41}E_x yz + 2r_{41}E_y xz + 2r_{63}E_z xy = 1 \qquad (9.1\text{-}6)$$

---

[2]That is, a rotation by $2\pi/4$ about this axis leaves the crystal structure invariant.

[3]The significance of the symmetry group symbols and a listing of most known crystals and their symmetry groups is to be found in any basic book on crystallography.

where the constants involved in the first three terms do not depend on the field and, since the crystal is uniaxial, are taken as $n_x = n_y = n_o$, $n_z = n_e$. We thus find that the application of an electric field causes the appearance of "mixed" terms in the equation of the index ellipsoid. These are the terms with xy, xz, and yz. This means that the major axes of the ellipsoid, with a field applied, are no longer parallel to the x, y, and z axes. It becomes necessary, then, to find the directions and magnitudes of the new axes, in the presence of **E**, so that we may determine the effect of the field on the propagation. To be specific we choose the direction of the applied field parallel to the z axis, so (9.1-6) becomes

$$\frac{x^2 + y^2}{n_o^2} + \frac{z^2}{n_e^2} + 2r_{63}E_z xy = 1 \tag{9.1-7}$$

The problem is one of finding a new coordinate system—x', y', z'—in which the equation of the ellipsoid (9.1-7) contains no mixed terms; that is, it is of the form

$$\frac{x'^2}{n_{x'}^2} + \frac{y'^2}{n_{y'}^2} + \frac{z'^2}{n_{z'}^2} = 1 \tag{9.1-8}$$

x', y', and z' are then the directions of the major axes of the ellipsoid in the presence of an external field applied parallel to z. The length of the major axes of the ellipsoid is, according to (9.1-8), $2n_{x'}$, $2n_{y'}$, and $2n_{z'}$, and these will, in general, depend on the applied field.

In the case of (9.1-7) it is clear from inspection that in order to put the equation in a diagonal form we need to choose a coordinate system x', y', z' where z' is parallel to z, and because of the symmetry of (9.1-7) in x and y, x' and y' are related to x and y by a 45° rotation, as shown in Figure 9-1. The transformation relations from x, y to x', y' are thus

$$x = x' \cos 45° + y' \sin 45°$$

$$y = -x' \sin 45° + y' \cos 45°$$

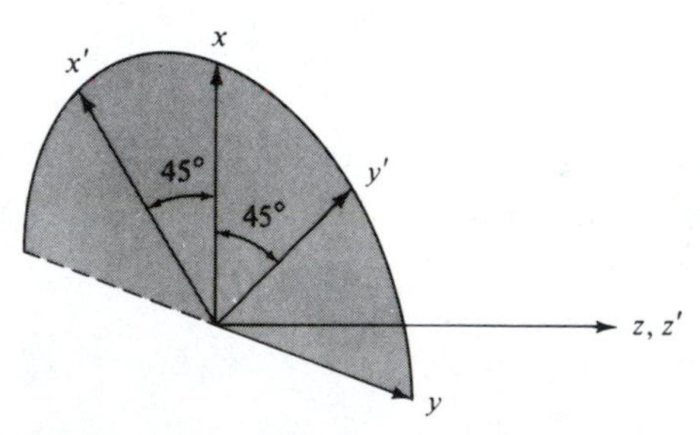

Figure 9-1 The x, y, and z axes of $\bar{4}2m$ crystals (such as KH_2PO_4) and the x', y', and z' axes, where z is the fourfold optic axis and x and y are the twofold axes of crystals with $\bar{4}2m$ symmetry.

which, upon substitution in (9.1-7), yield

$$\left(\frac{1}{n_o^2} - r_{63}E_z\right) x'^2 + \left(\frac{1}{n_o^2} + r_{63}E_z\right) y'^2 + \frac{z^2}{n_e^2} = 1 \qquad (9.1\text{-}9)$$

Equation (9.1-9) shows that x', y', and z are indeed the principal axes of the ellipsoid when a field is applied along the z direction. According to (9.1-9), the length of the x' axis of the ellipsoid is $2n_{x'}$, where

$$\frac{1}{n_{x'}^2} = \frac{1}{n_o^2} - r_{63}E_z$$

which, assuming $r_{63}E_z \ll n_o^{-2}$ and using the differential relation $dn = -(n^3/2)\, d(1/n^2)$, gives for the change in $n_{x'}$, $dn_{x'} = -(n_o^3/2)r_{63}E_z$ so that

$$n_{x'} = n_o + \frac{n_o^3}{2} r_{63}E_z \qquad (9.1\text{-}10)$$

and, similarly,

$$n_{y'} = n_o - \frac{n_o^3}{2} r_{63}E_z \qquad (9.1\text{-}11)$$

$$n_z = n_e \qquad (9.1\text{-}12)$$

The electrooptic effect in the practical imortant $\bar{4}3m$ crystal class (GaAs, InP, ZnS) is treated in detail in Appendix B.

The General Solution

We now consider the problem of optical propagation in a crystal in the presence of an external dc field along an arbitrary direction.

The index ellipsoid with the dc field on is given by (9.1-2), which we reexpress in the quadratic form

$$S_{ij}x_ix_j = 1 \qquad (9.1\text{-}13)$$

so that $S_{11} = (1/n^2)_1$, $S_{32} = S_{23} = (1/n^2)_4$, and so on. We also use the convention of summation over repeated indices. Our problem consists of finding the directions and magnitudes of the principal axes of the ellipsoid (9.1-13).

Before proceeding we need remind ourselves of one basic result of vector calculus. If the vector from the origin to a point (x_1, x_2, x_3) on the ellipsoid (9.1-13) is denoted by $\mathbf{R}(x_1, x_2, x_3)$, then the vector $\mathbf{N}$ with components

$$N_i = S_{ij}x_j \qquad (9.1\text{-}14)$$

is normal to the ellipsoid at $\mathbf{R}$.

We next apply the last result to determine the directions and magnitudes of the principal axes of the ellipsoid (9.1-13). Since the principal axes are normal to the

Table 9-2 Linear Electrooptic Coefficients of Some Commonly Used Crystals

Substance	Symmetry	Wavelength λ (μm)	Electrooptic Coefficients r_{lk} (10^{-12} m/V)	Index of Refraction n_i	n^3r (10^{-12} m/V)	Dielectric Constant* $\epsilon_i(\epsilon_0)$
CdTe (See App. B)	$\bar{4}3m$	1.0	$(T)\ r_{41} = 4.5$	$n = 2.84$	103	$(S)\ \epsilon = 9.4$
		3.39	$(T)\ r_{41} = 6.8$			
		10.6	$(T)\ r_{41} = 6.8$	$n = 2.60$	120	
		23.35	$(T)\ r_{41} = 5.47$	$n = 2.58$	94	
		27.95	$(T)\ r_{41} = 5.04$	$n = 2.53$	82	
GaAs (See App. B)	$\bar{4}3m$	0.9	$r_{41} = 1.1$	$n = 3.60$	51	$(S)\ \epsilon = 13.2$
		1.15	$(T)\ r_{41} = 1.43$	$n = 3.43$	58	$(T)\ \epsilon = 12.3$
		3.39	$(T)\ r_{41} = 1.24$	$n = 3.3$	45	
		10.6	$(T)\ r_{41} = 1.51$	$n = 3.3$	54	
GaP (See App. B)	$\bar{4}3m$	0.55–1.3	$(T)\ r_{41} = -1.0$	$n = 3.66–3.08$		$(S)\ \epsilon = 10$
		0.633	$(S)\ r_{41} = -0.97$	$n = 3.32$	35	
		1.15	$(S)\ r_{41} = -1.10$	$n = 3.10$	33	
		3.39	$(S)\ r_{41} = -0.97$	$n = 3.02$	27	
β-ZnS (sphalerite) (See App. B)	$\bar{4}3m$	0.4	$(T)\ r_{41} = 1.1$	$n = 2.52$	18	$(T)\ \epsilon = 16$
		0.5	$(T)\ r_{41} = 1.81$	$n = 2.42$		$(S)\ \epsilon = 12.5$
		0.6	$(T)\ r_{41} = 2.1$	$n = 2.36$		
		0.633	$(S)\ r_{41} = -1.6$	$n = 2.35$		
		3.39	$(S)\ r_{41} = -1.4$			
ZnSe (See App. B)	$\bar{4}3m$	0.548	$(T)\ r_{41} = 2.0$	$n = 2.66$		$(T)\ \epsilon = 9.1$
		0.633	$(S)\ r_{41} = 2.0$	$n = 2.60$	35	$(S)\ \epsilon = 9.1$
		10.6	$(T)\ r_{41} = 2.2$	$n = 2.39$		
ZnTe (See App. B)	$\bar{4}3m$	0.589	$(T)\ r_{41} = 4.51$	$n = 3.06$		$(T)\ \epsilon = 10.1$
		0.616	$(T)\ r_{41} = 4.27$	$n = 3.01$		$(S)\ \epsilon = 10.1$
		0.633	$(T)\ r_{41} = 4.04$	$n = 2.99$	108	
			$(S)\ r_{41} = 4.3$			
		0.690	$(T)\ r_{41} = 3.97$	$n = 2.93$		
		3.41	$(T)\ r_{41} = 4.2$	$n = 2.70$	83	
		10.6	$(T)\ r_{41} = 3.9$	$n = 2.70$	77	

$Bi_{12}SiO_{20}$	23	0.633	$r_{41} = 5.0$		$n = 2.54$	82
CdSe	6 mm	3.39	$(S)\ r_{13} = 1.8$		$n_o = 2.452$	$(T) \epsilon_1 = 9.70$
						$(T)\ \epsilon_3 = 10.65$
			$(T)\ r_{33} = 4.3$		$n_e = 2.471$	$(S)\ \epsilon_1 = 9.33$
						$(S)\ \epsilon_3 = 10.20$
α-ZnS	6 mm	0.633	$(S)\ r_{13} = 0.9$		$n_o = 2.347$	$(T)\ \epsilon_1 = \epsilon_2 = 8.7$
(wurtzite)			$(S)\ r_{33} = 1.8$		$n_e = 2.360$	$(S)\ \epsilon_1 = 8.7$
$Pb_{0.814}La_{0.214}$-$(Ti_{0.6}Zr_{0.4})O_3$ (PLZT)	$\infty\ m$	0.546	$n_e^3 r_{33} - n_o^3 r_{13} = 2320$		$n_o = 2.55$	
$LiIO_3$	6	0.633	$(S)\ r_{13} = 4.1$	$(S)\ r_{33} = 6.4$	$n_o = 1.8830$	
			$(S)\ r_{41} = 1.4$	$(S)\ r_{51} = 3.3$	$n_o = 1.7367$	
Ag_3AsS_3	$3m$	0.633	$(S)\ n_e^3 r_e = 70$		$n_o = 3.019$	
			$(S)\ n_o^3 r_{22} = 29$		$n_e = 2.739$	
$LiNbO_3$	$3m$	0.633	$(T_4)\ r_{13} = 9.6$	$(S)\ r_{13} = 8.6$	$n_o = {}_4 2.286$	$(T)\ \epsilon_1 = \epsilon_2 = 78$
($T_c = 1230°C$)			$(T)\ r_{22} = 6.8$	$(S)\ r_{22} = 3.4$	$n_e = 2.200$	$(T)\ \epsilon_2 = 32$
			$(T)\ r_{33} = 30.9$	$(S)\ r_{33} = 30.8$		$(S)\ \epsilon_1 = \epsilon_2 = 43$
			$(T)\ r_{51} = 32.6$	$(S)\ r_{51} = 28$		$(S)\ \epsilon_3 = 28$
			$(T)\ r_c = 21.1$			
		1.15	$(T)\ r_{22} = 5.4$		$n_o = 2.229$	
			$(T)\ r_c = 19$		$n_e = 2.150$	
		3.39	$(T)\ r_{22} = 3.1$	$(S)\ r_{33} = 28$	$n_o = 2.136$	
			$(T)\ r_c = 18$	$(S)\ r_{22} = 3.1$	$n_e = 2.073$	
				$(S)\ r_{13} = 6.5$		
				$(S)\ r_{51} = 23$		

Table 9-2 *(continued)*

Substance	Symmetry	Wavelength λ (μm)	Electrooptic Coefficients r_{lk} (10^{-12} m/V)		Index of Refraction n_i	n^3r (10^{-12} m/V)	Dielectric Constant* $\epsilon_i(\epsilon_0)$
$LiTaO_3$	$3m$	0.633	$(T)\ r_{13} = 8.4$	$(S)\ r_{13} = 7.5$	$n_o = 2.176$		$(T)\ \epsilon_1 = \epsilon_2 = 51$
			$(T)\ r_{33} = 30.5$	$(S)\ r_{33} = 33$	$n_e = 2.180$		$(T)\ \epsilon_3 = 45$
			$(T)\ r_{22} = -0.2$	$(S)\ r_{51} = 20$			$(S)\ \epsilon_1 = \epsilon_2 = 41$
			$(T)\ r_c = 22$	$(S)\ r_{22} = 1$			$(S)\ \epsilon_3 = 43$
		3.39	$(S)\ r_{33} = 27$		$n_o = 2.060$		
			$(S)\ r_{13} = 4.5$		$n_e = 2.065$		
			$(S)\ r_{51} = 15$				
			$(S)\ r_{22} = 0.3$				
$AgGaS_2$	$\bar{4}2m$	0.633	$(T)\ r_{41} = 4.0$		$n_o = 2.553$		
			$(T)\ r_{63} = 3.0$		$n_e = 2.507$		
CsH_2AsO_4	$\bar{4}2m$	0.55	$(T)\ r_{41} = 14.8$		$n_o = 1.572$		
(CDA)			$(T)\ r_{63} = 18.2$		$n_e = 1.550$		
KH_2PO_4	$\bar{4}2m$	0.546	$(T)\ r_{41} = 8.77$		$n_o = 1.5115$		$(T)\ \epsilon_1 = \epsilon_2 = 42$
(KDP)			$(T)\ r_{63} = 10.3$		$n_e = 1.4698$		$(T)\ \epsilon_3 = 21$
		0.633	$(T)\ r_{41} = 8$		$n_o = 1.5074$		$(S)\ \epsilon_1 = \epsilon_2 = 44$
			$(T)\ r_{63} = 11$		$n_e = 1.4669$		$(S)\ \epsilon_3 = 21$
		3.39	$(T)\ r_{63} = 9.7$				
			$(T)\ n_o^3 r_{63} = 33$				
KD_2PO_4	$\bar{4}2m$	0.546	$(T)\ r_{63} = 26.8$		$n_o = 1.5079$		$(T)\ \epsilon_3 = 50$
(KD*P)			$(T)\ r_{41} = 8.8$		$n_e = 1.4683$		$(S)\ \epsilon_1 = \epsilon_2 = 58$
		0.633	$(T)\ r_{63} = 24.1$		$n_o = 1.502$		$(S)\ \epsilon_3 = 48$
					$n_e = 1.462$		
$(NH_4)H_2PO_4$	$\bar{4}2m$	0.546	$(T)\ r_{41} = 23.76$		$n_o = 1.5266$		$(T)\ \epsilon_1 = \epsilon_2 = 56$
(ADP)			$(T)\ r_{63} = 8.56$		$n_e = 1.4808$		$(T)\ \epsilon_3 = 15$
		0.633	$(T)\ r_{41} = 23.41$		$n_o = 1.5220$		$(S)\ \epsilon_1 = \epsilon_2 = 58$
			$(T)\ n_o^3 r_{63} = 27.6$		$n_e = 1.4773$		$(S)\ \epsilon_3 = 14$

$(NH_4)D_2PO_4$	$\bar{4}2m$	0.633	$(T)\ r_{41} = 40$		$n_o = 1.516$	
(AD*P)			$(T)\ r_{63} = 10$		$n_e = 1.475$	
$BaTiO_3$	4 mm	0.546	$(T)\ r_{51} = 1640$	$(S)\ r_{51} = 820$	$n_o = 2.437$	$(T)\ \epsilon_1 = \epsilon_2 = 3600$
$(T_c = 395$ K)			$(T)\ r_c = 108$	$(S)\ r_c = 23$	$n_e = 2.365$	$(T)\ \epsilon_3 = 135$
$(KTa_xNb_{1-x}O_3$		0.633	$(T)\ r_{51} = 8000(T_c - 28)$		$n_o = 2.318$	
(KTN),						
$x = 0.35$			$(T)\ r_c = 500(T_c - 28)$		$n_e = 2.277$	
$(T_c = 40$–60°C)			$(T)\ r_{51} = 3000(T_c - 16)$		$n_o = 2.318$	
			$(T)\ r_c = 700(T_c - 16)$		$n_e = 2.281$	
$Ba_{0.25}Sr_{0.75}Nb_2O_6$	4 mm	0.633	$(T)\ r_{13} = 67$	$(T)\ r_{51} = 42$	$n_o = 2.3117$	$\epsilon_3 = 3400$
$(T_c = 395$ K)			$(T)\ r_{33} = 1340$	$(S)\ r_c = 1090$	$n_e = 2.2987$	(15 MHz)
α-HIO_3	222	0.633	$(T)\ r_{41} = 6.6$	$(S)\ r_{41} = 2.3$	$n_1 = 1.8365$	
			$(T)\ r_{52} = 7.0$	$(S)\ r_{52} = 2.6$	$n_2 = 1.984$	
			$(T)\ r_{63} = 6.0$	$(S)\ r_{63} = 4.3$	$n_3 = 1.960$	
$KNbO_3$	2 mm	0.633	$(T)\ r_{13} = 28$	$(T)\ r_{23} = 1.3$ $(T)\ r_{33} = 64$	$n_1 = 2.280$	
			$(T)\ r_{42} = 380$	$(S)\ r_{42} = 270$	$n_2 = 2.329$	
			$(T)\ r_{51} = 105$		$n_3 = 2.169$	
KIO_3	1	0.500	$r_{62} = 90$		$n_1 = 1.700$	
					$n_2 = 1.828$ (5893 Å)	
					$n_3 = 1.832$	

*"(T)" = low frequency from dc through audio range: (S) = high frequency.

surface, we can determine their points of intersection (x_1, x_2, x_3) with the ellipsoid by requiring that at such points the radius vector be parallel to the normal, that is,

$$S_{ij}x_j = Sx_i \tag{9.1-15}$$

where S is a constant independent of i.

Writing out (9.1-15) in component form for $i = 1, 2, 3$ gives

$$\begin{aligned}(S_{11} - S)x_1 + S_{12}x_2 + S_{13}x_3 &= 0\\ S_{21}x_1 + (S_{22} - S)x_2 + S_{23}x_3 &= 0\\ S_{31}x_1 + S_{32}x_2 + (S_{33} - S)x_3 &= 0\end{aligned} \tag{9.1-16}$$

(9.1-16) constitutes a system of three homogeneous equations for the unknowns x_1, x_2, and x_3. The condition for a nontrivial solution is that the determinant of the coefficients vanishes, that is,

$$\det[S_{ij} - S\delta_{ij}] = 0 \tag{9.1-17}$$

This is a cubic equation in S. For real S_{ij}, which is the case with lossless crystals, the three roots S', S'', and S''' of (9.1-17) are real numbers. Having solved (9.1-17) we use the three roots, one at a time, in (9.1-16) to solve, to within a multiplicative constant, for the radius vector (x_1, x_2, x_3) to the point of intersection of the principal axis with the ellipsoid. The first vector, obtained by using S', is denoted by $\mathbf{X}'(x_1', x_2', x_3')$, the second by $\mathbf{X}''(x_1'', x_2'', x_3'')$, and the third, obtained from S''', is $\mathbf{X}'''(x_1''', x_2''', x_3''')$. Since the vectors satisfy (9.1-15), we have

$$S_{ij}x_j' = S'x_i' \tag{9.1-18}$$

with a similar relation applying to x_i'' and x_i'''.

It is an easy task to prove that the three principal axis vectors $\mathbf{X}'$, $\mathbf{X}''$, $\mathbf{X}'''$ are mutually orthogonal.

So far we have solved for the directions of the principal axes. Next we obtain their magnitudes. We multiply (9.1-18) by x_i'

$$S_{ij}x_i'x_j' = S'x_i'x_i' = S'|\mathbf{X}'|^2 \tag{9.1-19}$$

But the left side of (9.1-19) is, according to (9.1-13), equal to unity since the point (x_1', x_2', x_3') is on the ellipsoid (9.1-13). We can thus write

$$|\mathbf{X}'| = \frac{1}{\sqrt{S'}}$$

with similar results for $\mathbf{X}''$ and $\mathbf{X}'''$. The lengths of the principal axes of the index ellipsoid are thus $2(S')^{-1/2}$, $2(S'')^{-1/2}$, and $2(S''')^{-1/2}$. If we then express the equation of the index ellipsoid in terms of a Cartesian coordinate system whose axes are parallel to $\mathbf{X}'$, $\mathbf{X}''$, and $\mathbf{X}'''$, it becomes

$$S'x'^2 + S''y'^2 + S'''z'^2 = 1 \tag{9.1-20}$$

where the unit vectors $\mathbf{x}'$, $\mathbf{y}'$, and $\mathbf{z}'$ here are taken as parallel to $\mathbf{X}'$, $\mathbf{X}''$, and $\mathbf{X}'''$, respectively.

The bit of mathematics starting with (9.1-13) is referred to as the transformation of a quadratic form to a principal coordinate system. An equivalent description of this transformation is by the term matrix diagonalization. The original matrix being the ordered array of the coefficients S_{ij}

$$\mathbf{S} \equiv \begin{vmatrix} S_{11} & S_{12} & S_{13} \\ S_{21} & S_{22} & S_{23} \\ S_{31} & S_{32} & S_{33} \end{vmatrix} \tag{9.1-21}$$

The set of S', S'', and S''', which are the roots of (9.1-17), are the *eigenvalues* of the matrix $\mathbf{S}$, while the vectors $\mathbf{X}'$, $\mathbf{X}''$, and $\mathbf{X}'''$ are its eigenvectors. The term matrix diagonalization follows from the fact that if we express the quadric surface

$$S_{ij}x_ix_j = 1$$

whose coefficients form the matrix $\mathbf{S}$ of (9.1-21), in terms of a Cartesian coordinate system whose axes are $\mathbf{X}'$, $\mathbf{X}''$, and $\mathbf{X}'''$, it assumes the form (9.1-20) with the diagonal form of the matrix $\mathbf{S}$ as

$$\mathbf{S} = \begin{bmatrix} S' & 0 & 0 \\ 0 & S'' & 0 \\ 0 & 0 & S''' \end{bmatrix} \tag{9.1-22}$$

Example: Electrooptic Field in KH_2PO_4

To illustrate the method of matrix diagonalization, we use the example of KH_2PO_4(KDP) with a dc field along the crystal z axis, which was solved above in a somewhat less formal fashion.

The index ellipsoid is given by (9.1-7) as

$$\frac{x^2}{n_0^2} + \frac{y^2}{n_0^2} + \frac{z^2}{n_e^2} + 2r_{63}E_zxy = 1 \tag{9.1-23}$$

The S_{ij} matrix is thus

$$S_{ij} = \begin{bmatrix} \frac{1}{n_0^2} & r_{63}E_z & 0 \\ r_{63}E_z & \frac{1}{n_0^2} & 0 \\ 0 & 0 & \frac{1}{n_e^2} \end{bmatrix} \tag{9.1-24}$$

The eigenvalues are given according to (9.1-17) as the roots of the equation

$$\det \begin{vmatrix} \dfrac{1}{n_0^2} - S & r_{63}E_z & 0 \\ r_{63}E_z & \dfrac{1}{n_0^2} - S & 0 \\ 0 & 0 & \dfrac{1}{n_e^2} - S \end{vmatrix} \tag{9.1-25}$$

which upon evaluation is

$$\left(\frac{1}{n_e^2} - S\right)\left[\left(\frac{1}{n_0^2} - S\right)^2 - (r_{63}E_z)^2\right] = 0$$

The roots are

$$\begin{aligned} S' &= \frac{1}{n_e^2} \\ S'' &= \frac{1}{n_0^2} + r_{63}E_z \\ S''' &= \frac{1}{n_0^2} - r_{63}E_z \end{aligned} \tag{9.1-26}$$

in agreement with (9.1-9). These roots are used, one at a time, in the equation

$$S_{ij}x_j = Sx_i \qquad i = 1, 2, 3 \tag{9.1-27}$$

to obtain the eigenvectors. Starting with S' we have

$$\begin{aligned} \left(\frac{1}{n_0^2} - \frac{1}{n_e^2}\right) x_1' + r_{63}E_z x_2' &= 0 \\ r_{63}E_z x_1' + \left(\frac{1}{n_0^2} - \frac{1}{n_e^2}\right) x_2' &= 0 \\ \left(\frac{1}{n_e^2} - \frac{1}{n_e^2}\right) x_3' &= 0 \end{aligned} \tag{9.1-28}$$

The first two equations above are satisfied by $x_1' = 0$ and $x_2' = 0$, while the third is satisfied by any value of x_3'. The eigenvector $\mathbf{X}'$ corresponding to $S'(=1/n_e^2)$ is thus parallel to the z axis. In a like fashion we substitute the value of S'' into (9.1-27) and find that the corresponding eigenvector $\mathbf{X}''$ is parallel to the direction $\mathbf{x} + \mathbf{y}$ while using S''' shows that $\mathbf{X}'''$ is parallel to $\mathbf{x} - \mathbf{y}$. Referring to the last two eigenvector directions as x' and y', we can rewrite the equation of the index ellipsoid in the x', y', z (principal) coordinate system as

$$\left(\frac{1}{n_0^2} - r_{63}E_z\right) x'^2 + \left(\frac{1}{n_0^2} + r_{63}E_z\right) y'^2 + \frac{z^2}{n_e^2} = 1 \tag{9.1-29}$$

where the quantities in parentheses are the eigenvalues given by (9.1-26). Equation (9.1-29) is the same as (9.1-9).

9.2 ELECTROOPTIC RETARDATION

The index ellipsoid for KDP with **E** applied parallel to z is shown in Figure 9-2. If we consider propagation along the z direction, then according to the procedure described in Section 1.4 we need to determine the ellipse formed by the intersection of the plane $z = 0$ (in general, the plane that contains the origin and is normal to the propagation direction) and the ellipsoid. The equation of this ellipse is obtained from (9.1-9) by putting $z = 0$ and is

$$\left(\frac{1}{n_o^2} - r_{63}E_z\right) x'^2 + \left(\frac{1}{n_o^2} + r_{63}E_z\right) y'^2 = 1 \tag{9.2-1}$$

One quadrant of the ellipse is shown shaded in Figure 9-2, along with its minor and major axes, which in this case coincide with x' and y', respectively. It follows from Section 1.4 that the two allowed directions of polarization are x' and y' and that their indices of refraction are $n_{x'}$ and $n_{y'}$, which are given by (9.1-10) and (9.1-11).

We are now in a position to take up the concept of retardation. We consider an optical field that is incident normally on the $x'y'$ plane with its **E** vector along the x direction. We can resolve the optical field at $z = 0$ (input plane) into two mutually orthogonal components polarized along x' and y'. The x' component propagates as

$$e_{x'} = Ae^{i[\omega t - (\omega/c)n_{x'}z]}$$

which, using (9.1-10), becomes

$$e_{x'} = Ae^{i\{\omega t - (\omega/c)[n_o + (n_o^3/2)r_{63}E_z]z\}} \tag{9.2-2}$$

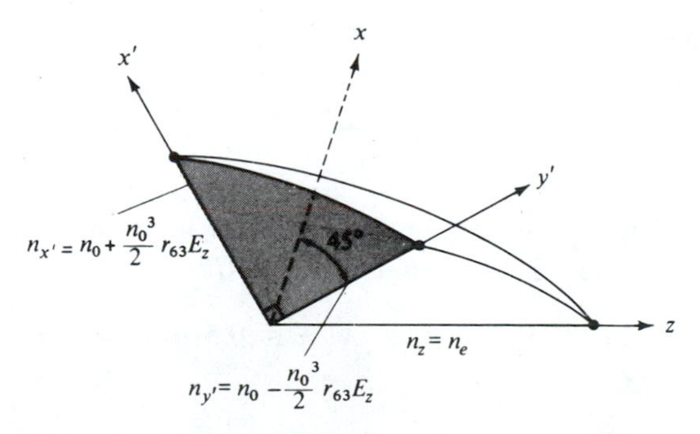

Figure 9-2 A section of the index ellipsoid of KDP, showing the principal dielectric axes x', y', and z due to an electric field applied along the z axis. The directions x' and y' are defined by Figure 9-1.

while the y' component is given by

$$e_{y'} = Ae^{i\{\omega t - (\omega/c)[n_o - (n_o^3/2)r_{63}E_z]z\}} \tag{9.2-3}$$

The phase difference at the output plane $z = l$ between the two components is called the *retardation.* It is given by the difference of the exponents in (9.2-2) and (9.2-3) and is equal to

$$\Gamma = \phi_{x'} - \phi_{y'} = \frac{\omega n_o^3 r_{63} V}{c} \tag{9.2-4}$$

where $V = E_z l$ and $\phi_{x'} = (\omega n_{x'}/c)l$.

Figure 9-3 shows $e_{x'}(z)$ and $e_{y'}(z)$ at some moment in time. Also shown are the curves traversed by the tip of the optical field vector at various points along the path. At $z = 0$, the retardation is $\Gamma = 0$ and the field is linearly polarized along x. At point e, $\Gamma = \pi/2$; thus, omitting a common phase factor, we have

$$e_{x'} = A \cos\left(\omega t - \frac{\pi}{2}\right) = A \sin \omega t$$

$$e_{y'} = A \cos \omega t \tag{9.2-5}$$

and the electric field vector is circularly polarized in the counterclockwise sense as shown in the figure. At point i, $\Gamma = \pi$ and thus

$$e_{x'} = A \cos(\omega t - \pi) = -A \cos \omega t$$

$$e_{y'} = A \cos \omega t$$

and the radiation is again linearly polarized, but this time along the y axis—that is, at 90° to its input direction of polarization.

The retardation as given by (9.2-4) can also be written as

$$\Gamma = \pi \frac{E_z l}{V_\pi} = \pi \frac{V}{V_\pi} \tag{9.2-6}$$

where V_π, the voltage yielding a retardation $\Gamma = \pi$,[4] is

$$V_\pi = \frac{\lambda}{2n_o^3 r_{63}} \tag{9.2-7}$$

where $\lambda = 2\pi c/\omega$ is the free space wavelength. Using, as an example, the value of r_{63} for ADP as given in Table 9-2, we obtain from (9.2-7)

$$(V_\pi)_{\text{ADP}} \approx 10{,}000 \text{ volts} \quad \text{at } \lambda = 0.5\ \mu\text{m}$$

[4] V_π is referred to as the "half-wave" voltage since, as can be seen in Figure 9-3c(i), it causes the two waves that are polarized along x' and y' to acquire a relative spatial displacement of $\Delta z = \lambda/2$, where λ is the optical wavelength.

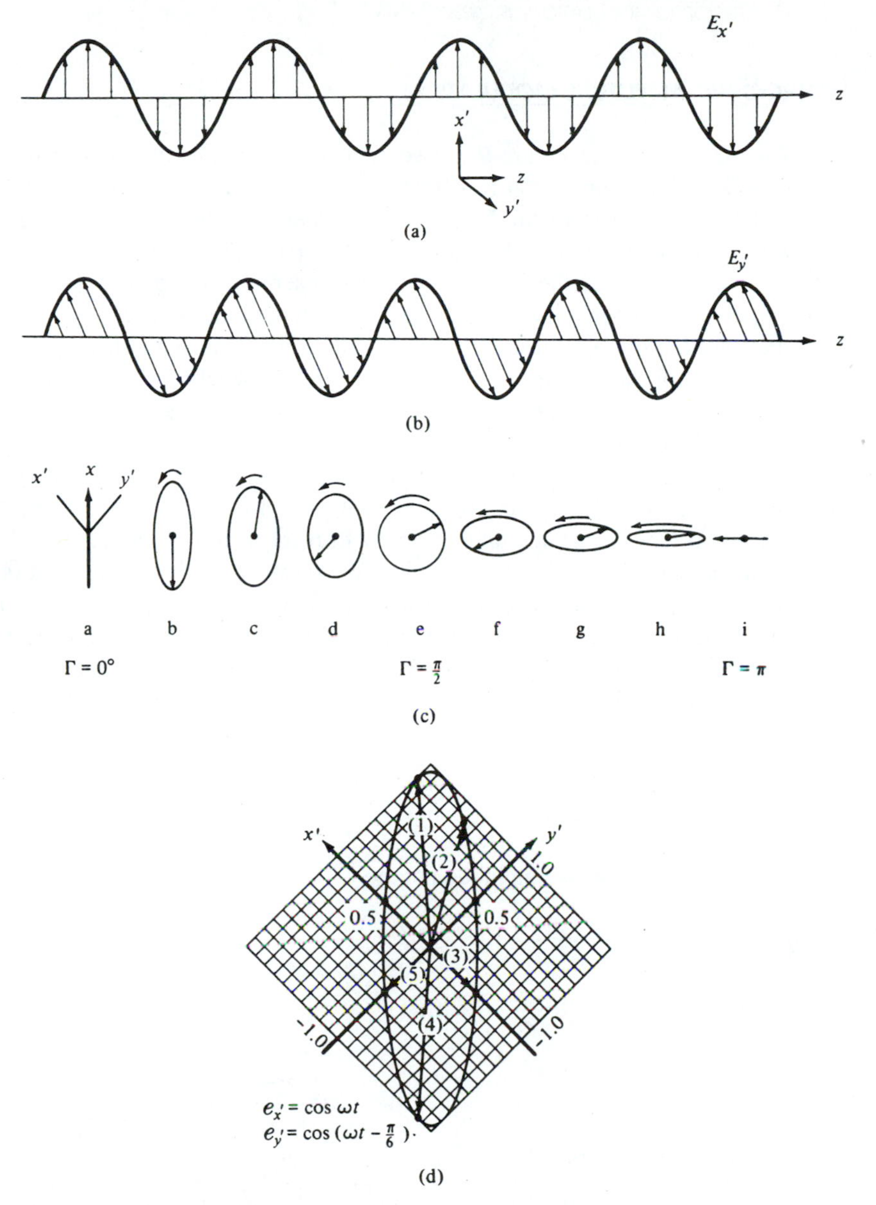

Figure 9-3 An optical field that is linearly polarized along x is incident on an electrooptic crystal having its electrically induced principal axes along x' and y'. (This is the case in KH_2PO_4 when an electric field is applied along its z axis.) (a) The component $e_{x'}$ at some time t as a function of the position z along the crystal. (b) $e_{y'}$ as a function of z at the same value of t as in (a). (c) The ellipses in the x'-y' plane traversed by the tip of the optical electric field at various points (a through i) along the crystal during one optical cycle. The arrow shows the instantaneous field vector at time t, while the curved arrow gives the sense in which the ellipse is traversed. (d) A plot of the polarization ellipse due to two orthogonal components $e_{x'} = \cos \omega t$ and $e_{y'} = \cos (\omega t - \pi/6)$. Also shown are the instantaneous field vectors at (1) $\omega t = 0°$, (2) $\omega t = 60°$, (3) $\omega t = 120°$, (4) $\omega t = 210°$, and (5) $\omega t = 270°$.

9.3 ELECTROOPTIC AMPLITUDE MODULATION

An examination of Figure 9-3 reveals that the electrically induced birefringence causes a wave launched at $z = 0$ with its polarization along x to acquire a y polarization, which grows with distance at the expense of the x component until at point i, at which $\Gamma = \pi$, the polarization becomes parallel to y. If point i corresponds to the output plane of the crystal and if one inserts at this point a polarizer at right angles to the input polarization—that is, one that allows only E_y to pass—then with the field on, the optical beam passes through unattenuated, whereas with the field off ($\Gamma = 0$), the output beam is blocked off completely by the crossed output polarizer. This control of the optical energy flow serves as the basis of the electrooptic amplitude modulation of light.

A typical arrangement of an electrooptic amplitude modulator is shown in Figure 9-4. It consists of an electrooptic crystal placed between two crossed polarizers, which, in turn, are at an angle of 45° with respect to the electrically induced birefringent axes x' and y'. To be specific, we show how this arrangement is achieved using a KDP crystal. Also included in the optical path is a naturally birefringent crystal that introduces a fixed retardation, so the total retardation Γ is the sum of the retardation due to this crystal and the electrically induced one. The incident field is parallel to x at the input face of the crystal, thus having equal-in-phase components along x' and y' that we take as

$$e_{x'} = A \cos \omega t$$

$$e_{y'} = A \cos \omega t$$

or, using the complex amplitude notation,

$$E_{x'}(0) = A$$

$$E_{y'}(0) = A$$

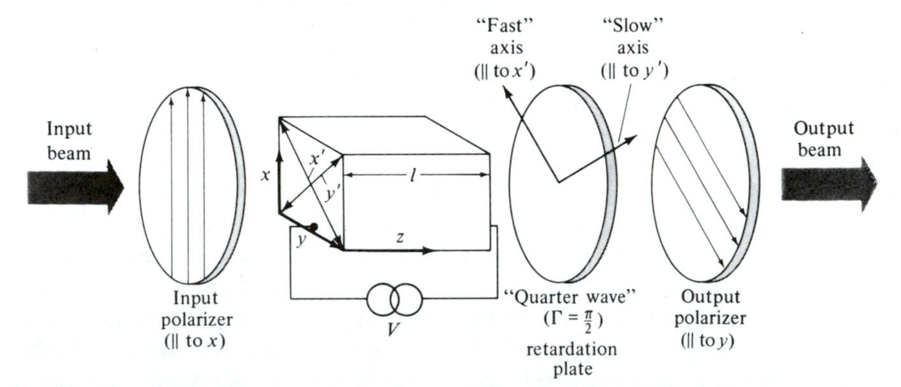

Figure 9-4 A typical electrooptic amplitude modulator. The total retardation Γ is the sum of the fixed retardation bias ($\Gamma_B = \pi/2$) introduced by the quarter-wave plate and that is attributable to the electrooptic crystal.

The incident intensity is thus[5]

$$I_i \propto \mathbf{E} \cdot \mathbf{E}^* = |E_{x'}(0)|^2 + |E_{y'}(0)|^2 = 2A^2 \tag{9.3-1}$$

Upon emerging from the output face $z = l$, the x' and y' components have acquired, according to (9.2-4), a relative phase shift (retardation) of Γ radians, so we may take them as

$$E_{x'}(l) = Ae^{-i\Gamma}$$
$$E_{y'}(l) = A \tag{9.3-2}$$

The total (complex) field emerging from the output polarizer is the sum of the y components of $E_{x'}(l)$ and $E_{y'}(l)$

$$(E_y)_o = \frac{-A}{\sqrt{2}}(e^{-i\Gamma} - 1) \tag{9.3-3}$$

which corresponds to an output intensity

$$I_o \propto [(E_y)_o(E_y^*)_o]$$
$$= \frac{A^2}{2}[(e^{-i\Gamma} - 1)(e^{i\Gamma} - 1)] = 2A^2 \sin^2 \frac{\Gamma}{2}$$

where the proportionality constant is the same as in (9.3-1). The ratio of the output intensity to the input is thus

$$\frac{I_o}{I_i} = \sin^2 \frac{\Gamma}{2} = \sin^2 \left[\left(\frac{\pi}{2}\right)\frac{V}{V_\pi}\right] \tag{9.3-4}$$

The second equality in (9.3-4) was obtained from (9.2-6). The transmission factor (I_o/I_i) is plotted in Figure 9-5 against the applied voltage.

The process of amplitude modulation of an optical signal is also illustrated in Figure 9-5. The modulator is usually biased[6] with a fixed retardation $\Gamma = \pi/2$ to the 50 percent transmission point. A small sinusoidal modulation voltage would then cause a nearly sinusoidal modulation of the transmitted intensity as shown.

To treat the situation depicted by Figure 9-5 mathematically, we take

$$\Gamma = \frac{\pi}{2} + \Gamma_m \sin \omega_m t \tag{9.3-5}$$

where the retardation bias is taken as $\pi/2$, and Γ_m is related to the amplitude V_m of the modulation voltage $V_m \sin \omega_m t$ by (9.2-6); thus, $\Gamma_m = \pi(V_m/V_\pi)$.

---

[5]We recall here that the time average of the product of two harmonic fields Re $[Be^{i\omega t}]$ and Re $[Ce^{i\omega t}]$ is equal to $\frac{1}{2}$ Re $[BC^*]$.

[6]This bias can be achieved by applying a voltage $V = V_\pi/2$ or, more conveniently, by using a naturally birefringent crystal as in Figure 9-4 to introduce a phase difference (retardation) of $\pi/2$ between the x' and y' components.

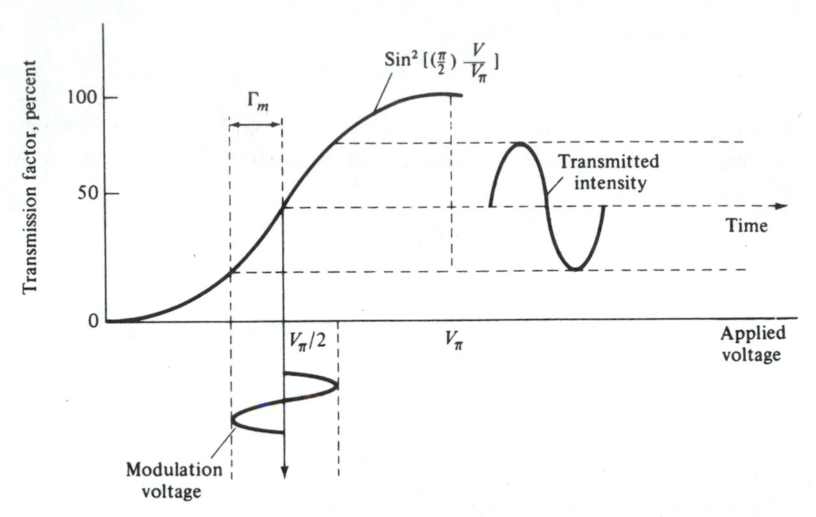

Figure 9-5 Transmission factor of a cross-polarized electrooptic modulator as a function of an applied voltage. The modulator is biased to the point $\Gamma = \pi/2$, which results in a 50 percent intensity transmission. A small applied sinusoidal voltage modulates the transmitted intensity about the bias point.

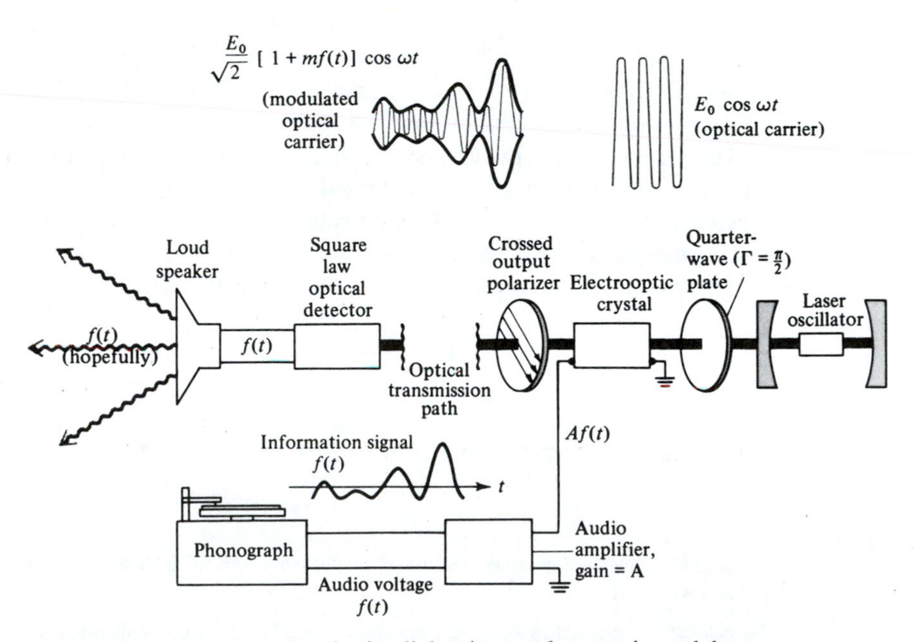

Figure 9-6 An optical communication link using an electrooptic modulator.

Using (9.3-4) we obtain

$$\frac{I_o}{I_i} = \sin^2\left(\frac{\pi}{4} + \frac{\Gamma_m}{2}\sin\omega_m t\right) \tag{9.3-6}$$

$$= \tfrac{1}{2}[1 + \sin(\Gamma_m \sin\omega_m t)] \tag{9.3-7}$$

which, for $\Gamma_m \ll 1$, becomes

$$\frac{I_o}{I_i} \simeq \frac{1}{2}(1 + \Gamma_m \sin\omega_m t) \tag{9.3-8}$$

so that the intensity modulation is a linear replica of the modulating voltage $V_m \sin\omega_m t$. If the condition $\Gamma_m \ll 1$ is not fulfilled, it follows from Figure 9-5 or from (9.3-7) that the intensity variation is distorted and will contain an appreciable amount of the higher (odd) harmonics. The dependence of the distortion of Γ_m is discussed further in Problem 9.3.

In Figure 9-6 we show how some information signal $f(t)$ (the electric output of a phonograph stylus in this case) can be impressed electrooptically as an amplitude modulation on a laser beam and subsequently be recovered by an optical detector. The details of the optical detection are considered in Chapter 11.

9.4 PHASE MODULATION OF LIGHT

In the preceding section we saw how the modulation of the state of polarization, from linear to elliptic, of an optical beam by means of the electrooptic effect can be converted, using polarizers, to intensity modulation. Here we consider the situation depicted by Figure 9-7, in which, instead of there being equal components along the induced birefringent axes (x' and y' in Figure 9-4), the incident beam is polarized parallel to one of them, x' say. In this case the application of the electric field does not change the state of polarization, but merely changes the output phase by

$$\Delta\phi_{x'} = -\frac{\omega l}{c}\Delta n_{x'}$$

where, from (9.1-10),

$$\Delta\phi_{x'} = -\frac{\omega n_o^3 r_{63}}{2c} E_z l \tag{9.4-1}$$

If the bias field is sinusoidal and is taken as

$$E_z = E_m \sin\omega_m t \tag{9.4-2}$$

then an incident optical field, which at the input ($z = 0$) face of the crystal is given by $e_{\text{in}} = A\exp(i\omega t)$, will emerge according to (9.2-2) as

$$e_{\text{out}} = A\exp\left\{i\left[\omega t - \frac{\omega}{c}\left(n_o + \frac{n_o^3}{2} r_{63} E_m \sin\omega_m t\right)l\right]\right\}$$

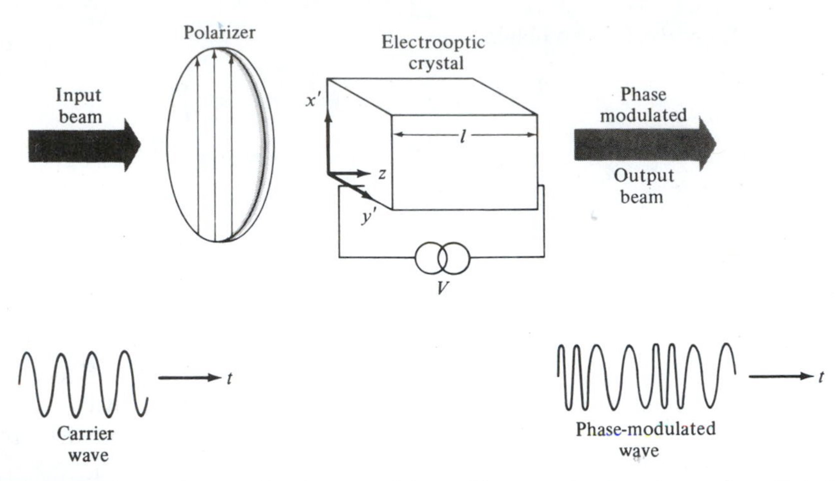

Figure 9-7 An electrooptic phase modulator. The crystal orientation and applied directions are appropriate to KDP. The optical polarization is parallel to an electrically induced principal dielectric axis (x').

where l is the length of the crystal. Dropping the constant phase factor, which is of no consequence here, we rewrite the last equation as

$$e_{\text{out}} = A \exp[i(\omega t + \delta \sin \omega_m t)] \tag{9.4-3}$$

where

$$\delta = \frac{-\omega n_o^3 r_{63} E_m l}{2c} = \frac{-\pi n_o^3 r_{63} E_m l}{\lambda} \tag{9.4-4}$$

is referred to as the phase modulation index. The optical field is thus phase-modulated with a modulation index δ. If we use the Bessel function identity

$$\exp(i\delta \sin \omega_m t) = \sum_{n=-\infty}^{\infty} J_n(\delta)\exp(in\omega_m t) \tag{9.4-5}$$

we can rewrite (9.4-3) as

$$e_{\text{out}} = A \sum_{n=-\infty}^{\infty} J_n(\delta)\, e^{i(\omega + n\omega_m)t} \tag{9.4-6}$$

which form gives the distribution of energy in the sidebands as a function of the modulation index δ. We note that, for $\delta = 0$, $J_0(0) = 1$ and $J_n(\delta) = 0$, $n \neq 0$. Another point of interest is that the phase modulation index δ as given by (9.4-4) is one half the retardation Γ as given by (9.2-4).

9.5 TRANSVERSE ELECTROOPTIC MODULATORS

In the examples of electrooptic retardation discussed in the two preceding sections, the electric field was applied along the direction of light propagation. This is the so-

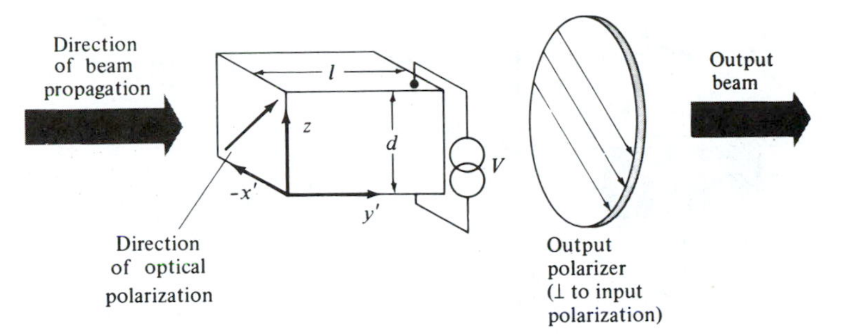

Figure 9-8 A transverse electrooptic amplitude modulator using a KH_2PO_4 (KDP) crystal in which the field is applied normal to the direction of propagation.

called longitudinal mode of modulation. A more desirable mode of operation is the transverse one, in which the field is applied normal to the direction of propagation. The reason is that in this case the field electrodes do not interfere with the optical beam, and the retardation, being proportional to the product of the field times the crystal length, can be increased by the use of longer crystals. In the longitudinal case the retardation, according to (9.2-4), is proportional to $E_z l = V$ and is independent of the crystal length l. Figures 9-1 and 9-2 suggest how transverse retardation can be obtained using a KDP crystal with the actual arrangement shown in Figure 9-8. The light propagates along y' and its polarization is in the $x'-z$ plane at 45° from the z axis. The retardation, with a field applied along z, is, from (9.1-10) and (9.1-12),

$$\Gamma = \phi_{x'} - \phi_z = \frac{\omega l}{c}\left[(n_o - n_e) + \frac{n_o^3}{2} r_{63}\left(\frac{V}{d}\right)\right] \tag{9.5-1}$$

where d is the crystal dimension along the direction of the applied field. We note that Γ contains a term that does not depend on the applied voltage. This point will be discussed in Problem 9-2. A detailed example of transverse electrooptic modulation using $\bar{4}3m$, cubic zinc-blende type crystals is given in Appendix C.

9.6 HIGH-FREQUENCY MODULATION CONSIDERATIONS

In the examples considered in the three preceding sections, we derived expressions for the retardation caused by electric fields of low frequencies. In many practical situations the modulation signal is often at very high frequencies and, in order to utilize the wide frequency spectrum available with lasers, may occupy a large bandwidth. In this section we consider some of the basic factors limiting the highest usable modulation frequencies in a number of typical experimental situations.

Consider first the situation described by Figure 9-9. The electrooptic crystal is placed between two electrodes with a modulation field containing frequencies near

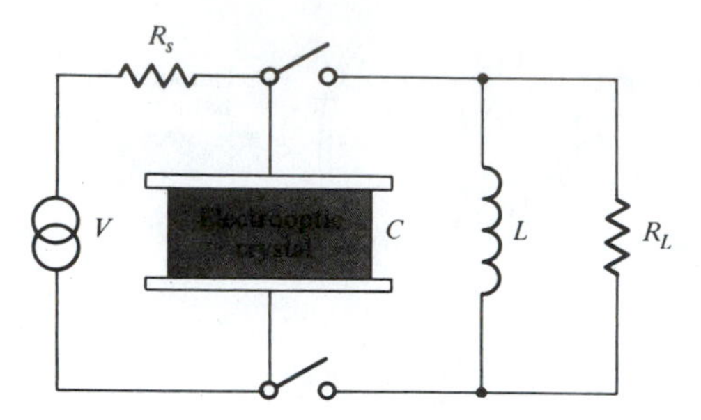

Figure 9-9 Equivalent circuit of an electrooptic modulation crystal in a parallel-plate configuration.

$\omega_0/2\pi$ applied to it. R_s is the internal resistance of the modulation source and C represents the parallel-plate capacitance due to the electrooptic crystal. If $R_s > (\omega_0 C)^{-1}$, most of the modulation voltage drop is across R_s and is thus wasted, since it does not contribute to the retardation. This can be remedied by resonating the crystal capacitance with an inductance L, where $\omega_0^2 = (LC)^{-1}$, as shown in Figure 9-9. In addition, a shunting resistance R_L is used so that at $\omega = \omega_0$ the impedance of the parallel RLC circuit is R_L, which is chosen to be larger than R_s so most of the modulation voltage appears across the crystal. The resonant circuit has a finite bandwidth—that is, its impedance is high only over a frequency interval $\Delta\omega/2\pi \simeq 1/2\pi R_L C$ (centered on ω_0). Therefore, the maximum modulation bandwidth (the frequency spectrum occupied by the modulation signal) must be less than

$$\frac{\Delta\omega}{2\pi} \simeq \frac{1}{2\pi R_L C} \tag{9.6-1}$$

if the modulation field is to be a faithful replica of the modulation signal.

In practice, the size of the modulation bandwidth $\Delta\omega/2\pi$ is dictated by the specific application. In addition, one requires a certain peak retardation Γ_m. Using (9.2-4) to relate Γ_m to the peak modulation voltage $V_m = (E_z)_m l$, we can show, with the aid of (9.6-1), that the power $V_m^2/2R_L$ needed in KDP-type crystals to obtain a peak retardation Γ_m is related to the modulation bandwidth $\Delta\nu = \Delta\omega/2\pi$ as

$$P = \frac{\Gamma_m^2 \lambda^2 A \epsilon \Delta\nu}{4\pi l n_0^6 r_{63}^2} \tag{9.6-2}$$

where $n_0 l$ is the length of the optical path in the crystal, A is the cross-sectional area of the crystal normal to l, and ϵ is the dielectric constant at the modulation frequency ω_0.

Transit-Time Limitations to High-Frequency Electrooptic Modulation

According to (9.2-4) the electrooptic retardation due to a field E can be written as

$$\Gamma = aEl \tag{9.6-3}$$

where $a = \omega n_0^3 r_{63}/c$ and l is the length of the optical path in the crystal. If the field E changes appreciably during the transit time $\tau_d = nl/c$ of light through the crystal, we must replace (9.6-3) by

$$\Gamma(t) = a \int_0^l e(z)\, dz = a \frac{c}{n} \int_{t-\tau_d}^{t} e(t')\, dt' \tag{9.6-4}$$

where c is the velocity of light and $e(t')$ is the instantaneous electric field. In the second integral we replace integration over z by integration over time, recognizing that the portion of the wave that reaches the output face $z = l$ at time t entered the crystal at time $t - \tau_d$. We also assumed that at any given moment the field $e(t)$ has the same value throughout the crystal.

Taking $e(t')$ as a sinusoid

$$e(t') = E_m e^{i\omega_m t'}$$

we obtain from (9.6-4)

$$\Gamma(t) = a \frac{c}{n} E_m \int_{t-\tau_d}^{t} e^{i\omega_m t'}\, dt'$$
$$= \Gamma_0 \left[\frac{1 - e^{-i\omega_m \tau_d}}{i\omega_m \tau_d}\right] e^{i\omega_m t} \tag{9.6-5}$$

where $\Gamma_0 = a(c/n)\tau_d E_m = alE_m$ is the peak retardation, which obtains when $\omega_m \tau_d \ll 1$. The factor

$$r = \frac{1 - e^{-i\omega_m \tau_d}}{i\omega_m \tau_d} \tag{9.6-6}$$

gives the decrease in peak retardation resulting from the finite transit time. For $r \simeq 1$ (that is, no reduction), the condition $\omega_m \tau_d \ll 1$ must be satisfied, so the transit time must be small compared to the shortest modulation period. The factor r is plotted in Figure 11-17.

If, somewhat arbitrarily, we take the highest useful modulation frequency as that for which $\omega_m \tau_d = \pi/2$ (at this point, according to Figure 11-17, $|r| = 0.9$) and we use the relation $\tau_d = nl/c$, we obtain

$$(\nu_m)_{\max} = \frac{c}{4nl} \tag{9.6-7}$$

which, using a KDP crystal ($n \simeq 1.5$) and a length $l = 1$ cm, yields $(\nu_m)_{\max} = 5 \times 10^9$ Hz.

Traveling-Wave Modulators

One method that can, in principle, overcome the transit-time limitation, involves applying the modulation signal in the form of a traveling wave [3], as shown in Figure 9-10. If the optical and modulation field phase velocities are equal to each other, then a portion of an optical wavefront will exercise the same instantaneous

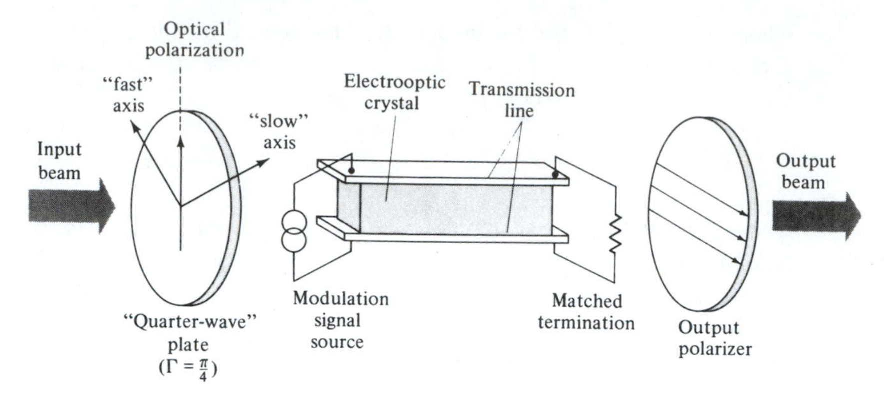

Figure 9-10 A traveling-wave electrooptic modulator.

electric field, which corresponds to the field it encounters at the entrance face, as it propagates through the crystal and the transit-time problem discussed above is eliminated. This form of modulation can be used only in the transverse geometry that was discussed in the preceding section, since the RF field in most propagating structures is predominantly transverse.

Consider an element of the optical wavefront that *enters* the crystal at $z = 0$ at time t. The position z of this element at some later time t' is

$$z(t') = \frac{c}{n}(t' - t) \tag{9.6-8}$$

where c/n is the optical phase velocity. The retardation exercised by this element is given similarly to (9.6-4) by

$$\Gamma(t) = \frac{ac}{n}\int_t^{t+\tau_d} e[t', z(t')]\, dt' \tag{9.6-9}$$

where $e[t', z(t')]$ is the instantaneous modulation field as seen by an observer traveling with the phase front. Taking the traveling modulation field as

$$e(t', z) = E_m e^{i[\omega_m t' - k_m z]}$$

we obtain, using (9.6-8),

$$e[t', z(t')] = E_m e^{i[\omega_m t' - k_m (c/n)(t' - t)]} \tag{9.6-10}$$

Recalling that $k_m = \omega_m/c_m$, where c_m is the phase velocity of the modulation field, we substitute (9.6-10) in (9.6-9) and, carrying out the simple integration, obtain

$$\Gamma(t) = \Gamma_0 e^{i\omega_m t}\left[\frac{e^{i\omega_m \tau_d(1 - c/nc_m)} - 1}{i\omega_m \tau_d(1 - c/nc_m)}\right] \tag{9.6-11}$$

where $\Gamma_0 = alE_m = a(c/n)\tau_d E_m$ is the retardation that would result from a dc field equal to E_m.

The reduction factor

$$r = \frac{e^{i\omega_m \tau_d(1 - c/nc_m)} - 1}{i\omega_m \tau_d(1 - c/nc_m)} \tag{9.6-12}$$

is of the same form as that of the lumped-constant modulator (9.6-6) except that τ_d is replaced by $\tau_d(1 - c/nc_m)$. If the two phase velocities are made equal so that $c/n = c_m$, then $r = 1$ and maximum retardation is obtained *regardless* of the crystal length.

The maximum useful modulation frequency is taken, as in the treatment leading to (9.6-7), as that for which $\omega_m \tau_d(1 - c/nc_m) = \pi/2$, yielding

$$(\nu_m)_{\max} = \frac{c}{4ln(1 - c/nc_m)} \tag{9.6-13}$$

which, upon comparison with (9.6-7), shows an increase in the frequency limit or useful crystal length of $(1 - c/nc_m)^{-1}$. The problem of designing traveling wave electrooptic modulators is considered in References [4–6].

For a more detailed treatment of electrooptic modulation including the traveling-wave and high-frequency cases, the student should consult Reference [11] as well as the treatment of Section 9.9.

9.7 ELECTROOPTIC BEAM DEFLECTION

The electrooptic effect is also used to deflect light beams [7]. The operation of such a beam deflector is shown in Figure 9-11. Imagine an optical wavefront incident on a crystal in which the optical path length depends on the transverse position x. This could be achieved by having the velocity of propagation—that is, the index of refraction n—depend on x, as in Figure 9-11. Taking the index variation to be a linear function of x, the upper ray A "sees" an index $n + \Delta n$ and hence traverses the crystal in a time

$$T_A = \frac{l}{c}(n + \Delta n)$$

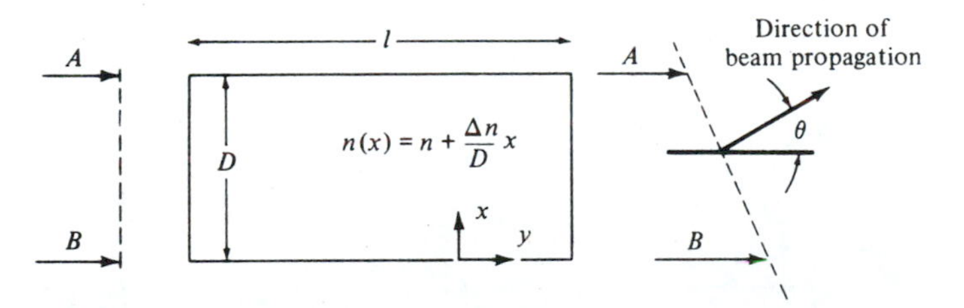

Figure 9-11 Schematic diagram of a beam deflector. The index of refraction varies linearly in the x direction as $n(x) = n_o + ax$. Ray B "gains" on ray A in passing through the crystal axis, thus causing a tilting of the wavefront by θ.

The lower portion of the wavefront (that is, ray B) "sees" an index n and has a transit time

$$T_B = \frac{l}{c} n$$

The difference in transit times results in a lag of ray A with respect to B of

$$\Delta y = \frac{c}{n}(T_A - T_B) = l \frac{\Delta n}{n}$$

which corresponds to a deflection of the beam-propagation axis, as measured inside the crystal, at the output face of

$$\theta' = -\frac{\Delta y}{D} = -\frac{l\Delta n}{Dn} = -\frac{l}{n}\frac{dn}{dx} \tag{9.7-1}$$

where we replaced $\Delta n/D$ by dn/dx. The external deflection angle θ, measured with respect to the horizontal axis, is related to θ' by Snell's law

$$\frac{\sin \theta}{\sin \theta'} = n$$

which, using (9.7-1) and assuming $\sin \theta \simeq \theta \ll 1$ yields

$$\theta = \theta' n = -l \frac{\Delta n}{D} = -l \frac{dn}{dx} \tag{9.7-2}$$

A simple realization of such a deflector using a KH_2PO_4(KDP) crystal is shown in Figure 9-12. It consists of two KDP prisms with edges along the x', y', and z

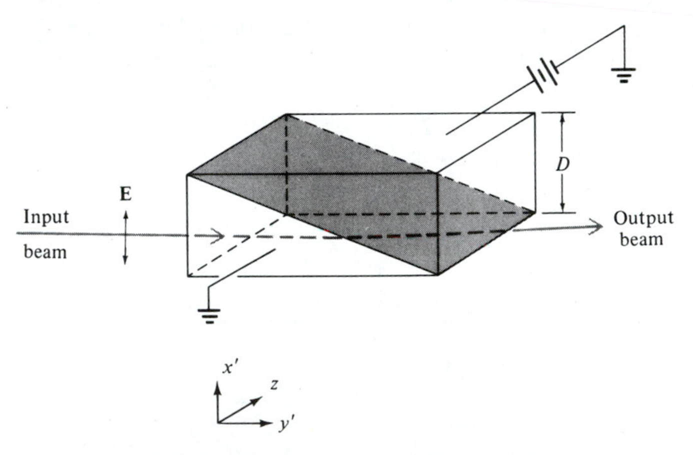

Figure 9-12 Double-prism KDP beam deflector. Upper and lower prisms have their z axes reversed with respect to each other. The deflection field is applied parallel to z.

directions.[7] The two prisms have their z axes opposite to one another, but are otherwise similarly oriented. The electric field is applied parallel to the z direction, and the light propagates in the y' direction with its polarization along x'. For this case the index of refraction "seen" by ray A, which propagates entirely in the upper prism, is given by (9.1-10) as

$$n_A = n_o + \frac{n_o^3}{2} r_{63} E_z$$

while in the lower prism the sign of the electric field with respect to the z axis is reversed so that

$$n_B = n_o - \frac{n_o^3}{2} r_{63} E_z$$

Using (9.7-2) with $\Delta n = n_A - n_B$, the deflection angle is given by

$$\theta = \frac{l}{D} n_o^3 r_{63} E_z \tag{9.7-3}$$

According to (2.5-18), every optical beam has a finite, far-field divergence angle that we call θ_{beam}. It is clear that a fundamental figure of merit for the deflector is not the angle of deflection θ that can be changed by a lens, but the factor N by which θ exceeds θ_{beam}. If one were, as an example, to focus the output beam, then N would correspond to the number of resolvable spots that can be displayed in the focal plane using fields with a magnitude up to E_z.

To get an expression for N we assume that the crystal is placed astride the "waist" of a Gaussian (fundamental) beam with a spot size ω_0. According to (2.5-18) the far-field diffraction angle in air is

$$\theta_{\text{beam}} = \frac{\lambda}{\pi \omega_0}$$

Such a beam can be passed through a crystal with height $D = 2\omega_0$ so that, using (9.7-3), the number of resolvable spots is

$$N = \left| \frac{\theta}{\theta_{\text{beam}}} \right| = \frac{\pi l n_o^3 r_{63}}{2\lambda} E_z \tag{9.7-4}$$

It follows directly from (9.7-4), the details being left as a problem, that an electric field that induces a birefringent retardation (in a distance l) $\Delta\Gamma = \pi$ will yield $N \simeq 1$. Therefore, fundamentally, the electrooptic extinction of a beam, which according to (9.3-4) requires $\Gamma = \pi$, is equivalent to a deflection by one spot diameter.

[7]These are the principal axes of the index ellipsoid when an electric field is applied along the z direction as described in Section 9.1 and in the example of Section 9.5.

The deflection of an optical beam by diffraction from a sound wave is discussed in Chapter 12. Electrooptic modulation in thin dielectric waveguides [8–10] is discussed in Chapter 13.

9.8 ELECTROOPTIC MODULATION—COUPLED WAVE ANALYSIS

The analysis of electrooptic modulation in Section 9.3 was based on an approach that requires one to determine the propagating electromagnetic eigenmodes of the crystal in the presence of the (low-frequency) electric field. In Section 9.2, we found as a special case the two eigenmodes and their indices in the presence of a dc field $\hat{\boldsymbol{E}}_m = \hat{\boldsymbol{e}}_z E_z$ for propagation along the z axis of KH_2PO_4.

The formalism of this section uses a different point of view, that of the coupled modes approach, in which the modulation field is viewed as a perturbation that causes exchange of power, coupling, between the eigenmodes of the crystal in the absence of an applied modulating field. The two approaches are, of course, formally equivalent as will be shown in Section 13.9. The author, however, prefers the coupled-mode approach since it dispenses with the need to diagonalize, as in (9.1-25), the permeability tensor. In addition, it lends itself more easily to an accurate analysis in situations in which the modulation field is a high-frequency field and the transit time of the optical field is not small compared to the modulation period.

We start with Equation (1.2-9), which using (1.2-3) can be written as

$$\mathbf{e} \cdot \mathbf{i} = -\nabla \cdot (\mathbf{e} \times \mathbf{h}) - \mathbf{h} \cdot \frac{\partial \mathbf{b}}{\partial t} - \mathbf{e} \cdot \frac{\partial \mathbf{d}}{\partial t}$$

Using the Gauss theorem (1.2-10a) to integrate the last equation over an arbitrary volume V

$$\begin{aligned} -\int_V \nabla \cdot (\mathbf{e} \times \mathbf{h})\, dV &= -\int_s (\mathbf{e} \times \mathbf{h}) \cdot \mathbf{n}\, da \\ &= \int_V (\mathbf{e} \cdot \mathbf{i})\, dV + \int_V \left(\mathbf{h} \cdot \frac{\partial \mathbf{b}}{\partial t} + \mathbf{e} \cdot \frac{\partial \mathbf{d}}{\partial t} \right) dV \end{aligned}$$

where S is the surface bounding V. If we further assume that the material medium is linear (i.e., ϵ and μ are independent of the field strengths), we can rewrite the last result as

$$-\int_s (\mathbf{e} \times \mathbf{h}) \cdot \mathbf{n}\, da = \int_V (\mathbf{e} \cdot \mathbf{i})\, dV + \frac{d}{dt} \int_V \frac{1}{2} (\mathbf{e} \cdot \mathbf{d} + \mathbf{h} \cdot \mathbf{b}) dV \qquad \textbf{(9.8-1)}$$

The usual interpretation of (9.8-1) is that the left side represents the electromagnetic power flowing into V, the term $(\mathbf{e} \cdot \mathbf{i})$ represents the losses due to conduction (ohmic losses), while the last term represents the temporal rate of change of the electromagnetic energy stored in V. It follows that the energy density (J/m^3) due to the electric field is

$$\omega_e = \frac{1}{2} \mathbf{E} \cdot \mathbf{D} \qquad \textbf{(9.8-2)}$$

where to conform with the notation of this section we changed $\mathbf{e} \rightarrow \mathbf{E}$, $\mathbf{h} \rightarrow \mathbf{H}$. In a linear (but not necessarily isotropic) medium we can write

$$D_i = \epsilon_{ij}E_j, \qquad E_i = \eta_{ij}D_j, \qquad \overline{\overline{\boldsymbol{\eta}}} \equiv \overline{\overline{\boldsymbol{\epsilon}}}^{-1} \tag{9.8-3}$$

so that

$$2\omega_e = \eta_{ij}D_jD_i \tag{9.8-4}$$

Let

$$\frac{D_i}{\sqrt{2\omega_e\epsilon_0}} \longrightarrow x_i$$

so that the last equation becomes

$$\epsilon_0\eta_{ij}x_ix_j = 1$$

which, after adopting the Voigt notation

$$11 \rightarrow 1,\ 22 \rightarrow 2,\ 33 \rightarrow 3,\ 23 \rightarrow 4,\ 13 \rightarrow 5,\ 12 \rightarrow 6 \tag{9.8-5}$$

becomes

$$\epsilon_0\,[\eta_1x^2 + \eta_2y^2 + \eta_3z^2 + 2\eta_4yz + 2\eta_5xz + 2\eta_6xy] = 1 \tag{9.8-6}$$

This equation is identical to that of the optical indicatrix (9.1-2), provided we associate

$$\left(\frac{1}{n^2}\right)_i \longrightarrow \epsilon_0\eta_i \tag{9.8-7}$$

So that the definition (9.1-3) of the linear electrooptic effect can be written as $\epsilon_0\Delta\eta_i = r_{ik}E_k^{(0)}$ or, restoring the full subscript labeling,

$$\epsilon_0\Delta\eta_{ij} = r_{ijk}E_k^{(0)} \tag{9.8-8}$$

where $E_k^{(0)}$ is a dc or low-frequency field and summation over repeated indices is assumed.

Returning for a moment to (9.8-3), we will assume that $\overline{\overline{\boldsymbol{\epsilon}}}$ and $\overline{\overline{\boldsymbol{\eta}}}$ are expressed initially in the principal dielectric coordinate system where, with no applied field, $E_k^{(0)} = 0$, ϵ_{ij}, η_{ij} are zero when $i \neq j$, i.e., $\overline{\overline{\boldsymbol{\epsilon}}}$ and $\overline{\overline{\boldsymbol{\eta}}}$ are diagonal. We use the rule for matrix inversion to obtain

$$\begin{aligned} \eta_{ii} &= (\epsilon_{ii})^{-1} \\ \underset{i\neq j}{\eta_{ij}} &= \frac{-\epsilon_{ji}}{\epsilon_{ii}\epsilon_{jj}} = \frac{-\epsilon_{ij}}{\epsilon_{ii}\epsilon_{jj}} \end{aligned} \tag{9.8-9}$$

and after combining (9.8-8) and (9.8-9)

$$\begin{aligned} \Delta\epsilon_{ji} &= -\frac{\epsilon_{ii}\epsilon_{jj}}{\epsilon_0}\,r_{ijk}E_k^{(0)} \\ (r_{ijk} &= r_{jik}) \end{aligned} \tag{9.8-10}$$

which is our principal result. It expresses the linear electrooptic effect as the first-order perturbation in the elements of the dielectric tensor.

Now consider the effect of applying a low-frequency field $E_k^{(0)}$ to a crystal in the presence of an optical field $E_j^{(\omega)}$ $(\mathbf{r}, t)$ using

$$D_i^{(\omega)} = \epsilon_{ij} E_j^{(\omega)} = \epsilon_0 E_i^{(\omega)} + P_i^{(\omega)}$$

or

$$P_i^{(\omega)} = (\epsilon_{ij} - \epsilon_0 \delta_{ij}) E_j^{(\omega)} \tag{9.8-11}$$

so that a perturbation $\Delta\epsilon_{ij}$ causes a perturbation in the medium optical polarization of

$$\begin{aligned} \Delta P_i^{(\omega)} &= (\Delta\epsilon_{ij}) E_j^{(\omega)} \\ &= -\frac{\epsilon_{ii}\epsilon_{jj}}{\epsilon_0} r_{ijk} E_j^{(\omega)} E_k^{(0)} \end{aligned} \tag{9.8-12}$$

The effect of the dc (low-frequency) field $E_k^{(0)}$ on the optical field propagating through the crystal is thus represented by an effective optical polarization $\Delta P_i^{(\omega)}$ of (9.8-12). The latter can now be used in Maxwell's equations to investigate the effect of the dc field on optical propagation. This will be done next.

The Wave Equation

Starting with Maxwell's equations as in Section 8.2, we have

$$\nabla^2 \mathbf{E}(\mathbf{r}, t) - \mu_0 \epsilon_0 \frac{\partial^2}{\partial t^2} \mathbf{E}(\mathbf{r}, t) - \mu_0 \frac{\partial^2}{\partial t^2} \mathbf{P} = 0 \tag{9.8-13}$$

The polarization vector $\mathbf{P}$ is taken as the sum

$$\mathbf{P} = \epsilon_0 \overline{\overline{\chi}} \mathbf{E} + \Delta\mathbf{P}$$

where $\overline{\overline{\chi}}$ is the second-rank linear susceptibility tensor and $\Delta\mathbf{P}$ the perturbation in $\mathbf{P}$ as given by (9.8-12). Using the last expression in (9.8-13) leads to

$$\begin{aligned} &\nabla^2 \mathbf{E} - \mu_0 \frac{\partial^2}{\partial t^2} (\overline{\overline{\mathbf{1}}} + \overline{\overline{\chi}}) \mathbf{E} - \mu_0 \frac{\partial^2}{\partial t^2} (\Delta\mathbf{P}) \\ &\qquad = \nabla^2 \mathbf{E} - \mu_0 \frac{\partial^2}{\partial t^2} \overline{\overline{\epsilon}} \mathbf{E} - \mu_0 \frac{\partial^2}{\partial t^2} (\Delta\mathbf{P}) \end{aligned} \tag{9.8-14}$$

defining the dielectric tensor

$$\overline{\overline{\epsilon}} = \epsilon_0 (\overline{\overline{\mathbf{1}}} + \overline{\overline{\chi}})$$

the wave equation (9.8-13) becomes

$$\nabla^2 \mathbf{E} - \mu_0 \frac{\partial^2}{\partial t^2} \overline{\overline{\epsilon}} E - \mu_0 \frac{\partial^2}{\partial t^2} (\Delta\mathbf{P}) = 0 \tag{9.8-15}$$

Most of the scenarios of electrooptic modulation may be described as an exchange of power between two optical eigenfields.[8] These can be two mutually orthogonal transverse polarizations of a field propagating along, say, the $\hat{\zeta}$ direction. As a result we take the total field as

$$\begin{aligned} \mathrm{E}(r, t) &= \tfrac{1}{2}\,\hat{\mathbf{e}}_1 A_1(\mathbf{r}, t) e^{i(\omega t - k_1\zeta)} \\ &\quad + \tfrac{1}{2}\,\hat{\mathbf{e}}_2 A_2(\mathbf{r}, t) e^{i(\omega t - k_2\zeta)} + \text{c.c.} \end{aligned} \tag{9.8-16}$$

where A_1 and A_2 are the "slowly" varying complex amplitudes of the fields that are polarized along the $\hat{\mathbf{e}}_1$ and $\hat{\mathbf{e}}_2$ directions normal to $\hat{\zeta}$ and

$$k_{1,2} = (\omega/c)n_{1,2}$$

Substitution of the last equation in (9.8-15) gives

$$\begin{aligned} &\hat{\mathbf{e}}_1 e^{i(\omega t - k_1\zeta)} \left[\left(\frac{\partial^2}{\partial\zeta^2} - 2ik_1\frac{\partial}{\partial\zeta} - k_1^2\right) - \mu_0\epsilon_{11}\left(\frac{\partial^2}{\partial t^2} + 2i\omega\frac{\partial}{\partial t} - \omega^2\right)\right]\frac{A_1(\zeta, t)}{2} \\ &\quad + \hat{\mathbf{e}}_2 e^{i(\omega t - k_2\zeta)} \left[\left(\frac{\partial}{\partial\zeta^2} - 2ik_2\frac{\partial}{\partial\zeta} - k_2^2\right) - \mu_0\epsilon_{22} \right.\\ &\quad\quad \left. \times \left(\frac{\partial^2}{\partial t^2} + 2i\omega\frac{\partial}{\partial t} - \omega^2\right)\right]\frac{A_2(\zeta, t)}{2} + \text{c.c.} \\ &\quad\quad\quad = i\mu_0\frac{\partial^2}{\partial t^2}\Delta\boldsymbol{P}(\mathbf{r}, t) \end{aligned} \tag{9.8-17}$$

where we assume that $\hat{\mathbf{e}}_1$ and $\hat{\mathbf{e}}_2$ are unit vectors along the principal dielectric axes (this is the case in nearly all experimental situations) and used

$$\epsilon_{11} = \epsilon_0(1 + \chi_{11}) \qquad \epsilon_{22} = \epsilon_0(1 + \chi_{22})$$

Recognizing that $k_1^2 = \omega_1^2\mu_0\epsilon_{11}$, $n_i = \sqrt{\epsilon_{ii}/\epsilon_0}$ $(i = 1, 2)$, and making the slowly varying envelope approximation $\partial^2/\partial\zeta^2 \ll k\partial/\partial\zeta$ and $\partial^2/\partial t^2 \ll \omega\partial/\partial t$, we obtain

$$\left(\frac{\partial}{\partial\zeta} + \frac{n_1}{c}\frac{\partial}{\partial t}\right) A_1(\zeta, t) = \frac{i\mu_0}{k_1} e^{-i(\omega t - k_1\zeta)}\frac{\partial^2}{\partial t^2}[\Delta P(\mathbf{r}, t)]_1 \tag{9.8-18a}$$

The subscript 1 on the right side signifies that we need only consider that part of $\Delta\mathbf{P}$ that contains the factor $\exp[i(\omega t - k_1\zeta)]$. The rest average out to zero.

$$\left(\frac{\partial}{\partial\zeta} + \frac{n_2}{c}\frac{\partial}{\partial t}\right) A_2(\zeta, t) = \frac{i\mu_0}{k_2} e^{-i(\omega t - k_2\zeta)}\frac{\partial^2}{\partial t^2}[\Delta P(\mathbf{r}, t)]_2 \tag{9.8-18b}$$

These are our basic working equations. They can be used to analyze most of the situations arising in electrooptic or acoustooptic modulation.

---

[8] We use the term *eigen (self) field* in the sense of the eigen modes of quantum mechanics. It is used here to describe a propagating monochromatic field, say, along z, that except for a propagation delay does not depend on z.

We will next show how Equations (9.8-18) are used in the important cases of electrooptic phase and amplitude modulation with a traveling modulation field.

9.9 PHASE MODULATION

In this case we have a traveling modulation field with some polarization, say, k, so that in (9.8-12) we take

$$E_k^{(0)} = E_{mk} \sin(\omega_m t - k_m \zeta) \tag{9.9-1}$$

For pure phase modulation of, say, wave 1, it is necessary that the perturbation polarization $(\Delta P)_1$ involve only A_1 and not A_2. This polarization is then given by (9.8-12) as

$$\Delta P_1(\zeta, t) = -\frac{(\epsilon_{11})^2}{2\epsilon_0} r_{11k} A_1(\zeta, t) e^{i(\omega t - k_1 \zeta)} E_{mk} \sin(\omega_m t - k_m \zeta) + \text{c.c.} \tag{9.9-2}$$

So that Equation (9.8-18a) becomes

$$\left(\frac{\partial}{\partial \zeta} + \frac{n_1}{c}\frac{\partial}{\partial t}\right) A_1(\zeta, t) = i\beta \sin(\omega_m t - k_m \zeta) A_1(\zeta, t) \tag{9.9-3}$$

$$\begin{aligned} \beta &\equiv +\frac{k_1 n_1^2}{2} r_{11k} E_{mk} \\ &= +\frac{\omega}{2c} n_1^3 r_{11k} E_{mk} \end{aligned} \tag{9.9-4}$$

In deriving (9.9-3) we took advantage of the fact that $\omega \ggg \omega_m$ (typically $\omega \sim 10^{15}$, $\omega_m < 10^{11}$), and replaced $\partial^2/\partial t^2$ on the right side of (9.8-17) by $-\omega^2$.

Equation (9.9-3) is a first-order linear partial differential equation and can be integrated by a change of variables

$$\begin{aligned} u &= \zeta + \frac{c}{n} t \\ v &= \zeta - \frac{c}{n} t \\ &(n \equiv n_1) \end{aligned} \tag{9.9-5}$$

Using

$$\begin{aligned} \frac{\partial}{\partial \zeta} &= \frac{\partial u}{\partial \zeta}\frac{\partial}{\partial u} + \frac{\partial v}{\partial \zeta}\frac{\partial}{\partial v} = \frac{\partial}{\partial u} + \frac{\partial}{\partial v} \\ \frac{\partial}{\partial t} &= \frac{\partial u}{\partial t}\frac{\partial}{\partial u} + \frac{\partial v}{\partial t}\frac{\partial}{\partial v} = \frac{c}{n}\left(\frac{\partial}{\partial u} - \frac{\partial}{\partial v}\right) \end{aligned} \tag{9.9-6}$$

Equation (9.9-3) becomes ($A_1 \equiv A$).

$$2\frac{\partial}{\partial u}A = i\beta \sin\left[\frac{n\omega_m}{2c}(u - v) - \frac{k_m}{2}(u + v)\right]A \qquad (9.9\text{-}7)$$

By treating u and v as independent variables, an integration of (9.9-7) yields

$$A(\zeta, t) = C\left(\zeta - \frac{c}{n}t\right)\exp\left[-i\frac{\beta c}{\omega_m(n - n_m)}\cos(\omega_m t - k_m\zeta)\right] \qquad (9.9\text{-}8)$$

where C is an arbitrary function and $n_m = ck_m/\omega_m$ is the index of refraction at ω_m. The boundary condition at the input ($\zeta = 0$) face of the crystal is

$$A(0, t) = A_0, \qquad (9.9\text{-}9)$$

where A_0 is an arbitrary constant. This condition requires that the function C be of the form

$$C\left(\zeta - \frac{c}{n}t\right) = A_0 \exp\left[i\frac{\beta c}{\omega_m(n - n_m)}\cos\left(\omega_m t - \frac{n}{c}\omega_m\zeta\right)\right] \qquad (9.9\text{-}10)$$

The mode amplitude $A(\zeta, t)$ is then, according to Equations (9.9-8) and (9.9-10), given by

$$A(\zeta, t) = A_0 \exp \times\left\{i\frac{\beta c}{\omega_m(n - n_m)}\left[\cos\left(\omega_m t - \frac{\omega_m}{c}n\zeta\right) - \cos\left(\omega_m t - \frac{\omega_m}{c}n_m\zeta\right)\right]\right\} \qquad (9.9\text{-}11)$$

By using the trigonometric identity

$$\cos\alpha - \cos\beta = -2\sin\tfrac{1}{2}(\alpha + \beta)\sin\tfrac{1}{2}(\alpha - \beta)$$

the amplitude at the output face ($\zeta = L$) of the crystal can be written as

$$A(L, t) = A_0 \exp[i\delta\sin(\omega_m t - \phi)] \qquad (9.9\text{-}12)$$

where

$$\delta = \beta L \frac{\sin\frac{\omega_m}{2c}(n_m - n)L}{\frac{\omega_m}{2c}(n_m - n)L} \qquad (9.9\text{-}13)$$

$$\phi = \frac{\omega_m}{2c}(n + n_m)L \qquad (9.9\text{-}14)$$

If there is no further perturbation beyond $\zeta = L$, then the emerging beam can be written (using 9.8-16) as

$$E_1(L, t) = A_0 \exp\{i[\omega t + \delta\sin(\omega_m t - \phi) - kL]\} \qquad (9.9\text{-}15)$$

$$k = \frac{\omega n}{c}$$

This is our main result. The output consists of a phase-modulated wave with a modulation index δ. The value of δ given by Equation (9.9-13) for this case is no longer proportional to the length of the crystal, L, and is reduced from its maximum value βL by a factor

$$\eta \equiv \frac{\sin \Delta L}{\Delta L}$$

where

$$\Delta \equiv \frac{\omega_m}{2c}(n_m - n) = \frac{\omega_m}{2}\left(\frac{1}{v_m} - \frac{1}{v_0}\right) \tag{9.9-16}$$

in which $v_0 = c/n_1$ and $v_m = c/n_m$ are the phase velocities of the light and modulating wave, respectively. Physically, this reduction factor is due to the mismatch of the phase velocities of the waves. In the event that the light wave and the modulating wave are traveling with the same phase velocities, the light wave will experience a constant modulating field as it propagates through the electrooptic crystal. The reduction factor η in this case is unity; that is, there is no reduction in the modulation index δ. The modulation index in this case is linearly proportional to the length of the crystal. When the phase velocities are not equal, δ becomes a periodic function

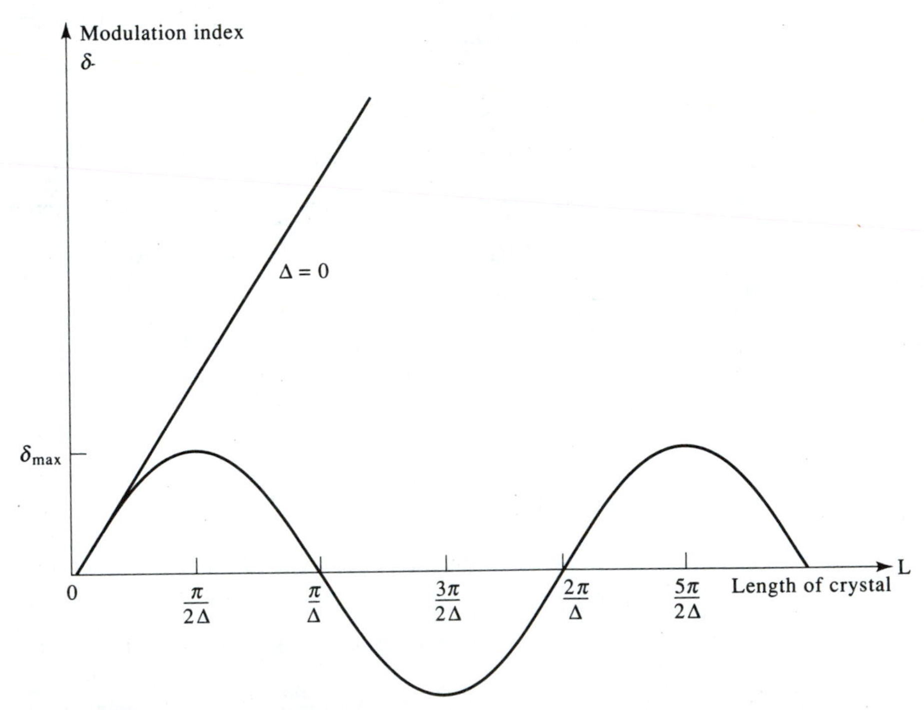

Figure 9-13 The modulation index δ versus the length L of the crystal.

of the length of the crystal. A plot of δ versus L is shown in Figure 9-13. The maximum δ occurs at the condition when

$$\frac{\omega_m}{2c}|n_m - n|L = \frac{\pi}{2} \tag{9.9-17}$$

with the maximum modulation index, $\delta_{\max}$, given by

$$\delta_{max} = \frac{\omega}{\omega_m}\frac{n^3}{|n - n_m|} r_{11k}E_{mk} \tag{9.9-18}$$

Example: $LiNbO_3$ Phase Modulator

Referring to Figure 9-14, we consider a rectangular $LiNbO_3$ rod with its input and output planes perpendicular to the y axis. The z direction is a principal dielectric axis of the crystal. An RF field with an **E** vector parallel to the z axis is applied to the crystal. Both the RF field and the optical beam are propagating in the y direction. An input polarizer in front of the input plane ensures that the light is polarized along the z direction of the crystal. The modulation index δ is given, according to Equation (9.9-13), by

$$\delta = \frac{\omega}{2c} n_e^3 r_{33} E_z L \frac{\sin \frac{\omega_m}{2c}(n_m - n_e)L}{\frac{\omega_m}{2c}(n_m - n_e)L} \tag{9.9-19}$$

where n_e is the extraordinary index of refraction of the crystal, L is the length of the crystal, and r_{33} is the relevant electrooptic coefficient. Let $\omega_m/2\pi = 6$ GHz, $n_m = 1.84$, and $n_e = 2.2$; then the maximum modulation occurs at $L = 6.8$ cm according to Equation (9.9-17). The symmetry group of $LiNbO_3$ is 3 m. From Table 9.2 it follows that the relevant electrooptic coefficient for the structure shown in Figure 9-14 is r_{33}.

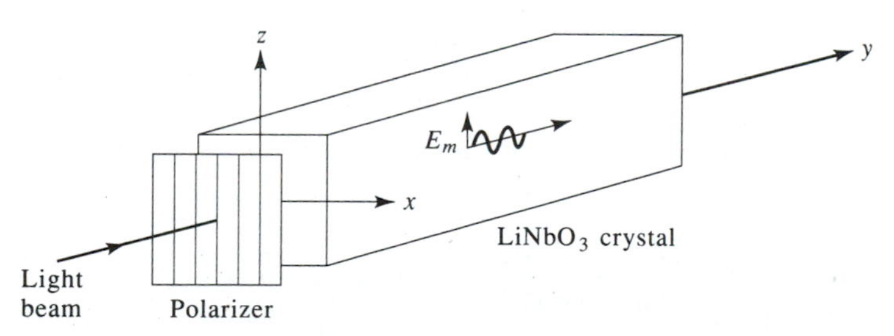

Figure 9-14 A $LiNbO_3$ rod used as the electrooptic crystal for phase modulation. The modulating RF field is polarized along the z axis and is propagating along the y axis.

Amplitude Modulation (advanced topic)

We now consider the case when the orthogonally polarized normal modes A_1 and A_2 are coupled (i.e., exchange power) by the applied modulation electric field. This occurs when the perturbation polarization (ΔP_1) in (9.8-18) is proportional to A_2 while, at the same time, ΔP_2 is proportional to A_1. In this case, electromagnetic energy is exchanged between the coupled modes as the wave propagates in the crystal. The magnitudes of the mode amplitudes are therefore functions of space and time. The mode amplitudes satisfy the coupled-mode equation (9.8-18). We will now consider a case of pure amplitude modulation. Again we assume that a traveling modulation wave $E_{mk}\sin(\omega_m t - k_m\zeta)$ yielding the coupled-mode equations (9.8-18). In our new language of mode coupling, amplitude modulation takes place when the presence of a modulation field

$$\mathbf{E}_{mod} = \hat{e}_k E_{mk} \sin(\omega_m t - k_m z) \tag{9.9-20}$$

causes power exchange between the modes 1 and 2. An inspection of the coupled-mode equations (9.8-18) shows that this coupling occurs when the presence of $\mathbf{E}_{\text{mod}}$ creates a perturbation $\Delta\epsilon_{ij}(i \neq j)$, since this, according to (9.8-13), will couple mode $i(1)$ to $j(2)$ and vice versa. Using (9.8-13), the condition for pure amplitude modulation is that the set of conditions

$$\begin{aligned} r_{11k}E_k^{(0)} &= r_{22k}E_k^{(0)} = 0 \\ r_{12k}E_k^{(0)} &= r_{21k}E_k^{(0)} \neq 0 \end{aligned} \tag{9.9-21}$$

be satisfied for some Cartesian direction k. If we take the modulation field as in (9.9-1), the coupled-mode equations (9.8-18) become

$$\begin{aligned} \left(\frac{\partial}{\partial\zeta} + \frac{n_1}{c}\frac{\partial}{\partial t}\right) A_1 &= i\kappa \sin(\omega_m t - k_m\zeta) A_2 e^{i(k_1-k_2)\zeta} \\ \left(\frac{\partial}{\partial\zeta} + \frac{n_2}{c}\frac{\partial}{\partial t}\right) A_2 &= i\kappa \sin(\omega_m t - k_m\zeta) A_1 e^{-i(k_1-k_2)\zeta} \end{aligned} \tag{9.9-22}$$

where

$$\kappa = \frac{n_1^2 n_2^2}{n_1 + n_2} \frac{\omega}{c} r_{12k} E_{mk} \tag{9.9-23}$$

and summation over k is assumed. In the case when $n_1 = n_2$, the coupled equations become

$$\begin{aligned} \left(\frac{\partial}{\partial\zeta} + \frac{n}{c}\frac{\partial}{\partial t}\right) A_1 &= i\kappa \sin(\omega_m t - k_m\zeta) A_2 \\ \left(\frac{\partial}{\partial\zeta} + \frac{n}{c}\frac{\partial}{\partial t}\right) A_2 &= i\kappa \sin(\omega_m t - k_m\zeta) A_1 \end{aligned} \tag{9.9-24}$$

with

$$\kappa = \frac{\omega n^3}{2c} r_{12k} E_{mk} \tag{9.9-25}$$

The solutions of the coupled mode-equations (9.9-22) for the general case ($n_1 \neq n_2$) are very complicated and will not be discussed here. For the case when $n_1 = n_2 = n$, the general solution of the coupled-mode equations is given by

$$\begin{aligned} A_1(\zeta, t) &= C_1\left(\zeta - \frac{c}{n} t\right) \cos\left[\frac{\kappa c}{\omega_m(n - n_m)} \cos(\omega_m t - k_m \zeta)\right] \\ &\quad + C_2\left(\zeta - \frac{c}{n} t\right) \sin\left[\frac{\kappa c}{\omega_m(n - n_m)} \cos(\omega_m t - k_m \zeta)\right] \\ A_2(\zeta, t) &= iC_2\left(\zeta - \frac{c}{n} t\right) \cos\left[\frac{\kappa c}{\omega_m(n - n_m)} \cos(\omega_m t - k_m \zeta)\right] \\ &\quad - iC_1\left(\zeta - \frac{c}{n} t\right) \sin\left[\frac{\kappa c}{\omega_m(n - n_m)} \cos(\omega_m t - k_m \zeta)\right] \end{aligned} \tag{9.9-26}$$

where C_1 and C_2 are arbitrary functions. Let the boundary condition at the input ($\zeta = 0$) face of the crystal be

$$\begin{aligned} A_1(0, t) &= A_0 \\ A_2(0, t) &= 0 \end{aligned} \tag{9.9-27}$$

These conditions correspond to the case where the input polarizer is parallel to $\hat{\mathbf{e}}_1$ (one of the unperturbed principal axes).

Let $\zeta = 0$ in Equation (9.9-26), then the boundary condition (9.9-27) becomes

$$\begin{aligned} &C_1\left(-\frac{c}{n} t\right) \cos\left[\frac{\kappa c}{\omega_m(n - n_m)} \cos \omega_m t\right] \\ &\quad + C_2\left(-\frac{c}{n} t\right) \sin\left[\frac{\kappa c}{\omega_m(n - n_m)} \cos \omega_m t\right] = A_0 \\ &C_1\left(-\frac{c}{n} t\right) \sin\left[\frac{\kappa c}{\omega_m(n - n_m)} \cos \omega_m t\right] \\ &\quad - C_2\left(-\frac{c}{n} t\right) \cos\left[\frac{\kappa c}{\omega_m(n - n_m)} \cos \omega_m t\right] = 0 \end{aligned} \tag{9.9-28}$$

This gives at $\zeta = 0$

$$\begin{aligned} C_1\left(-\frac{c}{n} t\right) &= A_0 \cos\left[\frac{\kappa c}{\omega_m(n - n_m)} \cos \omega_m t\right] \\ C_2\left(-\frac{c}{n} t\right) &= A_0 \sin\left[\frac{\kappa c}{\omega_m(n - n_m)} \cos \omega_m t\right] \end{aligned} \tag{9.9-29}$$

Equations (9.9-29) give the functions C_1 and C_2 at $\zeta = 0$. Since C_1 and C_2 are, in general, functions of $\zeta - (c/n)t$, they are given by

$$C_1\left(\zeta - \frac{c}{n}t\right) = A_0 \cos\left[\frac{\kappa c}{\omega_m(n - n_m)} \cos\left(\omega_m t - \frac{\omega_m}{c} n\zeta\right)\right]$$

$$C_2\left(\zeta - \frac{c}{n}t\right) = A_0 \sin\left[\frac{\kappa c}{\omega_m(n - n_m)} \cos\left(\omega_m t - \frac{\omega_m}{c} n\zeta\right)\right] \quad (9.9\text{-}30)$$

Substituting Equations (9.9-30) into Equations (9.9-26), the mode amplitudes become

$$A_1(\zeta, t) = A_0 \cos\left\{\frac{\kappa c}{\omega_m(n - n_m)}\left[\cos(\omega_m t - k_m\zeta) - \cos\left(\omega_m t - \frac{\omega_m}{c} n\zeta\right)\right]\right\}$$

$$A_2(\zeta, t) = iA_0 \sin\left\{\frac{\kappa c}{\omega_m(n - n_m)}\left[\cos\left(\omega_m t - \frac{\omega_m}{c} n\zeta\right) - \cos(\omega_m t - k_m\zeta)\right]\right\} \quad (9.9\text{-}31)$$

Using next the trigonometric identity

$$\cos\alpha - \cos\beta = -2 \sin\tfrac{1}{2}(\alpha + \beta) \sin\tfrac{1}{2}(\alpha - \beta)$$

and the relation $k_m = (\omega_m/c)n_m$, the mode amplitudes at the output plane ($\zeta = L$) of the crystal become

$$A_1(L, t) = A_0 \cos[\delta \sin(\omega_m t - \phi)]$$

$$A_2(L, t) = iA_0 \sin[\delta \sin(\omega_m t - \phi)] \quad (9.9\text{-}32)$$

where

$$\delta = \kappa L \frac{\sin \dfrac{\omega_m}{2c}(n - n_m)L}{\dfrac{\omega_m}{2c}(n - n_m)L} \quad (9.9\text{-}33)$$

and ϕ is given by Equation (9.9-14). We notice that δ in Equation (9.9-33) is identical to the phase modulation index (9.9-13) in its dependence on the length of the crystal L. Therefore, all the discussion of the phase-velocity matching for phase modulation can also be applied to amplitude modulation. In particular, maximum modulation occurs when condition (9.9-17) is satisfied and the maximum modulation depth is given by

$$\delta_{\max} = \frac{\omega}{\omega_m} \frac{n^3}{|n - n_m|} r_{12k} E_{mk} \quad (9.9\text{-}34)$$

It is interesting to compare the final result (9.9-32) with our previous formalism that led to (9.3-2) and (9.3-3). We associate the direction x of Figure 9-4 with the direction "1" of this section and y with "2." Using $\Gamma(t) = \Gamma_m \sin \omega_m t$ after reverting to real-time notation

$$E_x(L) = \frac{1}{\sqrt{2}} [(E_{x'}(L) + E_{y'}(L)] = \frac{A}{\sqrt{2}} \text{Re}[e^{i(\Gamma_m/2)\sin \omega_m t} + e^{-i(\Gamma_m/2)\sin \omega_m t}]$$

$$= \sqrt{2}\, A \cos\left(\frac{\Gamma_m}{2} \sin \omega_m t\right) \tag{9.9-35}$$

$$E_y(L) = \frac{1}{\sqrt{2}} [E_{x'}(L) - E_{y'}(L)] = i\sqrt{2}\, A \sin\left(\frac{\Gamma_m}{2} \sin \omega_m t\right) \tag{9.9-36}$$

If we use the definition (9.2-4), we have $\Gamma_m/2 = \kappa L$. It follows that Equations (9.9-35, 9.9-36) reduce to the form of (9.9-32) for the phase-matched case $n = n_m$, where $\delta = \kappa L = \Gamma_m/2$. Equations (9.9-35 and 9.9-36), however, account accurately for the important case when $n \neq n_m$, i.e., the phase velocity of the modulation field is different from that of the optical wave. The "exact" analysis also gives us the correct form of the phase delay ϕ.

Problems

9.1 Derive the equations of the ellipses traced during one period by the optical field vector as shown in Figure 9-3(c) for $\Gamma = 0, \frac{\pi}{4}, \frac{\pi}{2}, \frac{3\pi}{4}, \pi, \frac{5\pi}{4}$.

9.2 Discuss the consequence of the field-independent retardation $(\omega l/c_c)(n_0 - n_e)$ in Equation (9.5-1) on an amplitude modulator such as that shown in Figure 9-4.

9.3 Use the Bessel-function expansion of sin $[a \sin x]$ to express (9.3-7) in terms of the harmonics of the modulation frequency ω_m. Plot the ratio of the third harmonic $(3\omega_m)$ of the output intensity to the fundamental as a function of Γ_m. What is the maximum allowed Γ_m if this ratio is not to exceed 10^{-2}? (*Answer:* $\Gamma_m < 0.5$.)

9.4 Show that, if a phase-modulated optical wave is incident on a square-law detector, the output contains no alternating currents.

9.5 Using References [4] and [5], design a partially loaded KDP traveling wave phase modulator that operates at $\nu_m = 10^9$ Hz and yields a peak phase excursion of $\delta = \pi/3$. What is the modulation power?

9.6 Derive the expression [similar to Equation (9.6-2)] for the modulation power of a transverse $\bar{4}3m$ crystal electrooptic modulator of the type described in the Appendix B.

9.7 Derive an expression for the modulation power requirement [corresponding to Equation (9.6-2)] for a GaAs transverse modulator.

9.8 Show that if a ray propagates at an angle $\theta(\ll 1)$ to the z axis in the arrangement of Figure 9-4, it exercises a birefringent contribution to the retardation.

$$\Delta\Gamma_{\text{birefringent}} = \frac{\omega l}{2c} n_0 \left(\frac{n_0^2}{n_e^2} - 1\right) \theta^2$$

which corresponds to a change in index

$$n_0 - n_e(\theta) = \frac{n_0 \theta^2}{2} \left(\frac{n_0^2}{n_e^2} - 1\right)$$

9.9 Derive an approximate expression for the maximum allowable beam-spreading angle in Problem 9.8 for which $\Delta\Gamma_{\text{birefringent}}$ does not interfere with the operation of the modulator. *Answer:*

$$\theta < \left[\frac{\lambda}{4 l n_0 \{n_0^2/n_e^2 - 1\}}\right]^{1/2}$$

9.10 Consider the index ellipsoid S defined by

$$S_{ij} x_i x_j = 1$$

Show that the vector **N** defined by

$$N_i = S_{ij} x_j$$

is perpendicular to S at the point (x_1, x_2, x_3) on S.

9.11 Consider the case of a KH_2PO_4(KDP) crystal with an applied field along the x axis. Show that in the new principal dielectric axes coordinate system (x', y', z'), x' coincides with x while y' and z' are in the y–z plane, but rotated from their original positions by θ, where

$$\tan 2\theta = \frac{2 r_{41} E_x}{1/n_0^2 - 1/n_e^2}$$

Show that in the x, y', z' system the equation for the index ellipsoid is

$$\frac{x^2}{n_0^2} + \left(\frac{1}{n_0^2} + r_{41} E_x \tan\theta\right) y'^2 + \left(\frac{1}{n_e^2} - r_{41} E_x \tan\theta\right) z'^2 = 1$$

9.12 An optical beam with amplitude E_0 and frequency $\omega/2\pi$ is split, equally, in two. One of the beams is left as is, while the other is phase modulated according to

$$\Delta\phi = a + \delta \cos\omega_m t \ (\omega_m \ll \omega)$$

The two beams are then recombined coherently (the whole procedure can be accomplished by a Michelson–Morley, or a Mach–Zehnder, interferometer with a phase modulator placed in one arm).

a. Express the recombined field in the form

$$E_{\text{rec}} = f(t) e^{i(\alpha + \beta \cos\omega_m t)}$$

b. Show that for $a = \pi/2$, $\delta \ll 1$

$$E_{\rm rec} \approx E_0 \left(1 + \frac{\delta}{2} \cos\omega_m t\right) e^{i\left(\frac{\pi}{4} + \frac{\delta}{2}\cos\omega_m t\right)}$$

c. Obtain the (approximate) optical spectrum of the output beam.

d. Derive the intensity modulation characteristics

$$\frac{|E_{\rm rec}|^2}{|E_0|^2}$$

for the general case.

e. Using the results from **d**, determine how we can obtain a nearly linear modulation response in which the detected photocurrent, $I_{\rm det} \propto |E_{\rm rec}|^2$, is proportional to the modulation signal $\delta \cos\omega_m t$.

9.13 In Section 9.1 show that the three principal vectors X', X'', and X''' are perpendicular to each other.

9.14 Let x, y, z be the principal dielectric axes of a crystal with dielectric tensor elements ϵ_{xx}, ϵ_{yy}, ϵ_{zz}. Consider a new coordinate system ξ, η, z where ξ and η are rotated at an angle θ about the z axis (the z axis is the same in both systems). Show that the $\bar{\bar{\epsilon}}$ tensor in the new system is

	ξ	η	z
ξ	$\epsilon_{xx} + \delta \sin^2 \theta$	$\delta/2 \sin (2\theta)$	0
η	$\delta/2 \sin (2\theta)$	$\epsilon_{xx} + \delta \cos^2 \theta$	0
z	0	0	ϵ_{zz}

where $\delta \equiv \epsilon_{yy} - \epsilon_{xx}$.

9.15 Consider a crystal with principal dielectric axes x, y, z and corresponding ϵ_{xx}, ϵ_{yy}, ϵ_{zz}. Let the application of an electric field (or strain) cause an off-diagonal element ϵ_{zy} to appear.

a. Show the new principal dielectric axes are rotated about the x axis by an angle

$$\theta \simeq \frac{\epsilon_{zy}}{\epsilon_{yy} - \epsilon_{zz}} \quad (\epsilon_{zy} \ll \epsilon_{yy}, \epsilon_{zz})$$

b. Show that in KDP the application of a dc field $\mathbf{E} = \hat{\mathbf{e}}_x E_x$ causes a rotation β of the z and y principal axes about the x axis where

$$\beta = -\frac{n_e^2 n_0^2 r_{41} E_x}{n_0^2 - n_e^2}$$

9.16

a. Design an electrooptic waveguide modulator in a $LiTaO_3$ crystal as shown.

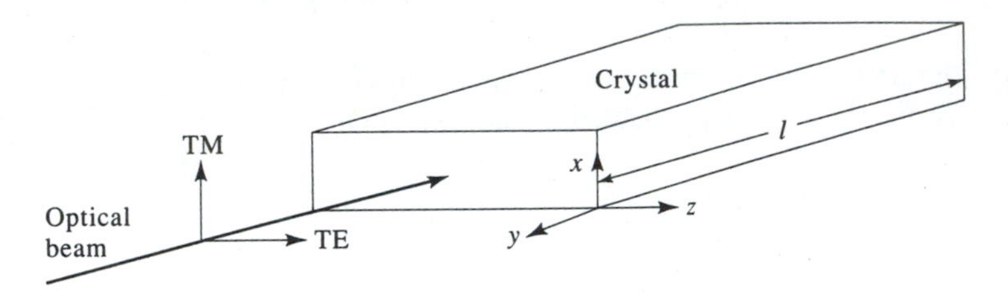

Show that the phase retardation $\Gamma \equiv \theta_{TE} - \theta_{TM}$ is given by

$$\Gamma = \frac{\omega l}{c} [n_0^3 r_{13} - n_e^3 r_{33}] E_z$$

b. Describe how you will use the waveguide as: (1) an amplitude modulator, (2) a phase modulator. Calculate the requisite modulation voltage assuming a width in the z direction of 5 μm and $\lambda = 0.6328$ μm.

9.17

a. Design a polarization switch (mode coupler) TE↔TM using $LiNbO_3$, the crystal geometry of problem 9.15, and a dc field parallel to the x axis. Show how you can overcome the velocity mismatch problem ($n_0 \neq n_e$) by using a spatially periodic dc field

$$E_x = E_0 \cos \frac{2\pi}{\Lambda} y$$

with a proper choice of the period Λ.

b. What is the value of Λ at $1 = 1.15$ μm? (See Table 9-2 for dispersion data.)

References

1. See, for example, J. F. Nye, *Physical Properties of Crystals.* New York: Oxford, 1957, p. 123.
2. See, for example, A. Yariv, *Quantum Electronics.* New York: Wiley, 1967, Chap. 14.
3. Peters, L. C., "Gigacycle bandwidth coherent light traveling-wave phase modulators," *Proc. IEEE* 51:147, 1963.
4. Rigrod, W. W., and I. P. Kaminow, "Wide-band microwave light modulation," *Proc. IEEE* 51:137, 1963.
5. Kaminow, I. P., and J. Lin, "Propagation characteristics of partially loaded two-conductor transmission lines for broadband light modulators," *Proc. IEEE* 51:132, 1963.
6. White, R. M., and C. E. Enderby, "Electro-optical modulators employing intermittent interaction," *Proc. IEEE* 51:214, 1963.

7. Fowler, V. J., and J. Schlafer, ''A survey of laser beam deflection techniques,'' *Proc. IEEE* 54:1437, 1966.
8. Hall, D., A. Yariv, and E. Garmire, ''Observation of propagation cutoff and its control in thin optical waveguides,'' *Appl. Phys. Lett.* 17:127, 1970.
9. Hall, D., A. Yariv, and E. Garmire, ''Optical guiding and electrooptic modulation in GaAs epitaxial layers,'' *Opt. Commun.* 1:403, 1970.
10. Hammer, J. M., and W. Phillips, ''Low-loss single mode optical waveguide and efficient high-speed modulators of $LiNb_xTa_{1-x}O_3$ on $LiTaO_3$,'' *Appl. Phys. Lett.* 24:545, 1974.
11. Yariv, A., and P. Yeh, *Optical Waves in Crystals.* New York: Wiley-Interscience, 1983, Chaps. 7 and 8.

Noise in Optical Detection and Generation

10.0 INTRODUCTION

In this chapter we study the effect of noise in a number of important physical processes. We will take the term noise to represent random electromagnetic fields occupying the same spectral region as that occupied by some "signal." The effect of noise will be considered in the following cases.

1. *Measurement of optical power.* In this case the noise causes fluctuations in the measurement, thus placing a lower limit on the smallest amount of power that can be measured.
2. *Linewidth of laser oscillators.* The presence of incoherent spontaneous emission power will be found to be the cause for a finite amount of spectral line broadening in the output of single-mode laser oscillators. This broadening manifests itself as a limited coherence time.
3. *Optical communication system.* We will consider the case of an optical communication system using a binary pulse code modulation in which the information is carried by means of a string of 1 and 0 pulses. The presence of noise will be shown to lead to a certain probability that any given pulse in the reconstructed train pulse is in error.

In this chapter we consider optical detectors utilizing light-generated charge carriers. These include the photomultiplier, the photoconductive detector, the *p-n* junction photodiode, and the avalanche photodiode. These detectors are the main ones used in the field of quantum electronics, because they combine high sensitivity with very short response times. Other types of detectors, such as bolometers, Golay cells, and thermocouples, whose operation depends on temperature changes induced by the absorbed radiation, will not be discussed.[1]

Two types of noise will be discussed in detail. The first type is thermal (Johnson) noise, which represents noise power generated by thermally agitated charge carriers. The expression for this noise will be derived by using the conventional thermodynamic treatment as well as by a statistical analysis of a particular model in which the physical origin of the noise is more apparent. The second type, shot noise (or generation-recombination noise in photoconductive detectors), is attributable to the random way in which electrons are emitted or generated in the process of interacting with a radiation field. This noise exists even at zero temperature, where thermal agitation or generation of carriers can be neglected. In this case it results from the randomness with which carriers are generated by the *very signal that is measured.* Detection in the limit of signal-generated shot noise is called quantum-limited detection, since the corresponding sensitivity is that allowed by the uncertainty principle in quantum mechanics. This point will be brought out in the next chapter.

A quantum optics treatment of noise and of squeezing of field fluctuations [25, 26, 27, 28] is given in Chapter 20.

10.1 LIMITATIONS DUE TO NOISE POWER

Measurement of Optical Power

Consider the problem of measuring an optical signal field

$$v_S(t) = V_S \cos \omega t \tag{10.1-1}$$

in the presence of a noise field. The instantaneous noise field that adds to that of the signal can be taken as the sum of an in-phase component and a quadrature component according to

$$v_N(t) = V_{NC}(t) \cos \omega t + V_{NS}(t) \sin \omega t \tag{10.1-2}$$

where $V_{NC}(t)$ and $V_{NS}(t)$ are slowly [compared to $\exp(i\omega t)$] varying random uncorrelated quantities with a zero mean. The total field at the detector $v(t) = v_S(t) + v_N(t)$ can be written as

$$v(t) = \mathrm{Re}\{[V_S + V_{NC}(t) - iV_{NS}(t)]e^{i\omega t}\} \tag{10.1-3}$$

$$\equiv \mathrm{Re}[V(t)e^{i\omega t}] \tag{10.1-4}$$

The total (signal plus noise) field phasor $V(t)$ is shown in Figure 10-1.

In most situations of interest to optical detection the sources of noise are due to the concerted action of a large number of independent agents. In this case the central

---

[1] The interested reader will find a good description of these devices in Reference [6].

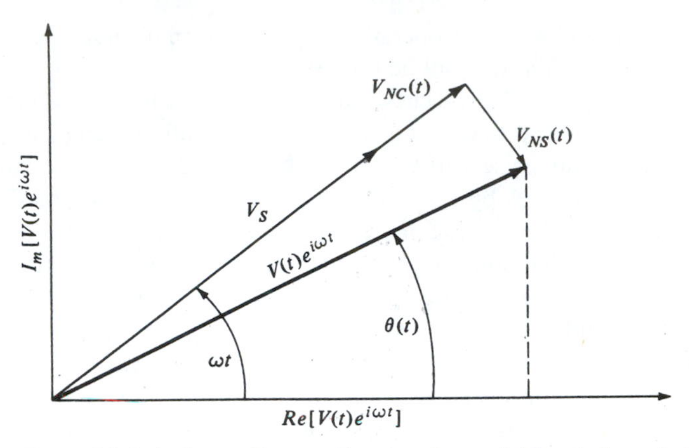

Figure 10-1 A phasor diagram showing the total (signal plus noise) field phasor $V(t)$ at time t. The instantaneous field is given by the horizontal projection of $V(t)\exp(i\omega t)$.

limit theorem of statistics [1] tells us that the probability function for finding $V_{NC}(t)$ at time t between V_{NC} and $V_{NC} + dV_{NC}$ is described by a Gaussian

$$p(V_{NC})\,dV_{NC} = \frac{1}{\sqrt{2\pi}\sigma} e^{-V_{NC}^2/2\sigma^2}\,dV_{NC} \tag{10.1-5}$$

and by a similar expression in which V_{NS} replaces V_{NC} for $p(V_{NS})$. Since $V_{NC}(t)$ has a unity probability of having some value between $-\infty$ and ∞, it follows that

$$\int_{-\infty}^{\infty} p(V_{NC})\,dV_{NC} = 1 \tag{10.1-6}$$

It follows from (10.1-5) that $\overline{V}_{NC}$, the ensemble average[2] (denoted by a horizontal bar) of V_{NC}, is zero,[3] whereas the mean square value is

$$\overline{V_{NC}^2} = \overline{V_{NS}^2} = \int_{-\infty}^{\infty} V_{NC}^2 p(V_{NC})\,dV_{NC} = \sigma^2 \tag{10.1-7}$$

---

[2]The ensemble average $\overline{A(t)}$ of a quantity $A(t)$ is obtained by measuring A simultaneously at time t in a very large number of systems that, *to the best of our knowledge*, are identical. Mathematically,

$$\overline{A(t)} = \lim_{N\to\infty} \left[\frac{1}{N}\sum_{n=1}^{N} A_n(t)\right]$$

where $A_n(t)$ denotes the observation in the nth system. In a truly random phenomenon, the time averaging and ensemble averaging lead to the same result, so the ensemble average is independent of the time t in which it is performed and can also be obtained from

$$\overline{A} = \int_{-\infty}^{\infty} Ap(A)\,dA$$

where $p(A)$ is the probability function, in the sense of (10.1-5), of the variable A.

[3]The reason for $\overline{V_{NC}(t)} = 0$ can be appreciated from Figure 10-1. $V_{NC}(t)$ has an equal probability of being in phase with V_S as of being out phase, thus averaging out to zero.

The "power" in $v(t)$ is obtained using (1.1-12) as

$$P(t) \equiv \tfrac{1}{2}[V(t)e^{i\omega t}][V^*(t)e^{-i\omega t}]$$
$$= \tfrac{1}{2}V_S^2 + 2V_S V_{NC} + V_{NC}^2 + V_{NS}^2 \qquad (10.1\text{-}8)$$

The ensemble average (or *long* time average) of $P(t)$ is

$$\overline{P} \equiv \overline{P(t)} = \tfrac{1}{2}(\overline{V_S^2} + \overline{V_{NC}^2} + \overline{V_{NS}^2}) = \tfrac{1}{2}(\overline{V_S^2} + 2\sigma^2) \qquad (10.1\text{-}8a)$$

where use has been made of the fact that $\overline{V}_{NC} = 0$ and of (10.1-7).

The physical significance of the time-varying power $P(t)$ and its long-time (or ensemble) average $\overline{P}$ is illustrated by Figure 10-2.

It is clear from the fluctuating nature of $P(t)$ that any measurement of this power is subject to an uncertainty due to the random nature of V_{NC} and V_{NS} in (10.1-8). As a measure of the uncertainty in power measurement, we may reasonably take the root mean square (rms) power deviation

$$\Delta P \equiv [\overline{(P(t) - \overline{P})^2}]^{1/2}$$

Using (10.1-8) and (10.1-8a), we obtain after some algebra

$$\Delta P = (4\overline{V_S^2 V_{NC}^2} + 2\overline{V_{NC}^4} - 2\overline{V_{NS}^2}\;\overline{V_{NC}^2})^{1/2} \qquad (10.1\text{-}9)$$

Using (10.1-5) we obtain

$$\overline{V_{NC}^4} = \int_{-\infty}^{\infty} V_{NC}^4 p(V_{NC})\, dV_{NC} = 3\sigma^4 \qquad (10.1\text{-}10)$$

so that using $\overline{V_{NC}^2} = \overline{V_{NS}^2} = \sigma^2$ in (10.1-9) results in

$$\Delta P = \sigma(\overline{V_S^2} + \sigma^2)^{1/2} = \sigma(2P_S + \sigma^2)^{1/2} \qquad (10.1\text{-}11)$$

where according to (10.1-8) we may associate $P_S = \overline{V_S^2}$ with the signal power that is, the power that would be measured if V_{NC} and V_{NS} were, hypothetically, rendered zero.

A question of practical importance involves the minimum signal power that can be measured in the presence of noise. We may, somewhat arbitrarily, take this power

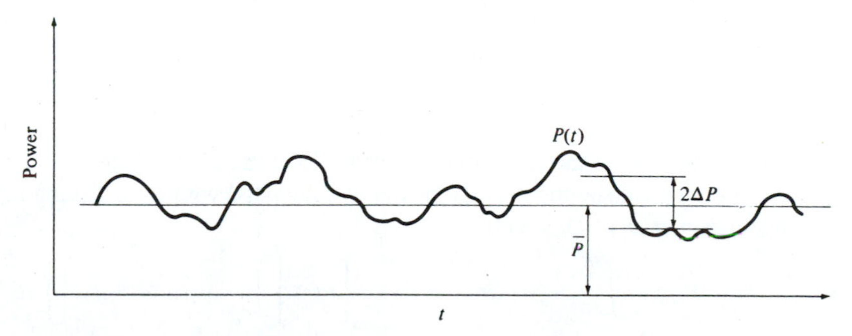

Figure 10-2 The intermingling of noise power with that of a signal causes the total power to fluctuate. The rms fluctuation ΔP limits the accuracy of power measurements.

P_{limit} to be that at which the uncertainty ΔP becomes equal to the signal power P_S. At this point we have from (10.1-11)

$$P_{\text{limit}} = \sigma(2P_{\text{limit}} + \sigma^2)^{1/2}$$

or, after solving for P_{limit},

$$P_{\text{limit}} = \sigma^2(1 + \sqrt{2}) = P_N(1 + \sqrt{2}) \qquad \textbf{(10.1-12)}$$

where $P_N = \sigma^2 = \frac{1}{2}(\overline{V_{NC}^2} + \overline{V_{NS}^2})$ is the noise power. Widespread convention chooses to define the minimum detectable signal power as equal to P_N instead of $2.414P_N$, as obtained above. This simplification is understandable, since our choice of the limit of detectability $\Delta P = P_S$ was somewhat arbitrary. In any case the main conclusion to remember is that near the limit of detectivity, the rms power fluctuation is comparable to the signal power. The next task, which will be taken up in this chapter and in Chapter 11, is to find out the main sources of noise power and consequently ways to minimize them. Before tackling this task, however, we need to develop some mathematical tools for dealing with random processes.

10.2 NOISE—BASIC DEFINITIONS AND THEOREMS

A real function $v(t)$ and its Fourier transform $V(\omega)$ are related by

$$V(\omega) = \frac{1}{2\pi}\int_{-\infty}^{\infty} v(t)e^{-i\omega t}\,dt \qquad \textbf{(10.2-1)}$$

and

$$v(t) = \int_{-\infty}^{\infty} V(\omega)e^{i\omega t}\,d\omega \qquad \textbf{(10.2-2)}$$

In the process of measuring a signal $v(t)$, we are not in a position to use the infinite time interval needed, according to (10.2-1), to evaluate $V(\omega)$. If the time duration of the measurement is T, we may consider the function $v(t)$ to be zero when $t \leq -T/2$ and $t \geq T/2$ and, instead of (10.2-1), get

$$V_T(\omega) = \frac{1}{2\pi}\int_{-T/2}^{T/2} v(t)e^{-i\omega t}\,dt \qquad \textbf{(10.2-3)}$$

Since $v(t)$ is real, it follows that

$$V_T(\omega) = V_T^*(-\omega) \qquad \textbf{(10.2-4)}$$

T is usually called the resolution or integration time of the system.

Let us evaluate the average power P associated with $v(t)$. Taking the instantaneous power as $v^2(t)$, we obtain[4]

$$P = \frac{1}{T}\int_{-T/2}^{T/2} v^2(t)\,dt = \frac{1}{T}\int_{-T/2}^{T/2}\left\{v(t)\left[\int_{-\infty}^{\infty} V_T(\omega)e^{i\omega t}\,d\omega\right]\right\}dt \qquad \textbf{(10.2-5)}$$

---

[4]It may be convenient for this purpose to think of $v(t)$ as the voltage across a one-ohm resistance.

Using (10.2-3) and (10.2-4) in the last equation and interchanging the order of integration leads to

$$P = \frac{2\pi}{T} \int_{-\infty}^{\infty} |V_T(\omega)|^2 \, d\omega \tag{10.2-6}$$

or

$$P = \frac{4\pi}{T} \int_{0}^{\infty} |V_T(\omega)|^2 \, d\omega \tag{10.2-7}$$

where we used

$$\lim_{T\to\infty} (2\pi)^{-1} \int_{-T/2}^{T/2} dt \exp\,[i(\omega + \omega')t] = \delta(\omega + \omega')$$

If we define the *spectral density function* $S_v(\omega)$ of $v(t)$ by

$$S_v(\omega) = \lim_{T\to\infty} \frac{4\pi |V_T(\omega)|^2}{T} \tag{10.2-8}$$

then, according to (10.2-7), $S_v(\omega) d\omega$ is the portion of the average power of $v(t)$ that is due to frequency components between ω and $\omega + d\omega$. According to this physical interpretation, we may measure $S_v(\omega)$ by separating the spectrum of $v(t)$ into its various frequency classes as shown in Figure 10-3 and then measuring the power output $S_v(\omega_i)\Delta\omega_i$ of each of the filters [2].

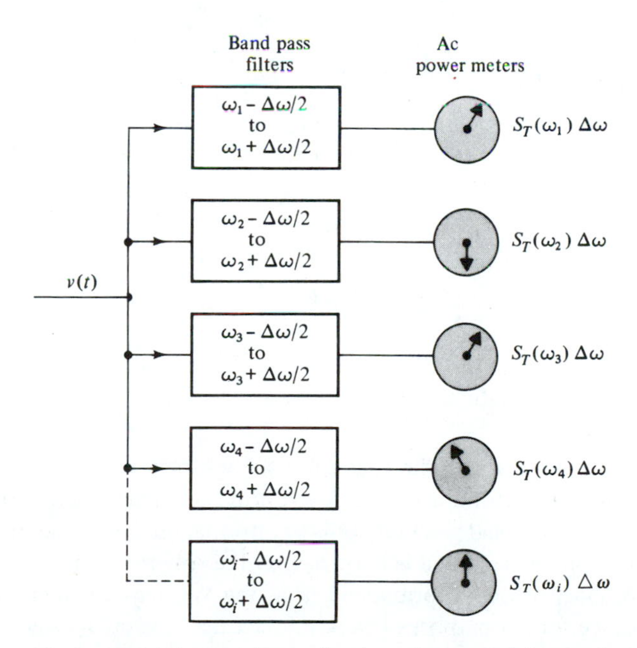

Figure 10-3 Diagram illustrating how the spectral density function $S_T(\omega)$ of a signal $v(t)$ can be obtained by measuring the power due to different frequency intervals.

Wiener–Khintchine Theorem

We will next derive another formal result involving the spectral density function.

Consider the time average of the product of some field quantity $v(t)$ with its delayed version $v(t + \tau)$

$$C_v(\tau) = \overline{v(t)v(t + \tau)} \tag{10.2-9}$$

The function $C_v(\tau)$ is termed the autocorrelation function of $v(t)$. We use (10.2-2) to carry out the integration indicated in (10.2-9)

$$\begin{aligned} C_v(\tau) &= \frac{1}{T}\int_{-T/2}^{T/2} v(t)v(t + \tau)\,dt \\ &= \frac{1}{T}\int_{-\infty}^{\infty}\int_{-\infty}^{\infty}\int_{-T/2}^{T/2} d\omega\,d\omega'\,dt\,V_T(\omega)V_T(\omega')e^{i(\omega+\omega')t}e^{i\omega\tau} \end{aligned} \tag{10.2-10}$$

In the limit $T \to \infty$,

$$\lim_{T\to\infty}\int_{-T/2}^{T/2} dt\,e^{i(\omega+\omega')t} = 2\pi\delta(\omega + \omega') \tag{10.2-11}$$

so that

$$\begin{aligned} C_v(\tau) &= \lim_{T\to\infty}\frac{2\pi}{T}\int_{-\infty}^{\infty}\int_{-\infty}^{\infty} V_T(\omega')V_T(\omega)\delta(\omega + \omega')e^{i\omega\tau}\,d\omega\,d\omega' \\ &= \lim_{T\to\infty}\frac{1}{2}\int_{-\infty}^{\infty}\frac{4\pi|V_T(\omega)|^2}{T}\,e^{i\omega\tau}\,d\omega \end{aligned} \tag{10.2-12}$$

The quantity $4\pi|V_T(\omega)|^2/T$ is, according to (10.2-8), the spectral density function of $S_v(\omega)$ of $v(t)$, so that

$$C_v(\tau) = \frac{1}{2}\int_{-\infty}^{\infty} S_v(\omega)e^{i\omega\tau}\,d\omega \tag{10.2-13}$$

so that using (10.2-1)

$$S_v(\omega) = \frac{1}{\pi}\int_{-\infty}^{\infty} C_v(\tau)e^{-i\omega\tau}\,d\tau \tag{10.2-14}$$

The last two equations state that the spectral density function $S_v(\omega)$ and the autocorrelation function $C_v(\tau)$ form a Fourier transform pair. This result is one of the more important theoretical and practical tools of information theory and of the mathematics of random processes, and it is known, after the American and Russian mathematicians who, independently, formulated it, as the Wiener–Khintchine theorem. Its main importance for our purposes lies in the fact that it is often easier to obtain, experimentally or theoretically, $C_v(\tau)$ rather than $S_v(\omega)$, so that $S_v(\omega)$ is derived by a Fourier transformation of $C_v(\tau)$.

10.3 THE SPECTRAL DENSITY FUNCTION OF A TRAIN OF RANDOMLY OCCURRING EVENTS

Consider a time-dependent random variable $i(t)$ made up of a very large number of individual events $f(t - t_i)$ that occur at random times t_i.[5] An observation of $i(t)$ during a period T will yield

$$i_T(t) = \sum_{i=1}^{N_T} f(t - t_i) \qquad 0 \le t \le T \tag{10.3-1}$$

where N_T is the total number of events occurring in T. Typical examples of a random function $i(t)$ are provided by the thermionic emission current from a hot cathode (under temperature-limited conditions), or the electron current caused by photoemission from a surface. In these cases $f(t - t_i)$ represents the current resulting from a single electron emission occurring at t_i.

The Fourier transform of $i_T(t)$ is given according to (10.2-3) by

$$I_T(\omega) = \sum_{i=1}^{N_T} F_i(\omega) \tag{10.3-2}$$

where $F_i(\omega)$ is the Fourier transform[6] of $f(t - t_i)$

$$\begin{aligned} F_i(\omega) &= \frac{1}{2\pi}\int_{-\infty}^{\infty} f(t - t_i)e^{-i\omega t}\, dt = \frac{e^{-i\omega t_i}}{2\pi}\int_{-\infty}^{\infty} f(t)\, e^{-i\omega t}\, dt \\ &= e^{-i\omega t_i}F(\omega) \end{aligned} \tag{10.3-3}$$

From (10.3-2) and (10.3-3) we obtain

$$\begin{aligned} |I_T(\omega)|^2 &= |F(\omega)|^2 \sum_{i=1}^{N_T}\sum_{j=1}^{N_T} e^{-i\omega(t_i - t_j)} \\ &= |F(\omega)|^2 \left(N_T + \sum_{i \ne j}^{N_T}\sum_{j}^{N_T} e^{i\omega(t_j - t_i)}\right) \end{aligned} \tag{10.3-4}$$

If we take the average of (10.3-4) over an ensemble of a very large number of physically identical systems, the second term on the right side of (10.3-4) can be neglected in comparison to N_T, since the times t_i are random. This results in

$$\overline{|I_T(\omega)|^2} = \overline{N}_T|F(\omega)|^2 \equiv \overline{N}T\,|F(\omega)|^2 \tag{10.3-5}$$

[5]This means that the *a priori* probability that a given event will occur in any time interval is distributed uniformly over the interval, or equivalently, that the probability $p(n)$ for n events to occur in an observation period T is given by the Poisson distribution function [2]

$$p(n) = \frac{(\overline{n})^n e^{-\overline{n}}}{n!}$$

where $\overline{n}$ is the average number of events occurring in T.

[6]We assume that the individual event $f(t - t_i)$ is over in a short time compared to the observation period T, so the integration limits can be taken as $-\infty$ to ∞ instead of 0 to T.

where the horizontal bar denotes ensemble averaging and where $\overline{N}$ is the average rate at which the events occur so that $\overline{N}_T = \overline{N}T$. The spectral density function $S_i(\omega)$ of the function $i_T(t)$ is given according to (10.2-8) and (10.3-5) as

$$S_i(\omega) = 4\pi\overline{N}|F(\omega)|^2 \tag{10.3-6}$$

In practice, one uses more often the spectral density function $S(\nu)$ defined so that the average power due to frequencies between ν and $\nu + d\nu$ is equal to $S(\nu)\,d\nu$. It follows then, that $S(\nu)\,d\nu = S(\omega)\,d\omega$; thus, since $\omega = 2\pi\nu$,

$$S_i(\nu) = 8\pi^2\overline{N}|F(2\pi\nu)|^2 \tag{10.3-7}$$

The last result is known as Carson's theorem and its usefulness will be demonstrated in the following sections where we employ it in deriving the spectral density function associated with a number of different physical processes related to optical detection.

Equation (10.3-7) was derived for the case in which the individual events $f(t - t_i)$ were displaced in time but were otherwise identical. There are physical situations in which the individual events may depend on one or more additional parameters. Denoting the parameter (or group of parameters) as α, we can clearly single out the subclass of events $f_\alpha(t - t_i)$ whose α is nearly the same and use (10.3-7) to obtain directly

$$S_\alpha(\nu) = 8\pi^2\overline{N}(\alpha)|F_\alpha(2\pi\nu)|^2\,\Delta\alpha \tag{10.3-8}$$

for the contribution of this subclass of events to $S(\nu)$. $F_\alpha(\omega)$ is the Fourier transform of $f_\alpha(t)$, and thus $\overline{N}(\alpha)\Delta\alpha$ is the average number of events per second whose α parameter falls between α and $\alpha + \Delta\alpha$.

$$\int_{-\infty}^{\infty} \overline{N}(\alpha)\,d\alpha = \overline{N}$$

The probability distribution function for α is $p(\alpha) = \overline{N}(\alpha)/\overline{N}$; therefore,

$$\int_{-\infty}^{\infty} p(\alpha)\,d\alpha = \frac{1}{\overline{N}}\int_{-\infty}^{\infty} \overline{N}(\alpha)\,d\alpha = 1 \tag{10.3-9}$$

Summing (10.3-8) over all classes α and weighting each class by the probability $p(\alpha)\,\Delta\alpha$ of its occurrence, we obtain

$$\begin{aligned} S_i(\nu) &= \sum_\alpha S_\alpha(\nu) = 8\pi^2 \sum_\alpha \overline{N}(\alpha)|F_\alpha(2\pi\nu)|^2\,\Delta\alpha \\ &= 8\pi^2\overline{N} \sum_\alpha |F_\alpha(2\pi\nu)|^2 p(\alpha)\,\Delta\alpha \\ &= 8\pi^2\overline{N}\int_{-\infty}^{\infty} |F_\alpha(2\pi\nu)|^2 p(\alpha)\,d\alpha = 8\pi^2\overline{N}\,\overline{|F(2\pi\nu)|^2} \end{aligned} \tag{10.3-10}$$

where the bar denotes averaging over α. Equation (10.3-10) is thus the extension of (10.3-7) to the case of events whose characterization involves, in addition to their time t_i, some added parameters. We will use it further in this chapter to derive the noise spectrum of photoconductive detectors in which case α is the lifetime of the excited photocarriers.

10.4 SHOT NOISE (3)

Let us consider the spectral density function of current arising from random generation and flow of mobile charge carriers. This current is identified with "shot noise." To be specific, we consider the case illustrated in Figure 10-4, in which electrons are released at random into the vacuum from electrode A to be collected at electrode B, which is maintained at a slight positive potential relative to A.

The average rate $\overline{N}$ of electron emission from A is $\overline{N} = \overline{I}/e$, where $\overline{I}$ is the average current and the electronic charge is taken as $-e$. The current pulse due to a single electron as observed in the external circuit is

$$i_e(t) = \frac{ev(t)}{d} \tag{10.4-1}$$

where $v(t)$ is the instantaneous velocity and d is the separation between A and B. To prove (10.4-1), consider the case in which the moving electron is replaced by a thin sheet of a very large area and of total charge $-e$ moving between the plates, as illustrated in Figure 10-5.

It is a simple matter to show (see Problem 10.1), using the relation $\nabla \cdot \mathbf{E} = \rho/\epsilon$, that the charge induced by the moving sheet on the left electrode is

$$Q_1 = \frac{e(d - x)}{d} \tag{10.4-2}$$

and that on the right electrode is

$$Q_2 = \frac{ex}{d} \tag{10.4-3}$$

where x is the position of the charged sheet measured from the left electrode. The current in the external circuit due to a single electron is thus

$$i_e(t) = \frac{dQ_2}{dt} = \frac{e}{d}\frac{dx}{dt} = \frac{e}{d}v(t) \tag{10.4-4}$$

in agreement with (10.4-1).

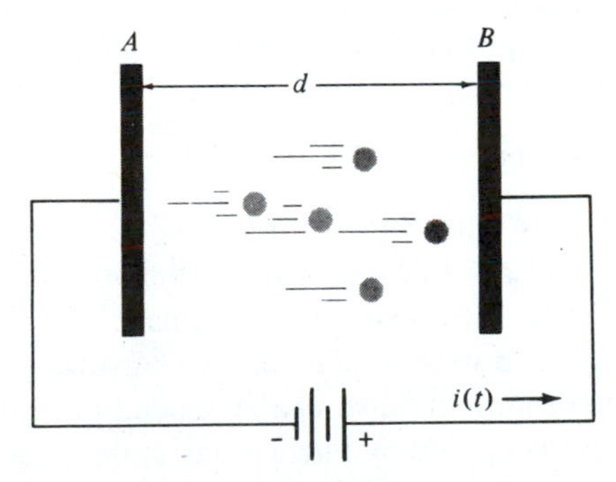

Figure 10-4 Random electron flow between two electrodes. This basic configuration is used in the derivation of shot noise.

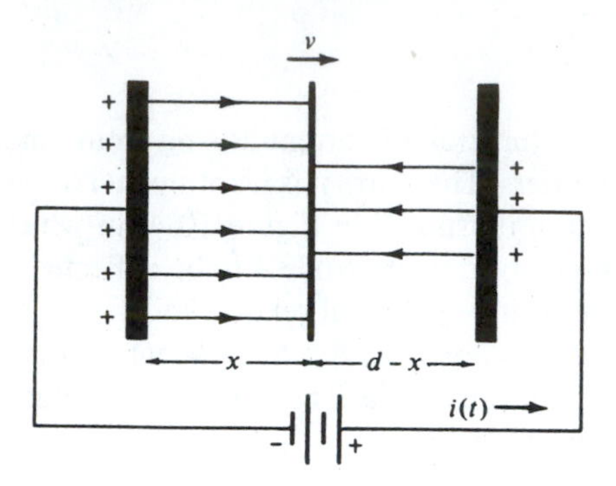

Figure 10-5 Induced charges and field lines due to a thin charge layer between the electrodes.

The Fourier transform of a single current pulse is

$$F(\omega) = \frac{e}{2\pi d} \int_0^{t_a} v(t) e^{-i\omega t} \, dt \tag{10.4-5}$$

where t_a is the arrival time of an electron emitted at $t = 0$. If the transit time of an electron is sufficiently small that, at the frequency of interest ω,

$$\omega t_a \ll 1 \tag{10.4-6}$$

i.e., $i_e(t) \propto \delta(t)$, we can replace $\exp(-i\omega t)$ in (10.4-5) by unity and obtain

$$F(\omega) = \frac{e}{2\pi d} \int_0^{t_a} \frac{dx}{dt} \, dt = \frac{e}{2\pi} \tag{10.4-7}$$

since $x(t_a)$ is, by definition, equal to d. Using (10.4-7) in (10.3-7) and recalling that $\bar{I} = e\bar{N}$ gives

$$S(\nu) = 8\pi^2 \bar{N} \left(\frac{e}{2\pi}\right)^2 = 2e\bar{I} \tag{10.4-8}$$

The power (in the sense of 10.2-5) in the frequency interval ν to $\nu + \Delta\nu$ associated with the current is, according to the discussion following (10.2-8), given by $S(\nu)$ $\Delta\nu$. It is convenient to represent this power by an *equivalent noise generator* at ν with a mean-square current amplitude

$$\overline{i_N^2}(\nu) \equiv S(\nu) \, \Delta\nu = 2e\bar{I} \, \Delta\nu \tag{10.4-9}$$

The noise mechanism described above is referred to as *shot noise*.

It is interesting to note that e in (10.4-9) is the charge of the particle responsible for the current flow. If, hypothetically, these carriers had a charge of $2e$, then at the *same average current* $\bar{I}$ the shot-noise power would double. Conversely, shot noise would disappear if the magnitude of an individual charge tended to zero. This is a reflection of the fact that shot noise is caused by fluctuations in the current that are due to the discreteness of the charge carriers and to the random electronic emission (for which the number of electrons emitted per unit time obey Poisson statistics [2]).

The ratio of the fluctuations to the average current decreases with increasing number of events.[7]

Another point to remember is that, in spite of the appearance of $\bar{I}$ on the right side of (10.4-9), $\overline{i_N^2}(\nu)$ represents an alternating current with frequencies near ν.

10.5 JOHNSON NOISE

Johnson, or *Nyquist noise* describes the fluctuations in the voltage across a dissipative circuit element; see References [4, 5]. These fluctuations are most often caused by the thermal motion of the charge carriers.[8] The charge neutrality of an electrical resistance is satisfied when we consider the whole volume, but locally the random thermal motion of the carriers sets up fluctuating charge gradients and, correspondingly, a fluctuating (ac) voltage. If we now connect a second resistance across the first one, the thermally induced voltage described above will give rise to a current and hence to a power transfer to the second resistor.[9] This is the so-called *Johnson noise*, whose derivation follows.

Consider the case illustrated in Figure 10-6 of a transmission line connected between two similar resistances R, which are maintained at the same temperature T.

[7]More precisely, for events obeying Poisson statistics we have (Reference [1] or derivable directly from footnote 5)

$$\frac{[\overline{(\Delta N)^2}]^{1/2}}{\overline{N}} = \frac{1}{(\overline{N})^{1/2}}$$

where N is the number of events in an observation time, $\overline{N}$ is the average value of N, and $\overline{(\Delta N)^2} \equiv \overline{(N - \overline{N})^2}$.

[8]We use the word ''carriers'' rather than ''electrons'' to include cases of ionic conduction or conduction by holes.

[9]The same argument applies to the second resistor, so at thermal equilibrium the net power leaving each resistor is zero.

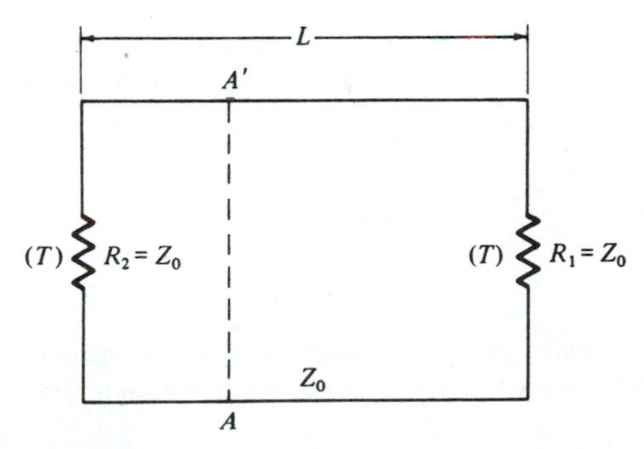

Figure 10-6 Lossless transmission line of characteristic impedance Z_0 connected between two matched loads ($R = Z_0$) at temperature T.

We choose the resistance R to be equal to the characteristic impedance Z_0 of the line, so that no reflection can take place at the ends. The transmission line can support traveling voltage waves of the form

$$v(t) = A \cos (\omega t \pm kz) \tag{10.5-1}$$

where $k = 2\pi/\lambda$ and the phase velocity is $c = \omega/k$.

For simplicity we require that the allowed solutions be periodic in the distance L,[10] so if we extend the solution outside the limits $0 \leq z \leq L$ we obtain

$$v(t) = A \cos [\omega t \pm k(z + L)] = A \cos (\omega t \pm kz)$$

This condition is fulfilled when

$$kL = 2m\pi \qquad m = 1, 2, 3, \ldots \tag{10.5-2}$$

Therefore, two adjacent modes differ in their value of k by

$$\Delta k = \frac{2\pi}{L} \tag{10.5-3}$$

and the number of modes having their k values somewhere between zero and $+k$ is[11]

$$N_k = \frac{kL}{2\pi} \tag{10.5-4}$$

or, using $k = 2\pi\nu/c$, we obtain

$$N(\nu) = \frac{\nu L}{c}$$

for the number of positively traveling modes with frequencies between zero and ν.

The number of modes per unit frequency interval is

$$p(\nu) = \frac{dN(\nu)}{d\nu} = \frac{L}{c} \tag{10.5-5}$$

Consider the power flowing in the $+z$ direction across some arbitrary plane, $A - A'$ say. It is clear that due to the lack of reflection this power must originate in R_2. Since the power is carried by the electromagnetic modes of the system, we have

$$\text{Power} = \frac{\text{energy}}{\text{distance}} \text{ (velocity of energy)}$$

---

[10]This seemingly arbitrary type of boundary condition is used extensively in similar situations in thermodynamics to derive the blackbody radiation density, or in solid-state physics to derive the density of electronic states in crystals.

[11]Negative k values correspond, according to (10.5-1), to waves traveling in the $-z$ direction. Our bookkeeping is thus limited to modes carrying power in the $+z$ direction.

We find, taking the velocity of light as c, that the power P due to frequencies between ν and $\nu + \Delta\nu$ is given by

$$P = \left(\frac{1}{L}\right)\left(\begin{array}{c}\text{number of modes between}\\ \nu \text{ and } \nu + \Delta\nu\end{array}\right)(\text{energy per mode})(c)$$
$$= \left(\frac{1}{L}\right)\left(\frac{L}{c}\,\Delta\nu\right)\left(\frac{h\nu}{e^{h\nu/kT} - 1}\right)(c)$$

or

$$P = \frac{h\nu\Delta\nu}{e^{h\nu/kT} - 1} \approx kT\Delta\nu \qquad (kT \gg h\nu) \tag{10.5-6}$$

where we used the fact that in thermal equilibrium the energy of a mode is given by [7]

$$\mathscr{E} = \frac{h\nu}{e^{h\nu/kT} - 1} \tag{10.5-7}$$

This result is also obtained in Appendix D from a different point of view. An equal amount of noise power is, of course, generated in the right resistor and is dissipated in the left one, so in thermal equilibrium the net power crossing any plane is zero.

The power given by (10.5-6) represents the maximum noise power available from the resistance, since it is delivered to a matched load. If the load connected across R has a resistance different from R, the noise power delivered is less than that given by (10.5-6). The noise-power bookkeeping is done correctly if the resistance R appearing in a circuit is replaced by either one of the following two equivalent circuits: a noise generator in series with R with mean-square voltage amplitude

$$\overline{v_N^2}(\nu) = \frac{4h\nu R\Delta\nu}{e^{h\nu/kT} - 1} \underset{kT \gg h\nu}{\simeq} 4kTR\Delta\nu \tag{10.5-8}$$

or a noise current generator of mean square value

$$\overline{i_N^2}(\nu) = \frac{4h\nu\Delta\nu}{R(e^{h\nu/kT} - 1)} \underset{kT \gg h\nu}{\simeq} \frac{4kT\Delta\nu}{R} \tag{10.5-9}$$

in parallel with R. The noise representations of the resistor are shown in Figure 10-7. There are numerous other derivations of the formula for Johnson noise. For deri-

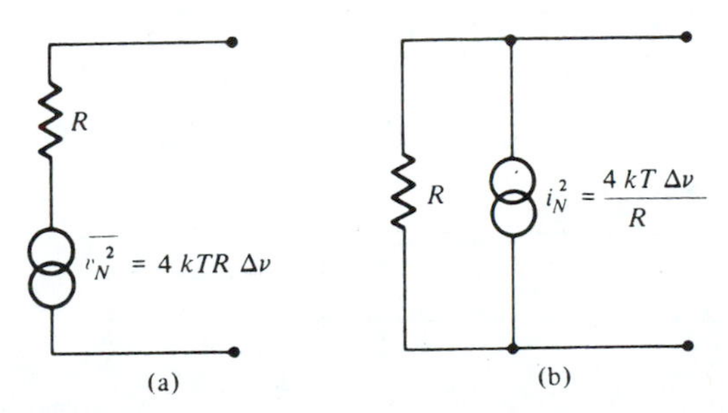

Figure 10-7 (a) Voltage and (b) current noise equivalent circuits of a resistance.

vations using lumped-circuit concepts and an antenna example, the reader is referred to References [6, 7], respectively.

Statistical Derivation of Johnson Noise

The derivation of Johnson noise leading to (10.5-6) leans heavily on thermodynamic and statistical mechanics considerations. It may be instructive to obtain this result using a physical model for a resistance and applying the mathematical tools developed in this chapter. The model used is shown in Figure 10-8.

The resistor consists of a medium of volume $V = Ad$, which contains N_e free electrons per unit volume. In addition, there are N_e positively charged ions, which preserve the (average) charge neutrality. The electrons move about randomly with an average kinetic energy per electron of

$$\overline{E} = \tfrac{3}{2}kT = \tfrac{1}{2}m(\overline{v_x^2} + \overline{v_y^2} + \overline{v_z^2}) \tag{10.5-10}$$

where $\overline{v_x^2} = \overline{v_y^2} = \overline{v_z^2}$ refer to thermal averages. A variety of scattering mechanisms including electron–electron, electron–ion, and electron–phonon collisions act to interrupt the electron motion at an average rate of τ_0^{-1} times per second. τ_0 is thus the mean scattering time. These scattering mechanisms are responsible for the electrical resistance and give rise to a dc conductivity[12]

$$\sigma = \frac{N_e e^2 \tau_0}{m} \tag{10.5-11}$$

where m is the mass of the electron.[13] The sample dc resistance is thus

$$R = \frac{d}{\sigma A} = \frac{md}{Ne^2\tau_0 A} \tag{10.5-12}$$

while its ac resistance $R(\omega)$ is $md(1 + \omega^2\tau_0^2)/Ne^2\tau_0 A$.

---

[12]The derivation of (10.5-11) can be found in any introductory book on solid-state physics.

[13]In a semiconductor we use the effective mass of the charge carrier.

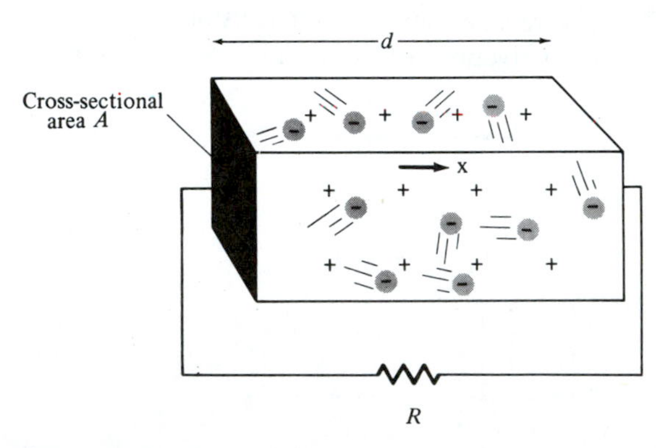

Figure 10-8 Model of a resistance used in deriving the Johnson-noise formula.

We apply next the results of Section 10.3 to the problem and choose as our basic single event the current pulse $i_e(t)$ in the external circuit due to the motion of *one* electron between two successive scattering events. Using (10.4-1), we write

$$i_e(t) = \begin{cases} \dfrac{ev_x}{d} & 0 \leq t \leq \tau \\ 0 & \text{otherwise} \end{cases} \tag{10.5-13}$$

where v_x is the x component of the velocity (assumed constant) and where τ is the scattering time of the electron under observation. Taking the Fourier transform of $i_e(t)$, we have

$$I_e(\omega, \tau, v_x) = \frac{1}{2\pi}\int_0^{\tau} i_e(t)e^{-i\omega t}\, dt = \frac{(1/2\pi)ev_x}{-i\omega d}\,[e^{-i\omega\tau} - 1] \tag{10.5-14}$$

from which

$$|I_e(\omega, \tau, v_x)|^2 = \frac{e^2 v_x^2}{4\pi^2\omega^2 d^2}\,[2 - e^{i\omega\tau} - e^{-i\omega\tau}] \tag{10.5-15}$$

According to (10.3-10) we need to average $|I_e(\omega, \tau, v_x)|^2$ over the parameters τ and v_x. We assume that τ and v_x are independent variables—that is, that the probability function

$$p(\alpha) = p(\tau, v_x) = g(\tau)f(v_x)$$

is the product of the individual probabilities [1]—and take $g(\tau)$ as[14]

$$g(\tau) = \frac{1}{\tau_0}\, e^{-\tau/\tau_0} \tag{10.5-16}$$

and, performing the averaging over τ, obtain

$$\overline{|I_e(\omega, v_x)|^2} = \int_0^{\infty} g(\tau)|I_e(\omega, v_x, \tau)|^2\, d\tau = \frac{2e^2 v_x^2 \tau_0^2}{4\pi^2 d^2(1 + \omega^2\tau_0^2)} \tag{10.5-17}$$

[14]If the collision probability per carrier per unit times is $1/\tau_0$ and $q(t)$ is the probability that an electron *has not* collided by time t, we have:

$$q'(t) = -q(t)\,\frac{1}{\tau_0} \Rightarrow q(t) = e^{-t/\tau_0}$$

Taking $g(t)\, dt$ as the probability that a collision will occur between τ and $t + dt$, it follows that

$$q(t) = 1 - \int_0^t g(t')\, dt'$$

and thus

$$g(t) = -\frac{dq}{dt} = \frac{1}{\tau_0}\, e^{-t/\tau_0}$$

as in (10.5-16).

The second averaging over v_x^2 is particularly simple, since it results in the replacement of v_x^2 in (10.5-17) by its average $\overline{v_x^2}$, which, for a sample at thermal equilibrium, is given according to (10.5-10) by $\overline{v_x^2} = kT/m$. The final result is then

$$\overline{|I_e(\omega)|^2} = \frac{2e^2\tau_0^2 kT}{4\pi^2 md^2(1 + \omega^2\tau_0^2)} \tag{10.5-18}$$

The average number of scattering events per second $\overline{N}$ is equal to the total number of electrons N_eV divided by the mean scattering time τ_0

$$\overline{N} = \frac{N_eV}{\tau_0} \tag{10.5-19}$$

thus, from (10.3-10), we obtain

$$S_i(\nu) = 8\pi^2\overline{N}\overline{|I_e(\omega)|^2} = \frac{4NVe^2\tau_0 kT}{md^2(1 + \omega^2\tau_0^2)}$$

and, after using (10.5-12) and limiting ourselves as in (10.4-6) to frequencies where $\omega\tau_0 \ll 1$, we get

$$\overline{i_N^2}(\nu) \equiv S_i(\nu)\Delta\nu = \frac{4kT\Delta\nu}{R(\nu)} \tag{10.5-20}$$

in agreement with (10.5-9).

10.6 SPONTANEOUS EMISSION NOISE IN LASER OSCILLATORS

Another type of noise that plays an important role in quantum electronics is that of spontaneous emission in laser oscillators and amplifiers. As shown in Chapter 5, a necessary condition for laser amplification is that the atomic population of a pair of levels 1 and 2 be inverted. If $E_2 > E_1$, gain occurs when $N_2 > N_1$. Assume that an optical wave with frequency $\nu \simeq (E_2 - E_1)/h$ is propagating through an inverted population medium. This wave will grow coherently due to the effect of stimulated emission. In addition, its radiation will be contaminated by noise radiation caused by spontaneous emission from level 2 to level 1. Some of the radiation emitted by the spontaneous emission will propagate very nearly along the same direction as that of the stimulated emission and cannot be separated from it. This has two main consequences. First, the laser output has a finite spectral width. This effect is described in this section. Second, the signal-to-noise ratio achievable at the output of laser amplifiers [7] is limited because of the intermingling of spontaneous emission noise power with that of the amplified signal. (See Figure 10-9 and Appendix C.)

Returning to the case of a laser oscillator, we represent it by an *RLC* circuit, as shown in Figure 10-10. The presence of the laser medium with negative loss (that is, gain) is accounted for by including a negative conductance $-G_m$ while the or-

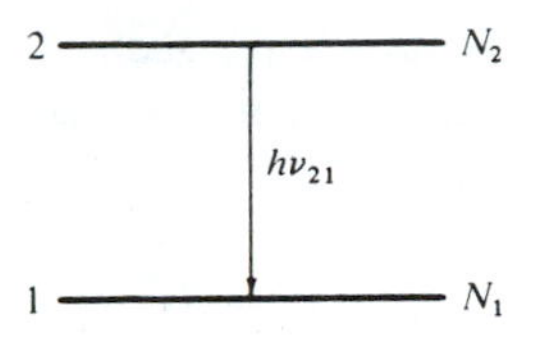

Figure 10-9 An atomic transition with $N_2 > N_1$ providing gain for laser oscillation.

dinary loss mechanisms described in Chapter 6 are represented by the positive conductance G_0. The noise generator associated with the losses G_0 is given according to (10.5-9) as

$$\overline{i_N^2} = \frac{4\hbar\omega G_0(\Delta\omega/2\pi)}{e^{\hbar\omega/kT} - 1}$$

where T is the actual temperature of the losses. Spontaneous emission is represented by a similar expression[15]

$$(\overline{i_N^2})_{\substack{\text{spont}\\\text{emission}}} = \frac{4\hbar\omega(-G_m)(\Delta\omega/2\pi)}{e^{\hbar\omega/kT_m} - 1} \tag{10.6-1}$$

where the term $(-G_m)$ represents negative losses and T_m is a temperature determined by the population ratio according to

$$\frac{N_2}{N_1} = e^{-\hbar\omega/kT_m} \tag{10.6-2}$$

Since $N_2 > N_1$, then $T_m < 0$, $(\overline{i_N^2})$ in (10.6-1) is positive definite.

[15]The 2π factor appearing in the denominators of $\overline{i_N^2}$ is due to the fact that here we use $\overline{i_N^2}(\omega)$ instead of $\overline{i_N^2}(\nu)$ with

$$\overline{i_N^2}(\omega)\Delta\omega = \overline{i_N^2}(\nu)\,\Delta\nu \qquad \Delta\omega = 2\pi\Delta\nu$$

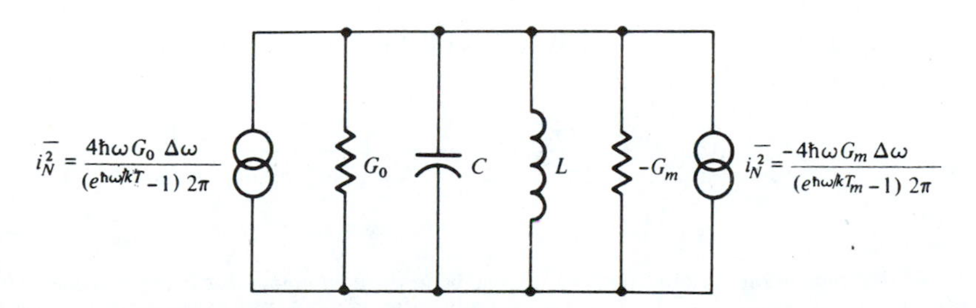

Figure 10-10 Equivalent circuit of a laser oscillator.

Although a detailed justification of (10.6-1) is outside the scope of the present treatment, a strong case for its plausibility can be made by noting that since $G_m \propto N_2 - N_1$, $\overline{(i_N^2)}$ in (10.6-1) can be written, using (10.6-2), as[16]

$$\overline{(i_N^2)}_{\substack{\text{spont}\\\text{emission}}} \propto \frac{-4\hbar\omega\Delta\omega(N_2 - N_1)}{(N_1/N_2) - 1} = 4\hbar\omega\Delta\omega N_2 \tag{10.6-3}$$

and is thus proportional to N_2. This makes sense, since spontaneous emission power is due to $2 \rightarrow 1$ transitions and should consequently be proportional to N_2.

Returning to the equivalent circuit, its quality factor Q is given by

$$Q^{-1} = \frac{G_0 - G_m}{\omega_0 C} = \frac{1}{Q_0} - \frac{1}{Q_m} \tag{10.6-4}$$

where $\omega_0^2 = (LC)^{-1}$. The circuit impedance is

$$\begin{aligned} Z(\omega) &= \frac{1}{(G_0 - G_m) + (1/i\omega L) + i\omega C} \\ &= \frac{i\omega}{C}\frac{1}{(i\omega\omega_0/Q) + (\omega_0^2 - \omega^2)} \end{aligned} \tag{10.6-5}$$

so the voltage across this impedance due to a current source with a complex amplitude $I(\omega)$ is

$$V(\omega) = \frac{i}{C}\frac{I(\omega)}{[(\omega_0^2 - \omega^2)/\omega] + (i\omega_0/Q)} \tag{10.6-6}$$

which, near $\omega = \omega_0$, becomes

$$\overline{|V(\omega)|^2} = \frac{1}{4C^2}\frac{\overline{|I(\omega)|^2}}{(\omega_0 - \omega)^2 + (\omega_0^2/4Q^2)} \tag{10.6-7}$$

The current sources driving the resonant circuit are those shown in Figure 10-10; since they are not correlated, we may take $|I(\omega)|^2$ as the sum of their mean-square values

$$\overline{|I(\omega)|^2} = 4\hbar\omega\left[\frac{G_m N_2}{N_2 - N_1} + \frac{G_0}{e^{\hbar\omega/kT} - 1}\right]\frac{d\omega}{2\pi} \tag{10.6-8}$$

where in the first term inside the square brackets we used (10.6-2). In the optical region, $\lambda = 1\ \mu$m say, and for $T = 300°$K we have $\hbar\omega/kT \simeq 50$; thus, since near oscillation $G_m \simeq G_0$, we may neglect the thermal (Johnson) noise term in (10.6-8), thereby obtaining

$$\overline{|V(\omega)|^2}_{\omega\simeq\omega_0} = \frac{\hbar G_m}{2\pi C^2}\left(\frac{N_2}{N_2 - N_1}\right)\frac{\omega\, d\omega}{(\omega_0 - \omega)^2 + (\omega_0^2/4Q^2)} \tag{10.6-9}$$

---

[16]The proportionality of G_m to $N_2 - N_1$ can be justified by noting that in the equivalent circuit (Figure 10-10) the stimulated emission power is given by $v^2 G_m$ where v is the voltage. Using the field approach, this power is proportional to $E^2(N_2 - N_1)$ where E is the field amplitude. Since v is proportional to E, G_m is proportional to $N_2 - N_1$.

Equation (10.6-9) represents the spectral distribution of the laser output. If we subject the output to high-resolution spectral analysis, we should, according to (10.6-9), measure a linewidth

$$\Delta\omega = \frac{\omega_0}{Q} \tag{10.6-10}$$

between the half-intensity points. The trouble is that, though correct, (10.6-10) is not of much use in practice. The reason is that according to (10.6-4), Q^{-1} is equal to the difference of two nearly equal quantities neither of which is known with high enough accuracy. We can avoid this difficulty by showing that Q is related to the laser power output, and thus $\Delta\omega$ may be expressed in terms of the power.

The total optical oscillation power extracted from the atoms comprising the laser is

$$\begin{aligned} P &= G_0 \int_0^\infty \frac{|V(\omega)|^2}{d\omega}\, d\omega \\ &= \frac{\hbar G_m G_0}{2\pi C^2}\left(\frac{N_2}{N_2 - N_1}\right)\int_0^\infty \frac{\omega\, d\omega}{(\omega_0 - \omega)^2 + (\omega_0/2Q)^2} \end{aligned} \tag{10.6-11}$$

Since the integrand peaks sharply near $\omega \simeq \omega_0$, we may replace ω in the numerator of (10.6-11) by ω_0 and after integration obtain

$$P = \frac{\hbar G_m G_0 Q}{C^2}\left(\frac{N_2}{N_2 - N_1}\right) \tag{10.6-12}$$

which is the desired result linking P to Q. In a laser oscillator the gain very nearly equals the loss, or in our notation, $G_m \simeq G_0$. Using this result in (10.6-12), we obtain

$$Q = \frac{C^2}{\hbar G_0^2}\left(\frac{N_2 - N_1}{N_2}\right) P$$

which, when substituted in (10.6-10), yields

$$\Delta\nu = \frac{2\pi h\nu_0(\Delta\nu_{1/2})^2}{P}\left(\frac{N_2}{N_2 - N_1}\right) \tag{10.6-13}$$

where $\Delta\nu_{1/2}$ is the full width of the passive cavity resonance given in (4.7-6) as $\Delta\nu_{1/2} = \nu_0/Q_0 = (1/2\pi)(G_0/C)$. It is worthwhile to recall here that $\Delta\nu$ represents, in the quantum limit, the laser field spectral width. The expression (10.6-13) is known as the Schawlow–Townes linewidth after the two American co-inventors of the laser [18] who first derived it.

Equation (10.6-13) does not predict an inverse dependence of $\Delta\nu$ on P, as may be deduced at a first glance, because of the dependence of N_2 on P. For very large powers, $P \to \infty$, N_2 is proportional to P, while $N_2 - N_1$ remains clamped at its threshold value. This leads to a residual power independent value of $\Delta\nu$. To appreciate this argument qualitatively, we note that unless the lifetime t_1 of the lower laser level is zero, as P increases, N_1 must increase since the increased (net)-induced transition rate into level 1 must equal in steady state N_1/t_1, the rate of emptying of level 1. This causes the population N_2 to increase in order to keep $N_2 - N_1$ and thus the gain, a constant. At sufficiently high values of P, N_2 becomes and stays proportional to P and the ratio N_2/P in (10.6-13) approaches a constant value, thus leading to a residual power independent linewidth.

To obtain the power dependence of the factor

$$\mu \equiv \frac{N_2}{(N_2 - N_1)_{\text{th}}}$$

we solve the rate equations for the atomic populations plus the equation for the photon number p(p = number of photons in the optical resonator)

$$\frac{dN_2}{dt} = R - \frac{N_2}{t_2} - (N_2 - N_1)W_i$$

$$\frac{dN_1}{dt} = -\frac{N_1}{t_1} + (N_2 - N_1)W_i + \frac{N_2}{t_2}$$

$$\frac{dp}{dt} = (N_2 - N_1)W_i - \frac{p}{t_c} \tag{10.6-14}$$

The first two equations are similar to (5.6-3) and (5.6-4) with $R_1 = 0$, $t_2 \to t_{\text{spont}}$, $R_2 \to R$, W_i is the induced transition rate and N_2, N_1, representing the total atomic populations of the laser transition levels 2 and 1, respectively. The third equation is a conservation equation for the total number of photons. W_i is the induced transition rate. The photon lifetime t_c is related to the cavity linewidth $\Delta\nu_{1/2}$ by $\Delta\nu_{1/2} = (2\pi t_c)^{-1}$. At equilibrium, $d/dt = 0$, we can solve (10.6-14) to obtain

$$N_2 - N_1 = \frac{R(t_2 - t_1)}{1 + W_i t_2}$$

$$N_2 = \frac{Rt_2(1 + W_i t_1)}{1 + W_i t_2} \tag{10.6-15}$$

so that

$$\frac{N_2}{(N_2 - N_1)_{\text{th}}} = \frac{t_2}{t_2 - t_1}(1 + W_i t_1) \tag{10.6-16}$$

where the subscript "th" indicates the value at threshold. The power output, including "wall losses" of the laser, is

$$P = (N_2 - N_1)_{\text{th}} W_i h\nu_0 \tag{10.6-17}$$

which, when used together with (10.6-16) in (10.6-13) gives

$$\Delta\nu_{\text{laser}} = \frac{2\pi h\nu_0(\Delta\nu_{1/2})^2}{P}\frac{t_2}{t_2 - t_1} + \frac{c\Delta\nu_{1/2}\lambda_0^2}{8\pi n^3 \Delta\nu_{\text{gain}} V}\frac{t_1}{t_2 - t_1} \tag{10.6-18}$$

where $\Delta\nu_{\text{gain}}$ is the linewidth of atomic transition responsible for the laser gain. V is the mode volume. In obtaining (10.6-18), we use

$$(N_2 - N_1)_{\text{th}} = \frac{8\pi\nu_0^2 n^3 \Delta\nu_{\text{gain}} V t_2}{c^3 t_c} \quad (t_2 = t_{\text{spont}}) \tag{10.6-19}$$

which is obtained from (6.1-11) if we put $\Delta\nu_{\text{gain}} = 1/g(\nu)$. The first term on the right-hand side of (10.6-18) is the conventional Schawlow–Townes expression containing the inverse P dependence. The second term is power independent and corresponds to a residual linewidth as $P \to \infty$.

To get an idea of the magnitudes involved, we consider the case of a 0.6328 μm He–Ne laser with mirror reflectivities of $R = 0.99$, a resonator length of $1 =$ 30 cm, and take $t_1/t_2 = 0.1$. We obtain

$$\Delta\nu_{1/2}(\text{Hz}) = \frac{(1 - R)c}{2\pi n l} = 1.6 \times 10^6$$

and

$$\Delta\nu_{\text{laser}}(\text{Hz}) \simeq \frac{10^{-3}}{P(\text{mW})} + 3.8 \times 10^{-4}$$

The residual linewidth thus dominates at power levels exceeding a few milliwatts.

10.7 PHASOR DERIVATION OF THE LASER LINEWIDTH

The derivation of the laser linewidth in Section 10.6 takes advantage of the highly sophisticated and efficient concepts and phenomena represented by the seemingly simple circuit model of a laser oscillator. The price we pay when taking this approach is a certain loss of physical insight into the mechanisms whereby spontaneous emission affects the laser linewidth.

In this section we will derive the expression (10.6-13) for the laser linewidth using a different approach. This is done not only for pedagogic purposes, but because some of the interim results involving phase fluctuations are useful in their own right.

The Phase Noise

An ideal monochromatic radiation field can be written as

$$\mathscr{E}(t) = \text{Re}[E_0 e^{i(\omega_0 t + \theta)}] \tag{10.7-1}$$

where ω_0 the radian frequency, E_0 the field amplitude, and θ are constants. A real field including that of lasers undergoes random phase and amplitude fluctuations that can be represented by writing

$$\mathscr{E}(t) = \text{Re}[E(t) e^{i[\omega_0 t + \theta(t)]}] \tag{10.7-2}$$

where $E(t)$ and $\theta(t)$ vary only "slightly" during one optical period.

There are many reasons in a practical laser for the random fluctuation in amplitude and phase. Most of these can be reduced, in theory, to inconsequence by various improvements such as ultrastabilization of the laser cavity length and the near elimination of microphonic and temperature variations. There remains, however, a basic source of noise that is quantum mechanical in origin. This is due to spontaneous emission that continually causes new power to be added to the laser oscillation field. The electromagnetic field represented by this new power, not being coherent with the old field, causes phase, as well as amplitude, fluctuations. These are responsible ultimately for the deviation of the evolution of the laser field from that of an ideal monochromatic field, i.e., for the quantum mechanical noise.

Let us consider the effect of one spontaneous emission event on the electromagnetic field of a single oscillating laser mode. A field such as (10.7-1) can be represented by a phasor of length E_0 rotating with an angular (radian) rate ω_0. In a frame rotating at ω_0 we would see a constant vector E_0. Since $E_0^2 \propto \bar{n}$, the average number of quanta in the mode, we shall represent the laser field phasor before a spontaneous emission event by a phasor of length $\sqrt{\bar{n}}$ as in Figure 10-11. The spontaneous emission adds *one* photon to the field, and this is represented, according to our conversion, by an incremental vector of unity length. Since this field increment is not correlated in phase with the original field, the angle ϕ is a random variable (i.e., it is distributed uniformly between zero and 2π). The resulting change $\Delta\theta$ of the field phase can be approximated for $\bar{n} \gg 1$ by

$$\Delta\theta_{\text{one emission}} = \frac{1}{\sqrt{\bar{n}}} \cos\phi \tag{10.7-3}$$

Next consider the effect of N spontaneous emissions on the phase of the laser field. The problem is one of random walk, since ϕ may assume with equal probability any value between 0 and 2π. We can then write

$$\langle[\Delta\theta(N)]^2\rangle = \langle(\Delta\theta_{\text{one emission}})^2\rangle\, N \tag{10.7-4}$$

and from (10.7-3)

$$\langle[\Delta\theta(N)]^2\rangle = \frac{1}{\bar{n}} \langle\cos^2\phi\rangle\, N$$

where $\langle\ \rangle$ denotes an ensemble average taken over a very large number of individual emission events.

Equation (10.7-4) is a statement of the fact that in a random walk problem the mean squared distance traversed after N steps is the square of the size of one step times N. The mean deviation $\langle\Delta\theta(N)\rangle$ after N spontaneous emissions is, of course, zero. Any one experiment, however, will yield a nonzero result. The mean squared deviation is thus nonzero and is a measure of the phase fluctuation. To obtain the root-mean-square (rms) phase deviation in a time t, we need to calculate the average number of spontaneous emission events $N(t)$ into a single laser mode in a time t.

The total number of spontaneous transitions per second into all modes is N_2/t_{spont}, where N_2 is the total number of atoms in the upper laser level 2 and t_{spont}

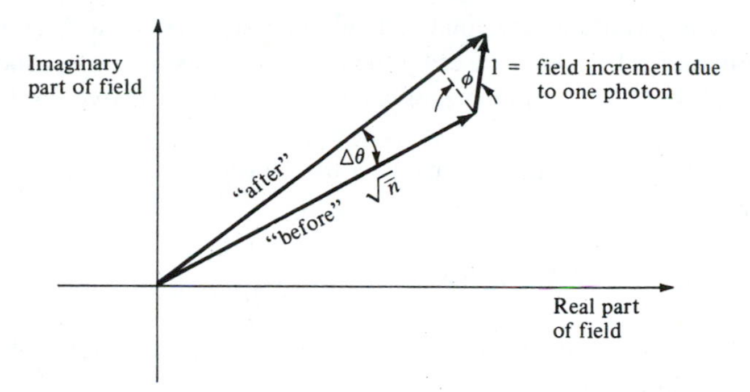

Figure 10-11 The phasor model for the effect of a single spontaneous emission event on the laser field phase.

is the spontaneous lifetime of an atom in 2. The total number of transitions per second into one mode is thus

$$\frac{N_{\text{spont}}}{\text{second-mode}} = \frac{N_2}{t_{\text{spont}} p} \tag{10.7-5}$$

where

$$p = \frac{8\pi\nu_0^2 \,\Delta\nu V n^3}{c^3} \tag{10.7-6}$$

is the number of modes interacting with the laser transition, i.e., partaking in the spontaneous emission. V is the mode volume, and $\Delta\nu$ is the linewidth of the atomic transition responsible for the laser gain. We can rewrite (10.7-5) as

$$\frac{N_{\text{spont}}}{\text{second-mode}} = \left(\frac{N_2}{\Delta N_t}\right) \frac{(\Delta N_t)}{t_{\text{spont}} p} \tag{10.7-7}$$

where ΔN_t is the population inversion $(N_2 - N_1)$ at threshold. Next we use the result (6.1-11)

$$\Delta N_t = \frac{p t_{\text{spont}}}{t_c}$$

where t_c is the photon lifetime in the resonator, and obtain

$$\frac{N_{\text{spont}}}{\text{second-mode}} = \frac{\mu}{t_c} \qquad \mu \equiv \frac{(N_2)_t}{\Delta N_t} = \frac{(N_2)_t}{(N_2 - N_1)_t} \tag{10.7-8}$$

The number of spontaneous transitions into a single mode in a time τ is thus

$$N(\tau) = \frac{\mu\tau}{t_c} \tag{10.7-9}$$

We recall here that in an ideal four-level laser $N_1 = 0$ and $\Delta N_t = N_2$, i.e., $\mu = 1$. In a three-level laser, on the other hand, μ can be appreciably larger than unity. In a ruby laser at room temperature, for example (see Section 7.2), $\mu \simeq 50$. This reflects the fact that for a given gain the total excited population N_2 of a three-level laser must exceed that of a four-level laser by the factor μ, since gain is proportional to $N_2 - N_1$. Equation (10.7-8) is also equivalent to stating that above threshold there are μ spontaneously emitted photons present in a laser mode.

Using (10.7-9) in (10.7-4), we obtain for the root-mean-square phase deviation after τ seconds

$$\Delta\theta(t) \equiv \langle[\Delta\theta(t)]^2\rangle^{1/2} = \sqrt{\frac{1}{2\bar{n}}\frac{\mu t}{t_c}}$$

The maximum time t available for such an experiment is the integration time T of the measuring apparatus so that

$$\Delta\theta(T) = \sqrt{\frac{1}{2\bar{n}}\frac{\mu T}{t_c}} \tag{10.7-10}$$

The rms frequency excursion caused by $\Delta\theta$ is

$$(\Delta\omega)_{\text{RMS}} = \frac{\Delta\theta(T)}{T} = \sqrt{\frac{\mu}{2\bar{n}t_c T}} \tag{10.7-11}$$

We can cast the last result in a more familiar form by using the relations

$$P_e = \frac{\bar{n}\hbar\omega_0}{t_c} \qquad B = \frac{1}{2T} \tag{10.7-11a}$$

Here P_e is the power emitted by the atoms (i.e., the sum of the useful power output plus any power lost by scattering and absorption), and B is the bandwidth in hertz of the phase-measuring apparatus. The result is

$$(\Delta\omega)_{RMS} = \sqrt{\frac{\mu\hbar\omega_0}{P_e t_c^2}B} \tag{10.7-12}$$

From the experimental point of view $(\Delta\omega)_{\text{RMS}}$ is the root-mean-square deviation of the reading of an instrument whose output is the frequency $\omega(t) \equiv d\theta/dt$. We will leave it as an exercise (Problem 10.11) for the student to design an experiment that measures $(\Delta\omega)_{\text{RMS}}$.

Ring laser gyroscopes sense rotation by comparing the oscillation frequencies of two counter-propagating modes in a rotating ring resonator. Their sensitivity, i.e., the smallest rotation rate that they can sense, is thus limited by any uncertainty $\Delta\omega$ in the laser frequency. Experiments have indeed demonstrated a rotation measuring sensitivity approaching the quantum limit as given by (10.7-12).

The Laser Field Spectrum

Next we address the case where one measures directly the spectrum of the optical field

$$\mathscr{E}(t) = \text{Re}[E(t)e^{i[\omega_0 t + \theta(t)]}] \tag{10.7-13}$$

using, say, a scanning Fabry–Perot etalon. If the etalon has a sufficiently high spectral resolution, the measurement should yield the spectral density function $S_{\mathscr{E}}(\omega)$ of the laser field. We will, consequently, proceed to obtain an expression for this quantity. We make use of the Wiener–Khintchine theorem (10.2-14) according to which $S_{\mathscr{E}}(\omega)$ is the Fourier integral transform of the field autocorrelation function $C_{\mathscr{E}}(\tau)$

$$S_{\mathscr{E}}(\omega) = \frac{1}{\pi}\int_{-\infty}^{\infty} C_{\mathscr{E}}(\tau)e^{-i\omega\tau}\,d\tau \tag{10.7-14}$$

$$S_{\mathscr{E}}(\omega) = \frac{4\pi}{T}|\mathscr{E}_T(\omega)|^2 \qquad \mathscr{E}_T(\omega) = \frac{1}{2\pi}\int_{-T/2}^{T/2} \mathscr{E}(t)e^{-i\omega t}\,dt$$

$$C_{\mathscr{E}}(\tau) \equiv \langle \mathscr{E}(t)\mathscr{E}(t+\tau)\rangle \tag{10.7-15}$$

where the symbol $\langle\ \rangle$ represents an ensemble, or time, average.

Using (10.7-13) we obtain

$$\begin{aligned} C_{\mathscr{E}}(\tau) = \frac{1}{4}\langle&[E(t)e^{i[\omega_0 t+\theta(t)]} + E^*(t)e^{-i[\omega_0 t+\theta(t)]}] \\ &\times [E(t+\tau)e^{i[\omega_0(t+\tau)+\theta(t+\tau)]} + E^*(t+\tau)e^{-i[\omega_0(t+\tau)+\theta(t+\tau)]}]\rangle \end{aligned} \tag{10.7-16}$$

Now, for example,

$$\langle E(t)E(t+\tau)e^{i[2\omega_0 t+\theta(t)+\theta(t+\tau)]}\rangle = 0$$

since it corresponds to averaging a signal oscillating at twice the optical frequency over many periods. So if we keep only the slowly varying terms in $C_{\mathscr{E}}(\tau)$, we obtain

$$\begin{aligned} C_{\mathscr{E}}(\tau) &= \frac{1}{4}\langle E(t)E^*(t+\tau)e^{i[-\omega_0\tau+\theta(t)-\theta(t+\tau)]} + E^*(t)E(t+\tau)e^{i[\omega_0\tau-\theta(t)+\theta(t+\tau)]}\rangle \\ &= \frac{1}{4}[I(\tau) + I^*(\tau)] \end{aligned} \tag{10.7-17}$$

$$I(\tau) = \langle E^*(t)E(t+\tau)e^{i[\Delta\theta(t,\tau)+\omega_0\tau]}\rangle \tag{10.7-18}$$

$$\Delta\theta(t,\tau) \equiv \theta(t+\tau) - \theta(t) \tag{10.7-19}$$

The main contributions to the laser noise are due to fluctuations of the phase $\theta(t)$ and not the amplitude $E(t)$, since the amplitude fluctuations are kept negligibly small by gain saturation. Taking advantage of this fact, we write $\langle E^*(t)E(t+\tau)\rangle = \langle E^2\rangle \approx$ constant so that

$$I(\tau) = \langle E^2\rangle e^{i\omega_0\tau}\langle e^{i\Delta\theta(t,\tau)}\rangle \tag{10.7-20}$$

Given a (normalized) probability distribution function for $\Delta\theta$, $g(\Delta\theta)$, the expectation value of exp $\{i\Delta\theta(t,\tau)\}$ is obtained from

$$\langle e^{i\Delta\theta(t,\tau)}\rangle = \int_{-\infty}^{\infty} e^{i\Delta\theta(t,\tau)}g(\Delta\theta)\,d(\Delta\theta) \tag{10.7-21}$$

Since the total phase excursion $\Delta\theta$ is the net result of many small and statistically independent (spontaneous transitions) excursions, the central limit theorem of statistics applies, and $g(\Delta\theta)$ is a Gaussian, which we write as

$$g(\Delta\theta) = \frac{1}{\sqrt{2\pi\langle(\Delta\theta)^2\rangle}}\,e^{-(\Delta\theta)^2/2\langle(\Delta\theta)^2\rangle} \tag{10.7-22}$$

where

$$\langle(\Delta\theta)^2\rangle = \int_{-\infty}^{\infty} (\Delta\theta)^2 g(\Delta\theta) d(\Delta\theta) \tag{10.7-23}$$

Using (10.7-22) in (10.7-21), we obtain

$$\langle e^{i\Delta\theta(t,\tau)}\rangle = e^{-\langle(\Delta\theta)^2\rangle/2} = e^{-\mu|\tau|/(4\bar{n}t_c)} \tag{10.7-24}$$

where in order to obtain the last result, we used (10.7-10) with $T = |\tau|$. Using (10.7-24) in (10.7-20),

$$C_{\mathscr{E}}(\tau) = \frac{1}{4}\langle E^2\rangle e^{-\mu|\tau|/4\bar{n}t_c}(e^{i\omega_0\tau} + e^{-i\omega_0\tau}) \tag{10.7-25}$$

The spectral density function of the laser field $S_{\mathscr{E}}(\omega)$, the quantity observed by a spectral analysis of the field, is given according to (10.7-14) and (10.7-25) by

$$S_{\mathscr{E}}(\omega) = \frac{\langle E^2\rangle}{4\pi}\int_{-\infty}^{\infty} e^{(-\mu|\tau|/4\bar{n}t_c)-i\omega\tau}\,(e^{i\omega_0\tau} + e^{-i\omega_0\tau}\, dt)\, d\tau \tag{10.7-26}$$

$$= \frac{\langle E^2\rangle}{2}\left(\frac{\mu/4\bar{n}t_c}{(\mu/4\bar{n}t_c)^2 + (\omega - \omega_0)^2} + \frac{\mu/4\bar{n}t_c}{(\mu/4\bar{n}t_c)^2 + (\omega + \omega_0)^2}\right) \tag{10.7-27}$$

We have defined in (10.2-7) the spectral density function in such a way that only positive frequencies need to be considered. For $\omega > 0$ the second term on the right side of (10.7-27) contributes negligibly so that

$$S_{\mathscr{E}}(\omega) = \frac{\langle E^2\rangle}{2\pi}\frac{\mu/4\bar{n}t_c}{(\mu/4\bar{n}t_c)^2 + (\omega - \omega_0)^2} \tag{10.7-28}$$

which corresponds to a Lorentzian-shaped function centered on the nominal laser frequency ω_0 with a full width at half-maximum of

$$(\Delta\omega)_{\text{laser}} = \frac{\mu}{2\bar{n}t_c} \tag{10.7-29}$$

Recalling that the total power emitted by the electrons is $P = \bar{n}\hbar\omega_0/t_c$ and defining the passive resonator linewidth $\Delta\nu_{1/2} = (2\pi t_c)^{-1}$, we can rewrite (10.7-29) using (10.7-11a) as

$$(\Delta\nu)_{\text{laser}} = \frac{(\Delta\omega)_{\text{laser}}}{2\pi} = \frac{2\pi h\nu_0(\Delta\nu_{1/2})^2\mu}{P} \tag{10.7-30}$$

which, recalling the definition (10.7-8) of μ, is half the result of the circuit model (10.6-13).[17]

Numerical Example: Linewidth of a He–Ne Laser and a Semiconductor Diode Laser

To obtain an order of magnitude estimate of the linewidth $(\Delta\nu)_{\text{laser}}$ predicted by (10.7-30), we will calculate it in the case of two largely different types of CW lasers: (1) a He–Ne laser and (2) a semiconductor GaInAsP laser.

[17]The discrepancy by a factor of 2 should not be taken too seriously considering the very different mathematical approaches employed by the two derivations.

1. *He–Ne laser.*

$$\nu = 4.741 \times 10^{14} \text{ Hz } (\lambda = 6328 \text{ Å})$$
$$l \text{ (distance between reflectors)} = 100 \text{ cm}$$
$$\text{Loss} = (1 - R) = 1\% \text{ per pass}$$

From these numbers we get

$$(\Delta\nu_{1/2}) = \frac{1}{2\pi t_c} \approx \frac{(1 - R)c}{2\pi nl} \approx 5 \times 10^5$$

(i.e., $t_c = 3.2 \times 10^{-7}$ s) and from (10.7-30), assuming $\mu = 1$ (i.e., $N_1 \ll N_2$),

$$(\Delta\nu)_{\text{laser}} \cong 2 \times 10^{-3} \text{ Hz}$$

at a power level $P = 1$ mW.

The predicted linewidth is thus so small as to be completely masked in almost all experimental situations by contributions due to extraneous causes, such as vibrations and temperature fluctuations.

2. *Semiconductor laser.* We use as a typical example the case of a GaInAsP ($\lambda = 1.55$ μm) laser with the following pertinent characteristics:

$$P = 3 \text{ mW}$$
$$\nu = 1.935 \times 10^{14} \ (\lambda_0 = 1.55 \ \mu\text{m})$$
$$\Delta\nu_{1/2} = \frac{(1 - R)c}{2\pi nl}$$

R (reflectivity) = 30%

$$l = 300 \ \mu\text{m}$$
$$n = 3.5$$
$$\mu = 3 \text{ (at } T = 300 \text{ K)}$$

This results in $\Delta\nu_{1/2} \sim 3 \times 10^{10}$ (i.e., $t_c = 1/(2\pi\Delta\nu_{1/2}) = 5 \times 10^{-12}$ s) and

$$(\Delta\nu)_{\text{laser}} = 0.817 \times 10^6 \text{ Hz}$$

The experimental curve of Figure 10-12 shows the predicted [Equation (10.7-30)] P^{-1} dependence of $(\Delta\nu)_{\text{laser}}$, but the measured values of the linewidth are larger by a factor of ~70 than those predicted by the analysis. This discrepancy has been studied by a number of investigators [20–22], who have shown that the analysis leading to (10.7-30) ignores the modulation of the index of refraction of the laser medium, which is due to fluctuations of the electron density caused by spontaneous emission. When this effect is included, the result is to multiply Equation (10.7-30) by the factor

$$1 + \left(\frac{\Delta n'}{\Delta n''}\right) \tag{10.7-31}$$

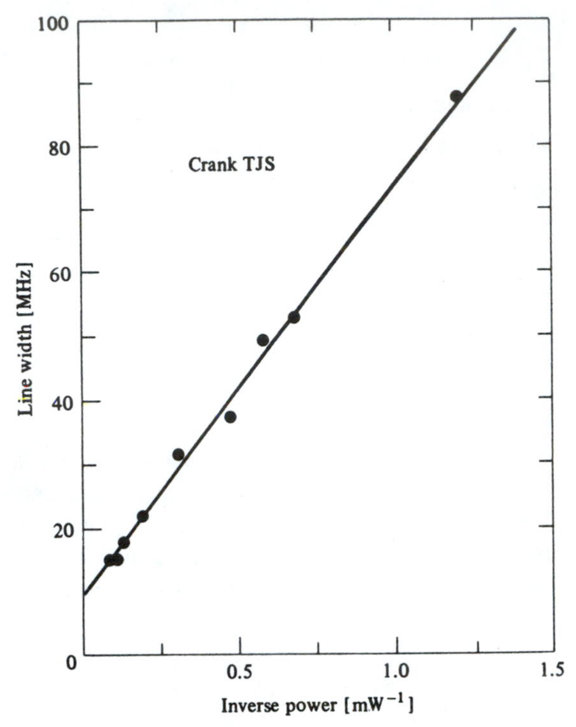

Figure 10-12 The measured dependence of the spectral linewidth of a semiconductor laser on the power output. (After Reference [19].)

where $\Delta n'$ and $\Delta n''$ are, respectively, the changes in the real and imaginary parts of the index of refraction "seen" by the laser field due to some change in the electron density. The factor $1 + (\Delta n'/\Delta n'')^2$ can be calculated from measured parameters of the laser or measured directly [6]. Its value is ~30 in typical cases, enough to reconcile the observed data of Figure 10-12 and the prediction of Equation (10.7-30).

The big difference, over nine orders of magnitude, between the limiting linewidth of conventional lasers, say gas lasers and semiconductor lasers, is due mostly to the very short photon lifetime t_c in semiconductor laser resonators. At a given power output we have from (10.7-30) $(\Delta\nu)_{\text{laser}} \propto (\Delta\nu_{1/2})^2 \propto t_c^{-2}$. In the above examples we obtained $t_c \simeq 3 \times 10^{-8}$ s in the case of the He–Ne laser, and $t_c \simeq 5 \times 10^{-12}$ s in the semiconductor laser. Since $t_c \sim ln/c(1 - R)$, the main hope for increasing t_c in a semiconductor laser, thus decreasing the linewidth $(\Delta\nu)_{\text{laser}}$, is to increase l by placing the laser in an external resonator and by using high reflectance mirrors $R \sim 1$. Semiconductor laser linewidths in the kilohertz regime are obtainable.

An actual (measured) GaAs/GaAlAs semiconductor laser, Lorentzian field spectrum is shown in Figure 10-13.

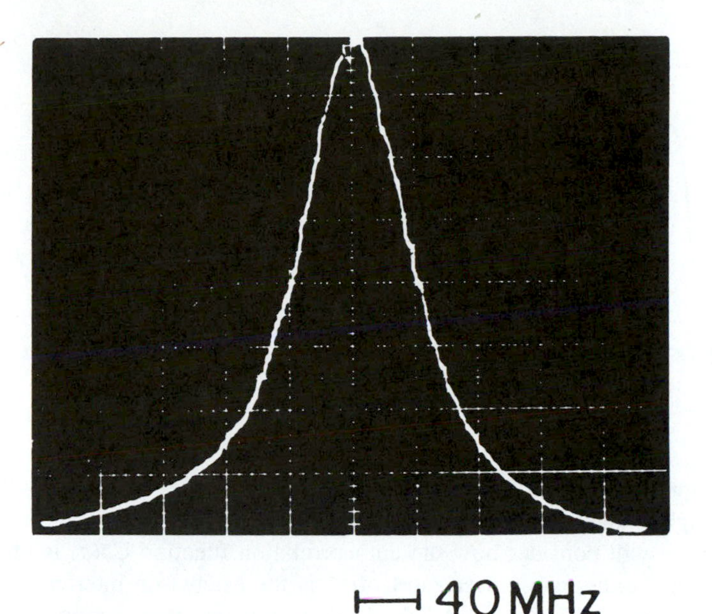

Figure 10-13 The measured Lorentzian field spectrum $S_{\mathscr{E}}(\omega)$ of a semiconductor laser. (After Reference [19].)

10.8 COHERENCE AND INTERFERENCE

In Section 10.7 [Equation (10.7-25)] we have derived the following expression for the autocorrelation function of the single-mode laser field

$$C_{\mathscr{E}}(\tau) \equiv \langle \mathscr{E}(t)\mathscr{E}(t+\tau)\rangle \propto \cos \omega_0 \tau e^{-\mu|\tau|/4\bar{n}_c}$$
$$= \cos \omega_0 \tau e^{-|\tau|/\tau_c}$$
$$\tau_c = \frac{4\bar{n}t_c}{\mu} \tag{10.8-1}$$

where $\bar{n}$ is the number of photons inside the resonator, $\mu = N_2/(N_2 - N_1)$ and t_c is the photon lifetime (the decay time constant for the mode optical energy if the gain mechanism were turned off).

The parameter τ_c is called the coherence time of the laser field. According to (10.7-29) it is equal to $2/(\Delta\omega)_{\text{laser}}$ where $(\Delta\omega)_{\text{laser}}$ is the laser output field linewidth. In practical terms it is the time duration during which we can count on the laser to act as a well-behaved sinusoidal oscillator with a well-defined phase. If we try and correlate (by means to be discussed below) the laser field with itself using a time delay exceeding τ_c, the result approaches zero. One form of a field $\mathscr{E}(t)$ that will display this behavior is shown in Figure 10-14. The field undergoes a phase memory loss on the average every τ_c seconds. It is intuitively clear that performing the au-

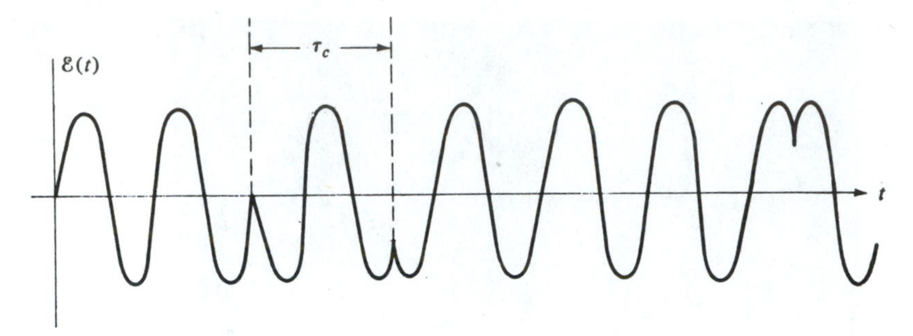

Figure 10-14 A sinusoidal field whose phase coherence is interrupted on the average every τ_c seconds.

tocorrelation operation as defined by the first equality of (10.8-1) will yield a result whose rough features agree with the form $(\cos \omega_0 \tau)e^{-|\tau|/\tau_c}$.

Next we will consider how the autocorrelation function $C_{\mathscr{E}}(\tau)$ is obtained in practice. The configuration used most often is the Michelson interferometer illustrated in Figure 10-15. An input field $\mathscr{E}_i(t)$ is split into two components. One of these fields is delayed relative to the second by a time delay

$$\tau = \frac{2(L_1 - L_2)}{c} \tag{10.8-2}$$

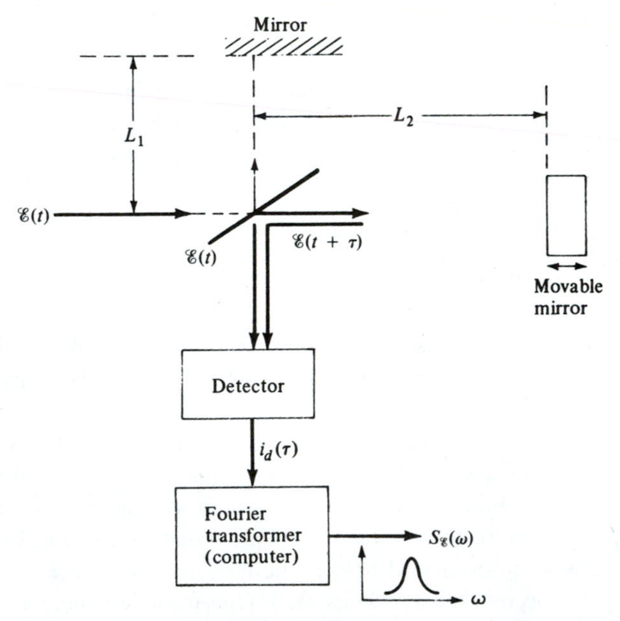

Figure 10-15 A Michelson interferometer "splits" an input beam into a two-component beam and then recombines them with a controlled time delay $\tau = 2(L_1 - L_2)/c$.

The two fields are then incident on a square-law detector whose current constitutes the useful output of the experiment.

Assuming equal division of power, the total optical field at the detector plane is

$$\mathscr{E}_d(t) = \mathscr{E}(t) + \mathscr{E}(t + \tau) \tag{10.8-3}$$

According to the discussion of Section 11.1, which the student is advised to preview at this point, the output current of the detector is

$$i_d = a\overline{\mathscr{E}_d^2(t)} \tag{10.8-4}$$

a is some constant that is irrelevant in the present discussion, and the bar indicates, as it does throughout this book, time-averaging. The duration of this averaging depends on the detector and its associated electrical circuitry and in the very fastest detectors may be as short as 10^{-11} s. It is thus *always* very long compared to the optical field period which is $\sim 10^{-15}$ s.

The detector output is then

$$\begin{aligned} i_d &= a[\overline{\mathscr{E}^2(t)} + \overline{\mathscr{E}^2(t + \tau)} + \overline{2\mathscr{E}(t)\mathscr{E}(t + \tau)}] \\ &= 2a[\overline{\mathscr{E}^2} + \overline{\mathscr{E}(t)\mathscr{E}(t + \tau)}] \end{aligned} \tag{10.8-5}$$

since $\overline{\mathscr{E}^2(t)} = \overline{\mathscr{E}^2(t + \tau)} \equiv \overline{\mathscr{E}^2}$. The output current from the detector is thus made up of a dc component $2a\overline{\mathscr{E}^2}$ and a component $2a\overline{\mathscr{E}(t)\mathscr{E}(t + \tau)}$. The ratio of these two current components is, according to (10.2-9), the (normalized) autocorrelation function of the optical field $\mathscr{E}(t)$.

$$\gamma(\tau) \equiv \frac{\tau_{\text{dependent part of } i_d}}{\tau_{\text{independent part}}} \propto C_{\mathscr{E}}(\tau) \tag{10.8-6}$$

The spectral density function $S_{\mathscr{E}}(\omega)$ is obtained, according to (10.2-14), by a Fourier transformation

$$S_{\mathscr{E}}(\omega) = \frac{1}{\pi}\int_{-\infty}^{\infty} C_{\mathscr{E}}(\tau)e^{-i\omega\tau}\,d\tau \tag{10.8-7}$$

The above scheme for obtaining the spectrum (spectral density function) of optical fields is termed Fourier transform spectroscopy, and the configuration of Figure 10-15 is representative of commercial instruments designed for this purpose. These instruments are popular especially in the far infrared (say $\lambda > 10\ \mu$m), since the relative inefficiency of detectors in this wavelength region can be compensated to some degree by a slow scanning rate (of τ) that allows for long integration times and better noise averaging.

A basic result of the Fourier integral transform relationships (10.2-13) and (10.2-14) between $C_{\mathscr{E}}(\tau)$ and $S_{\mathscr{E}}(\omega)$ is that in order to resolve $S_{\mathscr{E}}(\omega)$ to within, say, $\delta\omega$, i.e., to discern structure in $S_{\mathscr{E}}(\omega)$ on the scale of $\delta\omega$, we need to employ time delays $\tau > \pi/\delta\omega$. If we were, as an example, to employ interference spectroscopy to measure the output spectrum of a commercial semiconductor laser with a linewidth of $(\Delta\omega)_{\text{laser}} = 2\pi \times 10^6$ Hz, we would need a delay time τ that could be varied from 0 to 5×10^{-7} s.

In the case of lasers the finite spectral width of the optical field is due predominantly to phase, rather than amplitude, fluctuations. In this case a rather simple technique that involves mixing (heterodyning) the laser field with a delayed version of itself is sufficient to obtain the laser spectrum. This method, which employs a fixed delay instead of the variable delay of the Fourier transform method, is described next.

Delayed Self-Heterodyning of Laser Fields

Consider the configuration of Figure 10-16. An optical field is split into two components that, after a relative path delay t_d, are recombined at a detector. The spectrum of the resulting photocurrent is displayed by a spectrum analyzer. This detection method is referred to as *delayed self-heterodyning* since it involves a "mixing" of the field with a delayed version of itself.

Since the main fluctuation of laser fields is that of the phase and not the amplitude (see comment following Equation 10.7-19), we can approximate the field at the detector by the (complex) phasor

$$E_{\text{total}} = \frac{1}{4} E_0 e^{i\theta(t)} + \frac{1}{4} E_0 e^{i[\omega_0 t_d + \theta(t+t_d)]} \tag{10.8-8}$$

This field is illustrated in Figure 10-17. For delays t_d that are considerably shorter than the phase coherence time τ_c of the laser field (defined by Equation (10.7-24)), $\theta(t + t_d) \simeq \theta(t)$ and the magnitude of the total field phasor is a constant as shown in Figure 10-17. Although the phase angle $\theta(t)$ varies randomly, the angle α that determines the magnitude of E_{total} depends only on the difference $\theta(t + t_d) - \theta(t)$ and, in the limit $t_d \ll \tau_c$, does not change with time. The output current from the detector is constant, and nothing can be learned from it about the laser field spectrum. It is clear that we need to consider the case of $t_d \gg \tau_c$. In what follows we will consider the general case of arbitrary t_d.

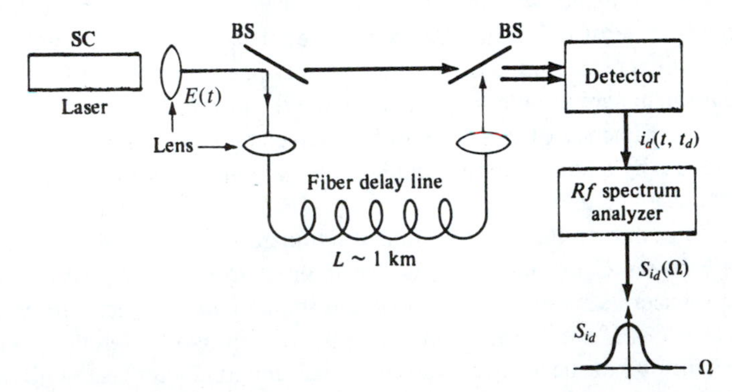

Figure 10-16 An interferometric arrangement employing a fiber delay for obtaining the spectrum $S_{\mathscr{E}}(\omega)$ of the laser field. (After Reference [23].)

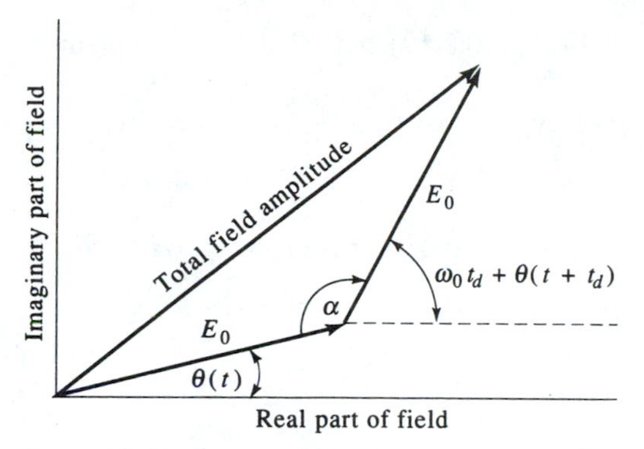

Figure 10-17 Construction showing the total optical field at the detector. For short delays, $t_d \ll \tau_c$, $\theta(t + t_d) \approx \theta(t)$ so that α and, consequently, the total field amplitude are constant.

The output current i_d is proportional to the time average of the square of the total optical field incident on the detector. It is thus proportional (see Equation 1.1-12) to the product of the complex amplitude of this field and its complex conjugate. Using (10.8-8) leads to

$$i_d = SE_0^2 \{e^{i\theta(t)} + e^{i[\omega_0 t_d + \theta(t+t_d)]}\} \times \{e^{-i\theta(t)} + e^{-i[\omega_0 t_d + \theta(t+t_d)]}\} \tag{10.8-9}$$

$$= SE_0^2 \{2 + e^{i[\theta(t) - \omega_0 t_d - \theta(t+t_d)]} + e^{-i[\theta(t) - \omega_0 t_d - \theta(t+t_d)]}\} \tag{10.8-10}$$

where S is a constant depending on the detector. We will derive the spectrum of i_d by employing the Wiener–Khintchine theorem (Equation [10.2-14]) so that first we need to obtain the autocorrelation function of $C_{i_d}(\tau)$ of the current i_d. Defining as in Equation (10.7-19)

$$\Delta\theta(t, \tau) \equiv \theta(t + \tau) - \theta(t) \tag{10.8-11}$$

We have reasoned in the last section (see discussion following Equation [10.7-21]) that $\Delta\theta(t, \tau)$ is a random Gaussian variable. It follows that the difference $\Delta\theta(t, \tau) - \Delta\theta(t + t_d, \tau)$ is also a Gaussian variable so that, in a manner identical to that used to derive Equation (10.7-24), we obtain

$$\langle e^{i[\Delta\theta(t,\tau) - \Delta\theta(t+t_d,\tau)]}\rangle = e^{-1/2\langle[\Delta\theta(t,\tau) - \Delta\theta(t+t_d,\tau)]^2\rangle} \tag{10.8-12}$$

Now

$$\langle[\Delta\theta(t, \tau) - \Delta\theta(t + t_d, \tau)]^2\rangle = 2\langle[\Delta\theta(\tau)]^2\rangle - 2\langle\Delta\theta(t, \tau)\Delta\theta(t + t_d, \tau)\rangle \tag{10.8-13}$$

where we used

$$\langle[\Delta\theta(t, \tau)]^2\rangle = \langle[\Delta\theta(t + t_d, \tau)]^2\rangle \equiv \langle[\Delta\theta(\tau)]^2\rangle$$

From the equation preceding (10.7-10) and putting $t = \tau$

$$\langle[\Delta\theta(\tau)]^2\rangle = \frac{\mu|\tau|}{2\bar{n}t_c} = \frac{2|\tau|}{\tau_c} \qquad \tau_c = 4\bar{n}t_c/\mu \tag{10.8-14}$$

Using (10.8-13) and (10.8-14) in (10.8-12) and (10.8-10), we obtain

$$C_{i_d}(\tau) \equiv \langle i_d(t) i_d(t + \tau)\rangle = S^2 E_0^4 \left[4 + 2e^{-\frac{|\tau|}{(\tau_c/2)}} e^{\langle \Delta\theta(t,\tau)\Delta\theta(t+t_d,\tau)\rangle}\right] \quad \textbf{(10.8-15)}$$

Special Case $t_d \gg \tau_c$

In the special, but important, long delay case $t_d \gg \tau_c$, we have

$$\lim_{t_d \to \infty} \langle \Delta\theta(t, \tau)\Delta\theta(t + t_d, \tau)\rangle \longrightarrow 0 \quad \textbf{(10.8-16)}$$

and

$$C_{i_d}(\tau)_{t_d \gg \tau_c} = S^2 E_0^4 \left(4 + e^{-\frac{|\tau|}{(\tau_c/2)}}\right) \quad \textbf{(10.8-17)}$$

Employing (10.7-14) or using directly the results of (10.7-28), we obtain the following expression for the spectral density of the current i_d

$$S_{i_d}(\Omega)_{t_d \gg \tau_c} = \frac{2S^2E_0^4}{\pi}\left[\frac{\left(\frac{4}{\tau_c}\right)}{\left(\frac{2}{\tau_c}\right)^2 + \Omega^2} + 4\pi\delta(\Omega)\right] \quad \textbf{(10.8-18)}$$

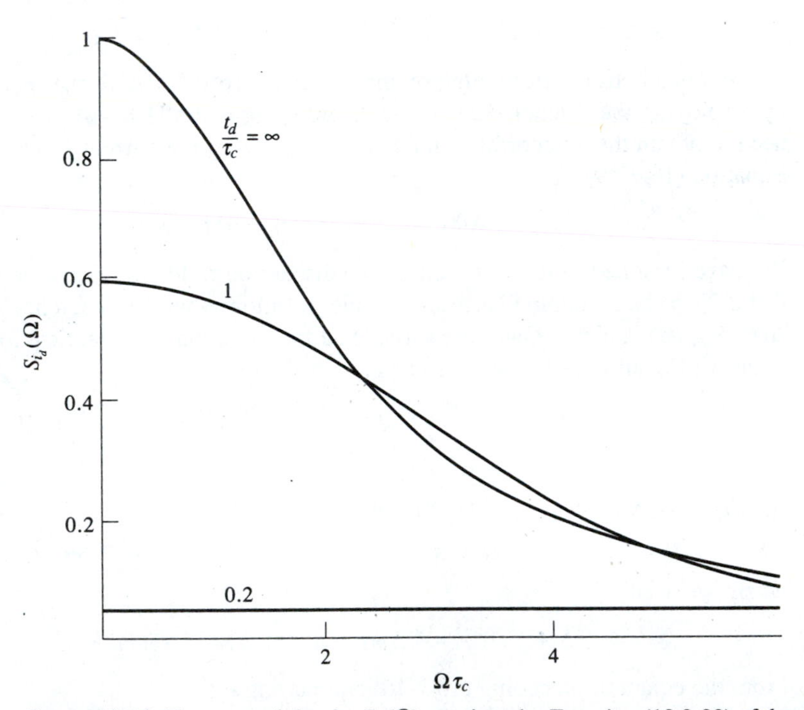

Figure 10-18 The spectral density $S_{i_d}(\Omega)$, as given by Equation (10.8-22) of the photocurrent in a delayed self-heterodyne detection of the output of a laser. The ratio of the delay time (t_d) to the laser field coherence time (τ_c) is a parameter. The frequency abcissa is in units of τ_c^{-1}, $\tau_c = (\Delta\nu)_{\text{laser}}^{-1}$.

The spectrum thus consists of a dc, $4\pi\delta(\Omega)$, term plus a Lorentzian distribution centered (if we count negative frequencies $\Omega < 0$) on $\Omega = 0$ with a full width at half maximum of

$$(\Delta\Omega)_{\text{FWHM}} = \frac{4}{\tau_c} = 2(\Delta\omega)_{\text{laser}} \tag{10.8-19}$$

The last equality, derived from (10.7-29) states that the width of the spectrum of the photo-detected current in the limit $t_d \gg \tau_c$ is twice that of the laser field.

The rigorous treatment of the general case involving arbitrary values of the delay t_d is beyond the scope of this book, since it requires a knowledge of the function $\langle \Delta\theta(t, \tau)\Delta\theta(t + t_d, \tau)\rangle$. The derivation of this function involves the solution of the nonlinear, noise-driven laser equation. The result is (see Reference [22])

$$\langle \Delta\theta(t, \tau)\Delta\theta(t + t_d, \tau)\rangle = \frac{2|\tau|}{\tau_c} - \frac{2}{\tau_c}\min(|\tau|, t_d) \tag{10.8-20}$$

where $\min(\tau, t_d)$ signifies the smallest of τ and t_d. The last result together with (10.8-13) and (10.8-20) when substituted in (10.8-15) give

$$C_{i_d}(\tau) \equiv \langle i_d(t)i_d(t + \tau)\rangle = S^2E_0^4\,[4 + 2e^{-(2|\tau|/\tau_c)\min(|\tau|,t_d)}] \tag{10.8-21}$$

$$\begin{aligned} S_{i_d}(\Omega) &= \frac{1}{\pi}\int_{-\infty}^{\infty} C_{i_d}(\tau)e^{-i\Omega\tau}\,d\tau \\ &= 8S^2E_0^4\,(1 + 0.5\,e^{-2t_d/\tau_c})\,\delta(\Omega) \\ &\quad + \left(\frac{8S^2E_0^4}{\pi\tau_c}\right)\frac{\left[1 - e^{-2t_d/\tau_c}\left(\cos\Omega t_d + \dfrac{2\sin\Omega t_d}{\Omega\tau_c}\right)\right]}{\left(\dfrac{2}{\tau_c}\right)^2 + \Omega^2} \end{aligned} \tag{10.8-22}$$

The integration leading to (10.8-22) is long but straightforward. Equation (10.8-22) reduces, as it should, to (10.8-18) when $t_d/\tau_c \to \infty$. In summation, we recall that only in the case $t_d/\tau_c \gg 1$, i.e., a long relative delay, is the spectrum $S_{i_d}(\Omega)$ a Lorentzian. A typical spectrum of a semiconductor laser obtained with a setup similar to that of Figure 10-16 is shown in Figure 10-13. A plot of the theoretical spectra of (10.8-22) for the cases $t_d/\tau_c = \infty$, 1, 0.2 is contained in Figure 10-18.

10.9 ERROR PROBABILITY IN A BINARY PULSE CODE MODULATION SYSTEM

The simplicity and reliability of digital processing by integrated electronic circuits has made it increasingly attractive to transmit information in the form of binary pulse trains. For optical communication systems, the analog data to be transmitted are coded into a train of 1 and 0 electrical pulses so that each pulse carries one bit of information. The electrical signal thus generated is impressed, say, by means of a modulator, on an optical beam, resulting in an optical train pulse. The optical signal having propagated through air or on optical fiber, is detected in the receiving end, thus yielding an electrical train of pulses.

Now, ideally, the reconstructed train of electrical pulses should be an exact replica of (or, more generally, constitute an exact analog of) the input train. The intermingling of noise at the detector output with the signal makes this perfect reconstruction impossible. A figure of merit used to describe the "quality" of the reconstructed signal is the *error probability*, *EP*, which is defined as the probability that any given pulse in the detected train does not agree with the corresponding pulse in the input train.

Figure 10-19 shows part of a pulse sequence containing three "1" pulses and two "0" pulses. An ideal noiseless detection should yield the sequence [Figure 10-19(a)] where the pulse height (say in amperes) is i_S. The presence of noise, however, introduces random fluctuations so that the detected signal may appear as in Figure 10-19(b).

A threshold decision circuit is usually employed that samples or integrates the signal [Figure 10-19(b)] once each period, yielding a 1 pulse if the sample exceeds a predetermined value ki_S $(k < 1)$ and a 0 pulse if the measured sample is smaller than ki_S [14]. In the case shown in Figure 10-19(b) the choice of the indicated threshold value will lead to a correct reconstruction of all pulses, except the last one, where a negative noise fluctuation has conspired to keep the pulse below the threshold value.

If a given pulse is a "1," then an erroneous reconstruction would result if during the sampling the noise current i_N is negative and such that

$$i_N < -i_S(1 - k)$$

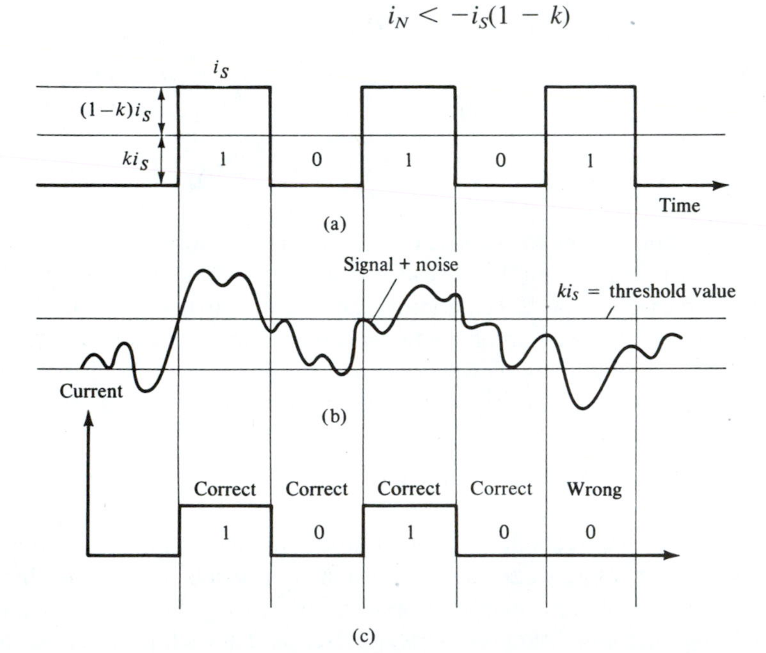

Figure 10-19 An ideal noiseless pulse train (a) is contaminated by noise as in (b). A reconstruction using a threshold decision level ki_S leads to (c). Note that the reconstruction of the last "1" pulse is in error because of a large negative noise fluctuation.

since in this case $i_S + i_N$ is smaller than the threshold value ki_S and "0" will result. In a like manner the reconstruction of a "0" pulse will be in error if

$$i_N > ki_S$$

On the average, half the pulses are 0 and half are 1 so that the probability of a wrong reconstruction of any given pulse is the bit error rate (BER)

$$\text{BER} = \tfrac{1}{2}[\text{probability that } i_N < -i_S(1-k) + \text{probability that } i_N > ki_S] \quad (10.9\text{-}1)$$

If the noise current i_N is a random Gaussian variable, which is the case in most applications, we can use (10.1-5) to evaluate the error probability. In this case σ is the root mean square (rms) value of the noise current i_N, so that $\sigma^2 = \overline{i_N^2}$ is the mean square noise current, as derived in Sections 10.4, 10.5, and 10.6. To simplify the result, let us choose $k = \frac{1}{2}$. Using the fact that, according to (10.1-5), $p(i_N) = p(-i_N)$ (10.9-1) becomes

$$\text{BER} = \text{probability that } i_N > \frac{i_S}{2}$$

$$= \int_{i_S/2}^{\infty} p(i_N)\, di_N = \frac{1}{\sigma\sqrt{2\pi}} \int_{i_S/2}^{\infty} e^{-(i_N^2/2\sigma^2)}\, di_N \quad (10.9\text{-}2)$$

$$= \frac{1}{\sqrt{\pi}} \int_{i_S/2\sqrt{2}\sigma}^{\infty} e^{-\xi^2}\, d\xi \quad (10.9\text{-}3)$$

Using the definition of the error function

$$\text{erf } z = \frac{2}{\sqrt{\pi}} \int_0^z e^{-\xi^2}\, d\xi \qquad \text{and} \qquad \int_0^{\infty} p(i_N)\, di_N = \tfrac{1}{2}$$

we can write (10.9-3) as

$$\text{BER} = \frac{1}{2}\left(1 - \text{erf}\, \frac{i_S}{2\sqrt{2}\sigma}\right)$$

$$\equiv \frac{1}{2}\, \text{erfc}\, \frac{i_S}{2\sqrt{2}\sigma} = \frac{1}{2}\, \text{erfc}\, \frac{i_S}{2\sqrt{2}\langle i_N\rangle} \quad (10.9\text{-}4)$$

where $\langle i_N\rangle \equiv \sigma\ (= (\overline{i_N^2})^{1/2})$ is the rms noise current.

A theoretical plot of BER as a function of the (peak) signal-to-noise ratio $i_S/\langle i_N\rangle$ is shown in Figure 10-20. We recall that $\overline{i_S^2}$ represents the electrical signal power at the detector output and not the optical power. It is interesting to note the extremely small error probabilities resulting from even moderate signal-to-noise power ratios. As an example, BER $= 10^{-9}$ when $i_S/\langle i_N\rangle = 11.89$ (21.5 db).

Experimental measurement of error probability in a detected optical pulse train is described by Reference [15]. Other pertinent discussions are to be found in References [16, 17].

A detailed example using the results of this section in designing a binary optical fiber communication system appears at the end of Chapter 11.

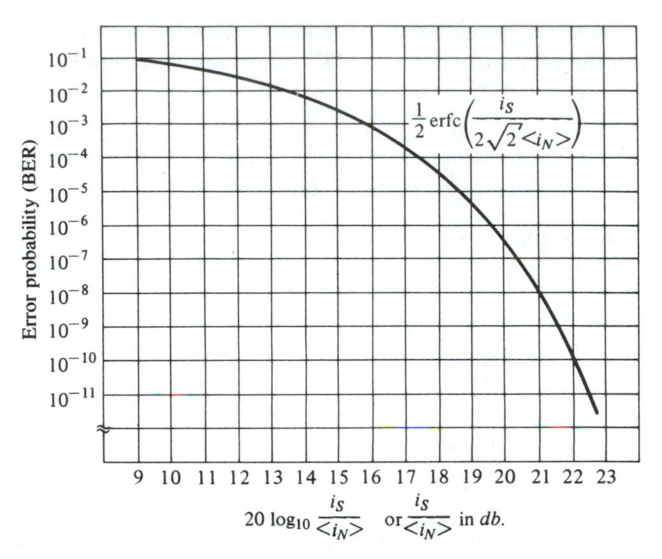

Figure 10-20 Plot of Equation (10.9-4) for the error probability (BER) as a function of the (peak) signal to noise current ratio at the detector output.

Problems

10.1 Derive Equations (10.4-2) and (10.4-3). [*Hint:* Apply the relation

$$\int_S \mathbf{D}\cdot\mathbf{n}\,ds = \int_v \rho\,dv$$

to a differential volume containing the charge sheet.]

10.2 Derive the shot-noise formula without making the restriction (Equation (10.4-6)) $\omega t_a \gg 1$. Assume the carriers move between the electrodes at a constant velocity.

10.3 Derive Equation (10.5-11).

10.4 Complete the missing steps in the derivation of Equation (10.5-20).

10.5 Estimate the scattering time τ_0 of carriers in copper at $T = 300°\text{K}$ using a tabulated value for its conductivity. At what frequencies is the condition $\omega\tau_0 \ll 1$ violated?

10.6 Repeat Problem 10.5 for a material with a carrier density of 10^{22} cm^{-3} and $\sigma = 10^{-5}$ (ohm-cm)$^{-1}$.

10.7 What is the change $\Delta\nu$ in the resonant frequency of a laser whose cavity length changes by Δl?

10.8

a. Estimate the frequency smearing $\Delta\nu$ of a laser in which fused-quartz rods are used to determine the length of the optical cavity in an environment where the temperature stability is ± 0.5 K. [*Caution*: Do not forget the dependence of n on T.]

b. What temperature stability is needed to reduce $\Delta\nu$ to less than 10^3 Hz?

10.9 Derive expression (10.5-9), $\overline{i_N^2}(\nu) = 4kT\Delta\nu/R$, for the Johnson noise by considering a high-Q parallel RLC circuit that is shunted by a current source of mean-square amplitude $\overline{i_N^2}(\nu)$. The magnitude of $\overline{i_N^2}(\nu)$ is to be chosen so that the resulting excitation of the circuit corresponds to a stored electromagnetic energy of kT. [*Hint:* Since the magnetic and electric energies are equal, then

$$kT = C\overline{v^2(t)} = C\int_0^\infty \frac{\overline{V_N^2(\nu)}}{\Delta\nu}\,d\nu$$

where $\overline{V_N^2(\nu)} = \overline{i_N^2(\nu)}|Z(\nu)|^2$. Also assume that $\overline{i_N^2(\nu)}/\Delta\nu$ is independent of frequency.]

10.10 Derive and plot the error probability as a function of $i_S/\langle i_N\rangle$ for (**a**) $k = 0.75$, (**b**) $k = 0.25$.

10.11 Design an experimental system for measuring the root-mean-square deviation of the laser frequency $(\Delta\omega)_{\text{RMS}} \equiv \langle(\omega(t) - \omega_0)^2\rangle^{1/2}$.

10.12

a. Please write a short report on the fundamentals of laser gyroscopes and of the Sagnac interferometer rotation sensor. You may, for example, look up the *Journal of Quantum Electronics* index section for listing of articles on "gyroscopes."

b. What is the minimum rotation rate detectable by each of the two types of interferometers when the laser field spectral purity is limited by the quantum effects discussed in Section 10.7?

References

1. The basic concepts of noise theory used in this chapter can be found, for example, in W. B. Davenport and W. L. Root, *An Introduction to the Theory of Random Signals and Noise.* New York: McGraw-Hill, 1958.
2. Bennett, W. R., "Methods of solving noise problems," *Proc. IRE* 44:609, 1956.
3. The classic reference to this topic is: S. O. Rice, "Mathematical analysis of random noise," *Bell Syst. Tech. J.* 23:282, 1944; 24:46, 1945.
4. Johnson, J. B., "Thermal agitation of electricity in conductors," *Phys. Rev.* 32:97, 1928.
5. Nyquist, H., "Thermal agitation of electric charge in conductors," *Phys. Rev.* 32:110, 1928.
6. Smith, R. A., F. A. Jones, and R. P. Chasmar, *The Detection and Measurement of Infrared Radiation.* New York: Oxford, 1968.
7. Yariv, A., *Quantum Electronics,* 2d ed. New York: Wiley, 1975.
8. Gordon, J. P., H. J. Zeiger, and C. H. Townes, "The maser—New type of microwave amplifier, frequency standard and spectrometer," *Phys. Rev.* 99:1264, 1955.

9. Gordon, E. I., ''Optical maser oscillators and noise,'' *Bell Syst. Tech. J.* 43:507, 1964.

10. Grivet, P. A., and A. Blaquiere, *Optical Masers.* New York: Polytechnic Press, 1963, p. 69.

11. Jaseja, T. J., Z. Javan, and C. H. Townes, ''Frequency stability of He–Ne masers and measurements of length,'' *Phys. Rev. Lett.* 10:165, 1963.

12. Egorov, Y. P., ''Measurements of natural line width of the emission of a gas laser with coupled modes,'' *JETP Lett.* 8:320, 1968.

13. Hinkley, E. D., and C. Freed, ''Direct observation of the Lorentzian lineshape as limited by quantum phase noise in a laser above threshold,'' *Phys. Rev. Lett.* 23:277, 1969.

14. Bennett, W. R., and J. R. Davey, *Data Transmission.* New York: McGraw-Hill, 1965, p. 100.

15. Goell, J. E., ''A 274 Mb/s optical repeater experiment employing a GaAs laser,'' *Proc. IEEE* 61:1504, 1973.

16. Personick, S. D., ''Receiver design for digital fiber optic communication systems,'' *Bell Syst. Tech. J.* 52:843, 1973.

17. Miller, S. E., T. Li, and E. A. J. Marcatili, ''Toward optic fiber transmission systems—devices and system considerations,'' *Proc. IEEE* 61:1726, 1973.

18. Schawlow, A. L., and C. H. Townes, *Phys. Rev.* 112:1940, 1958.

19. Courtesy of Kerry Vahala and Chris Harder of the California Institute of Technology.

20. Henry, C. H., ''Theory of the linewidth of semiconductor lasers,'' *IEEE J. Quant. Elect.* QE-18(2):259, February 1982.

21. Fleming, M., and A. Mooradian, ''Fundamental line broadening of single mode (GaAl)As diode lasers,'' *Appl. Phys. Lett.* 38:511, 1981.

22. Vahala, K., and A. Yariv, ''Semiclassical theory of noise in semiconductor lasers,'' *IEEE J. Quant. Elect.* QE-19:1096, 1983.

23. Yamamoto, Y., T. Mukai, and S. Saito, ''Quantum phase noise and linewidth of a semiconductor laser,'' *Elec. Lett.* 17:327, 1981.
Also: Okoshi, T., K. Kikuchi, and A. Nakayma, ''Novel method for high resolution measurement of laser output spectrum,'' *Elec. Lett.* 6:630, 1980.

24. Measurement of α parameter. Harder, C., Vahala, K., and A. Yariv, ''Measurement of the linewidth enhancement factor α in semiconductor lasers,'' *Appl. Phys. Lett.* 42:328–330, 1983.

25. Yuen, H. P., and J. H. Shapiro, ''Optical communication with two-photon coherent states,'' Part III *IEEE Trans. Inf. Theory* IT-26:78, 1980.

26. Kimble, H. J., and D. F. Walls, ''Squeezed states of electromagnetic fields,'' (special issue) *J. Opt. Soc.* B, 4:1353, 1987.

27. Slusher, R. E., L. W. Hollberg, B. Yurke, D. C. Mertz, and J. F. Valley, ''Observation of squeezed states generated by four-wave mixing in an optical cavity,'' *Phys. Rev. Lett.* 55:2409.

28. Haus, H. A., ''From classical to quantum noise,'' *J. Opt. Soc. Am. V* (in press).

Detection of Optical Radiation

11.0 INTRODUCTION

The detection of optical radiation is often accomplished by converting the radiant energy into an electric signal whose intensity is measured by conventional techniques. Some of the physical mechanisms that may be involved in this conversion include

1. The generation of mobile charge carriers in solid-state photoconductive detectors
2. Changing through absorption the temperature of thermocouples, thus causing a change in the junction voltage
3. The release by the photoelectric effect of free electrons from photoemissive surfaces

In this chapter we consider in some detail the operation of four of the most important detectors:

1. The photomultiplier
2. The photoconductive detector
3. The photodiode
4. The avalanche photodiode

The limiting sensitivity of each is discussed and compared to the theoretical limit. We will find that by use of the heterodyne mode of detection the theoretical limit of sensitivity may be approached.

11.1 OPTICALLY INDUCED TRANSITION RATES

A common feature of all the optical detection schemes discussed in this chapter is that the electric signal is proportional to the rate at which electrons are excited by the optical field. This excitation involves a transition of the electron from some initial bound state, say a, to a final state (or a group of states) b in which it is free to move and contribute to the current flow. For example, in an n-type photoconductive detector, state a corresponds to electrons in the filled valence band or localized donor impurity atoms, while state b corresponds to electrons in the conduction band. The two levels involved are shown schematically in Figure 11-1. A photon of energy $h\nu$ is absorbed in the process of exciting an electron from a "bound" state a to a "free" state b in which the electron can contribute to the current flow.

An important point to understand before proceeding with the analysis of different detection schemes is the manner of relating the transition rate per electron from state a to b to the intensity of the optical field. This rate is derived by quantum mechanical considerations.[1] In our case it can be stated in the following form: Given a nearly sinusoidal optical field[2]

$$e(t) = \tfrac{1}{2}[E(t)e^{i\omega_0 t} + E^*(t)e^{-i\omega_0 t}] \equiv \mathrm{Re}[V(t)] \qquad (11.1\text{-}1)$$

where $V(t) = E(t)\exp(i\omega_0 t)$,[3] the transition rate per electron induced by this field is proportional to $V(t)V^*(t)$. Denoting the transition rate as $W_{a\to b}$, we have

$$W_{a\to b} \propto V(t)V^*(t) \qquad (11.1\text{-}2)$$

We can easily show that $V(t)V^*(t)$ is equal to twice the average value of $e^2(t)$, where the averaging is performed over a few optical periods.

[1]More specifically, from first order time-dependent perturbation theory; see, for example, Reference [1].

[2]By "nearly sinusoidal" we mean a field where $E(t)$ varies slowly compared to $\exp(i\omega_0 t)$ or, equivalently, where the Fourier spectrum of $E(t)$ occupies a bandwidth that is small compared to ω_0. Under these conditions the variation of the amplitude $E(t)$ during a few optical periods can be neglected.

[3]$V(t)$ is referred to as the "analytic signal" of $e(t)$. See Problem 1.1.

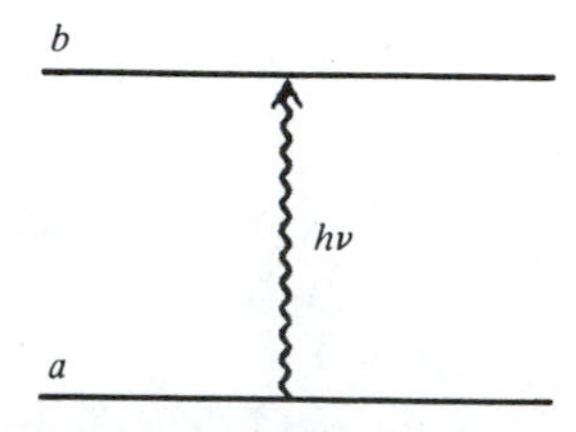

Figure 11-1 Most high-speed optical detectors depend on absorption of photons of energy $h\nu$ accompanied by a simultaneous transition of an electron (or hole) from a quantum state of low mobility (a) to one of higher mobility (b).

To illustrate the power of this seemingly simple result, consider the problem of determining the transition rate due to a field

$$e(t) = E_0 \cos(\omega_0 t + \phi_0) + E_1 \cos(\omega_1 t + \phi_1) \quad \textbf{(11.1-3)}$$

taking E_0 and E_1 real and $\omega_1 - \omega_0 \equiv \omega \ll \omega_0$. We can rewrite (11.1-3) as

$$\begin{aligned} e(t) &= \mathrm{Re}(E_0\, e^{i(\omega_0 t + \phi_0)} + E_1\, e^{i(\omega_1 t + \phi_1)}) \\ &= \mathrm{Re}[(E_0\, e^{i\phi_0} + E_1\, e^{i(\omega t + \phi_1)})e^{i\omega_0 t}] \end{aligned} \quad \textbf{(11.1-4)}$$

and, using (11.1-1), identify $V(t)$ as

$$V(t) = [E_0\, e^{i\phi_0} + E_1\, e^{i(\omega t + \phi_1)}]\, e^{i\omega_0 t}$$

thus, using (11.1-2), we obtain

$$\begin{aligned} W_{a\to b} &\propto (E_0\, e^{i\phi_0} + E_1\, e^{i(\omega t + \phi_1)})(E_0\, e^{-i\phi_0} + E_1\, e^{-i(\omega t + \phi_1)}) \\ &= E_0^2 + E_1^2 + 2E_0E_1 \cos(\omega t + \phi_1 - \phi_0) \end{aligned} \quad \textbf{(11.1-5)}$$

This shows that the transition rate has, in addition to a constant term $E_0^2 + E_1^2$, a component oscillating at the difference frequency ω with a phase equal to the difference of the two original phases. This coherent ''beating'' effect forms the basis of the heterodyne detection scheme, which is discussed in detail in Section 11.4.

11.2 PHOTOMULTIPLIER

The photomultiplier, one of the most common optical detectors, is used to measure radiation in the near ultraviolet, visible, and near infrared regions of the spectrum. Because of its inherent high current amplification and low noise, the photomultiplier is one of the most sensitive instruments devised by man and under optimal operation—which involves long integration time, cooling of the photocathode, and pulse-height discrimination—has been used to detect power levels as low as about 10^{-19} watt [2].

A schematic diagram of a conventional photomultiplier is shown in Figure 11-2. It consists of a photocathode (C) and a series of electrodes, called dynodes, that are labeled 1 through 8. The dynodes are kept at progressively higher potentials with respect to the cathode, with a typical potential difference between adjacent dynodes of 100 volts. The last electrode (A), the anode, is used to collect the electrons. The whole assembly is contained within a vacuum envelope in order to reduce the possibility of electronic collisions with gas molecules.

The photocathode is the most crucial part of the photomultiplier, since it converts the incident optical radiation to electronic current and thus determines the wavelength-response characteristics of the detector and, as will be seen, its limiting sensitivity. The photocathode consists of materials with low surface work functions. Compounds involving Ag-O-Cs and Sb-Cs are often used; see References [2, 3]. These compounds possess work functions as low as 1.5 eV, as compared to 4.5 eV in typical metals. As can be seen in Figure 11-3, this makes it possible to detect photons with longer wavelengths. It follows from the figure that the low-frequency

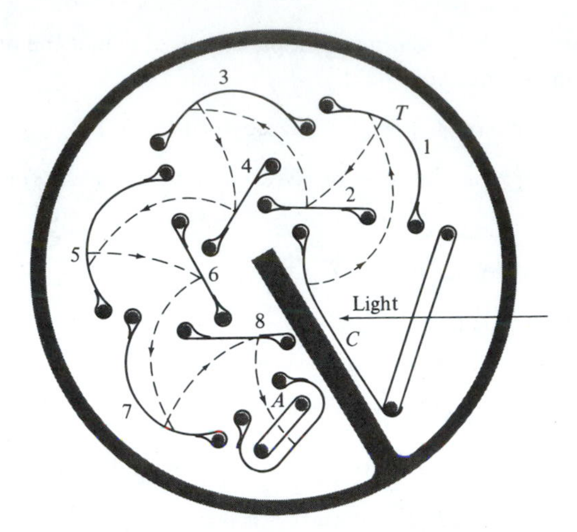

Figure 11-2 Photocathode and focusing dynode configuration of a typical commercial photomultiplier. C = cathode; 1–8 = secondary-emission dynodes; A = collecting anode. (After Reference [3].)

detection limit corresponds to $h\nu = \phi$. At present the lowest-work-function materials make possible photoemission at wavelengths as long as 1–1.1 μm.

Spectral response curves of a number of commercial photocathodes are shown in Figure 11-4. The quantum efficiency (or quantum yield as it is often called) is defined as the number of electrons released per incident photon.

The electrons that are emitted from the photocathode are focused electrostatically and accelerated toward the first dynode, arriving with a kinetic energy of, typically, about 100 eV. Secondary emission from dynode surfaces causes a multiplication of the initial current. This process repeats itself at each dynode until the initial current emitted by the photocathode is amplified by a very large factor. If the average secondary emission multiplication at each dynode is δ (that is, δ secondary

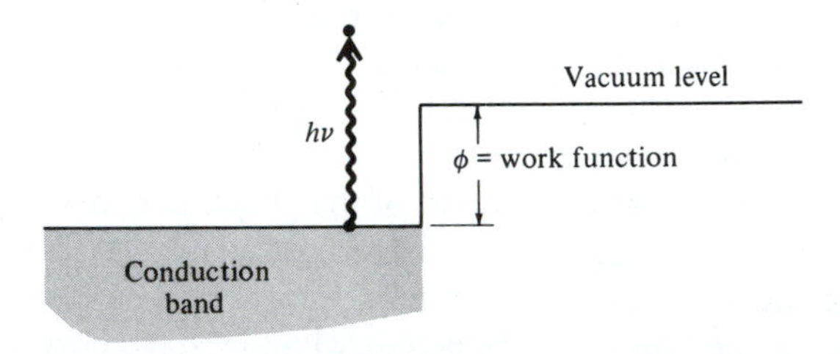

Figure 11-3 Photomultiplier photocathode. The vacuum level corresponds to the energy of an electron at rest at infinite distance from the cathode. The work function ϕ is the minimum energy required to lift an electron from the metal into the vacuum level, so only photons with $h\nu > \phi$ can be detected.

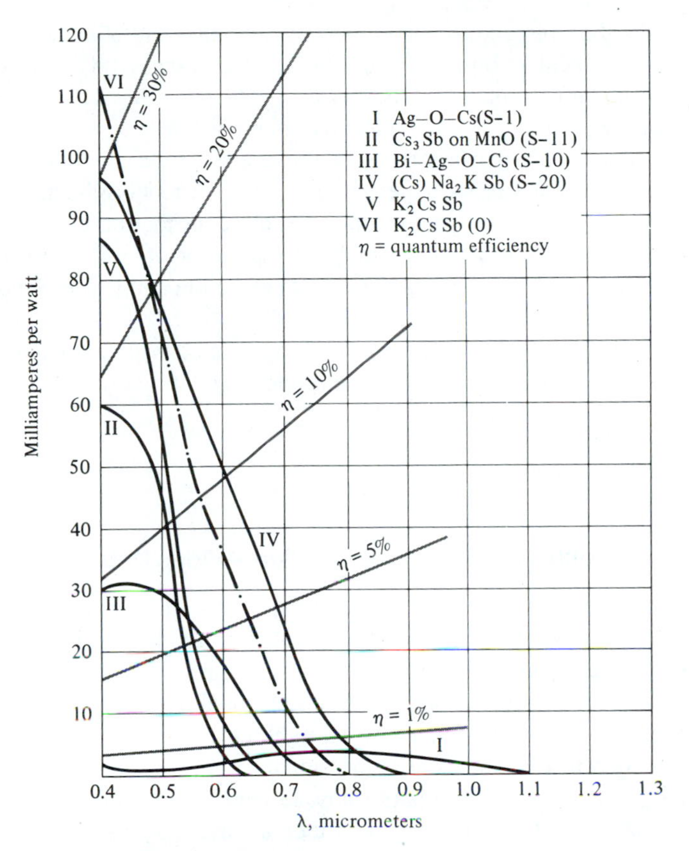

Figure 11-4 Photoresponse versus wavelength characteristics and quantum efficiency of a number of commercial photocathodes. (After Reference [3], p. 228.)

electrons for each incident one) and the number of dynodes is N, the total current multiplication between the cathode and anode is

$$G = \delta^N$$

which, for typical values[4] of $\delta = 5$ and $N = 9$, gives $G \simeq 2 \times 10^6$.

11.3 NOISE MECHANISMS IN PHOTOMULTIPLIERS

The random fluctuations observed in the photomultiplier output are due to

1. Cathode shot noise, given according to (10.4-9) by

$$\overline{(i_{N_1}^2)} = G^2 2\, e(\bar{i}_c + \bar{i}_d)\Delta\nu \tag{11.3-1}$$

[4]The value of δ depends on the voltage V between dynodes, and values of $\delta \simeq 10$ can be obtained (for $V \simeq 400$ volts). In commercial tubes, values of $\delta \simeq 5$, achievable with $V \simeq 100$ volts, are commonly used.

where $\bar{i}_c$ is the average current emitted by the photocathode due to the signal power that is incident on it. The current i_d is the so-called dark current, which is due to random thermal excitation of electrons from the surface as well as to excitation by cosmic rays and radioactive bombardment.

2. Dynode shot noise, which is the shot noise due to the random nature of the secondary emission process at the dynodes. Since current originating at a dynode does not exercise the full gain of the tube, the contribution of all the dynodes to the total shot noise output is smaller by a factor of $\sim\delta^{-1}$ than that of the cathode; since $\delta \simeq 5$ it amounts to a small correction and will be ignored in the following.
3. Johnson noise, which is the thermal noise associated with the output resistance R connected across the anode. Its magnitude is given by (10.5-9) as

$$\overline{(i_{N_2}^2)} = \frac{4kT\Delta\nu}{R} \tag{11.3-2}$$

Minimum Detectable Power in Photomultipliers—Video Detection

Photomultipliers are used primarily in one of two ways. In the first, the optical wave to be detected is modulated at some low frequency ω_m before impinging on the photocathode. The signal consists then, of an output current oscillating at ω_m, which, as will be shown below, has an amplitude proportional to the optical intensity. This mode of operation is known as *video*, or straight, detection.

In the second mode of operation, the signal to be detected, whose optical frequency is ω_s, is combined at the photocathode with a much stronger optical wave of frequency $\omega_s + \omega$. The output signal is then a current at the offset frequency ω. This scheme, known as *heterodyne* detection, will be considered in detail in Section 11-4.

The optical signal in the case of video detection may be taken as

$$\begin{aligned} e_s(t) &= E_s(1 + m \cos \omega_m t) \cos \omega_s t \\ &= \mathrm{Re}[E_s(1 + m \cos \omega_m t)e^{i\omega_s t}] \end{aligned} \tag{11.3-3}$$

where the factor $(1 + m \cos \omega_m t)$ represents amplitude modulation of the carrier.[5] The photocathode current is given, according to (11.1-2), by

$$\begin{aligned} i_c(t) &\propto [E_s(1 + m \cos \omega_m t)]^2 \\ &= E_s^2 \left[\left(1 + \frac{m^2}{2}\right) + 2m \cos \omega_m t + \frac{m^2}{2} \cos 2\omega_m t \right] \end{aligned} \tag{11.3-4}$$

[5]The amplitude modulation can be due to the information carried by the optical wave or, as an example, to chopping before detection.

To determine the proportionality constant involved in (11.3-4), consider the case of $m = 0$. The average photocathode current due to the signal is then[6]

$$\bar{i}_c = \frac{P e\eta}{h\nu_s} \tag{11.3-5}$$

where $\nu_s = \omega_s/2\pi$, P is the average optical power, and η (the quantum efficiency) is the average number of electrons emitted from the photocathode per incident photon. This number depends on the photon frequency, the photocathode surface, and in practice (see Figure 11-4) is found to approach 0.3. Using (11.3-5), we rewrite (11.3-4) as

$$i_c(t) = \frac{P e\eta}{h\nu_s}\left[\left(1 + \frac{m^2}{2}\right) + 2m \cos \omega_m t + \frac{m^2}{2} \cos 2\omega_m t\right] \tag{11.3-6}$$

The signal output current at ω_m is

$$i_s = \frac{GP e\eta}{h\nu_s}(2m) \cos \omega_m t \tag{11.3-7}$$

If the output of the detector is limited by filtering to a bandwidth $\Delta\nu$ centered on ω_m, it contains a shot-noise current, which, according to (11.3-1), has a mean-squared amplitude

$$\overline{(i_{N_1}^2)} = 2G^2 e(\bar{i}_c + i_d)\,\Delta\nu \tag{11.3-8}$$

where $\bar{i}_c$ is the average signal current and i_d is the dark current.

The noise and signal equivalent circuit is shown in Figure 11-5, where for the sake of definiteness we took the modulation index $m = 1$. R represents the output load of the photomultiplier. T_e is chosen so that the term $4kT_e\Delta\nu/R$ accounts for the thermal noise of R as well as for the noise generated by the amplifier that follows the photomultiplier.

[6] $P/h\nu_s$ is the rate of photon incidence on the photocathode; thus, if it takes $1/\eta$ photons to generate one electron, the average current is given by (11.3-5).

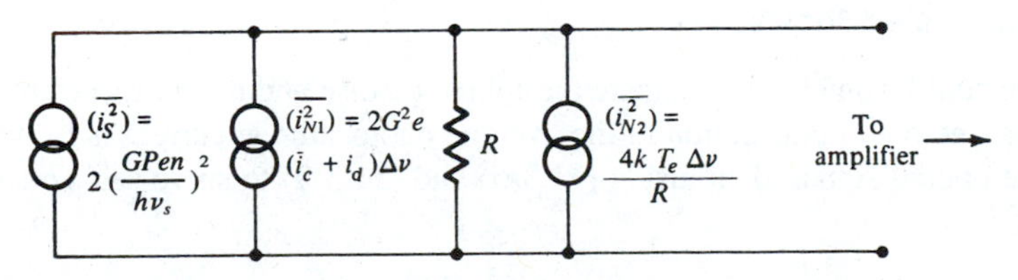

Figure 11-5 Equivalent circuit of a photomultiplier.

The signal-to-noise power ratio at the output is thus

$$\frac{S}{N} = \frac{\overline{i_s^2}}{(\overline{i_{N_1}^2}) + (\overline{i_{N_2}^2})}$$

$$= \frac{2(Pe\eta/h\nu_s)^2 G^2}{2G^2 e(\bar{i}_c + i_d)\Delta\nu + (4kT_e\Delta\nu/R)} \tag{11.3-9}$$

Due to the large current gain ($G \simeq 10^6$), the first term in the denominator of (11.3-9), which represents amplified cathode shot noise, is much larger than the thermal and amplifier noise term $4kT_e\Delta\nu/R$. Neglecting the term $4kT_e\Delta\nu/R$, assuming $i_d \gg \bar{i}_c$, and setting $S/N = 1$, we can solve for the minimum detectable optical power as

$$P_{\min} = \frac{h\nu_s(i_d\Delta\nu)^{1/2}}{\eta e^{1/2}} \tag{11.3-10}$$

Numerical Example: Sensitivity of Photomultiplier

Consider a typical case of detecting an optical signal under the following conditions:

$\nu_s = 6 \times 10^{14}$ Hz ($\lambda = 0.5$ μm)

$\eta = 10$ percent

$\Delta\nu = 1$ Hz

$i_d = 10^{-15}$ ampere (a typical value of the dark photocathode current)

Substitution in (11.3-10) gives

$$P_{\min} = 3 \times 10^{-16} \text{ watt}$$

The corresponding cathode signal current is $\bar{i}_c \sim 2.4 \times 10^{-17}$ ampere, so the assumption $i_d \gg \bar{i}_c$ is justified.

Signal-Limited Shot Noise

If one could, somehow, eliminate the Johnson noise and the dark current altogether, so that the only contribution to the average photocathode current is $\bar{i}_c$, which is due to the optical signal, then, using (11.3-5) and (11.3-9) to solve self-consistently for $P_{\min}$,

$$P_{\min} \simeq \frac{h\nu_s\Delta\nu}{\eta} \tag{11.3-11}$$

This corresponds to the quantum limit of optical detection. Its significance will be discussed in the next section. The practical achievement of this limit in video detection is nearly impossible since it depends on near total suppression of the dark current and other extraneous noise sources such as background radiation reaching the photocathode and causing shot noise.

The quantum detection limit (11.3-11) can, however, be achieved in the heterodyne mode of optical detection. This is discussed in the next section.

11.4 HETERODYNE DETECTION WITH PHOTOMULTIPLIERS

In the heterodyne mode of optical detection, the signal to be detected $E_s \cos \omega_s t$ is combined with a second optical field, referred to as the local-oscillator field, $E_L \cos(\omega_s + \omega)t$, shifted in frequency by $\omega(\omega \ll \omega_s)$. The total field incident on the photocathode is therefore given by

$$e(t) = \mathrm{Re}[E_L\, e^{i(\omega_s+\omega)t} + E_s\, e^{i\omega_s t}] \equiv \mathrm{Re}[V(t)] \tag{11.4-1}$$

The local-oscillator field originates usually at a laser at the receiving end, so that it can be made very large compared to the signal to be detected. In the following we will assume that

$$E_L \gg E_s \tag{11.4-2}$$

A schematic diagram of a heterodyne detection scheme is shown in Figure 11-6. The current emitted by the photocathode is given, according to (11.1-2) and (11.4-1), by

$$i_c(t) \propto V(t)V^*(t) = E_L^2 + E_s^2 + 2E_L E_s \cos \omega t$$

which, using (11.4-2) can be written as

$$i_c(t) = aE_L^2\left(1 + \frac{2E_s}{E_L}\cos \omega t\right) = aE_L^2\left(1 + 2\sqrt{\frac{P_s}{P_L}}\cos \omega t\right) \tag{11.4-3}$$

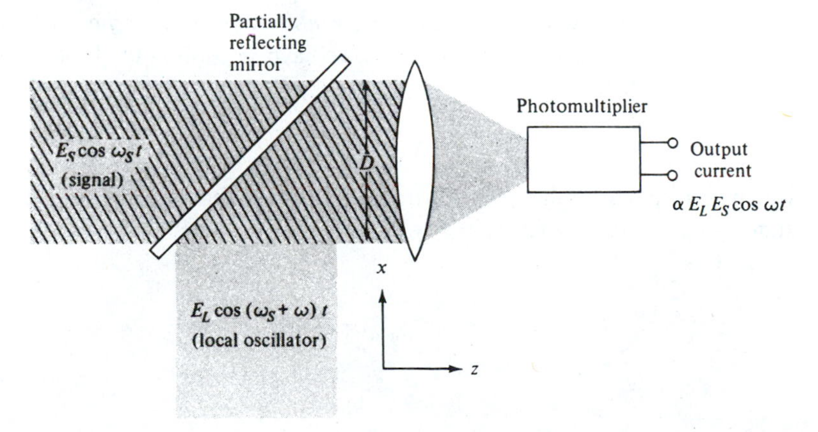

Figure 11-6 Schematic diagram of a heterodyne detector using a photomultiplier.

where P_s and P_L are the signal and local-oscillator powers, respectively. The proportionality constant a in (11.4-3) can be determined as in (11.3-6) by requiring that when $E_s = 0$ the direct current be related to the local-oscillator power P_L by $\bar{i}_c = P_L \eta e/h\nu_L$,[7] so taking $\nu \approx \nu_L \approx \nu_s$

$$i_c(t) = \frac{P_L e\eta}{h\nu}\left(1 + 2\sqrt{\frac{P_s}{P_L}}\cos\omega t\right) \tag{11.4-4}$$

The total cathode shot noise is thus

$$\overline{(i_{N_1}^2)} = 2e\left(i_d + \frac{P_L e\eta}{h\nu}\right)\Delta\nu \tag{11.4-5}$$

where i_d is the average dark current while $P_L e\eta/h\nu$ is the dc cathode current due to the strong local-oscillator field. The shot-noise current is amplified by G, resulting in an output noise

$$\overline{(i_N^2)}_{\text{anode}} = G^2 2e\left(i_d + \frac{P_L e\eta}{h\nu}\right)\Delta\nu \tag{11.4-6}$$

The mean-square signal current at the output is, according to (11.4-4),

$$\overline{(i_s^2)}_{\text{anode}} = 2G^2\left(\frac{P_s}{P_L}\right)\left(\frac{P_L e\eta}{h\nu}\right)^2 \tag{11.4-7}$$

The signal-to-noise power ratio at the output is given by

$$\frac{S}{N} = \frac{2G^2(P_s P_L)(e\eta/h\nu)^2}{[G^2 2e(i_d + P_L e\eta/h\nu) + 4kT_e/R]\Delta\nu} \tag{11.4-8}$$

where, as in (11.3-9), the last term in the denominator represents the Johnson (thermal) noise generated in the output load, plus the effective input noise of the amplifier following the photomultiplier. The big advantage of the heterodyne detection scheme is now apparent. By increasing P_L the S/N ratio increases until the denominator is dominated by the term $G^2 2eP_L e\eta/h\nu$. This corresponds to the point at which the *shot noise produced by the local oscillator current dwarfs all the other noise contributions.* When this state of affairs prevails, we have, according to (11.4-8),

$$\frac{S}{N} \simeq \frac{P_s}{h\nu\,\Delta\nu/\eta} \tag{11.4-9}$$

which corresponds to the quantum-limited detection limit. The minimum detectable signal—that is, the signal input power leading to an output signal-to-noise ratio of 1—is thus

$$(P_s)_{\min} = \frac{h\nu\,\Delta\nu}{\eta} \tag{11.4-10}$$

[7]This is just a statement of the fact that each incident photon has a probability η of releasing an electron.

This power corresponds for $\eta = 1$ to a flux at a rate of one photon per $(\Delta\nu)^{-1}$ seconds—that is, one photon per resolution time of the system.[8]

Numerical Example: Minimum Detectable Power with a Heterodyne System

It is interesting to compare the minimum detectable power for the heterodyne system as given by (11.4-10) with that calculated in the example of Section 11.3 for the video system. Using the same data,

$$\nu = 6 \times 10^{14}\ \text{Hz}(\lambda = 0.5\ \mu\text{m})$$

$$\eta = 10\ \text{percent}$$

$$\Delta\nu = 1\ \text{Hz}$$

we obtain

$$(P_s)_{\min} \simeq 4 \times 10^{-18}\ \text{watt}$$

to be compared with $P_{\min} \simeq 3 \times 10^{-16}$ watt in the video case.

Limiting Sensitivity as a Result of the Particle Nature of Light

The quantum limit to optical detection sensitivity is given by (11.4-10) as

$$(P_s)_{\min} = \frac{h\nu\,\Delta\nu}{\eta} \tag{11.4-11}$$

This limit was shown to be due to the shot noise of the photoemitted current. We may alternatively attribute this noise to the granularity—that is, the particle nature—of light, according to which the minimum energy increment of an electromagnetic wave at frequency ν is $h\nu$. The average power P of an optical wave can be written as

$$P = \overline{N}h\nu \tag{11.4-12}$$

where $\overline{N}$ is the average number of photons arriving at the photocathode per second. Next assume a hypothetical noiseless photomultiplier in which *exactly* one electron is produced for each η^{-1} incident photons. The measurement of P is performed by counting the number of electrons produced during an observation period T and then averaging the result over a large number of similar observations.

---

[8] A detection system that is limited in bandwidth to $\Delta\nu$ cannot resolve events in time that are separated by less than $\sim(\Delta\nu)^{-1}$ second. Thus $(\Delta\nu)^{-1}$ is the resolution time of the system.

The average number of electrons emitted per observation period T is

$$\overline{N}_e = \overline{N}T\eta \tag{11.4-13}$$

If the photons arrive in a perfectly random manner, then the number of photons arriving during the fixed observation period obeys Poissonian statistics[9]. Since in our ideal example, the electrons that are emitted mimic the arriving photons, they obey the same statistical distribution law. This leads to a fluctuation

$$\overline{(\Delta N_e)^2} \equiv \overline{(N_e - \overline{N}_e)^2} = \overline{N}_e = \overline{N}T\eta$$

Defining the minimum detectable number of quanta as that for which the rms fluctuation in the number of emitted photoelectrons equals the average value, we get

$$(\overline{N}_{\min}T\eta)^{1/2} = \overline{N}_{\min}T\eta$$

or

$$(\overline{N})_{\min} = \frac{1}{T\eta} \tag{11.4-14}$$

If we convert the last result to power by multiplying it by $h\nu$ and recall that $T^{-1} \simeq \Delta\nu$, where $\Delta\nu$ is the bandwidth of the system, we get

$$(P_s)_{\min} = \frac{h\nu\,\Delta\nu}{\eta} \tag{11.4-15}$$

in agreement with (11.4-10).

The above discussion points to the fact that the noise (fluctuation) in the photo current can be blamed on the physical process that introduces the randomness. In the case of Poissonian photon arrival statistics (as is the case with ordinary lasers) and perfect photon emission ($\eta = 1$), the fluctuations are due to the photons. The opposite, hypothetical, case of no photon fluctuations but random photoemission ($\eta < 1$) corresponds to pure shot noise. The electrical measurement of noise power will yield the same result in either case and cannot distinguish between them.

---

[9]This follows from the assumption that the photon arrival is perfectly random, so the probability of having N photons arriving in a given time interval is given by the Poisson law

$$p(N) = \frac{(\overline{N})^N e^{-\overline{N}}}{N!}$$

The mean-square fluctuation is given by

$$\overline{(\Delta N)^2} = \sum_{N=0}^{\infty} p(N)(N - \overline{N})^2 = \overline{N}$$

where

$$\overline{N} = \sum_{0}^{\infty} Np(N)$$

is the average N.

11.5 PHOTOCONDUCTIVE DETECTORS

The operation of photoconductive detectors is illustrated in Figure 11-7. A semiconductor crystal is connected in series with a resistance R and a supply voltage V. The optical field to be detected is incident on and absorbed in the crystal, thereby exciting electrons into the conduction band (or, in p-type semiconductors, holes into the valence band). Such excitation results in a lowering of the resistance R_d of the semiconductor crystal and hence in an increase in the voltage drop across R, which, for $\Delta R_d/R_d \ll 1$, is proportional to the incident optical intensity.

To be specific, we show the energy levels involved in one of the more popular semiconductive detectors—mercury-doped germanium [7]. Mercury atoms enter germanium as acceptors with an ionization energy of 0.09 eV. It follows that it takes a photon energy of at least 0.09 eV (that is, a photon with a wavelength shorter than 14 μm) to lift an electron from the top of the valence band and have it trapped by the Hg (acceptor) atom. Usually the germanium crystal contains a smaller density N_D of donor atoms, which at low temperatures find it energetically profitable to lose their valence electrons to one of the far more numerous Hg acceptor atoms, thereby becoming positively ionized and ionizing (negatively) an equal number of acceptors.

Since the acceptor density $N_A \gg N_D$, most of the acceptor atoms remain neutrally charged.

An incident photon is absorbed and lifts an electron from the valence band onto an acceptor atom, as shown in process A in Figure 11-8. The electronic deficiency (that is, the hole) thus created is acted upon by the electric field, and its drift along the field direction gives rise to the signal current. The contribution of a given hole to the current ends when an electron drops from an ionized acceptor level back into the valence band, thus eliminating the hole as in B. This process is referred to as electron–hole recombination or trapping of a hole by an ionized acceptor atom.

By choosing impurities with lower ionization energies, even lower-energy photons can be detected, and, indeed, photoconductive detectors commonly operate at wavelengths up to $\lambda = 50$ μm. Cu, as an example, enters into Ge as an acceptor with an ionization energy of 0.04 eV, which would correspond to long-wavelength

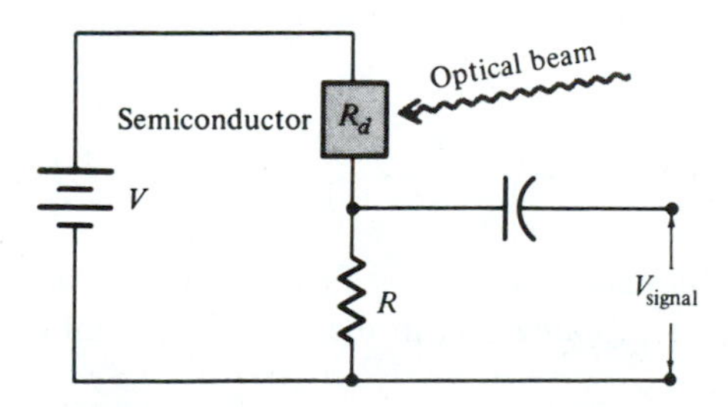

Figure 11-7 Typical biasing circuit of a photoconductive detector.

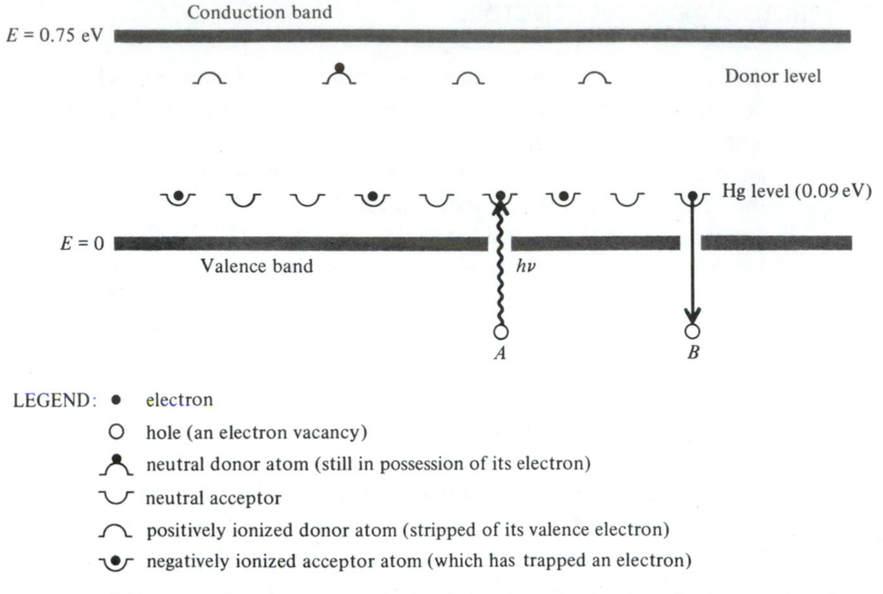

Figure 11-8 Donor and acceptor impurity levels involved in photoconductive semiconductors.

detection cutoff of $\lambda \simeq 32\ \mu$m. The response of a number of commercial photoconductive detectors is shown in Figure 11-9.

It is clear from this discussion that the main advantage of photoconductors compared to photomultipliers is their ability to detect long-wavelength radiation, since the creation of mobile carriers does not involve overcoming the large surface potential barrier. On the debit side we find the lack of current multiplication and the need to cool the semiconductor so that photoexcitation of carriers will not be masked by thermal excitation.

Consider an optical beam, of power P and frequency ν, that is incident on a photoconductive detector. Taking the probability for excitation of a carrier by an incident photon—the so-called quantum efficiency—as η, the carrier generation rate is $G = P\eta/h\nu$. If the carriers last on the average τ_0 seconds before recombining, the average number of carriers N_c is found by equating the generation rate to the recombination rate (N_c/τ_0), so

$$N_c = G\tau_0 = \frac{P\eta\tau_0}{h\nu} \tag{11.5-1}$$

Each one of these carriers drifts under the electric field influence[10] at a velocity $\bar{v}$ giving rise, according to (10.4-1), to a current in the external circuit of $i_e = e\bar{v}/d$,

[10] The drift velocity is equal to μE, where μ is the mobility and E is the electric field.

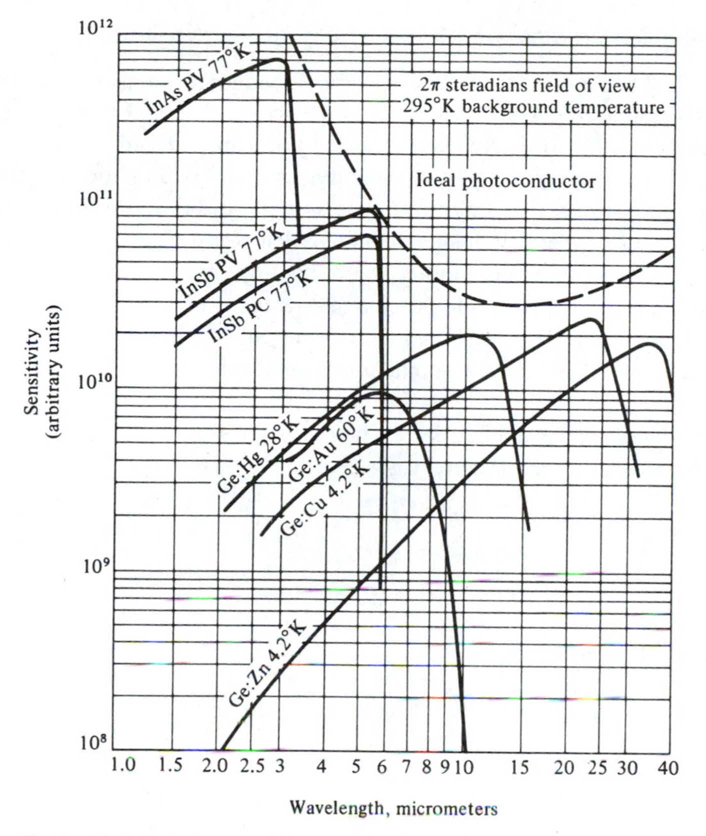

Figure 11-9 Relative sensitivity of a number of commercial photoconductors. (Courtesy Santa Barbara Research Corp.)

where d is the length (between electrodes) of the semiconductor crystal. The total current is thus the product of i_e and the number of carriers present, or, using (11.5-1),

$$\bar{i} = N_c i_e = \frac{P\eta\tau_0 e\bar{v}}{h\nu d} = \frac{e\eta}{h\nu}\left(\frac{\tau_0}{\tau_d}\right)P \tag{11.5-2}$$

where $\tau_d = d/\bar{v}$ is the drift time for a carrier across the length d. The factor (τ_0/τ_d) is thus the fraction of the crystal length drifted by the average excited carrier before recombining.

Equation (11.5-2) describes the response of a photoconductive detector to a constant optical flux. Our main interest, however, is in the heterodyne mode of photoconductive detection, which, as has been shown in Section 11.4, allows detection sensitivities approaching the quantum limit. In order to determine the limiting sensitivity of photoconductive detectors, we need first to understand the noise contribution in these devices.

Generation Recombination Noise in Photoconductive Detectors

The principal noise mechanism in cooled photoconductive detectors reflects the randomness inherent in current flow. Even if the incident optical flux were constant in time, the generation of individual carriers by the flux would constitute a random process. This is exactly the type of randomness involved in photoemission, and we may expect, likewise, that the resulting noise will be shot noise. This is almost true except for the fact that in a photoconductive detector a photoexcited carrier lasts τ seconds[11] (its recombination lifetime) before being captured by an ionized impurity. The contribution of the carrier to the charge flow in the external circuit is thus $e(\tau/\tau_d)$, as is evident from inspection of (11.5-2). Since the lifetime τ is not a constant, but must be described statistically, another element of randomness is introduced into the current flow.

Consider a carrier excited by a photon absorption and lasting τ seconds. Its contribution to the external current is, according to (10.4-1)

$$i_e(t) = \begin{cases} \dfrac{e\overline{v}}{d} & 0 \leq t \leq \tau \\ 0 & \text{otherwise} \end{cases} \tag{11.5-3}$$

which has a Fourier transform

$$I_e(\omega, \tau) = \frac{e\overline{v}}{2\pi d} \int_0^\tau e^{-i\omega t}\, dt = \frac{-ie\overline{v}}{2\pi\omega d}\,[1 - e^{-i\omega\tau}] \tag{11.5-4}$$

so that

$$|I_e(\omega, \tau)|^2 = \frac{e^2\overline{v}^2}{4\pi^2\omega^2 d^2}\,[2 - e^{-i\omega\tau} - e^{i\omega\tau}] \tag{11.5-5}$$

According to (10.3-10) we need to average $|I_e(\omega, \tau)|^2$ over τ. This is done in a manner similar to the procedure used in Section 10.5. Taking the probability function[12] $g(\tau) = \tau_0^{-1} \exp(-\tau/\tau_0)$, we average (11.5-5) over all the possible values of τ according to

$$\begin{aligned} \overline{|I_e(\omega)|^2} &= \int_0^\infty |I_e(\omega, \tau)|^2 g(\tau)\, d\tau \\ &= \frac{2e^2\overline{v}^2\tau_0^2}{4\pi^2 d^2(1 + \omega^2\tau_0^2)} \end{aligned} \tag{11.5-6}$$

The spectral density function of the current fluctuations is obtained using Carson's theorem (10.3-10) as

$$S(\nu) = 2\overline{N}\,\frac{2e^2(\tau_0^2/\tau_d^2)}{1 + \omega^2\tau_0^2} \tag{11.5-7}$$

[11]The parameter τ_0 appearing in (11.5-2) is the value of τ averaged over a large number of carriers.

[12]$g(\tau)\,d\tau$ is the probability that a carrier lasts between τ and $\tau + d\tau$ seconds before recombining.

where we used $\tau_d = d/\bar{v}$ and where $\overline{N}$, the average number of carriers generated per second, can be expressed in terms of the average current $\bar{I}$ by use of the relation[13]

$$\bar{I} = \overline{N} \frac{\tau_0}{\tau_d} e \tag{11.5-8}$$

leading to

$$S(\nu) = \frac{4e\bar{I}(\tau_0/\tau_d)}{1 + 4\pi^2\nu^2\tau_0^2}$$

Therefore, the mean-square current representing the noise power in a frequency interval ν to $\nu + \Delta\nu$ is

$$\overline{i_N^2} \equiv S(\nu)\,\Delta\nu = \frac{4e\bar{I}(\tau_0/\tau_d)\Delta\nu}{1 + 4\pi^2\nu^2\tau_0^2} \tag{11.5-9}$$

which is the basic result for generation–recombination noise.

Numerical Example: Generation Recombination Noise in Hg Doped Germanium Photoconductive Detector

To better appreciate the kind of numbers involved in the expression for $\overline{i_N^2}$ we may consider a typical mercury-doped germanium detector operating at 20 K with the following characteristics:

$$d = 10^{-1} \text{ cm}$$

$$\tau_0 = 10^{-9} \text{ s}$$

$$V \text{ (across the length } d) = 10 \text{ volts} \Rightarrow E = 10^2 \text{ V/cm}$$

$$\mu = 3 \times 10^4 \text{ cm}^2/\text{V-s}$$

The drift velocity is $\bar{v} = \mu E = 3 \times 10^6$ cm/s and $\tau_d = d/\bar{v} \simeq 3.3 \times 10^{-8}$ second, and therefore $\tau_0/\tau_d = 3 \times 10^{-2}$. Thus, on the average, a carrier traverses only 3 percent of the length ($d = 1$ mm) of the sample before recombining. Comparing (11.5-9) to the shot-noise result (10.4-9), we find that for a given average current $\bar{I}$ the generation recombination noise is reduced from the shot-noise value by a factor

$$\frac{\overline{(i_N^2)}_{\text{generation-recombination}}}{\overline{(i_N^2)}_{\text{shot noise}}} \underset{\omega\tau_0 \ll 1}{=} 2\left(\frac{\tau_0}{\tau_d}\right) \tag{11.5-10}$$

[13]This relation follows from the fact that the average charge per carrier flowing through the external circuit is $e(\tau_0/\tau_d)$, which, when multiplied by the generation rate $\overline{N}$, gives the current.

which, in the foregoing example, has a value of about 1/15. Unfortunately, as will be shown subsequently, the reduced noise is accompanied by a reduction by a factor of (τ_0/τ_d) in the magnitude of the signal power, which wipes out the advantage of the lower noise.

Heterodyne Detection in Photoconductors

The situation here is similar to that described by Figure 11-6 in connection with heterodyne detection using photomultipliers. The signal field

$$e_s(t) = E_s \cos \omega_s t$$

is combined with a strong local-oscillator field

$$e_L(t) = E_L \cos(\omega + \omega_s)t \qquad E_L \gg E_s$$

so the total field incident on the photoconductor is

$$e(t) = \mathrm{Re}(E_s\, e^{i\omega_s t} + E_L\, e^{i(\omega_s+\omega)t}) \equiv \mathrm{Re}[V(t)] \tag{11.5-11}$$

The rate at which carriers are generated is taken, following (11.1-2), as $aV(t)V^*(t)$ where a is a constant to be determined. The equation describing the number of excited carriers N_c is thus

$$\frac{dN_c}{dt} = aVV^* - \frac{N_c}{\tau_0} \tag{11.5-12}$$

where τ_0 is the average carrier lifetime, so N_c/τ_0 corresponds to the carrier's decay rate. We assume a solution for $N_c(t)$ that consists of the sum of dc and a sinusoidal component in the form of

$$N_c(t) = N_0 + (N_1\, e^{i\omega t} + \text{c.c.}) \tag{11.5-13}$$

where c.c. stands for "complex conjugate."

Substitution in (11.5-12) gives

$$N_c(t) = a\tau_0(E_s^2 + E_L^2) + a\tau_0 \left(\frac{E_s E_L\, e^{i\omega t}}{1 + i\omega\tau_0} + \text{c.c.} \right) \tag{11.5-14}$$

where we took E_s and E_L as real. The current through the sample is given by the number of carriers per unit length N_c/d times $e\bar{v}$, where $\bar{v}$ is the drift velocity

$$i(t) = \frac{N_c(t)e\bar{v}}{d} \tag{11.5-15}$$

which, using (11.5-14), gives

$$i(t) = \frac{e\bar{v}a\tau_0}{d} \left(E_s^2 + E_L^2 + \frac{2E_s E_L \cos(\omega t - \phi)}{\sqrt{1 + \omega^2\tau_0^2}} \right) \tag{11.5-16}$$

where $\phi = \tan^{-1}(\omega\tau_0)$.

The current is thus seen to contain a signal component that oscillates at ω and is proportional to E_s. The constant a in (11.5-16) can be determined by requiring that, when $P_s = 0$, the expression for the direct current predicted by (11.5-16) agree with (11.5-2). This condition is satisfied if we rewrite (11.5-16) as

$$i(t) = \frac{e\eta}{h\nu}\left(\frac{\tau_0}{\tau_d}\right)\left[P_s + P_L + \frac{2\sqrt{P_s P_L}}{\sqrt{1 + \omega^2\tau_0^2}}\cos(\omega t - \phi)\right] \tag{11.5-17}$$

where P_s and P_L refer, respectively, to the incident-signal and local-oscillator powers and $\nu = \nu_s = \omega_s/2\pi$ and η, the quantum efficiency, is the number of carriers excited per incident photon. The signal current is thus

$$i_s(t) = \frac{2e\eta}{h\nu}\left(\frac{\tau_0}{\tau_d}\right)\frac{\sqrt{P_s P_L}}{\sqrt{1 + \omega^2\tau_0^2}}\cos(\omega t - \phi) \tag{11.5-18}$$

while the dc (average) current is

$$\bar{I} = \frac{e\eta}{h\nu}\left(\frac{\tau_0}{\tau_d}\right)(P_s + P_L) \tag{11.5-19}$$

Since the average current $\bar{I}$ appearing in the expression (11.5-9) for the generation recombination noise is given in this case by

$$\bar{I} = \left(\frac{e\eta}{h\nu}\right)\left(\frac{\tau_0}{\tau_d}\right)P_L \qquad P_L \gg P_S$$

we can, by increasing P_L, increase the noise power $\overline{i_N^2}$ and at the same time, according to (11.5-18), the signal i_s^2 until the generation recombination noise (11.5-9) is by far the largest contribution to the total output noise. When this condition is satisfied, the signal-to-noise ratio can be written, using (11.5-9), (11.5-18), and (11.5-19) and taking $P_L \gg P_s$, as

$$\frac{S}{N} = \frac{\overline{i_s^2}}{\overline{i_N^2}} = \frac{2(e\eta\tau_0/h\nu\tau_d)^2 P_s P_L/(1 + \omega^2\tau_0^2)}{4e^2\eta(\tau_0/\tau_d)^2 P_L\,\Delta\nu/(1 + \omega^2\tau_0^2)h\nu} = \frac{P_s\eta}{2h\nu\,\Delta\nu} \tag{11.5-20}$$

The minimum detectable signal—that which leads to a signal-to-noise ratio of unity—is found by setting the left side of (11.5-20) equal to unity and solving for P_s. It is

$$(P_s)_{\min} = \frac{2h\nu\,\Delta\nu}{\eta} \tag{11.5-21}$$

which, for the same η, is twice that of the photomultiplier heterodyne detection as given by (11.4-10). In practice, however, η in photoconductive detectors can approach unity, whereas in the best photomultipliers $\eta \simeq 30$ percent.

Numerical Example: Minimum Detectable Power of a Heterodyne Receiver Using a Photoconductor at 10.6 μm

Assume the following:

$$\lambda = 10.6\ \mu\text{m}$$

$$\Delta\nu = 1\ \text{Hz}$$

$$\eta \simeq 1$$

Substitution in (11.5-21) gives a minimum detectable power of

$$(P_s)_{\min} \simeq 10^{-19}\ \text{watt}$$

Experiments ([8, 9]) have demonstrated that the theoretical signal-to-noise ratio as given by (11.5-20) can be realized quite closely in practice; see Figure 11-10.

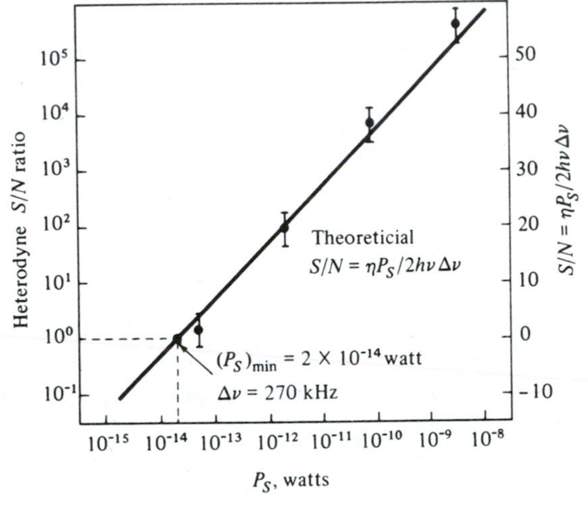

Figure 11-10 Signal-to-noise ratio of heterodyne signal to Ge:Cu detector at a heterodyne frequency of 70 MHz. Data points represent observed values. (After Reference [8].)

11.6 THE *p-n* JUNCTION

Before embarking on a description of the *p-n* diode detector, we need to understand the operation of the semiconductor *p-n* junction. Consider the junction illustrated in Figure 11-11. It consists of an abrupt transition from a donor-doped (that is, *n*-type) region of a semiconductor, where the charge carriers are predominantly electrons, to an acceptor-doped (*p*-type) region, where the carriers are holes. The doping profile—that is, the density of excess donor (in the *n* region) atoms or acceptor atoms

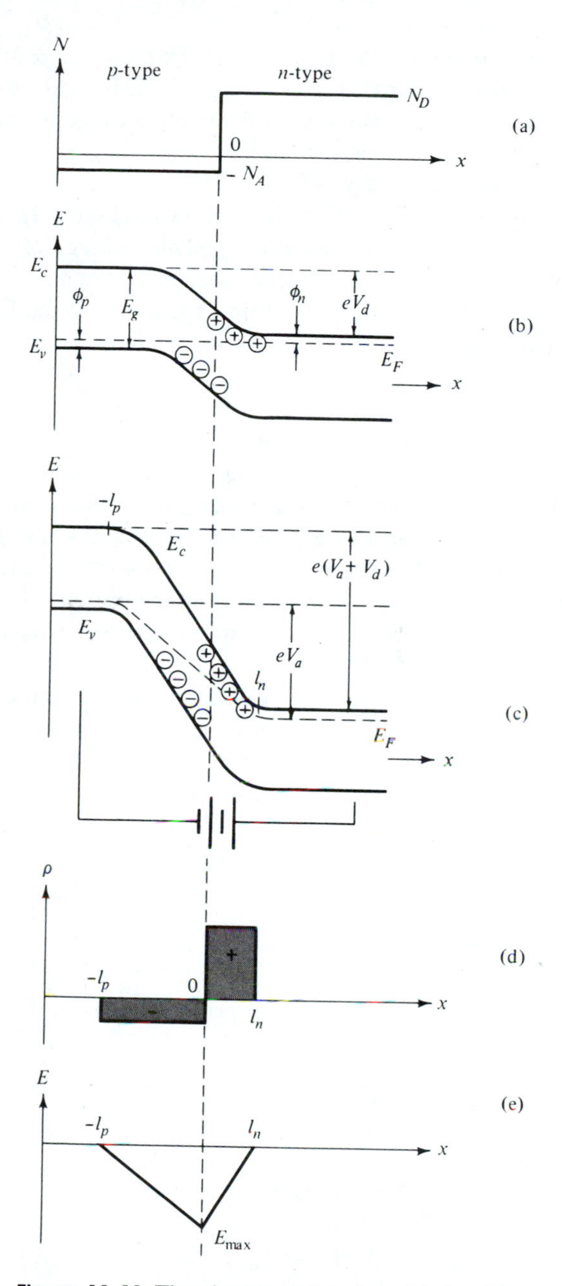

Figure 11-11 The abrupt *p-n* junction. (a) Impurity profile. (b) Energy-band diagram with zero applied bias. (c) Energy-band diagram with reverse applied bias. (d) Net charge density in the depletion layer. (e) The electric field. The circles in (b) and (c) represent ionized impurity atoms in the depletion layer.

(in the p region)—is shown in Figure 11-11(a). This abrupt transition results usually from diffusing suitable impurity atoms into a substrate of a semiconductor with the opposite type of conductivity. In our slightly idealized abrupt junction we assume that the n region ($x > 0$) has a constant (net) donor density N_D and the p region ($x < 0$) has a constant acceptor density N_A.

The energy-band diagram at zero applied bias is shown in Figure 11-11(b). The top (or bottom) curve can be taken to represent the potential energy of an electron as a function of position x, so the minimum energy needed to take an electron from the n to the p side of the junction is eV_d. Taking the separations of the Fermi level from the respective band edges as ϕ_n and ϕ_p as shown, we have

$$eV_d = E_g - (\phi_n + \phi_p)$$

V_d is referred to as the "built-in" junction potential.

Figure 11-11(c) shows the energy band diagram in the junction with an applied reverse bias of magnitude V_a. This leads to a separation of eV_a between the Fermi levels in the p and n regions and causes the potential barrier across the junction to increase from eV_d to $e(V_d + V_a)$. The change of potential between the p and n regions is due to a sweeping of the mobile charge carriers from the region $-l_p < x < l_n$, giving rise to a charge double layer of stationary (ionized) impurity atoms, as shown in Figure 11-11(d).

In the analytical treatment of the problem we assume that in the depletion layer ($-l_p < x < l_n$) the excess impurity atoms are fully ionized and thus, using $\nabla \cdot \mathbf{E} = \rho/\epsilon$ and $\mathbf{E} = -\nabla V$, where V is the potential, we have

$$\frac{d^2V}{dx^2} = \frac{eN_A}{\epsilon} \qquad \text{for } -l_p < x < 0 \tag{11.6-1}$$

and

$$\frac{d^2V}{dx^2} = -\frac{eN_D}{\epsilon} \qquad 0 < x < l_n \tag{11.6-2}$$

where the charge of the electron is $-e$ and the permittivity is ϵ. The boundary conditions are

$$E = -\frac{dV}{dx} = 0 \text{ at } x = -l_p \quad \text{and } x = +l_n \tag{11.6-3}$$

$$V \text{ and } \frac{dV}{dx} \text{ are continuous at } x = 0 \tag{11.6-4}$$

$$V(l_n) - V(-l_p) = V_d + V_a \tag{11.6-5}$$

The solutions of (11.6-1) and (11.6-2) conforming with the arbitrary choice of $V(0) = 0$ are

$$V = \frac{e}{2\epsilon} N_A(x^2 + 2l_p x) \qquad \text{for } -l_p < x < 0 \tag{11.6-6}$$

$$V = -\frac{e}{2\epsilon} N_D(x^2 - 2l_n x) \qquad 0 < x < l_n \tag{11.6-7}$$

which, using (11.6-4), gives

$$N_A l_p = N_D l_n \tag{11.6-8}$$

so the double layer contains an equal amount of positive and negative charge.

Condition (11.6-5) gives

$$V_d + V_a = \frac{e}{2\epsilon}(N_D l_n^2 + N_A l_p^2) \tag{11.6-9}$$

which, together with (11.6-8) leads to

$$l_p = (V_d + V_a)^{1/2}\left(\frac{2\epsilon}{e}\right)^{1/2}\left(\frac{N_D}{N_A(N_A + N_D)}\right)^{1/2} \tag{11.6-10}$$

$$l_n = (V_d + V_a)^{1/2}\left(\frac{2\epsilon}{e}\right)^{1/2}\left(\frac{N_A}{N_D(N_A + N_D)}\right)^{1/2} \tag{11.6-11}$$

and, therefore, as before,

$$\frac{l_p}{l_n} = \frac{N_D}{N_A} \tag{11.6-12}$$

Differentiation of (11.6-6) and (11.6-7) yields

$$E = -\frac{e}{\epsilon}N_A(x + l_p) \qquad \text{for } -l_p < x < 0$$

$$E = -\frac{e}{\epsilon}N_D(l_n - x) \qquad 0 < x < l_n \tag{11.6-13}$$

The field distribution of (11.6-13) is shown in Figure 11-11(e). The maximum field occurs at $x = 0$ and is given by

$$\begin{aligned} E_{\max} &= -2(V_d + V_a)^{1/2}\left(\frac{e}{2\epsilon}\right)^{1/2}\left(\frac{N_D N_A}{N_A + N_D}\right)^{1/2} \\ &= -\frac{2(V_d + V_a)}{l_p + l_n} \end{aligned} \tag{11.6-14}$$

The presence of a charge $Q = -eN_A l_p$ per unit junction area on the p side and an equal and opposite charge on the n side leads to a junction capacitance. The reason is that l_p and l_n depend, according to (11.6-10) and (11.6-11), on the applied voltage V_a, so a change in voltage leads to a change in the charge $eN_A l_p = eN_D l_n$ and hence to a differential capacitance per unit area,[14] given by

$$\begin{aligned} \frac{C_d}{\text{area}} &\equiv \frac{dQ}{dV_a} = eN_A\frac{dl_p}{dV_a} \\ &= \left(\frac{\epsilon e}{2}\right)^{1/2}\left(\frac{N_A N_D}{N_A + N_D}\right)^{1/2}\left(\frac{1}{V_a + V_d}\right)^{1/2} \end{aligned} \tag{11.6-15}$$

---

[14]The capacitance is defined by $C = Q/V_a$, whereas the differential capacitance $C_d = dQ/dV_a$ is the capacitance "seen" by a small ac voltage when the applied bias is V_a.

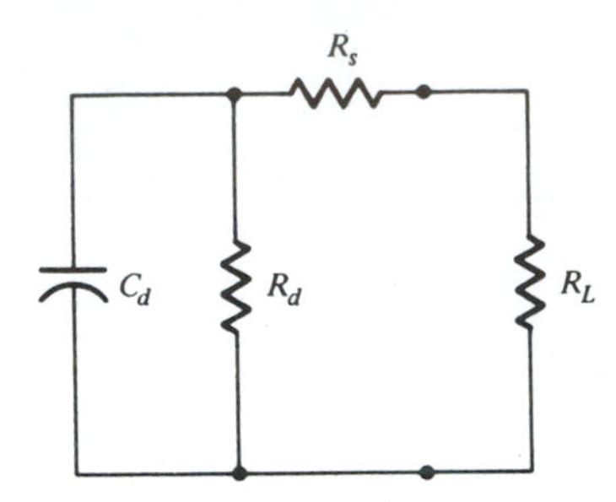

Figure 11-12 Equivalent circuit of a *p-n* junction. In typical back-biased diodes, $R_d \gg R_s$ and R_L, and $R_L \gg R_s$, so the resistance across the junction can be taken as equal to the load resistance R_L.

which, using (11.6-10) and (11.6-11), can be shown to be equal to

$$\frac{C_d}{\text{area}} = \frac{\epsilon}{l_p + l_n} \tag{11.6-16}$$

as appropriate to a parallel-plate capacitance of separation $l = l_p + l_n$. The equivalent circuit of a *p-n* junction is shown in Figure 11-12. The capacitance C_d was discussed above. The diode shunt resistance R_d in back-biased junctions is usually very large ($>10^6$ ohms) compared to the load impedance R_L and can be neglected. The resistance R_s represents ohmic losses in the bulk *p* and *n* regions adjacent to the junction.

11.7 SEMICONDUCTOR PHOTODIODES

Semiconductor *p-n* junctions are used widely for optical detection: see References [10–12]. In this role they are referred to as junction photodiodes. The main physical mechanisms involved in junction photodetection are illustrated in Figure 11-13. At *A*, an incoming photon is absorbed in the *p* side creating a hole and a free electron. If this takes place within a diffusion length (the distance in which an excess minority concentration is reduced to e^{-1} of its peak value, or in physical terms, the average distance a minority carrier traverses before recombining with a carrier of the opposite type) of the depletion layer, the electron will, with high probability, reach the layer boundary and will drift under the field influence across it. An electron traversing the junction contributes a charge *e* to the current flow in the external circuit, as described in Section 10.4. If the photon is absorbed near the *n* side of the depletion layer, as shown at *C*, the resulting hole will diffuse to the junction and then drift across it again, giving rise to a flow of charge *e* in the external load. The photon may also be absorbed in the depletion layer as at *B*, in which case both the hole and electron that are created drift (in opposite directions) under the field until they reach the *p* and *n* sides, respectively. Since in this case each carrier traverses a distance that is less than the full junction width, the contribution of this process to charge flow in the external circuit is, according to (10.4-1) and (10.4-7), *e*. In practice this last process is the most desirable, since each absorption gives rise to a charge *e*, and

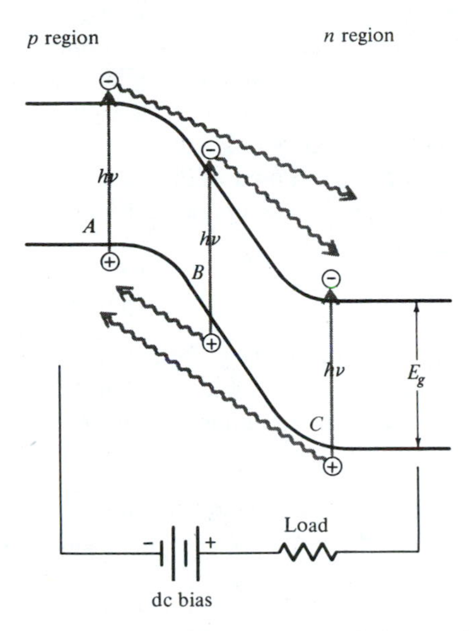

Figure 11-13 The three types of electron–hole pair creation by absorbed photons that contribute to current flow in a *p-n* photodiode.

delayed current response caused by finite diffusion time is avoided. As a result, photodiodes often use a *p-i-n* structure in which an intrinsic high resistivity (*i*) layer is sandwiched between the *p* and *n* regions. The potential drop occurs mostly across this layer, which can be made long enough to ensure that most of the incident photons are absorbed within it. Typical construction of a *p-i-n* photodiode is shown in Figure 11-14.

It is clear from Figure 11-13 that a photodiode is capable of detecting only radiation with photon energy $h\nu > E_g$, where E_g is the energy gap of the semiconductor. If, on the other hand, $h\nu \gg E_g$, the absorption, which in a semiconductor increases strongly with frequency, will take place entirely near the input face (in the

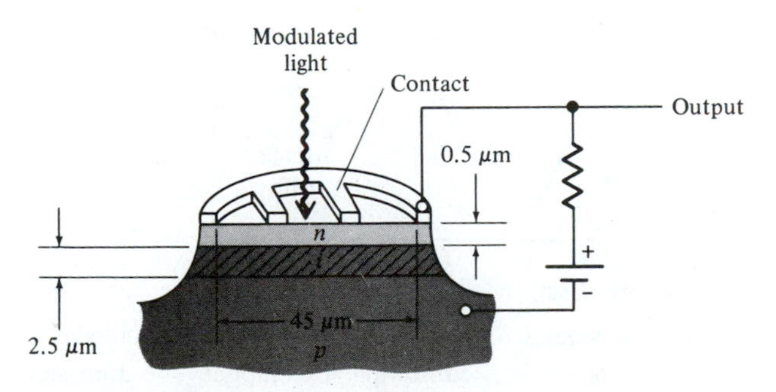

Figure 11-14 A *p-i-n* photodiode. (After Reference [13].)

n region of Figure 11-14) and the minority carriers generated by absorbed photons will recombine with majority carriers before diffusing to the depletion layer. This event does not contribute to the current flow and, as far as the signal is concerned, is wasted. This is why the photoresponse of diodes drops off when $h\nu > E_g$. Typical frequency response curves of photodiodes are shown in Figure 11-15. The number of carriers flowing in the external circuit per incident photon, the so-called quantum efficiency, is seen to approach 50 percent in Ge.

Frequency Response of Photodiodes

One of the major considerations in optical detectors is their frequency response—that is, the ability to respond to variations in the incident intensity such as those caused by high-frequency modulation. The three main mechanisms limiting the frequency response in photodiodes are:

1. The finite diffusion time of carriers produced in the p and n regions. This factor was described in the last section, and its effect can be minimized by a proper choice of the length of the depletion layer.
2. The shunting effect of the signal current by the junction capacitance C_d shown in Figure 11-12. This places an upper limit of

$$\omega_m \simeq \frac{1}{R_e C_d} \tag{11.7-1}$$

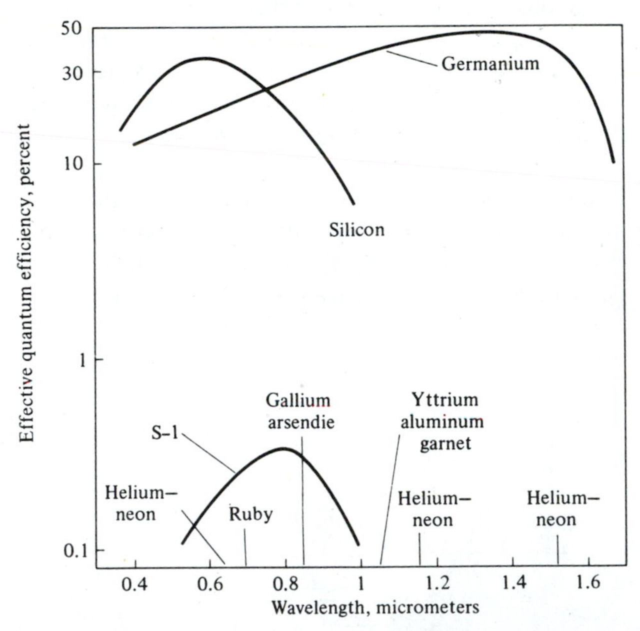

Figure 11-15 Quantum efficiencies for silicon and germanium photodiodes compared with the efficiency of the S-1 photodiode used in a photomultiplier tube. Emission wavelengths for various lasers are also indicated. (After Reference [13].)

on the intensity modulation frequency where R_e is the equivalent resistance in parallel with the capacitance C_d.

3. The finite transit time of the carriers drifting across the depletion layer.

To analyze first the limitation due to transit time, we assume the slightly idealized case in which the carriers are generated in a *single* plane, say point A in Figure 11-13, and then drift the full width of the depletion layer at a constant velocity v. For high enough electric fields, the drift velocity of carriers in semiconductors tends to saturate, so the constant velocity assumption is not very far from reality even for a nonuniform field distribution, such as that shown in Figure 11-11(e), provided the field exceeds its saturation value over most of the depletion layer length. The saturation of the whole velocity in germanium, as an example, is illustrated by the data of Figure 11-16.

The incident optical field is taken as

$$\begin{aligned} e(t) &= E_s(1 + m \cos \omega_m t) \cos \omega t \\ &\equiv \text{Re}[V(t)] \end{aligned} \tag{11.7-2}$$

where

$$V(t) \equiv E_s(1 + m \cos \omega_m t)\, e^{i\omega t} \tag{11.7-3}$$

Thus, the amplitude is modulated at a frequency $\omega_m/2\pi$. Following the discussion of Section 11.1 we take the generation rate $G(t)$; that is, the number of carriers generated per second, as proportional to the average of $e^2(t)$ over a time long com-

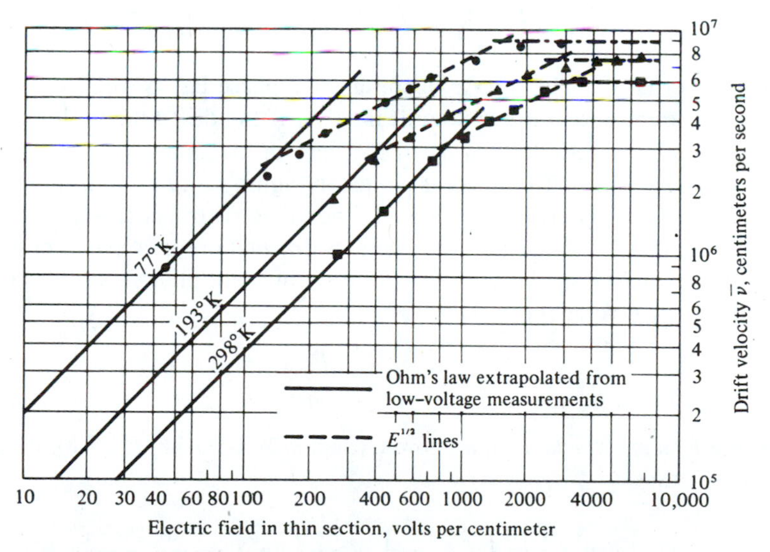

Figure 11-16 Experimental data showing the saturation of the drift velocity of holes in germanium at high electric fields. (After Reference [14].)

pared to the optical period $2\pi/\omega$. This average is equal to $\frac{1}{2}V(t)V^*(t)$, so the generation rate is taken as

$$G(t) = aE_s^2\left[\left(1 + \frac{m^2}{2}\right) + 2m \cos \omega_m t + \frac{m^2}{2} \cos 2\omega_m t\right] \tag{11.7-4}$$

where a is a proportionality constant to be determined. Dropping the term involving $\cos 2\omega_m t$ and using complex notation, we rewrite $G(t)$ as

$$G(t) = aE_s^2\left[1 + \frac{m^2}{2} + 2me^{i\omega_m t}\right] \tag{11.7-5}$$

A single carrier drifting at a velocity $\overline{v}$ contributes, according to (10.4-1), an instantaneous current

$$i = \frac{e\overline{v}}{d} \tag{11.7-6}$$

to the external circuit, where d is the width of the depletion layer. The current due to carriers generated between t' and $t' + dt'$ is $(e\overline{v}/d)G(t')\,dt'$ but, since each carrier spends a time $\tau_d = d/\overline{v}$ in transit, the instantaneous current at time t is the sum of contributions of carriers generated between t and $t - \tau_d$

$$i(t) = \frac{e\overline{v}}{d}\int_{t-\tau_d}^{t} G(t')\,dt' = \frac{e\overline{v}aE_s^2}{d}\int_{t-\tau_d}^{t}\left(1 + \frac{m^2}{2} + 2m\,e^{i\omega_m t'}\right)dt'$$

and, after integration,

$$i(t) = \left(1 + \frac{m^2}{2}\right)eaE_s^2 + 2meaE_s^2\left(\frac{1 - e^{-i\omega_m\tau_d}}{i\omega_m\tau_d}\right)e^{i\omega_m t} \tag{11.7-7}$$

The factor $(1 - e^{-i\omega_m\tau_d})/i\omega_m\tau_d$ represents the phase lag as well as the reduction in signal current due to the finite drift time τ_d. If the drift time is short compared to the modulation period, so $\omega_m\tau_d \ll 1$, it has its maximum value of unity, and the signal is maximum. This factor is plotted in Figure 11-17 as a function of the transit phase angle $\omega_m\tau_d$. We can determine the value of the constant a in (11.7-7) by requiring that (11.7-7) agree with the experimental observation according to which in the absence of modulation, $m = 0$, each incident photon will create η carriers. Thus the dc (average) current is

$$\bar{I} = \frac{Pe\eta}{h\nu} \tag{11.7-8}$$

where P is the optical (signal) power when $m = 0$. Using (11.7-8), we can rewrite (11.7-7) as

$$i(t) = \frac{Pe\eta}{h\nu}\left(1 + \frac{m^2}{2}\right) + \frac{Pe\eta}{h\nu}2m\left(\frac{1 - e^{-i\omega_m\tau_d}}{i\omega_m\tau_d}\right)e^{i\omega_m t} \tag{11.7-9}$$

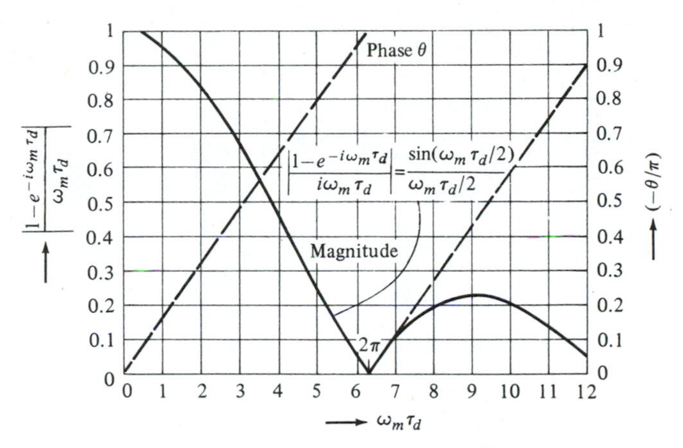

Figure 11-17 Phase and magnitude of the transit-time reduction factor $(1 - e^{-i\omega_m\tau_d})/\omega_m\tau_d$.

To evaluate the effect of the other limiting factors on the modulation frequency response of a photodiode, we refer to the diode equivalent ac circuit in Figure 11-18. Here R_d is the diode incremental (ac) resistance, C_d the junction capacitance, R_s represents the contact and series resistance, L_p the parasitic inductance associated mostly with the contact leads, and C_p the parasitic capacitance due to the contact leads and the contact pads.

Recent advances [20–22, 24] have resulted in metal–GaAs (Schottky) diodes with frequency response extending up to 10^{11} Hz. Figure 11-19 shows a schematic diagram of such a diode. This high-frequency limit was achieved by using a very small area (5 μm $\times$ 5 μm) that minimizes C_d, by using extremely short contact leads to reduce R_s and L_p, by fabricating the diode on semi-insulating GaAs substrate [20] to reduce C_p, and by using a thin (0.3 μm) n^-GaAs drift region to reduce the transit time. The resulting measured frequency response is shown in Figure 11-20. The measurement of the frequency response up to 100 GHz is by itself a considerable achievement. This was accomplished by first obtaining the impulse response of the

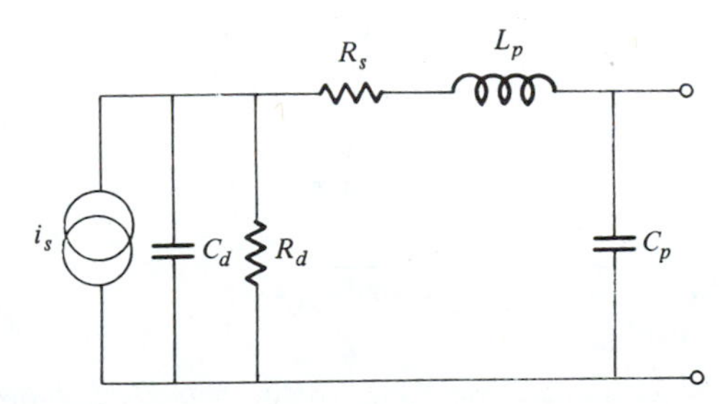

Figure 11-18 The equivalent high-frequency circuit of a semiconductor photodiode.

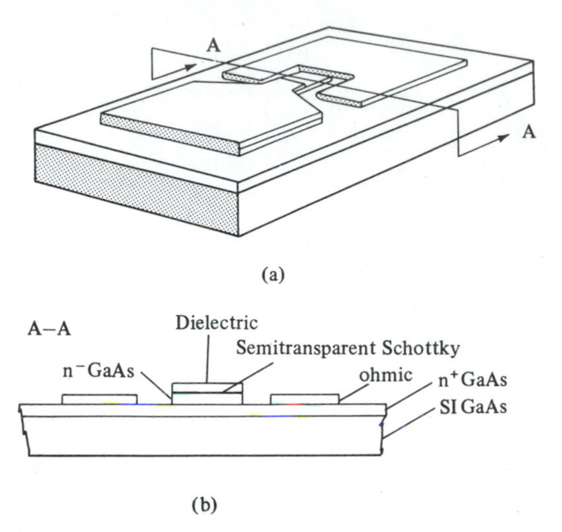

Figure 11-19 (a) Planar GaAs Schottky photodiode. (b) Cross section along A–A. The n^-GaAs layer (0.3 μm thick) and the n^+GaAs (0.4 μm thick) are grown by liquid-phase epitaxy on semi-insulating GaAs substrate. The semitransparent Schottky consists of 100 Å of Pt (After Reference [22].)

photodiode by exciting it with picosecond pulses (which, for the range of frequencies of interest, may be considered as delta functions) from a mode-locked laser [21]. The diode response, which is only a few picoseconds long, is measured by a new electrooptic sampling technique [23, 24]. The frequency response, as plotted in Figure 11-20, is obtained by taking the Fourier transform of the measured impulse response.

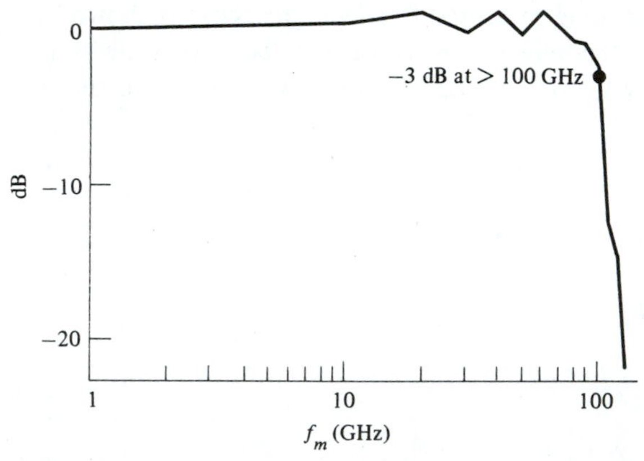

Figure 11-20 The modulation frequency response of the Schottky photodiode shown in Figure 11-19. (After Reference [22].)

Numerical Example: Modulation Response of a GaAs *p-n* Junction Photodiode

Let us calculate the upper limit on the frequency response of the diode shown in Figure 11-19. The following data apply.

$$\text{Area} = 5\ \mu\text{m} \times 5\ \mu\text{m}$$

$$\epsilon = 12.25\epsilon_0$$

$$d = 0.3\ \mu\text{m}\ (= \text{thickness of drift region})$$

$$\bar{v} = 10^7\ \text{cm/s (saturation velocity of electron in GaAs)}$$

$$R_s \approx 10\ \text{ohms}$$

The transit-time limit f_m is obtained from the condition $2\pi f_m \tau_d = 2$. This, according to Figure 11-17, is the frequency where the response is down to 84 percent of its maximum (zero-frequency) value. The result is

$$f_m \sim \frac{\bar{v}}{\pi d} \sim 1.06 \times 10^{11}\ \text{Hz}$$

The junction capacitance, based on the above data, is $\sim 10^{-14}$ farad. The parasitic capacitance can be kept in this case to $\sim 10^{-13}$ farad. Since the resistance R_d of the reverse-biased junction is very large, it is usually neglected.

The circuit limit to the frequency response is $f_m \sim 1/(2\pi R_s C_p) = 1.59 \times 10^{11}$ Hz. Since this value is larger than the transit-time limit, we conclude that the frequency response is transit-time limited to a value $\sim 10^{11}$ Hz, which is in agreement with the value obtained from Figure 11-20.

Detection Sensitivity of Photodiodes

We assume that the modulation frequency of the light to be detected is low enough that the transit time factor is unity and that the condition

$$\omega_m \ll \frac{1}{R_e C_d} \tag{11.7-10}$$

is fulfilled and, therefore, according to (11.7-1), the shunting of signal current by the diode capacitance C_d can be neglected. The diode current is given by (11.7-9) as

$$i(t) = \frac{Pe\eta}{h\nu}\left(1 + \frac{m^2}{2}\right) + \frac{Pe\eta}{h\nu}\, 2m\, e^{i\omega_m t} \tag{11.7-11}$$

The noise equivalent circuit of a diode connected to a load resistance R_L is shown in Figure 11-21. The signal power is proportional to the mean-square value of the sinusoidal current component, which, for $m = 1$, is

$$\overline{i_s^2} = 2\left(\frac{Pe\eta}{h\nu}\right)^2 \tag{11.7-12}$$

Two noise sources are shown. The first is the shot noise associated with the random generation of carriers. Using (10.4-9), this is represented by a noise generator $\overline{i_{N_1}^2} = 2e\bar{I}\,\Delta\nu$, where $\bar{I}$ is the average current as given by the first term on the right side of (11.7-11). Taking $m = 1$, we obtain

$$\overline{i_{N_1}^2} = \frac{3e^2(P + P_B)\eta\,\Delta\nu}{h\nu} + 2ei_d\Delta\nu \tag{11.7-13}$$

where P_B is the background optical power entering the detector (in addition to the signal power) and i_d is the "dark" direct current that exists even when $P_s = P_B = 0$. The second noise contribution is the thermal (Johnson noise) generated by the output load, which, using (10.5-9), is given by

$$\overline{i_{N_2}^2} = \frac{4kT_e\Delta\nu}{R_L} \tag{11.7-14}$$

where T_e is chosen to include the equivalent input noise power of the amplifier following the diode.[15] The signal-to-noise power ratio at the amplifier output is thus

$$\frac{S}{N} = \frac{\overline{i_s^2}}{\overline{i_{N_1}^2} + \overline{i_{N_2}^2}} = \frac{2(Pe\eta/h\nu)^2}{3e^2(P + P_B)\eta\Delta\nu/h\nu + 2ei_d\Delta\nu + 4kT_e\Delta\nu/R_L} \tag{11.7-15}$$

---

[15] In practice it is imperative that the signal-to-noise ratio take account of the noise power contributed by the amplifier. This is done by characterizing the "noisiness" of the amplifier by an effective input noise "temperature" T_A. The amplifier noise power measured at its output is taken as $GkT_A\Delta\nu$, where G is the power gain. (A hypothetical noiseless amplifier will thus be characterized by $T_A = 0$.) This power can be referred to the input by dividing by G, thus becoming $kT_A\Delta\nu$. The total effective noise power at the amplifier input is the sum of this power and the Johnson noise $kT\Delta\nu$ due to the diode load resistance; that is, $k(T + T_A)\Delta\nu \equiv kT_e\Delta\nu$. The amplifier noise temperature T_A is related to its "noise figure" F by the definition

$$F = 1 + \frac{T_A}{290}$$

It follows that the noise power generated within the amplifier and measured at its output is

$$N_A = GkT_A\Delta\nu = G(F - 1)kT_0\Delta\nu$$

where $T_0 = 290$. The ratio of the signal-to-noise power ratio at the input of the amplifier to the same ratio at the output is thus

$$\frac{(S/N)_{\text{in}}}{(S/N)_{\text{out}}} = \frac{S_{\text{in}}[G(F - 1)kT_0\Delta\nu + GkT\Delta\nu]}{kT\Delta\nu GS_{\text{in}}}$$

This ratio becomes equal to the "noise figure" F when the temperature T of the detector output load is equal to T_0. (Note that the choice $T_0 = 290$ is a matter of, universal, convention.)

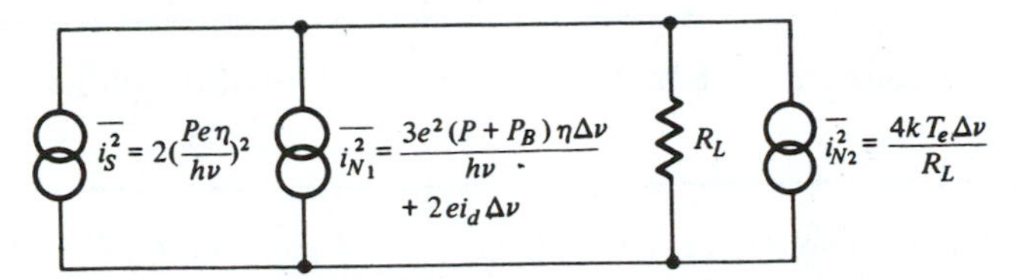

Figure 11-21 Noise equivalent circuit of a photodiode operating in the direct (video) mode. The modulation index m is taken as unity, and it is assumed that the modulation frequency is low enough that the junction capacitance and transit-time effects can be neglected. The resistance R_L is assumed to be much smaller than the shunt resistance R_d of the diode, so the latter is neglected. Also neglected is the series diode resistance, which is assumed small compared with R_L.

In most practical systems the need to satisfy Equation (11.7-10) forces one to use small values of load resistance R_L. Under these conditions and for values of P that are near the detectability limit ($S/N = 1$), the noise term (11.7-14) is much larger than the shot noise (11.7-13) and the detector is consequently not operating near its quantum limit. Under these conditions we have

$$\frac{S}{N} \simeq \frac{2(Pe\eta/h\nu)^2}{4kT_e\Delta\nu/R_L} \tag{11.7-16}$$

The ''minimum detectable optical power'' is by definition that yielding $S/N = 1$ and is, from (11.7-16),

$$(P)_{\min} = \frac{h\nu}{e\eta}\sqrt{\frac{2kT_e\Delta\nu}{R_L}} \tag{11.7-17}$$

which is to be compared to the theoretical limit of $h\nu\,\Delta\nu/\eta$, which, according to (11.3-11), obtains when the signal shot-noise term predominates. In practice, the value of R_L is related to the desired modulation bandwidth $\Delta\nu$ and the junction capacitance C_d by

$$\Delta\nu \simeq \frac{1}{2\pi R_L C_d} \tag{11.7-18}$$

which, when used in (11.7-16), gives

$$P_{\min} \simeq 2\sqrt{\pi}\,\frac{h\nu\Delta\nu}{e\eta}\sqrt{kT_e C_d} \tag{11.7-19}$$

This shows that sensitive detection requires the use of small area junctions so that C_d will be at a minimum.

Numerical Example: Minimum Detectable Power in the Case of Amplifier Limited Detection

Assume a typical Ge photodiode operating at $\lambda = 1.4\ \mu\text{m}$ with $C_d = 1$ pF, $\Delta\nu = 1$ GHz, and $\eta = 50$ percent. Let the amplifier following the diode have an effective noise temperature $T_e = 1200 + 290 = 1490$ K (see footnote 15) [14–15]. Substitution in (11.7-19) gives

$$P_{\min} \simeq 3.34 \times 10^{-7} \text{ watt}$$

for the minimum detectable signal power.

11.8 THE AVALANCHE PHOTODIODE

By increasing the reverse bias across a p-n junction, the field in the depletion layer can increase to a point at which carriers (electrons or holes) that are accelerated across the depletion layer can gain enough kinetic energy to "kick" new electrons from the valence to the conduction band, while still traversing the layer. This process, illustrated in Figure 11-22, is referred to as avalanche multiplication. An absorbed photon (A) creates an electron–hole pair. The electron is accelerated until at point C it has gained sufficient energy to excite an electron from the valence to the conduction band, thus creating a new electron–hole pair. The newly generated carriers drift in turn in opposite directions. The hole (F) can also cause carrier multiplication as in G. The result is a dramatic increase (avalanche) in junction current that sets in when the electric field becomes high enough. This effect, discovered first in gaseous

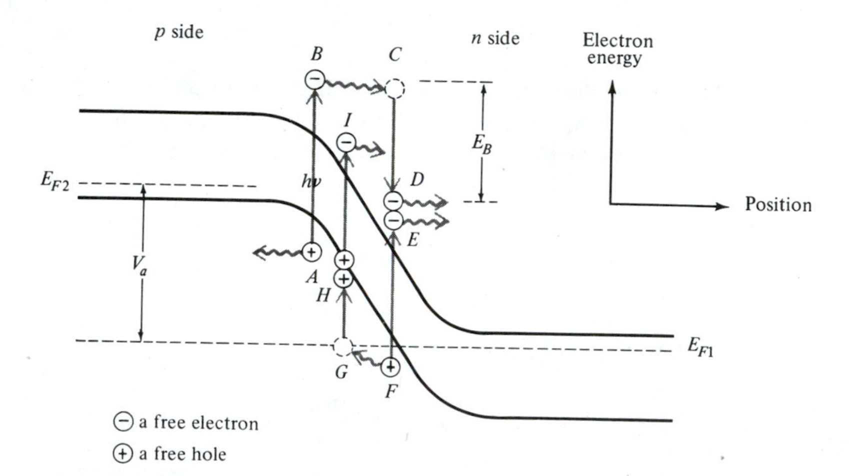

Figure 11-22 Energy-position diagram showing the carrier multiplication following a photon absorption in a reverse-biased avalanche photodiode.

plasmas and more recently in *p-n* junctions (References [15, 16]), gives rise to a multiplication of the current over its value in an ordinary (nonavalanching) photodiode. An experimental plot of the current gain M as a function of the junction field is shown in Figure 11-23.[16]

Avalanche photodiodes are similar in their construction to ordinary photodiodes except that, because of the steep dependence of M on the applied field in the avalanche region, special care must be exercised to obtain very uniform junctions. A sketch of an avalanche photodiode is shown in Figure 11-24.

Since an avalanche photodiode is basically similar to a photodiode, its equivalent circuit elements are given by expressions similar to those given above for the photodiode. Its frequency response is similarly limited by diffusion, drift across the depletion layer, and capacitive loading, as discussed in Section 11.7.

A multiplication by a factor M of the photocurrent leads to an increase by M^2

[16]If the probability that a photo-excited electron–hole pair will create another pair during its drift is denoted by p, the current multiplication is

$$M = (1 + p + p^2 + p^3 + \cdots) = \frac{1}{1 - p}$$

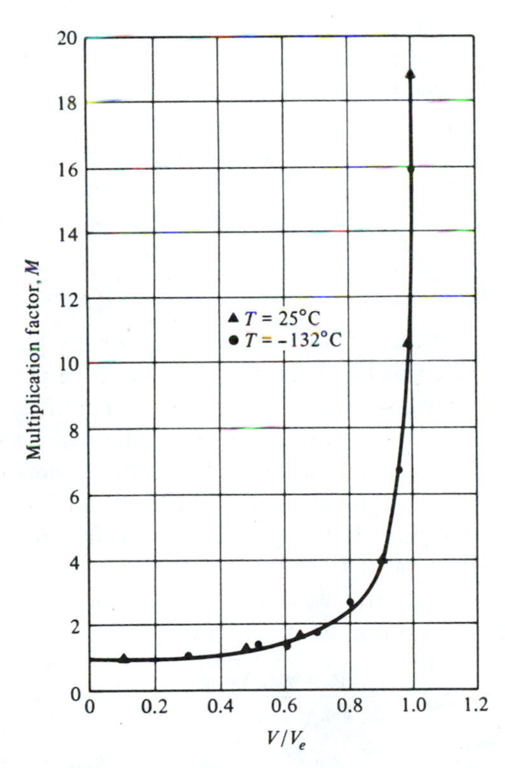

Figure 11-23 Current multiplication factor in an avalanche diode as a function of the electric field. (After Reference [16].)

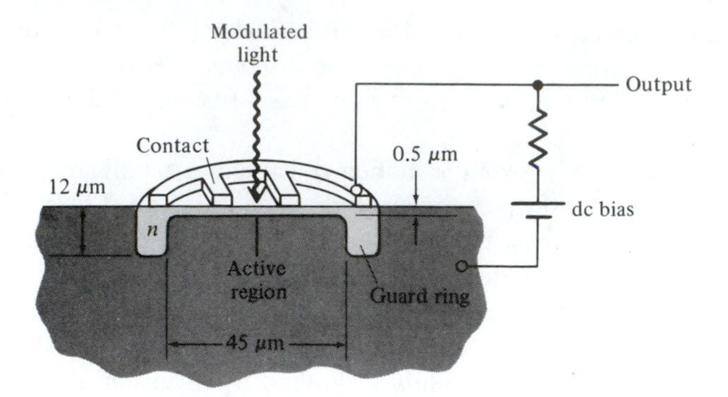

Figure 11-24 Planar avalanche photodiode. (After Reference [13].)

of the signal power S over that which is available from a photodiode so that, using (11.7-12), we get

$$S \propto \overline{i_s^2} = 2M^2\left(\frac{Pe\eta}{h\nu}\right)^2 \tag{11.8-1}$$

where P is the optical power incident on the diode. This result is reminiscent of the signal power from a photomultiplier as given by the numerator of (11.3-9), where the avalanche gain M plays the role of the secondary electron multiplication gain G. We may expect that, similarly, the shot-noise power will also increase by M^2. The shot noise, however, is observed to increase as M^n, where $2 < n < 3$.[17] Experimental observation of a near ideal $M^{2.1}$ behavior is shown in Figure 11-25.

The signal-to-noise power ratio at the output of the diode is thus given, following (11.7-15), by

$$\frac{S}{N} = \frac{2M^2(Pe\eta/h\nu)^2}{[3e^2(P + P_B)\eta\Delta\nu/h\nu]M^n + 2ei_d\Delta\nu M^n + 4kT_e\Delta\nu/R_L} \tag{11.8-2}$$

The advantage of using an avalanche photodiode over an ordinary photodiode is now apparent. When $M = 1$, the situation is identical to that at the photodiode as described by (11.7-15). Under these conditions the thermal term $4kT_e\Delta\nu/R_L$ in the denominator of (11.8-2) is typically much larger than the shot-noise terms. This causes S/N to increase with M. This improvement continues until the shot-noise terms become comparable with $4kT_e\Delta\nu/R_L$. Further increases in M result in a reduction of S/N since $n > 2$, and the denominator of (11.8-2) grows faster than the numerator. If we assume that M is adjusted optimally so that the denominator of (11.8-2) is equal to twice the thermal term $4kT_e\Delta\nu/R_L$, we can solve for the minimum detectable power (that is, the power input for which $S/N = 1$) obtaining

$$P_{\min} = \frac{2h\nu}{M'e\eta}\sqrt{\frac{kT_e\Delta\nu}{R_L}} \tag{11.8-3}$$

[17]A theoretical study by McIntyre [17] predicts that if the multiplication is due to either holes or electrons, $n = 2$, whereas if both carriers are equally effective in producing electron–hole pairs, $n = 3$.

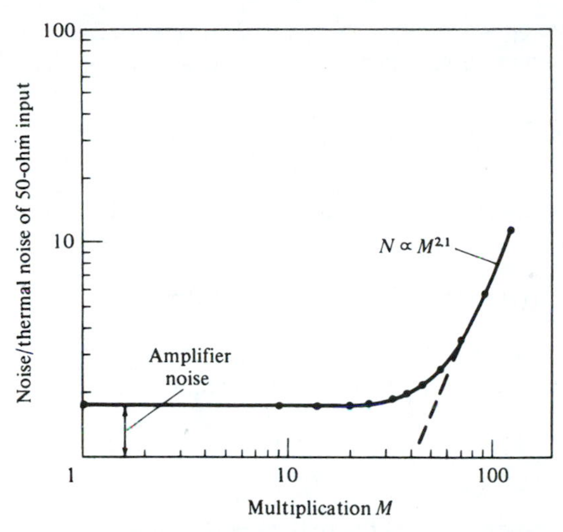

Figure 11-25 Noise power (measured at 30 MHz) as a function of photocurrent multiplication for an avalanche Schottky-barrier photodiode. (After Reference [18].)

where M' is the optimum value of M as discussed previously. The improvement in sensitivity over the photodiode result (11.7-17) is thus approximately M'. Values of M' between 30 and 100 are commonly employed, so the use of avalanche photodiodes affords considerable improvement in sensitivity over that available from photodiodes.

11.9 POWER FLUCTUATION NOISE IN LASERS

The power output from lasers is ever fluctuating. This fluctuation may be due to temperature variations, acoustic vibrations, and other man-made causes. Even if all of these extraneous effects are eliminated, there remains a basic (quantum mechanical) contribution that is due to spontaneous emission of radiation into the laser mode by atoms dropping from the upper transition level into the lower level. The field due to this spontaneous emission is not coherent with that of the laser mode, thus causing phase and amplitude fluctuations [1]. Since these fluctuations are random, they are described and quantified in terms of the statistical noise tools developed earlier in this chapter and in Chapter 10.

Let the power output of the laser be

$$P(t) = P_0 + \Delta P(t) \tag{11.9-1}$$

where the time-averaging value of the fluctuation is zero.

$$\overline{\Delta P}(\mathbf{t}) = 0$$

so that P_0 is the average optical power. Using (10.2-6) and (10.2-8) we characterize the ''power'' of the fluctuation via the mean of the squared deviation[18]

$$\overline{(P(t) - P_0)^2} \equiv \overline{(\Delta P(t))^2} = \int_0^\infty S_{\Delta P}(f)df \tag{11.9-2}$$

$S_{\Delta P}(f)$ is related to the spectral density function $S_{\Delta P}(\omega)$, defined by (10.2-8) and (10.2-14), by

$$S_{\Delta P}(f) = 2\pi S_{\Delta P}(\omega) \qquad (\omega = 2\pi f) \tag{11.9-3}$$

If an optical field at frequency ν with a power $P(t)$ is incident on a detector whose quantum efficiency (electrons per photon) is η, the output current is

$$i(t) = \frac{e\eta P(t)}{h\nu}$$

so that according to Equation (11.9-1) the optical power fluctuation $\Delta P(t)$ causes a fluctuating current component $\Delta i(t) = e\eta\Delta P(t)/h\nu$ with a mean square

$$\overline{i_{NL}^2}(t) \equiv \overline{[\Delta i(t)]^2} = \frac{e^2\eta^2}{(h\nu)^2}\overline{[\Delta P(t)]^2} = \frac{e^2\eta^2}{(h\nu)^2} S_{\Delta P}(f)\Delta f \tag{11.9-4}$$

where Δf is the bandwidth of the electronic detection circuit.

The relative intensity noise (RIN), is defined as the relative fluctuation ''power'' in a $\Delta f = 1\text{Hz}$ bandwidth

$$\text{RIN} \equiv \frac{S_{\Delta P}\Delta f(=1\text{Hz})}{P_0^2} \tag{11.9-5}$$

A single-mode semiconductor laser might possess a value of RIN $\approx 10^{-16}$ (or -160 db). Assuming that the detector circuit has a bandwidth of, say, $\Delta f = 10^9$ Hz, the relative mean-squared fluctuation in the detected current is

$$\frac{\overline{(\Delta i_d)^2}}{i_{d0}^2} = \frac{\overline{(\Delta P)^2}}{P_0^2} = \frac{S_{\Delta P}(f)\ \Delta f}{P_0^2} = 10^{-16} \times 10^9 = 10^{-7}$$

The RMS value of the power fluctuation is thus

$$\frac{\{\overline{[\Delta P(t)]^2}\}^{1/2}}{P_0} = 3.16 \times 10^{-4}$$

The mean-squared noise current in the output of the detector due to these fluctuations is given by (11.9-4)

$$\overline{i_{NL}^2}(t) = \frac{e^2\eta^2}{(h\nu)^2}(\text{RIN})P_0^2\Delta f \tag{11.9-6}$$

[18]In this section, we will use f to denote ''low'' (RF) frequencies and ν for optical frequencies.

Assuming as an example that $\lambda = 1.3\ \mu$m, $P_0 = 3$ mW, RIN $= 10^{-16}$ Hz^{-1}, $\Delta f = 10^9$ Hz, and $\eta = 0.6$, we obtain

$$(\overline{i_{NL}^2})^{1/2} = 5.95 \times 10^{-7}\ \text{A}$$

Example: Optical Fiber Link Design

Our task here is to determine the maximum allowed repeater spacing for an optical fiber communication link. We will assume that the optical source is a 1.3 μm GaInAsP laser ($\nu = c/\lambda = 2.31 \times 10^{14}$ Hz) and that the fiber possesses an attentuation of 0.3 db/km (corresponding to an attenuation constant $\alpha = 0.3/4.343 = 0.0691$ (km)$^{-1}$). The optical power launched into the fiber is $P_0 = 3$ mW. The channel is to transmit 10^9 bits/s so that the bandwidth of the detector circuit is taken as $\Delta f = 1/\text{period} = 10^9$ Hz. The system considerations dictate that the bit error probability at the detector output not exceed 10^{-10}. The detector output impedance is $R_L = 1{,}000\ \Omega$, and the amplifier (following the detector) noise figure is 6 db, i.e., $F = 4$ (see footnote 15).

From Figure 10-20 we determine that the signal-to-noise power ratio at the amplifier output must exceed 22 db to assure a bit error probability upon detection that is smaller than 10^{-10}. Our task is thus to calculate the signal power $\overline{i_s^2}$ and the total noise power $\overline{i_N^2}$ at the output of the detector as a function of the length L of the link.

The signal power is obtained from (11.7-11), assuming a modulation index $m = 0.5$, as

$$\overline{i_s^2} = \frac{e^2\eta^2P_0^2e^{-2\alpha L}}{2(h\nu)^2} \tag{11.9-7}$$

The total noise power at the output of the amplifier referred to its input is

$$\overline{i_N^2} = \overline{i_{NL}^2} + \overline{i_{NS}^2} + \overline{i_{NA}^2} \underset{\text{(in order)}}{=} \frac{\eta^2e^2}{(h\nu)^2}(\text{RIN})P_0^2e^{-2\alpha L}\Delta f + \frac{2\eta e^2}{h\nu}P_0e^{-\alpha L}\Delta f + \frac{4kT_e}{R_L}\Delta f \tag{11.9-8}$$

The first noise term is that due to power fluctuation (11.9-6); the second is the shot noise associated with the average current at the output of the detector $I_{d0} = \eta P_0 e \exp(-\alpha L)/(h\nu)$. The third term represents, as in (11.7-15), both the Johnson noise of the output resistor R_L as well as the amplifier output noise power (referred to its input, see footnote 15). If the temperature of the output resistor R_L is $T = 290$ K, $T_E = T + (F - 1)290 = 1160$ K.

Figure 11-26(a) shows the main elements of an optical fiber link. Figure 11-26(b) shows a plot of $\overline{i_s^2}$, $\overline{i_{NL}^2}$, $\overline{i_{NS}^2}$, and $\overline{i_{NA}^2}$ as well as the total noise power as a function of the link length L. The important thing to note is the relative change of the various powers with distance. The distance L_0, where the detected signal-to-noise power ratio is down to 22 db, is read off as $L_0 = 87$ km.[19] This distance is thus chosen as the link length. Notice, as an example, that the dominant noise contribution at $L > 33$ km is the amplifier-detector noise $\overline{i_{NA}^2}$. If the latter were reduced by, say, 3 db, the link length could be increased by 5 km, as indicated by the dashed line.

The signal-to-noise power ratio of a *p-n* diode detector is given by (11.7-16) in the case where the dominant contributions to the noise power are the amplifier noise and the Johnson (thermal) noise of the load resistance R_L in the diode output circuit. The mean-square noise current is then

$$\overline{i_N^2} \approx \frac{4kT_e\Delta f}{R_L} \tag{11.9-9}$$

The signal peak current is given by (11.7-8) for the case $m = \frac{1}{2}$ is

$$i_S = \frac{P_S e\eta}{h\nu} \tag{11.9-10}$$

where P_s is the peak pulsed optical power incident on the detector. The signal-to-noise current ratio at the amplifier output is thus

$$\frac{i_S}{(\overline{i_N^2})^{1/2}} = \frac{P_s e\eta/h\nu}{(4kT_e\Delta f/R_L)^{1/2}} \tag{11.9-11}$$

Our next problem is that of finding the minimum value of the signal power P_s so that $i_S/(\overline{i_N^2})^{1/2}$ in (11.9-11) exceeds the needed value of 12.59. We thus need to know T_e, R_L, and Δf. T_e is obtained from the given value of the amplifier noise figure ($F = 6$ dB). Taking $T = 290$ K, we obtain, using footnote 15, $T_e = 290 +$

[19]That is, $10 \log (\overline{i_s^2}/\overline{i_N^2}) = 22$.

Figure 11-26 (a) An optical fiber communication link consisting of a laser, an optical coupling system c, a fiber L/(km long), a detector D, an output resistance R_L and an amplifier A with a current gain G and a noise figure F. (b) The signal ($\overline{i_s^2}$), laser fluctuation ($\overline{i_{NL}^2}$), detector shot noise ($\overline{i_{NS}^2}$), combined Johnson-amplifier noise ($\overline{i_{NA}^2}$), and the total noise

$$\overline{i_N^2} = \overline{i_{NL}^2} + \overline{i_{NS}^2} + \overline{i_{NA}^2}$$

currents as a function of the link length L. The currents are referred to the amplifier input plane S, i.e., they correspond to output currents divided by the current gain G of the output amplifier.

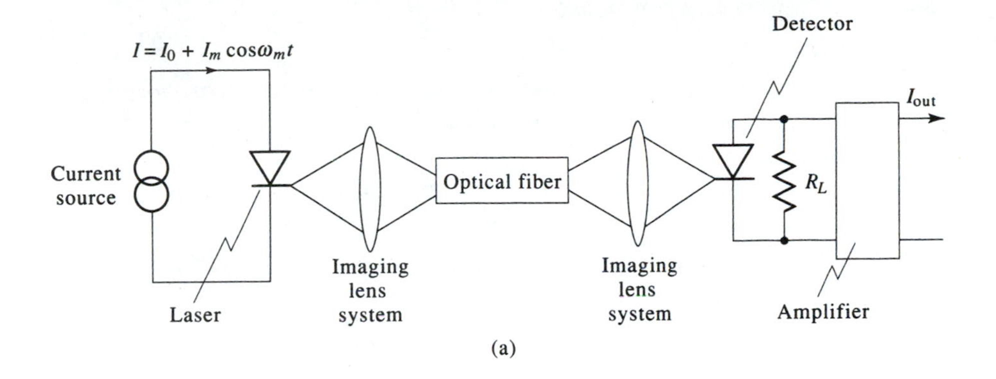

$I = I_0 + I_m \cos\omega_m t$
Detector
I_{out}
Current source
Optical fiber
R_L
Laser
Imaging lens system
Imaging lens system
Amplifier

(a)

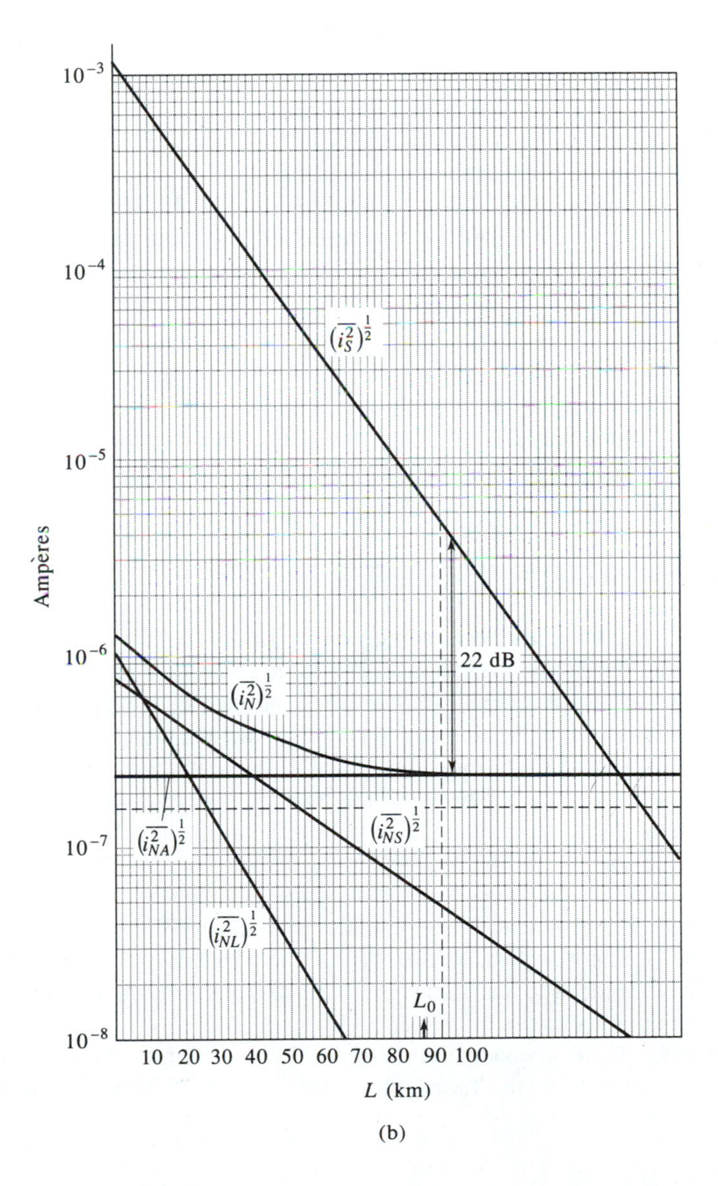

10^{-3}
10^{-4}
10^{-5}
10^{-6}
10^{-7}
10^{-8}
Ampères
$(\overline{i_S^2})^{\frac{1}{2}}$
22 dB
$(\overline{i_N^2})^{\frac{1}{2}}$
$(\overline{i_{NA}^2})^{\frac{1}{2}}$
$(\overline{i_{NS}^2})^{\frac{1}{2}}$
$(\overline{i_{NL}^2})^{\frac{1}{2}}$
L_0
10 20 30 40 50 60 70 80 90 100
L (km)

(b)

$(4 - 1)290 = 1160$. In order to achieve this bandwidth, the load resistance R_L must not exceed (see 11.7-18) the value

$$R_L = \frac{1}{2\pi \Delta f C} \tag{11.9-12}$$

where C is the total output capacitance given as 3×10^{-12} F. Using the above value of Δf and C, we obtain

$$R_L \leqslant 53 \text{ ohms}$$

We return now to (11.9-11), which, using $\eta = 0.5$, $\lambda = 1.35$ μm, $i_S/(\overline{i_N^2})^{1/2} = 12.59$, yields

$$P_S \cong 2.55 \times 10^{-5} \text{ watt}$$

for the minimum power input to the photodiode.

The total transmission loss in the 50 km fiber is 20 dB. We will assume that an additional 4 dB loss is caused by coupling the laser output to the fiber and at the fiber output so that the total loss is 24 dB (that is, 251). The laser power output must thus exceed

$$P_{\text{laser}} = 6.4 \times 10^{-3} \text{ watt}$$

which is a reasonable power level for CW diode lasers.

If the fiber had been substantially lossier than in the above example, we could still have met our design specifications by using an avalanche photodiode.

11.10 INFRARED IMAGING AND BACKGROUND-LIMITED DETECTION (25–28)

Arrays of cooled infrared detectors based mostly on photoconductive semiconductors such as mercury cadmium telluride (HgCdTe) have become increasingly important elements in the fast developing technology of infrared imaging and detection. The application areas served by this new technology include tumor detection, the mapping of earth resources by orbiting satellites, ''spy'' satellites, and nighttime ''seeing.'' We will not concern ourselves here with the system aspects of these applications but rather with the basic noise physics of a single element that is prerequisite to system considerations. The concepts involved here are the same as those we have encountered in the early sections of this chapter, but the operational considerations merit a dedicated treatment. To be specific, we will focus our discussion to doped, say, *n*-type, photoconductors, such as HgCdTe, in which the optical input field causes excitation of electrons to the conduction band so that the signal current is due to the drift of the excited carriers.

Consider the photoconductive detector shown in Figure 11-27. The detected radiation is incident on the ''face'' whose area is A. The thickness of the detector is

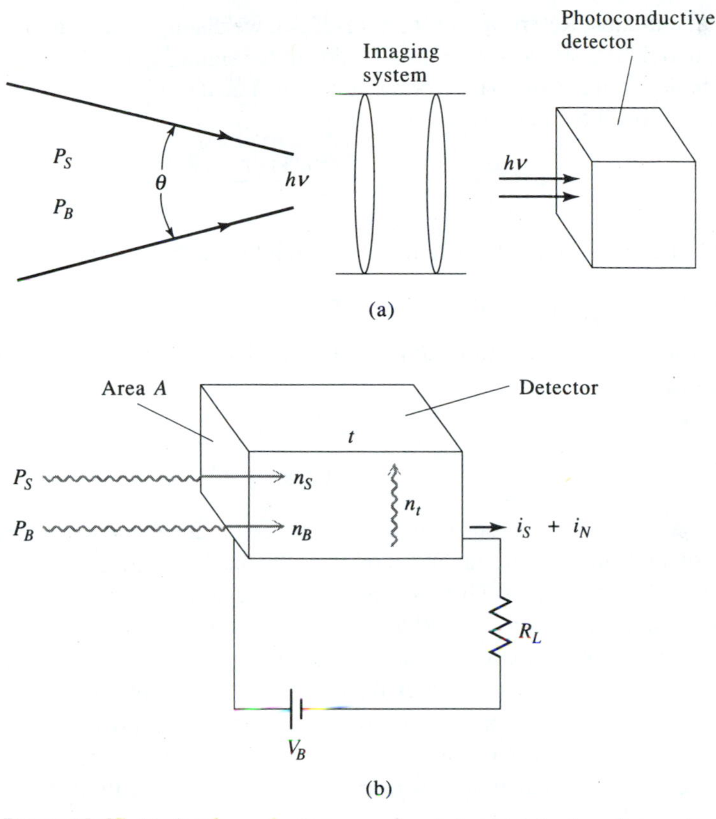

Figure 11-27 (a) A schematic diagram of an infrared detecting element intercepting radiation from an acceptance angle θ. (b) A detailed view of the photoconductor.

t. The optical signal power input to the detector is P_s. This power gives rise to a signal current given according to (11.5-2) by

$$i_s = \frac{\eta P_s e}{h\nu_{\text{opt}}} \left(\frac{\tau_0}{\tau_d} \right) \tag{11.10-1}$$

where τ_0 is the lifetime of the photo-excited carriers, $\tau_d = \frac{t}{\bar{v}}$ is the transit time for a carrier with a velocity $\bar{v}$. ν_{opt} is the frequency of the optical beam. η is the fraction of the incoming photons that are usefully absorbed in the photoconductor. τ_0 is the average lifetime of a photo-excited carrier. In addition to the signal current, we have two other major sources of noise currents that are not related to the signal.

The first is the shot noise associated with the drift under the influence of the applied external field of thermally excited (minority) carriers, while the second is the shot noise due to carriers excited by the ever present incoming background optical radiation. Referring to Figure 11-27(b), we distinguish in our photoconductor three populations of carriers: n_s (cm^{-3}) due to the signal, n_t due to thermal excitation, and n_B due to incoming background radiation. The density n_B of carriers excited by the background radiation is

$$n_B = \frac{(P_B/A)\eta\tau_0}{h\nu_{\text{opt}}t} \tag{11.10-2}$$

This last expression can be obtained by setting the rate of photon absorption per unit volume $\frac{(P_B\eta)}{(h\nu_{\text{opt}}At)}$ equal to the rate $\frac{n_B}{\tau_0}$ of minority carrier recombination. The incident background power P_B is most often that of the background blackbody radiation, in which case

$$\mathcal{I}_B \equiv \frac{P_B}{A} = \frac{2\pi h\nu_{\text{opt}}^3\Delta\nu(\sin^2\theta/2)}{C^2(e^{h\nu_{\text{opt}}/kT_B}-1)} \tag{11.10-3}$$

where A is the cross-sectional area of the detector, $\Delta\nu$ is the optical bandwidth of the radiation allowed into the detector, T_B is the background temperature, while θ is the angle [see Figure 11-27(a)] within which radiation is accepted by the detector. In practice the acceptance angle θ and the background temperature T_B are dictated by the application. A reasonable strategy in such a case is to cool the detector to the point where $n_t < n_B$. This renders the contribution of n_t to the shot noise equal to that of n_B so that additional cooling will not materially improve the signal-to-noise ratio at the output. Since the signal-to-noise ratio under this condition is determined by the background radiation, it is referred to as background limited infrared performance (BLIP).

Let us assume that a BLIP condition has been achieved and proceed to calculate the resulting detector performance. From (11.5-9) the mean-squared output noise current is (in the limit $\nu\tau_0 \ll 1$)

$$\begin{aligned}\overline{i_{NB}^2} &= 4e\bar{I}_B\left(\frac{\tau_0}{\tau_d}\right)\Delta f \\ &= 4e(n_Be\bar{v}A)\frac{\tau_0}{\tau_d}\Delta f\end{aligned} \tag{11.10-4}$$

Where $\bar{I}_B = n_Be\bar{v}A$ is the average current due to the background radiation–excited carriers and Δf is the bandwidth of the (electronic) detection circuit. Substituting for n_B from (11.10-2) gives

$$\overline{i_{NB}^2} = \frac{4e^2P_B\eta}{h\nu_{\text{opt}}}\left(\frac{\tau_0}{\tau_d}\right)^2\Delta f \tag{11.10-5}$$

The minimum detectable signal, also known as the noise equivalent power (NEP) of the detector, is that value of the signal power P_s for which

$$\overline{i_s^2} = \overline{i_N^2}$$

Using (11.10-1) and (11.10-5) we can solve for the minimum detectable power when the main noise contribution is due to the background radiation, i.e., $n_t < n_B$

$$(\text{NEP})_B = P_s(\overline{i_s^2} = \overline{i_{NB}^2}) = 2\sqrt{\frac{A\mathcal{I}_B h\nu_{\text{opt}}\Delta f}{\eta}} \tag{11.10-6}$$

A common figure of merit used in the infrared imaging community to describe detector sensitivity is the specific peak detectivity D* (''Dee'' star) defined as

$$\text{D}^* = \frac{\sqrt{A\Delta f}}{\text{NEP}}$$

When the detector is cooled sufficiently so that it is background limited, D* becomes

$$\text{D}_B^* = \frac{\sqrt{A\Delta f}}{(\text{NEP})_B} = \frac{1}{2}\sqrt{\frac{\eta}{h\nu_{\text{opt}}\mathcal{I}_B}} \tag{11.10-6a}$$

where, to remind us, the B subscript stands for the background limited condition and Δf is bandwidth of the detection circuit including the photoconductive element.

In a detector limited by thermal excitation of carriers, i.e., one where $n_t > n_B$ we have

$$\bar{I}_t = n_t e\bar{v}A \tag{11.10-7}$$

$$\overline{i_{Nt}^2} = 4e\bar{I}_t\left(\frac{\tau_0}{\tau_d}\right)\Delta f = 4e(n_t e\bar{v}A)\left(\frac{\tau_0}{\tau_d}\right)\Delta f \tag{11.10-8}$$

Equating $\overline{i_{Nt}^2}$ to $\overline{i_s^2}$ as in (11.10-6), we obtain

$$(\text{NEP})_t = \frac{2h\nu_{\text{opt}}}{\eta}\sqrt{\left(\frac{\tau_d}{\tau_0}\right)n_t\bar{v}A\Delta f}$$

and

$$\text{D}_t^* = \frac{\sqrt{A\Delta f}}{(\text{NEP})_t} = \frac{\eta}{2h\nu_{\text{opt}}}\sqrt{\frac{\tau_0/\tau_d}{n_t\bar{v}}} = \frac{\eta}{2h\nu_{\text{opt}}}\sqrt{\frac{\tau_0}{n_t t}} \tag{11.10-9}$$

where, in the last expression, we used $\tau_d \equiv t/\bar{v}$.

It is obvious that the condition $n_t \leq n_B$ is equivalent to

$$(\text{NEP})_t \leq (\text{NEP})_B \quad \text{or} \quad \text{D}_t^* \geq \text{D}_B^* \tag{11.10-10}$$

A key issue in infrared detection is to determine to what temperature a detector element need be cooled to be background limited. To answer this question we need to know n_B and the dependence of n_t on the material parameters and the temperature.

As an example consider an infrared detector as shown in Figure 11-28(a) in which the photoconductive medium is a GaAs/GaAlAs superlattice [29]. It is based on excitation of electrons from a confined ''quantum-well'' state (see Section 16.1) to continuum (unconfined) states where they are free to conduct. These wells consist of thin layers (~100Å) of crystalline GaAs layers sandwiched between higher energy gap $Ga_{1-x}Al_xAs$ crystalline layers.

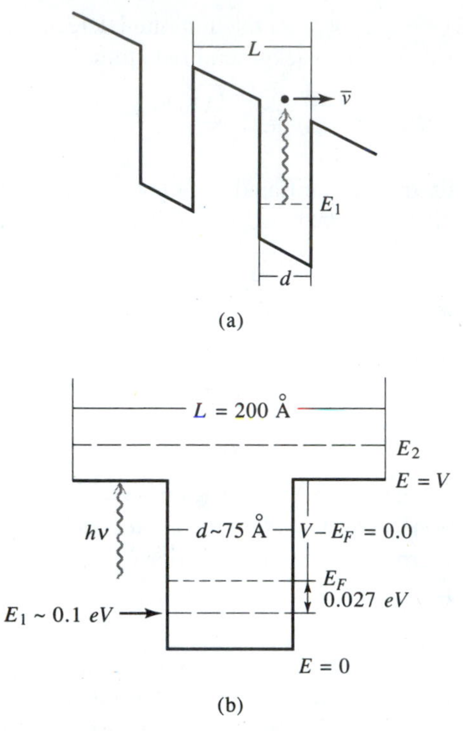

Figure 11-28 Schematic illustration of a typical quantum well detector structure (only two wells are shown) with the relavant energies. In the upper part of the figure (a), the quantum well detector under bias is shown. The second part of the figure (b) shows the relevant distances and energies. The sub-band levels are given by E_1 and E_2.

In this case [25],

$$n_t = \frac{m^*kT}{\pi\hbar^2 L} e^{-(V-E_F)/kT} \tag{11.10-11}$$

where L is the width of the unit cell in Figure 11-28(a), m^* is the carrier (electron) effective mass, E_F the Fermi energy of the material, and V the depth of the quantum well. Using the data of Figure 11-28(b) and $m^* = 0.067m_e$, $\frac{\tau_0}{\tau_d} = 0.5$, $\lambda = \frac{c}{\nu_{\text{opt}}} =$ 10 μm, we can use (11.10-9) to plot D_t^* vs. the detector temperature T. The result is shown in Figure 11-29(a).

To use this curve we need first to obtain a value for D_B^* using (11.10-3) and (11.10-6a). We then find the temperature in Figure 11-29(a) where $D_t^* = D_B^*$. As an example, given $D_B^* = 10^{12}$ we find from the figure that the quantum-well detector needs to be cooled to $T < 48$ K in order to become background limited.[20] Figure

[20]In practice cooling the detector below the liquid N_2 temperature, 77.7°K, is expensive and is reserved to very demanding applications.

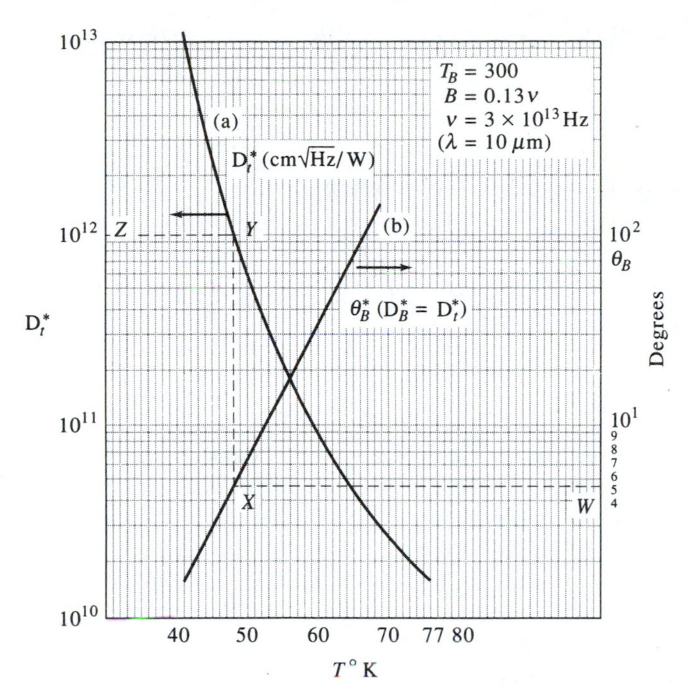

Figure 11-29 Curve (a) is a plot of the thermal-excitation-limited detectivity D_t^* at $\lambda = 10$ μm vs. temperature of the GaAs quantum well detector. Curve (b) is a plot of the acceptance angle θ_B which results in $D_B^* = D_t^*$ (BLIP) for the background conditions stated in the figure.

11-29 curve (b) is a plot based on Equations (11.10-3, 11.10-6, 11.10-9) of the acceptance angle θ_B for which D_B^* is equal to D_t^*. Further increases of θ_B will thus cause the detector to be background limited. In the figure we used a background temperature $T_B = 300$ K, $\nu_{\text{opt}} = 3 \times 10^{13}$ ($\lambda = 10$ μm), $\Delta\nu = 0.13\nu_{\text{opt}}$. As an example, if our detector has a D_t^* of 10^{12}, then the BLIP condition obtains when $\theta_B = 4.6°$.

We can also use the curves of Figure 11-29 in reverse to find D_B^* for a given θ_B. For instance, for $\theta_B = 4.6°$ we follow the sequence $W \rightarrow X \rightarrow Y \rightarrow Z$ to find $D_B^* = 10^{12}$; in the process we also learn that the detector temperature must be below 48 K in order to be background limited.

The most widely used material for infrared imaging near $\lambda = 10$ μm is the semiconductor HgCdTe [28] (MCT) whose composition can be adjusted to yield an energy of a photon with $\lambda \sim 10$ μm. The photoconduction in this case is due to excitation across the energy gap of the semiconductor. Typical MCT photoconductive detectors use N-type material so that the carriers responsible for the signal (and noise) are the (minority) holes. Our theoretical discussion up to this point applies if we merely take n_B, n_t, and n_s, respectively, as the density of holes excited by the background radiation, thermal process, and the "signal" radiation.

The most important task that confronts the infrared detector scientist is to develop materials that enable background-limited performance at the *highest possible temperature.* The background limit condition $n_t = n_B$ can be written using (11.10-2) as

$$\frac{\mathcal{I}_B \eta \tau_0}{h\nu_{\text{opt}} t} = n_t \tag{11.10-12}$$

Since n_t invariably increases with T, it follows from (11.10-12) that *the temperature for background-limited operation increases with the carrier lifetime* τ_0. In the example given above, the temperature T for background-limited detection condition $n_t = n_B$ is given, according to (11.10-11), by the condition $n_B = n_t$, i.e., by the value of T satisfying

$$\frac{\mathcal{I}_B \eta \tau_0}{h\nu_{\text{opt}} t} = \frac{m^* kT}{\pi \hbar^2 L} e^{-(V-E_F)/kT} \tag{11.10-13}$$

In a typical HgCdTe at 77 K, the carrier lifetime is $\tau_0 \sim 10^{-6}$ s while in our quantum well detector $\tau_0 \sim 10^{-11}$ s [the time for an excited carrier to drop in energy below the top of the well thus becoming immobile (trapped). This happens after the emission of only a few optical-branch phonons by the excited carrier.]

It follows that HgCdTe is background limited and thus has an NEP, described by (11.10-6), at a higher temperature than a GaAs/GaAlAs detector used in the above example. To illustrate this point, we show in Figure 11-30 a plot of the thermal generation current that is the rate of decay (per unit of incidence area) of thermally excited carriers[21]

$$I_{t-g} = \frac{n_t t}{\tau_0} \; [\text{photons}/(\text{cm}^2 - \text{s})] \tag{11.10-14}$$

where n_t is the density of thermally excited carriers (for HgCdTe it is the minority carrier density [26–28]). I_{t-g} is commonly used in the infrared imaging community to compare different materials since at the background limit (BLIP) it is equal to the rate $\eta \mathcal{I}_B / h\nu_{\text{opt}}$ of (absorbed) background photons (per cm^2) incident on the detector. Since the latter rate is determined by system considerations (see Equation (11.10-3) and the following discussion), given the background absorbed photon flux $\eta \mathcal{I}_B / h\nu_{\text{opt}}$ we can determine at a glance the temperature to which our detector needs to be cooled to achieve BLIP condition or, equivalently, the temperature to which the detector needs to be cooled for BLIP operation at a given background photon flux.

We note by comparing Figure 11-30(a) to 11-30(b) that at given T, I_{t-g} in HgCdTe is ~6 orders of magnitudes smaller than in the GaAs/GaAlAs detector, reflecting mainly the difference in carrier lifetime τ_0. It is ~10^{-11} s in GaAs/GaAlAs [25] and ~10^{-6} s in HgCdTe. As an example consider a system subject to a background photon flux near $\lambda = 10$ μm of 3×10^{13} cm^{-2} − s^{-1}. The HgCdTe detector becomes, according to Figure 11-30(a), background limited at ~80 K while the GaAs/GaAlAs detector [Figure (b)] needs to be cooled to ~45 K. If we calculate the D_t^* corresponding to this incoming background flux, we obtain using, for example, Figure 11-29(a) at 45 K, $D_t^* = 2.2 \times 10^{13}$ cm-Hz$^{1/2}$-W^{-1}.

[21] This rate is equal, at thermal equilibrium, to the rate at which the "thermal" carriers are generated.

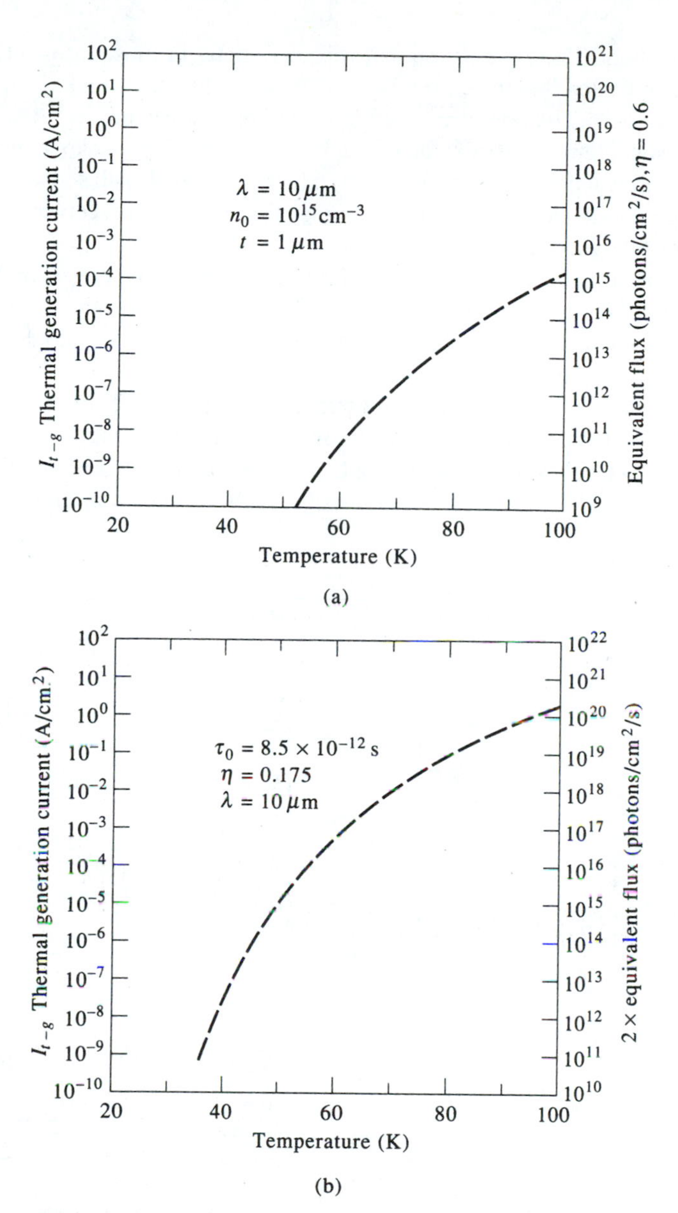

Figure 11-30 (a) Thermal equivalent current for HgCdTe IR devices. On the left ordinate it is expressed in A/cm^2 while on the right ordinate it is expressed in the equivalent arrival rate of (10 μm) photons/cm^2. (b) Thermal equivalent generation current for GaAs/AlGaAs.

11.11 OPTICAL AMPLIFICATION IN FIBER LINKS

Optical amplification in fiber links [31–33] has recently been recognized as having major system implications for very long distance transmission of information (>1000 km) using optical fibers and for distribution systems involving a large number of subscribers. These purely optical repeaters may, in most cases, obviate the need for the repeater stations currently used that involve detection, electronic amplification, and remodulation of a (new) launched optical beam.

The raison d'être for the optical amplifiers is that they make it possible to maintain the optical power at sufficiently high levels along the path so that the signal-to-noise ratio (SNR) degradation due to signal shot noise and receiver noise is reduced to practical inconsequence.

A new and dominant noise source, amplified spontaneous emission, however, is introduced by the optical amplifier [35], and its effect on the signal-to-noise ratio (SNR) of the detected signal current will be considered below. Before doing so, we will briefly review the relevant physics of the amplifier.

The most common amplifier uses a transition at $\lambda = 1.535\ \mu m$ in an E_r^{3+} ion introduced as a dopant into a silica fiber [34, 35]. This wavelength has assumed a very exalted status because of the low optical losses of silica fibers, in this spectral region (see Figure 3-19). The pertinent energy levels are shown in Figure 11-31(a).

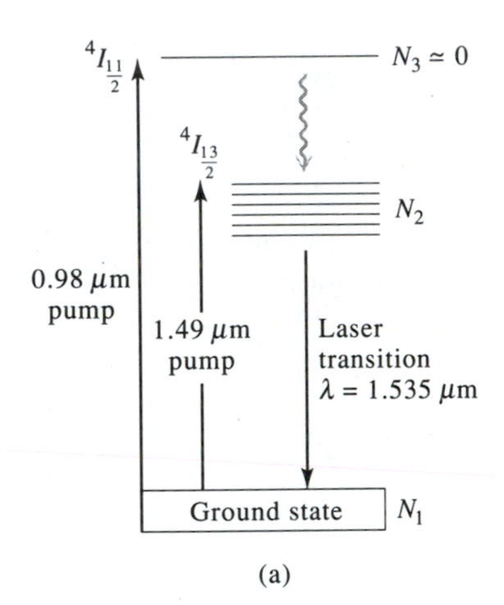

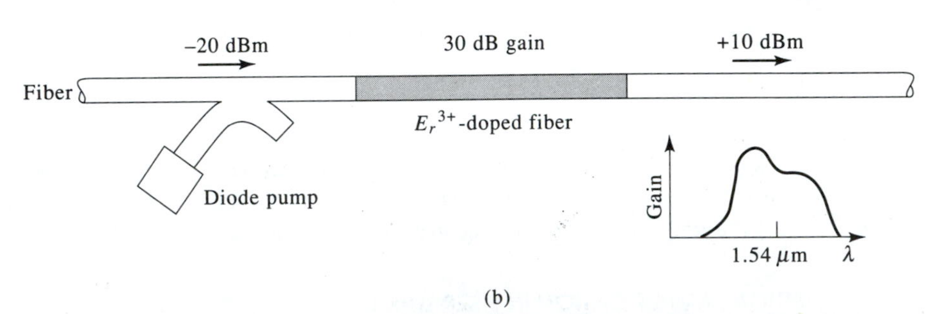

Figure 11-31 (a) The pertinent energy level diagram of E_r^{3+} in silica for pumping at $\lambda = .98$ μm (preferred). (b) A schematic diagram showing the amplifying fiber spliced into the transmission fiber and the method for coupling the pump radiation into the fiber.

The laser transition can be pumped by radiation at $\lambda \sim 0.98$ μm or $\lambda \sim 1.49$ μm as shown. This pumping field is usually obtained from semiconductor lasers and is coupled into the amplifying fiber whose length is typically between a few meters and a few tens of meters. A schematic diagram of the amplifier configuration is shown in Figure 11-31(b). The fiber amplifier section can be spliced smoothly into the fiber. A plot of the gain vs. signal wavelength is shown in Figure 11-32(a).

The main effect of the optical amplifier on the SNR of the detected signal is to add, upon detection, a noise current component, at frequencies near that of the signal current. This noise is due to beating between the amplified (optical) spontaneous emission (ASE) power of the amplifier and the signal optical field.

Figure 11-32(b) shows the two spectral windows of the amplified output spontaneous emission power that beat (at the detector) with the optical signal field S_0 at ω_0 to generate an output noise current at some arbitrary frequency ω_m. This mechanism thus gives rise to a spectral continuum of RF noise current extending from dc to approximately $\Delta\omega_{\text{gain}}$, the width of the (amplified) spontaneous emission spectrum. To estimate this current we first need to obtain an expression for the optical spontaneous emission power at the output of an optical amplifier. This topic is the subject of Appendix C. The main result, Equation C-8, is that the (amplified) spontaneous emission power in a *single* mode within a spectral bandwidth $\Delta\nu_{\text{opt}}$ at the output of an optical amplifier is [36]

$$F_0 = \mu h\nu\, \Delta\nu_{\text{opt}}(G - 1) \tag{11.11-1}$$

where $G = \exp(\gamma l)$ is the power gain of the optical amplifier with a distributed gain γ and length l and

$$\mu = \frac{N_2}{N_2 - N_1\dfrac{g_2}{g_1}} \tag{11.11-2}$$

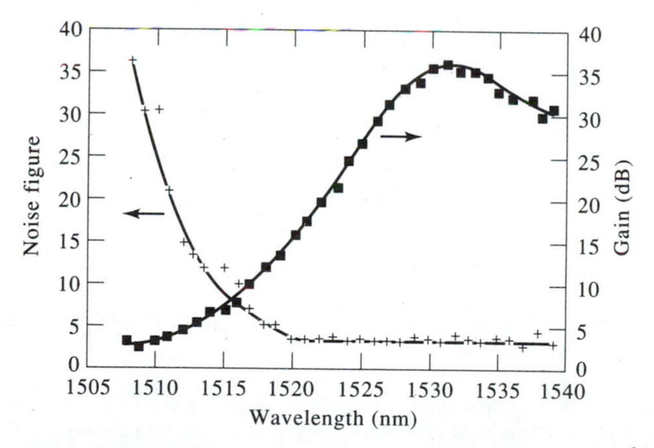

Figure 11-32(a) Noise factor and gain spectrum of the silica E_r^{3+} fiber amplifier for a constant pump power of 34.2 mW at 0.98 μm. (After Reference [35].)

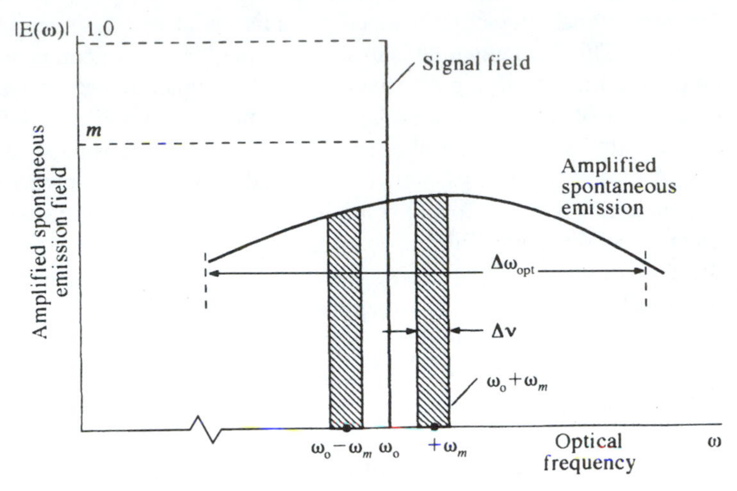

Figure 11-32 (b) At the output of the laser, the amplified spontaneous emission fields near $\omega_0 + \omega_m$ and near $\omega_0 - \omega_m$ will, each, beat (mix) at the detector with the amplified optical signal field at ω_0 to yield radio frequency (RF) currents with frequencies near ω_m. These currents, which occupy a spectral width of $\Delta\omega_{\text{opt}}$, cannot be separated from those due to intentional (signal) modulation of the intensity at ω_m and thus constitute RF noise.

is the atomic inversion factor of the transition. It accounts for the larger value of N_2, and hence larger spontaneous emission power, in atomic (amplifier) systems in which $N_1 \neq 0$.[22]

If we denote the output optical power at ω_2 as S and that of the spontaneous emission at ω_1 as F, then the beat current component with a frequency $\omega_{\text{m}} = \omega_2 - \omega_1$ is (see Equation 11.4-4)

$$i = \frac{Se\eta}{h\nu}\left[1 + \frac{F}{S} + 2\sqrt{\frac{F}{S}}\cos(\omega_{\text{m}}t + \Phi_{\text{ASE}} - \Phi_s)\right] \tag{11.11-3}$$

where ϕ_{ASE} and ϕ_S are the phases of the ASE field and the signal optical field, respectively. The mean-squared beat current is then

$$(\overline{i^2})_{\text{ASE-signal}} = 2FS\left(\frac{e\eta}{h\nu}\right)^2 \tag{11.11-4}$$

which, using (11.11-1) and putting $\Delta\nu_{\text{opt}} = \Delta\omega_{\text{opt}}/2\pi = 2\Delta f_{\text{sig}}$, yields[23]

$$(\overline{i^2})_{\text{ASE-signal}} = \frac{4e^2\eta^2 S(G-1)\mu\Delta f_{\text{sig}}}{h\nu} \tag{11.11-5}$$

In the remainder we will drop the subscript "sig" and use Δf only.

[22]In a laser the gain per pass is given by $G = \exp[a(N_2 - N_1)L_{\text{amp}}]$ where L_{amp} is the length and a is a constant depending on the atoms. A large N_1 thus causes a larger N_2 for a given gain. The SE power is proportional to N_2.

[23]Two ASE frequency bands, each with a width $\Delta\nu_{\text{sig}}$, one above and one below the signal frequency contribute incoherently to the beat power so that the effective $\Delta\nu_{\text{opt}} = 2\Delta f_{\text{sig}}$.

Consider an optical in-line amplifier as shown in Figure 11-33. The input signal power is S_0, and it enters the amplifier in a *single* transverse (usually the fundamental) fiber mode. The amplified output signal is GS_0, while F_0, as given by (11.11-1), represents the (optical) amplified spontaneous emission power at the output, which is generated within the amplifier in a band $\Delta\nu$. If we were to detect the signal at the input to the amplifier, the main noise contribution would, in an ideal case, i.e., a noiseless receiver, be that of the signal shot noise so that the signal-to-noise power ratio (SNR) at the input to the amplifier is

$$\mathrm{SNR}_{\mathrm{in}} = \frac{\left(\dfrac{S_0 e}{h\nu}\right)^2}{2e^2 \dfrac{S_0}{h\nu} \Delta f} = \frac{S_0}{2h\nu\Delta f} \tag{11.11-6}$$

where we assume 100% detection efficiency $\eta = 1$ for simplicity.

The detected signal "power"[24] at the output is

$$(\overline{i^2})_{\mathrm{out}} = \left(\frac{GS_0 e}{h\nu}\right)^2 \tag{11.11-7}$$

while the noise power is that of the ASE-signal noise (11.11-5) and the shot noise

$$(\overline{i^2_{\mathrm{shot}}})_{\mathrm{out}} = \frac{2e^2 GS_0}{h\nu} \Delta f \tag{11.11-8}$$

The noise current component that is due to beating of ASE frequencies with themselves is proportional to F_0^2 and can be made to be negligible compared to the ASE-signal current if the signal power $S(z)$ is not allowed to drop too far and/or by optical filtering. We have neglected for similar reasons the shot noise due to the ASE. The (S/N) ratio at the output of the amplifier is thus

$$\mathrm{SNR}_{\mathrm{out}} = \frac{\left(\dfrac{GS_0 e}{h\nu}\right)^2}{\dfrac{2e^2 GS_0}{h\nu} \Delta f + \dfrac{4e^2 G(G-1) S_0 \mu \Delta f}{h\nu}} \tag{11.11-9}$$

[24]The "power" everywhere is taken as the mean square of the current. Since our final results involve only (signal-to-noise) power ratios, this procedure is justified.

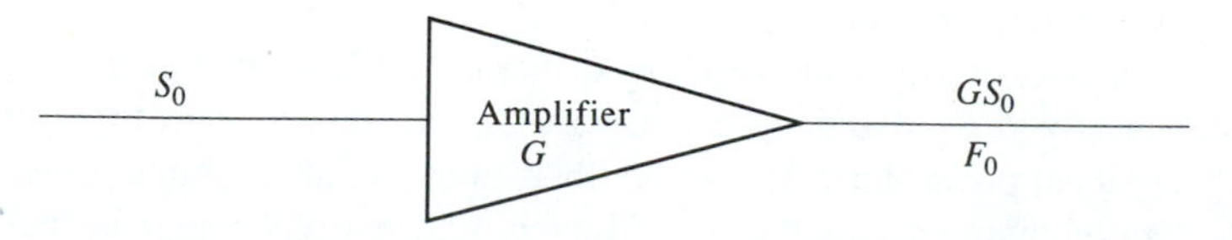

Figure 11-33 An optical amplifier with a power gain G and an input signal power S_0. F_0 is the total power of the amplified spontaneous emission (ASE) at the output of the amplifier in the appropriate bandwidth $\Delta\nu$.

where we assumed a 100 percent detector quantum efficiency. For large gain $G \gg 1$, the second term in the denominator of (11.11-9), dominates, and

$$\mathrm{SNR}_{\mathrm{out}} \approx \frac{S_0}{4\mu h\nu\Delta f} \tag{11.11-10}$$

The ratio of the input (SNR) to the output value is thus

$$\frac{\mathrm{SNR}_{\mathrm{in}}}{\mathrm{SNR}_{\mathrm{out}}} \approx 2\mu$$

which in an ideal, four-level ($N_1 = 0$, $\mu = 1$) amplifier is equal to 2. The single high-gain optical amplifier will thus degrade the SNR of the detected output by a factor of 2 (3 db). We recall that this degradation is tolerated only in order to save the signal from the, far worse, fate of succumbing, in its attenuated state, to the noise of the receiver. An experimental verification of the 3 db limit is shown in Figure 11-32(a).

In a very long (100 km) fiber link, we will need to amplify the signal a number of times. We will consequently develop in what follows a formalism for treating systematically cascades of amplifiers.

A generalization of the expression (11.11-9) for the SNR of the detected signal at an arbitrary point z along the link is to write

$$\mathrm{SNR}(z) = \frac{\left[\dfrac{eS(z)}{h\nu}\right]^2}{\dfrac{2e^2S(z)\Delta f}{h\nu} + \dfrac{4e^2F(z)S(z)}{(h\nu)^2} + \dfrac{4kT_e\Delta f}{R}} \tag{11.11-11}$$

where the last term in the denominator represents the mean-squared thermal noise current of the receiver (at point z) whose effective noise temperature is T_e. R is the output impedance of the detector including the receiver's input impedance. Equation (11.11-11) neglects, again, the shot noise due to the ASE, the ASE-ASE beat noise, and intensity fluctuation noise of the source laser. If the signal power $S(z)$ can be maintained above a certain level by repeated amplification, we can neglect the receiver noise term. Under these realistic circumstances, the SNR expression (11.11-11) becomes

$$\mathrm{SNR}(z) = \frac{S^2(z)}{2S(z)h\nu\Delta f + 4S(z)F(z)} \tag{11.11-12}$$

$S(z)$ is the signal power at z, while $F(z)$ is the total ASE power at z originating in *all* the preceding amplifiers ($z' < z$).

Let us next consider the realistic scenario of a long fiber with amplifiers employed serially at fixed and equal intervals (z_0), as illustrated in Figure 11-34.

The signal power level $S(z)$ at the fiber input and at the output of each amplifier is S_0. The signal is attenuated by a factor of $L \equiv \exp(-\alpha z_0)$ in the distance z_0 between amplifiers and is boosted back up by the gain $G = L^{-1} = e^{\alpha z_0}$ at each amplifier to the initial level S_0. The spontaneous emission power $F(z)$ is attenuated

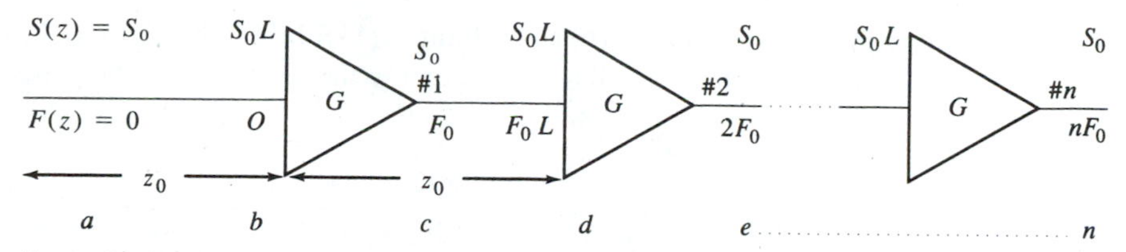

Figure 11-34 A fiber link with periodic amplification. The spontaneous emission power F(z) and the signal power S(z) at the amplifiers' input and output planes are indicated.

by a factor L between two neighboring amplifiers and increases by an increment of F_0 at the output of each amplifier. We employ Equation (11.11-1) to calculate the SNR of the detected current at the output of the nth amplifier. Assuming $G \gg 1$, the result is

$$\mathrm{SNR}_n = \frac{S_0}{2h\nu\Delta f[1 + 2n\mu(e^{\alpha z_0} - 1)]} \quad \text{(11.11-13)}$$

where, because of the high short noise and ASE levels, we neglected the thermal receiver noise. When $\exp(\alpha z_0) = G \gg 1$, we find a z^{-1} (more exactly an n^{-1}) dependence of the SNR rather than the $\exp(-\alpha z)$ dependence of a fiber without amplification in which the main noise mechanism is shot noise. The physical reason for this difference is that the repeated amplification keeps the signal level high as well as the level of the signal-ASE beat noise. The latter is kept well above the signal shot noise. A fixed amount of beat noise power is thus added at each stage leading to the inverse distance dependence of the SNR.

Equation (11.11-13) suggests that the SNR at z can be improved by reducing z_0, i.e., by using smaller intervals between the amplifiers which, of course, entails reducing the gain $G = \exp(\alpha z_0)$ of each. Let us take the limit of Equation (11.11-13) as $z_0 \to 0$, i.e., the separation between amplifiers tends to zero. In this limit the whole length of the fiber acts as a distributed amplifier with a gain constant $g = \alpha$, just enough to maintain the signal at a constant value. Since $S(z)$ is a constant, we need only evaluate the ASE optical power $F(z)$ in order to obtain, using (11.11-12), an expression for the SNR at z. To find how much noise power is added by the amplifying fiber, we consider a differential length dz. It may be viewed as a discrete amplifier with a gain of $\exp(gdz)$ so that its contribution to $F(z)$ is given by (11.11-1) as

$$dF = (e^{g(dz)} - 1)\mu h\nu\Delta f \quad \text{(11.11-14)}$$

or

$$\frac{dF}{dz} = g\mu h\nu\Delta f, \; F(z) = g\mu h\nu\Delta f z \quad \text{(11.11-15)}$$

where, since no spontaneous emission is present at the input, we used $F(0) = 0$. Using (11.11-15) in (11.11-11) and taking $S(z) = S_0$, $g = \alpha$ results in

$$\mathrm{SNR}(z) = \frac{S_0}{2[1 + 2\mu\alpha z]h\nu\Delta f} \quad \text{(11.11-16)}$$

We can also obtain (11.11-16) as the limit of (11.11-13) when $z_0 \to 0$. It is interesting to compare the (ideal) distributed amplifier to the discrete amplifier case of Equation (11.11-13)

$$(\text{SNR})(z) = \frac{S_0}{2[1 + 2(z/z_0)\mu(e^{\alpha z_0} - 1)]\, h\nu\Delta f} \qquad \textbf{(11.11-17)}$$

where we used $G = \exp(\alpha z_0)$ and $n = z/z_0$.

Figure 11-35 shows plots of the ideal continuous amplification case described by Equation (11.11-16) as well as two cases of discrete amplifier cascades [Equation (11.11-13)]. The advantage of continuous amplification compared to, say, amplification every α^{-1} is seen to be less than 2 db so that the latter may be taken as a practical optimum configuration. In a low-loss optical fiber, say with $\alpha = 0.2$ db/km, the distance between amplifiers that are placed every α^{-1} km would be 21.7 km. Figure 11-36 shows the SNR of the detected signal along a realistic link for the case of (a) continuous amplification; (b) discrete amplifiers spaced by $z_0 = \alpha_0^{-1}$; and (c) for the case of no amplification at all. The launched power is $P_0 = 5$ mW, $\lambda = 1.55\ \mu$m, $\Delta f = 10^9$ Hz, and $\alpha = 0.2$ db/km. Curve (b) is to be read only at multiples of $z = \alpha^{-1} = 21.7$ km, which are the output planes of the optical amplifiers. Curve

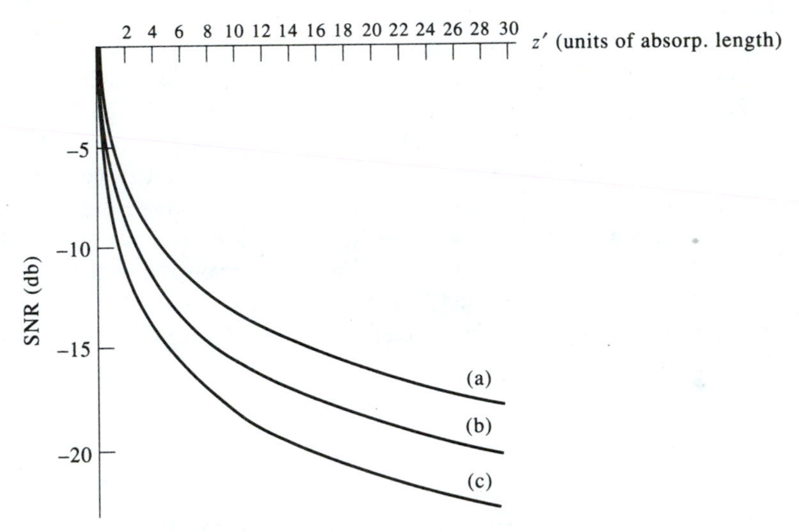

Figure 11-35 A universal plot of the degradation of the SNR compared to the initial ($z = 0$) value in the cases of (a) continuous amplification ($g = \alpha$), ($\mu = 1$); (b) periodic amplification every $z_0 = \alpha^{-1}$ ($z' = 1, 2, 3, \ldots$), ($\mu = 1$), (curve is to be read only at $z' = 1, 2, 3 \ldots$); and (c) periodic amplification every $z_0 = 2\alpha^{-1}$ ($z' = 2, 4, 6, \ldots$), ($\mu = 1$), (curve is to be read only at $z' = 2, 4, 6, \ldots$).

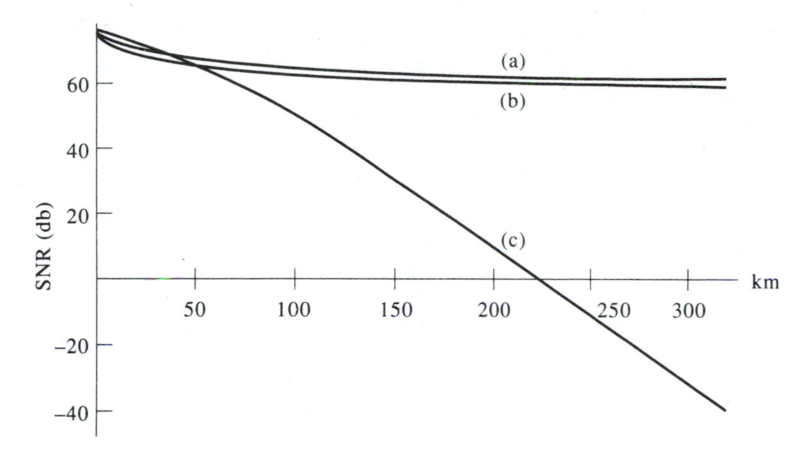

Figure 11-36 SNR of detected signal in a fiber link with (a) a continuous amplifier $g = \alpha$, ($\mu = 1$); (b) discrete amplifiers employed every absorption length $\alpha^{-1} = 21.7$ km (0.2 db/km fiber loss), ($\mu = 1$) (curve is to be read only at multiples of 21.7 km); and (c) no optical amplification and detection with a receiver with a noise figure of 4 db. The power launched into the fiber is 5 mW, the fiber loss is 0.2 db/km, $\lambda = 1.55$ μm, the detection bandwidth is $\Delta f = 10^9$, and the detector load impedance is 1000 ohms.

(c) assumes detection with a receiver with $T_e = 725$ K ($F = 4$ db) and an input impedance of 1000 Ω.

We note that if, for example, we need to maintain a SNR exceeding 50 db, we must use a fiber link shorter than 100 km if no amplifier is used, but if laser amplifiers are used every, say, $z_0 = \alpha^{-1}$ (=21.7 km), fiber length in excess of 1000 km can be employed.

Serious consideration has also been devoted to the use of semiconductor (SC) laser amplifiers [31]. These are identical in their construction to semiconductor laser oscillators, which are discussed in Chapter 15, except that the facets are coated with antireflection layers to reduce optical feedback and thus prevent oscillation from taking place. The main advantage is the possibility of very large gains (> 20 db) in a short (< ~400 μm) semiconductor chip. The main disadvantages of the SC amplifier compared to the fiber amplifier is the presence of residual reflections and the resulting need for optical isolators. The presence of even minute reflection ($R < 10^{-5}$) can give rise to instabilities and excess noise in the source laser oscillator. Impressive results, however, have been demonstrated [36].

The above discussion centers on the use of optical amplification in long distance transmission of data. A second class of applications, no less important, is that of distribution systems with a very large number of subscribers. The use of optical amplifiers makes it possible to maintain the power arriving at a subscriber's premises at sufficiently high levels so as not to be degraded by the receiver noise. The number of subscribers that can thus be served by a single laser can be increased by anywhere from 1 to 3 orders of magnitude. This topic is the subject of Problem 11.13.

Problems

11.1 Show that the total output shot-noise power in a photomultiplier including that originating in the dynodes is given by

$$\overline{(i_N^2)} = G^2 2e(\bar{i}_c + i_d)\Delta f \frac{1 - \delta^{-N}}{1 - \delta^{-1}}$$

where δ is the secondary-emission multiplication factor and N is the number of stages.

11.2 Calculate the minimum power that can be detected by a photoconductor in the presence of a strong optical background power P_B. *Answer:*

$$(P_s)_{\min} = 2\left(\frac{P_B h\nu \Delta f}{\eta}\right)^{1/2}$$

11.3 Derive the expression for the minimum detectable power using a photoconductor in the video mode (that is, no local-oscillator power) and assuming that the main noise contribution is the generation–recombination noise. The optical field is given by $e(t) = E(1 + \cos \omega_m t) \cos \omega t$, and the signal is taken as the component of the photocurrent at ω_m.

11.4 Derive the minimum detectable power of a Ge:Hg detector with characteristics similar to those described in Section 11.7 when the average current is due mostly to blackbody radiation incident on the photocathode. Assume $T = 295$ K, an acceptance solid angle $\Omega = \pi$ and a photocathode area of 1 mm^2. Assume that the quantum yield η for blackbody radiation at $\lambda < 14$ μm is unity and that for $\lambda > 14$ μm, $\eta = 0$. [*Hint:* Find the flux of photons with wavelengths 14 μm $> \lambda > 0$ using blackbody radiation formulas or, more easily, tables or a blackbody "slide rule."]

11.5 Find the minimum detectable power in Problem 11.4 when the input field of view is at $T = 4.2$ K.

11.6 Derive Equations (11.6-15) and (11.6-16).

11.7 Show that the transit time reduction factor $(1 - e^{-i\omega_m \tau_d})/i\omega_m \tau_d$ in Equation (11.7-7) can be written as

$$\alpha - i\beta$$

where

$$\alpha = \frac{\sin \omega_m \tau_d}{\omega_m \tau_d} \qquad \beta = \frac{1 - \cos \omega_m \tau_d}{\omega_m \tau_d}$$

Plot α and β as functions of $\omega_m \tau_d$.

11.8 Derive the minimum detectable optical power for a photodiode operated in the heterodyne mode. (*Answer:* $P_{\min} = h\nu\, \Delta\nu/\eta$.)

11.9 Discuss the limiting sensitivity of an avalanche photodiode in which the noise increases as M^2. Compare it with that of a photomultiplier. What is the minimum detectable power in the limit of $M \gg 1$, and of zero background radiation and no dark current?

11.10 Derive an expression for the magnitude of the output current in a heterodyne detection scheme as a function of the angle θ between the signal and local-oscillator propagation directions. Taking the aperture diameter (see Figure 11-6) as D, show that if the output is to remain near its maximum ($\theta = 0°$) value, θ should not exceed λ/D. [*Hint:* You may replace the lens in Figure 11-6 by the photoemissive surface.] Show that instead of Equation (11.4-4) the current from an element $dx\, dy$ of the detector is

$$di(x, t) = \frac{P_L e\eta}{h\nu(\pi D^2/4)}\left[1 + 2\sqrt{\frac{P_s}{P_L}}\cos(\omega t + kx \sin\theta)\right] dx\, dy$$

The propagation directions lie in the z-x plane. The contribution of $dx\, dy$ to the (complex) signal current is thus

$$dI_s(x, t) = \frac{2\sqrt{P_s P_L}}{h\nu(\pi D^2/4)}\, e^{ikx\sin\theta}\, dx\, dy$$

11.11 Show that for a Poisson distribution (footnote 9) $\overline{(\Delta N)^2} = \overline{N}$.

11.12 Calculate the smallest temperature increment that can be measured by an infrared detector "looking" at an object at $T = 350$ K with a background temperature of $T = 300$ K. The detector has a $D^*_\lambda = 10^{11}$ cm $(\text{Hz})^{1/2}$/W and responds to $\Delta\lambda \sim 0.1\lambda$ centered on $\lambda = 10\ \mu$m. The output circuit bandwidth is $\Delta f = 10^3$ Hz.

11.13 Assume a fiber distribution network fed by a single semiconductor laser at $\lambda = 1.55\ \mu$m with a power output $P_0 = 10$ mW. The power is divided into N branches, amplified by an optical fiber amplifier (in each branch) and then divided again into M branches.

Determine the maximum number of "subscribers" NM that can be serviced by the system assuming: $\Delta f = 10^9$ Hz; R (receiver input impedance) is 10^3 ohms, $T_e = 1000$ K; and a minimum SNR at the subscriber of 42 db. The maximum power level at the output of the amplifiers is 10 mW.

References

1. Yariv, A., *Quantum Electronics*, 3d ed. New York: Wiley, 1988, p. 54.
2. Engstrom, R. W., "Multiplier phototube characteristics: Application to low light levels," *J. Opt. Soc. Am.* 37:420, 1947.
3. Sommer, A. H., *Photo-Emissive Materials.* New York: Wiley, 1968.
4. Forrester, A. T., "Photoelectric mixing as a spectroscopic tool," *J. Opt. Soc. Am.* 51:253, 1961.
5. Siegman, A. E., S. E. Harris, and B. J. McMurtry, "Optical heterodyning and optical demodulation at microwave frequencies." In *Optical Masers*, J. Fox, ed. New York: Wiley, 1963, p. 511.

6. Mandel, L., ''Heterodyne detection of a weak light beam,'' *J. Opt. Soc. Am.* 56:1200, 1966.
7. Chapman, R. A., and W. G. Hutchinson, ''Excitation spectra and photoionization of neutral mercury centers in germanium,'' *Phys. Rev.* 157:615, 1967.
8. Teich, M. C., ''Infrared heterodyne detection,'' *Proc. IEEE* 56:37, 1968.
9. Buczek, C., and G. Picus, ''Heterodyne performance of mercury doped germanium,'' *Appl. Phys. Lett.* 11:125, 1967.
10. Lucovsky, G., M. E. Lasser, and R. B. Emmons, ''Coherent light detection in solid-state photodiodes,'' *Proc. IEEE* 51:166, 1963.
11. Riesz, R. P., ''High speed semiconductor photodiodes,'' *Rev. Sci. Instr.* 33:994, 1962.
12. Anderson, L. K., and B. J. McMurtry, ''High speed photodetectors,'' *Appl. Opt.* 5:1573, 1966.
13. D'Asaro, L. A., and L. K. Anderson, ''At the end of the laser beam, a more sensitive photodiode,'' *Electronics*, May 30, 1966, p. 94.
14. Shockley, W., ''Hot electrons in germanium and Ohm's law,'' *Bell Syst. Tech. J.* 30:990, 1951.
15. McKay, K. G., and K. B. McAfee, ''Electron multiplication in silicon and germanium,'' *Phys. Rev.* 91:1079, 1953.
16. McKay, K. G., ''Avalanche breakdown in silicon,'' *Phys. Rev.* 94:877, 1954.
17. McIntyre, R., ''Multiplication noise in uniform avalanche diodes,'' *IEEE Trans. Elect. Devices* ED-13:164, 1966.
18. Lindley, W. T., R. J. Phelan, C. M. Wolfe, and A. J. Foyt, ''GaAs Schottky barrier avalanche photodiodes,'' *Appl. Phys. Lett.* 14:197, 1969.
19. Nahory, R. E., M. A. Pollack, E. D. Beebe, and J. C. DeWinter, ''Continuous operation of a 1.0 μm wavelength $GaAs_{1-x}Sb_x/Al_yGa_{1-y}$-$As_{1-x}Sb_x$ double-heterostructure injection laser,'' *Appl. Phys. Lett.* 28:19, 1976.
20. Bar-Chaim, N., K. Y. Lau, I. Ury, and A. Yariv, ''High speed GaAlAs/GaAs photodiode on a semi-insulating GaAs substrate,'' *Appl. Phys. Lett.* 43:261, 1983.
21. Wang, S. Y., D. M. Bloom, and D. M. Collins, ''20-GHz bandwidth GaAs photodiode,'' *Appl. Phys. Lett.* 42:190, 1983.
22. Wang, S. Y., and D. M. Bloom, ''100 GHz bandwidth planar GaAs Schottky photodiode,'' *Elec. Lett.* 19:554, 1983.
23. Valdmanis, J. A., G. Mourou, and C. W. Gabel, ''Picosecond electro-optic sampling system,'' *Appl. Phys. Lett.* 41:211, 1982.
24. Kolner, B. H., D. M. Bloom, and P. S. Cross, ''Characterization of high speed GaAs photodiodes using a 100-GHz electro-optic sampling system,'' 1983 Conference on Lasers and Electro-optics, paper ThGl.
25. Kinch, M. A. and A. Yariv, ''Performance limitations of GaAs/GaAlAs infrared superlattices,'' *Appl. Phys. Lett.* 55:2093, 1990.
26. R. E. Burgess, ''Fluctuations in the number of electrons and holes in semiconductors.'' *Proc. Phys. Soc. B* 68:661, 1955.
27. Van Der Ziel, A., *Fluctuation Phenomena in Semiconductors* Ch. 4, New York: Academic Press, 1959.

28. Long, D. ''On generation-recombination noise in infrared detector materials,'' *Infrared Phys.* 7:167, Pergamon Press, 1967.

29. L. C. Chiu, J. S. Smith, S. Margalit, A. Yariv, and A. Y. Cho, ''Application of internal photoemission from quantum well heterojunction superlattices to infrared detectors'' *Infrared Phys.* 23(2):93, 1983.

30. B. F. Levine, C. G. Bethea, G. Hasnain, J. Walker, and R. J. Malek, ''High detectivity $D^* = 10^{10}$ cm$\sqrt{\text{Hz}}$/W GaAs/AlGaAs multiquantum well $\lambda = 8.3$ μm Infrared Detector,'' *Appl. Phys. Lett.* 53:2196, 1988.

31. Simon J. C., ''Semiconductor laser amplifier for single mode optical fiber communications,'' *J. Opt. Commun.* 4:51, 1983.

32. Mears, R. J., L. Reekie, I. M. Jauncey, and D. N. Payne, ''Low noise erbium-doped fiber amplifier operating at 1.54 mm,'' *Elec. Lett.* 23:1026, 1987.

33. Hagimoto, K., et al. ''A 212 Km non-repeatered transmission experiment at 1.8 Gb/s using LD pumped Er^{3+}-doped fiber amplifiers in an Im/direct-detection repeater system,'' in Proceedings of the Optical Fiber Conference, Houston, TX, postdeadline Paper PD15, 1989.

34. Olshansky, R., ''Noise figure for Er-doped optical fibre amplifiers,'' *Elec. Letts.* 24:1363, 1988.

35. Payne, David N., ''Tutorial session abstracts,'' Optical Fiber Communication (OFC 1990) Conference, San Francisco 1990. Published by Opt. Soc. of Am., Washington, D.C.

36. See, for example, Eisenstein, G., U. Koren, G. Raybon, T. L. Koch, M. Wiesenfeld, M. Wegener, R. S. Tucker, and B. I. Miller, ''Large-signal and small-signal gain characteristics of 1.5 mm quantum well optical amplifiers,'' *Appl. Phys. Lett.* 56:201, 1990.

Supplementary Reference

37. Boyd, R. W., *Radiometry and the Detection of Optical Radiation.* New York: Wiley, 1983.

Interaction of Light and Sound

12.0 INTRODUCTION

Diffraction of light by sound[1] waves was predicted by Brillouin in 1922 [1] and demonstrated experimentally some ten years later [2]. Recent developments in high-frequency acoustics [3] and in lasers caused a renewed interest in this field because the scattering of light from sound affords a convenient means of controlling the frequency, intensity, and direction of an optical beam. This type of control enables a large number of applications involving the transmission, display, and processing of information [4].

12.1 SCATTERING OF LIGHT BY SOUND

A sound wave consists of a sinusoidal perturbation of the density of the material, or strain, that travels at the sound velocity v_s, as shown in Figure 12-1. A change in the density of the medium causes a change in its index of refraction, which, to first order, is proportional to it.[2] We can, consequently, represent the sound wave shown in Figure 12-1 by

$$\Delta n(z, t) = \Delta n \sin(\omega_s t - k_s z) \tag{12.1-1}$$

where $\omega_s/k_s = v_s$.

---

[1]In this chapter we use the word *sound* to describe acoustic waves with frequencies that in practice may range through the microwave region ($f \simeq 10^{10}$ Hz).

[2]This is easily understood in the case where each atom (or molecule) contributes a constant amount to the index, n, so the latter is proportional to the material density.

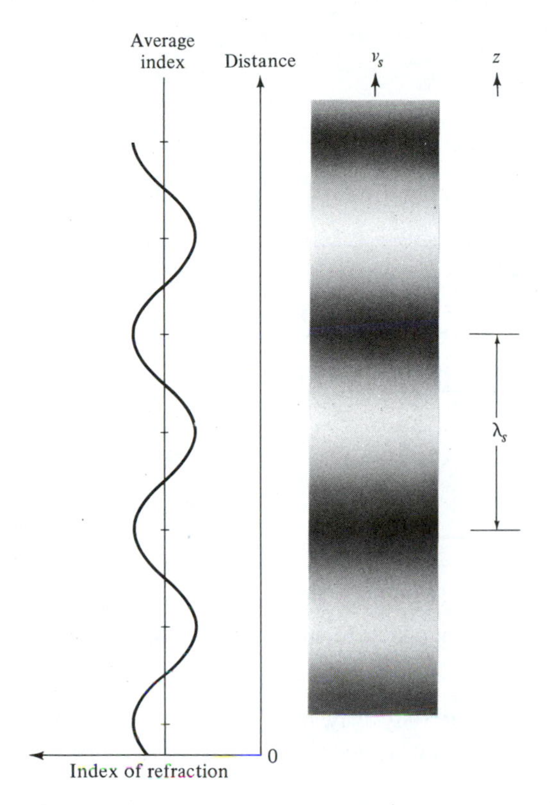

Figure 12-1 Traveling sound wave "frozen" at some instant of time. It consists of alternating regions of compression (dark) and rarefaction (white), which travel at the sound velocity v_s. Also shown is the instantaneous spatial variation of the index of refraction that accompanies the sound wave.

Next consider an optical beam incident on a sound wave at an angle θ_i as in Figure 12-2. For the purpose of the immediate discussion, we can characterize the sound wave as a series of partially reflecting mirrors,[3] separated by the sound wavelength λ_s, that are moving at a velocity v_s. Ignoring, for the moment, the motion of the mirrors, let us consider the diffracted wave and take the diffraction angle as θ_r. A necessary condition for diffraction in a given direction is that *all the points on a given mirror contribute in phase* to the diffraction along this direction. Considering the diffraction from two points, such as C and B in Figure 12-2, it is then necessary that the optical path difference $AC - BD$ be some multiple of the optical wavelength λ/n for diffraction along θ_r to occur. This condition takes the form

$$x(\cos \theta_i - \cos \theta_r) = \frac{m\lambda}{n} \qquad \textbf{(12.1-2)}$$

[3]This is due to the fact that the index of refraction is higher in the compressed portions of the sound wave and lower in the rarefied regions. Since a change in index causes reflection, the mirrors' analogy follows.

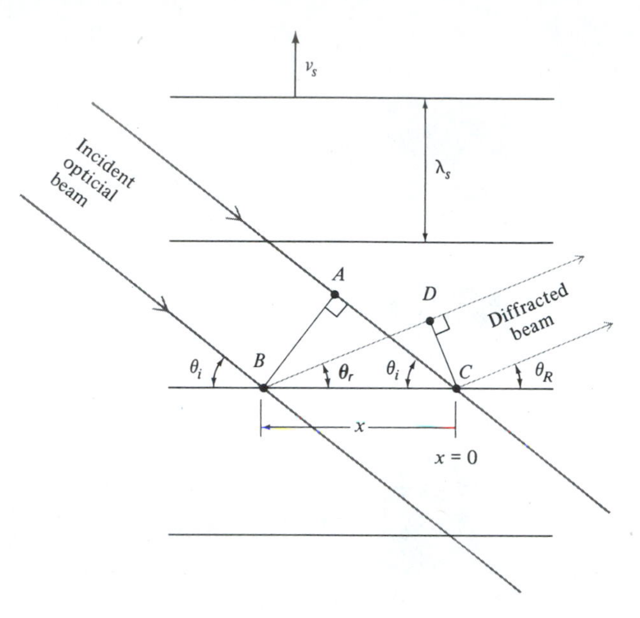

Figure 12-2 Diffraction of an incident optical beam from an array of equispaced reflectors.

where $m = 0, \pm 1, \pm 2, \ldots$. The only way in which (12.1-2) can be satisfied simultaneously *for all points* x along a *given* reflector is if $m = 0$, from which it follows that

$$\theta_i = \theta_r \tag{12.1-3}$$

In addition to the requirement that the different parts of a given acoustic phase front interfere constructively, which leads to (12.1-3), we require that the diffraction from any two acoustic phase fronts add up in phase along the direction of the reflected beam. The path difference, $AO + OB$ shown in Figure 12-3, of a given optical wavefront resulting from reflection from two equivalent acoustic wavefronts (that is, planes separated by λ_s) must thus be equal to the optical wavelength λ. Using (12.1-3) and Figure 12-3 we find that this condition can be written as[4]

$$2\lambda_s \sin\theta = \lambda/n \tag{12.1-4}$$

where $\theta_i = \theta_r = \theta$.

The diffraction of light that satisfies (12.1-4) is known as Bragg diffraction after a similar law applying in X-ray diffraction from crystals. To get an idea of the order of magnitude of the angle θ, consider the case of diffraction of light with $\lambda/n = 0.5$ μm from a 500-MHz sound wave. Taking the sound velocity as $v_s = 3 \times 10^5$ cm/sec, we have $\lambda_s = v_s/\nu_s = 6 \times 10^{-4}$ cm and, from (12.1-4),

$$\theta \simeq 4 \times 10^{-2} \text{ rad} \simeq 3.5°$$

[4]The reader may justly wonder why path differences of $2\lambda/n$, $3\lambda/n$, and so on, do not lead to maximum diffraction as well as a path difference of λ/n. This point is considered in Problem 12.6.

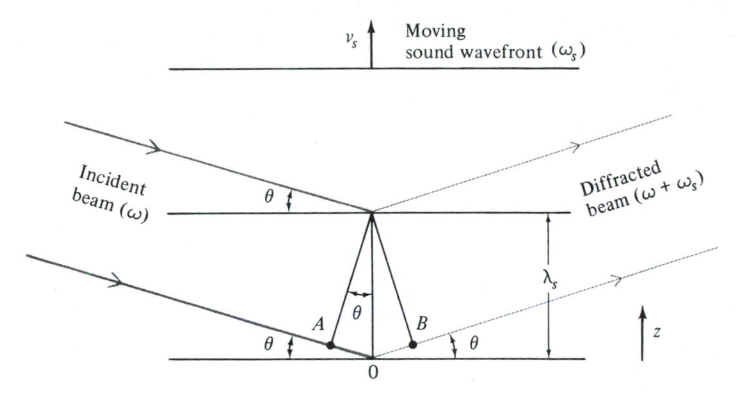

Figure 12-3 The reflections from two equivalent planes in the sound beam (that is, planes separated by the sound wavelength λ_s), which add up in phase along the direction θ if the optical path difference $AO + OB$ is equal to one optical wavelength.

12.2 PARTICLE PICTURE OF BRAGG DIFFRACTION OF LIGHT BY SOUND

Many of the features of Bragg diffraction of light by sound can be deduced if we take advantage of the dual particle-wave nature of light and of sound. According to this picture a light beam with a propagation vector $\mathbf{k}$[5] and frequency ω can be considered to consist of a stream of particles (photons) with momentum $\hbar\mathbf{k}$ and energy $\hbar\omega$. The sound wave, likewise, can be thought of as made up of particles (phonons) with momentum $\hbar\mathbf{k}_s$ and energy $\hbar\omega_s$. The diffraction of light by an *approaching* sound beam illustrated in Figure 12-3 can be described as a series of collisions, each of which involves an annihilation of *one* incident photon at ω_i and *one* phonon and a simultaneous creation of a new (diffracted) photon at a frequency ω_d, which propagates along the direction of the scattered beam. The conservation of momentum requires that the momentum $\hbar(\mathbf{k}_s + \mathbf{k}_i)$ of the colliding particles be equal to the momentum $\hbar\mathbf{k}_d$ of the scattered photon, so

$$\mathbf{k}_d = \mathbf{k}_s + \mathbf{k}_i \tag{12.2-1}$$

The conservation of energy takes the form

$$\omega_d = \omega_i + \omega_s \tag{12.2-2}$$

From (12.2-2) we learn that the diffracted beam is shifted in frequency by an amount equal to the sound frequency. Since the interaction involves the annihilation of a phonon, conservation of energy decrees that the shift in frequency is such that $\omega_d > \omega_i$ and the phonon energy is *added* to that of the annihilated photon to form a new photon. Using this argument it follows that if the direction of the sound beam

[5]The beam is of the form $\cos(\omega t - \mathbf{k} \cdot \mathbf{r})$, so it propagates in a direction parallel to $\mathbf{k}$ with a wavelength $2\pi/k$.

in Figure 12-3 were reversed so that it was receding from the incident optical wave, the scattering process could be considered as one in which a new photon (diffracted photon) and a *new* phonon are generated while the incident photon is annihilated. In this case, the conservation-of-energy principle yields

$$\omega_d = \omega_i - \omega_s$$

The relation between the sign of the frequency change and the sound propagation direction will become clearer using Doppler-shift arguments, as is done at the end of this section.

The conservation-of-momentum condition (12.2-1) is equivalent to the Bragg condition (12.1-4). To show why this is true, consider Figure 12-4. Since the sound frequencies of interest are below 10^{10} Hz and those of the optical beams are usually above 10^{13} Hz, we have

$$\omega_d = \omega_i + \omega_s \simeq \omega_i, \qquad \text{so } k_d \simeq k_i$$

and the magnitude of the two optical wave vectors is taken as k (see also Problem 12.4). The magnitude of the sound wave vector is thus

$$k_s = 2k \sin \theta \tag{12.2-3}$$

Using $k_s = 2\pi/\lambda_s$, this equation becomes

$$2\lambda_s \sin \theta = \lambda/n \tag{12.2-4}$$

which is the same as the Bragg-diffraction condition (12.1-4).

Doppler Derivation of the Frequency Shift

The frequency-shift condition (12.2-2) can also be derived by considering the Doppler shift exercised by an optical beam incident on a mirror moving at the sound velocity v_s at an angle satisfying the Bragg condition (12.1-4). The formula for the Doppler frequency shift of a wave reflected from a moving object is

$$\Delta\omega = 2\omega\frac{v}{c/n}$$

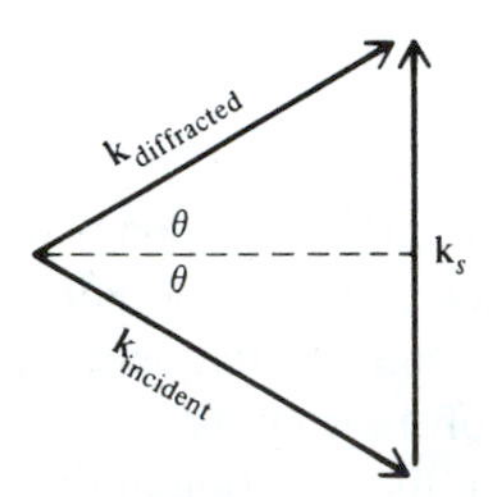

Figure 12-4 The momentum-conservation relation, Equation (12.2-1), used to derive the Bragg condition $2\lambda_s \sin \theta = \lambda/n$, for an optical beam that is diffracted by an approaching sound wave. θ is the angle between the incident or diffracted beam and the acoustic wavefront.

where ω is the optical frequency and v is the component of the object velocity that is parallel to the wave propagation direction. From Figure 12-3 we have $v = v_s \sin \theta$, and thus

$$\Delta\omega = 2\omega \frac{v_s \sin \theta}{c/n} \tag{12.2-5}$$

Using (12.1-4) for $\sin \theta$ we obtain

$$\Delta\omega \equiv \frac{2\pi v_s}{\lambda_s} = \omega_s \tag{12.2-6}$$

and, therefore, $\omega_d = \omega + \omega_s$.

If the direction of propagation of the sound beam is reversed so that, in Figure 12-3, the sound recedes from the optical beam, the Doppler shift changes sign and the diffracted beam has a frequency $\omega - \omega_s$.

12.3 BRAGG DIFFRACTION OF LIGHT BY ACOUSTIC WAVES—ANALYSIS

In treating the diffraction of light by acoustic waves, we assume a long interaction path so that higher diffraction orders [5] are missing and the only two waves coupled by the sound are the incident wave at ω_i and a diffracted wave at $\omega_d = \omega_i + \omega_s$ or at $\omega_i - \omega_s$, depending on the direction of the Doppler shift as discussed in Section 12.2.

According to the discussion in Section 12.1, the sound wave causes a traveling modulation of the index of refraction given by

$$\Delta n(\mathbf{r}, t) = \Delta n \cos(\omega_s t - \mathbf{k}_s \cdot \mathbf{r})$$

This modulation interacts with the fields at ω_i and ω_d to give rise to additional electric polarization in the medium, which is given by[6]

$$\Delta\mathbf{p}(\mathbf{r}, t) = 2\sqrt{\epsilon\epsilon_0}\, \Delta n(\mathbf{r}, t)\mathbf{e}(\mathbf{r}, t) \tag{12.3-1}$$

where $\mathbf{e}(\mathbf{r}, t)$ is the sum of the fields at ω_i and ω_d. The polarization term $\Delta n\mathbf{e}$ in (12.3-1) will be shown, in what follows, to cause exchange of power between the fields at ω_i and ω_d.

We start with the wave equation (8.2-5) modified for the case of no losses $(\sigma = 0)$.

$$\nabla^2\mathbf{e}(\mathbf{r}, t) = \mu\epsilon \frac{\partial^2 \mathbf{e}}{\partial t^2} + \mu \frac{\partial^2}{\partial t^2} \mathbf{p}_{NL}(\mathbf{r}, t) \tag{12.3-2}$$

[6]Equation (12.3-1) can be derived using the relations $\mathbf{d} = \epsilon_0\mathbf{e} + \mathbf{p} = \epsilon\mathbf{e}$, $\mathbf{p} \equiv \epsilon_0\chi\mathbf{e}$, and $n^2 \equiv \epsilon/\epsilon_0$. This leads to $\mathbf{p} = \epsilon_0(n^2 - 1)\mathbf{e}$, whence Equation (12.3-1).

$\mathbf{p}_{NL}(\mathbf{r}, t)$ is given in our case by $\Delta\mathbf{p}(\mathbf{r}, t)$. Equation (12.3-2) must be satisfied separately for the fields at ω_i and ω_d. Writing it for the former case and assuming that both the incident and diffracted fields are linearly polarized results in

$$\nabla^2 e_i = \mu\epsilon \frac{\partial^2 e_i}{\partial t^2} + \mu \frac{\partial^2}{\partial t^2} (\Delta p)_i \tag{12.3-3}$$

where e_i is the magnitude of the vector $\mathbf{e}_i$ and $(\Delta p)_i$ is the component of $\Delta\mathbf{p}(\mathbf{r}, t)$ parallel to $\mathbf{e}_i$, which oscillates at a frequency ω_i. The polarization components oscillating at other frequencies are nonsynchronous, and their contribution to e_i averages out to zero. The total field $\mathbf{e}(\mathbf{r}, t)$ is taken as the sum of two traveling waves

$$e_i(\mathbf{r}, t) = \tfrac{1}{2}E_i(r_i)\, e^{i(\omega_i t - \mathbf{k}_i\cdot\mathbf{r})} + \text{c.c.}$$
$$e_d(\mathbf{r}, t) = \tfrac{1}{2}E_d(r_d)\, e^{i(\omega_d t - \mathbf{k}_d\cdot\mathbf{r})} + \text{c.c.} \tag{12.3-4}$$

where $\mathbf{k}_i$ and $\mathbf{k}_d$ are parallel to the direction of propagation of the incident and diffracted waves, respectively. Two differentiations of the first of Equations (12.3-4) lead to

$$\nabla^2 e_i(\mathbf{r}, t) = -\frac{1}{2}\left[k_i^2 E_i + 2ik_i \frac{dE_i}{dr_i} - \nabla^2 E_i\right] e^{i(\omega_i t - \mathbf{k}_i\cdot\mathbf{r})}$$

Assuming "slow" variation of $E_i(r_i)$ so that $\nabla^2 E_i \ll k_i dE_i/dr_i$, we combine (12.3-3) with the last equation and, recalling that $k_i^2 = \omega_i^2 \mu\epsilon$, obtain

$$k_i \frac{dE_i}{dr_i} = i\mu\left[\frac{\partial^2}{\partial t^2} (\Delta p)_i\right] e^{-i(\omega_i t - \mathbf{k}_i\cdot\mathbf{r})} \tag{12.3-5}$$

Using the relation $\Delta\mathbf{p} = 2\sqrt{\epsilon\epsilon_0}\, \Delta n(\mathbf{r}, t) \cdot [\mathbf{e}_i(\mathbf{r}, t) + \mathbf{e}_d(\mathbf{r}, t)]$, $(\Delta p)_i$ is given by

$$[\Delta p(\mathbf{r}, t)]_i = \tfrac{1}{2}\sqrt{\epsilon\epsilon_0}\, \Delta n E_d\{e^{i[(\omega_s+\omega_d)t - (\mathbf{k}_s+\mathbf{k}_d)\cdot\mathbf{r}]}\} + \text{c.c.} \tag{12.3-6}$$

Note that in taking the product $\Delta n(\mathbf{r}, t)e(\mathbf{r}, t)$ we assumed that $\omega_i = \omega_s + \omega_d$ and therefore neglected nonsynchronous terms with frequencies $\omega_d - \omega_s$ and $\omega_i \pm \omega_s$. Substituting (12.3-6) for $(\Delta p)_i$ in (12.3-5) leads to

$$\frac{dE_d}{dr_d} = -i\eta_d E_i\, e^{-i(\mathbf{k}_i - \mathbf{k}_s - \mathbf{k}_d)\cdot\mathbf{r}}$$

and similarly

$$\frac{dE_i}{dr_i} = -i\eta_i E_d\, e^{i(\mathbf{k}_i - \mathbf{k}_s - \mathbf{k}_d)\cdot\mathbf{r}} \tag{12.3-7}$$

with

$$\eta_{i,d} = \tfrac{1}{2}\omega_{i,d}\sqrt{\mu\epsilon_0}\, \Delta n = \frac{\omega_{i,d}\, \Delta n}{2c} \tag{12.3-8}$$

where c is the velocity of light in vacuum. An inspection of (12.3-7) reveals that a prerequisite for continuous cumulative interaction between the incident field (E_i) and the diffracted field (E_d) is that

$$\mathbf{k}_i = \mathbf{k}_s + \mathbf{k}_d \tag{12.3-9}$$

Otherwise, it follows from (12.3-7) that contributions to E_i, as an example, from different path elements do not add in phase and no sustained spatial growth of E_i is possible.

Equation (12.3-9) is, as shown in Section 12.2, the Bragg condition for scattering of light by sound. The difference between (12.3-9) and (12.2-1) is due to the fact that the latter was derived for the case of diffraction from an approaching sound beam so that $\omega_d = \omega_i + \omega_s$ resulting in a "momentum" condition $\mathbf{k}_d = \mathbf{k}_i + \mathbf{k}_s$, while in the treatment leading to (12.3-9) we recall that the sound wave is taken as receding from the incident field so that $\omega_d = \omega_i - \omega_s$.

Assuming that the Bragg condition (12.3-9) is satisfied, (12.3-7) becomes

$$\begin{aligned} \frac{dE_i}{dr_i} &= -i\eta E_d \\ \frac{dE_d}{dr_d} &= -i\eta E_i \end{aligned} \tag{12.3-10}$$

where, since $\omega_i \approx \omega_d$, we took $\eta_i = \eta_d \equiv \eta$.

Equations (12.3-10) are our main result. An apparent difficulty in solving (12.3-10) is the fact that they involve two different spatial coordinates r_i and r_d measured along the two respective ray directions. This difficulty can be resolved by transforming to a coordinate ζ measured along the bisector of the angle formed between $\mathbf{k}_i$ and $\mathbf{k}_d$, as shown in Figure 12-5. Defining the values of r_d and r_i, which correspond to a given ζ as the respective projections of ζ along $\mathbf{k}_d$ and $\mathbf{k}_i$, we have

$$r_i = \zeta \cos \theta \qquad r_d = \zeta \cos \theta \tag{12.3-11}$$

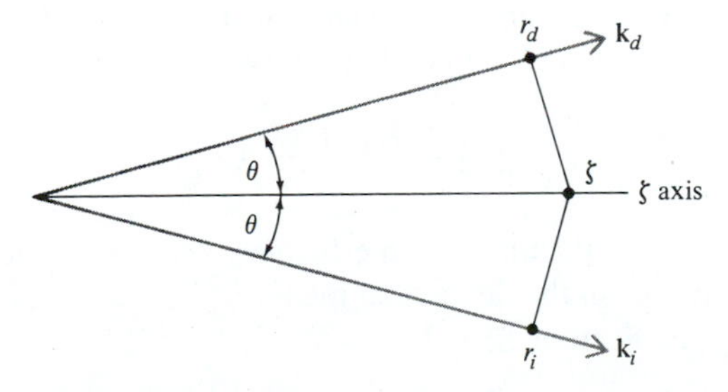

Figure 12-5 The directions and angles appearing in the diffraction equations (12.3-12).

so that (12.3-10) become

$$\frac{dE_i}{d\zeta} = \frac{dE_i}{dr_i}\cos\theta = -i\eta E_d \cos\theta$$

$$\frac{dE_d}{d\zeta} = -i\eta E_i \cos\theta \tag{12.3-12}$$

whose solutions are

$$E_i(\zeta) = E_i(0)\cos(\eta\zeta\cos\theta) - iE_d(0)\sin(\eta\zeta\cos\theta)$$

$$E_d(\zeta) = E_d(0)\cos(\eta\zeta\cos\theta) - iE_i(0)\sin(\eta\zeta\cos\theta)$$

Using the correspondence between ζ, r_i, and r_d defined above, we can rewrite the solutions as

$$E_i(r_i) = E_i(0)\cos(\eta r_i) - iE_d(0)\sin(\eta r_i)$$

$$E_d(r_d) = E_d(0)\cos(\eta r_d) - iE_i(0)\sin(\eta r_d) \tag{12.3-13}$$

which is the desired result. It is of sufficient generality to describe the interaction between two input fields at ω_i and ω_d with arbitrary phases [$E_i(0)$ and $E_d(0)$ are complex], and arbitrary amplitudes as long as the Bragg condition (12.3-9) and the frequency condition $\omega_i = \omega_s + \omega_d$ are fulfilled. In the special case of a single frequency input at ω_i, $E_d(0) = 0$, and

$$E_i(r_i) = E_i(0)\cos(\eta r_i)$$

$$E_d(r_d) = -iE_i(0)\sin(\eta r_d) \tag{12.3-14}$$

we note that

$$|E_i(r_i)|^2 + |E_d(r_d = r_i)|^2 = |E_i(0)|^2 \tag{12.3-15}$$

so that the total optical power carried by both waves is conserved.

If the interaction distance between the two beams is such that $\eta r_i = \eta r_d = \pi/2$, the total power of the incident beam is transferred into the diffracted beam. Since this process is used in a large number of technological and scientific applications, it may be worthwhile to gain some appreciation for the diffraction efficiencies possible using known acoustic media and conveniently available acoustic power levels.

The fraction of the power of the incident beam transferred in a distance l into the diffracted beam is given, using (12.3-8) and (12.3-14), by

$$\frac{I_{\text{diffracted}}}{I_{\text{incident}}} = \frac{E^2_{\text{diffracted}}}{E_i^2(0)} = \sin^2\left(\frac{\omega l}{2c}\Delta n\right) \tag{12.3-16}$$

It is advantageous to express the diffraction efficiency (12.3-16) in terms of the acoustic intensity $I_{\text{acoustic}}(\text{W/m}^2)$ in the diffraction medium. First we relate the index change Δn to the strain s (see Section 12.1) by [4–5]

$$\Delta n \equiv -\frac{n^3 p}{2}s \tag{12.3-17}$$

where p, the photoelastic constant of the medium,[7] is defined by (12.3-17). The strain s is related to the acoustic intensity I_{acoustic} by[8]

$$s = \sqrt{\frac{2I_{\text{acoustic}}}{\rho v_s^3}} \tag{12.3-18}$$

where v_s is the velocity of sound in the medium and ρ is the mass density (kg/m^3). Combining (12.3-17) and (12.3-18) in (12.3-16) we obtain

$$\frac{I_{\text{diffracted}}}{I_{\text{incident}}} = \sin^2\left[\frac{\pi l}{\sqrt{2}\lambda}\sqrt{\frac{n^6p^2}{\rho v_s^3}I_{\text{acoustic}}}\right] \tag{12.3-19}$$

and using the following definition for the diffraction figure of merit

$$M \equiv \frac{n^6p^2}{\rho v_s^3} \tag{12.3-20}$$

(12.3-19) becomes

$$\frac{I_{\text{diffracted}}}{I_{\text{incident}}} = \sin^2\left(\frac{\pi l}{\sqrt{2}\lambda}\sqrt{MI_{\text{acoustic}}}\right) \tag{12.3-21}$$

Taking water as an example, an optical wavelength of $\lambda = 0.6328\ \mu$m, and the constants (taken from Table 12-1)

$$n = 1.33$$

$$p = 0.31$$

$$v_s = 1.5 \times 10^3 \text{ m/s}$$

$$\rho = 1000 \text{ kg/m}^3$$

[7]In the case of interactions using crystals, (12.3-17) becomes a tensor relation and p becomes a fourth rank tensor. In this case we can often simplify the problem in such a way that only one tensor element is important so that (12.3-17) can be used.

[8]The (elastic) potential energy per unit volume due to an instantaneous strain $s(t)$ is

$$\tfrac{1}{2}Ts^2(t)$$

where T is the bulk modulus (elastic stiffness constant). The time averaged energy per unit volume due to the propagation of a sound wave with a strain amplitude s is the sum of the (equal) average potential and kinetic energy densities

$$\frac{\mathscr{E}}{\text{vol}} = 2(\tfrac{1}{2})T\overline{s^2}(t) = \tfrac{1}{2}Ts^2$$

since $\overline{s^2}(t) = \frac{1}{2}s^2$, the bar denoting time-averaging. Using the relation $I_{\text{acoustic}} = v_s\mathscr{E}/\text{vol}$ and $T/\rho = v_s^2$ where ρ is the mass density and v_s the velocity of sound, we get

$$I_{\text{acoustic}} = \tfrac{1}{2}\rho v_s^3 s^2$$

or

$$s = \sqrt{\frac{2I_{\text{acoustic}}}{\rho v_s^3}}$$

which is the result stated in (12.3-18).

Equation (12.3-21) gives

$$\left(\frac{I_{\text{diffracted}}}{I_{\text{incident}}}\right)_{\substack{H_2O \\ \text{at } \lambda=0.6328\ \mu m}} = \sin^2\left(1.4 l \sqrt{I_{\text{acoustic}}}\right) \tag{12.3-22}$$

For other materials and at other wavelengths we can combine the last two equations to obtain a convenient working formula

$$\frac{I_{\text{diffracted}}}{I_{\text{incident}}} = \sin^2\left(1.4\,\frac{0.6328}{\lambda\ \mu m}\, l\sqrt{M_\omega I_{\text{acoustic}}}\right) \tag{12.3-23}$$

where $M_\omega = M_{\text{material}}/M_{H_2O}$ is the diffraction figure of merit of the material relative to water. Values of M and M_ω for some common materials are listed in Tables 12-1 and 12-2.

According to (12.3-19), at small diffraction efficiencies, the diffracted light intensity is proportional to the acoustic intensity. This fact is used in acoustic modulation of optical radiation. The information signal is used to modulate the intensity of the acoustic beam. This modulation is then transferred, according to (12.3-19), as intensity modulation onto the diffracted optical beam.

Table 12-1 A List of Some Materials Commonly Used in the Diffraction of Light by Sound and Some of Their Relevant Properties. ρ Is the Density, v_s the Velocity of Sound, n the Index of Refraction, p the Photoelastic Constant as Defined by Equation (12.3-7), and M_ω Is the Relative Diffraction Constant Defined Above (After Reference [4].)

Material	ρ (mg/m^3)	v_s (km/s)	n	p	M_ω
Water	1.0	1.5	1.33	0.31	1.0
Extra-dense flint glass	6.3	3.1	1.92	0.25	0.12
Fused quartz (SiO_2)	2.2	5.97	1.46	0.20	0.006
Polystyrene	1.06	2.35	1.59	0.31	0.8
KRS-5	7.4	2.11	2.60	0.21	1.6
Lithium niobate ($LiNbO_3$)	4.7	7.40	2.25	0.15	0.012
Lithium fluoride (LiF)	2.6	6.00	1.39	0.13	0.001
Rutile (TiO_2)	4.26	10.30	2.60	0.05	0.001
Sapphire (Al_2O_3)	4.0	11.00	1.76	0.17	0.001
Lead molybdate ($PbMO_4$)	6.95	3.75	2.30	0.28	0.22
Alpha iodic acid (HIO_3)	4.63	2.44	1.90	0.41	0.5
Tellurium dioxice (TeO_2) (Slow shear wave)	5.99	0.617	2.35	0.09	5.0

Table 12-2 A List of Materials Commonly Used in Acoustooptic Interactions and Some of Their Relevant Properties. $M = n^6p^2/\rho v_s^3$ Is the Figure of Merit, Defined by (12.3-20) and Is Given in MKS Units (After Reference [6].)

Material	$\lambda(\mu m)$	n	$\rho(g/cm^3)$	Acoustic Wave Polarization and Direction	$v_s(10^5$ cm/s)	Optical Wave Polarization and Direction*	$M = n^6p^2/\rho v_s^3$
Fused quartz	0.63	1.46	2.2	long.	5.95	$\perp$	1.51×10^{-15}
Fused quartz	0.63			trans.	3.76	$\parallel$ or $\perp$	0.467
GaP	0.63	3.31	4.13	long. in [110]	6.32	$\parallel$	44.6
GaP	0.63			trans. in [100]	4.13	$\parallel$ or $\perp$in [010]	24.1
GaAs	1.15	3.37	5.34	long. in [110]	5.15	$\parallel$	104
GaAs	1.15			trans. in [100]	3.32	$\parallel$ or $\perp$in [010]	46.3
TiO_2	0.63	2.58	4.6	long. in [11–20]	7.86	$\perp$in [001]	3.93
$LiNbO_3$	0.63	2.20	4.7	long. in [11–20]	6.57		6.99
YAG	0.63	1.83	4.2	long. in [100]	8.53	$\parallel$	0.073
YAG	0.63			long. in [110]	8.60	$\perp$	0.012
YIG	1.15	2.22	5.17	long. in [100]	7.21	$\perp$	0.33
$LiTaO_3$	0.63	2.18	7.45	long. in [001]	6.19	$\parallel$	1.37
As_2S_3	0.63	2.61	3.20	long.	2.6	$\perp$	433
As_2S_3	1.15	2.46		long.		$\parallel$	347
SF-4	0.63	1.616	3.59	long.	3.63	$\perp$	4.51
β-ZnS	0.63	2.35	4.10	long. in [110]	5.51	$\parallel$ in [001]	3.41
β-ZnS	0.63			trans. in [110]	2.165	$\parallel$ or $\perp$in [001]	0.57
α-Al_2O_3	0.63	1.76	4.0	long. in [001]	11.15	$\parallel$ in [11–20]	0.34
CdS	0.63	2.44	4.82	long. in [11–20]	4.17	$\parallel$	12.1
ADP	0.63	1.58	1.803	long. in [100]	6.15	$\parallel$ in [010]	2.78
ADP	0.63			trans. in [100]	1.83	$\parallel$ or $\perp$in [001]	6.43
KDP	0.63	1.51	2.34	long. in [100]	5.50	$\parallel$ in [010]	1.91
KDP	0.63			trans. in [100]		$\parallel$ or $\perp$in [001]	3.83
H_2O	0.63	1.33	1.0	long.	1.5		160
Te	10.6	4.8	6.24	long. in [11–20]	2.2	$\parallel$ in [001]	4400
$PbMO_4$ [14]	0.63	2.4		long. $\parallel$ c axis	3.75	$\parallel$ or $\perp$	73

*The optical-beam direction actually differs from that indicated by the magnitude of the Bragg angle. The polarization is defined as parallel or perpendicular to the scattering plane formed by the acoustic and optical k vectors.

Numerical Example: Scattering in Fused Quartz

Calculate the fraction of 0.633 μm light that is diffracted under Bragg conditions from a sound wave in $PbMO_4$ with the following characteristics

Acoustic power = 1 watt

Acoustic beam cross section = 1 mm × 1 mm

l = optical path in acoustic beam = 1 mm

M_ω (from Table 12-1) = 0.22

Substituting these data into (12.3-23) yields

$$\frac{I_{\text{diffracted}}}{I_{\text{incident}}} \simeq 37\%$$

12.4 DEFLECTION OF LIGHT BY SOUND

One of the most important applications of acoustooptic interactions is in the deflection of optical beams. This can be achieved by changing the sound frequency while operating near the Bragg-diffraction condition. The situation is depicted in Figure 12-6 and can be understood using Figure 12-7. Let us assume first that the Bragg condition (12.1-4) is satisfied. The momentum vector diagram originally introduced in Figure 12-4 is closed, and the beam is diffracted along the direction θ as given by (12.1-4). Now let the sound frequency change from ν_s to $\nu_s + \Delta\nu_s$. Since $k_s = 2\pi\nu_s/v_s$, this causes a change of $\Delta k_s = 2\pi(\Delta\nu_s)/v_s$ in the magnitude of the sound wave vector as shown. Since the angle of incidence remains θ and the magnitude of the diffracted k vector is unchanged,[9] so its tip is constrained to the circle locus shown in Figure 12-7, we can no longer close the momentum diagram and thus momentum is no longer strictly conserved. The beam will be diffracted along the direction that least violates the momentum conservation.[10] This takes place along

[9]The small change in the diffracted wave vector that is attributable to the frequency change is typically about $\Delta k/k \simeq 10^{-7}$ and is neglected. (See Problem 12.4.)

[10]The violation of momentum conservation is equivalent to destructive interference in the diffracted beam, so the beam intensity will be less than under Bragg condition, where momentum is conserved. The diffracted beam will thus have its maximum value along the direction in which the destructive interference is smallest. This corresponds to the direction that minimizes the momentum mismatch, as shown in Figure 12-7.

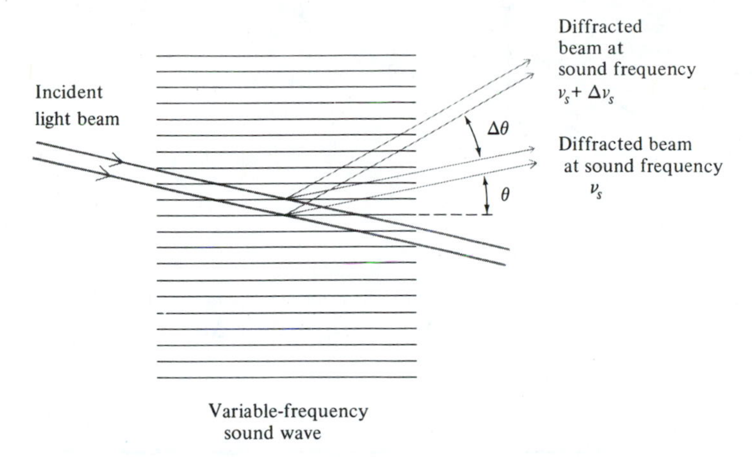

Figure 12-6 A change of frequency of the sound wave from ν_s to $\nu_s + \Delta\nu_s$ causes a change $\Delta\theta$ in the direction of the diffracted beam, according to Equation (12.4-1).

the direction OB, causing a deflection of the beam by $\Delta\theta$. Recalling that the angles θ and $\Delta\theta$ are all small and that $k_s = 2\pi\nu_s/v_s$, we obtain

$$\Delta\theta = \frac{\Delta k_s}{k} = \frac{\lambda}{nv_s}\Delta\nu_s \tag{12.4-1}$$

so that the deflection angle is proportional to the change of the sound frequency.

As in the case of electrooptic deflection, we are not interested so much in the absolute deflection $\Delta\theta$ as we are in the number of resolvable spots—that is, the factor by which $\Delta\theta$ exceeds the beam divergence angle. If we take the diffraction angle as $\sim\lambda/nD$, where D is the beam diameter,[11] the number of resolvable spots is

$$\begin{aligned} N &= \frac{\Delta\theta}{\theta_{\text{diffracted}}} = \left(\frac{\lambda}{v_s}\right)\frac{\Delta\nu_s}{\lambda/D} \\ &= \Delta\nu_s\left(\frac{D}{v_s}\right) = \Delta\nu_s\tau \end{aligned} \tag{12.4-2}$$

where $\tau = D/v_s$ is the time it takes the sound to cross the optical-beam diameter.

---

[11] According to (2.5-18), $\theta_{\text{beam}} = \lambda/\pi n\omega_0$ is the half-apex diffraction angle, so the full-beam diffraction angle can be taken as

$$\theta_{\text{diffraction}} = 2\theta_{\text{beam}} = \frac{4\lambda}{\pi nD}$$

where $D = 2\omega_0$ is the Gaussian spot diameter.

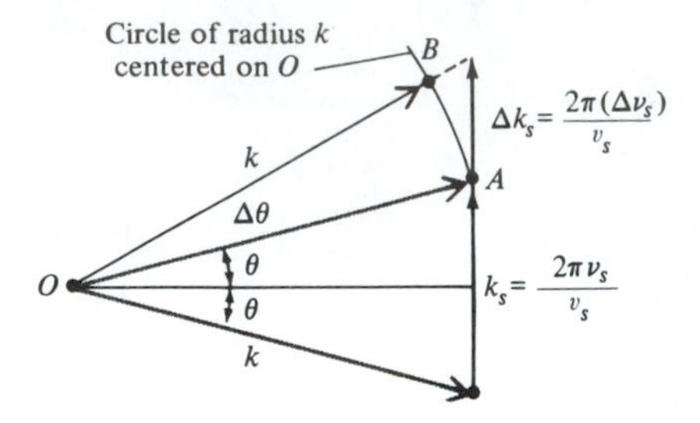

Figure 12-7 Momentum diagram, illustrating how the change in sound frequency from ν_s to $\nu_s + \Delta\nu_s$ deflects the diffracted light beam from θ to $\theta + \Delta\theta$.

Numerical Example: Beam Deflection

Consider a deflection system using flint glass and a sound beam that can be varied in frequency from 80 MHz to 120 MHz; thus, $\Delta\nu_s = 40$ MHz. Let the optical beam diameter be $D = 1$ cm. From Table 12-1 we obtain $v_s = 3.1 \times 10^5$ cm/s; therefore, $\tau = D/v_s = 3.23 \times 10^{-6}$ seconds and the number of resolvable spots is $N = \Delta\nu_s\tau \simeq 130$.

Bragg interactions have recently been demonstrated [15] between surface acoustic waves and optical modes confined in thin film dielectric waveguides. Since the

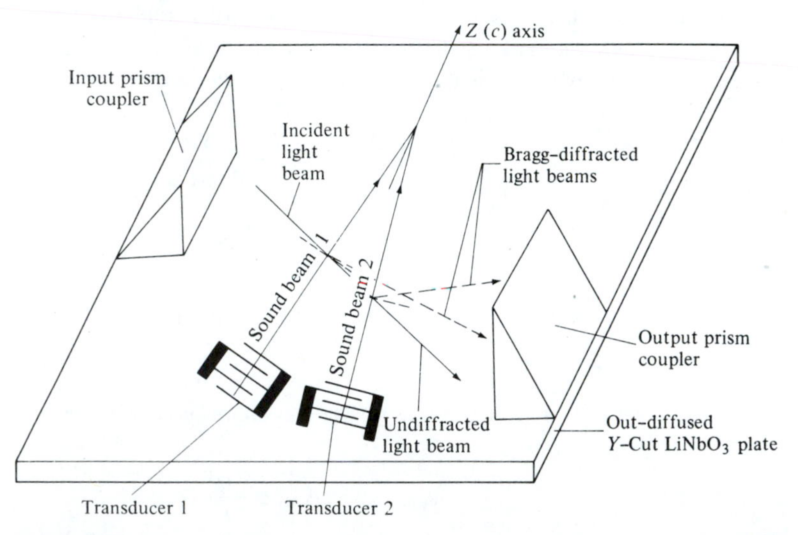

Figure 12-8 Guided-wave acoustooptic Bragg diffraction from two tilted surface acoustic waves. (Reference [16].)

modulation efficiency depends, according to (12.3-19), on the acoustic intensity, the confinement of the acoustic power near the surface (to a distance $\sim \lambda_s$) leads to low modulation or switching power.

Figure 12-8 shows an experimental setup in which both the acoustic surface wave and the optical wave are guided in a single crystal of $LiNbO_3$. The dielectric waveguide is produced by out-diffusion from a layer of $\sim$10 μm near the surface, which raises the index of refraction.

For a more advanced treatment of the subject of light and sound interaction in crystals and for some new devices that are based upon it, the student should consult Reference [17].

Problems

12.1 Derive the expression of the frequency shift, under Bragg conditions, from a receding sound wave.

12.2 Design an acoustic modulation system for transferring the output of a magnetic-cartridge phonograph onto an optical beam with $\lambda_0 = 0.6328$ μm and $I_{\text{incident}} = 10^{-3}$ watt. Specify the power levels involved and the essential characteristics of all the key components. [*Hint:* Use the audio output of the cartridge to modulate a high-frequency (100 MHz, say) carrier, which is then used to transduce an acoustic beam.]

12.3 What happens in Bragg diffraction of light from a standing sound wave? Describe the frequency shifts and direction of diffraction.

12.4 Using Figure (12-4) show that under Bragg conditions the change in wave vector of the diffracted wave is

$$\frac{k_{\text{diffracted}} - k}{k} = 2 \sin \theta \frac{v_s}{c}$$

12.5 Consult the literature (see References [4, 5], for example) and describe the difference between Bragg diffraction and Debye-Sears diffraction. Under what conditions is each observed?

12.6 Bragg's law for diffraction of X-rays in crystals is [7]

$$2d \sin \theta = m \frac{\lambda}{n} \qquad m = 1, 2, 3, \ldots$$

where d is the distance between equivalent atomic planes, θ is the angle of incidence, and λ/n is the wavelength of the diffracted radiation. Bragg diffraction of light from sound [see Equation (12.1-4)] takes place when

$$2\lambda_s \sin \theta = \frac{\lambda}{n}$$

Thus, if we compare it to the X-ray result and take $\lambda_s = d$, only the case of $m = 1$ is allowed. Explain the difference. Why don't we get light diffracted along directions

θ corresponding to $m = 2, 3, \ldots$? [*Hint:* The diffraction of X-rays takes place at discrete atomic planes, which can be idealized as infinitely thin sheets, whereas the sound wave is continuous in z; see Figure 12-3.]

12.7 Design an acoustic deflection system using $LiTaO_3$ to be used in scanning an optical beam in a manner compatible with that of commercial television receivers.

References

1. Brillouin, L., ''Diffusion de la lumière et des rayons X par un corps transparent homogène,'' *Ann. Physique* 17:88, 1922.
2. Debye, P., and F. W. Sears, ''On the scattering of light by supersonic waves,'' *Proc. Nat. Acad. Sci. U.S.* 18:409, 1932.
3. Dransfeld, K., ''Kilomegacycle ultrasonics,'' *Sci. Am.* 208:60, 1963.
4. See, for example, Robert Adler, ''Interaction between light and sound,'' *IEEE Spectrum* 4, May 1967:42.
5. Born, M., and E. Wolf, *Principles of Optics.* New York: Pergamon, 1965, Chap. 12.
6. Dixon, R. W., ''Photoelastic properties of selected materials and their relevance for applications to acoustic light modulators and scanners,'' *J. Appl. Phys.* 38:5149, 1967.
7. Kittel, C., *Introduction to Solid State Physics*, 3d ed. New York: Wiley, 1967, p. 38.
8. Quate, C. F., C. D. W. Wilkinson, and D. K. Winslow, ''Interactions of light and microwave sound,'' *Proc. IEEE* 53:1604, 1965.
9. Cohen, M. G., and E. I. Gordon, ''Acoustic beam probing using optical frequencies,'' *Bell Syst. Tech. J.* 44:693, 1965.
10. Cummings, H. Z., and N. Knable, ''Single sideband modulation of coherent light by Bragg reflection from acoustical waves,'' *Proc. IEEE* 51:1246, 1963.
11. Yariv, A., *Quantum Electronics.* New York: Wiley, 1967, Equation (25.4-14).
12. Gordon, E. I., ''A review of acousto-optical deflection and modulation devices,'' *Proc. IEEE* 54:1391, 1966.
13. Gordon, E. I., ''Measurement of light-sound interaction efficiencies in solids,'' *IEEE J. Quant. Elect.* QE-1:283, 1965.
14. D. A. Pinnow, L. G. Van Uitert, A. W. Warner, and W. A. Bonner. ''$PbMO_4$: A melt grown crystal with a high figure of merit for acoustooptic device applications,'' *Appl. Phys. Lett.* 15:83, 1969.
15. Kuhn, L. M., L. Dakss, F. P. Heidrich, and B. A. Scott, ''Deflection of optical guided waves by a surface acoustic wave,'' *Appl. Phys. Lett.* 17:265, 1970.
16. C. S. Tsai, Le T. Nguyen, S. K. Yao, and M. H. Alhaider, ''High performance acousto-optic guided light beam device using two tilting surface acoustic waves,'' *Appl. Phys. Lett.* 26:140, 1975.
17. Yariv, A., and P. Yeh, *Light Propagation in Crystals.* New York: Wiley-Interscience, 1980, Chaps. 9 and 10.

13 Propagation and Coupling of Modes in Optical Dielectric Waveguides—Periodic Waveguides

13.0 INTRODUCTION

In this chapter we discuss a number of topics that involve propagation of optical modes in dielectric films with thicknesses comparable to the wavelength.

The ability to generate, guide, modulate, and detect light in such thin film configurations [1–3] opens up new possibilities for monolithic "optical circuits" [4]—an endeavor going under the name of integrated optics [5].

We will first consider the basic problem of TE and TM mode propagation in slab dielectric waveguides. A coupled-mode formalism is then developed to describe situations involving exchange of power between modes. These include (a) periodic (corrugated) optical waveguides and filters, (b) distributed feedback lasers, (c) electrooptic mode coupling, and (d) directional couplers.

13.1 WAVEGUIDE MODES—A GENERAL DISCUSSION

A prerequisite to understanding guided wave interactions is a knowledge of the properties of the guided modes. A mode of a dielectric waveguide at a (radian) frequency ω is a solution of the wave equation

$$\nabla^2\mathbf{E}(\mathbf{r}) + k_0^2 n^2(\mathbf{r})\mathbf{E}(\mathbf{r}) = 0 \tag{13.1-1}$$

where $k_0^2 \equiv \omega^2\mu\varepsilon_0 = (2\pi/\lambda)^2$ and n is the index of refraction. The solutions are subject to the continuity of the tangential components of $\mathbf{E}$ and $\mathbf{H}$ at the dielectric interfaces. In (13.1-1) the form of the field is taken as

$$\mathbf{E}(\mathbf{r}, t) = \mathbf{E}(x, y)e^{i(\omega t-\beta z)} \tag{13.1-2}$$

so that (13.1-1) becomes

$$\left(\frac{\partial^2}{\partial x^2} + \frac{\partial^2}{\partial y^2}\right)\mathbf{E}(x, y) + [k_0^2 n^2(\mathbf{r}) - \beta^2]\mathbf{E}(x, y) = 0 \tag{13.1-3}$$

The basic features of the behavior of dielectric waveguides can be elucidated with the help of a slab (planar) model in which no variation exists in one (for example, y) dimension. Channel waveguides, in which the waveguide dimensions are finite in both the x and y directions, approach the behavior of the planar guide when one dimension is considerably larger than the other [6, 7]. Even when this is not the case, most of the phenomena of interest are only modified in a simple quantitative way when going from a planar to a channel waveguide. Because of the immense mathematical simplification that results, we will limit most of the following treatment to planar waveguides such as the one shown in Figure 13-1.

Putting $\partial/\partial y = 0$ in (13.1-3) and writing it separately for regions I, II, and III yields

Region I $$\frac{\partial^2}{\partial x^2}E(x, y) + (k_0^2 n_1^2 - \beta^2)E(x, y) = 0 \tag{13.1-4a}$$

Region II $$\frac{\partial^2}{\partial x^2}E(x, y) + (k_0^2 n_2^2 - \beta^2)E(x, y) = 0 \tag{13.1-4b}$$

Region III $$\frac{\partial^2}{\partial x^2}E(x, y) + (k_0^2 n_3^2 - \beta^2)E(x, y) = 0 \tag{13.1-4c}$$

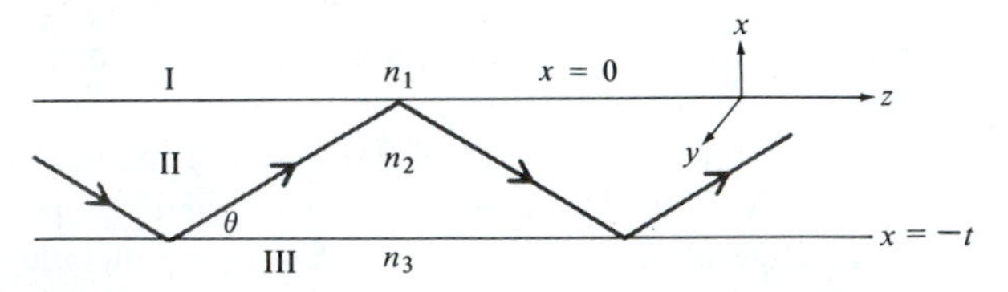

Figure 13-1 A slab ($\partial/\partial y = 0$) dielectric waveguide.

where $E(x, y)$ is a Cartesian component of $\mathbf{E}(x, y)$. Before embarking on a formal solution of (13.1-4), we may learn a great deal about the physical nature of the solutions by simple arguments. Let us consider the nature of the solutions as a function of the propagation constant β at a *fixed* frequency ω. Let us assume that $n_2 > n_3 > n_1$. For $\beta > k_0 n_2$ [that is, regime (a) in Figure 13-2], it follows directly from (13.1-4) that $(1/E)(\partial^2 E/\partial x^2) > 0$ everywhere, and $E(x)$ is exponential in all three layers (I, II, III) of the waveguides. Because of the need to match both $E(x)$ and its derivatives at the two interfaces, the resulting field distribution is as shown in Figure 13-2(a). The field increases without bound away from the waveguide so that the solution is not *physically realizable* and thus does not correspond to a real wave.

For $k_0 n_3 < \beta < k_0 n_2$, as in points (b) and (c), it follows from (13.1-4) that the solution is sinusoidal in region II, since $(1/E)(\partial^2 E/\partial x^2) < 0$, but is exponential in regions I and III. This makes it possible to have a solution $E(x)$ that satisfies the boundary conditions while *decaying* exponentially in regions I and III. Two such solutions are shown in Figure 13-2(b) and (c). The energy carried by these modes is confined to the vicinity of the guiding layer II, and we will, consequently, refer to them as confined, or guided, modes. From the above discussion it follows that a necessary condition for their existence is that $k_0 n_1, k_0 n_3 < \beta < k_0 n_2$ so that confined modes are possible only when $n_2 > n_1, n_3$; that is, the inner layer possesses the highest index of refraction.

Mode solutions for $k_0 n_1 < \beta < k_0 n_3$, regime (d), correspond according to (13.1-4) to exponential behavior in region I and to sinusoidal behavior in regions II and III as illustrated in Figure 13-2(d). We will refer to these modes as substrate radiation modes. For $0 < \beta < k_0 n_1$, as in (e), the solution for $E(x)$ becomes sinusoidal in all three regions. These are the so-called radiation modes of the waveguides.

A solution of (13.1-4) subject to the boundary conditions at the interfaces given in the next section shows that while in regimes (d) and (e) β is a continuous variable, the values of allowed β in the propagation regime $k_0 n_3 < \beta < k_0 n_2$ are *discrete.* The number of confined modes depends on the width, t, the frequency, and the indices of refraction n_1, n_2, n_3. At a given wavelength the number of confined modes increases from 0 with increasing t. At some t, the mode TE_1 becomes confined. Further increases in t will allow TE_2 to exist as well, and so on.

A useful point of view is one of considering the wave propagation in the inner layer 2 as that of a plane wave propagating at some angle θ to the horizontal axis and undergoing a series of total internal reflections at the interfaces II–I and II–III. This is based on (13.1-4b). Assuming a solution in the form of $E \propto \sin(hx + \alpha) \exp(-i\beta z)$, we obtain

$$\beta^2 + h^2 = k_0^2 n_2^2 \tag{13.1-5}$$

The resulting right-angle triangles with sides β, h, and $k_0 n_2$ are shown in Figure 13-2. Note that since the frequency is constant, $k_0 n_2 \equiv (\omega/c) n_2$ is the same for cases (b), (c), (d), and (e). The propagation can thus be considered formally as that of a plane wave along the direction of the hypotenuse with a *constant* propagation constant $k_0 n_2$. As β decreases, θ increases until, at $\beta = k_0 n_3$, the wave ceases to be

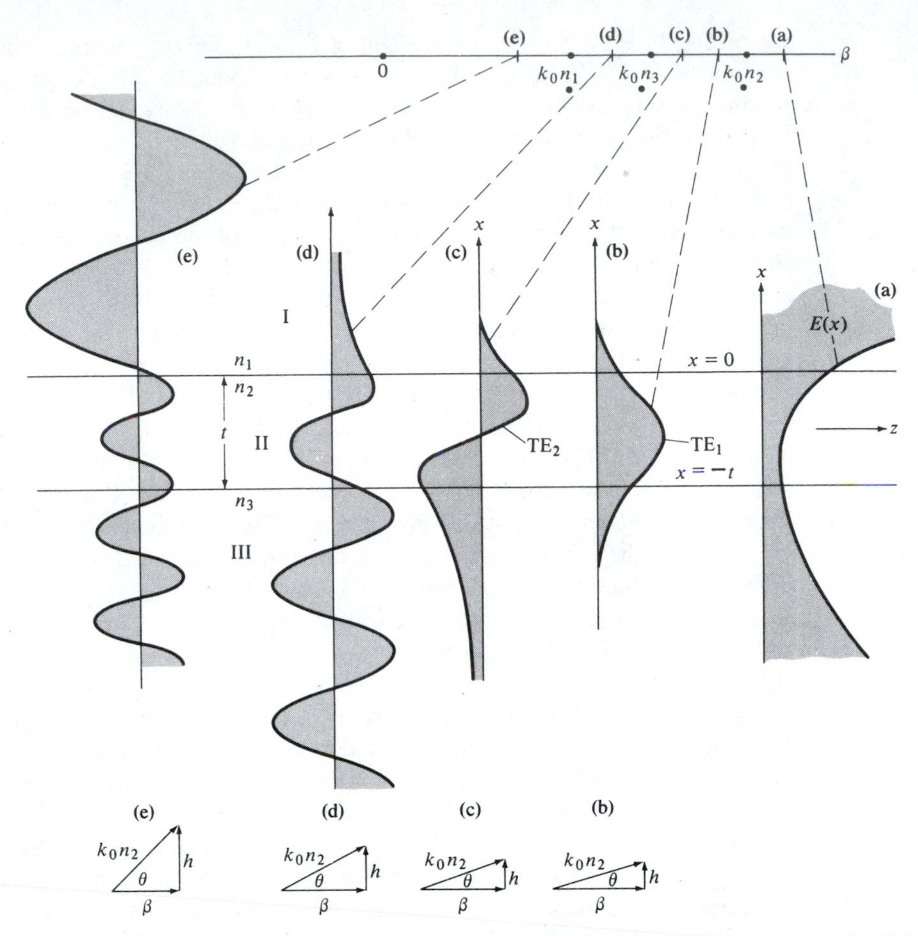

Figure 13-2 (*top*) The different regimes (a, b, c, d, e) of the propagation constant, β, of the waveguide shown in Figure 13-1. (*middle*) The field distributions corresponding to the different value of β. (*bottom*) The propagation triangles corresponding to the different propagation regimes.

totally internally reflected at the interface III–II. This follows from the fact that the guiding condition $\beta > k_0 n_3$ leads, using $\beta = k_0 n_2 \cos\theta$, to $\theta < \cos^{-1}(n_3/n_2) = \theta_c$, where θ_c is the total internal reflection angle at the interface between layers II–III. Since $n_3 > n_1$, total internal reflection at the II–III interface guarantees total internal reflection at the I–II interface.

Confined Modes in a Symmetric Slab Waveguide

Before considering the more general, and difficult, case of asymmetric waveguides, we will solve in some detail the case of the symmetric slab waveguide where $n_1 = n_3$. The guiding layer of index $n_2 > n_1$ occupies the region $-d < x < d$, as in Figure 13-3.

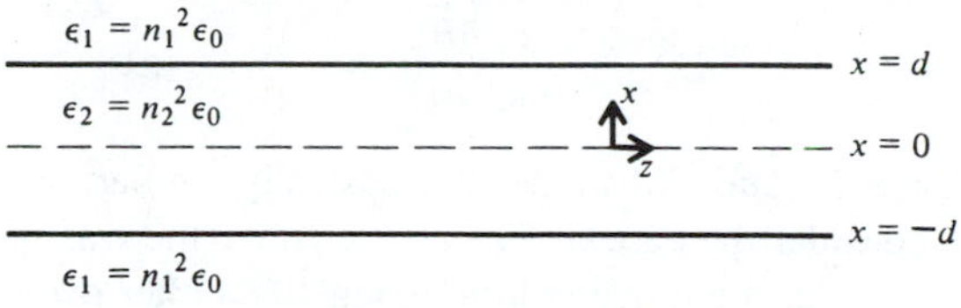

$\epsilon_1 = n_1^2 \epsilon_0$ $x = d$

$\epsilon_2 = n_2^2 \epsilon_0$ x $x = 0$ z

$x = -d$

$\epsilon_1 = n_1^2 \epsilon_0$

Figure 13-3 A symmetric slab waveguide.

We consider the case of harmonic time behavior in the form of exp $(i\omega t)$ and in infinite slab geometry so that there is no variation in the y directions $(\partial/\partial y = 0)$. Maxwell's curl equations (1.2-1) and (1.2-2) become

$$\frac{\partial E_y}{\partial z} = i\omega\mu H_x \tag{13.1-6a}$$

$$\frac{\partial E_x}{\partial z} - \frac{\partial E_z}{\partial x} = -i\omega\mu H_y \tag{13.1-6b}$$

$$\frac{\partial E_y}{\partial x} = -i\omega\mu H_z \tag{13.1-6c}$$

$$\frac{\partial H_y}{\partial z} = -i\omega\varepsilon E_x \tag{13.1-6d}$$

$$\frac{\partial H_x}{\partial z} - \frac{\partial H_z}{\partial x} = i\omega\varepsilon E_y \tag{13.1-6e}$$

$$\frac{\partial H_y}{\partial x} = i\omega\varepsilon E_z \tag{13.1-6f}$$

Next we assume that the modes propagate in the z direction with the z dependence in the form of exp $(-i\beta z)$ so that in (13.1-6) we can replace $\partial/\partial z$ by $-i\beta$. An inspection of (13.1-6) reveals that we may obtain two self-consistent types of solutions. The first contains only E_y, H_x, and H_z and is referred to as transverse electric (TE) modes, since the electric field (E_y) is restricted to the transverse (that is, normal to the direction of propagation) plane. Maxwell's equations (13.1-6) for the TE modes reduce to

$$E_y = -\frac{\omega\mu}{\beta} H_x \tag{13.1-7a}$$

$$\frac{\partial E_y}{\partial x} = -i\omega\mu H_z \tag{13.1-7b}$$

The second type of mode is transverse magnetic (TM) and involves H_y, E_x, and E_z. These are related, according to (13.1-6), by

$$H_y = \frac{\omega\varepsilon}{\beta} E_x \tag{13.1-8a}$$

$$E_z = -\frac{i}{\omega\varepsilon}\frac{\partial H_y}{\partial x} \tag{13.1-8b}$$

The solutions for the TE and TM modes are basically similar, so that in order to be specific we will consider the case of TE modes. Since the waveguide is symmetric about the plane $x = 0$, the mode solutions must be either even or odd in x, that is,

$$E_y(x, z, t) = E_y(-x, z, t)$$

in the case of even modes, and

$$E_y(x, z, t) = -E_y(-x, z, t)$$

for the odd modes. The solution for the even modes is taken in the form

$$E_y = A \exp[-p(|x| - d) - i\beta z] \qquad |x| \geq d \tag{13.1-9}$$

and

$$E_y = B \cos(hx) \exp(-i\beta z) \qquad |x| \leq d \tag{13.1-10}$$

where p and h are positive real constants to be determined. From (13.1-7b) we obtain

$$H_z = \mp\frac{ipA}{\omega\mu} \exp[-p(|x| - d) - i\beta z] \qquad |x| \geq d \tag{13.1-11}$$

The $(-)$ sign is used with $x \geq d$ and $(+)$ for $x \leq -d$

$$H_z = -\frac{ihB}{\omega\mu} \sin(hx) \exp(-i\beta z) \qquad |x| \leq d \tag{13.1-12}$$

Next we require that the tangential field components E_y and H_z be continuous across the interfaces.[1] The continuity of E_y at $x = \pm d$ leads according to (13.1-9) and (13.1-10) to

$$A = B \cos(hd) \tag{13.1-13}$$

while the continuity of H_z results in

$$pA = hB \sin(hd) \tag{13.1-14}$$

From (13.1-13) and (13.1-14) it follows that

$$pd = hd \tan(hd) \tag{13.1-15}$$

Since the field solutions (13.1-9) and (13.1-10) must satisfy the wave equation (13.1-4a,b), the following relations are obeyed

$$\beta^2 = k_0^2 n_2^2 - h^2$$
$$\beta^2 = k_0^2 n_1^2 + p^2 \tag{13.1-16}$$

[1]The reasons for these conditions are discussed in any elementary text on electromagnetic theory.

The last two equations can be combined to give

$$(pd)^2 + (hd)^2 = (n_2^2 - n_1^2)k_0^2 d^2 \qquad (13.1\text{-}17)$$

The propagation constants p and h of a given mode need to satisfy, simultaneously, (13.1-15) and (13.1-17). A straightforward graphical solution is illustrated by Figure 13-4 and consists of finding the intersections in the pd-hd plane of the circle $(pd)^2 + (hd)^2 = (n_2^2 - n_1^2)k_0^2 d^2$ with the curve $pd = hd \tan(hd)$. Each intersection with a $p > 0$ corresponds to a confined mode. The propagation constant β of a given mode can be obtained, once p and h are given, from (13.1-16).

To appreciate the nature of the solutions, let us consider what happens in a given waveguide (that is, fixed n_1, n_2, and d) as the frequency increases gradually from zero. Since $k_0 = \omega/c$, the effect of increasing the frequency is to increase the radius of the circle $(pd)^2 + (hd)^2 = (n_2^2 - n_1^2)k_0^2 d^2$. At low frequencies such that

$$0 < \sqrt{n_2^2 - n_1^2}\, k_0 d < \pi \qquad (13.1\text{-}18)$$

only one intersection (point A) exists between the circle and the curve $pd = hd \tan(hd)$ with $p > 0$. This is evident from an inspection of Figure 13-4. The mode is designated as TE_1 and has a tranverse h parameter falling within the range

$$0 < h_1 d < \frac{\pi}{2} \qquad (13.1\text{-}19)$$

so that it has no zero crossings in the interior of the slab $|x| \leq d$.

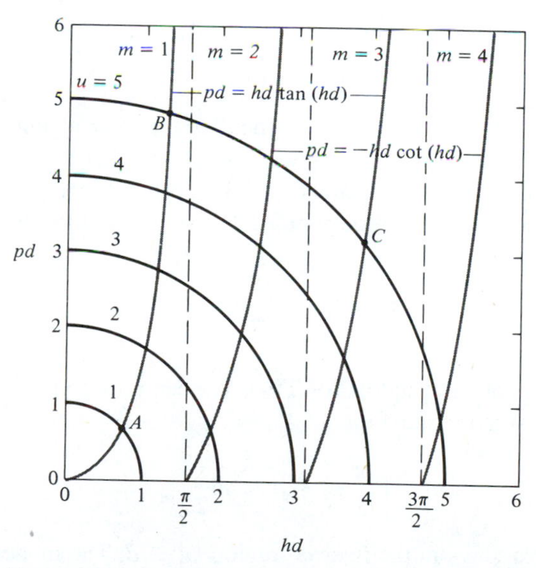

Figure 13-4 Plot of eigenvalue equations $pd = hd \tan(hd)$ for even TE modes, $pd = -hd \cot(hd)$ for odd TE modes, and the supplementary relationship $(pd)^2 + (hd)^2 = (n_2^2 - n_1^2)(k_0 d)^2 \equiv u^2$. (After Reference [8].)

When the parameter $u \equiv \sqrt{n_2^2 - n_1^2}k_0 d$ falls within the range

$$\pi < \sqrt{n_2^2 - n_1^2}k_0 d < 2\pi \tag{13.1-20}$$

we obtain two intersections with $p > 0$. One (point B) corresponds to a value of $hd < \pi/2$ and is thus that of the lowest order TE_1 mode. In the second mode (point C)

$$\pi < h_3 d < \frac{3\pi}{2} \tag{13.1-21}$$

and consequently this mode has two zero crossings (that is, points where $E_y = 0$) in the region $|x| < d$. This is the so-called TE_3 mode ($m = 3$ in Figure 13-4). Both of these modes correspond to the same frequency and can thus be excited simultaneously by the same input field. We notice, however, that the TE_1 mode has a larger value of p (that is, $p_1 > p_3$) and is therefore more highly confined to the interior slab. It also follows from (13.1-16) that $\beta_1 > \beta_3$, so that the phase velocity $v_1 = \omega/\beta_1$ of the TE_1 mode is smaller than that of the TE_3 mode.

From (13.1-19) and (13.1-21) one would conclude that no mode exists with $\pi/2 < hd < \pi$. This is due to the fact that up to this point, we considered only modes with even x symmetry as in (13.1-10). Another family of modes—the odd TE modes—exists and is described by

$$E_y = \begin{cases} A \exp[-p(x - d) - i\beta z] & x \geq d \\ -A \exp[p(x + d) - i\beta z] & x \leq -d \end{cases} \tag{13.1-22}$$

$$E_y = B \sin(hx) \exp(-i\beta z) \qquad |x| \leq d \tag{13.1-23}$$

Applying the continuity conditions at $|x| = d$ leads to

$$pd = -hd \cot(hd) \tag{13.1-24}$$

instead of (13.1-15). The mode solutions correspond to the intersection of (13.1-24) with the circle (13.1-17). Reference to Figure 13-4 shows that the corresponding values of h do indeed fill the gaps ''avoided'' by the even TE modes.

The lowest order odd TE mode is designated TE_2 ($m = 2$), since its h parameter h_2 satisfies

$$\frac{\pi}{2} < h_2 d < \pi \tag{13.1-25}$$

thus falling between h_1 (of TE_1) and h_3 (of TE_3). We can now generalize and state that the mth (TE or TM) mode satisfies

$$(m - 1)\frac{\pi}{2} < h_m d < m\frac{\pi}{2} \tag{13.1-26}$$

and has $m - 1$ zero crossings in the internal region $|x| \leq d$. The modes with m 1, 3, 5, . . . are even symmetric, while those with $m = 2, 4, 6$ . . . are odd. We note that all the modes except the fundamental ($m = 1$) can exist (that is, are confined) only above a ''cutoff'' frequency. The higher the mode index, the higher its cutoff frequency. The fundamental mode can exist at any frequency, as is evident from Figure 13-4. If the dielectric waveguide is asymmetric ($n_1 \neq n_3$), the lowest order ($m = 1$) modes also possess a cutoff frequency. This case will be taken up in the next section.

The general features of TM modes are similar to those of TE modes except that the corresponding values of p are somewhat smaller, indicating a lesser degree of confinement. A larger fraction of the total TM mode power thus propagates in the outer media compared to a TE mode of the same order. This point is taken up in Problem 13.7.

13.2 TE AND TM MODES IN AN ASYMMETRIC SLAB WAVEGUIDE

In this section we will derive the mode solutions for the general asymmetric ($n_1 \neq n_3$) slab waveguide shown in Figure 13-1. We limit the derivation to the guided modes that according to Figure 13-2 have propagation constants β

$$k_0 n_3 < \beta < k_0 n_2$$

where, to be specific, $n_3 > n_1$.

TE modes

The field component E_y of the TE modes obeys the wave equation

$$\nabla^2 E_{yi}(x, y, z) + \omega^2 \mu \varepsilon_0 n_i^2 E_{yi} = 0 \qquad i = 1, 2, 3 \tag{13.2-1}$$

where i refers to the layer and the (real) electric field is given by

$$E_{yi}(x, y, z, t) = \text{Re}[E_{yi}(x, y, z) e^{i\omega t}]$$

For waves propagating along the z direction and for $\partial/\partial y = 0$ we have

$$E_{yi}(x, y, z) = \mathscr{E}_{yi}(x) e^{-i\beta z} \tag{13.2-2}$$

The transverse function $\mathscr{E}_{yi}(x)$ is taken as

$$\mathscr{E}_y = \begin{cases} C \exp(-qx) & 0 \leq x < \infty \\ C\left(\cos hx - \dfrac{q}{h} \sin hx\right) & -t \leq x \leq 0 \\ C\left(\cos ht + \dfrac{q}{h} \sin ht\right) \exp[p(x + t)] & -\infty < x \leq -t \end{cases} \tag{13.2-3}$$

Applying (13.2-1) to (13.2-3) results in

$$h = (n_2^2 k_0^2 - \beta^2)^{1/2}$$

$$q = (\beta^2 - n_1^2 k_0^2)^{1/2}$$

$$p = (\beta^2 - n_3^2 k_0^2)^{1/2}$$

$$k_0 \equiv \frac{\omega}{c} \tag{13.2-4}$$

The acceptable solutions for $\mathscr{E}_y$ and $\mathscr{H}_z = (i/\omega\mu)(\partial\mathscr{E}_y/\partial x)$ should be continuous at both $x = 0$ and $x = -t$. The choice of coefficients in (13.2-3) is such as to make $\mathscr{E}_y$ continuous at both interfaces as well as $(\partial\mathscr{E}_y/\partial x)$ at $x = 0$. By imposing the continuity requirement on $\partial\mathscr{E}_y/\partial x$ at $x = -t$, we get from (13.2-3)

$$h \sin ht - q \cos ht = p\left(\cos ht + \frac{q}{h} \sin ht\right)$$

or

$$\tan ht = \frac{p + q}{h(1 - pq/h^2)} \tag{13.2-5}$$

In the symmetric case ($n_1 = n_3$) the field (13.2-3) must be odd or even about the midplane $x = -t/2$. This special case was treated in Section 13.1 and leads to the eigenvalue equations (13.1-15) and (13.1-24). The last equation in conjunction with (13.2-4) is used to obtain the eigenvalues β for the confined TE modes. An example of such a solution is shown in Figure 13-5.

The constant, C, appearing in (13.2-3) is arbitrary, yet for many applications, especially those in which propagation and exchange of power involve more than one mode, it is advantageous to define C in such a way that it is simply related to total power in the mode. This point will become clear in Section 13.3. We choose C so that the field $\mathscr{E}_y(x)$ in (13.2-3) corresponds to a power flow of one watt (per unit

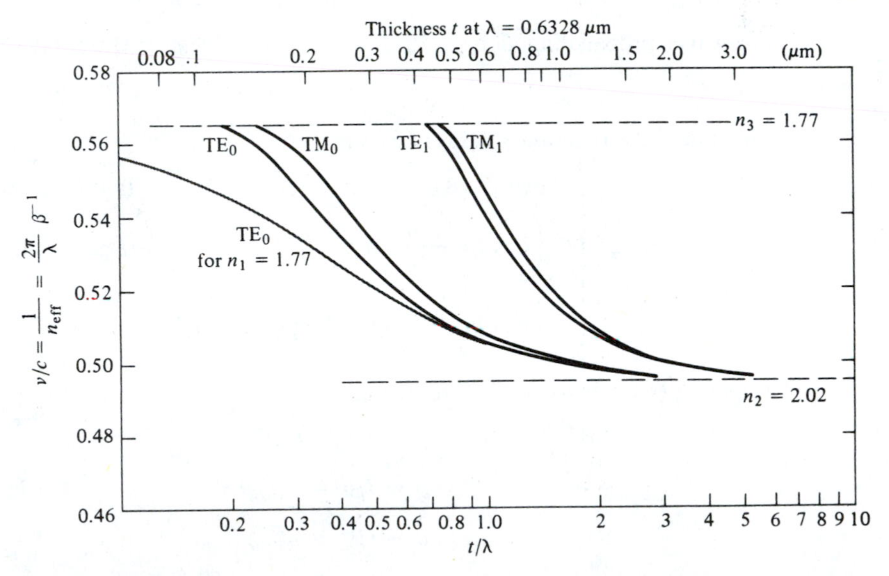

Figure 13-5 Dispersion curves for the confined modes of ZnO on sapphire waveguide $n_1 = 1$. (After Reference [10].)

width in y direction) in the mode. A mode for which $E_y = A\mathscr{E}_y(x)$ will thus correspond to a power flow of $|A|^2$ watts/m. The normalization condition becomes

$$-\frac{1}{2}\int_{-\infty}^{\infty} E_y H_x^* \, dx = \frac{\beta_m}{2\omega\mu}\int_{-\infty}^{\infty} [\mathscr{E}_y^{(m)}(x)]^2 \, dx = 1 \qquad (13.2\text{-}6)$$

where the symbol m denotes the mth confined TE mode [corresponding to the mth eigenvalue of (13.2-5)] and $H_x = -i(\omega\mu)^{-1} \, \partial E_y/\partial z$.

Using (13.2-3) in (13.2-6) leads, after substantial but straightforward calculation, to

$$C_m = 2h_m\left(\frac{\omega\mu}{|\beta_m|[t + (1/q_m) + (1/p_m)](h_m^2 + q_m^2)}\right)^{1/2} \qquad (13.2\text{-}7)$$

Since the modes $\mathscr{E}_y^{(m)}$ are orthogonal (see Problem 13.6), we have

$$\int_{-\infty}^{\infty} \mathscr{E}_y^{(l)}\mathscr{E}_y^{(m)} \, dx = \frac{2\omega\mu}{\beta_m}\,\delta_{l,m} \qquad (13.2\text{-}8)$$

TM modes

The derivation of the confined TM modes is similar in principle to that of the TE modes. Using (13.1-6) the field components are

$$\begin{aligned} H_y(x, z, t) &= \mathscr{H}_y(x)e^{i(\omega t - i\beta z)} \\ E_x(x, z, t) &= \frac{i}{\omega\varepsilon}\frac{\partial H_y}{\partial z} = \frac{\beta}{\omega\varepsilon}\mathscr{H}_y(x)e^{i(\omega t - \beta z)} \\ E_z(x, z, t) &= -\frac{i}{\omega\varepsilon}\frac{\partial H_y}{\partial x} \end{aligned} \qquad (13.2\text{-}9)$$

The transverse function, $\mathscr{H}_y(x)$, is taken as

$$\mathscr{H}_y(x) = \begin{cases} -C\left(\dfrac{h}{\overline{q}}\cos ht + \sin ht\right)e^{p(x+t)} & x \le -t \\ C\left(-\dfrac{h}{\overline{q}}\cos hx + \sin hx\right) & -t \le x \le 0 \\ -\dfrac{h}{\overline{q}}Ce^{-qx} & x \ge 0 \end{cases} \qquad (13.2\text{-}10)$$

The continuity of H_y and E_z at the two interfaces leads, in a manner similar to (13.2-5), to the eigenvalue equation

$$\tan ht = \frac{h(\overline{p} + \overline{q})}{h^2 - \overline{pq}} \qquad (13.2\text{-}11)$$

where

$$\overline{p} \equiv \frac{n_2^2}{n_3^2}p \qquad \overline{q} = \frac{n_2^2}{n_1^2}q$$

The normalization constant, C, is chosen so that the field represented by (13.2-9) and (13.2-10) carries *one* watt per unit width in the y direction

$$\frac{1}{2}\int_{-\infty}^{\infty} H_y E_x^* \, dx = \frac{\beta}{2\omega}\int_{-\infty}^{\infty} \frac{\mathcal{H}_y^2(x)}{\varepsilon(x)}\, dx = 1$$

or, using $n_i^2 \equiv \varepsilon_i/\varepsilon_0$,

$$\int_{-\infty}^{\infty} \frac{[\mathcal{H}_y^{(m)}(x)]^2}{n^2(x)}\, dx = \frac{2\omega\varepsilon_0}{\beta_m} \tag{13.2-12}$$

Carrying out the integration using (13.2-10) gives

$$C_m = 2\sqrt{\frac{\omega\varepsilon_0}{\beta_m t_{\text{eff}}}}$$

$$t_{\text{eff}} \equiv \frac{\overline{q}^2 + h^2}{\overline{q}^2}\left(\frac{t}{n_2^2} + \frac{q^2 + h^2}{\overline{q}^2 + h^2}\frac{1}{n_1^2 q} + \frac{p^2 + h^2}{\overline{p}^2 + h^2}\frac{1}{n_3^2 p}\right) \tag{13.2-13}$$

The general properties of the TE and TM mode solutions are illustrated in Figure 13-5. In general a mode becomes confined above a certain (cutoff) value of t/λ. At the cutoff value $p = 0$, and the mode extends to $x = -\infty$. For increasing values of t/λ, $p > 0$, and the mode becomes increasingly confined to layer 2. This is reflected in the effective mode index $\beta\lambda/2\pi$ that, at cutoff, is equal to n_3, and which, for large t/λ, approaches n_2. In a symmetric waveguide ($n_1 = n_3$) the lowest order mode TE_0 has no cutoff and is confined for all values of t/λ. The selective excitation of waveguide modes by means of prism couplers and a determination of their propagation constants β_m are described in Reference [11].

13.3 A PERTURBATION THEORY OF COUPLED MODES IN DIELECTRIC OPTICAL WAVEGUIDES

In Section 13.2 we obtained solutions for the confined modes supported by a slab dielectric waveguide such as that shown in Figure 13-1. An increasingly large number of experiments and devices involve coupling between such modes [12, 13, 17]. Typical examples are TM-to-TE mode conversion by the electrooptic or acoustooptic effect [12] or coupling of forward-to-backward modes by means of a corrugation in one of the waveguides interfaces [15, 16]. In this section we will develop a formalism for describing such coupling.

We start with the wave equation in the form

$$\nabla^2\mathbf{E}(\mathbf{r}, t) = \mu\epsilon_0\frac{\partial^2\mathbf{E}(\mathbf{r}, t)}{\partial t^2} + \mu\frac{\partial^2}{\partial t^2}\mathbf{P}(\mathbf{r}, t) \tag{13.3-1}$$

The total medium polarization can be taken as the sum

$$\mathbf{P}(\mathbf{r}, t) = \mathbf{P}_0(\mathbf{r}, t) + \mathbf{P}_{\text{pert}}(\mathbf{r}, t) \tag{13.3-2}$$

where

$$\mathbf{P}_0(\mathbf{r}, t) = [\varepsilon(\mathbf{r}) - \varepsilon_0]\mathbf{E}(\mathbf{r}, t) \tag{13.3-3}$$

is the polarization induced by $\mathbf{E}(\mathbf{r}, t)$ in the *unperturbed* waveguide whose dielectric constant is $\varepsilon(\mathbf{r})$. The perturbation polarization $\mathbf{P}_{\text{pert}}(\mathbf{r}, t)$ is then defined by (13.3-2) and represents any deviation of the polarization from that of the unperturbed waveguide. Using (13.3-2) and (13.3-3) in (13.3-1) gives

$$\nabla^2 E_y - \mu\varepsilon(\mathbf{r})\frac{\partial^2 E_y}{\partial t^2} = \mu\frac{\partial^2}{\partial t^2}[P_{\text{pert}}(\mathbf{r}, t)]_y \tag{13.3-4}$$

and similar expressions for E_x and E_z.

Ignoring the possibility of coupling to the continuum of radiation modes, regimes *d* and *e* in Figure 13-2, we expand the total field in the "perturbed" waveguide as a superposition of confined modes

$$E_y(\mathbf{r}, t) = \tfrac{1}{2}\sum_m A_m(z)\mathscr{E}_y^{(m)}(x)e^{i(\omega t-\beta_m z)} + \text{c.c.} \tag{13.3-5}$$

where m indicates the mth discrete eigenmode of (13.2-5), which satisfies

$$\left(\frac{\partial^2}{\partial x^2} - \beta_m^2\right)\mathscr{E}_y^{(m)}(\mathbf{r}) + \omega^2\mu\varepsilon(\mathbf{r})\mathscr{E}_y^{(m)}(\mathbf{r}) = 0 \tag{13.3-6}$$

where $\varepsilon(\mathbf{r}) = \varepsilon_0 n^2(\mathbf{r})$.

Substitution of (13.3-5) in (13.3-4) leads to

$$\begin{aligned} e^{i\omega t}\sum_m \Bigg[\frac{A_m}{2}\left(-\beta_m^2\mathscr{E}_y^{(m)} + \frac{\partial^2\mathscr{E}_y^{(m)}}{\partial x^2} + \omega^2\mu\varepsilon(\mathbf{r})\mathscr{E}_y^{(m)}\right)e^{-i\beta_m z} \\ +\frac{1}{2}\left(-2i\beta_m\frac{dA_m}{dz} + \frac{d^2A_m}{dz^2}\right)\mathscr{E}_y^{(m)}\,e^{-i\beta_m z}\Bigg] + \text{c.c.} \\ = \mu\frac{\partial^2}{\partial t^2}[P_{\text{pert}}(\mathbf{r}, t)]_y \end{aligned} \tag{13.3-7}$$

First we note that in view of (13.3-6) the sum of the first three terms in (13.3-7) is zero. We assume "slow" variation so that

$$\left|\frac{d^2A_m}{dz^2}\right| \ll \beta_m\left|\frac{dA_m}{dz}\right|$$

and obtain from (13.3-7)

$$\sum_m - i\beta_m\frac{dA_m}{dz}\mathscr{E}_y^{(m)}\,e^{i(\omega t-\beta_m z)} + \text{c.c.} = \mu\frac{\partial^2}{\partial t^2}[P_{\text{pert}}(\mathbf{r}, t)]_y \tag{13.3-8}$$

We take the product of (13.3-8) with $\mathscr{E}_y^{(s)}(x)$ and integrate from $-\infty$ to ∞. The result, using (13.2-8), is

$$\frac{dA_s^{(-)}}{dz} e^{i(\omega t+\beta_s z)} - \frac{dA_s^{(+)}}{dz} e^{i(\omega t-\beta_s z)} - \text{c.c.}$$
$$= -\frac{i}{2\omega}\frac{\partial^2}{\partial t^2}\int_{-\infty}^{\infty} [P_{\text{pert}}(\mathbf{r}, t)]_y \mathscr{E}_y^{(s)}(x)\, dx \qquad (13.3\text{-}9)$$

The presence of two terms on the left side of Equation (13.3-9) is due to the fact that the summation over m in (13.3-8) contains two terms involving $\mathscr{E}_y^{(m)}(x)$ for each value of m—one, designated as $(-)$, traveling in the $-z$ direction, and the other $(+)$, traveling in the $+z$ direction.

Equation 13.3-9 can be used to treat a large variety of mode interactions [12]. Each physical example involves, in general, a different perturbation polarization $\mathbf{P}_{\text{pert}}$ $(\mathbf{r}, t)$. They all, however, lead to the same set of "coupled mode" equations of the form of Equation (13.5-1). Some important examples are considered in the following sections.

13.4 PERIODIC WAVEGUIDE

Consider a periodic dielectric waveguide in which the periodicity is due to a corrugation of one of the interfaces as shown in Figure 13-6. Such periodic waveguides are used for optical filtering [16] as well as in the distributed feedback laser [17–19]. These two applications will be described further below.

The corrugation is described by the dielectric perturbation $\Delta\varepsilon(\mathbf{r}) \equiv \varepsilon_0 \Delta n^2(\mathbf{r})$ such that the total dielectric constant is

$$\varepsilon'(\mathbf{r}) = \varepsilon(\mathbf{r}) + \Delta\varepsilon(\mathbf{r})$$

The perturbation polarization is from (13.3-2) and (13.3-3)

$$\mathbf{P}_{\text{pert}}(\mathbf{r}, t) = \Delta\varepsilon(\mathbf{r})\mathbf{E}(\mathbf{r}, t) = \Delta n^2(\mathbf{r})\varepsilon_0 \mathbf{E}(\mathbf{r}, t) \qquad (13.4\text{-}1)$$

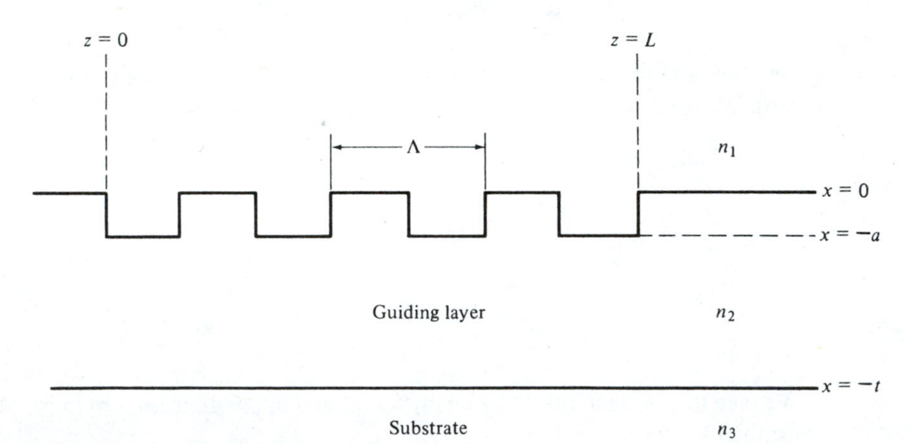

Figure 13-6 A corrugated periodic waveguide.

Since $\Delta n^2(\mathbf{r})$ is a scalar, it follows, from (13.3-4), that the corrugation couples only TE to TE modes and TM to TM, but not TE to TM.

To be specific consider TE mode propagation. Using (13.3-5) in (13.4-1) gives

$$[P_{\text{pert}}(\mathbf{r}, t)]_y = \frac{\Delta n^2(\mathbf{r})\varepsilon_0}{2} \sum_m [A_m \mathscr{E}_y^{(m)}(x) e^{i(\omega t - \beta_m z)} + \text{c.c.}] \tag{13.4-2}$$

which, when used in (13.3-9), leads to

$$\frac{dA_s^{(-)}}{dz} e^{i(\omega t + \beta_s z)} - \frac{dA_s^{(+)}}{dz} e^{i(\omega t - \beta_s z)} - \text{c.c.}$$

$$= -\frac{i\varepsilon_0}{4\omega} \frac{\partial^2}{\partial t^2} \sum_m \left[A_m \int_{-\infty}^{\infty} \Delta n^2(x, z) \mathscr{E}_y^{(m)}(x) \mathscr{E}_y^{(s)}(x)\, dx e^{i(\omega t - \beta_m z)} + \text{c.c.} \right] \tag{13.4-3}$$

We may consider the right side of (13.4-3) as a source wave term driving the forward wave $A_s^{(+)} \exp[i(\omega t - \beta_s z)]$ and the backward wave $A_s^{(-)} \exp[i(\omega t + \beta_s z)]$ on the left side. In order for a wave to be driven by a source, both source wave and driven wave must have the same frequency so that the interaction will not average out to zero over a long time (long compared to a period of their difference frequency). Equally important: Both source and wave need to have nearly the same phase dependence $\exp(i\beta z)$ so that the interaction does not average out to zero with distance of propagation z. If, for example, it is desired that the forward wave $A_s^{(+)} \exp[i(\omega t - \beta_s z)]$ be excited, it is necessary that at least one term on the right side of (13.4-3), say the lth one, vary as $\exp[i(\omega t - \beta z)]$ with $\beta \approx \beta_s$. If no other terms on the right side of (13.4-3) satisfy this condition, we simplify the equation by keeping only the forward wave on the left side and the lth on the right. We describe this situation by saying that the perturbation $\Delta n^2(x, z)$ couples the forward $(+s)$ mode to the lth mode and vice versa.

To be specific, let us assume that the period Λ in the z direction of the perturbation $\Delta n^2(x, z)$ is so chosen that $l\pi/\Lambda \approx \beta_s$ for some integer l. We can expand $\Delta n^2(x, z)$ of a square wave perturbation as a Fourier series

$$\Delta n^2(x, z) = \Delta n^2(x) \sum_{q=-\infty}^{\infty} a_q e^{i(2q\pi/\Lambda)z} \tag{13.4-4}$$

The right side of (13.4-3) now contains a term $(q = l,\ m = s)$ proportional to $A_s^{(+)} \exp[i(2l\pi/\Lambda - \beta_s)z]$. But

$$\frac{2l\pi}{\Lambda} - \beta_s \approx \beta_s$$

so that this term is capable of driving synchronously the amplitude $A_s^{(-)} \exp(i\beta_s z)$ on the left side of (13.4-3) with the result

$$\frac{dA_s^{(-)}}{dz} = \frac{i\omega\varepsilon_0}{4} A_s^{(+)} \int_{-\infty}^{\infty} \Delta n^2(x)[\mathscr{E}_y^{(s)}(x)]^2\, dx a_l e^{i[(2l\pi/\Lambda) - 2\beta_s]z} \tag{13.4-5}$$

The coupling between the backward $A_s^{(-)}$ and the forward $A_s^{(+)}$ by the lth harmonic of $\Delta n^2(x, z)$ can thus be described by

$$\frac{dA_s^{(-)}}{dz} = \kappa A_s^{(+)} e^{-i2(\Delta\beta)z} \tag{13.4-6}$$

and reciprocally

$$\frac{dA_s^{(+)}}{dz} = \kappa^* A_s^{(-)} e^{i2(\Delta\beta)z}$$

where

$$\kappa = \frac{i\omega\varepsilon_0 a_l}{4} \int_{-\infty}^{\infty} \Delta n^2(x)[\mathscr{E}_y^{(s)}(x)]^2 \, dx \tag{13.4-7}$$

$$\Delta\beta \equiv \beta_s - \frac{l\pi}{\Lambda} \equiv \beta_s - \beta_0 \tag{13.4-8}$$

Some General Properties of the Coupled Mode Equations

The coupled mode equations (13.4-6), first encountered here, play a major role in guided wave optics [12]. They describe not only coupling between modes due to a spatially periodic index perturbation but also, as we shall show later, many other situations, including electrooptic and acoustooptic coupling between modes in neighboring waveguides and within the same waveguide. It is thus worthwhile to pause and consider some of the basic properties of these equations, which are independent of the specific application.

To simplify our notation, we replace the amplitude $A_s^{(-)}$ of the backward propagating mode by A, and that of the forward mode by B. The coupled mode equations (13.4-6) now read

$$\begin{aligned} \frac{dA}{dz} &= \kappa_{ab} B \exp[-i2(\Delta\beta)z] \\ \frac{dB}{dz} &= \kappa_{ba} A \exp[i2(\Delta\beta)z] \end{aligned} \tag{13.4-9}$$

$$\begin{aligned} 2(\Delta\beta) &\equiv |\beta_b| + |\beta_a| - \ell \frac{2\pi}{\Lambda} \qquad \ell = 1, 2, 3 \ldots \\ \kappa_{ba} &= \kappa_{ab}^* \end{aligned} \tag{13.4-9a}$$

for modes carrying power in *opposite* directions. The phase of κ_{ab} is immaterial since it depends, according to Equation (13.4-4, 5), on the arbitrary choice of our $z = 0$ reference plane. It follows straightforwardly from (13.4-9) that

$$\frac{d}{dz}(|B(z)|^2 - |A(z)|^2) = 0 \tag{13.4-9b}$$

so that the total electromagnetic power carried by the two modes is conserved. A more detailed example of this type of coupling is treated in Section 13.5.

If the two modes A and B carry power in the same direction, say z, the coupling coefficients in Equations 13.4-9 must obey

$$\kappa_{ab} = -\kappa_{ba}^* \tag{13.4-9c}$$

$$2\Delta\beta \equiv |\beta_b| - |\beta_a| - \ell\frac{2\pi}{\Lambda} \qquad \ell = 0, \pm 1, \pm 2, \ldots \tag{13.4-9d}$$

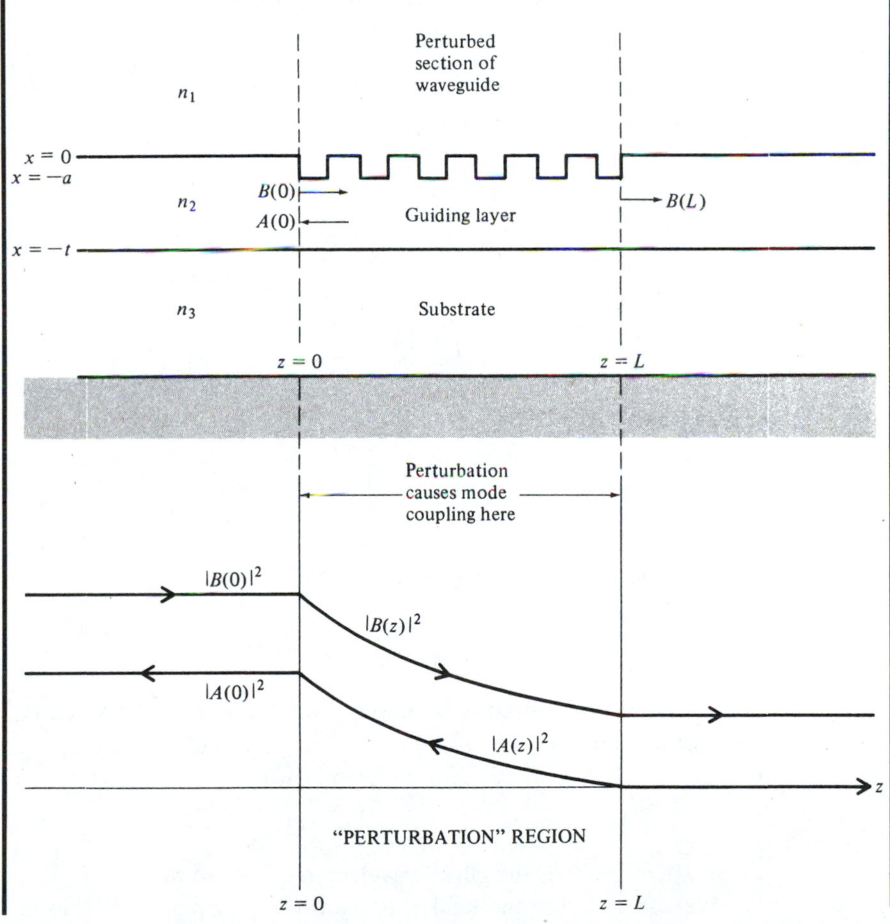

Figure 13-7 (*upper*) A corrugated section of a dielectric waveguide. (*lower*) The incident and reflected intensities inside the corrugated section.

The power conservation takes the form of

$$\frac{d}{dz}(|A(z)|^2 + |B(z)|^2) = 0 \tag{13.4-9e}$$

An example of this type of codirectional coupling is discussed in Section 13.7. A comparison of Figure 13-7 with Figure 13-11 reveals the fundamental difference in the nature of codirectional and contradirectional coupling. While the first case involves a spatially periodic exchange of power between A and B, the second one admits a continuous, one-sense exchange.

Let us consider the specific "square-wave" corrugation of Figure 13-6. In this case the periodicity (period = Λ) in the z direction is accounted for by taking

$$\begin{aligned}\Delta n^2(x, z) &= \Delta n^2(x)\left[\frac{1}{2} + \frac{2}{\pi}\left(\sin \eta z + \frac{1}{3}\sin 3\eta z + \cdots\right)\right] \\ &= \Delta n^2(x)\sum_l a_l e^{i\eta l z} \qquad l = 1, 3, 5, \ldots\end{aligned} \tag{13.4-10}$$

where

$$\Delta n^2(x) = \begin{cases} n_1^2 - n_2^2 & -a \le x \le 0 \\ 0 & \text{elsewhere}\end{cases} \tag{13.4-11}$$

$$\eta \equiv \frac{2\pi}{\Lambda}$$

so that

$$a_l = \begin{cases} \dfrac{-i}{\pi l} & l \text{ odd} \\ 0 & l \text{ even} \\ \frac{1}{2} & l = 0\end{cases}$$

and for l odd we obtain from (13.4-7) and (13.4-10)

$$\kappa = \frac{+\omega\varepsilon_0}{4\pi l}\int_{-\infty}^{\infty} \Delta n^2(x)[\mathscr{E}_y^{(s)}(x)]^2\, dx \tag{13.4-12}$$

In practice the period Λ is chosen so that, for some particular l, $\Delta\beta \approx 0$. We note that for $\Delta\beta = 0$

$$\Lambda = l\frac{\lambda_g^{(s)}}{2} \tag{13.4-13}$$

where $\lambda_g^{(s)} = 2\pi/\beta_s$ is the guide wavelength of the sth mode.

We can now use the field expansion (13.2-3) plus (13.4-11) to perform the integration of (13.4-12).

$$\int_{-\infty}^{\infty} \Delta n^2(x)[\mathscr{E}_y^{(s)}(x)]^2\, dx = (n_1^2 - n_2^2) \int_{-a}^{0} [\mathscr{E}_y^{(s)}(x)]^2\, dx$$
$$= (n_1^2 - n_2^2)C_s^2 \int_{-a}^{0} \left[\cos h_s x - \frac{q_s}{h_s} \sin h_s x\right]^2 dx \tag{13.4-14}$$

Although the integral can be calculated exactly using (13.2-3) and (13.2-5), an especially simple result follows if we consider that operation is sufficiently above propagation cutoff, $t(n_2 - n_3)/s\lambda \gg 1$ so that from (13.2-4) and (13.2-5)

$$\beta_s \approx n_2 k_0$$

$$h_s \to \frac{\pi s}{t} \qquad s = 1, 2, \ldots = \text{transverse mode number}$$

$$\frac{q_s}{h_s} \approx (n_2^2 - n_1^2)^{1/2} \left(\frac{2t}{s\lambda}\right) \tag{13.4-15}$$

The results can be verified using (13.2-4) and (13.2-5). In addition since $q_s \gg h_s$, we have, from (13.2-7),

$$C_s^2 = \frac{4h_s^2 \omega\mu}{\beta_s t q_s^2} \tag{13.4-16}$$

in the well-confined regime and for $h_s a \ll 1$ the integral of (13.4-14) becomes

$$(n_1^2 - n_2^2) \int_{-a}^{0} [\mathscr{E}_y^{(s)}(x)]^2\, dx = (n_1^2 - n_2^2) \frac{4\pi^2 \omega\mu}{3n_2 k_0} \left(\frac{a}{t}\right)^3 s^2 \left(1 + \frac{3}{q_s a} + \frac{3}{q_s^2 a^2}\right)$$

and, using (13.4-15),

$$\kappa_s \approx \frac{2\pi^2 s^2}{3l\lambda} \frac{(n_2^2 - n_1^2)}{n_2} \left(\frac{a}{t}\right)^3 \left[1 + \frac{3}{2\pi} \frac{\lambda/a}{(n_2^2 - n_1^2)^{1/2}} + \frac{3}{4\pi^2} \frac{(\lambda/a)^2}{(n_2^2 - n_1^2)}\right] \tag{13.4-17}$$

The problem of two-wave coupling by a corrugation has thus been reduced to a pair of coupled differential equations (13.4-6) and an expression (13.4-17) for the coupling constant.

13.5 COUPLED-MODE SOLUTIONS

Let us return to the coupled-mode equations (13.4-6). For simplicity let us put $A_s^{(-)} \equiv A$, $A_s^{(+)} \equiv B$ and write them as

$$\frac{dA}{dz} = \kappa_{ab} B e^{-i2(\Delta\beta)z}$$

$$\frac{dB}{dz} = \kappa_{ab}^* A e^{+i2(\Delta\beta)z} \tag{13.5-1}$$

Consider a waveguide with a corrugated section of length L as in Figure 13-6. A wave with an amplitude $B(0)$ is incident from the left on the corrugated section.

The solution of (13.5-1) for this case subject to $A(L) = 0$ is

$$A(z)e^{i\beta z} = B(0)\frac{i\kappa_{ab}e^{i\beta_0 z}}{-\Delta\beta \sinh SL + iS\cosh SL}\sinh[S(z-L)]$$

$$B(z)e^{-i\beta z} = B(0)\frac{e^{-i\beta_0 z}}{-\Delta\beta\sinh SL + iS\cosh SL} \times \{\Delta\beta\sinh[S(z-L)] + iS\cosh[S(z-L)]\} \qquad \textbf{(13.5-2)}$$

where

$$S = \sqrt{\kappa^2 - (\Delta\beta)^2}$$

$$\kappa \equiv |\kappa_{ab}| \qquad \textbf{(13.5-3)}$$

Under the matching condition $\Delta\beta = 0$, we have

$$A(z) = B(0)\frac{\kappa_{ab}}{\kappa}\frac{\sinh[\kappa(z-L)]}{\cosh\kappa L}$$

$$B(z) = B(0)\frac{\cosh[\kappa(z-L)]}{\cosh\kappa L} \qquad \textbf{(13.5-4)}$$

A plot of the mode powers $|B(z)|^2$ and $|A(z)|^2$ for this case is shown in Figure 13-7. For sufficiently large arguments of the cosh and sinh functions in (13.5-4), the incident mode power drops off exponentially along the perturbation region. This behavior, however, is due not to absorption but to *reflection* of power into the backward traveling mode, A.

From (13.3-5) and (13.5-2) we find that the z-dependent parts of the wave solutions in the periodic waveguide are exponentials with propagation constants

$$\beta' = \beta_0 \pm iS = \frac{l\pi}{\Lambda} \pm i\sqrt{\kappa^2 - [\beta(\omega) - \beta_0]^2} \qquad \textbf{(13.5-5)}$$

where we used $\Delta\beta \equiv \beta - \beta_0$, $\beta_0 \equiv \pi l/\Lambda$.

We note that for a range of frequencies such that $\Delta\beta(\omega) < \kappa$, β' has an imaginary part. This is the so-called "forbidden" region in which the evanescence behavior shown in Figure 13-7 occurs and which is formally analogous to the energy gap in semiconductors where the periodic crystal potential causes the electron propagation constants to become complex. Note that for each value of l, $l = 1, 2, 3 \ldots$, there exists a gap whose center frequency ω_{0l} satisfies $\beta(\omega_{0l}) = l\pi/\Lambda$. The exceptions are values of l for which κ is zero. We can approximate $\beta(\omega)$ near its Bragg value $(\pi l/\Lambda)$ by $\beta(\omega) \approx (\omega/c)n_{\text{eff}}$ (n_{eff} is an effective index of refraction). The result is

$$\beta' \cong \frac{l\pi}{\Lambda} \pm i\left[\kappa^2 - \left(\frac{n_{\text{eff}}}{c}\right)^2(\omega - \omega_0)^2\right]^{1/2} \qquad \textbf{(13.5-6)}$$

where ω_0, the midgap frequency, is the value of ω for which the unperturbed β is equal to $\beta_0 \equiv l\pi/\Lambda$.

A plot of Re β' and Im β' (for $l = 1$) versus ω, based on (13.5-6), is shown in Figure 13-8. We note that the height of the ''forbidden'' frequency zone is

$$(\Delta\omega)_{\text{gap}} = \frac{2\kappa c}{n_{\text{eff}}} \tag{13.5-7}$$

where κ is according to (13.4-17) a function of the integer l. It follows from (13.5-6) that

$$(\text{Im } \beta')_{\max} = \kappa \tag{13.5-8}$$

In solid-state physics, it is well known that the behavior of electrons is described by means of electron wave functions that are of the form

$$\Psi_i = u_i(\mathbf{r})\exp\left(-i\frac{E_i t}{\hbar} + ik_i \cdot \mathbf{r}\right)$$

There exist regions of electron energy E_i where the propagation constant $_ik$ is complex independently of the direction of k_i in complete formal analogy with (13.5-6). These are the so-called *forbidden energy gaps* of the crystal. Recent proposals and experiments [16,22] suggest that it should be possible to engineer ''optical crystals'' that have a three-dimensional periodicity that will possess a forbidden frequency gap for which optical propagation will be evanescent (i.e., with a complex propagation constant that is a three-dimensional generalization of (13.5-6)).

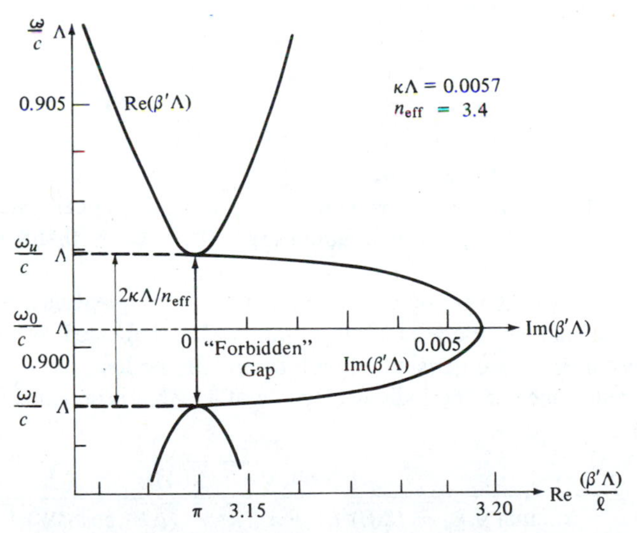

Figure 13-8 Dependence of the real and imaginary parts of the mode propagation constant, β', of the modes in a periodic waveguide. At frequencies $\omega_1 < \omega < \omega_u$, Im$(\beta') \neq 0$ and the modes are evanescent. At these frequencies, Re $\beta' = l\pi/\Lambda$.

Numerical Example

To appreciate the significance of $\Delta n \lesssim 10^{-3}$, which is achievable in a periodic index optical fiber in which the index perturbation is caused by ultraviolet exposure, we will calculate the coupling coefficient κ that results from an index perturbation

$$n^2(x,y,z) = (n_0 + \Delta n)^2 \approx n_0^2 + 2n_0\Delta n \sin \frac{2\pi}{\Lambda} z$$

where n_0 is the average index of refraction of the fiber. We use this result in (13.4-7) with $\Delta n^2(x) = 2n_0\Delta n$ and $a_\ell(\ell = 1) = 2/\pi$. Using the normalization integral (13.2-8) and the approximation $\beta \cong \omega\sqrt{\mu\epsilon_0}n_0$ leads to

$$\kappa = \frac{2\Delta n}{\lambda} \tag{13.5-9}$$

If the filter is to be used at the communication wavelength of $\lambda = 1.55$ μm in a fiber with $\Delta n = 10^{-3}$, the result is $\kappa = (2 \times 10^{-3})/(1.55 \times 10^{-4}) = 12.9$ cm^{-1}. A length of a periodic index fiber 3 mm long can thus result in a reflection of

$$|r(\omega)|^2 = |\tanh(\kappa L)|^2 = 0.998$$

13.6 PERIODIC WAVEGUIDES AS OPTICAL FILTERS AND REFLECTORS—PERIODIC FIBERS

One of the most interesting and potentially important applications of spatially periodic optical waveguides is their use as optical reflectors [16] and filters. These applications owe much of their impetus to advances in fabricating high-efficiency index gratings in spatially doped (and treated) silica fibers by means of exposure to standing wave patterns of ultraviolet light [17, 18, 19, 20, 21]. This is illustrated in Figure 13-9. The index perturbation is in the form

$$\Delta n(x,y,z) = \Delta n_0 \sin \frac{2\pi}{\Lambda} z \tag{13.6-1}$$

$$\Lambda = \frac{\lambda}{2 \sin \theta} \tag{13.6-2}$$

where θ is the incidence angle of the two interfering beams. Practical systems employ excimer lasers ($\lambda = 0.244$ μm) or doubled argon laser ($\lambda = 0.488$ μm) as the radiation source.

The basic feature of the periodic waveguide is that at frequencies near the Bragg frequency ω_0, an incident mode is strongly reflected as indicated by Figure 13-7. Frequencies not near ω_0 are transmitted with essentially no loss.

The field reflectance of a periodic waveguide of length L is obtained from Equations (13.5-2).

$$r(\omega) = \frac{A(0)}{B(0)} = \frac{-i\kappa\sinh(\sqrt{\kappa^2 - (\Delta\beta)^2}L)}{-\Delta\beta\sinh(\sqrt{\kappa^2 - (\Delta\beta)^2}L) + i\sqrt{\kappa^2 - (\Delta\beta)^2}\cosh(\sqrt{\kappa^2 - (\Delta\beta)^2}L)}$$

$$\Delta\beta \equiv \beta(\omega) - \beta_0 = \frac{\omega - \omega_0}{c} n_{\text{eff}}, \qquad \beta_0 = l\frac{\pi}{\Omega}, \; l = 1, 2, 3, \ldots \tag{13.6-3}$$

where n_{eff} is the effective mode index of refraction.

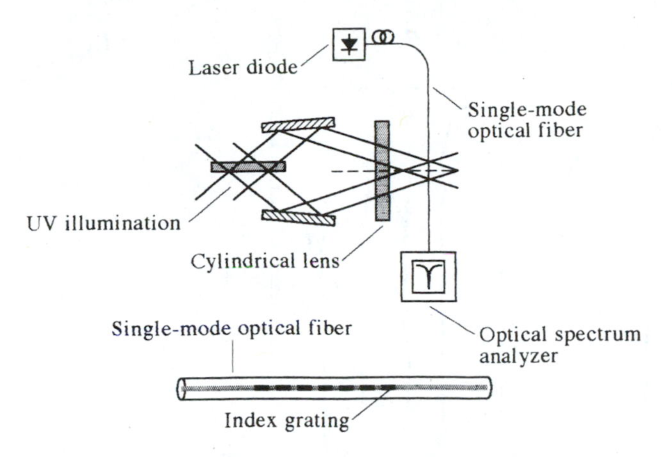

Figure 13-9 The irradiation of a silica fiber by the interferometric standing wave pattern of an ultraviolet light beam causes a periodic index perturbation in the fiber. The period is $\Lambda = \lambda/2 \sin \theta$. Doping with phosphorus and molecular hydrogen-loading of the fiber increase the sensitivity of the fiber to the ultraviolet radiation. The dip shown in the transmitted optical spectrum is due to the strong reflection for frequencies within the "forbidden gap."

Figure 13-10(a) shows a calculated plot of the power reflectivity $|r(\omega)|^2$ vs. frequency detuning $(\omega - \omega_0)/2\pi$. We note the extremely high mode reflectivities ($|r|^2 = .93$ in this example) that are attainable. The filter selectivity indicates a resolution of ~10 GHz. Figure 13-10(b) shows a plot of the phase shift of the reflected wave. This (phase) information is of importance in applications such as vertical cavity lasers, which will be discussed in Chapter 16, and in calculating the effect of the filter on incident optical pulses. A useful technique for analyzing periodic structures employs a transfer matrix approach and is outlined in Problem 13. The frequency dependence of $r(\omega)$ is due to the dependence of $\Delta\beta$ on ω (Eq. 13.6-3). We can use the separation between the two nearest zero crossings of $r(\omega)$, one on each side of ω_0, as a measure of the width of the reflectivity peak. These zero crossings occur when $\Delta\beta = \kappa$. Using (13.6-3) gives

$$\Delta\omega_{\text{filter}} = \frac{2\kappa c}{n_{\text{eff}}} \tag{13.6-4}$$

The value of $\Delta\omega_{\text{filter}}$ is equal to the forbidden "frequency gap" shown in Figure 13-8. This is not surprising since the evanescent exponential decay, i.e., $k Im \beta' \neq 0$, behavior characteristic of the gap is due to strong reflection of the wave at these frequencies.

DISTRIBUTED FEEDBACK LASERS

If a periodic medium is provided with sufficient gain at frequencies near the Bragg frequency ω_0 (where $l\pi/\Lambda \approx \beta$), oscillation can result without the benefit of end reflectors. The feedback is now provided by the continuous coherent backscattering from the periodic perturbation. In the following discussion we will consider two generic cases: (1) the bulk properties of a medium are perturbed periodically [17];

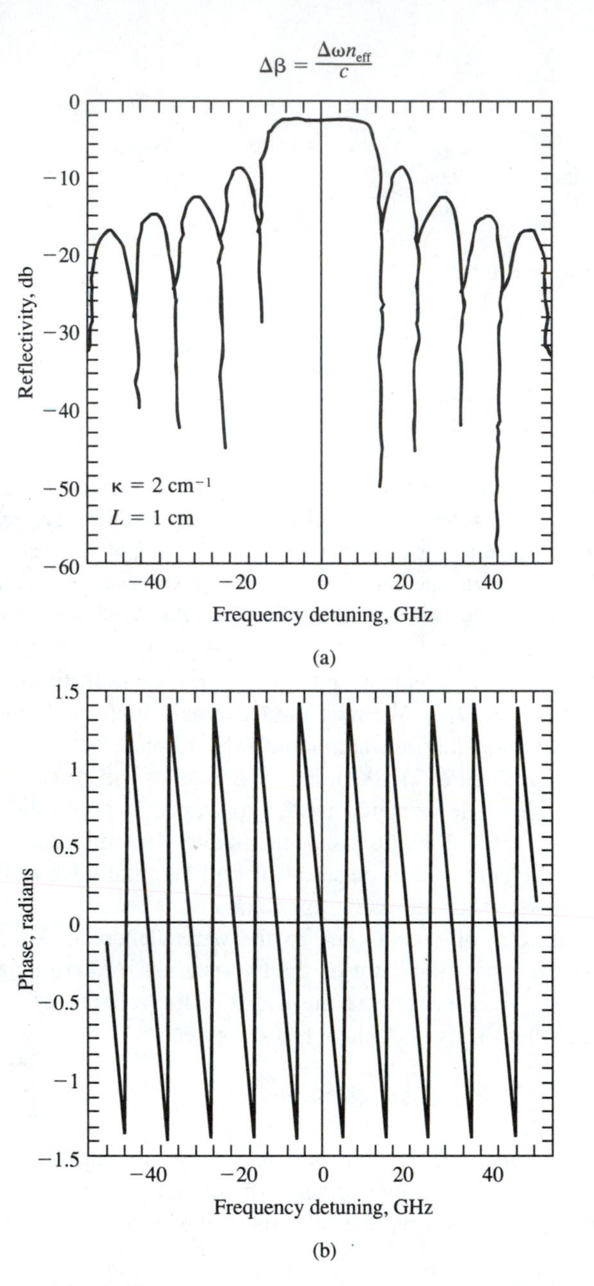

Figure 13-10 The reflection characteristics of a periodic waveguide. (a) The reflectivity vs. frequency deviation from the Bragg frequency. The plot is in db ($= 10 \log |r|^2$). (b) The phase of the reflected wave (phase of r) vs. frequency deviation. The center frequency is $\nu_0 = 1.9354 \times 10^{14}$ Hz ($\lambda_0 = 1.55\ \mu$m), $k = 2\ \text{cm}^{-1}$, $L = 1$ cm. [Courtesy of A. S. Kewitsch.]

(2) the boundary of a waveguide laser is perturbed periodically [18]. Both cases will be found to lead to the same set of equations.

Semiconductor lasers with built-in monolithic gratings, the so-called *distributed feedback lasers*, are the topic of Chapter 16.

13.7 ELECTROOPTIC MODULATION AND MODE COUPLING IN DIELECTRIC WAVEGUIDES

One of the most important applications of thin film waveguiding is in optical modulation and switching. The reason is twofold: (1) The confinement of the optical radiation to dimensions comparable to λ makes it possible to achieve the magnitude of electric fields that is necessary for modulation (see Section 9.5) with relatively small applied voltages. This leads to smaller modulation powers. (2) The absence of diffraction in a guided optical beam makes it possible to use longer modulation paths.

The main principle of electrooptic modulation in dielectric waveguides involves the diversion of all or part of the power from an input TE (or TM) mode to an output TM (or TE) mode that is caused by an applied dc (that is, low-frequency) field. To be specific we consider next the case of TM $\rightarrow$ TE mode conversion. This coupling is due to a perturbation polarization[2]

$$[P_{\text{pert}}(t)]_y \propto rE^{(0)}E_x^{(\omega)}(x)e^{i(\omega t-\beta_{\text{TM}}z)} \tag{13.7-1}$$

caused by the TM mode at ω whose field is $E_x^{(\omega)}(x)$ in the presence of the dc field $E^{(0)}$. The symbol r is an appropriate linear combination of electrooptic coefficients, which will be discussed below. This polarization, acting as a source, can excite, according to (13.3-9), a TE wave $A_s^{(+)}$. The application of a dc field thus causes a TM $\rightarrow$ TE power transfer.

The complex amplitude of the y polarization at ω produced by the TM field

$$E_x^{(\omega)}(x)e^{i(\omega t-\beta_{\text{TM}}z)} \tag{13.7-2}$$

in the presence of a dc field $E^{(0)}$ is[3]

$$[P_{\text{pert}}^{(\omega)}]_y = -\frac{\varepsilon^2 rE^{(0)}}{\varepsilon_0}E_x^{(\omega)}(x)e^{-i\beta_{\text{TM}}z} \tag{13.7-3}$$

---

[2]This follows from the wave equation (13.3-1) or (13.3-4), according to which a TE mode with a field E_y can be excited by P_y. An input TM mode with a field E_x can thus excite the TE wave ($E_y \neq 0$) if the medium has an off-diagonal ε_{yx} dielectric tensor component that generates $P_y = \varepsilon_{yx}E_x$. In the electrooptic case ε_{yx} is induced by and is proportional to the applied dc field $E^{(0)}$ as in (13.7-1).

[3]The origin of relation (13.7-3) is as follows: From the basic definition of the electrooptic tensor elements $(1/n^2)_{ij}$ in (9.1-3) it follows that a change in the indicatrix constant $(1/n^2)_{ij}$ is related to the corresponding change in the elements of the dielectric tensor ε_{ij} by

$$\Delta\varepsilon_{ij} = -\frac{\varepsilon_{ii}\varepsilon_{jj}}{\varepsilon_0}\Delta\left(\frac{1}{n^2}\right)_{ij}$$

Using (13.2-9) we take the TM input field $E_x^{(\omega)}(x)\exp[i(\omega t - \beta_{\text{TM}}z)]$ as that of the lth mode

$$E_x^l(\mathbf{r}, t) = \frac{\beta_l}{2\omega\varepsilon(x)} B_l \mathcal{H}_y^{(l)}(x)e^{i(\omega t - \beta_l^{\text{TM}}z)} + \text{c.c.} \tag{13.7-4}$$

where $\mathcal{H}_y^{(l)}(x)$ is given by (13.2-10) and $|B_l|^2$ is the mode power per unit width in the y direction. The polarization (13.7-3) can thus be written as

$$[P_{\text{pert}}(\mathbf{r}, t)]_y = -\frac{\varepsilon(x)r(x, z)E^{(0)}}{2\omega\varepsilon_0} \beta_l B_l \mathcal{H}_y^{(l)}(x)e^{i(\omega t - \beta_l^{\text{TM}}z)} + \text{c.c.} \tag{13.7-5}$$

Substitution of (13.7-5) into the wave equation (13.3-9) leads to

$$\frac{dA_m^{(+)}}{dz} \exp(-i\beta_m^{\text{TE}}z) - \frac{dA_m^{(-)}}{dz} \exp(i\beta_m^{\text{TE}}z)$$
$$= -\frac{i}{4}\int_{-\infty}^{\infty} \frac{\varepsilon(x)r(x, z)E^{(0)}(x, z)}{\varepsilon_0} \beta_l B_l \mathcal{H}_y^{(l)}(x)\mathcal{E}_y^{(m)}(x)\, dx \exp(-i\beta_l^{\text{TM}}z) \tag{13.7-6}$$

If $\beta_l^{\text{TM}} \approx \beta_m^{\text{TE}}$, the coupling excites only the $A_m^{(+)}$ wave, that is, it is codirectional. Dropping the plus and minus superscripts we can rewrite (13.7-6) as

$$\frac{dA_m}{dz} = -i\kappa_{ml}(z)B_l e^{-i(\beta_l^{\text{TM}} - \beta_m^{\text{TE}})z} \tag{13.7-7}$$

$$\kappa_{ml} = \frac{\beta_l}{4}\int_{-\infty}^{\infty} \frac{\varepsilon(x)r(x, z)E^{(0)}(x, z)}{\varepsilon_0} \mathcal{H}_y^{(l)}(x)\mathcal{E}_y^{(m)}(x)\, dx \tag{13.7-8}$$

Equation (13.7-8) is general enough to apply to a large variety of cases. The dependence of $E^{(0)}$ and $r(x, z)$ on x accounts for coupling by electrooptic material in the guiding or in the bounding layers. The z dependence allows for situations where $E^{(0)}$ or r depends on the longitudinal position. To be specific, we consider first the case where the guiding layer $-t < x < 0$ is uniformly electrooptic and where $E^{(0)}$ is

---

Using (9.1-3) we thus relate $\Delta\varepsilon_{ij}$ to an applied dc field $E_k^{(0)}$ by

$$\Delta\varepsilon_{ij} = -\frac{\varepsilon_{ii}\varepsilon_{jj}}{\varepsilon_0} r_{ijk}E_k^{(0)}$$

where r_{ijk} is the electrooptic tensor and where we sum over repeated indices. From the relation $D_i = \varepsilon_{ij}E_j + P_i$ we obtain

$$[P_{\text{pert}}^{(\omega)}]_i = \Delta\varepsilon_{ij}E_j^{(\omega)} = -\frac{\varepsilon_{ii}\varepsilon_{jj}}{\varepsilon_0} r_{ijk}E_j^{(\omega)}E_k^{(0)}$$

for the change in the ith component of the complex amplitude of the polarization at ω induced by an optical field with amplitude $E_j^{(\omega)}$ at ω in the presence of a dc field $E_k^{(0)}$. This last relation appears above in the form of (13.7-3) where the z and x dependence of $E_j^{(\omega)}$ are expressed explicitly and where

$$\varepsilon^2 rE^{(0)} \longrightarrow \varepsilon_{ii}\varepsilon_{jj}r_{ijk}E_k^{(0)}$$

uniform over the same region so that the integration in (13.7-8) is from $-t$ to 0. In that case, the overlap integral of (13.7-8) is maximum when the TE(m) and TM(l) modes are well confined and of the *same* order so that $l = m$. Under well-confined conditions, $p, q \gg h$ and the expressions (13.2-3), (13.2-7) for $\mathscr{E}_y^{(m)}(x)$, (13.2-10) and (13.2-13) for $\mathscr{H}_y^{(m)}(x)$ in the guiding layer become

$$\mathscr{E}_y^{(m)}(x) \longrightarrow \left(\frac{4\omega\mu}{t\beta_m^{\text{TE}}}\right)^{1/2} \sin\frac{m\pi x}{t}$$

$$\mathscr{H}_y^{(m)}(x) \longrightarrow \left(\frac{4\omega\varepsilon_0 n_2^2}{t\beta_m^{\text{TM}}}\right)^{1/2} \sin\frac{m\pi x}{t}$$

where for well-confined mode $\beta_l^{\text{TM}} \simeq \beta_m^{\text{TE}} \equiv \beta \approx k_0 n_2$. In this case the overlap integral in (13.7-8) becomes

$$\int_{-t}^{0} \mathscr{H}_y^{(m)}(x)\mathscr{E}_y^{(m)}(x)\,dx = \frac{4\omega\sqrt{\mu\varepsilon_2}}{t\beta}\int_{-t}^{0}\sin^2\frac{m\pi x}{t}\,dx \approx \frac{2\omega\sqrt{\mu\varepsilon_2}}{\beta}$$

since the integral approaches $t/2$. The coupling coefficient (13.7-8) achieves a maximum value of

$$\kappa \longrightarrow \frac{n_2^3 k_0 r E^{(0)}}{2} \tag{13.7-9}$$

The coupling is thus described by

$$\frac{dA_m}{dz} = -i\kappa B_m e^{-i(\beta_m^{\text{TM}}-\beta_m^{\text{TE}})z}$$

and (13.7-10)

$$\frac{dB_m}{dz} = -i\kappa A_m e^{i(\beta_m^{\text{TM}}-\beta_m^{\text{TE}})z}$$

The second equation of (13.7-10) can be obtained by a process similar to that leading to the first equation or by invoking the conservation of total power [12], which shows that the above expression for dB_m/dz is needed to satisfy

$$\frac{d}{dz}\left(|A_m|^2 + |B_m|^2\right) = 0$$

For the phase-matched condition $\beta_m^{\text{TM}} = \beta_m^{\text{TE}}$ the solution of (13.7-10) in the case of a single input ($B_m(0) \equiv B_0$, $A_m(0) = 0$) is

$$B_m(z) = B_0 \cos(\kappa z)$$

$$A_m(z) = -iB_0 \sin(\kappa z) \tag{13.7-11}$$

Using (13.7-9) we can show that the field length product $E^{(0)}L$ for which $\kappa L = \pi/2$, which is necessary to effect a complete TM $\leftrightarrow$ TE power transfer in a distance L, is the same as that needed to go from "on" to "off" in the bulk modulator shown in Figure 9-4. This result applies only in the limit of tight confinement. In general the coupling coefficient κ is smaller than the value given by (13.7-9), and the $E^{(0)}L$ product needed to achieve a complete power transfer is correspondingly larger.

When $\beta_m^{\text{TM}} \neq \beta_m^{\text{TE}}$, the solution of (13.7-10), subject to boundary conditions $B_m(0) = B_0$, $A_m(0) = 0$, is

$$B_m(z) = B_0 e^{i\delta z}\left[\cos(sz) - \frac{i\delta}{s}\sin(sz)\right]$$

$$A_m(z) = -iB_0 e^{-i\delta z}\frac{\kappa}{s}\sin(sz) \tag{13.7-12}$$

where

$$s^2 \equiv \kappa^2 + \delta^2 \qquad 2\delta \equiv \beta_m^{\text{TM}} - \beta_m^{\text{TE}} = \beta^B - \beta^A \tag{13.7-13}$$

In contrast to the phase-matched case (13.7-11), the maximum fraction of the power that can be coupled from the input mode, B_m, to A_m is

$$\text{Fraction of power exchanged} = \frac{\kappa^2}{\kappa^2 + \delta^2} \tag{13.7-14}$$

and becomes negligible once $\delta \gg \kappa$.

A plot of the mode power for the phase-matched ($\delta = 0$) and $\delta \neq 0$ case is shown in Figure 13-11.

A deliberate periodic variation of $E^{(0)}(z)$ or $r(z)$, in this case, with a spatial period $|2\pi/(\beta_m^{\text{TE}} - \beta_m^{\text{TM}})|$ can be used, according to (13.7-8), to compensate for the mismatch factor $\exp[-i(\beta_m^{\text{TM}} - \beta_m^{\text{TE}})z]$ in (13.7-10), thus leading again to a phase-matched operation.

Example: GaAs Thin-Film Modulator at $\lambda = 1\ \mu$m

To appreciate the order of magnitude of the coupling, consider a case where the guiding layer is GaAs and $\lambda = 1\ \mu$m. In this case (see Table 9-2)

$$n_2 \simeq 3.5 \qquad n^3 r \cong 60 \times 10^{-12}\ \frac{\text{m}}{\text{volt}}$$

Assuming an applied field $E^{(0)} = 10^6$ volt/m, we obtain, from (13.7-9),

$$\kappa = 1.88\ \text{cm}^{-1}$$

$$l \equiv \frac{\pi}{2\kappa} = 0.83\ \text{cm}$$

for the coupling constant and the power-exchange distance, respectively.

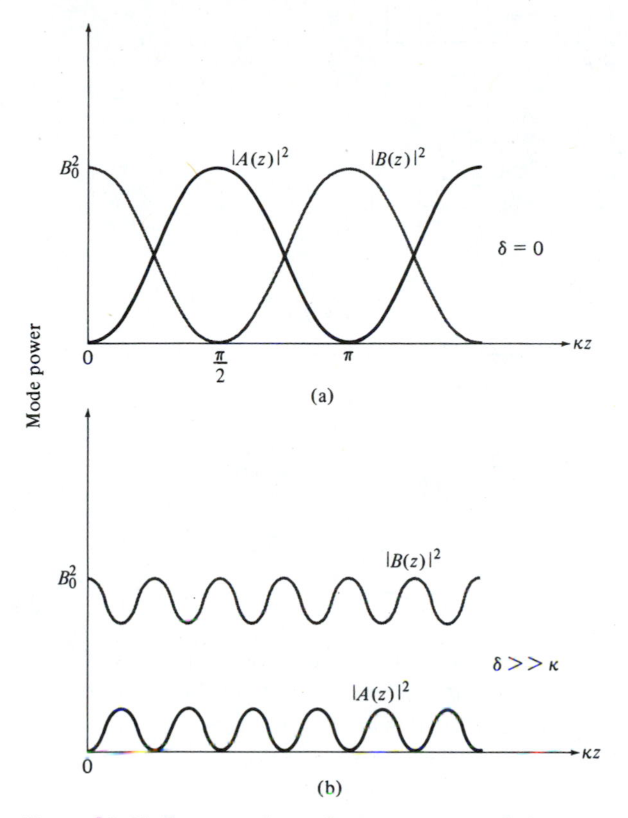

Figure 13-11 Power exchange between two coupled modes under (a) phase-matched conditions ($\beta_m^{\text{TM}} = \beta_m^{\text{TE}}$) as described by Equation (13.7- 11); (b) $\beta_m^{\text{TM}} \neq \beta_m^{\text{TE}}$, Equation (13.7-12).

An experimental setup used in one of the earliest demonstrations [3] of electrooptic thin film modulation is depicted in Figure 13-12. The modulation scheme is identical to that illustrated in Figure 9-8 and depends on an electrooptic induced phase retardation (9.5-1)

$$\begin{aligned}\Gamma &= (\beta_{\text{TM}} - \beta_{\text{TE}})l \\ &\approx \frac{\pi n^3 r_{41} l}{\lambda t} V\end{aligned} \tag{13.7-15}$$

where V is the applied voltage, t and l are the height and length of the waveguide, respectively.

The ratio of the transmitted to the input intensities is given by (9.3-4), or, equivalently by (13.7-11) [note that Γ of (9.2-4) is the same as κ in (13.7-9)] as

$$\frac{I_0}{I_i} = \sin^2 \frac{\Gamma}{2} \tag{13.7-16}$$

An experimental transmission plot is shown in Figure 13-13.

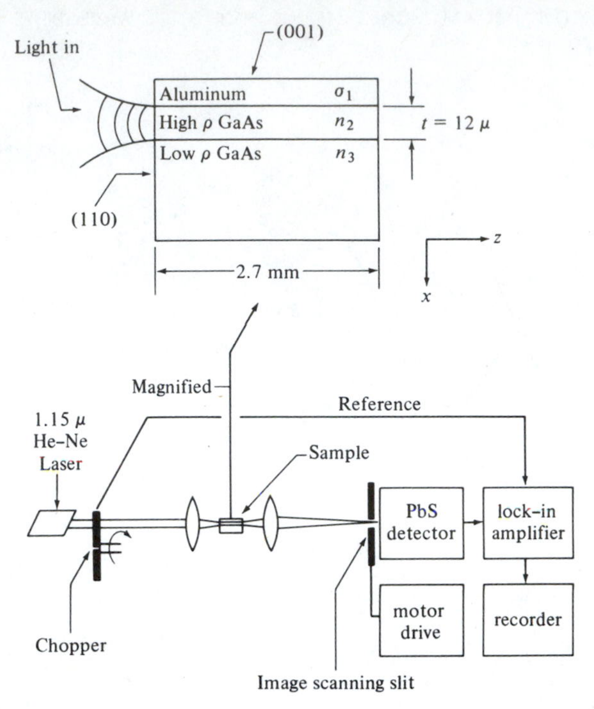

Figure 13-12 The first demonstration of electrooptic modulation in a dielectric waveguide. An electrooptic modulator in a GaAs epitaxial film. The modulation field is due to a reverse-bias voltage applied to the metal semiconductor junction [3].

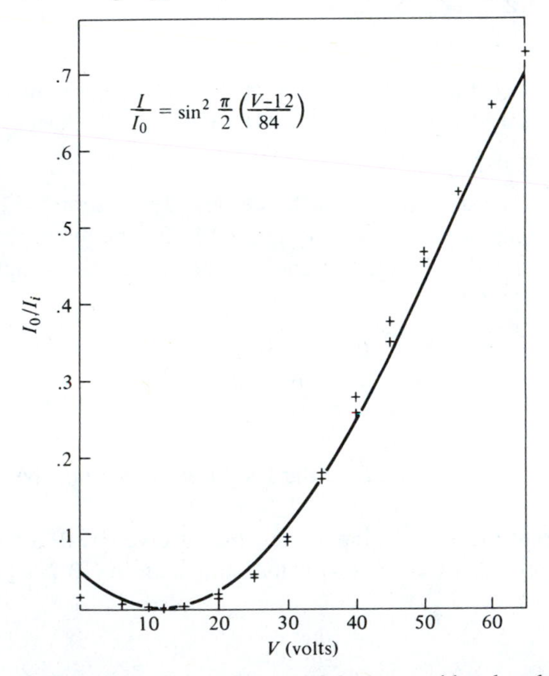

Figure 13-13 Transmittance of the waveguide, placed between crossed polarizers, as a function of the applied reverse voltage [3].

13.8 DIRECTIONAL COUPLING

Exchange of power between guided modes of adjacent waveguides is known as directional coupling. Waveguide directional couplers perform a number of useful functions in thin-film devices, including power division, modulation, switching, frequency selection, and polarization selection.

Waveguide coupling can be treated by the coupled-mode theory. Consider the case of the two planar waveguides illustrated in Figure 13-14. Refractive index distributions for the two guides in the absence of coupling are given by $n_a(x)$ and $n_b(x)$. The transverse electric field distribution for a particular guided mode of waveguide **a** alone and a particular mode of waveguide **b** alone will be denoted by $\mathscr{E}_y^{(a)}(x)$ and $\mathscr{E}_y^{(b)}(x)$, and their propagation constants by β_a and β_b. The field in the coupled-guide structure with an index $n_c(x)$ (for propagation in the positive z direction) is approximated by the sum of the unperturbed fields

$$E_y = A(z)\mathscr{E}_y^{(a)}(x)e^{i(\omega t-\beta_a z)} + B(z)\mathscr{E}_y^{(b)}(x)e^{i(\omega t-\beta_b z)} \quad \textbf{(13.8-1)}$$

In the absence of coupling—that is, if the distance between guides a and b were infinite—$A(z)$ and $B(z)$ do not depend on z and will be independent of each other, since each of the two terms on the right side of (13.8-1) satisfies the wave equation (13.3-1) separately.

The perturbation polarization responsible for the coupling is calculated by substituting (13.8-1) into (13.3-2) and (13.3-3). The result is

$$P_{\text{pert}} = e^{i\omega t}\varepsilon_0[\mathscr{E}_y^{(a)}A(z)(n_c^2(x) - n_a^2(x))e^{-i\beta_a z} + \mathscr{E}_y^{(b)}B(z)(n_c^2(x) - n_b^2(x))e^{-i\beta_b z}] \quad \textbf{(13.8-2)}$$

where $n_c(x)$ is the index profile of the two-guide structure. Substituting (13.8-2) in (13.3-9) and integrating over x gives

$$\frac{dA}{dz} = -i\kappa_{ab}Be^{-i(\beta_b-\beta_a)z} - iM_aA$$

$$\frac{dB}{dz} = -i\kappa_{ba}Ae^{-i(\beta_a-\beta_b)z} - iM_bB \quad \textbf{(13.8-3)}$$

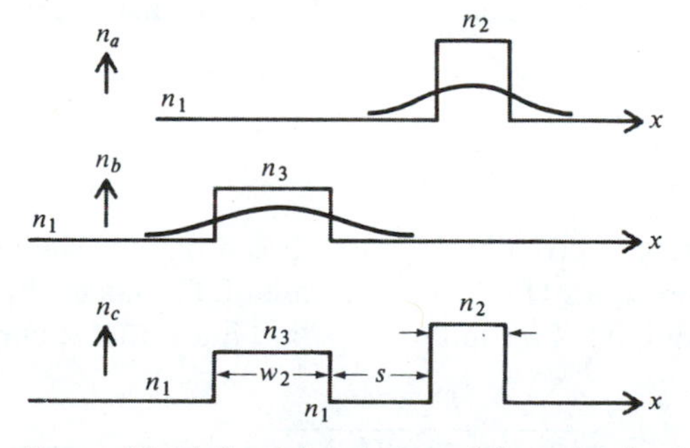

Figure 13-14 Spatial variation of the refractive index for uncoupled waveguides $n_a(x)$ and $n_b(x)$ and for a parallel waveguide structure $n_c(x)$.

where

$$\kappa_{ba,ab} = \frac{\omega\varepsilon_0}{4}\int_{-\infty}^{\infty} [n_c^2(x) - n_{(a,b)}^2(x)]\mathscr{E}_y^{(a)}\mathscr{E}_y^{(b)}\, dx \tag{13.8-4}$$

$$M_{(a,b)} = \frac{\omega\varepsilon_0}{4}\int_{-\infty}^{\infty} [n_c^2(x) - n_{(a,b)}^2(x)](\mathscr{E}_y^{(a,b)})^2\, dx \tag{13.8-5}$$

The terms M_a and M_b represent a small correction to the propagation constants β_a and β_b, respectively, due to the presence of the second guide. So if we take the total field as

$$E_y = A(z)\mathscr{E}_y^{(a)}e^{i[\omega t-(\beta_a+M_a)z]} + B(z)\mathscr{E}_y^{(b)}e^{i[\omega t-(\beta_b+M_b)z]}$$

instead of (13.8-1), Equations (13.8-3) become

$$\frac{dA}{dz} = -i\kappa_{ab}Be^{-i2\delta z}$$

$$\frac{dB}{dz} = -i\kappa_{ba}Ae^{i2\delta z} \tag{13.8-6}$$

where

$$2\delta = (\beta_b + M_b) - (\beta_a + M_a)$$

The solution of (13.8-6) subject to a single input at guide b ($B(0) = B_0$, $A(0) = 0$), and assuming $\kappa_{ab} = \kappa i_{ba}$, is given by (13.7-12). In terms of powers $P_a = AA^*$ and $P_b = BB^*$ in the two guides, the solution becomes

$$P_a(z) = P_0\frac{\kappa^2}{\kappa^2 + \delta^2}\sin^2 [(\kappa^2 + \delta^2)^{1/2}z]$$

$$P_b(z) = P_0 - P_a(z) \tag{13.8-7}$$

where $P_0 = |B(0)|^2$ is the input power to guide b. Complete power transfer from b to a occurs in a distance $L = \pi/2\kappa$ provided $\delta = 0$ (that is, equal phase velocities in both modes). For $\delta \neq 0$, the maximum fraction of power that can be transferred is from (13.8-7)

$$\frac{\kappa^2}{\kappa^2 + \delta^2} \tag{13.8-8}$$

The coupling constant κ is given by (13.8-4). It can be evaluated straightforwardly using the field expressions (13.2-3) in the case of TE modes. In the special case of identical waveguides, $h_1 = h_2$ and $p_1 = p_2$ in Figure 13-14, one obtains

$$\kappa = \frac{2h^2pe^{-ps}}{\beta(w + 2/p)(h^2 + p^2)} \tag{13.8-9}$$

The extension to channel waveguide couplers that are confined in the y, as well as in the x, direction is straightforward. In the well-confined case $w \gg 2/p$ and (13.8-9) simplifies to

$$\kappa = \frac{2h^2 p\, e^{-ps}}{\beta w(h^2 + p^2)} \tag{13.8-10}$$

A typical value of κ obtained at $\lambda \sim 1\ \mu$m with $w, s \sim 3\ \mu$m, and $\Delta n \sim 5 \times 10^{-3}$ is $\kappa \sim 5\ \text{cm}^{-1}$ so that coupling distances are of the order of magnitude of $\kappa^{-1} \approx 2$ mm.

A form of an electrooptic switch based on directional coupling [23] is as follows.

The length L of the coupler is chosen so that for $\delta = 0$ (that is, synchronous case) $\kappa L = \pi/2$. From (13.8-7) it follows that all the input power to guide b exits from guide a at $z = L$. The switching is achieved by applying an electric field to guide a (or b) in such a way as to change its propagation constant until

$$\delta L = \tfrac{1}{2}(\beta_a - \beta_b)L = \frac{\sqrt{3}}{2}\pi \tag{13.8-11}$$

that is, $\delta = \sqrt{3}\kappa$. It follows from (13.8-7) that at this value of δ

$$P_a = 0 \qquad P_b = P_0$$

that is, the power reappears at the output of guide b. A control of δ can thus be used to achieve any division of the powers between the outputs of guides a and b.

In practice a convenient way to control δ is to fabricate the directional coupler in an electrooptic crystal. In this case, according to (9.1-11), the application of an electric field E across one of the two waveguides will cause the index of refraction to change by

$$\Delta n \propto n^3 rE$$

where r is the appropriate electrooptic tensor element. The change Δn will give rise to a change in propagation constant

$$\delta \sim \frac{\omega}{c}\Delta n \sim \frac{\omega}{c} n^3 rE$$

The control of the power output from both arms of a directional coupler by means of an applied voltage is illustrated in Figure 13-15. The electrode geometry for applying a field to the waveguide is illustrated in Figure 13-16.

One of the interesting applications for electrooptically switched directional couplers is in the area of very high-frequency ($>5 \times 10^9$ Hz) sampling and of multiplexing and demultiplexing of optical binary pulse trains. An example of the latter is demonstrated by Figure 13-16. Two independent, but synchronized, data pulse trains A and B are fed into legs a and b, respectively, of a directional coupler. The length of the coupling section satisfies the power transfer condition $\kappa L = \pi/2$. The phase mismatch δ between the two waveguides is controlled, as discussed above, by

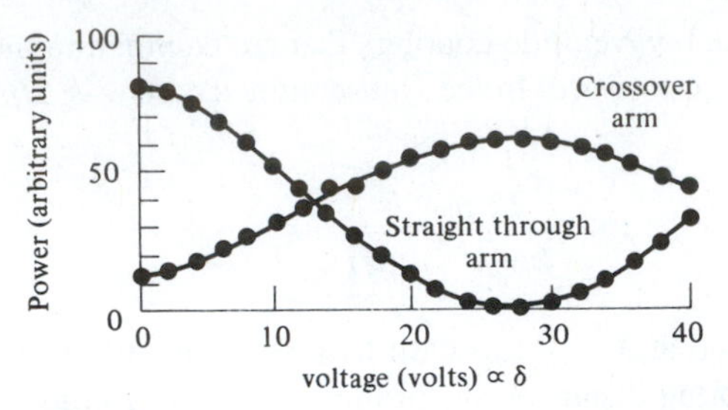

Figure 13-15 The dependence of the power output from the arms of a directional coupler on the (voltage-controlled) phase constant mismatch δ. (After Reference [24].)

an electric field applied across one of the waveguides. This electric field is due to a microwave signal at a frequency ω_m. The resulting peak phase constant mismatch, which occurs at the maxima and minima of the applied voltage, satisfies the condition (13.8-11)

$$|\delta_{\max}| = \frac{\sqrt{3}\pi}{2L} \tag{13.8-12}$$

so that the B pulses, which are synchronized to arrive during the extrema of the microwave signal, exit from arm b. Pulses A, on the other hand, arrive when the applied field, hence δ, is zero, and since $\kappa L = \pi/2$, cross over and exit from guide b. The result is that both pulse trains A and B are interleaved, or in the electrical engineering parlance, multiplexed in the output of guide b. The (combined) output

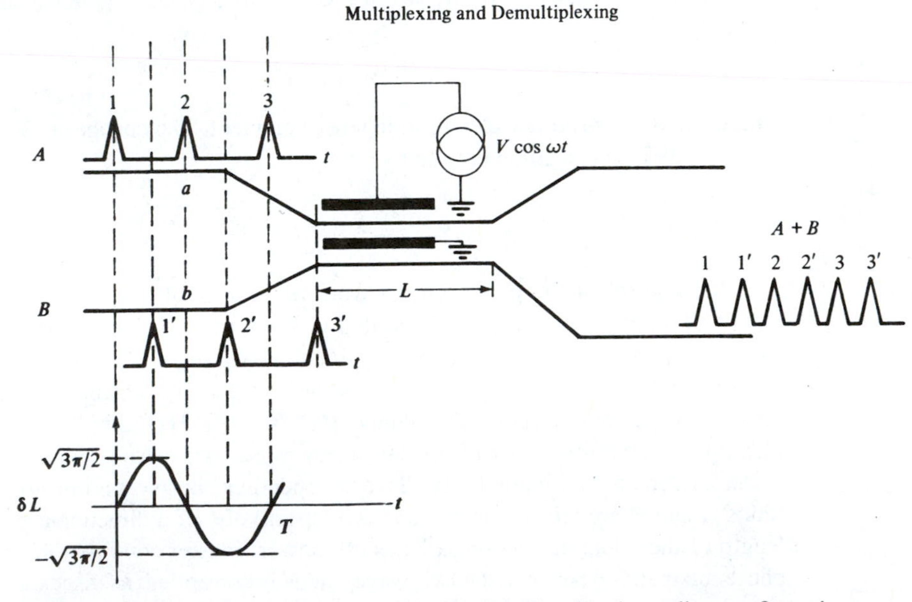

Figure 13-16 Multiplexing and demultiplexing in a directional coupling configuration.

from b can be fed into the input of a second directional coupler fed with a signal at $2\omega_m$ and multiplexed thereby with a second data train, and so on.

The device can, of course, be operated in reverse, right to left in the figure, and act as a demultiplexer for separating the dense bit train $A + B$ entering b into the individual trains A and B.

A multiguide directional coupler such as that shown in Figure 13-17 is described by a set of equations

$$\frac{dA_n}{dz} = -i\kappa A_{n-1} - i\kappa A_{n+1} \tag{13.8-13}$$

which is an obvious extension of (13.8-6) to the multimode synchronous case ($\delta = 0$) when only adjacent channels couple to each other. The solution of (13.8-13) in the case of a single input, that is, $A_n(0) = 1$, $n = 0$, $A_n(0) = 0$ $n \neq 0$ is [23]

$$A_n(z) = (-i)^n J_n(2\kappa z) \tag{13.8-14}$$

where J_n is the Bessel function of order n.

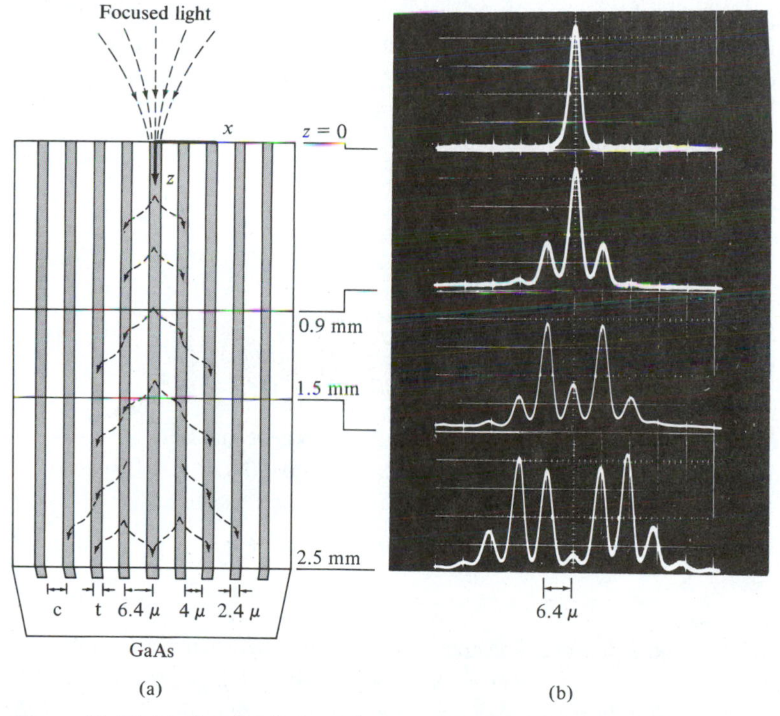

Figure 13-17 (a) Sketch of channel optical waveguide directional coupler showing flow of light energy into adjacent channels. (b) Measured guided-light intensity profiles at various lengths. The profiles have been displayed relative to the sketch at the proper value of z. Intensity scale is arbitrary. The guides were produced by proton implantation into p^+-GaAs crystal. (After Reference [23].)

A directional coupler based on this principle is shown in Figure 13-17(a). The predicted Bessel function distribution of the intensity at various propagation distances is shown in Figure 13-17(b).

13.9 THE EIGENMODES OF A COUPLED WAVEGUIDE SYSTEM (supermodes)

In the preceding section we treated the important case of directional coupling between two parallel waveguides (Figure 13-14) by means of the coupled-mode formalism. The same problem may be approached from a different, and equivalent, point of view that is better suited for the treatment of certain important classes of experimental and device configurations. In this new point of view, we seek to obtain the propagating eigenmodes of the two-waveguide (in general, the multiwaveguide) system shown in Figure 13-14. The eigenmode is, by definition, that propagating field solution of the waveguiding structure that, except for a propagation delay factor, does not depend on the propagation coordinate z. We can obtain these mode solutions by a straightforward extension of the formalism of Section 13.1 to the more complex waveguide whose index profile $n_c(x)$ is given at the bottom of Figure 13-14. This procedure, although exact, is laborious and does not submit itself readily to the intuitive understanding that characterizes the method of solution that starts with the coupled-mode equations. The following analysis follows closely that of Reference [12].

We recall that, according to (13.8-6), the normalized, individual waveguide, mode amplitudes obey the coupled-mode equations

$$\frac{dA}{dz} = \kappa B e^{-i2\delta z}$$

$$\frac{dB}{dz} = -\kappa^* A e^{i2\delta z} \tag{13.9-1}$$

$$2\delta = (\beta_b - \beta_a) + (M_b - M_a)$$

These equations are equivalent to (13.8-6) except that here $\kappa \equiv -i\kappa_{ab}$. These equations subject to boundary conditions $A(0)$ and $B(0)$ at $z = 0$ specify as in (13.8-1), the total field in terms of the individual waveguides' fields.

$$a(x, z, t) = A(z)\mathscr{E}_y^{(m)}(x)e^{i[\omega t-(\beta_a+M_a)z]} \qquad \text{in guide } a$$

$$b(x, z, t) = B(z)\mathscr{E}_y^{(l)}(x)e^{i[\omega t-(\beta_b+M_b)z]} \qquad \text{in guide } b \tag{13.9-2}$$

with m and l denoting the transverse mode order. Since the individual mode field profiles $\mathscr{E}_y^{(m)}(x)$ and $\mathscr{E}_y^{(l)}(x)$ are known as well as β_a, β_b, M_a, M_b, and the frequency ω, *the total field is specified once the (complex) amplitudes $A(z)$ and $B(z)$ are given.* We thus may *uniquely* describe the field at z by means of a column vector[4]

[4]The term *vector* is due merely to the fact that $\mathbf{E}(z)$ is specified by two numbers that can be viewed as two components and can thus be viewed formally as a vector in a two-dimensional space.

$$\mathbf{E}(z) \equiv \begin{vmatrix} B(z)e^{-i\beta'_b z} \\ A(z)e^{-i\beta'_a z} \end{vmatrix} \equiv \begin{vmatrix} E_1(z) \\ E_2(z) \end{vmatrix} \tag{13.9-3}$$

with

$$\beta'_{\substack{a\\b}} \equiv \beta_{\substack{a\\b}} + M_{\substack{a\\b}}$$

The evolution of $\mathbf{E}(z)$ is obtained from equations (13.9-1) as

$$\frac{d\mathbf{E}}{dz} = \tilde{\mathbf{C}}\mathbf{E} \tag{13.9-4}$$

with the matrix $\tilde{\mathbf{C}}$ given by

$$\tilde{\mathbf{C}} = \begin{vmatrix} -i\beta'_b & -\kappa^* \\ \kappa & -i\beta'_a \end{vmatrix} \tag{13.9-5}$$

Since an eigenmode depends on z only through a propagation phase factor, we postulate a solution

$$\mathbf{E}(z) = \mathbf{E}(0)e^{i\gamma z} \tag{13.9-6}$$

Combining (13.9-4) and (13.9-6) results in

$$\tilde{\mathbf{C}}\mathbf{E} = i\gamma\mathbf{E} \tag{13.9-7}$$

This is a standard matrix algebra eigenvalue problem [note similarity to Problem 2.2 and to Equation (2.1-12)], where $\mathbf{E}$ is the eigenvector and $i\gamma$ is the eigenvalue of the matrix $\tilde{\mathbf{C}}$. To determine $\mathbf{E}$ and γ, we write out the two equations represented by (13.9-7)

$$\begin{aligned} -i(\beta'_b + \gamma)E_1 - \kappa^* E_2 &= 0 \\ \kappa E_1 - i(\beta'_a + \gamma)E_2 &= 0 \end{aligned} \tag{13.9-8}$$

The condition for nontrivial solutions for E_1 and E_2 is the vanishing of the determinant of the coefficients in (13.9-8). The solution of the resulting quadratic equation yields the eigenvalues

$$\gamma_{1,2} = -\frac{\beta'_a + \beta'_b}{2} \pm \frac{1}{2}\sqrt{(\beta'_a - \beta'_b)^2 + 4\kappa^2} = -\overline{\beta} \pm S \tag{13.9-9}$$

$$\overline{\beta} \equiv \frac{1}{2}(\beta'_a + \beta'_b) \qquad S \equiv \sqrt{\delta^2 + \kappa^2} \qquad \delta = \beta'_b - \beta'_a \tag{13.9-10}$$

The two values γ_1 and γ_2 are substituted, one at a time, in (13.9-8) to obtain, to within an arbitrary constant, the corresponding eigenvectors. The result is

$$\mathbf{E}_1(z) = \begin{vmatrix} \dfrac{i\kappa^*}{\delta + S} \\ 1 \end{vmatrix} e^{-i(\overline{\beta} - S)z} \tag{13.9-11a}$$

$$\mathbf{E}_2(z) = \begin{vmatrix} \dfrac{i\kappa^*}{\delta - S} \\ 1 \end{vmatrix} e^{-i(\bar{\beta}+S)z} \tag{13.9-11b}$$

We note that, as expected, $\mathbf{E}_1 \cdot \mathbf{E}_2^* = 0$, i.e., the eigenmodes are orthogonal. The mode norms $\mathbf{E}_{\substack{1\\2}} \cdot \mathbf{E}_{\substack{1\\2}}^* = 1 + |\kappa|^2/(\delta \pm S)^2$ are proportional to the respective (eigen) mode powers and are thus a constant. The two components $i\kappa^*/(\delta \pm S)$ and 1 of each eigenvector represent, respectively, the normalized amplitudes of the individual waveguide modes, $\mathscr{E}_y^{(l)}(x)$ in guide b and $\mathscr{E}_y^{(m)}(x)$ in guide a, which together make up the eigenmode of the two-waveguide system. The ratio of the power in waveguides b and a in these two "supermodes" is thus $|\kappa|^2/(\delta \pm S)^2$. In the limit $\kappa/\delta \to 0$, the "velocity mismatch" limit, $\mathbf{E}_1$ and $\mathbf{E}_2$ become

$$\begin{aligned} \underset{(\kappa \ll \delta)}{\mathbf{E}_1} &\to \begin{vmatrix} 0 \\ 1 \end{vmatrix} e^{-i\beta'_a z} \\ \underset{(\kappa \ll \delta)}{\mathbf{E}_2} &\to \begin{vmatrix} 1 \\ 0 \end{vmatrix} e^{-i\beta'_b z} \end{aligned} \tag{13.9-12}$$

to within a multiplicative constant, i.e., the super-(eigen) modes become the uncoupled single-waveguide modes.

Another important situation occurs when the two individual waveguide modes have the same phase velocity, i.e., $\delta = 0$. In this case

$$\begin{aligned} \underset{(\delta=0)}{\mathbf{E}_1(z)} &= \begin{vmatrix} i\dfrac{\kappa^*}{|\kappa|} \\ 1 \end{vmatrix} e^{-i(\beta-|\kappa|)z} \\ & \qquad\qquad \beta \equiv \beta_1 = \beta_2 \\ \underset{(\delta=0)}{\mathbf{E}_2(z)} &= \begin{vmatrix} -i\dfrac{\kappa^*}{|\kappa|} \\ 1 \end{vmatrix} e^{-i(\beta+|\kappa|)z} \end{aligned} \tag{13.9-13}$$

The admixture is 50–50 percent and each waveguide carries half of the total power. In the case of identical waveguides at $\delta = 0$ and for $l = m$ (i.e., same order modes), the coupling constant κ is a negative imaginary number [see Equations (13.8-3) and (13.8-4)] so that the two eigenvectors [of (13.9-13)] take the form

$$\begin{aligned} \underset{(\delta=0)}{\mathbf{E}_1(z)} &= \begin{vmatrix} -1 \\ 1 \end{vmatrix} e^{-i(\beta-|\kappa|)z} \\ \underset{(\delta=0)}{\mathbf{E}_2(z)} &= \begin{vmatrix} 1 \\ 1 \end{vmatrix} e^{-i(\beta+|\kappa|)z} \end{aligned} \tag{13.9-14}$$

$\mathbf{E}_1(z)$ is thus the odd symmetric mode while $\mathbf{E}_2(z)$ is even symmetric as depicted in Figure 13-18.

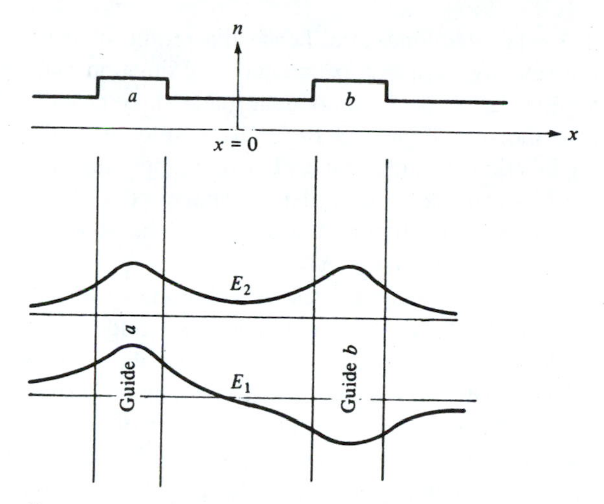

Figure 13-18 The transverse (x) field distribution of the two supermodes at the phase velocity matching ($\delta = 0$) condition of the parallel two-guide structure whose index of refraction profile is shown at the top.

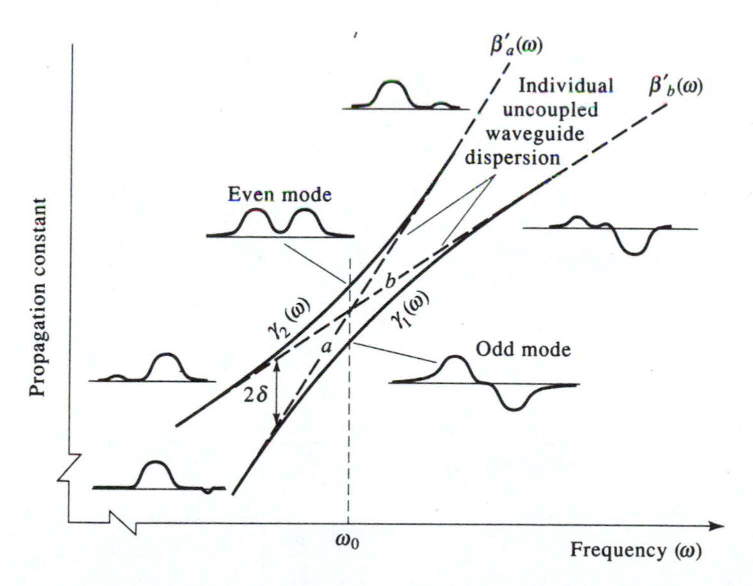

Figure 13-19 Some field distributions and the dispersion curves of the two lowest order array modes ("supermodes") of a two-waveguide configuration near the phase-matching frequency (ω_0). The dashed curves correspond to the guide's dispersion in the absence of coupling.

According to (13.9-11) the admixture, hence the profile of the supermodes, depends on the phase velocity mismatch parameter δ. A situation may exist where the uncoupled dispersion curves of the individual guides may cross each other at some frequency ω_0, as shown in Figure 13-19. In the vicinity of ω_0 the supermode profile is thus a strong function of δ and, hence, of ω. If we approximate the mismatch parameter near ω_0 by $\delta = \text{const}\ (\omega - \omega_0)$, then the supermodes' dispersion curves $\gamma_1(\omega)$ and $\gamma_2(\omega)$ are as shown in the figure. Also shown are the (super) mode profiles at ω_0 and at a frequency ω where $\delta(\omega) \gg \kappa$.

Since the supermode (i.e., eigenmode) description of the waveguide problem is formally equivalent to one that is based on individual waveguide coupled modes, it is instructive to consider how we might, for example, describe the phenomenon of directional coupling [see (13.8-7)] by means of our new point of view. To simplify matters, we assume $\delta = 0$ (phase-matching) and, referring to Figure 13-18, consider the case where at $z = 0$ power is fed into guide a (on the left) only. This boundary condition can be satisfied by expanding the total field at $z = 0$ as an equal admixture of the two supermodes (13.9-14) taken

$$\mathbf{E}_{\text{tot}}(0) = \mathbf{E}_1(0) + \mathbf{E}_2(0) = \begin{vmatrix} 0 \\ 2 \end{vmatrix} \tag{13.9-15}$$

It is also clear that if one were to add algebraically the fields of Figure 13-18, they would reinforce each other on the left and largely cancel each other on the right leading to the column vector in (13.9-15). Having established in (13.9-15) the proper admixture that satisfies the boundary condition at $z = 0$, we can determine the field at any z by simply inserting the z-dependence of each supermode.

$$\begin{aligned}\mathbf{E}_{\text{tot}}(z) &= \mathbf{E}_1(0)e^{-i(\beta-|\kappa|)z} + \mathbf{E}_2(0)e^{-i(\beta+|\kappa|)z} \\ &= \begin{vmatrix} \dfrac{i\kappa^*}{|\kappa|} \\ 1 \end{vmatrix} e^{-i(\beta-|\kappa|)z} + \begin{vmatrix} -\dfrac{i\kappa^*}{|\kappa|} \\ 1 \end{vmatrix} e^{-i(\beta+|\kappa|)z} \\ &= e^{-i(\beta-|\kappa|)z}[\mathbf{E}_1(0) + \mathbf{E}_2(0)e^{-i2|\kappa|z}]\end{aligned} \tag{13.9-16}$$

At a distance z where

$$|\kappa|z = \frac{\pi}{2}$$

the total field is

$$\begin{aligned}\mathbf{E}_{\text{tot}}\left(z = \frac{\pi}{2|\kappa|}\right) &= e^{-i(\beta-|\kappa|)z}[\mathbf{E}_1(0) - \mathbf{E}_2(0)] \\ &= e^{-i(\beta-|\kappa|)z} \begin{vmatrix} \dfrac{2i\kappa^*}{|\kappa|} \\ 0 \end{vmatrix}\end{aligned} \tag{13.9-17}$$

so that the power is completely in the right waveguide. This exchange of power between the two guides takes place every $\Delta z = \pi/2|\kappa|$ just as predicted by the coupled-mode solution (13.8-7). Using the present point of view, however, we attribute the sloshing of power between guides to the difference of the phase constants $\gamma_2 - \gamma_1 = 2|\kappa|$ of the two supermodes leading to the factor $\exp(-i2|\kappa|z)$ in the last term of (13.9-16).

13.10 LASER ARRAYS (25, 26)

In Section 13.9 we showed that starting with the coupled-mode equations (13.8-6) for two adjacent and interacting waveguides we can obtain the supermodes, i.e., the modes of the two-waveguide system. The same approach can be applied to the problem of finding the modes of an N-waveguide system. Such semiconductor laser arrays [25, 26] are now widely used to increase the power output relative to that which is available from a single waveguide laser.

Consider a structure consisting of N adjacent waveguides as shown in Figure 13-20. For simplicity we assume that each waveguide can support one mode only. If we label the waveguides by 1, 2, . . . , N and take the normalized electric field mode solution of the isolated mth waveguide as $A_m \mathscr{E}_m(x, y)e^{-i\beta_m z}$, we can describe the total field at some arbitrary plane z as

$$\mathscr{E}(z) = \sum_m A_m(z)\mathscr{E}_m(x, y)e^{-i\beta_m z} \tag{13.10-1}$$

where the z dependence of A_m reflects the possible amplitude and phase coupling between the "individual" waveguide modes. If the waveguides were perfectly isolated from each other, the A_m's would be constant and *independent* of each other. The individual waveguide modes are subject to a normalization condition (13.2-6)

$$\frac{\beta_m}{2\omega\mu} \int_{-\infty}^{\infty} |\mathscr{E}_m(x, y)|^2 dxdy = 1 \tag{13.10-2}$$

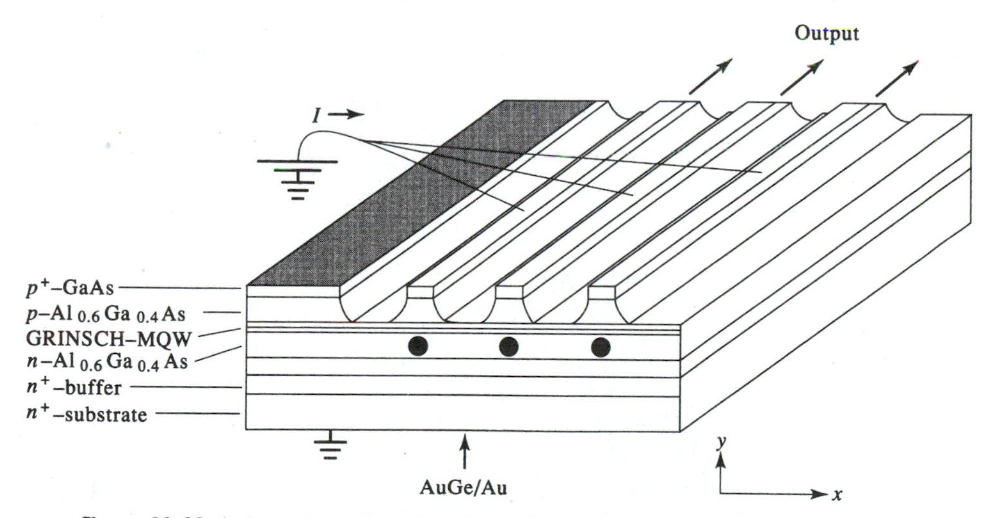

Figure 13-20 A three-channel semiconductor laser array. The active amplifying regions are GaAs quantum wells. The elliptical spots correspond to the near-field intensity pattern. (After Reference [27].)

so that the total power in the mth guide is $|A_m|^2$. The total field at z can be represented uniquely by the column vector

$$\mathbf{E}(z) \equiv \begin{vmatrix} A_1(z)e^{-i\beta_1 z} \\ A_2(z)e^{-i\beta_2 z} \\ \cdot \\ \cdot \\ \cdot \\ \cdot \\ \cdot \\ A_N(z)e^{-i\beta_N z} \end{vmatrix} \equiv \begin{vmatrix} E_1(z) \\ E_2(z) \\ \cdot \\ \cdot \\ \cdot \\ \cdot \\ \cdot \\ E_N(z) \end{vmatrix} \tag{13.10-3}$$

If we assume that each mode couples only to its immediate neighbors, then, in the manner of (13.9-1), the mode amplitudes $A_m(z)$ obey

$$\frac{dA_m}{dz} = \kappa_{m,m+1}A_{m+1}e^{i(\beta_m-\beta_{m+1})z} + \kappa_{m,m-1}A_{m-1}e^{i(\beta_m-\beta_{m-1})z} \tag{13.10-4}$$

From its definition in (13.10-3) and from (13.10-4), it follows that the component $E_m(z)$ of the vector $\mathbf{E}(z)$ obeys

$$\begin{aligned} \frac{dE_m}{dz} &= -i\beta_m A_m(z)e^{-i\beta_m z} + \frac{dA_m}{dz}e^{-i\beta_m z} \\ &= -i\beta_m A_m(z)e^{-i\beta_m z} + \kappa_{m,m+1}A_{m+1}e^{-i\beta_{m+1}z} \\ &\quad + \kappa_{m,m-1}A_{m-1}e^{-i\beta_{m-1}z} \end{aligned} \tag{13.10-5}$$

where, according to (13.9-1) and (13.8-3),

$$\kappa_{m,m+1} = -\kappa^*_{m+1,m} \tag{13.10-6}$$

The set of coupled equations (13.10-5) can be expressed, using the vector definition of Equation (13.10-3), as

$$\frac{d\mathbf{E}}{dz} = \tilde{\mathbf{C}}E \tag{13.10-7}$$

$$\tilde{\mathbf{C}} = \begin{vmatrix} -i\beta_1 & \kappa_{1,2} & 0 & 0 & \ldots 0 & 0 \\ \kappa_{2,1} & -i\beta_2 & \kappa_{2,3} & 0 & \ldots 0 & 0 \\ 0 & \kappa_{3,2} & -i\beta_3 & \kappa_{3,4} & \ldots 0 & 0 \\ \cdot & \cdot & \cdot & \cdot & \cdot & \cdot \\ \cdot & \cdot & \cdot & \cdot & \cdot & \cdot \\ \cdot & \cdot & \cdot & \cdot & \cdot & \cdot \\ \cdot & \cdot & \cdot & \cdot & \cdot & \cdot \\ 0 & 0 & 0 & 0 & \kappa_{N,N-1} & -i\beta_N \end{vmatrix} \tag{13.10-8}$$

A propagating supermode, by definition, is a field solution that except for a phase factor $\exp(i\gamma z)$ to be determined, is independent of the propagation coordinate z. We can express it as

$$\mathbf{E}(z) = \mathbf{E}(0)e^{i\gamma z} \tag{13.10-9}$$

$$\frac{d\mathbf{E}}{dz} = i\gamma\mathbf{E}$$

which combined with (13.10-7) results in

$$(\tilde{\mathbf{C}} - i\gamma\tilde{\mathbf{I}})\mathbf{E} = 0 \tag{13.10-10}$$

where $\tilde{\mathbf{I}}$ is the identity $N \times N$ matrix. Equation (13.10-10) is a generalization of (13.9-7). If written out in detail, it becomes

$$\begin{array}{lllll}
(C_{11} - i\gamma)E_1 & + \; C_{12}E_2 & + \cdots + \; C_{1N}E_N = 0 \\
C_{21}E_1 & + \; (C_{22} - i\gamma)E_2 & + \cdots + \; C_{2N}E_N = 0 \\
\cdot & & \\
\cdot & & \\
\cdot & & \\
\cdot & & \\
\cdot & & \\
\cdot & & \\
C_{N1}E_1 & + \; C_{N2}E_2 & + \cdots + \; (C_{NN} - i\gamma)E_N = 0
\end{array} \tag{13.10-11}$$

These are N homogeneous equations with N unknowns $(E_1, \ldots, E_N)$ similar to Equation (13.9-8). The method of solution involves first finding the roots of the determinantal equation

$$\det|\tilde{\mathbf{C}} - i\gamma\tilde{\mathbf{I}}| = 0 \tag{13.10-12}$$

for the N eigenvalues $i\gamma_1, \ldots, i\gamma_N$. Each eigenvalue, say, $i\gamma_\nu$, is then used in (13.10-11) to obtain the corresponding eigenvector $\mathbf{E}^\nu$, which thus satisfies

$$(\tilde{\mathbf{C}} - i\gamma_\nu\tilde{\mathbf{I}})\mathbf{E}^\nu = 0 \tag{13.10-13}$$

The νth supermode (sometimes called the *array mode*) thus consists of a unique linear superposition of the individual waveguide modes $\mathscr{E}_m(x, y)$ with a fixed relative phase between any two of them. This combination propagates together with a *single* phase factor $\exp(i\gamma_\nu z)$. The total field of the νth array mode (supermode) can be written using (13.10-1), (13.10-3), and (13.10-9) as

$$\mathscr{E}^\nu(x,y,z) = \left[\sum_m E_m^\nu \mathscr{E}_m(x, y)\right] e^{i\gamma_\nu z} \tag{13.10-14}$$

The simple case of identical waveguides that are equally spaced with nearest neighbor coupling deserves some special attention. In this case we have $\beta_1 = \beta_2 \cdots = \beta_N \equiv \beta$. We also have from (13.8-6)

$$\kappa_{m,m+1} = -\kappa_{m+1,m} = -i\kappa \tag{13.10-15}$$

Using the last relation we obtain directly from (13.10-12) and (13.10-11)

$$E_\ell^\nu = \sin\left(\ell \frac{\pi\nu}{N+1}\right) \qquad \begin{array}{l} \ell = 1, 2, \ldots, N \\ \nu = 1, 2, \ldots, N \end{array} \tag{13.10-16}$$

$$\gamma_\nu = -\beta - 2\kappa \cos\left(\frac{\pi\nu}{N+1}\right) \tag{13.10-17}$$

where, we recall, E_ℓ^ν is the complex field amplitude in waveguide ℓ corresponding to supermode ν as in Equation (13.10-14). $-\gamma_\nu$ is, according to (13.10-9) the propagation constant of supermode ν.

Theoretical plots of a four-waveguide supermodes are shown in Figure 13-21. We note that only the fundamental $\nu = 0$ mode, in which all the channel fields possess the same phase, has a far-field consisting (mostly) of a single lobe centered on $\theta = 0^0$.

Semiconductor laser arrays of the type illustrated in Figure 13-20 have received a great deal of attention during the last few years due to their potential for combining coherently the power outputs of many lasers [25, 26]. Many different approaches to "encourage" the semiconductor laser array to oscillate only in its desirable (in-phase) fundamental mode have not been entirely successful [26].

To appreciate the connection between the near-field, say, at the output facet, and the far-field, we recall that according to (13.10-14) we can write the near field of the νth supermode measured at the output plane $z = L$ in the form of a phase-locked superposition of individual waveguide modes

$$\mathscr{E}^\nu(x, y, z) = \sum_{\ell=1}^{N} E_\ell^\nu \mathscr{E}_\ell(x, y) \exp(i\gamma_\nu L) \tag{13.10-18}$$

The amplitudes $E_\ell^\nu = E_\ell^\nu(0)$ are the solution of (13.10-11) for the νth mode. Let $E_\ell(\theta)$ be the far field of the ℓth channel by itself when $E_\ell^\nu = 1$ while $E_m^\nu = 0$ and $m \neq \ell$. With θ measured from the normal to the exit plane in the x-z plane, it follows from (13.10-18) by superposition that the far field of the νth supermode is

$$F^\nu(\theta) = \sum_{\ell=1}^{N} E_\ell^\nu E_\ell(\theta) \tag{13.10-19}$$

In the, special but important, case of identical channels, we have

$$E_\ell(\theta) = E_{\ell+1}(\theta) e^{-ik_0 S \sin\theta} = E_0(\theta) e^{ik_0 \ell S \sin\theta}$$

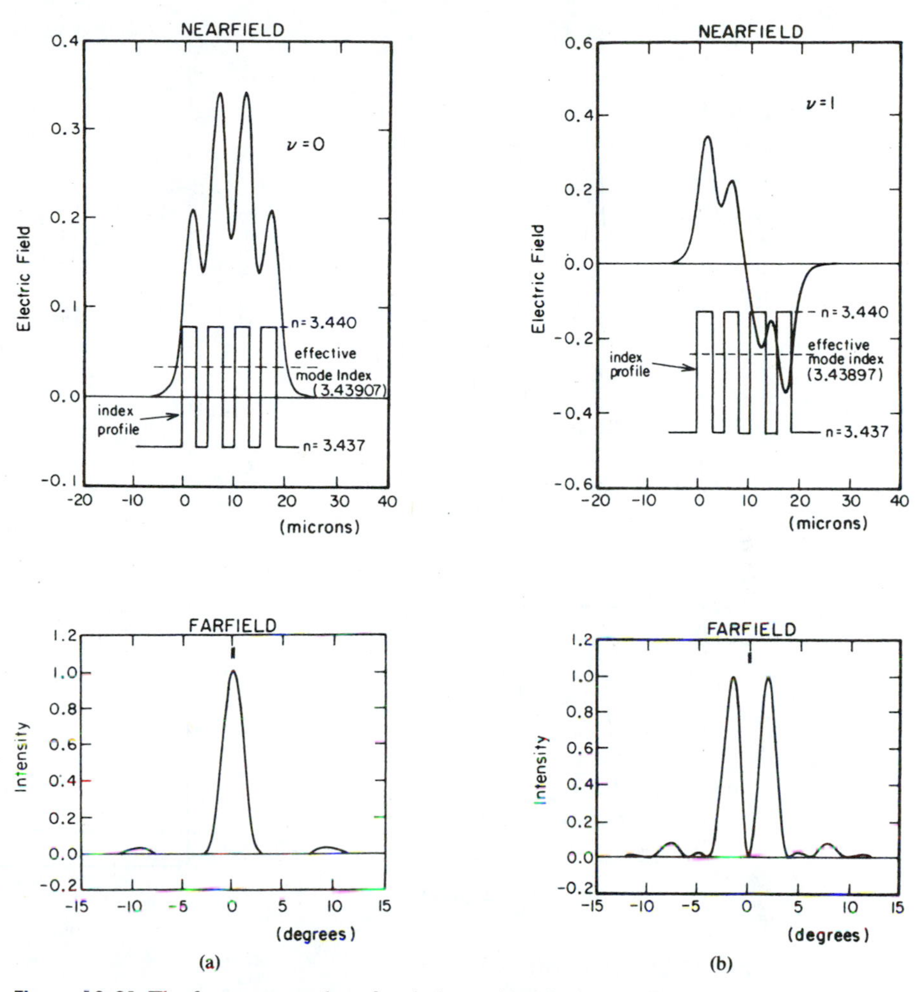

Figure 13-21 The four supermodes of an index-guided four-channel waveguide: (a) The lowest order (highest $|\gamma_\nu|$) $\nu = 0$, mode. (b), (c), (d) The next three modes in decreasing order of β. The upper figure in each case shows the near field in relation to the channel array. The lower figure is that of the far-field intensity. The individual channel waveguides can support a single mode (each) only. (Courtesy of C. Lindsey)

where S is the separation between two channels so that $k_0 S \sin\theta$ is the extra phase delay in the far field between the fields arriving from two adjacent channels. From the last two equations

$$F^\nu(\theta) = E_0(\theta) \sum_{\ell=0}^{N-1} E_\ell^\nu e^{ik_0 \ell S \sin\theta} \qquad (13.10\text{-}20)$$

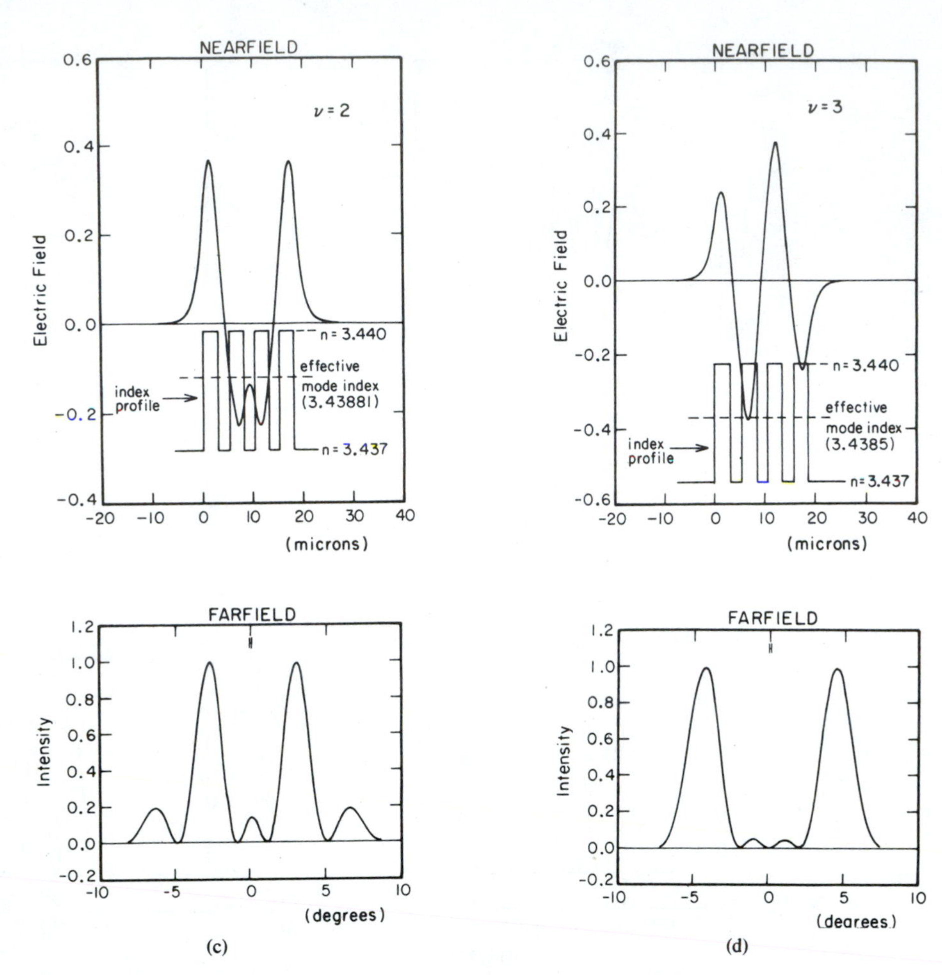

Figure 13-21 *(continued)*

The far-field intensity pattern is

$$|F^{\nu}(\theta)|^2 = |E_0(\theta)|^2 G^{\nu}(\theta) \tag{13-10.21}$$

with

$$G^{\nu}(\theta) \equiv \left| \sum_{\ell=0}^{N-1} E_{\ell}^{\nu} e^{ik_0 \ell S \sin\theta} \right|^2 \tag{13.10-22}$$

The "grating function" $G^{\nu}(\theta)$ is periodic in $\sin\theta$ with a period $\Delta(\sin\theta) = 2\pi/(k_0 S)$. It corresponds physically to the radiation pattern in the far field of N point (i.e., "δ-

function'') sources spaced by S whose relative amplitudes are proportional to E_ℓ^ν. The grating function $G^0(\theta)$ of the fundamental ($\nu = 0$, i.e., $+++++$) mode of a five-element ($N = 5$) array is shown in Figure 13-22(a). The solid curve is based on amplitudes E_ℓ^0 obtained from a solution of (13.10-11). The dashed curve is a plot of $G^\nu(\theta)$ with $E_\ell^\nu = 1$, i.e., equal and in-phase amplitudes. The remaining curves show $G^\nu(\theta)$ of the high order supermodes. The limited angular spread of the optical beam as described by $F^\nu(\theta)$ is due to the finite angular extent of the single aperture radiation pattern $|E_0(\theta)|^2$ which is a factor in $F^\nu(\theta)$. In Problem 13.12 we show that to obtain a single-lobed, far-field pattern it is sufficient to have an aperture whose width satisfies $W \gtrsim S/2$.

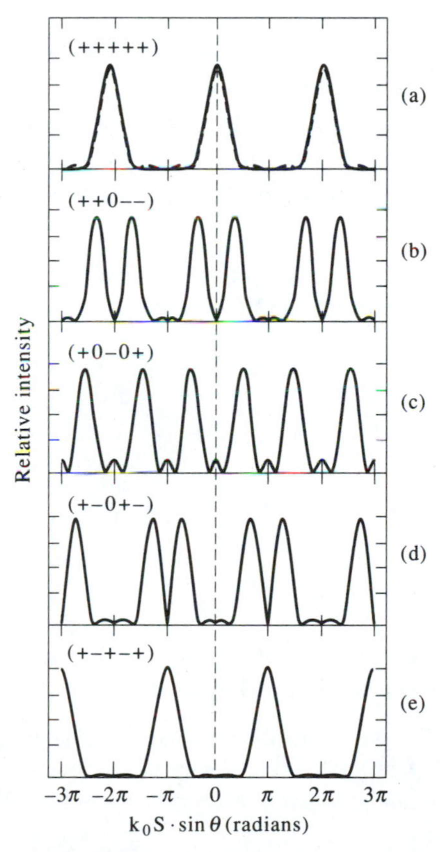

Figure 13-22 The grating functions G^ν of the five supermodes of a five-channel array. The dashed curve in (a) corresponds to an array of five equal amplitude ($E_i = 1$) radiators. (After Reference [25].)

Problems

13.1 Derive Equation (13.2-7).

13.2 Show that the form of Equation (13.7-10) is consistent with the conservation of the modes' power.

13.3 Derive the equations in (13.7-12).

13.4 Derive an expression for the modulation power of a transverse electrooptic waveguide modulator of length L and cross section $2\lambda \times 2\lambda$ (λ is the vacuum wavelength of the light). Compare to the bulk result ([9], p. 547). Estimate the power requirement for a $LiNbO_3$ modulator at $\lambda = 1\ \mu m$, L = 5 mm.

13.5 Derive Equation (13.2-13).

13.6 Prove the orthogonality relation Equation (13.2-8). What happens to this relation if ε is complex?

13.7 Compare the ratio of the power propagating in the regions with an index n_1 to that of the total power for a TE_m and TM_m mode.

13.8 Show that in the case of electrooptic mode coupling in which $\beta_m^{TM} \neq \beta_m^{TE}$ [see Equation (13.7-10)], one can use a z periodic electrooptic constant or electric field to obtain phase matched operation. How would you accomplish this in practice? (Be bold and invent freely.)

13.9 Show that in the case of coupling between modes that carry power in opposite directions the conservation of total power condition becomes

$$\frac{d}{dz}(|A|^2 - |B|^2) = 0$$

which can be satisfied if instead of (13.7-10) we have

$$\frac{dA}{dz} = \kappa_{ab} B e^{-i(\beta_B - \beta_A)z}$$

$$\frac{dB}{dz} = \kappa_{ab}^* A e^{i(\beta_B - \beta_A)z}$$

13.10 Find the eigenmodes, as in (13.9-11), of a two-channel (waveguide) system with individual exponential gain constants γ_a and γ_b in each guide.

13.11 Referring to Figure 13-20 assume that the regions between the optical waveguides are highly absorbing. Explain qualitatively why the desirable $(++ \cdots +)$ supermode has a smaller modal gain at a given injection current than the $(+-+- \cdots +)$ supermode.

13.12

a. Plot the array function $G^0(\theta)$ of a six-channel in-phase $(++++++)$ supermode taking $E_\ell^0 = 1$.

b. Using the diffraction result Equation (4.8-2), derive the far-field distribution $E_0(\theta)$ due to a single channel, say channel 0, whose near field is

$$\mathscr{E}_0(x, y) = \begin{cases} \cos\left(\dfrac{\pi x}{W}\right) & -W/2 \leq x \leq W/2 \\ 0 & \text{elsewhere} \end{cases}$$

c. Plot the far-field intensity distribution function $|F^0(\theta)|^2$ of the mode assuming $k_0S = 6\pi$ and (1) $W = 0.05S$, (2) $W = 0.2S$, and (3) $W = 0.6S$. (S is the distance between the centers of two nearest-neighbor channels.) Obtain an approximate relation between the number of lobes in the far field and the ratio W/S.

13.13 Repeat Problem 13.12 for the $(+ - + - +)$ mode, i.e., $E_0 = E_2 = E_{-2} = 1$, $E_1 = E_{-1} = -1$.

References

1. Yariv, A., and R.C.C. Leite, "Dielectric waveguide mode of light propagation in *p-n* junctions," *Appl. Phys. Lett.* 2:55, 1963.
2. Osterberg, H., and L. W. Smith, "Transmission of optical energy along surfaces," *J. Opt. Soc. Am.* 54:1073, 1964.
3. Hall, D., A. Yariv, and E. Garmire, "Optical guiding and electrooptic modulation in GaAs epitaxial layers," *Opt. Comm.* 1:403, 1970.
4. Shubert, R., and J. H. Harris, "Optical surface waves on thin films and their application to integrated data processors," *IEEE Trans. Microwave Theory Tech.* (1968 Symp. issue) MTT-16:1048–1054, Dec. 1968.
5. Miller, S. E., "Integrated optics, an introduction," *Bell Syst. Tech. J.* 48:2059, 1969.
6. Goell, J. E., "A circular harmonic computer analysis for rectangular dielectric waveguides," *Bell Syst. Tech. J.* 48:2133, 1968.
7. Marcatili, E.A.J., "Dielectric rectangular waveguide and directional couplers for integrated optics," *Bell Syst. Tech. J.* 48:2071, 1969.
8. Lotspeich, J. F., "Explicit general eigenvalue solutions for dielectric slab waveguides," *Appl. Opt.* 14:327, 1975.
9. Chapter 14, this book.
10. Hammer, J. M., D. J. Channin, and M. T. Duffy, "High speed electrooptic grating modulators," *RCA Technical Report,* unpublished.
11. Tien, P. K., R. Ulrich, and R. J. Martin, "Modes of propagating light in thin deposited semiconductor films," *Appl. Phys. Lett.* 144:291–293, 1969.
12. Yariv, A., "Coupled mode theory for guided wave optics," *IEEE J. Quant. Elec.* 9:919, 1973.
13. Kuhn, L., M. L. Dakss, P. F. Heidrich, and B. A. Scott, "Deflection of optical guided waves by a surface acoustic wave," *Appl. Phys. Lett.* 17:265, 1970.
14. Dixon, R. W., "The photoelastic properties of selected materials and their relevance to acoustic light modulators and scanners," *J. Appl. Phys.* 38:5149, 1967.

15. Stoll, H., and A. Yariv, ''Coupled mode analysis of periodic dielectric waveguides,'' *Opt. Commun.* 8:5, 1973.
16. Yablonovitch, E. ''Inhibited spontaneous emission in solid state physics and electronics,'' *Phys. Rev. Lett. 58*:2059, 1987.
17. K. O. Hill, Y. Fujii, D. C. Johnson, and B. Kawasaki, ''Photosensitivity in optical fiber waveguides: Application to reflection filter fabrications,'' *Appl. Phys. Lett.* 32:646–649, 1978.
18. G. Meltz, W. W. Morey, and W. H. Glenn, ''Formation of Bragg gratings in optical fibers by a transverse holographic method,'' *Opt. Lett.* 14:823–825, 1989.
19. K. O. Hill, B. Malo, F. Bilodeau, and D. C. Johnson, ''Photosensitivity in optical fibers,'' *Ann. Rev. Mater. Sci.* 23:125–157, 1993
20. J.-L. Archambault, L. Reekie, and P. St. J. Russell, ''100% reflectivity Bragg reflectors produced in optical fibres by single excimer laser pulses,'' *Electr. Lett.* 29:453–454, 1993
21. W. W. Morey, G. A. Ball, and G. Meltz, ''Photoinduced Bragg gratings in optical fibers,'' *Optics & Photonics News*, February 1994, 8–14.
22. Joanopoulos, J. D., R. D. Meade, and J. N. Winn, ''Photonic crystals: molding in the flow of light.'' Princeton, N.J.: Princeton Univ. Press, 1995.
23. Somekh, S., E. Garmire, A. Yariv, H. L. Garvin, and R. G. Hunsperger, ''Channel optical waveguide directional couplers,'' *Appl. Phys. Lett.* 27:327–329, 1975.
24. Campbell, J. C., F. A. Blum, D. W. Shaw, and K. L. Lawley, ''GaAs electro-optic directional-coupler switch,'' *Appl. Phys. Lett.* 27:202–205, 1975.
25. Kapon, E., J. Katz, and A. Yariv, ''Supermode analysis of phase-locked arrays of semiconductor lasers,'' *Opt. Lett.* 9:125, 1984.
26. Scifres, D. R., R. D. Burnham, and W. Streifer, ''Arrays of semiconductor lasers,'' *Appl. Phys. Lett.* 33:1015, 1978.
27. Kapon, E., C. Lindsey, J. Katz, S. Margalit, and A. Yariv, ''Coupling mechanism of gain guided laser arrays,'' *Appl. Phys. Lett.* 44:389, 1984.

Holography and Optical Data Storage

14.0 INTRODUCTION

This chapter takes up the basic concepts and some key applications of the remarkable field of holography [1–8]. This field traces its beginning to a 1948 paper by D. Gabor [1] and to a major improvement by Leith and Upatnieks [2] who solved a major problem of the original proposal. Holography is an imaging technique in which the recording is accomplished by an interference in the recording medium of two, usually mutually coherent waves: the image-bearing "picture wave" and the, usually, plane wave or spherical, "reference" wave. The intensity pattern due to this interference "burns" itself into the volume (or surface) of the recording medium in a proportionate manner by modifying its index of refraction or gain. This pattern—the hologram—clearly contains *both phase and amplitude* information of the "picture." Viewing (reconstruction) is achieved when a wave identical and usually from the same direction as the original reference wave is incident on the hologram. The wave created by the diffraction of this wave from the hologram is identical in all essential respects to the original picture wave so that the viewer perceives the three-dimensional aspects of the originally photographed objects.

In volume holograms, it is possible to store a large number of pictures and to view each one of them selectively, with negligible crosstalk from the other pictures.

14.1 THE MATHEMATICAL BASIS OF HOLOGRAPHY[1]

Figure 14-1 illustrates the experimental setup used in making a simple hologram. A plane-parallel light beam illuminates the object whose hologram is desired. Part of the same beam is reflected from a mirror (at this point we refer to it as the reference beam) and is made to interfere within the *volume* of the photosensitive medium with the beam reflected diffusely from the object (object beam). The photosensitive medium is then developed and forms the hologram.

The image reconstruction process is illustrated in Figure 14-2. It is performed by illuminating the hologram with the same wavelength laser beam and in the same relative orientation that existed between the reference beam and the photosensitive medium when the hologram was made. An observer facing the far side (B) of the hologram will now see a three-dimensional image occupying the same spatial position as the original object. The image is, ideally, indistinguishable from the direct image of the laser-illuminated object.

The Holographic Process Viewed as Bragg Diffraction

To illustrate the basic process involved in holographic wavefront reconstruction, consider the simple case in which the two beams reaching the photosensitive medium

[1]Chapters 17 and 18 deal with the more advanced topic of dynamic holography in nonlinear optical media. The treatment of this section is mostly kinetic. It tells us when and how things happen but does not address the magnitudes involved. It is meant as an introduction to the major concepts of holography.

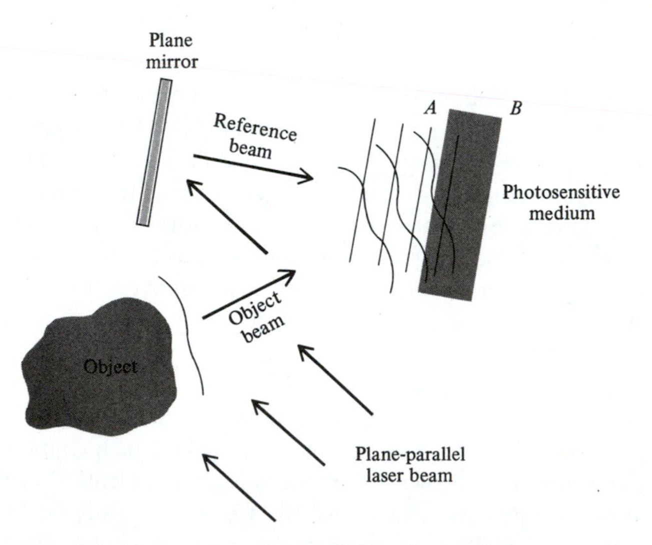

Figure 14-1 A hologram of an object can be made by exposing a photosensitive medium at the same time to coherent light, which is reflected diffusely from the object, and a plane-parallel reference beam, which is part of the same beam that is used to illuminate the object.

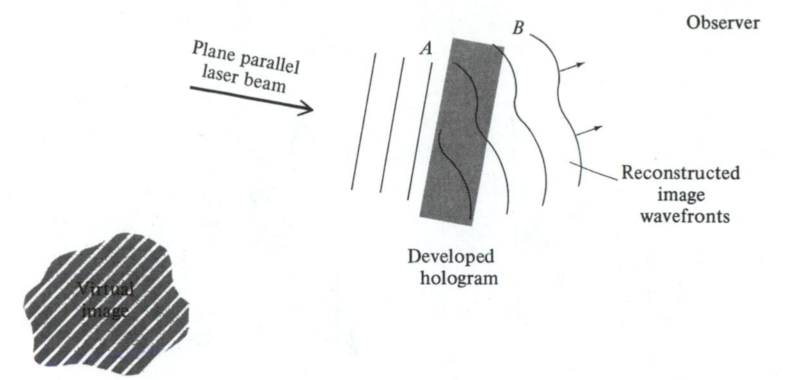

Figure 14-2 Wavefront reconstruction of the original image is usually achieved by illuminating the hologram with a laser beam of the same wavelength and relative orientation as the reference beam making it. An observer on the far side (B) sees a virtual image occupying the same space as the original subject.

in Figure 14-1 are plane waves. The situation is depicted in Figure 14-3. We choose the z axis as the direction of the bisector of the angle formed between the two propagation directions $\mathbf{k}_1$ and $\mathbf{k}_2$ of the reference and object plane waves inside the photosensitive layer. The x axis is contained in the plane of the paper. The electric fields of the two beams are taken as

$$e_{\text{object}}(\mathbf{r}, t) = E_1 e^{i(\mathbf{k}_1\cdot\mathbf{r}-\omega t)}$$

$$e_{\text{reference}}(\mathbf{r}, t) = E_2 e^{i(\mathbf{k}_2\cdot\mathbf{r}-\omega t)} \qquad \textbf{(14.1-1)}$$

From Figure 14-3 and the fact that $|\mathbf{k}_1| = |\mathbf{k}_2| = k$, we have

$$\mathbf{k}_1 = \mathbf{a}_x k \sin\theta + \mathbf{a}_z k \cos\theta$$

$$\mathbf{k}_2 = -\mathbf{a}_x k \sin\theta + \mathbf{a}_z k \cos\theta \qquad \textbf{(14.1-2)}$$

where $k = 2\pi/\lambda$, and $\mathbf{a}_x$ and $\mathbf{a}_z$ are unit vectors parallel to x and z, respectively.

The total complex field amplitude is the sum of the complex amplitudes of the two beams, which, using (14.1-1) and (14.1-2), can be written as

$$E(x, z) = E_1 e^{ik(x\sin\theta + z\cos\theta)} + E_2 e^{ik(-x\sin\theta + z\cos\theta)} \qquad \textbf{(14.1-3)}$$

If the photosensitive medium were a photographic emulsion, the exposure to the two beams and subsequent development would result in silver atoms developed out at each point in the emulsion in direct proportion to the time average of the square of the optical field. The density of silver in the developed hologram is thus proportional to $E(x, z)E^*(x, z)$, which, using (14.1-3), assuming E_1 and E_2 real, becomes

$$E(x, z)E^*(x, z) = E_1^2 + E_2^2 + 2E_1E_2 \cos(2kx\sin\theta) \qquad \textbf{(14.1-4)}$$

The hologram is thus seen to consist of a sinusoidal modulation of the silver density. The planes x = constant (that is, planes containing the bisector and normal to the

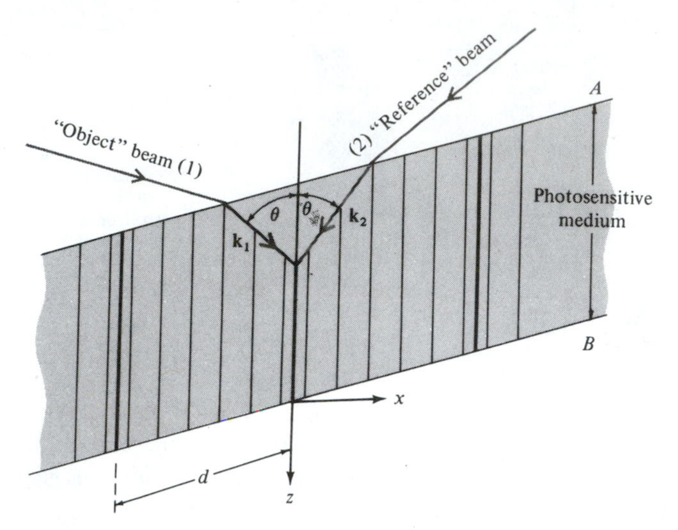

Figure 14-3 A sinusoidal "diffraction grating," produced by the interference of two plane waves inside a photographic emulsion. The density of black lines represents the exposure and hence the silver-atom density. The z direction is chosen as that of the bisector of the angle formed between the directions of propagation *inside* the photographic emulsion. It is not necessarily perpendicular to the surface of the hologram.

plane of Figure 14-3) correspond to equidensity planes. The distance between two adjacent peaks of this spatial modulation pattern is, according to (14.1-4),

$$d = \frac{\pi}{k \sin \theta} = \frac{\lambda/n}{2 \sin \theta} \tag{14.1-5}$$

In the process of wavefront reconstruction, the hologram is illuminated with a coherent laser beam. Since the hologram consists of a three-dimensional sinusoidal diffraction grating, the situation is directly analogous to the diffraction of light from sound waves, which was analyzed in Section 12.1. Applying the results of Bragg diffraction and denoting the wavelength of the light used in reconstruction (that is, in viewing the hologram) as λ_R, a diffracted beam exists *only* when the Bragg condition (12.1-4)

$$2d \sin \theta_B = \frac{\lambda_R}{n_R} \tag{14.1-6}$$

is fulfilled, where θ_B is the angle of incidence and of diffraction as shown in Figure 14-4 and n_R is the index of refraction. Substituting for d its value according to (14.1-5), we obtain

$$\sin \theta_B = \left(\frac{n}{n_R}\right) \frac{\lambda_R}{\lambda} \sin \theta \tag{14.1-7}$$

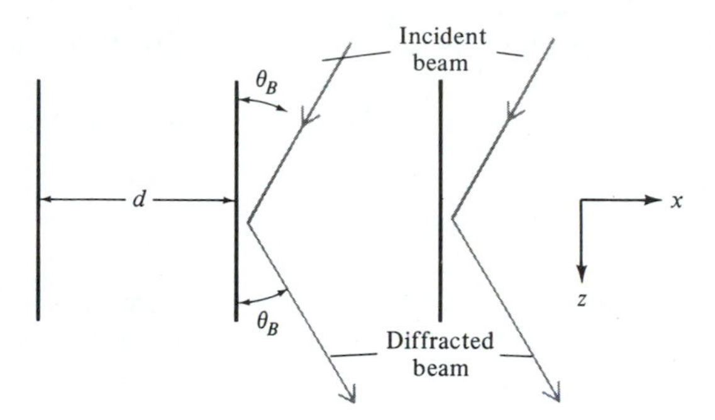

Figure 14-4 Bragg diffraction from a sinusoidal volume grating. The grating periodic distance d is the distance in which the grating structure repeats itself. In the case of a hologram we may consider the vertical lines in the figure as an edge-on view of planes of maximum silver density.

In the special case when $\lambda_R = \lambda$—that is to say, when the hologram is viewed with the same laser wavelength as that used in producing it—we have

$$\theta_B = \theta$$

so that wavefront reconstruction (that is, diffraction) results only when the beam used to view the hologram is incident on the diffracting planes at the same angle as the beam used to make the hologram. The diffracted beam emerges along the same direction ($\mathbf{k}_1$) as the original "object" beam, thus constituting a reconstruction of the latter.

We can view the complex beam reflected from the object toward the photographic emulsion when the hologram is made, as consisting of a "bundle" of plane waves each having a slightly different direction. Each one of these waves interferes with the reference beam, creating, after development, its own diffraction grating, which is displaced slightly in angle from that of the other gratings. During reconstruction the illuminating laser beam is chosen so as to nearly satisfy the Bragg condition (14.1-6) for these gratings. Each grating gives rise to a diffracted beam along the same direction as that of the object plane wave that produced it, so the total field on the far side of the hologram (B) is identical to that of the object field.

Basic Holography Formalism

The point of view introduced above, according to which a hologram may be viewed as a volume diffraction grating, is extremely useful in demonstrating the basic physical principles. A slightly different approach is to take the total field incident on the photosensitive medium as

$$A(\mathbf{r}) = A_1(\mathbf{r}) + A_2(\mathbf{r}) \tag{14.1-8}$$

where $A_1(\mathbf{r})$ may represent the complex amplitude of the diffusely reflected wave from the object while $A_2(\mathbf{r})$ is the complex amplitude of the reference beam. $A_2(\mathbf{r})$

is not necessarily limited to plane waves and may correspond to more complex wavefronts.

The intensity of the total radiation field can be taken, as in (14.1-4), to be proportional to

$$AA^* = A_1A_1^* + A_2A_2^* + A_1A_2^* + A_1^*A_2 \tag{14.1-9}$$

The first term $A_1A_1^*$ is the intensity I_1 of the light arriving from the object. If the object is a diffuse reflector, its unfocused intensity I_1 can be regarded as essentially uniform over the hologram's volume. $A_2A_2^*$ is the intensity I_2 of the reference beam. The change in the amplitude transmittance of the hologram ΔT can be taken as proportional to the exposure density so that

$$\Delta T \propto I_1 + I_2 + A_1A_2^* + A_1^*A_2$$

The reconstruction is performed by illuminating the hologram with the reference beam A_2 in the *same* relative orientation as that used during the exposure. Limiting ourselves to the portion of the transmitted wave modified by the exposure, we have

$$R = A_2\Delta T \propto (I_1 + I_2)A_2 + A_1^*A_2A_2 + I_2A_1 \tag{14.1-10}$$

The first term corresponds to a wavefront proportional to the reference beam. The second term, not being proportional to A_1, may be regarded as undesirable "noise." Since I_2 is a constant, the third term I_2A_1 corresponds to a transmitted wave that is proportional to A_1 and is thus a reconstruction of the object wavefront.

14.2 THE COUPLED WAVE ANALYSIS OF VOLUME HOLOGRAMS

In this section, we will extend the qualitative dynamic arguments of Section 14.1 and obtain analytic expressions that will be useful in analyzing certain holographic applications. We will start with a description of the recording process, followed by a coupled wave analysis of the reconstruction of the hologram. We will simplify the problem by limiting it to two plane waves. One, A_1, is the "picture" field; A_{2r} is the reference wave. The results can be extended to more complex picture fields.

The total field during the recording phase is thus taken as

$$E(\mathbf{r}) = \mathrm{Re}[(A_1e^{-i\mathbf{k}_1'\cdot\mathbf{r}} + A_{2r}e^{-i\mathbf{k}_2'\cdot\mathbf{r}})e^{i\omega t}]$$

where the subscript r denotes the reference wave. We assume that the index of refraction (rather than the absorption) of the holographic medium changes by an amount that is proportional to the optical intensity $I(\mathbf{r})$

$$\begin{aligned} I(\mathbf{r}) &\propto [(A_1e^{-i\mathbf{k}_1'\cdot\mathbf{r}} + A_{2r}e^{-\mathbf{k}_2'\cdot\mathbf{r}})(\text{c.c.})] \\ &= |A_1|^2 + |A_{2r}|^2 + A_1A_{2r}^*e^{-i(\mathbf{k}_1'-\mathbf{k}_2')\cdot\mathbf{r}} + A_1^*A_{2r}e^{i(\mathbf{k}_1'-\mathbf{k}_2')\cdot\mathbf{r}} \end{aligned} \tag{14.2-1}$$

We can thus take the index of refraction distrubution of a hologram as

$$\begin{aligned} n(\mathbf{r}) &= n_0 + n_1\cos(\mathbf{K}\cdot\mathbf{r} + \phi) \\ n_1 &\propto |A_1 A_{2r}| \propto \sqrt{I_1 I_2} \end{aligned} \tag{14.2-2}$$

$$\mathbf{K} = \mathbf{k}_2' - \mathbf{k}_1' \tag{14.2-3}$$

A graphical demonstration of $n_1(\mathbf{r})$—the hologram—is shown in Figure 14-5.

During the reconstruction of the hologram, it is illuminated by a reference wave $A_2 e^{-i\mathbf{k}_2\cdot\mathbf{r}}$ propagating along $\mathbf{k}_2$. Our task is to obtain an expression for the "picture" wave A_1 that results. We know in advance that if the Bragg condition $\mathbf{k}_2 - \mathbf{k}_1 = \mathbf{K}$

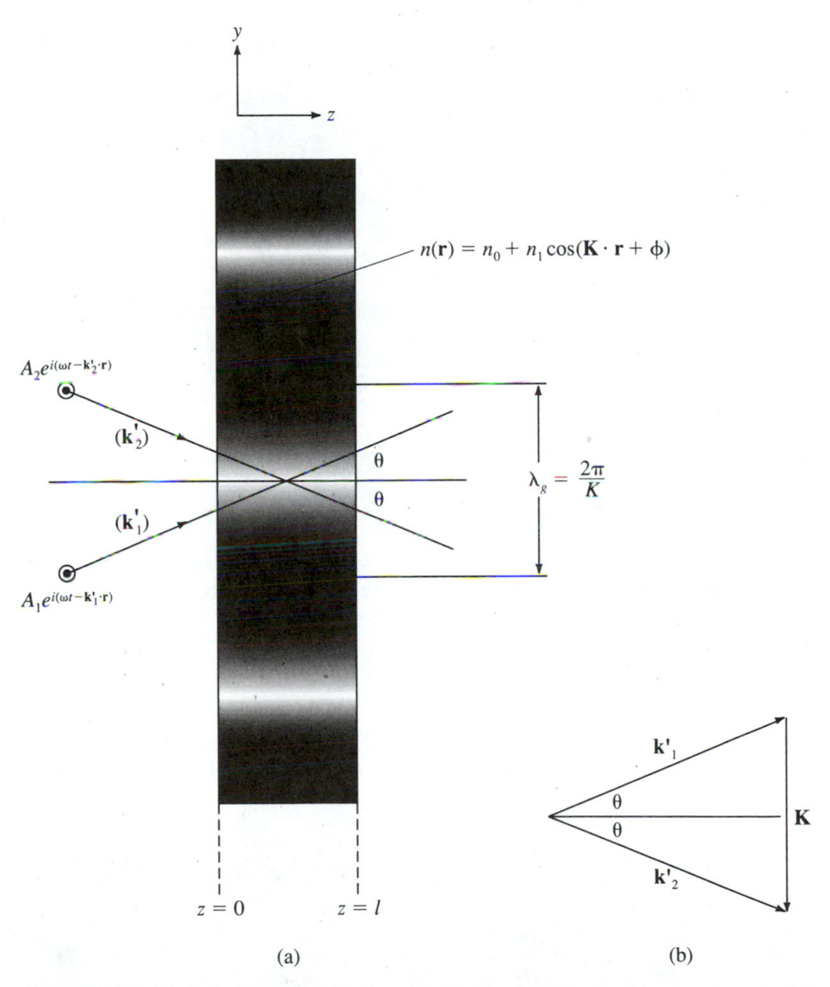

Figure 14-5 (a) A holograph $n(\mathbf{r})$ is recorded in a photosensitive medium by the standing wave pattern produced by two coherent beams that intersect in the medium. The hologram grating vector is $\mathbf{K} = \mathbf{k}_2' - \mathbf{k}_1'$. (b) The Bragg condition diagram $\mathbf{K} = \mathbf{k}_2 - \mathbf{k}_1$.

is satisfied, where $\mathbf{k}_1$ is the propagation vector of the diffracted wave, then the interaction involves exclusively waves A_1 and A_2 so that we can write for the total field in a long medium hologram

$$\mathbf{E}(\mathbf{r}) = \frac{1}{2}\mathbf{A}_1(\mathbf{r})e^{-i\mathbf{k}_1\cdot\mathbf{r}} + \frac{1}{2}\mathbf{A}_2(\mathbf{r})e^{-i\mathbf{k}_2\cdot\mathbf{r}} + \text{c.c.} \tag{14.2-4}$$

These two waves "see" a medium with a modulated index of refraction, the hologram, given by (14.2-2). The total field thus obeys the Helmholtz equation

$$\nabla^2\,\mathbf{E}(\mathbf{r}) + \omega^2\mu\epsilon(\mathbf{r})\mathbf{E} = 0 \tag{14.2-5}$$

$$\epsilon(r) = \epsilon_0 n^2(r) \cong \epsilon_0[n_0^2 + (n_0 n_1 e^{-i(\mathbf{K}\cdot\mathbf{r}+\phi)} + \text{c.c.})] \tag{14.2-6}$$

Substituting (14.2-4) and (14.2-6) in (14.2-5) leads to

$$\begin{aligned}
&\frac{1}{2}\left(-2ik_1\frac{dA_1}{dr_1} - k_1^2A_1\right)e^{-i\mathbf{k}_1\cdot\mathbf{r}} + \text{cc.}\\
&+\frac{1}{2}\left(-2ik_2\frac{dA_2}{dr_2} - k_2^2A_2\right)e^{-i\mathbf{k}_2\cdot\mathbf{r}} + \text{c.c.}\\
&+\omega^2\mu\epsilon_0[n_0^2 + (n_0n_1e^{-i\phi}e^{-i\mathbf{K}\cdot\mathbf{r}} + \text{c.c.})]\\
&\times\left[\frac{A_1}{2}e^{-i\mathbf{k}_1\cdot\mathbf{r}} + \frac{A_2}{2}e^{-i\mathbf{k}_2\cdot\mathbf{r}} + \text{c.c.}\right] = 0
\end{aligned} \tag{14.2-7}$$

where we neglected

$$\frac{d^2A}{dr^2} \ll k\frac{dA}{dr}$$

We observe by inspection that spatially cumulative exchange of power takes place when the (Bragg) condition

$$\mathbf{k}_2 - \mathbf{k}_1 = \mathbf{K} \tag{14.2-8}$$

is satisfied.[2] Keeping only synchronous terms (terms with similar exponents) and recalling that in an isotropic medium $k_1 = k_2 = \omega\sqrt{\mu\epsilon_0}n_0$ helps us simplify (14.2-7) to

$$\begin{aligned}
\cos\theta\frac{dA_1}{dz} &= -\frac{\alpha}{2}A_1 + i\frac{\pi n_1}{\lambda}e^{i\phi}A_2e^{i(\mathbf{k}_1-\mathbf{k}_2+\mathbf{K})\cdot\mathbf{r}}\\
\cos\theta\frac{dA_2}{dz} &= -\frac{\alpha}{2}A_2 + i\frac{\pi n_1}{\lambda}e^{-i\phi}A_1e^{-i(\mathbf{k}_1-\mathbf{k}_2+\mathbf{K})\cdot\mathbf{r}}
\end{aligned} \tag{14.2-9}$$

[2]When condition (14.2-8) is not satisfied the power exchange reverses sign every "coherence length" $L_c \equiv \pi/(|\mathbf{k}_2 - \mathbf{k}_1 - \mathbf{K}|)$ and averages out to near zero over distances $\gg L_c$.

where loss terms $-(\alpha/2)A_{1,2}$ were added phenomenologically to account for absorption, and $\lambda = 2\pi/(\omega\sqrt{\mu\epsilon_0})$ is the free space wavelength. 2θ is the angle between $\mathbf{k_1}$ and $\mathbf{k_2}$, and z is the distance measured along the bisector, so that $z = r_{1,2} \cos \theta$. Expressing the amplitudes in terms of magnitudes and phases by using the definition $A_j \equiv \sqrt{I_j} \exp(-i\phi_j)$ leads to (in what follows, we take $\mathbf{k_1} - \mathbf{k_2} + \mathbf{K} = 0$, i.e., the Bragg condition is satisfied)

$$\cos\theta \frac{dI_1}{dz} = -\alpha I_1 - \frac{2\pi n_1}{\lambda}\sqrt{I_1 I_2}\sin(\phi_1 - \phi_2 + \phi)$$

$$\cos\theta \frac{dI_2}{dz} = -\alpha I_2 + \frac{2\pi n_1}{\lambda}\sqrt{I_1 I_2}\sin(\phi_1 - \phi_2 + \phi) \tag{14.2-10}$$

Note that the coupling at point $\mathbf{r}$ depends on the local phase $\psi \equiv (\phi_1 - \phi_2 + \phi)$. If the phase $\psi = \pm\pi/2$, the exchange is maximum. The case $\psi = \pm\pi/2$ corresponds, according to Equations (14.2-4) and (14.2-5), to a grating that is displaced by a quarter period with respect to the intensity interference pattern of waves 1 and 2. In the most common scenario, a single wave, say 1, is incident on the grating, and wave 2 is the diffracted wave. In this case, it follows from the second equation of (14.2-10) that wave 2 is generated with a phase $\phi_2 = \phi_1 + \phi + \pi/2$, i.e., $\psi = -\pi/2$, which results, according to the second equation of (14.2-10), in a *maximum* positive value for the power exchange dI_2/dz.

The solution of (14.2-10) in the case of $\psi = +\pi/2$ and $I_2(0) = 0$ becomes [9]

$$I_1(z) = I_1(0)e^{-\left(\frac{\alpha z}{\cos\theta}\right)}\cos^2\left(\frac{\pi n_1 z}{\lambda\cos\theta}\right)$$

$$I_2(z) = I_1(0)e^{-\left(\frac{\alpha z}{\cos\theta}\right)}\sin^2\left(\frac{\pi n_1 z}{\lambda\cos\theta}\right) \tag{14.2-11}$$

so that in a grating of length ℓ, the diffraction efficiency is

$$\eta = \frac{I_2(\ell)}{I_1(0)} = \exp\left(-\frac{\alpha\ell}{\cos\theta}\right)\sin^2\left(\frac{\pi n_1 \ell}{\lambda\cos\theta}\right) \tag{14.2-12}$$

This formula is very useful in interpreting a large variety of experimental data involving fixed volume gratings and holograms.

Multihologram Recording and Readout—Crosstalk

It is possible in a volume hologram to record simultaneously a large number of holograms. The fundamental reason is that in a large volume, with dimensions $\gg L_c$, a reconstructed image results only when the Bragg condition, $\mathbf{k}_2 - \mathbf{k}_1 = \mathbf{K}$, is satisfied (see footnote 2). Here $\mathbf{k}_2$ and $\mathbf{k}_1$ are the propagation vectors of the incident and diffracted wave during the reconstruction, and $\mathbf{K}$ represents the hologram, as in (14.2-2). If the $\mathbf{K}$ vectors representing the different holograms are sufficiently different, it is possible to read one specific hologram with negligible contributions—crosstalk—from the others, since, for all the other holograms, the Bragg condition is strongly violated. To consider this problem quantitatively, consider the situation depicted in Figure 14-6, where two holograms $n_1(\mathbf{r})$ and $n_2(\mathbf{r})$ are recorded in the same volume using two different reference directions $k_2^{(1)}$ and $k_2^{(2)}$ but the same picture wave direction $\mathbf{k}_1$.

$$n_1^{(1)}(\mathbf{r}) \propto \sqrt{I_1^{(1)}I_2^{(1)}} \sin (\mathbf{K}^{(1)}\cdot\mathbf{r} + \phi^{(1)})$$
$$n_1^{(2)}(\mathbf{r}) \propto \sqrt{I_1^{(2)}I_2^{(2)}} \sin (\mathbf{K}^{(2)}\cdot\mathbf{r} + \phi^{(2)}) \qquad (14.2\text{-}13)$$

where

$$\mathbf{K}^{(1)} = \mathbf{k}_2^{(1)} - \mathbf{k}_1$$
$$\mathbf{K}^{(2)} = \mathbf{k}_2^{(2)} - \mathbf{k}_1 \qquad (14.2\text{-}14)$$

If we wish to reconstruct picture 1, we illuminate the hologram with the corresponding reference wave $\mathbf{k}_2^{(1)}$ (i.e., the same reference wave used to record it), as discussed above. This reference wave will encounter in the crystal, not only the desired hologram $n_1^{(2)}(\mathbf{r})$ but also hologram $n_1^{(2)}(\mathbf{r})$. Any light scattered from hologram $n_1^{(2)}(\mathbf{r})$ in the direction of $\mathbf{k}_1$ thus constitutes (noisy) crosstalk, which degrades the information contents of picture 1. This crosstalk places a fundamental limit on the number of holograms and their stored information contents. To quantify this argument, we will derive an expression for the power radiated along $\mathbf{k}_1$ due to the undesirable scattering of the reference beam employed ($\mathbf{k}_2^{(1)}$) off the "wrong" hologram of picture 2 $- \, n_1^{(2)}(\mathbf{r})$. The equations describing this process were derived in (14.2-9) and are reproduced here for the incident (A_2) and the diffracted (A_1) beams

$$\frac{dA_1}{dz} = i\frac{\pi n_1^{(2)}}{\lambda \cos\theta} A_2 e^{i(\mathbf{k}_1 - \mathbf{k}_2^{(1)} + \mathbf{k}_2^{(2)} - \mathbf{k}_1)\cdot\mathbf{r}}$$
$$\frac{dA_2}{dz} = i\frac{\pi n_1^{(2)}}{\lambda \cos\theta} A_1 e^{-i(\mathbf{k}_1 - \mathbf{k}_2^{(1)} + \mathbf{k}_2^{(2)} - \mathbf{k}_1)\cdot\mathbf{r}} \qquad (14.2\text{-}15)$$

where the grating vector $\mathbf{K}^{(2)} = \mathbf{k}_2^{(2)} - \mathbf{k}_1$ is that of hologram 2 and we took $\phi = 0$. The direction $\mathbf{k}_1$ is, according to Figure 14-6, the same for both $n_1^{(1)}(\mathbf{r})$ and $n_1^{(2)}(\mathbf{r})$, since the "picture" direction is the same for all the recorded holograms. We rewrite Equations (14.2-15) as

$$\frac{dA_1}{dz} = i\kappa A_2 e^{-i\delta z}$$
$$\frac{dA_2}{dz} = i\kappa A_1 e^{-i\delta z} \qquad (14.2\text{-}16)$$
$$\kappa = \frac{\pi n_1^{(2)}}{\lambda\cos\theta} \qquad \delta \equiv k_{2z}^{(2)} - k_{2z}^{(1)}$$

where, for the sake of simplicity, we assumed no optical losses ($\alpha = 0$).

We note that if we take $\mathbf{k}_2^{(1)} = \mathbf{k}_2^{(2)}$, Equations (14.2-16) lead to the solution (14.2-11) for the Bragg matched condition. Here, however, we are interested in the

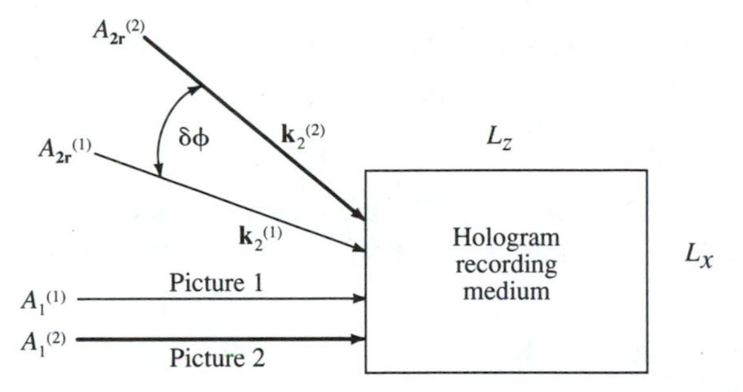

Figure 14-6 Two pictures $A_1^{(1)*}$ and $A_1^{(2)}$ are recorded, separately as volume holograms each with its angularly unique reference wave. In practice, we can record many pictures in the same volume. This procedure can be repeated, resulting in the recording of hundreds (or even thousands) of holograms in the same volume.

crosstalk case, $\mathbf{k}_2^{(1)} \neq \mathbf{k}_2^{(2)}$. In this case, $\delta \neq 0$ and the formal solution of (14.2-16) at the output of the hologram $z = L$ is

$$A_2(L) = A_2(0)e^{-i\delta L}\left[\cos(sL) + \frac{i\delta}{s}\sin(sL)\right]$$

$$A_1(L) = iA_2(0)e^{i\delta L}\frac{\kappa}{s}\sin(sL) \tag{14.2-17}$$

$$s^2 = \kappa^2 + \delta^2 \tag{14.2-18}$$

where we used the boundary condition $A_1(0) = 0$ so that only the reference wave $A_2(0)$ is present at the input. The output field $A_1(L)$ constitutes our crosstalk. The fraction of the incident power thus scattered is given by (14.2-17) as

$$\left|\frac{A_1(L)}{A_2(0)}\right|^2 \approx (\kappa L)^2 \frac{\sin^2(\delta L)}{(\delta L)^2}, \qquad \delta \gg \kappa \tag{14.2-19}$$

We can thus reduce this power to an acceptable level by using a sufficiently large δ. Using the definition $\delta = k_{2z}^{(2)} - k_{2z}^{(1)}$, we find that by employing a sufficiently large angular separation $\delta\phi$ between the two reference waves $\mathbf{k}_2^{(1)}$ and $\mathbf{k}_2^{(2)}$, we can reduce the crosstalk. Choosing, somewhat arbitrarily, the location of the second zero of (14.2-19) as a measure of the necessary selectivity, we obtain

$$\delta = k_{2z}^{(2)} - k_{2z}^{(1)} = K_z^{(2)} - K_z^{(1)} \geq \frac{2\pi}{L} \tag{14.2-20}$$

The strategy for volumetric data storage is thus to fill $\mathbf{K}$ space with $\mathbf{K}^{(i)}$ vectors, each representing a single hologram, which are separated by

$$\Delta K_x \geqq \frac{2\pi}{L_x},\ \Delta K_y \geqq \frac{2\pi}{L_y},\ \Delta K_z \geqq \frac{2\pi}{L_z}$$

where L_x, L_y, L_z are the dimensions of the hologram.

We may thus need allocate to each hologram a volume

$$d^3\mathbf{K} = \frac{8\pi^3}{L_x L_y L_z} = \frac{8\pi^3}{V_{\text{holog}}} \tag{14.2-21}$$

in **K** space in order to avoid crosstalk. The total number of holograms that can be stored thus corresponds to packing **K** space with nonoverlapping holograms whose total number is

$$N_{\text{holog}} = \frac{\text{Total volume in } \mathbf{K} \text{ space}}{\text{Volume per hologram}} \approx \left(\frac{1}{2}\right) \frac{4\pi k^3 V_{\text{holog}}}{3 \times 8\pi^3} = \left(\frac{2\pi}{3}\right) \frac{V_{\text{holog}}}{\lambda^3}$$

where we took $|\mathbf{K}| \sim k$. (Since $\mathbf{K} = \mathbf{k_2} - \mathbf{k_2}$, $|\mathbf{K}|$ can vary between zero and $2k$, so on the average $|\mathbf{K}| \sim k$.) The factor of 1/2 accounts for the fact that holograms with $\mathbf{K}$ and $-\mathbf{K}$ are not independent. λ is the wavelength in the recording medium.

In our example of plane wave holograms, each hologram carries one bit of information. The hologram either exists (a "1") or does not (a "0"), so that the number of stored bits is equal to the number of holograms. The total number of bits that can be stored is thus

$$N_{\text{bits}} \sim \frac{V_{\text{holog}}}{\lambda^3} \tag{14.2-22}$$

According to (14.2-22), if we use $\lambda = 1\ \mu\text{m}$, the number of bits that can be stored exceeds $10^{12}/\text{cm}^3$. This storage density is intriguingly large and helps explain the interest in holographic data storage. Figure 14-8 shows a diagram of the experimental system used in recording some 1,000 angle-multiplexed diagrams in a $LiNbO_3$ crystal.

Wavelength Multiplexing (11)

Another method for multiplexing numerous holograms uses the geometry demonstrated in Figure 14-9. Here a given hologram is recorded with two oppositely traveling waves of the same wavelength. Each hologram is recorded with a different wavelength. The **K** space representation of this method is shown in Figure 14-9(a). The reconstruction of a given picture, say i, is accomplished by illuminating the hologram with the reference wave at λ_i, used to record it, as shown in Figure 14-9(b). This method of recording is called *wavelength multiplexing*. It makes the best use of **K** space and minimizes crosstalk, which results when the information contents of the holograms causes them to spread beyond their nominal **K** space address and encroach on the territory of adjacent holograms [11]. Figure 14-10 shows the reflection vs. λ of a large number of wavelength multiplexed holograms.

Crosstalk in Data-Bearing Holograms (12)

Up to this point we considered only plane wave holograms. We showed that in a recording volume with dimensions $L_i (i = 1, 2, 3)$, such holograms needed to be separated in **K** by $\Delta K_i = 2\pi/L_i$ to avoid crosstalk. If the picture wave, prior to

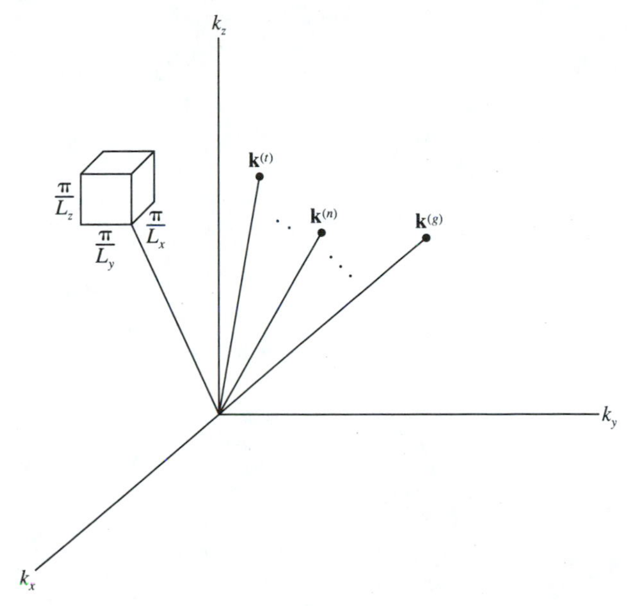

Figure 14-7 **K** space representation of holograms. Each hologram, $n_1^{(i)}(\mathbf{r}) \propto \sin(\mathbf{K_i} \cdot \mathbf{r} + \phi^{(i)})$, is represented by a fuzzy volume centered on $\mathbf{K_i}$. The fuzziness reflects the finite dimension L_x, L_y, L_z of the hologram. A plane wave hologram, such as, g, n, and t, is represented ideally as a point in **K** space. The small ''fuzzy'' volume shown represents the spread in **K** due to the finite volume of the hologram. The unit volume in **K** space associated with a single hologram is shown.

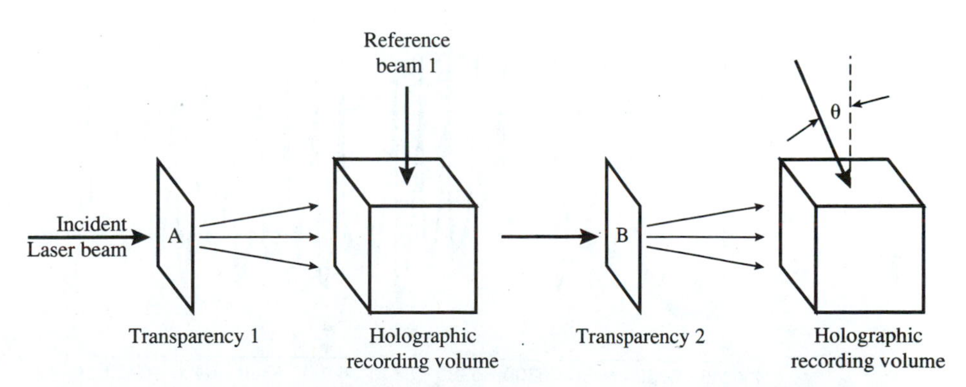

Figure 14-8 Generic schematic diagram of the optical setup for angular multiplexing of volume holograms. Two holograms are shown recorded in the *same* photosensitive volume but with each hologram using a different direction reference beam.

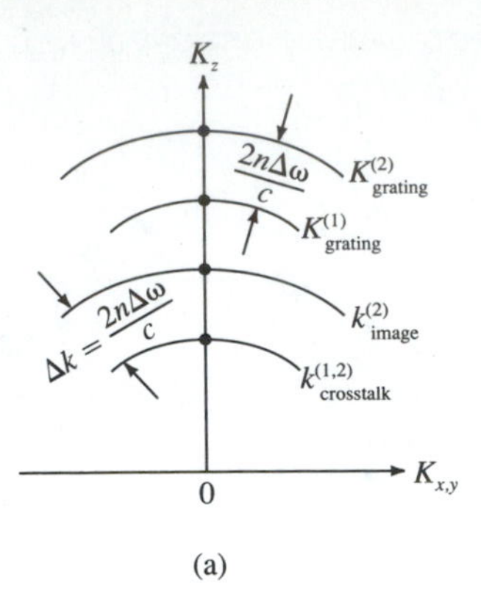

(a)

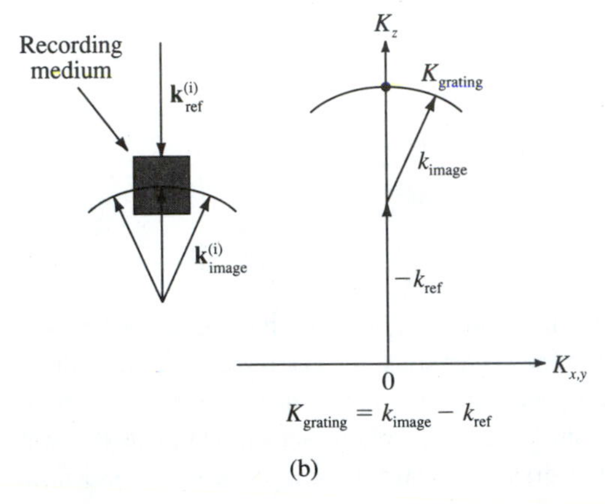

(b)

Figure 14-9 Holographic data storage with wavelength multiplexing. (a) The black dots on the K_z axis correspond to ''blank'' (no information) holograms. The transverse quasi-circular curves correspond to the loci of the tip of the $\mathbf{K}^{(i)}$ vector (recorded with λ_i) when the holograms bear pictorial information. (b) A construction illustrating how the angular fanning of the $k^{(i)}_{\text{image}}$ due to stored information spreads the loci of $\mathbf{K}^{(i)}$. (After Reference [11].)

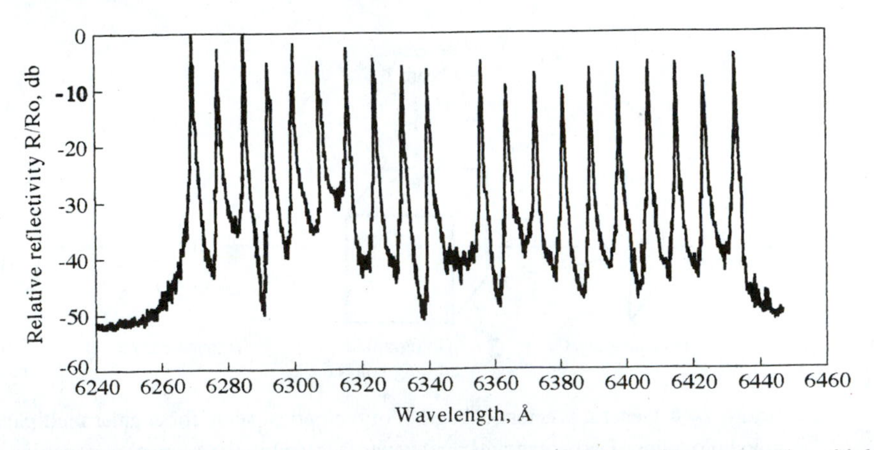

Figure 14-10 Diffraction (reflection) efficiency vs. λ in the case of 20 wavelength multiplexed hologram [Courtesy V. Leyva, G. Rakuljic-Accuwave Corp.]

reaching the crystal, passes through a transparency, or a spatial light modulator, it is modulated spatially and can no longer be represented by a simple plane wave. If the wave incident on the transparency is taken as $A_1 \exp(-ikz)$ and the transmittance of the transparency as $t(x, y)$, then the wave incident on the holographic medium can be taken as

$$\begin{aligned} E_{\text{picture}} &= A_1 e^{-ikz} t(x, y) \\ &= A_1 \iint \exp[i(k_x x + k_y y) + i\sqrt{k^2 - k_x^2 - k_y^2} z]\tilde{t}(k_x, k_y)\, dk_x dk_y \end{aligned} \tag{14.2-23}$$

where

$$t(x, y) = \iint \tilde{t}(k_x, k_y) \exp[i(k_x x + k_y y)]\, dk_x dk_y$$

with $\tilde{t}(k_x, k_y)$ the spatial Fourier transform of $t(x, y)$.

According to (14.2-23), the picture wave is no longer a single plane wave but a continuum of such waves, each with an amplitude $\tilde{t}(k_x, k_y)\, dk_x dk_y$ propagating along the direction

$$\hat{x}k_x + \hat{y}k_y + \hat{z}\sqrt{k^2 - k_x^2 - k_y^2}$$

as illustrated in Figure 14-11. If we plot the **K** vector of this hologram in **K** space as in Figure 14-9(b), the tip of the **K** vector now occupies a volume rather than a point. If the smallest feature size in the transparency is a, then we expect $\tilde{t}(k_x, k_y)$ to have appreciable value up to $\sim (k_x)_{\max} \sim \pi/a$, $(k_y)_{\max} \sim \pi/a$, so that, in general, the volume in **K** space needed to avoid overlap of holograms is now no longer $8\pi^3/L_x L_y L_z$ as in the case of plane wave (no information) holograms but becomes π^3/a^3. This reduces the number of holograms that can be recorded with negligible crosstalk from $\sim L_x L_y L_z/\lambda^3$ (see 14.2-22) to $\sim a^3/\lambda^3$. If we now associate a single

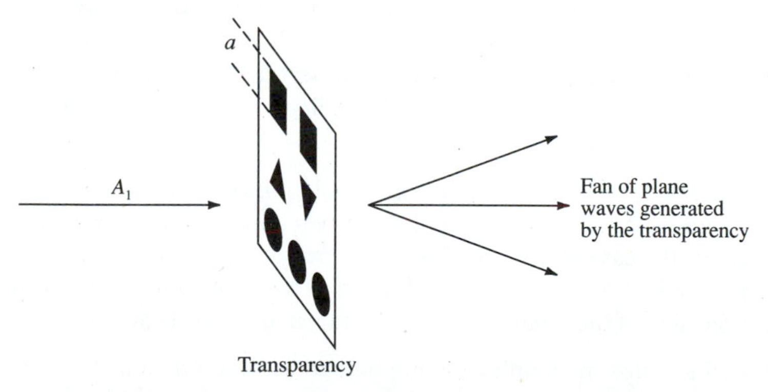

Figure 14-11 Passage of a plane wave through a transparency generates a continuum of plane waves on the output size.

bit with the volume a^3, the volume of the smallest feature, we find that the total number of bits that can be recorded and read with little crosstalk is

$$N_{\text{bits}} = N_{\text{holog}} \times \text{bits per hologram}$$
$$= \left(\frac{a}{\lambda}\right)^3 \left(\frac{V}{a^3}\right) = \left(\frac{V}{\lambda}\right)^3$$
$$(V = L_x L_y L_z \text{ is the hologram volume})$$

which is the same as the result of Equation (14.2-22) for the number of bits that can be stored in the case of (one bit each) plane wave holograms. This is just one manifestation of the fact that the basic limit $\sim V/\lambda^3$ to holographic data storage is insensitive to the spatial modulation format, i.e., single-bit plane holograms or holograms that store multibit pages.

Problems

14.1 Show that if a planar hologram is made using a wavelength λ but is reconstructed with a wavelength λ_R, the reconstructed image is magnified by a factor of λ_R/λ with respect to the original object.

14.2 A monochromaticwave $A(\mathbf{r})$ is propagating essentially in the $+z$ direction. Show that, if in any plane z we replace A by its complex conjugate (but leave the factor $\exp(i\omega t)$ unchanged), the result is a new wave propagating in the $-z$ direction, but possessing everywhere wavefronts (i.e., loci of constant phase) identical to those of the original wave. [*Hint*: The expansion (1.6-12) may be useful.]

14.3

a. Using the notation of section 14.2, show that if the hologram is illuminated with a plane wave A_2^* instead of A_2 the reconstructed image is A_1^* instead of A_1.
b. Show that the reconstructed image A_1^* is real—that is, A_1^* actually converges to an image. [*Hint:* Consider what happens to a bundle of rays originally emanating from a point on the object.]
c. Show that the reconstructed image A_1 observed when the hologram is illuminated by A_2 is virtual; that is, rays corresponding to a given image point do not cross unless imaged by a lens.

14.4 Consider the problem of making a hologram in which the reference and object beams are incident on the emulsion from two opposite sides. Draw the equidensity planes for the case where the beams are nearly antiparallel. Show that the viewing (reconstructing) of this beam is performed in the reflection mode: that is, the viewer faces the side of the emulsion that is illuminated by the beam.

14.5 Show that in an infinitely thin hologram both virtual and real images can be reconstructed simultaneously. [*Hint:* Consider the problems of light scattering from a surface grating (as opposed to a volume grating).]

14.6 Calculate the reconstruction angle sensitivity $d\theta_B/d\lambda_R$ for transmission holograms (as described in the text) and in reflection holograms (as described in Problem 14.4). θ_B is the Bragg angle, and λ_R is the wavelength used in reconstruction. Show that $d\theta_B/d\lambda_R$ is much larger in the case of the transmission hologram. Which hologram will yield better results when illuminated by white light?

14.7 Plot the locus in **K** space of holograms that contain pictorial information with dimensions a. (The holograms were recorded with plane wave passing through a transparency with feature size $\sim a$.)

a. Plot for the case of angular multiplexing.
b. Plot for the case of wavelength multiplexing.
c. Show that in case (a) the preferred geometry for minimizing crowding in **K** space is one where the reference and picture waves are at 90° to each other.

14.8 Calculate the wavelength selectivity of a hologram recorded with two oppositely traveling plane waves ($+z$ and $-z$) of wavelength λ. Specifically, plot the power reflectivity $R(\lambda) = |r(\lambda)|^2$ for a wave incident along the z direction.

References

1. Gabor, D., ''Microscopy by reconstructed wavefronts,'' *Proc. Roy. Soc. (London)*, ser. A, 197:454, 1949.
2. Leith, E. N., and J. Upatnieks, ''Wavefront reconstruction with diffused illumination and three-dimensional objects,'' *J. Opt. Soc. Am.* 54:1295, 1964.
3. Collier, R. J., ''Some current views on wavefront reconstruction,'' *IEEE Spectrum*, 3:67, July 1966.
4. Stroke, G. W., *An Introduction to Coherent Optics and Holography*, 2d ed. New York: Academic, 1969.
5. DeVelis, J. B., and G. O. Reynolds, *Theory and Applications of Holography*. Reading, MA: Addison-Wesley, 1967.
6. Smith, H. M., *Principles of Holography*. New York: Interscience, 1969.
7. Goodman, J. W., *Introduction to Fourier Optics*. New York: McGraw-Hill, 1968.
8. Yu, T. S. F., *Introduction to Diffraction Information Processing and Holography*. Cambridge, MA: MIT, 1973.
9. Kogelnik, H., ''Coupled wave theory for thick hologram gratings,'' *Bell Syst. Tech. J.* 48:2909, 1969.
10. Mok, H. F., ''Angle multiplexed storage of 5000 holograms in lithium niobate,'' *Opt. Lett.* 18:915, 1993.
11. Rakuljic, G., Leyva, V., and Yariv, A. ''Optical data storage by using wavelength multiplexed volume holograms,'' *Opt. Lett.* 17:1473, 1992.
12. Yariv, A. ''Interpage and interpixel crosstalk in wavelength multiplexed holograms,'' *Opt. Lett.* Vol. 18:652, 1993.

Semiconductor Lasers—Theory and Applications

15.0 INTRODUCTION

The semiconductor laser invented in 1961 [1–3] is the first laser to make the transition from a research topic and specialized applications to the mass consumer market. It is, by economic standards and the degree of its applications, the most important of all lasers.

The main features that distinguish the semiconductor laser are

1. Small physical size (300 μm $\times$ 10 μm $\times$ 50 μm) that enables it to be incorporated easily into other instruments.
2. Its direct pumping by low-power electric current (15 mA at 2 volts is typical), which makes it possible to drive it with conventional transistor circuitry.
3. Its efficiency in converting electric power to light. Actual operating efficiencies exceed 50 percent.
4. The ability to modulate its output by direct modulation of the pumping current at rates exceeding 20 GHz. This is of major importance in high-data-rate optical communication systems.
5. The possibility of integrating it *monolithically* with electronic field effect transistors, microwave oscillators, bipolar transistors, and optical components in III–V semiconductors to form integrated optoelectronic circuits.
6. The semiconductor-based manufacturing technology, which lends itself to mass production.
7. The compatibility of its output beam dimensions with those of typical silica-based optical fibers and the possibility of tailoring its output wavelength to the low-loss, low-dispersion region of such fibers.

From the pedagogic point of view, understanding how a modern semiconductor laser works requires, in addition to the basic theory of the interaction of radiation with electrons that was developed in Chapter 5, an understanding of dielectric waveguiding [4, 5] (Section 13.1) and elements of solid-state theory of semiconductors [6, 7]. The latter theory will be taken up in the next few sections.

15.1 SOME SEMICONDUCTOR PHYSICS BACKGROUND

In this section we will briefly develop some of the basic background material needed to understand semiconductor lasers. The student is urged to study the subject in more detail, using any of the numerous texts dealing with the wave mechanics of solids (Reference [6], for example).

The main difference between electrons in semiconductors and electrons in other laser media is that in semiconductors all the electrons occupy, thus share, the whole crystal volume, while in a conventional laser medium, ruby, for example, the Cr^{3+} electrons are localized to within 1 or 2 Å of their parent Cr^{3+} ion and electrons on a given ion, for the typical Cr doping levels used, do not communicate with those on other ions.

In a semiconductor, on the other hand, because of the spatial overlap of their wavefunctions, no two electrons in a crystal can be placed in the same quantum state, i.e., possess the same eigenfunction. This is the so-called *Pauli exclusion principle*, which is one of the more important axiomatic foundations of quantum mechanics. Each electron thus must possess a unique spatial wavefunction and an associated eigenenergy (the total energy associated with the state). If we plot a horizontal line, as in Figure 15-1, for each allowed electron energy (eigenenergy), we will discover that the energy levels cluster within bands that are separated by "energy gaps" ("forbidden" gaps). A schematic description of the energy level spectrum of electrons in a crystal is shown in Figure 15-1.

The manner in which the available energy states are occupied determines the conduction properties of the crystal. In an insulator the uppermost occupied band is filled up with electrons while the next highest band is completely empty. The gap between them is large enough, say, ~3 eV, so that thermal excitation across the gap is negligible. If we apply an electric field to such an idealized crystal, no current will flow, since the electronic motion in a filled band is completely balanced and for each electron moving with a velocity $\mathbf{v}$ there exists another one with $-\mathbf{v}$.

If the gap between the uppermost filled band—the valence band—and the next highest—the conduction band—is small, say, <2 eV, then thermal excitation causes partial transfer of electrons from the valence band to the conduction band and the crystal can conduct electricity. Such crystals are called *semiconductors*. Their degree of conductivity can be controlled not only by the temperature but also by "doping" them with impurity atoms.

The wavefunction of an electron in a given band, say, the valence band, is characterized by a vector $\mathbf{k}$ and a corresponding (Bloch) wavefunction

$$\psi_v(\mathbf{r}) = u_{v_\mathbf{k}}(\mathbf{r})\, e^{i\mathbf{k}\cdot\mathbf{r}} \qquad (15.1\text{-}1)$$

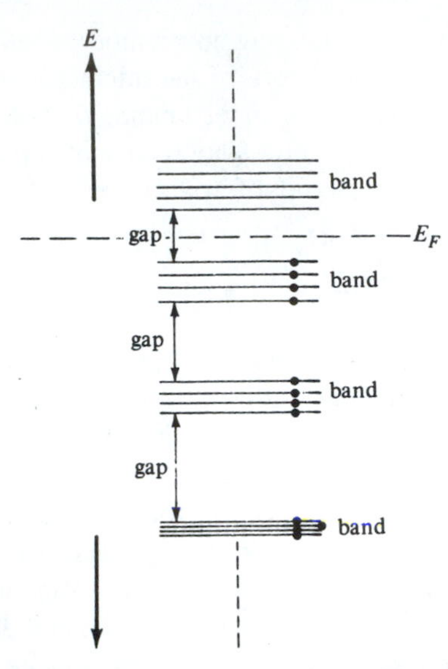

Figure 15-1 The energy levels of electrons in a crystal. In a given material these levels are usually occupied, in the ground state, up to some uppermost level. The energy E_F that marks in the limit of $T \to 0$, the transition from fully occupied electron states ($E < E_F$) to empty states ($E > E_F$), is called the *Fermi energy*. It does not, except accidentally, correspond to an eigenenergy of an electron in the crystal.

The function $u_{v_\mathbf{k}}$ possesses the same periodicity as the lattice. The factor $\exp(i\mathbf{k}\cdot\mathbf{r})$ is responsible for the wave nature of the electronic motion and is related to the de Broglie wavelength λ_e of the electron by[1]

$$\lambda_e = \frac{2\pi}{k} \tag{15.1-2}$$

The vector $\mathbf{k}$ can only possess a prescribed set of values (i.e., it is quantized), which is obtained by requiring that the total phase shift $\mathbf{k}\cdot\mathbf{r}$ across a crystal with dimensions L_x, L_y, L_z be some multiple integer of 2π.

$$k_i = \frac{2\pi}{L_i} s \qquad s = 1, 2, 3, \ldots \tag{15.1-3}$$

where $i = x, y, z$. We can thus divide the total volume in $\mathbf{k}$ space into cells each with a volume

$$\Delta V_k \equiv \Delta k_x \, \Delta k_y \, \Delta k_z = \frac{(2\pi)^3}{L_x L_y L_z} = \frac{(2\pi)^3}{V} \tag{15.1-4}$$

[1]This is true if the value of $\mathbf{k}$ is taken in the extended (i.e. not reduced) $\mathbf{k}$ space.

and associate with each such differential volume a quantum state (two states when we allow for the two intrinsic spin states of each electron). The number of such states within a spherical shell (in **k** space) of radial thickness dk and radius k is then given by the volume of the shell divided by the volume (15.1-4) ΔV_k per state

$$\rho(k)dk = \frac{k^2 V}{\pi^2}\, dk \tag{15.1-5}$$

so that $\rho(k)$ is the number of states per unit volume of **k** space. (A factor of 2 for spin was included to account for the fact that an electron in a given (spatial) state is also in a spin "up" or "down" state.)

The energy, measured from the bottom of the band, of an electron **k** in, say, the conduction band (indicated henceforth by a subscript c) is

$$E_c(\mathbf{k}) = \frac{\hbar^2 k^2}{2m_c} \tag{15.1-6}$$

where m_c is the effective mass of an electron in the conduction band. In the simplest and idealized case, which is the one we are considering here, the energy depends only on the magnitude k of the electron propagation vector and not its direction.

We often need to perform electron counting, not in **k** space but as a function of the energy. The density of states function $\rho(E)$ (the number of electronic states per unit energy interval per unit crystal volume) is determined from the conservation of states relation

$$\rho(E)\; dE = \frac{1}{V}\, \rho(k)\; dk$$

which with the use of (15.1-5) and (15.1-6) leads to

$$\rho_c(E) = \frac{1}{2\pi^2}\left(\frac{2m_c}{\hbar^2}\right)^{3/2} E^{1/2}$$

or

$$\rho_c(\omega) = \hbar\rho_c(E) = \frac{1}{2\pi^2}\left(\frac{2m_c}{\hbar}\right)^{3/2} \omega^{1/2} \tag{15.1-7}$$

where $\hbar\omega = E$. A similar expression but with m_c replaced by m_v, the effective mass in the valence band, applies to the valence band.

Figure 15-2 depicts the energy$-k$ relationship of a direct gap semiconductor, i.e., one where the conduction band minimum and the valence band maximum occur at the same value of **k**. The dots represent allowed (not necessarily occupied) electron energies. Note that, following (15.1-3), these states are spaced uniformly along the k axis.

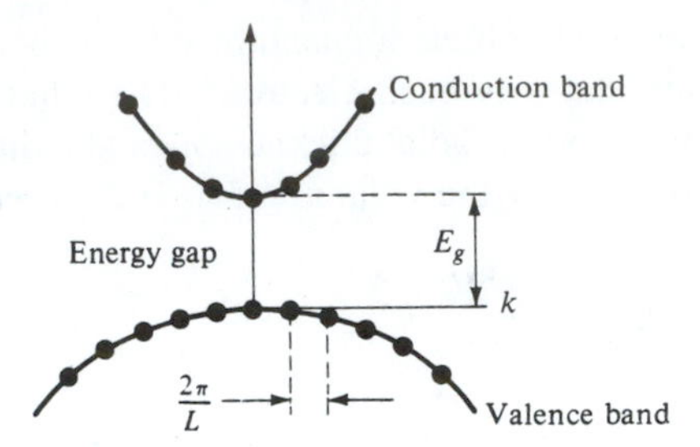

Figure 15-2 A typical energy band structure for a direct gap semiconductor with $m_c < m_v$. The uniformly spaced dots correspond to electron states.

The Fermi–Dirac Distribution Law

The probability that an electron state at energy E is occupied by an electron is given by the Fermi–Dirac law [6, 7]

$$f(E) = \frac{1}{e^{(E-E_F)/kT} + 1} \tag{15.1-8}$$

where E_F is the Fermi energy and T is the temperature. For electron energies well below the Fermi level such that $E_F - E \gg kT$, $f(E) \to 1$ and the electronic states are fully occupied, while well above the Fermi level $E - E_F \gg kT$, $f(E) \propto \exp(-E/kT)$ and approaches the Boltzmann distribution. At $T = 0$ $f(E) = 1$, for $E < E_F$, and $f(E) = 0$, for $E > E_F$ so that all levels below the Fermi level are occupied while those above it are empty. In thermal equilibrium a single Fermi energy applies to both the valence and conduction bands. Under conditions in which the thermal equilibrium is disturbed, such as in a *p-n* junction with a current flow or a bulk semiconductor in which a large population of conduction electrons and holes is created by photoexcitation, separate Fermi levels called *quasi-Fermi levels* are used for each of the bands. The concept of quasi-Fermi levels in excited systems is valid whenever the carrier scattering time within a band is much shorter than the equilibration time between bands. This is usually true at the large carrier densities used in *p-n* junction lasers.

In very highly doped semiconductors, the Fermi level is forced into either (1) the conduction band for donor impurity doping or (2) into the valence band for acceptor impurity doping. This situation is demonstrated by Figure 15-3. According to (15.1-8) at 0 K, all the states below E_F are filled while those above it are unoccupied as shown in the figure. In this respect the degenerate semiconductor behaves like a metal in which case the conductivity does not disappear at very low temperatures. The unoccupied states in the valence band [unshaded area in Figure 15-3(b)] are called *holes*, and they are treated exactly like electrons except that their charge, corresponding to an electron deficiency, is positive and their energy increases downward in the diagram. The number of holes in the semiconductor depicted by Figure 15-3(b) is the number of electron states falling within the unshaded area at the top of the valence band. The process of exciting an electron from state a to state b

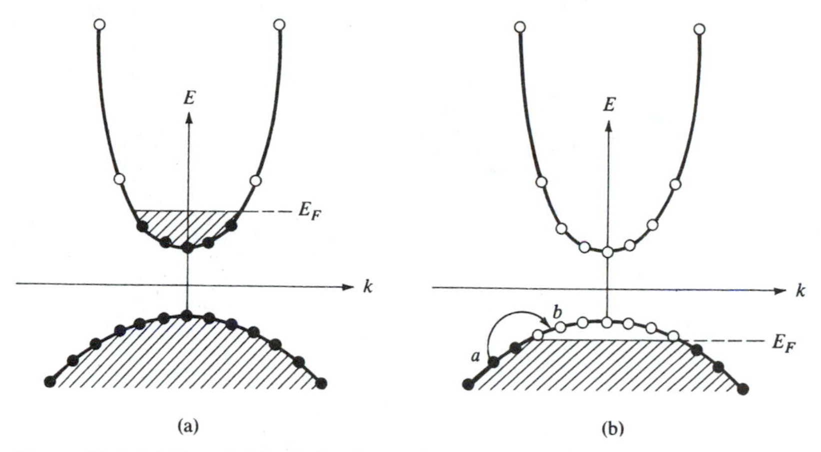

Figure 15-3 (a) Energy band of a degenerate *n*-type semiconductor at 0 K. (b) A degenerate *p*-type semiconductor at 0 K. The cross-hatching represents regions in which all the electron states are filled. Empty circles indicate unoccupied states (holes).

[Figure 15.3(b)] in the valence band can also be viewed as one whereby a hole is excited from b to a. The advantage of this point of view is the symmetry in the language and mathematical description that it brings to the discussions of current flow due to electrons in the conduction band and those in the valence band.

To better appreciate the role of the quasi-Fermi level, consider a nonthermal equilibrium situation in which electrons are excited into the conduction band of a degenerate *p*-type semiconductor at a very high rate. This can be done by injecting electrons into the *p* region across a *p-n* junction or by subjecting the semiconductor to an intense light beam with $h\nu > E_g + E_{Fc} + E_{Fv}$, so that for each absorbed photon an electron is excited into the conduction band from the valence band. This situation is depicted in Figure 15-4. Following this excitation, electrons relax, by emitting optical and acoustic phonons, to the bottom of the conduction band in times of $\sim 10^{-12}$ s while their relaxation across the gap back to the valence band—a process referred to as electron-hole recombination—is characterized by a time constant of

$$\tau \sim 3 - 4 \times 10^{-9} \text{ s}$$

It is important in analyzing the process of light amplification in semiconductors to determine the quasi-Fermi level E_{F_c} for a given rate of excitation. Assuming that the relaxation to the bottom of the band into which the carriers are excited is instantaneous, we have

$$\frac{N_c}{\tau} = \frac{I}{\text{eV}} \tag{15.1-9}$$

where N_c is the density (m^{-3}) of electrons in the conduction band, I the injection current (in amperes), τ is the electron relaxation time back to the valence band (electron-hole recombination time), and V is the volume into which the electrons are

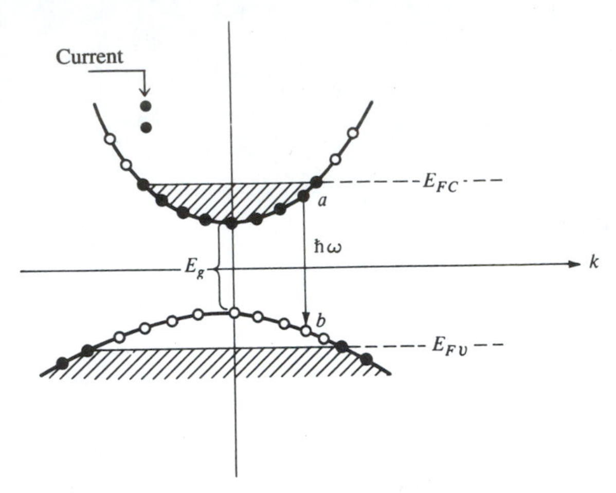

Figure 15-4 Electrons are injected at a rate of I/eV per unit volume (I = total current) into the conduction band of a semiconductor.

confined following injection. The density of electrons with energies between E and $E + dE$ is the product of $\rho_c(E)$—the density of allowed electron states—and the occupation probability $f_c(E)$ of these states.

Using (15.1-7) and (15.1-8),

$$N_c = \frac{I\tau}{\text{eV}} = \int_0^\infty \rho_c(E) f_c(E)\, dE$$
$$= \frac{1}{2\pi^2}\left(\frac{2m_c}{\hbar^2}\right)^{3/2} \int_0^\infty \frac{E^{1/2}}{e^{(E-E_{F_c})/kT} + 1}\, dE \qquad \textbf{(15.1-10)}$$

For a given injection current I the only unknown quantity in (15.1-10) is the conduction quasi-Fermi level E_{F_c}. We can thus invert, in practice by numerical methods, (15.1-10) and solve it for $E_{F_c}(T)$ as a function of I, or equivalently of N_c. We shall make use, later, of this fact. At $T = 0$ the integral is replaced by

$$\int_0^{E_{Fc}} E^{1/2}\, dE = \tfrac{2}{3} E_{Fc}^{3/2}$$

yielding

$$E_{Fc}(T = 0) = (3\pi^2)^{2/3} \frac{\hbar^2}{2m_c} N_c^{2/3} \qquad \textbf{(15.1-11)}$$

Another fact that we need before proceeding to the subject of optical gain in semiconductors is that when an electron makes a transition (induced or spontaneous) between a conduction band state and one in the valence band, the two states involved must have the same **k** vector. This is due to the fact that according to quantum mechanics the rate of such a transition is always proportional to an integral over the crystal volume that involves the product of the initial state wavefunction and the complex conjugate of that of the final state. Such an integral would, according to (15.1-1), be vanishingly small except when the condition

$$\mathbf{k}_f = \mathbf{k}_i \tag{15.1-12}$$

is satisfied. In band diagrams such as that of Figure 15-4, the transitions are consequently described by vertical arrows.

15.2 GAIN AND ABSORPTION IN SEMICONDUCTOR (laser) MEDIA

Consider the semiconductor material depicted in Figure 15-5 in which by virtue of electron pumping a nonthermal equilibrium steady state is obtained in which *simultaneously* large densities of electrons and holes coexist in the *same* space. These are characterized by quasi-Fermi levels E_{F_c} and E_{F_v}, respectively, as shown.

Let an optical beam at a (radian) frequency ω_0 travel through the crystal. This beam will induce downward $a \rightarrow b$ transitions that lead to amplification as well as $b \rightarrow a$ absorbing transitions. Net amplification of the beam results if the rate of $a \rightarrow b$ transitions exceeds that of $b \rightarrow a$.

As discussed in the previous section, only transitions in which the upper and lower electron states have the same **k** vector are allowed. The pair of levels a and b in Figure 15-5 are thus characterized by some **k** value. Let us consider a group of such levels with nearly the same **k** value and hence with nearly the same transition energy

$$\hbar\omega(\mathbf{k}) = E_g + \frac{\hbar^2 k^2}{2m_c} + \frac{\hbar^2 k^2}{2m_v} \tag{15.2-1}$$

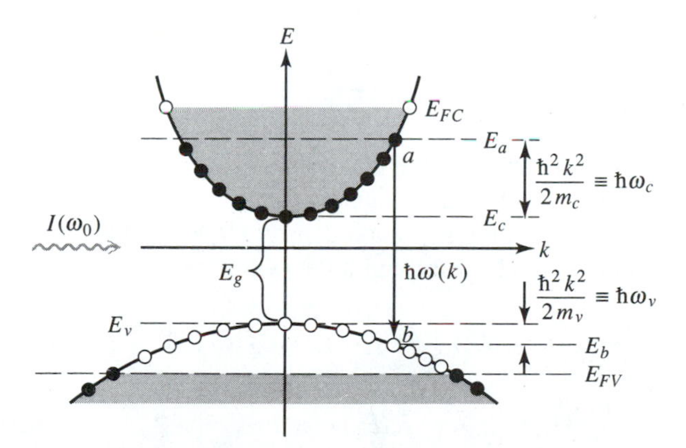

Figure 15-5 An optical beam at ω_0 with intensity $I(\omega_0)$ is incident on a pumped semiconductor medium characterized by quasi-Fermi levels E_{F_c} and E_{F_v}. A single level pair a-b with the same **k** value is shown. The induced transition $a \rightarrow b$ contributes one photon to the beam.

(In the following the **k** dependence of ω will be omitted but understood.) The density of such level pairs whose **k** values fall within a spherical shell of thickness dk is, according to (15.1-5), $\rho(k)\, dk/V$.

Before proceeding let us remind ourselves of some results developed in connection with conventional laser media. The gain constant $\gamma(\omega_0)$ is given by (5.5-7) as

$$\gamma(\omega_0) = -\frac{k}{n^2}\chi''(\omega_0) \qquad k = \frac{2\pi n}{\lambda} \tag{15.2-2}$$

where $\chi''(\omega_0)$, the imaginary part of the electric susceptibility, is

$$\chi''(\omega_0) = \frac{(N_1 - N_2)\lambda_0^3}{8\pi^3 t_{\text{spont}}\, \Delta\nu n} \frac{1}{1 + 4(\nu - \nu_0)^2/(\Delta\nu)^2} \tag{15.2-3}$$

Combining the last two equations and defining the "relaxation time" T_2 by $T_2 = (\pi\,\Delta\nu)^{-1}$ leads to

$$\gamma(\omega_0) = \frac{(N_2 - N_1)\lambda_0^2}{4n^2 t_{\text{spont}}} \frac{T_2}{\pi[1 + (\omega - \omega_0)^2 T_2^2]} \tag{15.2-4}$$

In semiconductors T_2 is the mean lifetime for coherent interaction of k electrons with a monochromatic field and is of the order of the phonon-electron collision time. Numerically $T_2 \sim 10^{-12}$ s. Given an electron in an upper state "a," the lower state "b" with the same **k** value may be occupied by another electron. The downward rate of transitions is thus proportional to

$$R_{a\to b} \propto f_c(E_a)[1 - f_v(E_b)]$$

i.e., to the product of the probabilities $f_c(E_a)$ that the upper (conduction) state is occupied and the probability $(1 - f_v)$ that the lower (valence) state is empty. The functions $f_{v,c}(E)$ are given, according to (15.1-8), by

$$f_c(E) = \frac{1}{e^{(E-E_{F_c})/kT} + 1} \tag{15.2-5}$$

$$f_v(E) = \frac{1}{e^{(E-E_{F_v})/kT} + 1} \tag{15.2-6}$$

allowing for the fact that under pumping conditions $E_{F_c} \neq E_{F_v}$.

In translating to the case of semiconductors the results that were developed for conventional lasers, the population inversion density $(N_2 - N_1)$ is thus replaced by the effective inversion due to electrons and holes within dk.

$$\begin{aligned} N_2 - N_1 &\to \frac{\rho(k)\, dk}{V}\{f_c(E_a)[1 - f_v(E_b)] - f_v(E_b)[1 - f_c(E_a)]\} \\ &= \frac{\rho(k)\, dk}{V}[f_c(E_a) - f_v(E_b)] \end{aligned} \tag{15.2-7}$$

$$E_a - E_b \equiv \hbar\omega = E_g + \frac{\hbar^2 k^2}{2m_c} + \frac{\hbar^2 k^2}{2m_v} \tag{15.2-8}$$

Equation (15.2-7) is of central importance and is a capsule statement of the difference between the population inversion in a conventional laser medium where the level occupation probability obeys Boltzmann statistics and that of a semiconductor medium governed by Fermi–Dirac statistics.

Returning to the gain expression (15.2-4), we use (15.2-7) to rewrite it as

$$d\gamma(\omega_0) = \frac{\rho(k)\,dk}{V}(f_c - f_v)\frac{\lambda_0^2}{4n^2\tau}\left(\frac{T_2}{\pi[1 + (\omega - \omega_0)^2 T_2^2]}\right)$$

where $\omega \equiv \omega(k)$ is the transition frequency at k as in (15-2-1). The differential designation $d\gamma(\omega_0)$ is to remind us that only electrons with **k** vectors within dk are included here. We have also replaced, to agree with popular usage, the term spontaneous lifetime (t_{spont}) by the recombination lifetime τ for an electron in the conduction band with a hole in the valence band. To obtain the gain constant, we must add up the contributions from all the electrons

$$\gamma(\omega_0) = \int_0^\infty \frac{dk\,\rho(k)}{V}[f_c(\omega) - f_v(\omega)]\frac{\lambda_0^2}{4n^2\tau}\left(\frac{T_2}{\pi[1 + (\omega - \omega_0)^2 T_2^2]}\right) \qquad (15.2\text{-}9)$$

We will find it easier to carry out the indicated integration in (15.2-9) in the ω domain [$\hbar\omega$ being the separation $E_a(\mathbf{k}) - E_b(\mathbf{k})$]. From (15.2-1)

$$\hbar\omega = E_g + \frac{\hbar^2}{2m_r}k^2 \qquad (15.2\text{-}10)$$

$$\frac{1}{m_r} = \frac{1}{m_v} + \frac{1}{m_c} \qquad (m_r \equiv \text{reduced effective mass}) \qquad (15.2\text{-}11)$$

Using the relations

$$d\omega = \frac{\hbar}{m_r}k\,dk$$

$$\text{k} = (\hbar\omega - E_g)^{1/2}\left(\frac{2m_r}{\hbar^2}\right)^{1/2}$$

the expression (15.2-9) for $\gamma(\omega_0)$ becomes

$$\gamma(\omega_0) = \int_0^\infty (\hbar\omega - E_g)^{1/2}\left(\frac{2m_r}{\hbar^2}\right)^{1/2} \frac{m_r\lambda_0^2 T_2[f_c(\omega) - f_v(\omega)]}{\pi^2\hbar 4n^2\tau\pi[1 + (\omega - \omega_0)^2 T_2^2]}\,d\omega \qquad (15.2\text{-}12)$$

In most situations we can replace the normalized function

$$\frac{T_2}{\pi[1 + (\omega - \omega_0)^2 T_2^2]} \to \delta(\omega - \omega_0)$$

which is merely a statement of the fact that its width $\Delta\omega \sim T_2^{-1}$ is narrower than other spectral features of interest. In this case the integration (15.2-12) leads to

$$\gamma(\omega_0) = \frac{\lambda_0^2}{8\pi^2 n^2\tau}\left(\frac{2m_c m_v}{\hbar(m_v + m_c)}\right)^{3/2}\left(\omega_0 - \frac{E_g}{\hbar}\right)^{1/2}[f_c(\omega_0) - f_v(\omega_0)] \qquad (15.2\text{-}13)$$

The condition for net gain $\gamma(\omega_0) > 0$ is thus

$$f_c(\omega_0) > f_v(\omega_0) \tag{15.2-14}$$

which is the equivalent, in a semiconductor, of the conventional inversion condition $N_2 > N_1$. Using (15.2-5) and (15.2-6), the gain condition (15.2-14) becomes

$$\frac{1}{e^{(E_a - E_{F_c})/kT} + 1} > \frac{1}{e^{(E_b - E_{F_v})/kT} + 1} \tag{15.2-15}$$

Recalling that $E_a - E_b = \hbar\omega_0$, (15.2-15) is satisfied provided

$$\hbar\omega_0 < E_{F_c} - E_{F_v} \tag{15.2-16}$$

so that only frequencies whose photon energies $\hbar\omega_0$ are smaller than the quasi-Fermi levels separation are amplified. Condition (15.2-16) was first derived by Basov, et al. [1], Bernard and Duraffourg [8]. The general features of the gain dependence $\gamma(\omega_0)$ on the frequency ω_0 are illustrated by Figure 15-6. The gain is zero at $\hbar\omega < E_g$, since no electronic transitions exist at these energies. The gain becomes zero

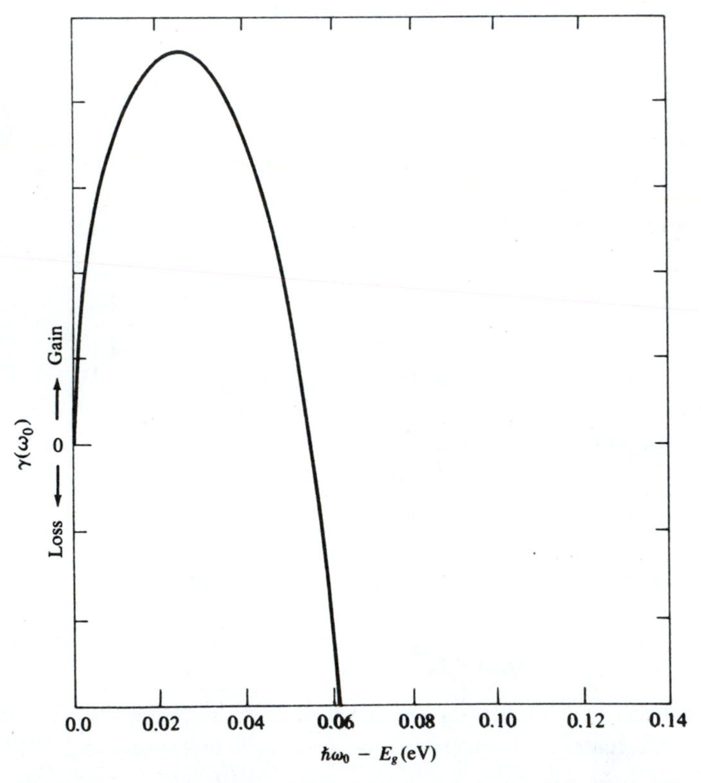

Figure 15-6 A typical plot of gain $\gamma(\omega_0)$ as a function of frequency for a fixed pumping level N. (After Reference [9].)

again at the frequency where $\hbar\omega_0 = E_{F_c} - E_{F_v}$. At higher frequencies the semiconductor absorbs.

Figure 15-7 shows calculated plots based on (15.2-12) with the density of the (injected) electrons as a parameter. The curves are based on the following physical constants of GaAs:$m_c = 0.067m_e$, $m_v = 0.48m_e$, $T_2 \sim 0.5$ ps, $\tau \simeq 3 \times 10^{-9}$ s, $E_g = 1.43$ eV. We note that the minimum density to achieve transparency ($\gamma = 0$) is $N_{tr} \sim 1.55 \times 10^{18}$ cm^{-3}. The peak gain corresponding to a given inversion density N_c is plotted in Figure 15-8.

It follows from Figure 15-8 that semiconductor media are capable of achieving very large gain ranging up to a few hundred cm^{-1}. In a laser the amount of gain that actually prevails is clamped by the phenomenon of saturation (see Section 5.6) to a value equal to the loss. In a typical semiconductor laser this works out to $20 < \gamma < 80$ cm^{-1}. In this region we can approximate the plot of Figure 15-8 by a linear relationship

$$\gamma_{max} = B(N - N_{tr}) \tag{15.2-17}$$

The constant B fitting the data of Figure 15-8 is $B \sim 1.5 \times 10^{-16}$ cm^2 and is typical of GaAs/GaAlAs lasers at 300 K. The gain constant B increases with the decrease of the temperature T. This is due to the narrowing of the transition regions of the Fermi functions $f_c(\omega)$ and $f_v(\omega)$ in (15.2-12). At 77 K, $B \sim 5 \times 10^{-16}$ cm^2. Figure 15-8 shows that the semiconductor diode is capable of producing extremely large incremental gains, with only moderate increases of the inversion density, hence the current, above the transparency value ($N_{tr} \sim 1.55 \times 10^{18}$ cm^{-3} in the figure). It is thus possible to obtain oscillation in a semiconductor laser with active regions that are only a few tens of microns long. Commercial diode lasers have typical lengths of ~250 μm.

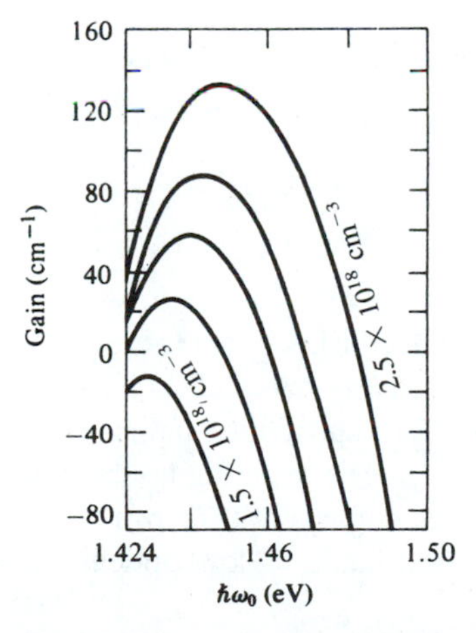

Figure 15-7 A plot based on (15.2-12) of the photon energy dependence of the optical gain (or loss = negative gain) of GaAs with the injected carrier density as a parameter. (After Reference [9].)

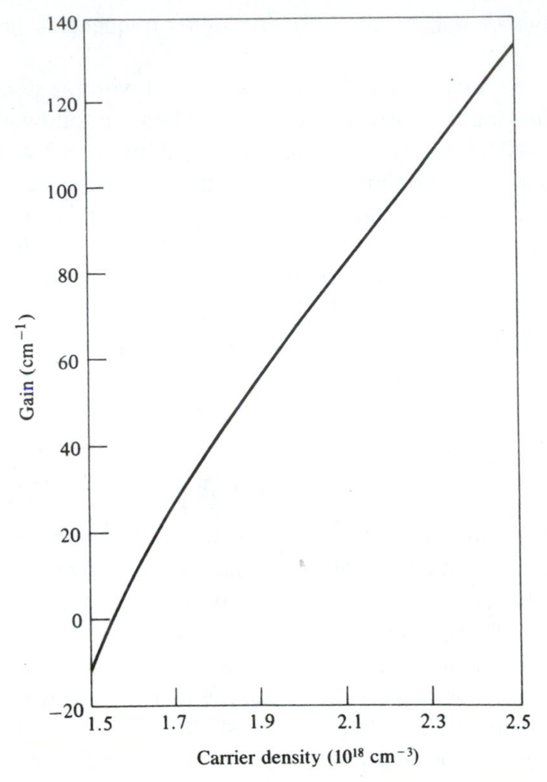

Figure 15-8 A plot of the peak gain γ_{max} of Figure 15-7 as a function of the inversion density at $T = 300$ K.

For additional background material on semiconductor lasers, the student is advised to consult References [10–12].

15.3 GaAs/$Ga_{1-x}Al_xAs$ LASERS

The two most important classes of semiconductor lasers are those that are based on III-V semiconductors. The first system is based on GaAs and $Ga_{1-x}Al_xAs$. The active region in this case is GaAs or $Ga_{1-x}Al_xAs$. The subscript x indicates the fraction of the Ga atoms in GaAs that are replaced by Al. The resulting lasers emit (depending on the active region molar fraction x and its doping) at $0.75\ \mu m < \lambda < 0.88\ \mu m$. This spectral region is convenient for the short-haul (<2 km) optical communication in silica fibers.

The second system has $Ga_{1-x}In_xAs_{1-y}P_y$ as its active region. The lasers emit in the $1.1\ \mu m < \lambda < 1.6\ \mu m$ depending on x and y. The region near 1.55 μm is especially favorable, since, as shown in Figure 3-19, optical fibers are available with losses as small as 0.15 dB/km at this wavelength, making it extremely desirable for long-distance optical communication.

In this section we will consider $GaAs/Ga_{1-x}Al_xAs$ lasers. A generic laser of this type, depicted in Figure 15-9, has a thin (0.1–0.2 μm) region of GaAs sandwiched between two regions of GaAlAs. It is consequently called a *double heterostructure* laser. The basic layered structure is grown epitaxially on a crystalline GaAs substrate so that it is uninterrupted crystalographically.

The favored crystal growth techniques are liquid-phase epitaxy and chemical vapor deposition using metallo-organic reagents (MOCVD) [11, 13, 14]. Another important technique—molecular beam epitaxy [11, 13, 15, 16]—uses atomic beams of the crystal constituents in ultra-high vacuum to achieve extremely fine thickness and doping control.

The thin active region is usually undoped while one of the bounding $Ga_{1-x}Al_xAs$ layers is doped heavily n-type and the other p-type. The difference

$$n_{\text{GaAs}} - n_{\text{Ga}_{1-x}\text{Al}_x\text{As}} \simeq 0.62x$$

between the indices of refraction of GaAs and the ternary crystal with a molar fraction x gives rise to a three-layered dielectric waveguide of the type illustrated in Figure 13-1. At this point the student should review the basic modal concepts discussed in Chapter 13. The lowest-order (fundamental) mode has its energy concentrated mostly in the GaAs (high index) layer. The index distribution and a typical modal intensity plot for the lowest-order mode are shown in Figure 15-10. When a positive bias is applied to the device, electrons are injected from the n-type $Ga_{1-x}Al_xAs$ into the active GaAs region while a density of holes equal to that of the electrons in the active region is caused by injection from the p side.

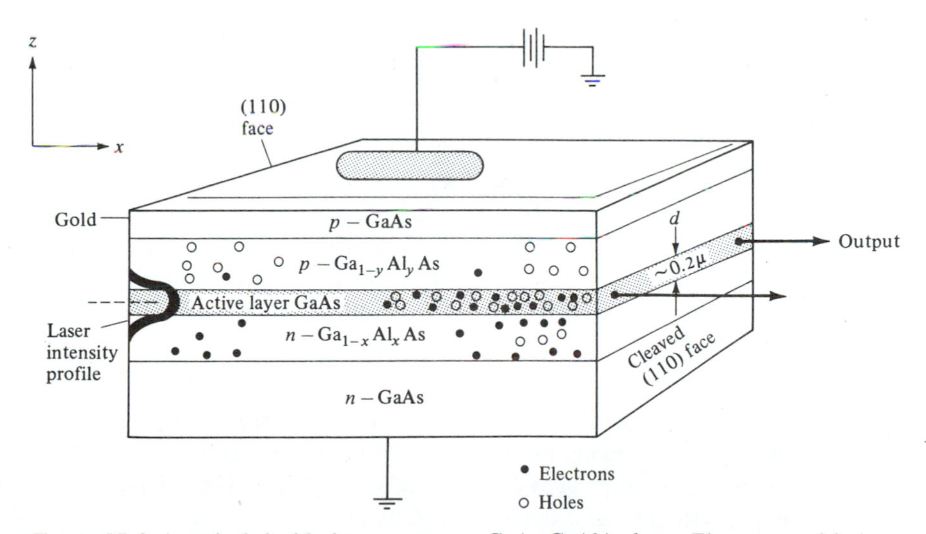

Figure 15-9 A typical double heterostructure GaAs-GaAlAs laser. Electrons and holes are injected into the active GaAs layer from the n and p GaAlAs. Frequencies near $\nu = E_g/h$ are amplified by stimulating electron–hole recombination.

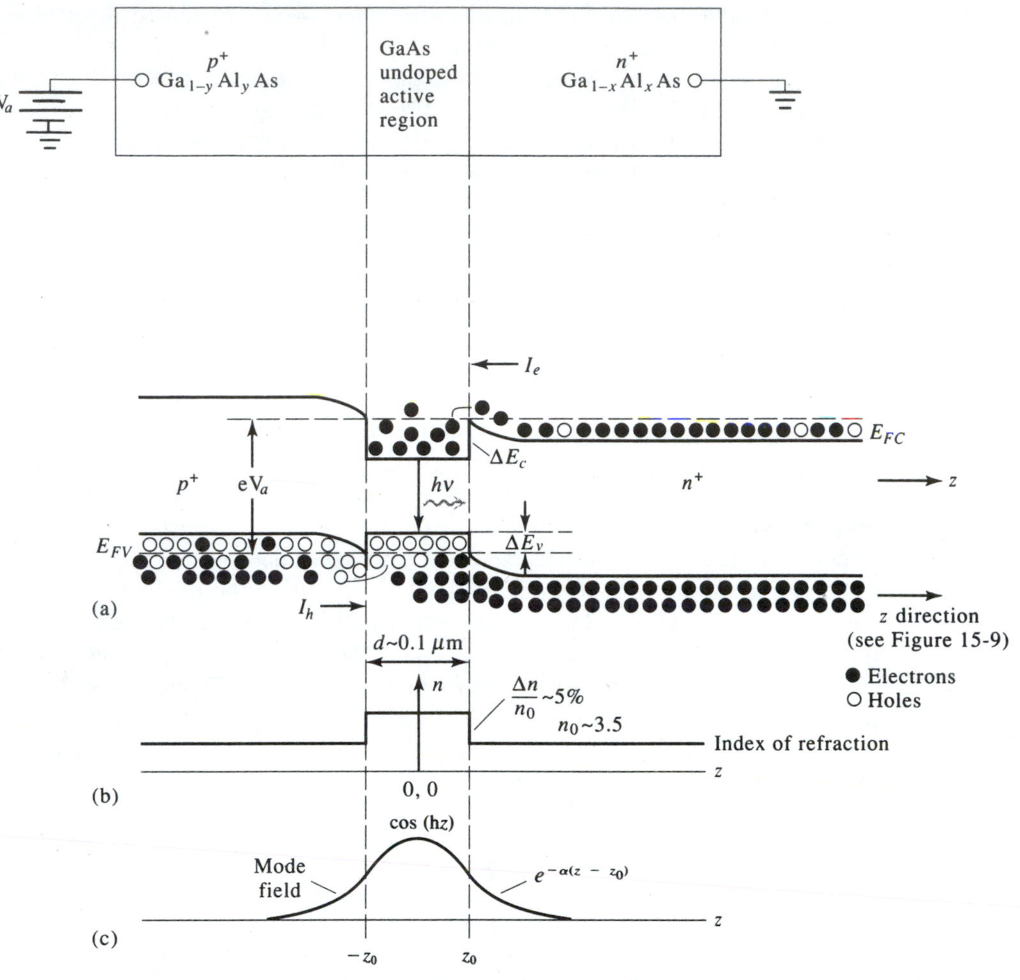

Figure 15-10 (a) The energy band edges of a strongly forward-biased (near-flattened) double heterostructure GaAs/GaAlAs laser diode. Note trapping of electrons (holes) in the potential well formed by the conduction (valence) band edge energy discontinuity $\Delta E_c(\Delta E_v)$. (b) The spatial (z) profle of the index of refraction which is responsible for dielectric waveguiding in the high index (GaAs) layer. (c) The intensity profle of the fundamental optical mode in a slab waveguide.

The electrons that are injected into the active region are prevented from diffusing out into the p region by means of the potential barrier due to the difference ΔE_g between the energy gaps of GaAs and $Ga_{1-x}Al_xAs$. The x-dependence of the energy gap of $Ga_{1-x}Al_xAs$ is approximated by [13]

$$E_g(x < 0.37) = (1.424 + 1.247x)\ \text{eV}$$

and is plotted in Figure 15-11.

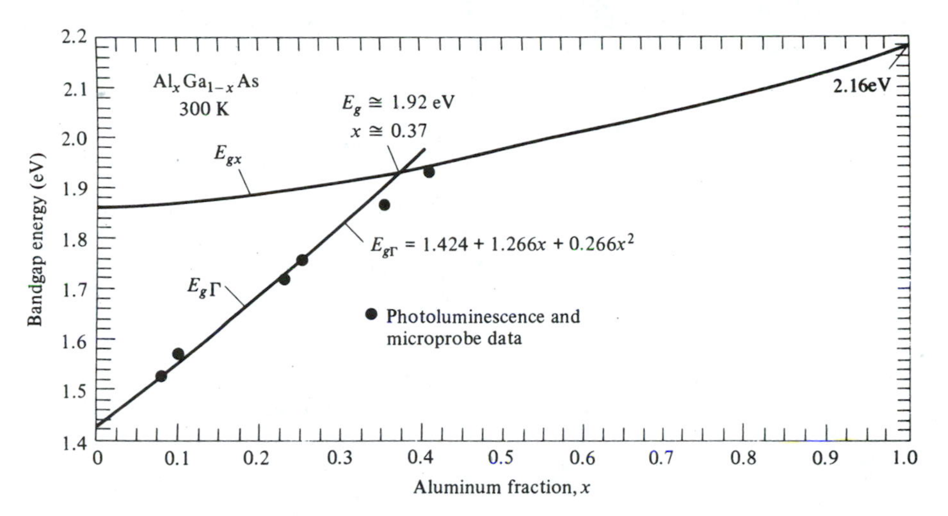

Figure 15-11 The magnitude of the energy gap in Ga$_{1-x}$Al$_x$As as a function of the molar fraction x. For $x > 0.37$ the gap is indirect. (After Reference [11].)

The total discontinuity ΔE_g of the energy gap at a GaAs/GaAlAs interface is taken up mostly (60 percent) by the conduction band edge, i.e., $\Delta E_c = 0.6\Delta E_g$, while 40 percent is left to the valence band, $\Delta E_v = 0.4\Delta E_g$, so that both holes and electrons are effectively confined to the active region. This double confinement of injected carriers as well as of the optical mode energy to the same region is probably the single most important factor responsible for the successful realization of low-threshold continuous semiconductor lasers [17–19]. Under these conditions we expect the gain experienced by the mode to vary as d^{-1}, where d is the thickness of the active (GaAs) layer, since at a given total current, the carrier density, hence the gain, will be proportional to d^{-1}. To quantify the last statement, we start with the basic definition of the modal gain

$$g = \frac{\text{power generated per unit length (in } x)}{\text{power carried by beam}}$$
$$= \frac{-\int_{-\infty}^{-d/2} \alpha_n |E|^2\, dz + \int_{-d/2}^{d/2} \gamma |E|^2\, dz - \int_{d/2}^{\infty} \alpha_p |E|^2\, dz}{\int_{-\infty}^{\infty} |E|^2\, dz} \qquad (15.3\text{-}1)$$

where γ is the gain constant experienced by a plane wave in a medium whose inversion density is equal to that of the active medium. γ is given by (15.2-12) and (15.2-17). α_n is the loss constant of the unpumped n-Ga$_{1-x}$Al$_x$As and is due mostly to free electron absorption. α_p is the loss (by free holes) in the bounding p-Ga$_{1-y}$Al$_y$As region. We note that as $d \to \infty$, $g \to \gamma$.

It is convenient to rewrite (15.3-1) as

$$g = \gamma\Gamma_a - \alpha_n\Gamma_n - \alpha_p\Gamma_p \tag{15.3-2}$$

$$\Gamma_a = \frac{\int_{-d/2}^{d/2} |E|^2\, dz}{\int_{-\infty}^{\infty} |E|^2\, dz} \tag{15.3-3a}$$

$$\Gamma_n = \frac{\int_{-\infty}^{-d/2} |E|^2\, dz}{\int_{-\infty}^{\infty} |E|^2\, dz} \tag{15.3-3b}$$

$$\Gamma_p = \frac{\int_{d/2}^{\infty} |E|^2\, dz}{\int_{-\infty}^{\infty} |E|^2\, dz} \tag{15.3-3c}$$

$$\Gamma_a + \Gamma_a + \Gamma_p = 1$$

Γ_a is very nearly the fraction of the mode power carried within the active GaAs layer, while Γ_n and Γ_p are, respectively, the fraction of the power in the n and p regions. As long as $\Gamma_a \sim 1$, i.e., most of the mode energy is in the active region, the gain g is inversely proportional to the active region thickness d since decreasing d, for example, increases the optical intensity for a given total beam power and, consequently, the rate of stimulated transitions. As d decreases, an increasing fraction

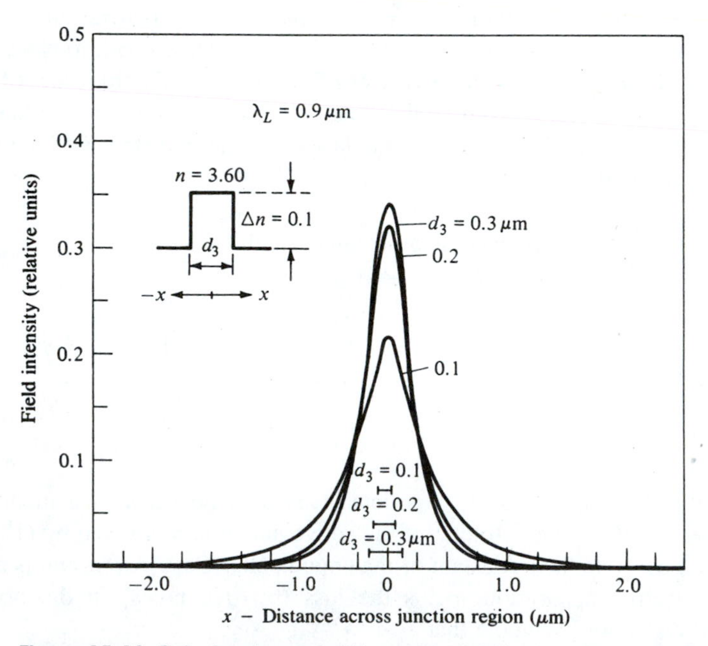

Figure 15-12 Calculated near field intensity distribution of the step discontinuity waveguide for various values of the guiding layer thickness. (After Reference [11].)

of the mode intensity is carried outside the active region as can be seen from the modal waveguide solution plotted in Figure 15-12 [11]. The resulting decrease of the confinement factor Γ_a eventually dominates over the d^{-1}-dependence and the gain begins to decrease with further decrease of d [22]. A plot of the threshold current dependence on d is depicted in Figure 15-13. The bottoming out and eventual increase of J_{th} for $d \lessapprox 0.1$ μm is due to the decrease of the confinement factor Γ_a and the increase of the relative role of the losses in the p and n GaAlAs bounding layer as Γ_n and Γ_p increase, i.e., as an increasing fraction of the mode intensity is carried within these lossy unpumped regions as shown in Figure 15-12.

Numerical Example: Threshold Current Density in Double Heterostructure Lasers

Consider the case of a GaAs/GaAlAs laser of the type illustrated in Figure 15-10. We will use the following parameters: $\tau \sim 4 \times 10^{-9}$ s, $L = 500$ μm. The threshold gain condition is (15.3-2)

$$\gamma\Gamma_a = \alpha_n\Gamma_n + \alpha_p\Gamma_p - \frac{1}{L}\ln R + \alpha_s \tag{15.3-4}$$

where the term α_s accounts for scattering losses (mostly at heterojunction interfacial imperfections). The largest loss term in lasers with uncoated faces is usually L^{-1} ln R. In our case, taking $R = 0.31$ as due to Fresnel reflectivity at a GaAs ($n = 3.5$) air interface, we obtain

$$-\frac{1}{L}\ln R = 23.4 \text{ cm}^{-1}$$

The rest of the loss terms are assumed to add up to $\sim$10 cm^{-1} so that taking $\Gamma_a \sim 1$ the total gain needed is 33.4 cm^{-1}. This requires, according to Figure 15-8, an injected carrier density of $N \sim 1.7 \times 10^{18}$ cm^{-3}. Under steady-state conditions the rate at which carriers are injected into the active region must equal the electron–hole recombination rate

$$\frac{J}{e} = \frac{Nd}{\tau}$$

Using the above data we obtain

$$\frac{J}{d} = \frac{eN}{\tau} \sim 6.8 \times 10^3 \text{ A/(cm}^2\text{-}\mu\text{m)}$$

This value of J/d is in reasonable agreement with the measured value of $\sim 5 \times 10^3$ in Figure 15-13. If we use this value to estimate the lowest threshold current density which from Figure 15-13 occurs when $d \sim 0.08$ μm, we obtain

$$J_{\text{min}} = 0.68 \times 10^4 \times 0.08 = 544 \text{ A/cm}^2$$

again, close to the range of observed values.

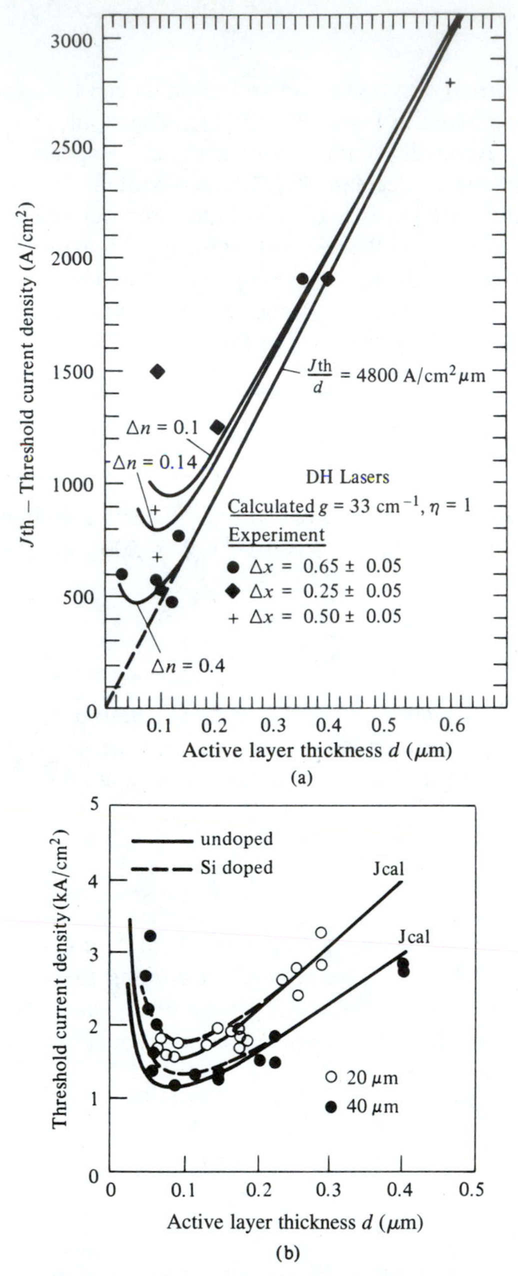

Figure 15-13 (a) Calculated and experimental values of the threshold current density as a function of the active layer thickness d for broad-area 500-μm-long AlGaAs DH diode lasers of "undoped" active layers. Notable exceptions are the experimental data for $\Delta x \simeq 0.25$, which were obtained from diodes with heavy-Ge-doped active layers. (After Reference [11].) (b) Calculated and experimental values of the threshold current density as a function of active layer thickness d for stripe-geometry (20- and 40-μm-wide stripe contacts) 300-μm-long AlGaAs DH diode lasers ($\Delta x = 0.25$) of "undoped" and low-Si-doped active layers ($x = 0.05$). (After Reference [21].)

The successful epitaxial growth of $Ga_{1-x}Al_xAs$ on top of GaAs (and vice versa), which is the main reason for the success of double heterostructure lasers, is due to the fact that their lattice constants are the same, to within a fraction of a percent, over the range $0 \leq x \leq 1$. This can be seen from the plot of Figure 15-14, which shows the lattice constant corresponding to various compositions of III-V semiconductors as a function of the band gap energy. We note that the line connecting the AlAs ($x = 1$) and the GaAs ($x = 0$) is nearly horizontal, which corresponds to a (very nearly) constant lattice constant over this compositional range.

15.4 SOME REAL LASER STRUCTURES

The double heterostructure lasers discussed in Section 15.3 lack the means for confining the current and the radiation in the lateral (y) direction. The outcome is that typical broad area lasers can support more than one transverse (y) mode, resulting in unacceptable mode hopping as well as spatial and temporal instabilities. To overcome these problems, modern semiconductor lasers employ some form of transverse optical and carrier confinement. A typical and successful example of this approach is the buried heterostructure laser [20] shown in Figure 15-15. To fabricate these lasers, the first three layers: n-$Ga_{1-x}Al_xAs$, GaAs, and p-$Ga_{1-y}Al_yAs$ are grown on an n-GaAs crystalline substrate by one of the epitaxial techniques described above. The structure is then etched through a mask down to the substrate level, leaving stand a thin (~3 μm) rectangular mesa composed of the original layers. A ''burying'' $Ga_{1-z}Al_zAs$ layer is then regrown on both sides of the mesa, resulting in the structure shown in Figure 15-15.

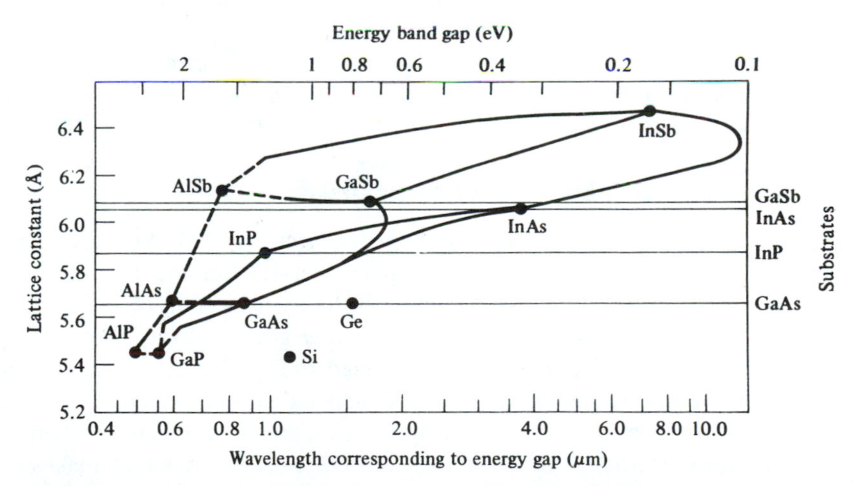

Figure 15-14 III-V compounds: Lattice constants versus energy band gaps and corresponding wavelengths. The solid lines correspond to direct-gap materials and the dashed lines to indirect-gap materials. The binary-compound substrates that can be used for lattice-matched growth are indicated on the right. [After Reference [11].)

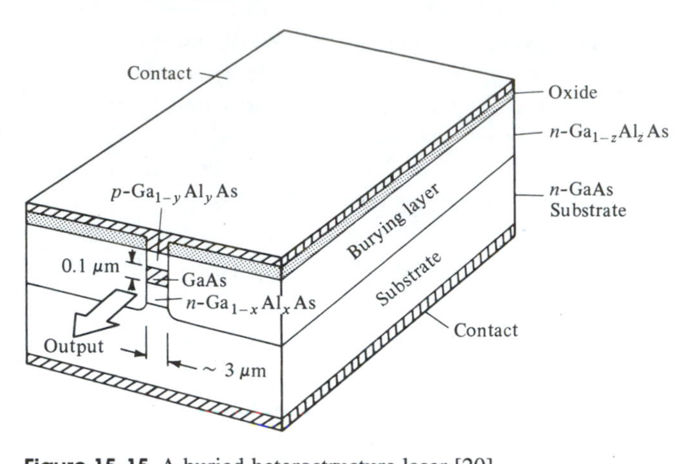

Figure 15-15 A buried heterostructure laser [20].

The most important feature of the buried heterostructure laser is that the active GaAs region is surrounded on *all* sides by the lower index GaAlAs, so that electromagnetically the structure is that of a rectangular dielectric waveguide. The transverse dimensions of the active region and the index discontinuities (i.e., the molar fractions x, y, and z) are so chosen that only the lowest-order transverse mode can propagate in the laser waveguide. Another important feature of this laser is the confinement of the injected carriers at the boundaries of the active region due to the energy band discontinuity at a GaAs/GaAlAs interface as discussed in the last section. These act as potential barriers inhibiting carrier escape out of the active region. GaAs semiconductor lasers utilizing this structure have been fabricated, see Chapter 16, with threshold currents of less than 1 milliampere [38]; more typical lasers have thresholds of ~20 milliamperes. Typical power vs. current plot of a commercial laser is shown in Figure 15-16, while the far field angular intensity distribution is shown in Figure 15-17.

Quaternary GaInAsP Semiconductor Lasers

Optical fiber communication over long distances (say > 10 km) uses, almost exclusively, lasers emitting in spectral regions near 1.3 μm and 1.55 μm. The 1.3 μm lasers are important because the group velocity dispersion of silica-based fibers at this wavelength is very small. The first-order group velocity dispersion parameter D, defined by Equation (3.4-12), is plotted in Figure 3-10 and is zero at $\lambda \approx 1.3$ μm, so that optical pulses at this wavelength undergo, according to (2.9-35), minimal spread with distance. The wavelength region around 1.55 μm is where the optical absorption coefficient of silica fibers reaches a minimum, which makes it a favorite for long-haul links. Lasers in these wavelength regions [24] are fabricated using active layers of $GA_{1-x}In_xAs_{1-y}P_y$. From Figure 15-14, we find that such lasers spanning the $0.9\ \mu m < \lambda < 1.7\ \mu m$ region can be lattice-matched to InP, which possesses a lower index of refraction to produce dielectric waveguides in which the InP epi-

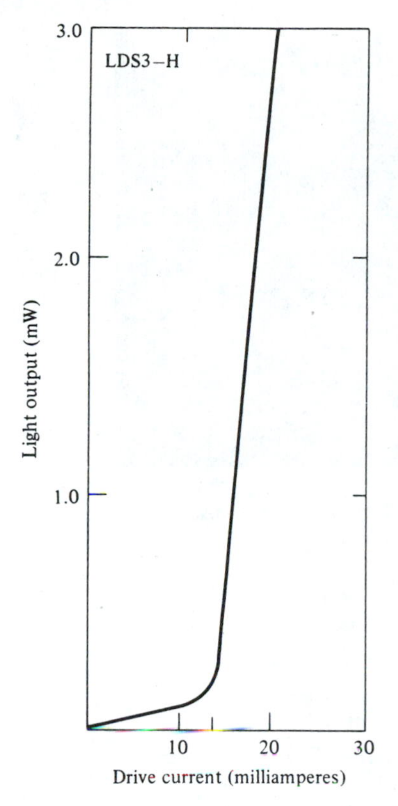

Figure 15-16 Power versus current plot of a low-threshold (~14 milliamperes) commercial DH GaAs/GaAlAs laser. (After Reference [23].)

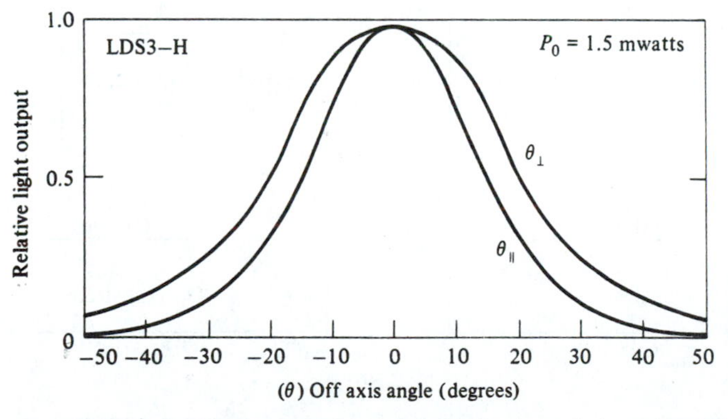

Figure 15-17 Far-field angular intensity distribution of a low-threshold commercial DH GaAs/GaAlAs laser. (After Reference [23].)

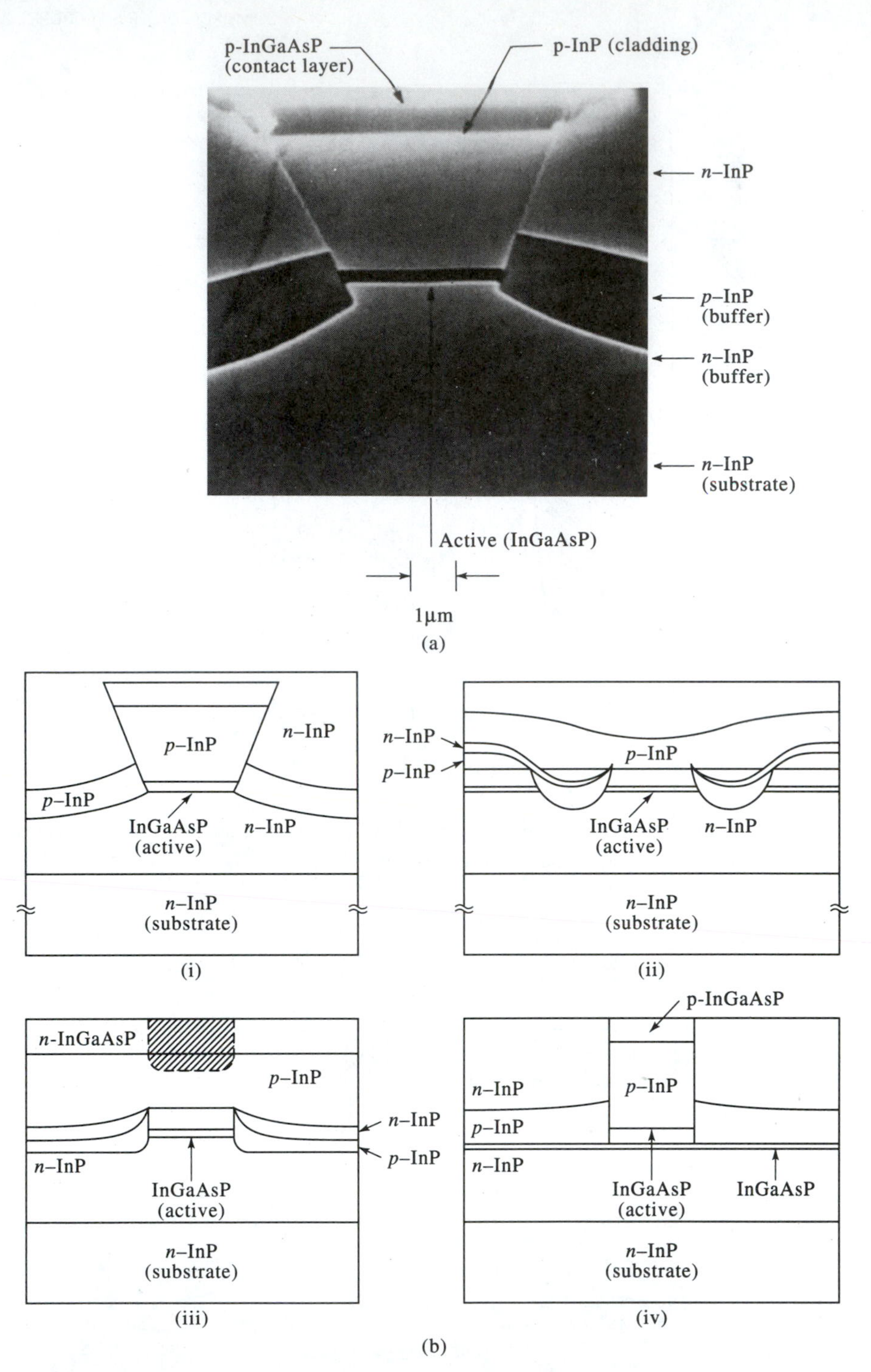

Figure 15-18 Typical structures of buried active region InP/GaInAsP diode lasers. (a) SEM [40, 41]. (b) Drawing of different structures [39].

taxial layers act as cladding layers to the quaternary $Ga_{1-x}In_xAs_{1-y}P_y$ active layer. The quaternary layer plays in this system the role played by GaAs in the GaAs/GaAlAs laser depicted in Figure 15-9. A typical quaternary laser structure is shown in Figure 15-18. Modern versions of these laser systems employ active regions with thicknesses in the 50Å → 100Å range. These are the so-called *quantum well lasers*. These lasers possess lower threshold currents and have a larger modulated bandwidth compared to earlier generations employing "thick" (~1000Å) active regions. They are discussed in detail in Chapter 16. Recent experiments [26, 39, 41, 42] have demonstrated propagation without repeaters at distances of ~ 150 km in optical fibers at 1.55 μm.

Power Output of Injection Lasers

The considerations of saturation and power output in an injection laser are basically the same as that of conventional lasers, which were described in Sections 5.6 and 6.4. As the injection current is increased above the threshold value, the laser oscillation intensity builds up. The resulting stimulated emission shortens the lifetime of the inverted carriers to the point where the magnitude of the inversion is clamped at its threshold value. Taking the probability that an injected carrier recombine radiatively within the active region as η_i,[2] we can write the following expression for the power emitted by stimulated emission:

$$P_e = \frac{(I - I_t)\eta_i}{e} h\nu \tag{15.4-1}$$

Part of this power is dissipated inside the laser resonator, and the rest is coupled out through the end reflectors. These two powers are, according to (15.3-4), proportional to the effective internal loss $\alpha \equiv \alpha_n \Gamma_n + \alpha_p \Gamma_p + \alpha_s$ and to $-L^{-1} \ln R$, respectively. We can thus write the output power as

$$P_0 = \frac{(I - I_t)\eta_i h\nu}{e} \frac{(1/L) \ln (1/R)}{\alpha + (1/L) \ln (1/R)} \tag{15.4-2}$$

The external differential quantum efficiency η_{ex} is defined as the ratio of the photon output rate that results from an increase in the injection rate (carriers per second) to the increase in the injection rate:

$$\eta_{ex} = \frac{d(P_0/h\nu)}{d[(I - I_t)/e]} \tag{15.4-3}$$

Using (15.4-2) we obtain

$$\eta_{ex}^{-1} = \eta_i^{-1} \left(\frac{\alpha L}{\ln (1/R)} + 1 \right) \tag{15.4-4}$$

---

[2]The reason for a quantum efficiency η_i that is less than unity is, mostly, the existence of a leakage current component that bypasses the active *p-n* junction region.

By plotting the dependence of η_{ex} on L we can determine η_i, which in GaAs is around 0.9–1.0.

Since the incremental efficiency of converting electrons into useful output photons is η_{ex}, the main remaining loss mechanisms degrading the conversion of electrical to optical power is the small discrepancy between the energy eV_{appl} supplied to each injected carrier and the photon energy $h\nu$. This discrepancy is due mostly to the series resistance of the laser diode. The efficiency of the laser in converting electrical power input to optical power is thus

$$\eta = \frac{P_0}{VI} = \eta_i \frac{I - I_t}{I} \frac{h\nu}{eV_{appl}} \frac{\ln(1/R)}{\alpha L + \ln(1/R)} \tag{15.4-5}$$

In practice $eV_{appl} \sim 1.4E_g$ and $h\nu \simeq E_g$. Values of $\eta \sim 30$ percent at 300 K have been achieved.

We conclude this section by showing in Figures 15-16 and 15-17 typical plots of the power output versus current and the far field of commercial low-threshold GaAs semiconductor lasers.

15.5 DIRECT-CURRENT MODULATION OF SEMICONDUCTOR LASERS

Since the main application of semiconductor lasers is as sources for optical communication systems, the problem of high-speed modulation of their output by the high-data-rate information is one of great technological importance.

A unique feature of semiconductor lasers is that, unlike other lasers that are modulated externally (see Chapter 9), the semiconductor laser can be modulated directly by modulating the excitation current. This is especially important in view of the possibility of monolithic integration of the laser and the modulation electronic circuit, as will be discussed in Section 15.7. The following treatment follows closely that of Reference [27].

If we denote the photon density inside the active region of a semiconductor laser by P and the injected electron (and hole) density by N, then we can write

$$\frac{dN}{dt} = \frac{I}{eV} - \frac{N}{\tau} - A(N - N_{tr})P$$

$$\frac{dP}{dt} = A(N - N_{tr})P\Gamma_a - \frac{P}{\tau_p} \tag{15.5-1}$$

where I is the total current, V the volume of the active region, τ the spontaneous recombination lifetime, τ_p the photon lifetime as limited by absorption in the bounding media, scattering and coupling through the output mirrors.

The term $A(N - N_{tr})P$ is the net rate per unit volume of induced transitions. N_{tr} is the inversion density needed to achieve transparency as defined by (15.2-17), and

A is a temporal growth constant that by definition is related to the constant B defined by (15.2-17) by the relation $A = Bc/n$. Γ_a is the filling factor defined by (15.3-3), and its presence here is to ensure that the total number, rather than the density variables used in (15.5-1), of electrons undergoing stimulated transitions is equal to the number of photons emitted. The contribution of spontaneous emission to the photon density is neglected since only a very small fraction ($\sim 10^{-4}$) of the spontaneously emitted power enters the lasing mode.

By setting the left side of (15.5-1) equal to zero, we obtain the steady-state solutions N_0 and P_0

$$0 = \frac{I_0}{eV} - \frac{N_0}{\tau} - A(N_0 - N_{tr})P_0$$

$$0 = A(N_0 - N_{tr})P_0\Gamma_a - \frac{P_0}{\tau_p} \tag{15.5-2}$$

We consider the case where the current is made up of dc and ac components

$$I = I_0 + i_1 e^{i\omega_m t} \tag{15.5-3}$$

and define the small-signal modulation response n_1 and p_1 by

$$N = N_0 + n_1 e^{i\omega_m t} \qquad P = P_0 + p_1 e^{i\omega_m t} \tag{15.5-4}$$

where N_0 and P_0 are the dc solutions of (15.5-2).

Using (15.5-3), (15.5-4), and the result $A(N_0 - N_{tr}) = (\tau_p\Gamma_a)^{-1}$ from (15.5-2) in (15.5-1) leads to the small-signal algebraic equations

$$-i\omega_m n_1 = -\frac{i_1}{eV} + \left(\frac{1}{\tau} + AP_0\right)n_1 + \frac{1}{\tau_p\Gamma_a}p_1$$

$$i\omega_m p_1 = AP_0\Gamma_a n_1 \tag{15.5-5}$$

Our main interest is in the modulation response $p_1(\omega_m)/i_1(\omega_m)$ so that from (15.5-5) we obtain

$$p_1(\omega_m) = \frac{-(i_1/eV)AP_0\Gamma_a}{\omega_m^2 - i\omega_m/\tau - i\omega_m AP_0 - AP_0/\tau_p} \tag{15.5-6}$$

A typical measurement of $p_1(\omega_m)$ is shown in Figure 15-19(b). The response curve is flat at small frequencies, peaks at the "relaxation resonance frequency" ω_R, and then drops steeply. The expression for the peak frequency is obtained by minimizing the magnitude of the denominator of (15.5-6)

$$\omega_R = \sqrt{\frac{AP_0}{\tau_p} - \frac{1}{2}\left(\frac{1}{\tau} + AP_0\right)^2} \tag{15.5-7}$$

In a typical semiconductor laser with $L = 300\ \mu$m, we have from (4.7-3) $\tau_p \simeq (n/c)(\alpha - (1/L)\ln R)^{-1} \sim 10^{-12}$ s, $\tau \sim 4 \times 10^{-9}$ s, and $AP_0 \sim 10^9\ \text{s}^{-1}$ so that to a very good accuracy

$$\omega_R = \sqrt{\frac{AP_0}{\tau_p}} \tag{15.5-8}$$

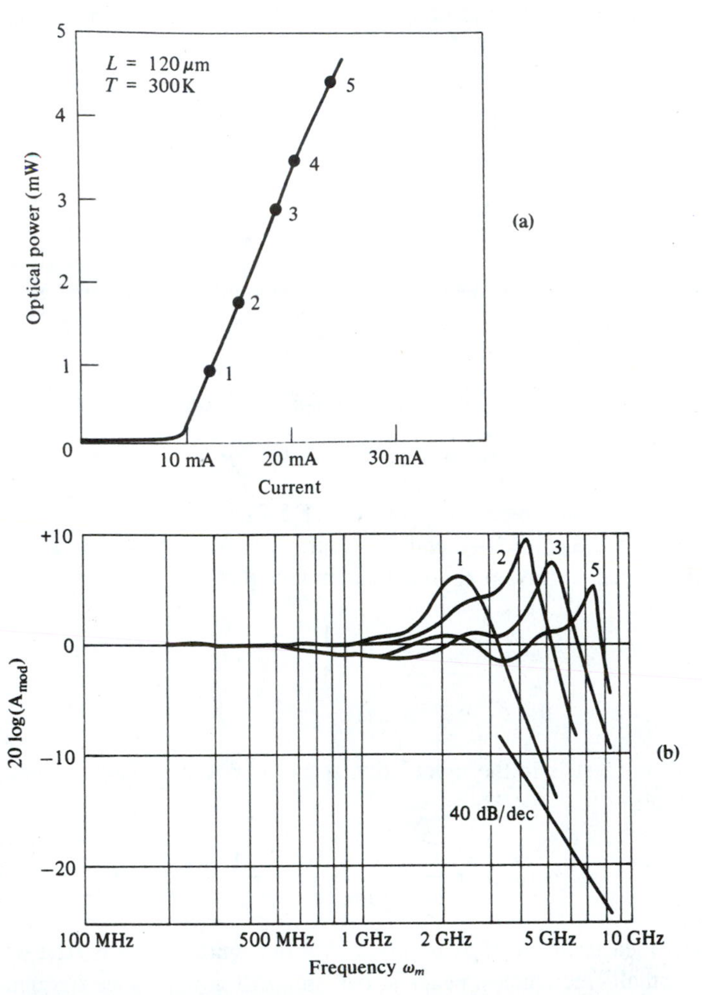

Figure 15-19 (a) *CW* light output power versus current characteristic of a laser of length = 120 μm. (b) Modulation characteristics of this laser at various bias points indicated in the plot. (c) Measured relaxation oscillation resonance frequency of lasers of various cavity lengths as a function of $\sqrt{P}$, where P is the cw output optical power. The points of catastrophic damage are indicated by downward pointing arrows. (After Reference [27].) (d) Current feed network for microwave modulation of high-speed lasers. (e) The corresponding frequency response (after Reference [41].)

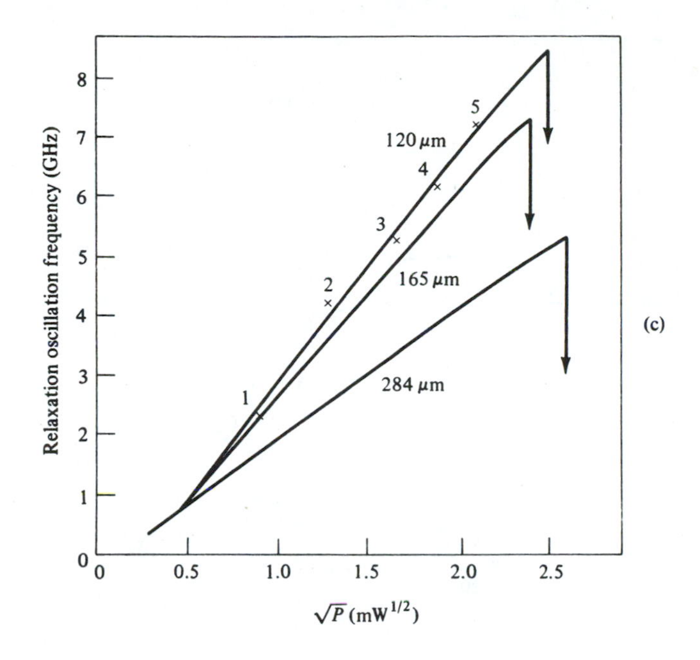
Relaxation oscillation frequency (GHz)
$\sqrt{P}$ (mW$^{1/2}$)
120 μm
165 μm
284 μm
0
0.5
1.0
1.5
2.0
2.5
1
2
3
4
5
6
7
8

(c)

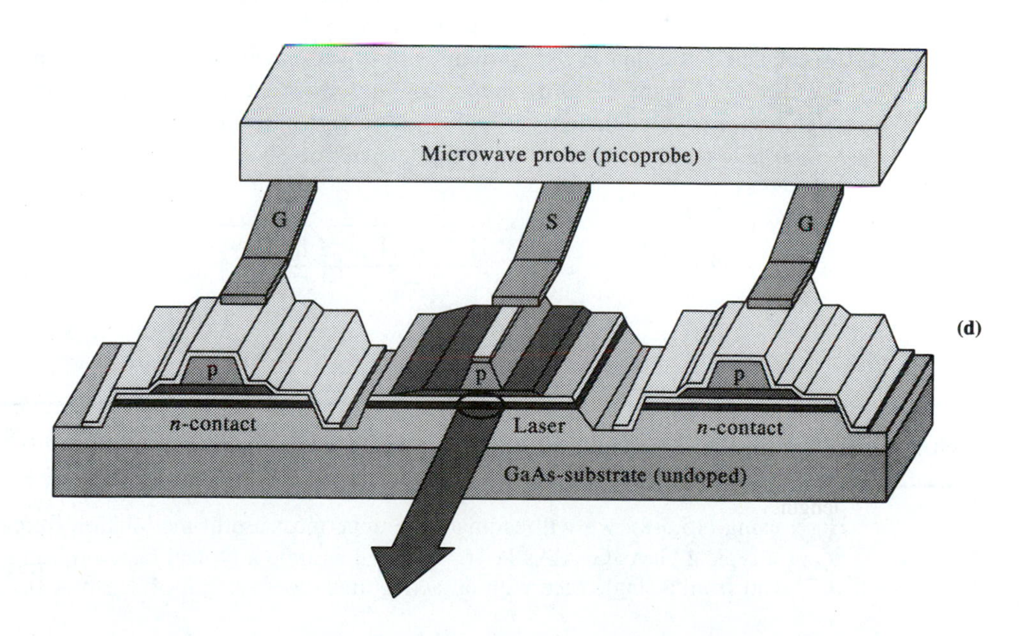
Microwave probe (picoprobe)
G
S
G
p
p
p
n-contact
Laser
n-contact
GaAs-substrate (undoped)

(d)

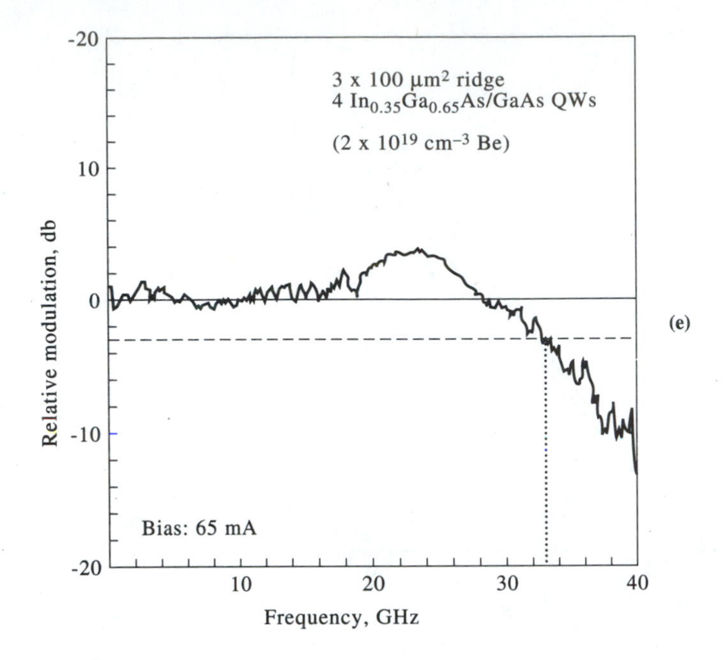

The last result is extremely useful, since it suggests that to increase ω_R and thus increase the useful linear region of the modulation response $p_1(\omega_m)/i_1(\omega_m)$, we need to increase the optical gain coefficient A, decrease the photon lifetime τ_p, and operate the laser at as high internal photon density P_0 as possible. The observed linear dependence of the modulation resonance frequency ω_R on the square root of the power output $p\sqrt{P}$ is demonstrated in Figure 15-19(c) for lasers of varying lengths. A detailed discussion of the optimum strategy for maximizing ω_R is given in Reference [27]. Figure 15-19(d) shows the microwave current feeding electrodes for high-frequency modulation, and 15-19(e) the corresponding frequency response.

It is somewhat tedious but straightforward to show that (15.5-8) can also be written as

$$\omega_R = \sqrt{\frac{1 + A\tau_p\Gamma_a N_{\text{tr}}}{\tau\tau_p}\left(\frac{I_0}{I_{\text{th}}} - 1\right)} \tag{15.5-9}$$

Numerical Example: Modulation Bandwidth in GaAs/GaAlAs Lasers

Here, using (15.5-8), we will estimate the uppermost useful modulation frequency ω_R of a typical GaAs/GaAlAs laser. We shall assume a typical laser emitting 5×10^{-3} watt from a single face with an active area cross section of 3 μm $\times$ 0.1 μm, a facet reflectivity of $R = 0.31$ and an index of refraction $n_0 = 3.5$. Solving for P_0 from the relationship

$$\frac{(1 - R)P_0 ch\nu}{n_0} = \frac{\text{power}}{\text{area}}$$

we obtain $P_0 = 1.21 \times 10^{15}$ photons/cm^3 for the photon density in the laser cavity. The constant A has a typical value of 2×10^{-6} cm^3/s. [This can be checked against the relationship $A = Bc/n_0$, where B is the spatial gain parameter of (15.2-17).] The photon lifetime τ_p is obtained from (4.7-3)

$$\tau_p = \frac{n_0}{c}\left(\alpha_{ab} - \frac{1}{L}\ln R\right)^{-1}$$

which for $L = 120\ \mu$m, $\alpha_{ab} = 10$ cm^{-1}, and $R = 0.31$ yields $\tau_p \sim 1.08 \times 10^{-12}$ s. Combining these results gives

$$\nu_R \equiv \frac{\omega_R}{2\pi} = \frac{1}{2\pi}\sqrt{\frac{AP_0}{\tau_p}} = \frac{1}{2\pi}\sqrt{\frac{2 \times 10^{-6} \times 1.2 \times 10^{15}}{1.08 \times 10^{-12}}}$$
$$= 7.53 \times 10^9 \text{ Hz}$$

This value is in the range of the experimental data shown in Figure 15-19, which was obtained on a laser with characteristics similar to that used in our example. The square root law dependence of ω_R on the photon density (or power output) predicted by (15.5-8) is verified by the data of Figure 15-19(c).

15.6 GAIN SUPPRESSION AND FREQUENCY CHIRP IN CURRENT-MODULATED SEMICONDUCTOR LASERS

In Section 15.5, we solved for the modulation of the power output (or, equivalently, the photon density, inside the laser resonator), which is due to a modulation of the current flowing through the laser. The current is taken as

$$I(t) = I_0 + i_1(\omega_m)\exp(i\omega_m t) \tag{15.6-1}$$

while the photon density inside the laser, which is proportional to the power output, is

$$P(t) = P_0 + p_1(\omega_m)\exp(i\omega_m t) \tag{15.6-2}$$

We also take the carrier density in the active regions (the inverted population) as

$$N(t) = N_0 + n_1(\omega_m)\exp(i\omega_m t) \tag{15.6-3}$$

Ideally we would like $p_1(\omega_m)/i_1(\omega_m)$, the frequency modulation response, to be a constant independent of ω_m and, above threshold, we will expect that $n_1(\omega_m) = 0$, indicating perfect gain clamping. As we shall find out, neither expectation is realized fully. As a matter of fact, if we solve Equations (15.5-5) for $n_1(\omega_m)$, the result is

$$n_1(\omega_m) = -i\left(\frac{i_1}{eV}\right)\frac{\omega_m}{\omega_m^2 - \dfrac{AP_0}{\tau_p} - i\omega_m\left(\dfrac{1}{\tau} + AP_0\right)} \tag{15.6-4}$$

We thus find that, under dynamic conditions, the carrier density, hence the gain, is not clamped at the threshold value N_0 but has an oscillating component whose amplitude $n_1(\omega_m)$ is given by (15.6-4). Its general features are depicted in Figure 15-20. We note a peak at ω_R, which is also the resonance frequency for the amplitude modulation response $p_1(\omega_m)/i_1(\omega_m)$, as given by (15.5-7). Since the index of refraction of a semiconductor medium depends on the carrier density, the modulation of the latter is accompanied by a modulation of the index of refraction, leading to a frequency modulation (FM) of the output optical field. This parasitic FM modulation, most of the time undesired, has a number of important consequences. The most important of these is the resulting spectral broadening of the laser field that, in dispersive, D $\neq$ 0, fibers, leads to an increase in the spreading of optical pulses with distance, as shown in Section 3.4.

Before embarking on the analysis of the parasitic frequency modulation, we need to introduce two new physical concepts: (1) the gain suppression effect and (2) the amplitude-phase coupling effect.

Gain suppression. The gain experienced in an inverted semiconductor laser medium by an optical wave is invariably lower the higher the optical intensity. This

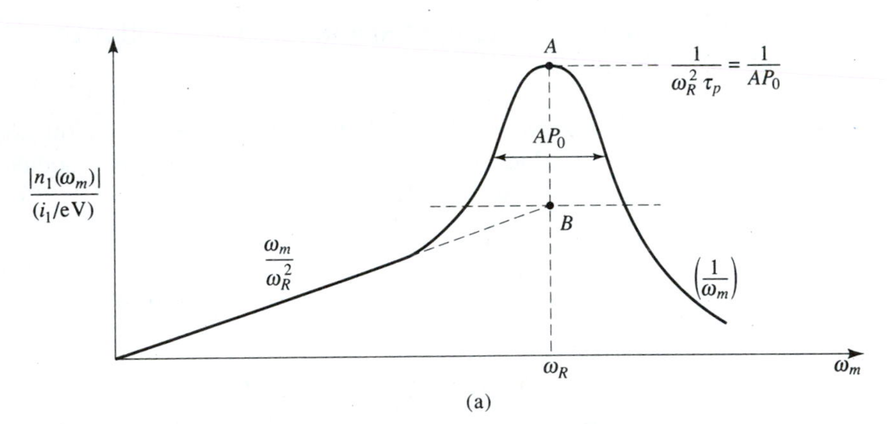

(a)

Figure 15-20 (a) A theoretical plot of the carrier density modulation n_1 as a function of the current modulation frequency ω_m. (b) (1) A scanning Fabry–Perot spectrum of a GaInAsP (λ = 1.31 μm) DFB laser with no current modulation. (2) The spectrum of the same laser when the current is modulated at f_m = 550 MHz, horiz. scale = 1 GHz/div. (Courtesy of H. Blauvelt, P. C. Chen, and N. Kwong of ORTEL Corporation, Alhambra, California)

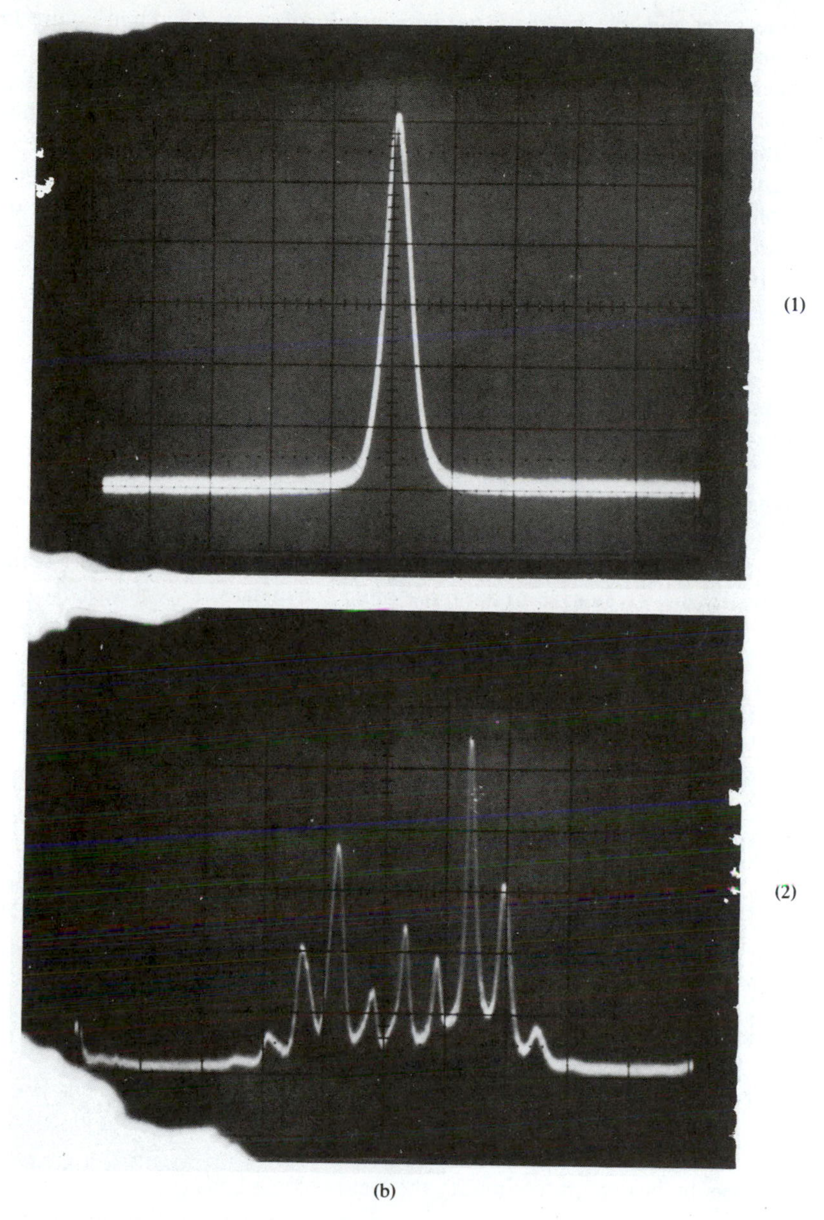

(b)

Figure 15-20 (*continued*)

is due partly to optical gain saturation and partly to gain suppression. The first effect reflects the drop of the total electron population density—N in (15.5-1)—with the increase in P. This effect is accounted for properly by the coupled rate equations (15.5-1). The second mechanism reducing the gain takes place even when the total density N is constant and reflects the reduction of the density of *resonant carriers* (electrons and holes) in the *immediate vicinity* of points a and b in Figure 15-5 which contribute to the gain. This is due both to spectral "hole burning," as discussed in Section 5.7 and illustrated in Figure 6-9(d), and to an increase in the effective electron temperature[3] by the optical field. Under dynamic nonthermal equilibrium conditions, this temperature may differ from the lattice temperature. Such an increase causes, according to the discussion in Sections 15.1 and 15.2, the electrons (and holes) to spread to higher energies, reducing in the process the density of resonant carriers that contribute to the gain. This effect asserts itself with a time constant of $<10^{-12}$ s, characteristic of electron-electron and electron-phonon collisions, and for system applications involving modulation at $\omega_m/2\pi < 3 \times 10^{10}$ Hz, it can be considered as responding instantaneously to the optical field.

Our main departure from the analyses of Section 15.5 is to take the optical gain constant as

$$G(N,P) = G(N)(1 - \epsilon P) \approx G(N_{th}) + A(N - N_{th}) - \epsilon G(N_{th})P \quad (15.6\text{-}5)$$

where gain suppression is represented by the factor $(1 - \epsilon P)$. The constant ϵ is called the *gain suppression factor*. We can view (15.6-5) as a Taylor expansion about the threshold point $N = N_{th}$, $P = 0$. The numerical value of ϵ can be estimated theoretically in a given system but more often is evaluated experimentally [42].

We now rewrite the rate equations (15.5-1) as

$$\frac{dN}{dt} = \frac{I}{eV} - \frac{N}{\tau} - G(N,P)P$$

$$\frac{dP}{dt} = \Gamma_a G(N,P)P - \frac{P}{\tau_p} \quad (15.6\text{-}6)$$

at steady state $d/dt = 0$

$$0 = \frac{I_0}{eV} - \frac{N_0}{\tau} - [G(N_{th}) + A(N_0 - N_{th}) - \epsilon G(N_{th})P_0]P_0$$

$$0 = \Gamma_a[G(N_{th}) + A(N_0 - N_{th}) - \epsilon G(N_{th})P_0]P_0 - \frac{P_0}{\tau_p}$$

The value of $G(N_{th})$ is obtained from the second equation evaluated at threshold ($N_0 = N_{th}$, $P_0 = 0$)

$$G(N_{th}) = \frac{1}{\Gamma_a \tau_p}$$

We use the last result to simplify the last two equations

$$0 = \frac{I_0}{eV} - \frac{N_0}{\tau} - \frac{P_0}{\Gamma_a \tau_p}$$

$$0 = A(N_0 - N_{th}) - \frac{\epsilon P_0}{\Gamma_a \tau_p} \quad (15.6\text{-}7)$$

[3]By electron "temperature" we mean the temperature used in the Fermi function (15.1-8).

Performing a "small signal" expansion as in Equations (15.5-3) and (15.5-4) leads to

$$i\omega_m n_1 = \frac{i_1}{eV} - \left(\frac{1}{\tau} + AP_0\right)n_1 - \frac{(1 - \epsilon P_0)}{\Gamma_a \tau_P} p_1$$

$$i\omega_m p_1 = \Gamma_a AP_0 n_1 - \frac{\epsilon P_0}{\tau_P} p_1 \tag{15.6-8}$$

We note that for the case $\epsilon = 0$, these equations reduce to Equations (15.5-5). Solving (15.6-8) for $p_1(\omega_m)$ yields

$$p_1(\omega_m) = \frac{-\Gamma_a AP_0\left(\dfrac{i_1}{eV}\right)}{\omega_m^2 - i\omega_m\left(\dfrac{1}{\tau} + AP_0 + \dfrac{\epsilon P_0}{\tau_P}\right) - \left(\dfrac{AP_0}{\tau_P} + \dfrac{\epsilon P_0}{\tau\tau_P}\right)} \tag{15.6-9}$$

Under typical conditions such as those of the numerical example above, we have

$$P_0 = 1.2 \times 10^{21} \text{ photons/m}^3$$

$$A = 2 \times 10^{-12} \text{ m}^3\text{/s} \qquad \tau = 4 \times 10^{-9} \text{ s}$$

$$\tau_P = 10^{-12} \text{ s} \qquad \epsilon = 10^{-23} \text{ m}^3$$

so that $\epsilon P_0/\tau_P = 1.2 \times 10^{10} \text{ s}^{-1} \gg AP_0$.

Using the inequalities $\epsilon P_0/\tau_P \gg AP_0$ and $\epsilon/\tau_P \gg A$, we find that the peak modulation response occurs at

$$\omega_R \approx \sqrt{\frac{AP_0}{\tau_P} - \frac{\epsilon^2 P_0^2}{2\tau_P^2}}$$

A comparison of this last result to (15.5-8) shows that when $\epsilon \neq 0$, i.e., gain suppression is considered, the modulation resonant frequency, ω_R, does not increase indefinitely with P_0 but reaches a maximum value at a photon density of

$$(P_0)_{\max} = \frac{A\tau_P}{\epsilon^2} \tag{15.6-10}$$

The maximum value of ω_R, which obtains when $P_0 = (P_0)_{\max}$, is

$$(\omega_R)_{\max} = \frac{A}{\sqrt{2}\epsilon} \tag{15.6-11}$$

Using the typical numerical constants given following (15.6-9) and in the numerical example of Section 15.5, we estimate

$$(P_0)_{\max} = \frac{A\tau_P}{\epsilon^2} = \frac{2 \times 10^{-12} \times 10^{-12}}{10^{-46}} = 1 \times 10^{22} \text{ photons/m}^3$$

which, using the expressions given in the example of Section 15.5, corresponds to a power output of ~80 mw per facet. The corresponding maximal resonant modulation frequency is

$$\left(\frac{\omega_R}{2\pi}\right)_{\max} \approx \left(\frac{1}{2\pi}\right)\frac{A}{\sqrt{2\epsilon}} = \frac{2 \times 10^{-12}}{2\pi\sqrt{2} \times 10^{-23}} = 2.24 \times 10^{10} \text{ Hz}$$

The last result points out the role of the gain suppression in placing a practical upper limit on the modulation bandwidth that can be achieved by current modulation of semiconductor lasers. To modulate at significantly higher frequencies, $\omega_m > \omega_R$, one needs to employ external modulators, such as these that are discussed in Sections 9.3 and 9.4.

Amplitude-phase coupling

Amplitude-phase coupling which causes a second effect that is attendant upon the carrier density modulation ($n_1(\omega_m) \neq 0$) is the frequency modulation ("chirp") of the laser output field. The amplitude $n_1(\omega_m)$ of the carrier density fluctuation is obtained by solving (15.6-8):

$$n_1(\omega_m) = \frac{-\left(i\omega_m + \dfrac{\epsilon P_0}{\tau_P}\right)\left(\dfrac{i_1}{eV}\right)}{\omega_m^2 - \left(\dfrac{AP_0}{\tau_P} + \dfrac{\epsilon P_0}{\tau\tau_P}\right) - i\omega_m\left(\dfrac{1}{\tau} + AP_0 + \dfrac{\epsilon P_0}{\tau_P}\right)} \tag{15.6-12}$$

A comparison of (15.6-12) to (15.6-9) shows that

$$n_1(\omega_m) = \frac{\left(i\omega_m + \dfrac{\epsilon P_0}{\tau_P}\right)}{\Gamma_a A P_0} p_1(\omega_m) \tag{15.6-13}$$

Since for our assumed $\exp(i\omega_m t)$ time dependence, $i\omega_m = d/dt$, we use (15.6-13) to write

$$\Delta N(t) = \frac{1}{\Gamma_a A}\left[\frac{1}{P_0}\frac{dP}{dt} + \frac{\epsilon}{\tau_p}\Delta P(t)\right] \tag{15.6-14}$$

where $N(t) = N_0 + \Delta N(t)$, $P(t) = P_0 + \Delta P(t)$. Equation (15.6-14) applies to any arbitrary photon density modulation, not necessarily harmonic.

Our next task is to find the effect of the carrier density modulation $\Delta N(t)$ on the laser frequency. The index of refraction of the gain medium is complex:

$$n_0(t) = n_0'(t) + in_0''(t) \tag{15.6-15}$$

where the time dependence reflects the dependence of the real index n_0' and the gain (or loss), which involves n_0'', on the time-modulated carrier density $\Delta N(t)$.

The imaginary part of the index n_0 is related to the spatial exponential gain constant of the laser medium since the spatial dependence of the field is

$$\mathscr{E}(z) \propto \mathscr{E}_0 \exp\left[-\frac{i\omega}{c}(n_0' + in_0'')z\right] = \mathscr{E}_0 \exp\left(-\frac{i\omega n_0'}{c} z\right) \exp\left(\frac{\omega n_0''}{c} z\right)$$

Since our rate equations (15.6-6) and (15.6-8) are in the time domain, we need to convert the spatial growth parameter $\omega n_0''/c$ to a temporal one. We use

$$\frac{d|\mathscr{E}|}{dt} = \frac{\partial|\mathscr{E}|}{\partial z}\frac{dz}{dt} \cong \left(\frac{\omega n_0''}{c}\right)\frac{c}{n_0'}|\mathscr{E}| = \frac{\omega n_0''}{n_0'}|\mathscr{E}|$$

It thus follows that the exponential gain constant $G(N,P)$ of (15.6-6) is related to n_0'' by

$$G = \frac{2\omega n_0''}{n_0'} = \frac{4\pi\nu n_0''}{n_0'} \tag{15.6-16}$$

where the factor of 2 accounts for the fact that G is the temporal growth constant of the photon density ($\propto$ optical intensity) i.e., of $|\mathscr{E}|^2$.

From the last equation, it follows that

$$\frac{\partial n_0''}{\partial N} = \frac{n_0'}{4\pi\nu}\frac{\partial G}{\partial N} = \frac{n_0'}{4\pi\nu} A \tag{15.6-17}$$

where we used (15.6-5) to substitute $\partial G/\partial N = A$. A modulation $\Delta N(t)$ of the carrier density causes, according to (15.6-17), a corresponding perturbation

$$\Delta n_0'' = \frac{n_0'}{4\pi\nu} A\ \Delta N(t) \tag{15.6-18}$$

Our next task is to obtain the dependence of the real index perturbation $\Delta n_0'$ on $\Delta N(t)$. Now $\Delta n_0'$ and $\Delta n_0''$ are related through Kramers-Kroning relations, which were discussed in Section 5.4. It has, however, proved useful to relate $\Delta n_0'$ and $\Delta n_0''$ by means of a parameter, the so-called phase-amplitude coupling constant, the α parameter [36, 43].

$$\alpha = \frac{\Delta n_0'}{\Delta n_0''} \tag{15.6-19}$$

The α parameter is a function of the carrier density N_0 of the semiconductor laser medium, and typical values for it range between 3 and 5 [36, 44]. Combining (15.6-19) and (15.6-18), results in

$$\Delta n_0'(t) = \frac{\alpha n_0' A}{4\pi\nu}\Delta N(t) \tag{15.6-20}$$

A perturbation $\Delta n_0'$ due to carrier density modulation causes a perturbation $\Delta\nu$ of the laser frequency

$$\frac{\Delta\nu}{\nu} = -\frac{\Delta n_0'}{n_0'}\Gamma_a = -\frac{\alpha\Gamma_a A}{4\pi\nu}\Delta N(t) \tag{15.6-21}$$

The first equality of (15.6-21) follows directly from the basic Fabry-Perot resonance frequency relation, Equation (6.2-2). The factor $\Gamma_a \approx V_{\text{active}}/V_{\text{mode}}$ accounts for (possibly) partial filling of the resonator by the active medium. Substitution in the last equation for $\Delta N(t)$ from (15.6-14) [36, 37] gives

$$\boxed{\Delta\nu(t) = -\frac{\alpha}{4\pi}\left[\frac{1}{P_0}\frac{dP}{dt} + \frac{\epsilon}{\tau_P}\Delta P(t)\right]} \tag{15.6-22}$$

for the laser frequency chirp. The contribution involving dP/dt is called the *transient chirp*, while that which is proportional to $\Delta P(t)$ is termed the *adiabatic chirp*. The latter involves the gain suppression parameter ϵ and is usually dominant at low ($\lesssim 10^8$ Hz) frequencies, while the transient term dominates at typical microwave ($> 10^9$ Hz) frequencies.

The Field Spectrum of a Chirping Laser (36, 37)

In the development leading to (15.6-22), we showed that any modulation of the power of a semiconductor laser by means of current modulation causes a frequency chirp. In the following, we will derive the spectrum of the output optical field of a laser with sinusoidal power modulation in order to obtain an appreciation of the expected order of magnitude of the effect, especially the amount of spectral spread. We take the optical field of the laser as

$$\mathscr{E}(t) = \left[E_0 + \frac{s}{2}E_0 \sin\omega_m t\right]\exp\{i[2\pi\nu_0 t + \phi(t)]\} \tag{15.6-23}$$

where ω_m is the modulation frequency, and ν_0 is the average optical frequency.

The phase $\phi(t)$ in (15.6-23) is related to the chirp $\Delta\nu(t)$ of (15.6-22) by the general result[4]

$$\phi(t) = 2\pi\int_0^t \Delta\nu(t')dt' = -\frac{\alpha}{2}\left[\frac{1}{P_0}P(t) + \frac{\epsilon}{\tau_P}\int_0^t \Delta P(t')dt'\right] \tag{15.6-24}$$

If $\phi(t)$ were a constant, which would be the case when $\alpha = 0$, the spectrum of the topical field $\mathscr{E}(t)$ would consist of a carrier of amplitude E_0 at ν_0 and two sidebands with amplitudes $s/4$. When $\phi(t)$ is not constant, additional sidebands appear and the optical spectrum broadens. This chirped-induced broadening is of spe-

---

[4]The basic definition of the frequency $\nu(t)$ of a field $A\exp[i\phi(t)]$ is $\nu(t) = (1/2\pi)(d\phi/dt)$ and corresponds to the number of optical cycles per second at time t.

cial concern in applications that involve high data rate communication in fibers, since, as shown in Section 2.9, the temporal broadening of pulses in dispersive fibers is directly proportional to the product of the spectral width of the light and the propagation length. Any increase in the laser's spectral width would thus limit the rate at which the data can be transmitted in such a fiber. In the case considered here, we can write the photon density $P(t)$ as

$$P(t) \propto \langle|\mathscr{E}(t)|^2\rangle = E_0^2\left(1 + \frac{s^2}{4}\right) + sE_0^2 \sin \omega_m t = P_0 + P_1 \sin \omega_m t \qquad (15.6\text{-}25)$$

where $\langle\ \rangle$ indicates averaging over a few optical periods. Substitution of the last expression (15.6-24) leads to

$$\phi(t) = -\frac{\alpha}{2}\left[\overset{\text{transient}}{\frac{P_1}{P_0}\sin \omega_m t} - \overset{\text{adiabatic}}{\frac{\epsilon P_1}{\omega_m \tau_P}\cos \omega_m t}\right]$$

where we left out time-independent terms that correspond to unimportant fixed phase shifts. At high modulation frequencies such that $\omega_m \gg \epsilon P_0/\tau_P$ (this corresponds, using the numerical data used earlier in this section, to $\omega_m/2\pi \gg 2 \times 10^9$ Hz), the first term ("transient") in the square brackets dominates, so that the optical phase can be taken as

$$\phi(t) = -\frac{m\alpha}{2}\sin \omega_m t$$

$$m \equiv \frac{P_1}{P_0} = \frac{s}{1 + s^2/4} \approx s \qquad (15.6\text{-}26)$$

when s (and m) $\ll 1$. The total optical field, Equation (15.6-23), assumes the form

$$\mathscr{E}(t) = \left[E_0 + \frac{m}{2}E_0 \sin(\omega_m t)\right]\exp\left\{i\left[\omega_0 t - \frac{m\alpha}{2}\sin(\omega_m t)\right]\right\} \qquad (15.6\text{-}27)$$

where we used $s \approx m(s \ll 1)$.

We can use the Bessel function identity

$$e^{i\delta \sin x} = \sum_{n=-\infty}^{\infty} J_n(\delta)e^{inx} \qquad (15.6\text{-}28)$$

to rewrite (15.6-27):

$$\frac{\mathscr{E}(t)}{E_0} = \sum_{n=-\infty}^{\infty} J_n(\delta)\exp[i(\omega_0 + n\omega_m)t]$$

$$- i\frac{m}{4}\sum_{n=-\infty}^{\infty} J_n(\delta)\exp\{i[\omega_0 + (n+1)\omega_m]t\}$$

$$+ i\frac{m}{4}\sum_{n=-\infty}^{\infty} J_n(\delta)\exp\{i[\omega_0 + (n-1)\omega_m]t\}$$

Some of the sidebands are:

$$
\begin{aligned}
&\text{at } \omega_0, && E_0\left[J_0(\delta) + i\frac{m}{2}J_1(\delta)\right]\exp(i\omega_0 t) \\
&\text{at } \omega_0 + \omega_m, && E_0\left[J_1(\delta) - i\frac{m}{4}J_0(\delta) + i\frac{m}{4}J_2(\delta)\right]\exp[i(\omega_0 + \omega_m)t] \\
&\text{at} \omega_0 - \omega_m, && E_0\left[-J_1(\delta) + i\frac{m}{4}J_0(\delta) - i\frac{m}{4}J_2(\delta)\right]\exp[i(\omega_0 - \omega_m)t] \\
&\text{at } \omega_0 + 2\omega_m, && E_0\left[J_2(\delta) - i\frac{m}{4}J_1(\delta) + i\frac{m}{4}J_3(\delta)\right]\exp[i(\omega_0 + 2\omega_m)t] \\
&\text{at } \omega_0 - 2\omega_m, && E_0\left[J_2(\delta) - i\frac{m}{4}J_1(\delta) + i\frac{m}{4}J_3(\delta)\right]\exp[i(\omega_0 - 2\omega_m)t]
\end{aligned}
\tag{15.6-29}
$$

where $\delta = -m\alpha/2$ is the phase modulation index of the optical field. The amplitudes of the sidebands at $\omega_0 \pm n\omega_m$ in this case have the same magnitude. This is a consequence of the form of (15.6-27) but is not generally true, so that the optical sidebands, in general, for m and δ not zero, are not symmetric about ω_0. This is considered in Problem 15.11. An experimental graph showing the spectrum of the output field of a laser whose current is modulated is given in Figure 15-20(b). The spectrum can be fit well with an adiabatic chirp spectrum (i.e., phase modulation 90° out of phase with current), corresponding to a field

$$
\mathscr{E}(t) = \left[E_0 + \frac{m}{2}E_0 \sin \omega_m t\right]\exp[i2\pi\nu_0 t + \delta \cos \omega_m t)] \tag{15.6-30}
$$

with $m = 0.2$ and $\delta = 3.3$.

In the transient case, we found [see (15.6-27)] that the phase modulation index δ, the amplitude of the phase excursion, is equal to $m\alpha/2$ and that it can be determined from a fit to the experimental sideband distribution. Since the intensity modulation index m can be determined straightforwardly from a spectral analysis of the laser intensity, the combination of the field spectrum, obtainable with a scanning Fabry-Perot etalon, and the intensity spectrum, obtained from a spectral analysis of the detected photocurrent, can be used to determine the amplitude-phase coupling constant α [36].

15.7 INTEGRATED OPTOELECTRONICS

In one of its rare moments of cooperative spirit, nature has endowed the III-V semiconductors based on GaAs/GaAlAs and InP/GaInAsP with a double gift. These are, as discussed above, the materials of choice for semiconductor lasers, but in addition it is possible to use them, especially In GaAs/GaAs and GaAs/GaAlAs, as base materials for electronic circuits in a manner similar to that in silicon.[5]

It was pointed out in 1971 [30] that it should be possible to bring together monolithically in a III-V semiconductor the two principal actors of the modern communication era—the transistor and the laser—in new integrated optoelectronic circuits. This new technology is now taking its first tentative steps from the laboratory to applications.

The basic philosophy, as well as an example of an integrated optoelectronic device, is shown in Figure 15-21, which shows a buried heterostructure GaAs/GaAlAs laser, similar to that illustrated in Figure 15-15, fabricated monolithically on the same crystal as a field-effect transistor (FET). The output current of the FET (see arrows) supplies the electron injection to the active region of the laser. This current and thus the laser power output can be controlled by a bias voltage applied to the gate electrode.

An example of a feasibility model of an integrated optoelectronic optical repeater, which incorporates a detector, a FET current preamplifier, a FET laser driver, and a laser, is shown in Figure 15-22. The main reason for the accelerating drive toward an integrated optoelectronic circuit technology [33] derives from the reduction of parasitic reactances that are always associated with conventional wire interconnections, plus the compatibility with the integrated electronic circuits technology that makes it possible to apply the advanced techniques of the latter to this new class

[5]A completely new electronic technology based on GaAs/GaAlAs is now emerging [29]. It takes advantage of the large mobility of electrons in GaAs for very high switching speeds.

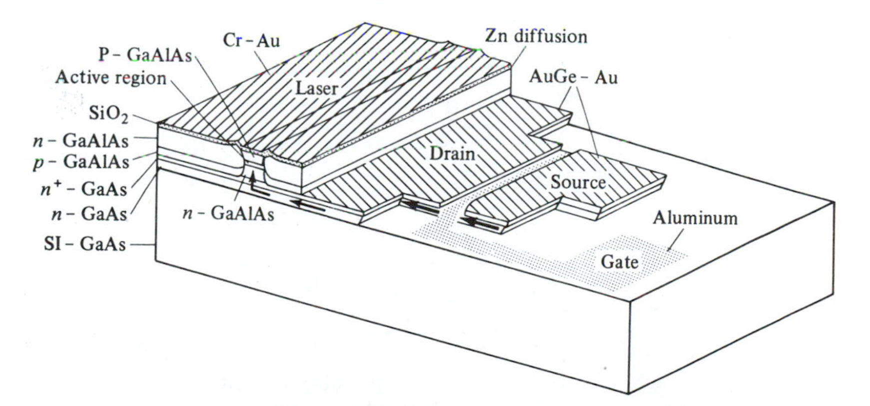

Figure 15-21 A GaAs *n*-channel field-effect transistor integrated monolithically with a buried heterostructure GaAs/GaAlAs laser. The application of a gate voltage is used to control the bias current of the laser. This voltage can oscillate and modulate the light at frequencies > 10 GHz. (After Reference [31].)

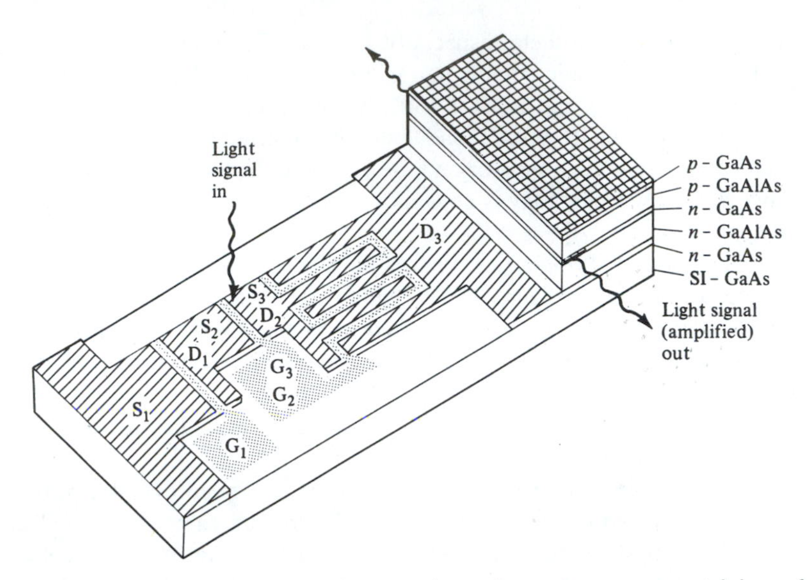

Figure 15-22 A monolithically integrated optoelectronic repeater containing a detector, transistor current source, a FET amplifier, and a laser on a single crystal GaAs substrate. (After Reference [32].)

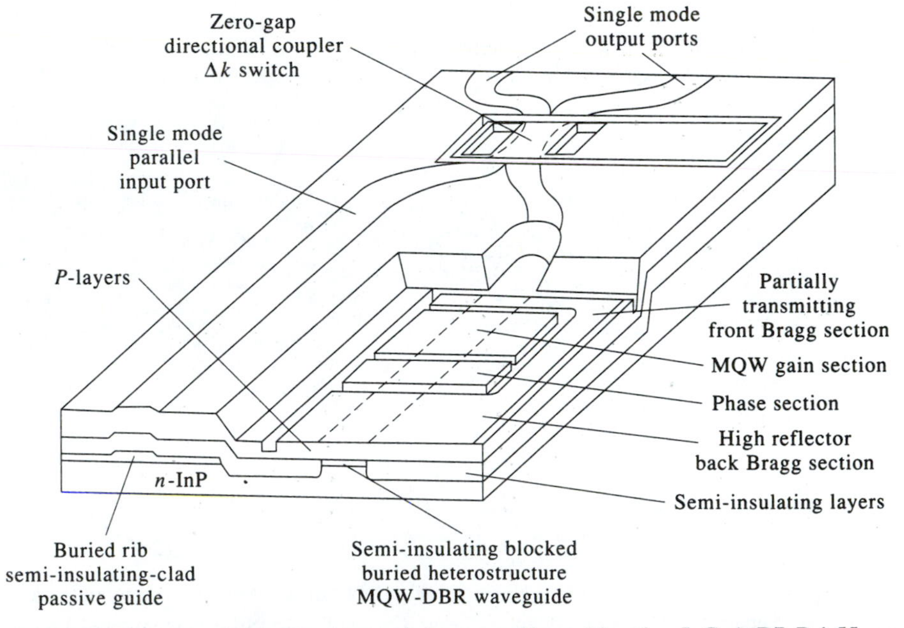

Figure 15-23 A monolithic circuit containing a tunable multisection InGaAsP/InP 1.55 μm laser employing multiquantum well gain section, a passive waveguide for an external input optical wave, and a directional coupler switch for combining the laser output field and that of the external input at the output ports. (After Reference [34].)

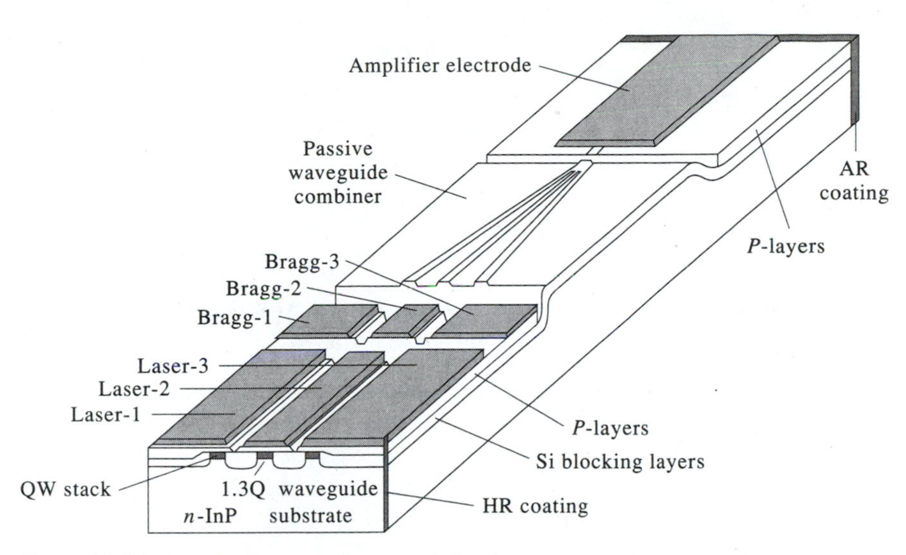

Figure 15-24 An optoelectronic integrated circuit composed of three ~1.5 μm InGaAs/InP distributed feedback lasers each tuned to a slightly different wavelength. The three wavelengths are fed into a single waveguide and amplified in a single amplifying section. (After Reference [35].)

of devices. More recent examples of optoelectronic integrated circuits are demonstrated in Figures 15-23 and 15-24.

Problems

15.1 Derive Equations (15.6-12).

15.2 Derive Equations 15.5-7 and 15.5-8.

15.3 Assume a fiber with $L = 10$ km and a group velocity dispersion parameter of 10 psec/nm-km (see Section 3.4). Calculate the maximum data rate through the fiber in bits/s if we use a semiconductor laser with characteristics similar to those used in the example of Section 15.6. For the purpose of this calculation, assume that a data rate of N bits/s is equivalent to a current modulation frequency of $\omega_m/2\pi = N$.

15.4 Find the frequency ω_R that maximizes $p_i(\omega_m)$ as given by equation (15.6-9) and, using the approximations given, derive (15.6-11).

15.5 Evaluate and plot:

(a) The gain $\gamma(\omega)$ of an inverted GaAs crystal under the following conditions:

$$N_{\text{elec}} = N_{\text{hole}} = 3 \times 10^{18} \text{ cm}^{-3}$$

$$m_c = 0.07\ m_{\text{electron}}$$

$$m_h = 0.4\ m_{\text{electron}}$$

$$T = 0\text{ K}$$

$$E_g = 1.45\text{ eV}$$

$$T_2 = \infty$$

(b) Comment qualitatively on the changes in $\gamma(\omega)$ as the temperature is raised.
(c) What is the effect of a finite T_2 on $\gamma(\omega)$?

15.6 Consider the effect on the modulation response $p_1(\omega_m)/i_1(\omega_m)$ of the inclusion of a nonlinear gain term bP in the rate equations (15.5-1)

$$\frac{dN}{dt} = \frac{I}{eV} - \frac{N}{\tau} - A(1 - bP)(N - N_{\text{tr}})P$$

$$\frac{dP}{dt} = A(1 - bP)(N - N_{\text{tr}})P\Gamma_a - \frac{P}{\tau_p}$$

where $bP \ll 1$. Show that the main effect is a damping of the resonance peak at ω_R.

15.7 Solve for the carrier density modulation $N = N_0 + N_1\, e^{i\omega_m t}$ in a semiconductor laser whose current is modulated at

$$I = I_0 + I_1 e^{i\omega_m t} \tag{1}$$

$$\omega_m = \text{modulation frequency} \ll \omega_{\text{opt}} \tag{2}$$

(See Section 15.5.)

15.8 Assume $\epsilon = \epsilon_0 - aN$, a is a constant and that the instantaneous frequency of the semiconductor laser obeys

$$\frac{\Delta\nu}{\nu} = -\frac{\Delta\epsilon}{\epsilon} \tag{3}$$

find the form of the laser optical field due to the current modulation. What is the (phase) modulating index of the field?

15.9 Using the data of Figure 15-7, what is the total current needed to render the active medium of a semiconductor laser transparent? Assume an active volume of $300 \times 2 \times 0.2$ (μm^3) and a recombination lifetime of $\tau = 3 \times 10^{-9}$ seconds.

15.10 If the thickness of the active region in Problem 15.9 were reduced to 100 Å, can we obtain enough gain from a semiconductor laser to overcome a distributed loss constant of $\alpha = 20\text{ cm}^{-1}$ and $R = 0.9$? What will be the transparency current? What will be the threshold current? Assume a mode height normal to the interfaces of $t = 4000$ A and

$$\Gamma_a \sim \frac{\text{d(active region)}}{\text{t(mode height)}}$$

15.11 Plot the optical spectrum of a wave with simultaneous AM and FM modulation

$$\mathscr{E}(t) = E_0\left(1 + \frac{m}{2} \sin \omega_m t\right) \exp\{i[\omega_0 t + \delta \sin(\omega_m t + \alpha)]\}$$

for $\alpha = 0$, $\pi/6$, $\pi/4$, $\pi/2$.

References

1. Basov, N. G., O. N. Krokhin, and Y. M. Popov, "Production of negative temperature states in *p-n* junctions of degenerate semiconductors," *J.E.T.P.* 40:1320, 1961.
2. Hall, R. N., G. E. Fenner, J. D. Kingsley, T. J. Soltys, and R. O. Carlson, "Coherent light emission from GaAs junctions," *Phys. Rev. Lett.* 9:366, 1962.
3. Nathan, M. I., W. P. Dumke, G. Burns, F. H. Dills, and G. Lasher, "Stimulated emission of radiation from GaAs *p-n* junctions," *Appl. Phys. Lett.* 1:62, 1962.
4. Yariv, A., and R. C. C. Leite, "Dielectric waveguide mode of light propagation in *p-n* junctions," *Appl. Phys. Lett.* 2:55, 1963.
5. Anderson, W. W., "Mode confinement in junction lasers," *IEEE J. Quant. Elec.* QE-1:228, 1965.
6. Kittel, C., *Introduction to Solid State Physics*, 5th ed. New York: Wiley, 1982.
7. Yariv, A., *Introduction to the Theory and Applications of Quantum Mechanics.* New York: Wiley, 1982.
8. Bernard, M. G., and G. Duraffourg, "Laser conditions in semiconductors," *Phys. Status Solidi* 1:699, 1961.
9. Vahala, K., L. C. Chiu, S. Margalit, and A. Yariv, "On the linewidth enhancement factor α in semiconductor injection lasers," *Appl. Phys. Lett.* 42:631, 1983.
10. Stern, F., "Semiconductor lasers: Theory," *Laser Handbook*, F. T. Arecchi and E. O. Schultz Du Bois, eds. Amsterdam: North Holland, 1972.
11. Kressel, H., and J. K. Butler, *Semiconductor Lasers and Heterojunction LEDS.* New York: Academic Press, 1977.
12. Yariv, A., *Quantum Electronics*, 2d ed. New York: Wiley, 1975, p. 219.
13. Casey, H. C., and M. B. Panish, *Heterostructure Lasers.* New York: Academic Press, 1978.
14. Dupuis, R. D., and P. D. Dapkus, *Appl. Phys. Lett.* 31:466, 1977.
15. Cho, A. Y., and J. R. Arthur, in *Progress in Solid State Physics*, J. O. McCaldin and G. Somoraj, eds., vol. 10. Elmsford, N.Y.: Pergamon Press, p. 157.
16. Tsang, W. T. and A. Y. Cho, "Growth of GaAs/GaAlAs by molecular beam epitoxy," *Appl. Phys. Lett.* 30:293, 1977.

17. Hayashi, J., M. B. Panish, and P. W. Foy, ''A low-threshold room-temperature injection laser,'' *IEEE J. Quant. Elec.* 5:211, 1969.
18. Kressel, H., and H. Nelson, ''Close confinement gallium arsenide *p-n* junction laser with reduced optical loss at room temperature,'' *RCA Rev.* 30:106, 1969.
19. Alferov, Zh. I., et al., *Sov. Phys.—Semicond.*, 4:1573, 1971.
20. Tsukada, T., ''GaAs-$Ga_{1-x}Al_xAs$ buried heterostructure injection lasers,'' *J. Appl. Phys.* 45:4899, 1974.
21. Chinone, N., H. Nakashima, I. Ikushima, and R. Ito, ''Semiconductor lasers with a thin active layer (<0.1 μm) for optical communication,'' *Appl. Opt.* 17:311, 1978.
22. Botez, D., and G. J. Herskowitz, ''Components for optical communication systems: A review,'' *Proc. IEEE* 68: 1980.
23. Ortel Corp., Alhambra, Calif. Product Data Sheets.
24. Hsieh, J. J., J. A. Rossi, and J. P. Donnelly, ''Room temperature CW operation of GaInAsP/InP double heterostructure diode lasers emitting at 1.1 μm,'' *Appl. Phys. Lett.* 28:709, 1976.
25. Yu, K. L., Koren, U., T. R. Chen, and A. Yariv, ''A Groove GaInAsP Laser on Semi-Insulating InP,'' *IEEE J. Quant. Elec.* QE-8:817, 1982.
26. Suematsu, Y., ''Long wavelength optical fiber communication,'' *Proc. IEEE.* 71:692, 1983.
27. Lau, K. T., N. Bar-Chaim, I. Ury, and A. Yariv, ''Direct amplitude modulation of semiconductor GaAs lasers up to X-band frequencies,'' *Appl. Phys. Lett.* 43:11, 1983.
28. Lau, K. Y., Ch. Harder, and A. Yariv, ''Direct modulation of semiconductor lasers at $f > 10$ GHz,'' *Appl. Phys. Lett.* 44:273–275, 1984.
29. See, for example, Bailbe, J. P., A. Marty, P. H. Hiep, and G. E. Rey, ''Design and fabrication of high speed GaAlAs/GaAs heterojunction transistors,'' *IEEE Trans. Elect. Dev.* ED-27:1160, 1980.
30. Yariv, A., ''Active integrated optics,'' in *Fundamental and Applied Laser Physics*, Proc. ESFAHAN Symposium, Aug. 29, 1971, M. S. Feld, A. Javan, N. Kurnit, eds. New York: Wiley, 1972.
31. Ury, I., K. Y. Lau, N. Bar-Chaim, and A. Yariv, ''Very high frequency GaAlAs laser-field effect transistor monolithic integrated circuit,'' *Appl. Phys. Lett.* 41:126, 1982.
32. Yust, M., et al., ''A monolithically integrated optical repeater,'' *Appl. Phys. Lett.* 10:795, 1979.
33. Bar-Chaim, N., I. Ury, and A. Yariv, ''Integrated optoelectronics,'' *IEEE Spectrum*, p. 38, May 1982.
34. Hernandez-Gil, F., T. L. Koch, U. Koren, R. P. Gnall, C. A. Burrus, ''Tunable MQW-DBR laser with a monolithically integrated InGaAsP/InP Directional coupler switch,'' paper PD 17, Conference on Laser Engineering and Optics (CLEO) 1989.
35. Koren, U., et al., ''Wavelength division multiplexing light source with integrated quantum well tunable lasers and optical amplifiers,'' *Appl. Phys. Lett.* 54:21 (May 1989).

36. Harder, C., K. Vahala, and A. Yariv, ''Measurement of the linewidth enhancement factor α of semiconductor lasers,'' *Appl. Phys. Lett.* 42:428 (1983).

37. Koch, T., and J. Bowers, ''Nature of wavelength chirping in directly modulated semiconductor lasers,'' *Electr. Lett.* 20:1038–1040, 1984.

38. Derry, P., et al., ''Ultra low threshold graded-index separate confinement single quantum well buried heterostructure (Al, Ga) as lasers with high reflectivity coatings,'' *Appl. Phys. Lett.* 50:1773, 1987.

39. A comprehensive book dealing with long wavelength lasers is G.P. Agrawal and N. K. Dutta, *Long Wavelength Semiconductor Lasers*, New York: Von Nostrand, 1986.

40. Hirao, M., S. Tsuji, K. Mizushi, A. Doi, and M. Nakamura, *J. Opt. Comm.* 1:10, 1980.

41. Ralston, J. D., S. Weisser, K. Eisele, R. E. Sah, E. C. Larkins, J. Rosenzweig, J. Fleissner, and K. Bender, ''Low-bias-current direct modulation up to 33 Ghz in InGaAs/AlGaAs pseudomorphic MQW Ridge-waveguide lasers,'' *IEEE Phot. Tech.* 6:1076, 1994.

42. Coldren, L. A., and S. W. Corzine, *Diode Lasers and Photonic Integrated Circuits,* New York: Wiley, 1995, p. 195.

43. Henry C. H., *Line Broadening of Semiconductor Lasers in Coherence Amplification and Quantum Effects,* in *Semiconductor Lasers* ed. Y. Yamamoto New York: Wiley-Interscience, 1991, Chap. 2.

44. Vahala, K., L. C. Chiu, S. Margalit, and A. Yariv, ''On the Linewidth Factor α in Semiconductor Injection Lasers,'' *Appl. Phys. Lett.* 42:631–633 (April 1983).

16 Advanced Semiconductor Lasers: Quantum Well Lasers Distributed Feedback Lasers Vertical Cavity Surface Emitting Lasers[1]

16.0 INTRODUCTION

During the last few years a new type of a semiconductor laser, the quantum well (QW) laser, has come to the fore [1]. It is similar in most respects to the conventional

[1] A basic, first-year knowledge of quantum mechanics is assumed in this chapter.

double heterostructure laser of the type shown in Figure 15-10 except for the thickness of the active layer. In the quantum well it is ~50–100 Å, while in conventional lasers it is ~1,000 Å. This feature leads to profound differences in performance. The main advantage to derive from the thinning of the active region is almost too obvious to state—a decrease in the threshold current that is nearly proportional to the thinning. This reduction can be appreciated directly from Figure 15-7. The carrier density in the active region needed to render the active region transparent is $\sim 10^{18}$ cm^{-3}. It follows that just to reach transparency we must maintain a total population of $N_{transp} \sim V_a \times 10^{18}$ electrons (holes) in the conduction (valence) band of the active region where $V_a(cm^3)$ is the volume of the active region. The injection current to sustain this population is approximately

$$I_{transp} \sim \frac{eV_a \times 10^{18}}{\tau} \tag{16.0-1}$$

and is proportional to the volume of the active region. A thinning of the active region thus reduces V_a and I_{transp} proportionately. In a properly designed laser, the sum of the free carrier, scattering, and mirror (output) coupling can be made small enough so that the increment of current, above the transparency value, needed to reach threshold is small in comparison to I_{transp}. The reduction of the transparency current that results from a small V_a thus leads to a small threshold current [4, 5, 6, 7, 8].

Quantum well active regions are also used as the amplifying medium in distributed feedback lasers and in vertical cavity surface emitting lasers. Both of these classes of important semiconductor lasers are discussed in separate sections of this chapter.

16.1 CARRIERS IN QUANTUM WELLS (Advanced Topic)

The essential difference between the gain of a pumped quantum well semiconductor medium and that of a bulk semiconductor laser has to do with the densities of states in both of these media. The density of states of a bulk semiconductor was derived in Section 15.1 and is given by Equation (15.1-7). It was used in (15.2-9) to derive an expression for the gain $\gamma(\omega_0)$ of that medium. In this section we will repeat this procedure using the density of states function of a quantum well. This is one of the few places in this book where we need to turn to quantum mechanics. The student without a quantum mechanical background but with a good electromagnetic preparation can simply think of the electron using the de Broglie picture as a wave obeying, not Maxwell's equations but the Schrödinger equation. The running wave solutions of this equation are of the form of modes $\psi_i(x, z) = u_i(\mathbf{r})\exp(-iE_it/\hbar)$ where $\hbar$ is Planck's constant divided by 2π, E_i is the quantized energy of an electron in the state i, while $u_i(\mathbf{r})$ is the eigen function.

We consider the electron in the conduction band of a QW to be free (with an effective mass m_c) to move in the x and y directions, but to be confined in the z (normal to the junction planes) as shown in Figure 16-1. The potential barrier ΔE_c was given in Section 15.3 for the GaAs/$Ga_{1-x}Al_xAs$ system as $\Delta E_c \sim 0.75 \times$ (eV).

For the sake of simplification, we shall take ΔE_c as infinite. (This is a close approximation for barriers $\gtrsim 100$ Å and $x \gtrsim 0.3$.) The wavefunction $u(\mathbf{r})$ of the electrons in this well obeys the time independent Schrödinger equation [2].

$$V(z)u(\mathbf{r}) - \frac{\hbar^2}{2m_c}\left(\frac{\partial^2}{\partial z^2} + \frac{\partial^2}{\partial x^2} + \frac{\partial^2}{\partial y^2}\right)u(\mathbf{r}) = Eu(\mathbf{r}) \tag{16.1-1}$$

E is the energy of the electron while $V(z) = E_c(z)$ is the potential energy function confining the electrons in the z direction. We will measure the energy relative to that of an electron at the bottom of the conduction band in the GaAs active region as shown in Figure 16-1. The eigenfunction $u(\mathbf{r})$ can be separated into a product

$$u(\mathbf{r}) = \psi_k(\mathbf{r}_\perp)u(z) \tag{16.1-2}$$

which, when substituted in (16.1-1), leads to

$$\left[V(z) - \frac{\hbar^2}{2m_c}\frac{\partial^2}{\partial z^2}\right]u(z) = E_z u(z) \tag{16.1-3}$$

where E_z is a separation constant to be determined. Since we agreed to take the height of $V(z)$ the well region as infinite, $u(z)$ must vanish at $z = \pm L_z/2$.

$$u_\ell(z) = \begin{cases} \cos \ell \dfrac{\pi}{L_z} z & \ell = 1, 3, 5, \ldots \\ \sin \ell \dfrac{\pi}{L_z} z & \ell = 2, 4, 6, \ldots \end{cases} \tag{16.1-4}$$

$$E_z = \ell^2 \frac{\hbar^2\pi^2}{2m_c L_z^2} = \ell^2 E_{1c} \equiv E_{\ell c} \qquad \ell = 1, 2, 3, \ldots \tag{16.1-5}$$

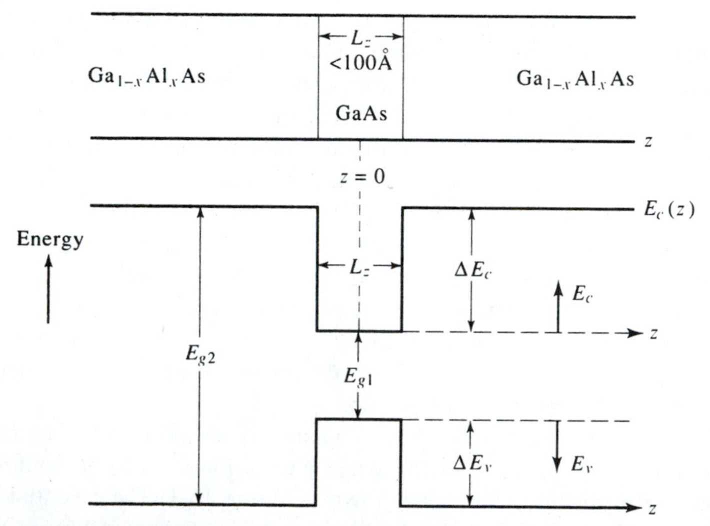

Figure 16-1 The layered structure and the band edges of a GaAlAs/GaAs/GaAlAs quantum well.

Using (16.1-3, 16.1-4, and 16.1-5) in (16.1-1) leads to

$$H(\mathbf{r}_\perp)\Psi(\mathbf{r}_\perp) = (E - E_z)\Psi(\mathbf{r}_\perp) \tag{16.1-6}$$

We can take $\psi(\mathbf{r}_\perp)$ as a two-dimensional Bloch wavefunction (see 15.1-1).

$$\psi(\mathbf{r}_\perp) = u_{\mathbf{k}_\perp}(\mathbf{r}_\perp)e^{(i\mathbf{k}_\perp \cdot \mathbf{r}_\perp)} \tag{16.1-7}$$

where $u_{\mathbf{k}_\perp}(\mathbf{r}_\perp)$ possesses the crystal periodicity. The wavefunction $\Psi(\mathbf{r}_\perp)$ obeys the Schrödinger equation

$$H(\mathbf{r}_\perp)\Psi(\mathbf{r}_\perp) = \frac{\hbar^2 k_\perp^2}{2m_c}\Psi(\mathbf{r}_\perp) \tag{16.1-8}$$

and from Equations [16.1-6, 7, 8]

$$E_c(\mathbf{k}_\perp, \ell) = \frac{\hbar^2 k_\perp^2}{2m_c} + \ell^2 \frac{\hbar^2 \pi^2}{2m_c L_z^2} = \frac{\hbar^2 k_\perp^2}{2m_c} + E_{\ell c} \qquad \ell = 1, 2, 3, \ldots \tag{16.1-9}$$

where the zero energy is taken as the bottom of the conduction band.

$u_{\mathbf{k}_\perp}(\mathbf{r}_\perp)$ possesses the lattice (two-dimensional) periodicity. Similar results with $m_c \rightarrow m_v$ apply to the holes in the valence band. We recall that the hole energy E_v is measured downward in our electronic energy diagrams so that

$$E_v(\mathbf{k}_\perp, l) = \frac{\hbar^2 k_\perp^2}{2m_v} + l^2 \frac{\hbar^2 \pi^2}{2m_v L_z^2} = \frac{\hbar^2 k_\perp^2}{2m_v} + E_{lv} \qquad l = 1, 2, 3, \ldots \tag{16.1-10}$$

measured (downward) from the top of the valence band. The complete wavefunctions are then

$$\psi_c(\mathbf{r}) = \sqrt{\frac{2}{L_z}}\,\Psi_{\mathbf{k}_{\perp c}}(\mathbf{r}_\perp) CS\left(\ell \frac{\pi}{L_z} z\right) \tag{16.1-11}$$

for electrons and

$$\psi_v(\mathbf{r}) = \sqrt{\frac{2}{L_z}}\,\Psi_{\mathbf{k}_{\perp v}}(\mathbf{r}_\perp) CS\left(\ell \frac{\pi}{L_z} z\right)$$

for holes. We defined $CS(x) \equiv \cos(x)$ or $\sin(x)$ in accordance with (16.1-4).

The lowest-lying electron and hole wavefunctions are

$$\psi_c(\mathbf{r})_{\text{ground state}} = \sqrt{\frac{2}{L_z}}\,\psi_{\mathbf{k}_{\perp c}}(\mathbf{r}_\perp)\cos\left(\frac{\pi}{L_z} z\right)$$

$$\psi_v(\mathbf{r})_{\text{ground state}} = \sqrt{\frac{2}{L_z}}\,\psi_{\mathbf{k}_{\perp v}}(\mathbf{r}_\perp)\cos\left(\frac{\pi}{L_z} z\right) \tag{16.1-12}$$

and are shown along with the next higher level in Figure 16-2. In a real semiconducting quantum well, the height ΔE_c of the confining well (see Figure 15-10) is finite, which causes the number of confined states, i.e., states with exponential decay in the z direction outside the well to be finite. The mathematical procedure for solving (16.1-3) is similar to that used in Section 13.1 to obtain the TE modes of a dielectric

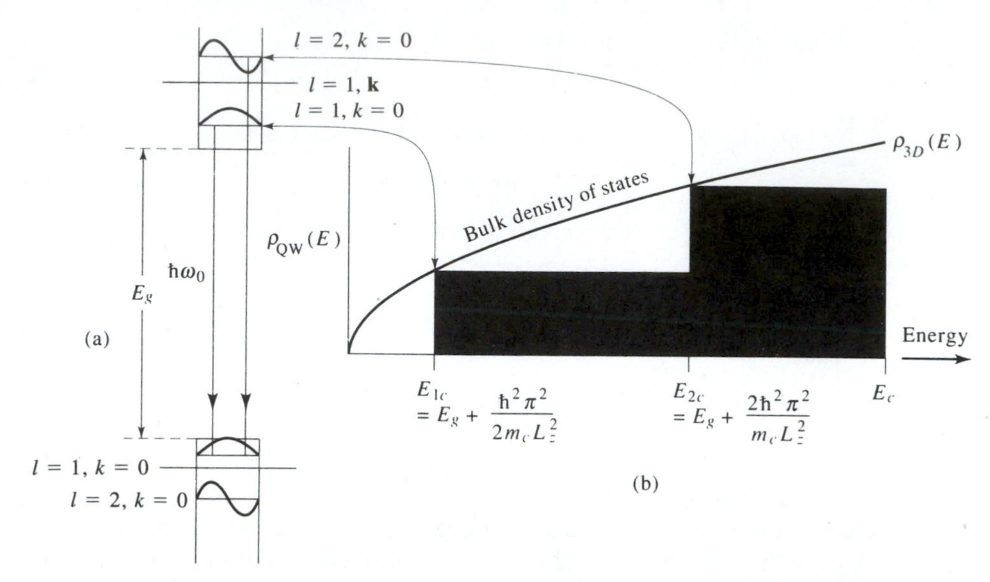

Figure 16-2 (a) The first two $l = 1$, $l = 2$ quantized electron and hole states and their eigenfunctions in an infinite potential well. (b) A plot of the volumetric density of states $(1/AL_z)/[dN(E)/dE]$ (i.e., the number of states per unit area (A) per unit energy divided by the thickness z of the active region) of electrons in a quantum well and of a bulk semiconductor. (Courtesy of M. Mittelstein, The California Institute of Technology, Pasadena, California)

waveguide that obeys a Schrödinger-like equation (13.3-1). As a matter of fact, to determine the number of confined eigenmodes as well as their eigenvalues we use a procedure identical to that of Figure 13-4.

The Density of States

The considerations applying here are similar to those of Section 15.1. Since the electron is "free" in the x and y directions, we apply two-dimensional quantization by assuming the electrons are confined to a rectangle L_xL_y. This leads, as in Equation (15.1-3), to a quantization of the components of the **k** vectors.

$$k_x = n\frac{\pi}{L_x} \qquad n = 1, 2, \ldots, \qquad k_y = m\frac{\pi}{L_y} \qquad m = 1, 2, \ldots$$

The area in $\mathbf{k}_\perp$ space per one eigenstate is thus $A_k = \pi^2/L_xL_y \equiv \pi^2/A_\perp$. We will drop the subnotation from now on so that $k \equiv k_\perp$. The number of states with transverse values of k' less than some given k is obtained by dividing the area $\pi k^2/4$ by A_k (the factor 1/4 is due to the fact that $\mathbf{k}_\perp$ and $-\mathbf{k}_\perp$ describe the same state). The result is

$$N(k) = \frac{k^2 A_\perp}{2\pi}$$

where a factor of two for the two spin orientations of each electron was included.

The number of states between k and $k + dk$ is

$$\rho(k)\, dk = \frac{dN(k)}{dk}\, dk = A_\perp \frac{k}{\pi}\, dk \qquad (16.1\text{-}13)$$

and is the same for the conduction or valence band. The total number of states with total energies between E and $E + dE$ is

$$\frac{dN(E)}{dE}\, dE = \frac{dN(k)}{dk}\frac{dk}{dE}\, dE \qquad (16.1\text{-}14)$$

The number of states per unit energy per unit area is thus

$$\frac{1}{A_\perp}\frac{dN(E)}{dE} = \frac{k}{\pi}\frac{dk}{dE}$$

from (16.1-9) with $\mathbf{k}_\perp \to \mathbf{k}$, $\ell = 1$, and limiting the discussion to the conduction band, the relation between the electron energy at the lowest state $\ell = 1$ and its $\mathbf{k}$ value is

$$k = \sqrt{\frac{2m_c}{\hbar^2}}\,(E - E_{1c})^{1/2}$$

so that the two-dimensional density of states (per unit energy and unit area) is

$$\rho_{\mathrm{QW}}(E) \equiv \frac{1}{A_\perp}\frac{dN(E)}{dE} = \frac{m_c}{\pi\hbar^2} \qquad (16.1\text{-}15)$$

Recall that this is the density of electron states. The actual density of electrons depends on the details of occupancy of these states as is discussed in the next section. An expression similar to (16.1-15) but with $m_c \to m_v$ applies to the valence band. In the reasoning leading to (16.1-15), we considered only one transverse $u(z)$ quantum state with a fixed ℓ quantum number (see Equation 16.1-15). But once $E > E_{2c}$, as an example, an electron of a given total energy E can be found in either $\ell = 1$ or $\ell = 2$ state so that the density of states at $E = E_{2c}$ doubles. At $E = E_{3c}$ it triples, and so on. This leads to a staircase density of states function. The total density of states thus increases by $m_c/\pi\hbar^2$ at each of the energies $E_{\ell c}$ of (16.1-5), which is expressed mathematically as

$$\rho_{\mathrm{QW}}(E) = \sum_{n=1}^{\text{all states}} \frac{m_c}{\pi\hbar^2} H(E - E_{nc}) \qquad (16.1\text{-}16)$$

where $H(x)$ is the Heaviside function that is equal to unity when $x > 0$ and is zero when $x < 0$.

The first two steps of the staircase density of states are shown in Figure 16-2. In the figure we plotted the volumetric density of states of the quantum well medium ρ_{QW}/L_z so that we can compare it to the bulk density of states in a conventional semiconductor medium. It is a straightforward exercise to show that in this case, the

QW volumetric density of states equals the bulk value $\rho_{3D}(E)$ at each of the steps, as shown in the figure.

The selection rules. Consider an amplifying electron transition from an occupied state in the conduction band to an unoccupied state in the valence band. The states $\ell = 1$ in the conduction band have the highest electron population. (The Fermi law Equation (15.1-8) shows how the electron occupation drops with energy.) The same argument shows that the highest population of holes is to be found in the $\ell = 1$ valence band state. It follows that, as far as populations are concerned, the highest optical gain will result from an $\ell = 1$ to $\ell = 1$ transition. The gain constant is also proportional to the (square of) the integral involving the initial and final states and the polarization direction x, y, or z of the optical field. Since the lowest lying electron and hole wavefunctions have, according to Equation (16.1-12), a z dependence that is proportional to $\cos(\pi z/L_z)$ and since

$$\int_{-L_z/2}^{L_z/2} z \cos^2 \frac{\pi z}{L_z}\, dz = 0$$

it follows that the optical field must be x or y polarized. The optical $\mathbf{E}$ vector thus must lie in the plane of the quantum well. A field polarized along the z direction does not stimulate any transitions between the two lowest lying levels and thus does not exercise gain (or loss).

16.2 GAIN IN QUANTUM WELL LASERS (3)

To obtain an expression for the gain of an optical wave confined (completely) within a quantum well medium, we follow a procedure identical to that employed in the case of a bulk semiconducting medium that was developed in Section 15.2. An amplifying transition at some frequency $\hbar\omega_0$ is shown in Figure 16-2. The upper electron state and the lower hole state (the unoccupied electron state in the valence band) have the same ℓ and $\mathbf{k}$ values (see discussion of selection rules in Section 16.1) so that the transition energy is

$$\begin{aligned}\hbar\omega = E_c - E_v &= E_g + E_c(\mathbf{k}, \ell) + E_v(\mathbf{k}, \ell)\\ &= E_g + \left(\frac{1}{m_c} + \frac{1}{m_v}\right)\frac{\hbar^2}{2}\left(k^2 + \ell^2\frac{\pi^2}{L_z^2}\right)\\ &= E_g + \frac{\hbar^2}{2m_r^*}\left(k^2 + \ell^2\frac{\pi^2}{L_z^2}\right)\end{aligned} \qquad (16.2\text{-}1)$$

$$m_r^* = \frac{m_c m_v}{m_c + m_v} \qquad (16.2\text{-}2)$$

$\ell = 1, 2, \ldots$ is the quantum number of the z dependent eigenfunction $u_\ell(z)$ as in Equation (16.1-4). We start again with (15.2-4) but this time in the correspondence of Equation (15.2-7) replace $\rho(k)/V$ by the equivalent quantum well $\rho_{QW}(\mathbf{k})/L_z$ volumetric carrier density

$$N_1(m^{-3}) \longrightarrow \frac{\rho_{QW}(\mathbf{k})}{L_z} dk f_v(E_v)[1 - f_c(E_c)] = \frac{k}{\pi L_z}(f_v - f_v f_c)dk$$

$$N_2(m^{-3}) \longrightarrow \frac{\rho_{QW}(\mathbf{k})}{L_z} dk f_c(E_c)[1 - f_v(E_v)] = \frac{k}{\pi L_z}(f_c - f_c f_v)dk$$

where $\rho_{QW}(\mathbf{k})$ is given by (16.1-15) and is independent of $\mathbf{k}$. The effective inversion population density due to carriers between k and $k + dk$ is thus

$$N_2 - N_1 \longrightarrow \frac{kdk}{\pi L_z}\left[f_c(E_c) - f_v(E_v)\right] \tag{16.2-3}$$

The division of ρ_{QW} by L_z is due to the need, in deriving the gain constant to use the *volumetric* density of inverted population consistent with the definition of N_1 and N_2 in (15.2-4). E_c and E_v are, respectively, the upper and lower energies of the carriers involved in a transition. We use (15.2-4) and (16.2-3) to write the contribution to the gain due to electrons within dk and in a single, say $\ell = 1$, sub-band as

$$d\gamma(\omega_0) = \frac{kdk}{\pi L_z}\left[f_c(E_c) - f_v(E_v)\right]\frac{\lambda_0^2}{4n^2\tau}\frac{T_2}{\pi[1 + (\omega - \omega_0)^2T_2^2]} \tag{16.2-4}$$

where T_2 is the coherence collision time of the electrons and τ is the electron-hole recombination lifetime assumed to be a constant. We find it more convenient to transform from the k variable to the transition frequency ω (see Equation 16.2-1). From (16.2-1) it follows that

$$dk = \frac{m_r^*}{\hbar k} d\omega$$

so that (16.2-4) becomes

$$\gamma_{n=1}(\omega_0) = \frac{m_r^*\lambda_0^2}{4\pi\hbar L_z n^2\tau}\int_0^\infty [f_c(\hbar\omega) - f_v(\hbar\omega)]\frac{T_2 d\omega}{\pi[1 + (\omega - \omega_0)^2T_2^2]} \tag{16.2-5}$$

where we used the convention that $f_c(\hbar\omega)$ is the Fermi function at the upper transition (electron) energy E_c, while $f_v(\hbar\omega)$ is the valence band Fermi function at the lower transition energy. To include, as we should, the contributions from all other sub-bands ($\ell = 2, 3, \ldots$) we replace, using (16.1-16)

$$\frac{m_r^*}{\pi\hbar^2} \longrightarrow \frac{m_r^*}{\pi\hbar^2}\sum_{\ell=1}^{\infty} H(\omega - \omega_\ell) \tag{16.2-6}$$

where $\hbar\omega_\ell$ is the energy difference between the bottom of the ℓ sub-band in the conduction band and the ℓ sub-band in the valence band.

$$\hbar\omega_\ell = E_g + \ell^2\frac{\hbar^2\pi^2}{2m_r^*L_z^2} \tag{16.2-7}$$

To get an analytic form for Equation (16.2-5) we will assume that the phase coherence "collision" time T_2 is long enough so that

$$\frac{T_2}{\pi[1 + (\omega - \omega_0)^2 T_2^2]} \longrightarrow \delta(\omega - \omega_0) \tag{16.2-8}$$

which simplifies (16.2-5) to

$$\gamma(\omega_0) = \frac{m_r \lambda_0^2}{4\pi\hbar L_z n^2 \tau} [f_c(\hbar\omega_0) - f_v(\hbar\omega_0)] \sum_{\ell=1}^{\infty} H(\hbar\omega_0 - \hbar\omega_\ell) \tag{16.2-9}$$

Equations 16.2-5 and 16.2-9 constitute our key result. They contain most of the basic physics of gain in quantum well media. Consider, first, the dependence of the gain on the Fermi functions f_c, f_v. An increase in the pumping current leads to an increase in the density of injected carriers in the active region and with it to an increase in the quasi-Fermi energies E_{F_c} and E_{F_v}. This leads to a larger region of ω_0 where the gain condition (Equation 15.2-14)

$$f_c(\hbar\omega_0) - f_v(\hbar\omega_0) > 0 \tag{16.2-10}$$

is satisfied. This situation is depicted in Figure 16-3. The solid curves (a), (b), and (c) show the modal gain of a typical GaAs quantum well laser at three successively increasing current densities. The modal gain is equal to the medium gain $\gamma(\omega_0)$ of Eq. (16.2-5) multiplied by the optical confinement factor $\Gamma_a \sim L_z/\alpha$. The dashed curve corresponds to the gain available at infinite current density ($f_v(\hbar\omega_0) = 0$, $f_c(\hbar\omega_0) = 1$) and thus, the gain in this case according to 16.2-9, is proportional to density-of-states function

$$\rho_{QW}(\hbar\omega_0) = \sum_{\ell=1}^{\infty} \frac{m_r^*}{\pi\hbar^2} H(\hbar\omega_0 - \hbar\omega_\ell) \tag{16.2-11}$$

The first frequency to experience transparency, then gain, as the current is increased, according to the idealized staircase density of states model, is ω_0 where

$$\hbar\omega_0 = E_g + E_{1c} + E_{1v} = E_g + \frac{\hbar^2\pi^2}{2m_r^* L_z^2} \tag{16.2-12}$$

$\hbar\omega_0$ is thus the energy difference between the $\ell = 1$ ($k = 0$) conduction band state and the $\ell = 1$ ($k = 0$) valence band state. The inversion factor $f_c(\hbar\omega_0) - f_v(\hbar\omega_0)$ is always larger at this frequency than at larger frequencies. As the current is increased, and with it the density of electrons (holes) in the conduction (valence) band, the quasi Fermi levels (E_{F_c}, E_{F_v}) move deeper into their respective bands. There now exists a range of frequencies between the value given by (16.2-12) and $\omega_0 = (E_g + E_{F_c} + E_{F_v})$ where the gain condition (16.2-10) is satisfied. At even higher pumping the contribution from the $\ell = 2$ sub-band [see Figure 16.3(b)] adds to that from $\ell = 1$ and the maximum available gain doubles to $2\gamma(\omega_0)$. Curve (d) in Figure 16-3 shows the gain of a conventional double heterostructure laser. We note that equal increments of current will yield larger increments of gain in the SQW case, that, at low currents, the SQW gain tends to saturate at a constant value γ_0, and that the width of the spectral region that experiences gain is much larger in SQW case compared to DH lasers.

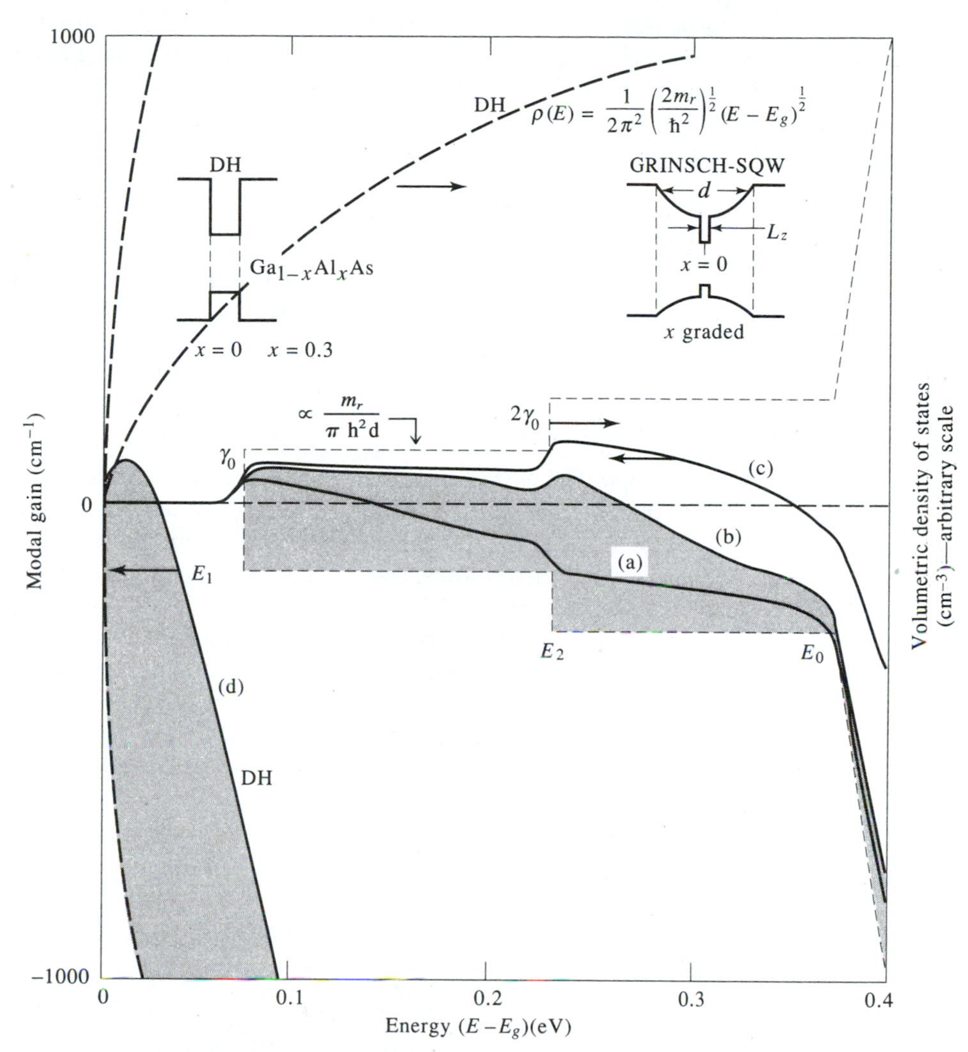

Figure 16-3 Gain (solid curves) and the joint density-of-states function (dashed lines) in a graded index, separate confinement heterostructure single quantum well laser (GRINSCH-SQW), and a conventional double heterostructure (DH) laser. The gain curves (a), (b), and (c) are for successively larger injection current densities, and curve (d) applies to the DH with the same current density as the QW laser curve (a). To meaningfully compare the density of states of a quantum well laser and the bulk (DH) laser we divided the former by the width $W = 4L_z$ of the optical confinement distance. This, in addition to rendering the dimensions identical, makes both curves proportional to the maximum (available) modal gain. (Courtesy of D. Mehuys, The California Institute of Technology)

The gain expression $\gamma(\omega_o)$ derived as (16.2-9) is that of the quantum well medium. It is the gain experienced by a wave that is completely confined to the quantum well. Since the quantum well thickness if typically $50\text{Å} < L_z < 100\text{Å}$, while the mode height is typically $d \sim 1000\text{Å}$, the actual gain experienced by the mode—modal gain—is given as in Equations (15.3-2, 3) by

$$g_{\text{modal}} = \gamma\Gamma_a \approx \gamma\frac{L_z}{d} \tag{16.2-13}$$

When we use the expression (16.2-9) for γ, we find that the modal gain at a given areal density of carriers is independent of the thickness L_z of the quantum well and is inversely dependent on the mode height d.

Multiquantum Well Laser

The small thickness of the quantum well relative to that of the mode height ($L_z/d \approx 2 \times 10^{-2}$ typically) makes it practical to employ more than one quantum well as the active region. To first approximation, the total electronic inversion is divided equally among the quantum wells, and the total modal gain is the sum of the individual modal gains of each well. The advantage of multiple-quantum well lasers is that, as shown in Figure 16-4, the gain from a single well tends to saturate with carrier density, hence with current, because of the flat-top nature of the density of states. The use of multiple-quantum wells enables each well to operate much within its linear gain-current region, thus extracting the maximum modal gain at a given total injected carrier density. This effect also results in a large differential gain $A \equiv$

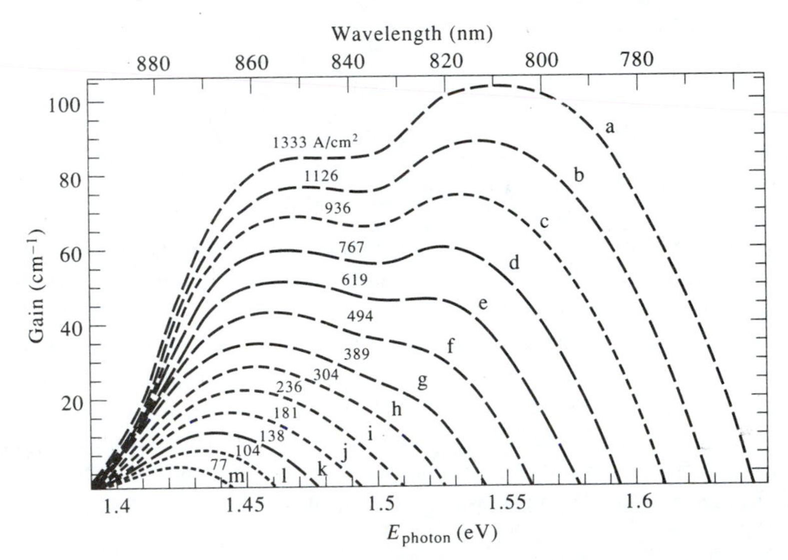

Figure 16-4 A theoretical plot of the exponential (modal) gain constant vs. wavelength of a quantum well laser. (Courtesy of Michael Mittelstein, The California Institute of Technology)

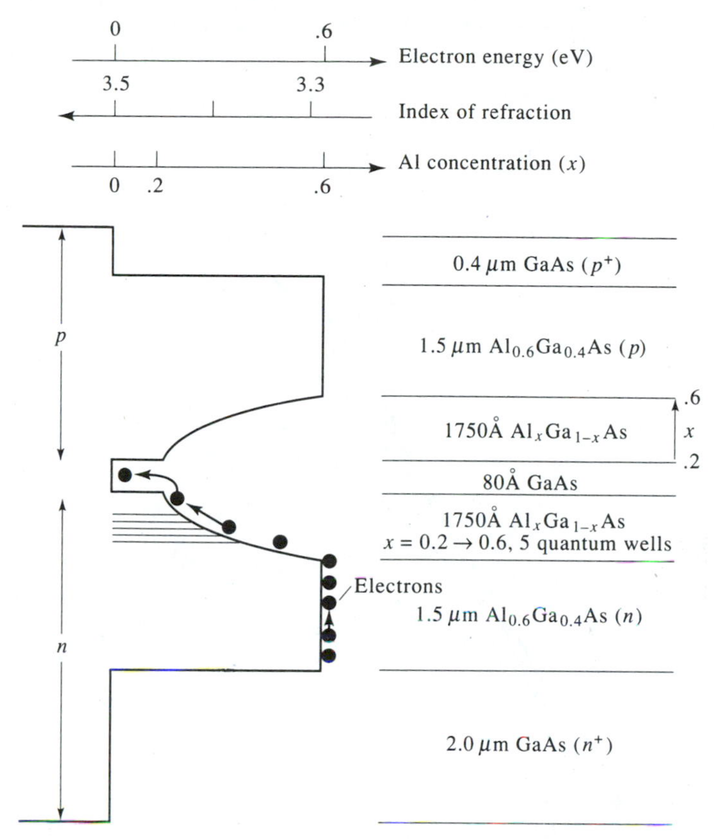

Figure 16-5 Schematic drawing of the conduction band edge and doping profile of a single quantum well, graded index separate confinement heterojunction lasers (GRINSCH). (Courtesy of H. Chen, The California Institute of Technology)

$\partial g/\partial N$, which, as shown in Section 15.5, leads to a higher laser modulation bandwidth. This behavior is illustrated by Figure 16-6. We see that the optimal number of wells in a given laser depends on the requisite modal gain which at oscillation, is equal to the laser losses. A laser with an effective loss constant of $d_{\text{eff}} = \alpha - 1/L$ ℓnR(α = loss constant, R is the mirrors' reflectivity) of, say, 10 cm^{-1}) will have, according to the figure, the lowest threshold current with one ($N = 1$) quantum well.

A theoretical plot of the exponential (modal) gain constant as a function of photon energy, or wavelength, is shown in Figure 16-4. The parameter is the injection current density. The interesting feature is the leveling off of the gain at the lower photon energies with increasing current and the appearance of a second peak at the higher current due to the population of the $\ell = 2$ well state. An experimental measurement of the gain vs. λ is discussed and illustrated in Section 16.3.

Figure 16-5 shows the layered structure of a single quantum well GaAs/GaAlAs laser. The 80 Å wide quantum well is bounded on each side with a graded

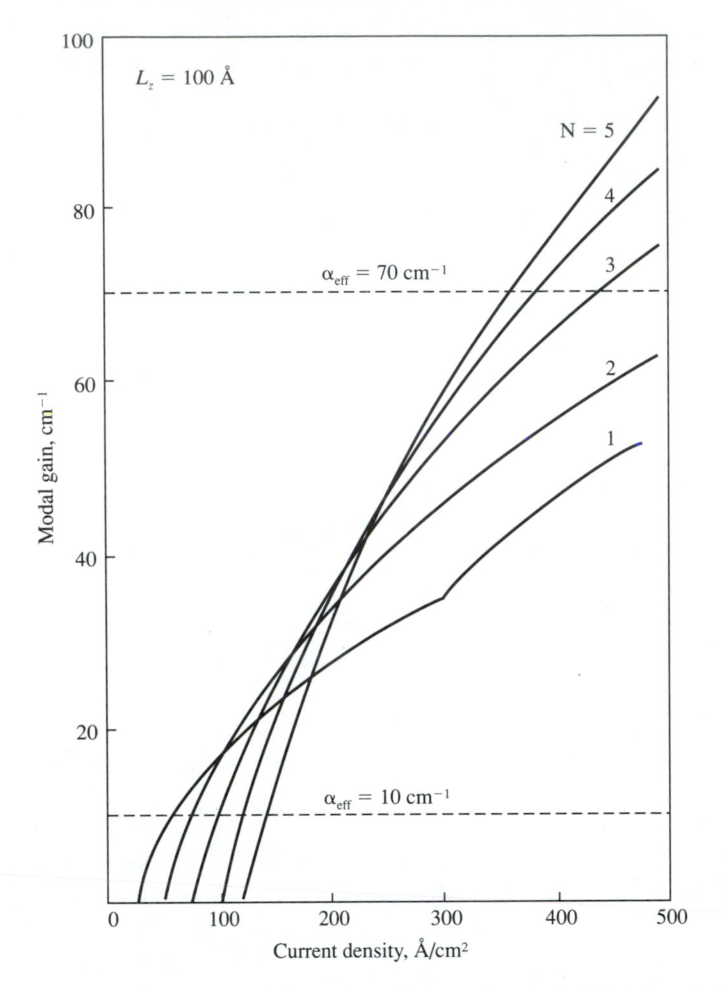

Figure 16-6 The modal gain $g_{mod}(= \Gamma g)$ as a function of the injected current with the number N of quantum wells as a parameter. (After Reference [3].)

index region. This graded index (and graded energy gap) region is grown by tapering the Al concentration from 0% to 60% in a gradual fashion as shown. The graded region functions as both a dielectric waveguide and as a funnel for the injected electrons and, not shown, the holes, herding them into the quantum well.

16.3 DISTRIBUTED FEEDBACK LASERS (9, 10, 11)

All laser oscillators employ optical feedback. By the word *feedback* we mean a means for ensuring that part of the optical field passing through a given point returns to the point repeatedly. If the delay is equal to an integral number of optical periods, this leads, in the presence of gain, to a sustained self-consistent oscillating mode where the field stimulated by atoms at any moment adds up coherently and *in phase* to those emitted earlier. In the laser resonators studied so far in this book, the feedback was provided by two oppositely facing reflectors. Feedback can also be achieved in a traveling wave folded-path geometry.

In distributed feedback (DFB) lasers, the reflection feedback of forward into backward waves, and vice versa, takes place not at the end reflectors but continuously throughout the length of the resonator. This coupling is due to a spatially periodic modulation of the index of refraction of the medium or of its optical gain. These lasers enjoy a wavelength stability that is far superior to those of ordinary Fabry–Perot lasers. This stability is due to the fact that the laser mode prefers to oscillate at a frequency such that the spatial period Λ of the index perturbation is equal to some (usually small) integer (ℓ) number of guide half wavelengths:

$$\Lambda = \ell \frac{\lambda_g}{2} \left(\lambda_g = \frac{2\pi}{\beta} \right) \qquad \ell = 1, 2, 3, \ldots$$

where β is the propagation constant of the optical field in the waveguide. This condition, which ensures that reflections from different unit cells of the periodic perturbation add up in phase, is referred to as the *Bragg condition*. This is in analogy with the, formally similar, phenomenon of x-ray diffraction from the periodic lattice of crystals. This enables the laser designer, through a choice of Λ, to "force" the laser to oscillate at any predetermined wavelength, provided that the amplifying medium is capable of providing sufficient gain at that wavelength. This property is especially important in semiconductor lasers used in optical fiber communication. Such lasers are often required to operate within narrow, prescribed wavelength regions to minimize pulse spreading by chromatic (group velocity) dispersion or to avoid crosstalk from other laser beams at different wavelengths sharing the same fiber.

We will start our treatment of the distributed feedback (DFB) laser with a derivation of the relevant coupled mode equations. The essence of the DFB laser is a spatially periodic waveguide *with gain*. It is thus described by the coupled-mode equations (13.5-1) with the addition of gain terms

$$\frac{dA}{dz} = \kappa_{ab} e^{-2i(\Delta\beta)z} - \gamma A$$

$$\frac{dB}{dz} = \kappa_{ab}^* e^{i2(\Delta\beta)z} + \gamma B \qquad \textbf{(16.3-1)}$$

$$\Delta\beta \equiv \beta - \beta_0 = \beta - \frac{\pi}{\Lambda} \qquad \text{for } \ell = 1 \qquad \textbf{(16.3-2)}$$

We shall choose, without loss of generality, a real κ so that $\kappa_{ab} = \kappa_{ba}^* \equiv \kappa$. The gain terms, $-\gamma A$ and γB, are chosen such that if, hypothetically, we eliminate the periodic perturbation ($\kappa = 0$), the waves A and B are uncoupled and grow exponentially, each along its direction of propagation as exp($\gamma \times$ distance). We could, of course, have derived Equations (16.3-1) by including *ab initio* gain in the derivation leading to Equations (13.5-1).[2] Equations (16.3-2) can be simplified by defining new complex amplitudes $A'(z)$ and $B'(z)$

$$A(z) = A'(z)\exp(-\gamma z)$$
$$B(z) = B'(z)\exp(\gamma z) \tag{16.3-4}$$

The result is

$$\frac{dA'}{dz} = \kappa_{ab}B'e^{-i2(\Delta\beta+i\gamma)z} \tag{16.3-5}$$

$$\frac{dB'}{dz} = \kappa_{ab}^{*}A'e^{i2(\Delta\beta+i\gamma)z} \tag{16.3-6}$$

These equations are identical in form to those of (13.5-1), provided we replace in the latter, $\Delta\beta \rightarrow \Delta\beta + i\gamma$. We can thus use the solution (13.5-2) of Equations (13.5-1) to write down directly the solutions for the complex amplitudes $E_i(z)$ and $E_r(z)$ of the incident and reflected waves, respectively, *inside* the *amplifying* periodic waveguide. We take the boundary conditions to be those of a *single* right-traveling wave incident from the left with an amplitude $B(0)$. The solution of Equations (16.3-5, 6) in this case is

$$E_i(z) = B'(z)e^{[(-i\beta+\gamma)z]}$$
$$= B(0)\frac{e^{-i\beta_0 z}\{(\gamma - i\Delta\beta)\sinh[S(L-z)] - S\cosh[S(L-z)]\}}{(\gamma - i\Delta\beta)\sinh(SL) - S\cosh(SL)} \tag{16.3-7}$$

$$E_r(z) = A'(z)e^{[(i\beta-\gamma)z]} = B(0)\frac{\kappa_{ab}e^{i\beta_0 z}\sinh[S(L-z)]}{(\gamma - i\Delta\beta)\sinh(SL) - S\cosh(SL)} \tag{16.3-8}$$

$$S^2 = |\kappa|^2 + (\gamma - i\Delta\beta)^2 \tag{16.3-9}$$

The fact that S now is complex makes for a major qualitative difference between the behavior of the passive periodic guide (13.5-2) and the periodic guide with gain. To demonstrate this difference, consider the case when the condition

$$(\gamma - i\Delta\beta)\sinh SL = S\cosh SL \tag{16.3-10}$$

is satisfied. It follows from (16.3-7, 8) that both the reflectance, $E_r(0)/E_i(0)$, and the transmittance, $E_i(L)/E_i(0)$, become infinite. *The device acts as an oscillator*, since it yields finite output fields $E_r(0)$ and $E_i(L)$ with no input [$E_i(0) = 0$]. Condition (16.3-10) is thus the oscillation condition for a distributed feedback laser. For the case of $\gamma = 0$, it follows, from (13.5-2), that $|E_i(L)/E_i(0)| < 1$, and $|E_r(0)/E_i(0)| < 1$ as appropriate to a passive device with no internal gain.

For frequencies very near the Bragg frequency $\omega_0(\Delta\beta \cong 0)$ and for sufficiently

[2]This can be done formally by replacing the real dielectric constant $\epsilon(\mathbf{r})$ in (13.3-3) by a complex $\epsilon_c(\mathbf{r})$: $\epsilon_c(\mathbf{r}) = \epsilon_r(\mathbf{r}) + i\epsilon_i(\mathbf{r})$. For the case of a uniform $\epsilon_i(\mathbf{r}) = \epsilon_i$, we obtain $\gamma = \omega\sqrt{\mu/\epsilon_r}\epsilon_i$. Otherwise γ involves a spatial integral (see Problem 16.6).

high-gain constant γ so that (16.3-10) is nearly satisfied, the guide acts as a high-gain amplifier. The amplified output is available either in reflection with a field gain

$$\frac{E_r(0)}{E_i(0)} = \frac{\kappa_{ab} \sinh SL}{(\gamma - i\Delta\beta)\sinh SL - S\cosh SL} \tag{16.3-11}$$

or in transmission with a gain

$$\frac{E_i(L)}{E_i(0)} = \frac{-Se^{i\beta_0 L}}{(\gamma - i\Delta\beta)\sinh SL - S\cosh SL} \tag{16.3-12}$$

The behavior of the incident and reflected field for a high-gain case is sketched in Figure 16-7. Note the qualitative difference between this case and the passive (no gain) one depicted in Figure 13-7.

The reflection power gain, $|E_r(0)/E_i(0)|^2$, and the transmission power gain $|E_i(L)/E_i(0)|^2$, are plotted in Figure 16.8 as a function of $\Delta\beta$ and γ. Each plot contains four infinite gain singularities at which the oscillation condition (16.3-10) is satisfied. These are four longitudinal laser modes. Higher orders exist but are not shown.

Oscillation Condition

The oscillation condition (16.3-10) can be written as

$$\frac{S - (\gamma - i\Delta\beta)}{S + (\gamma - i\Delta\beta)} e^{2SL} = -1 \tag{16.3-13}$$

In general, one has to resort to a numerical solution to obtain the threshold values of $\Delta\beta$ and γ for oscillation [17]. In some limiting cases, however, we can

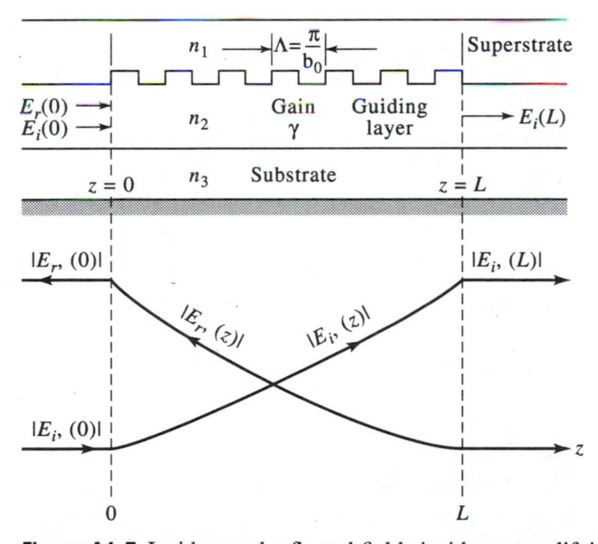

Figure 16-7 Incident and reflected fields inside an amplifying periodic waveguide.

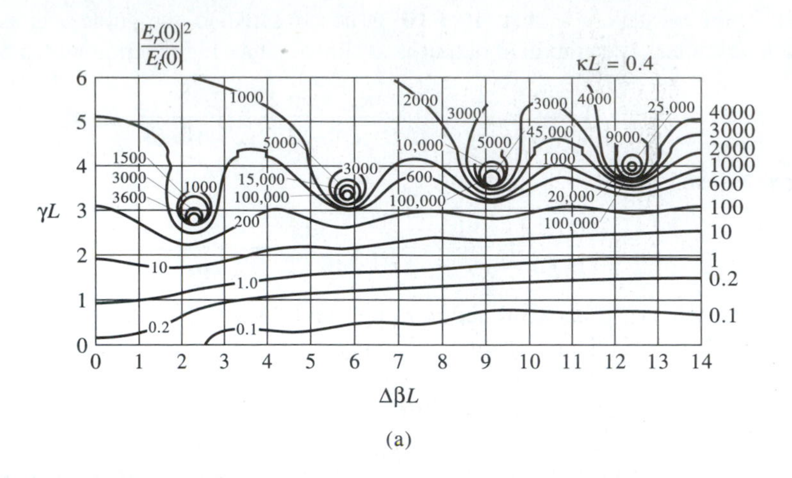

(a)

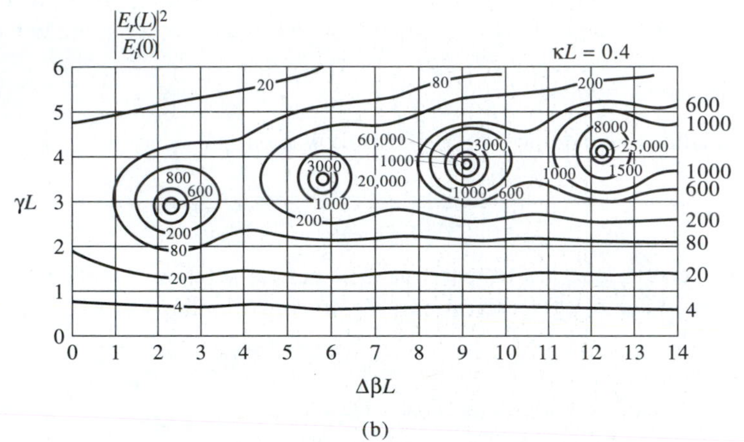

(b)

Figure 16-8 (a) Reflection gain contours in the $\Delta\beta L - \gamma L$ plane. $\Delta\beta$ is defined by (16.3-2) and is proportional to the deviation of the frequency $\omega_0 \equiv \pi c/\Lambda n$. The plots are symmetric about $\Delta\beta$, so that only one-half ($\Delta\beta > 0$) of the plots is shown. (b) Transmission gain. [Courtesy of H. W. Yen.]

obtain approximate solutions. In the high-gain $\gamma \gg \kappa$ case, we have from the definition of $S^2 = \kappa^2 + (\gamma - i\Delta\beta)^2$

$$S \approx -(\gamma - i\Delta\beta)\left(1 + \frac{\kappa^2}{2(\gamma - i\Delta\beta)^2}\right) \qquad \gamma \gg \kappa$$

so that

$$S - (\gamma - i\Delta\beta) \cong -2(\gamma - i\Delta\beta)$$

$$S + (\gamma - i\Delta\beta) \cong \frac{-\kappa^2}{2(\gamma - i\Delta\beta)}$$

and (16.3-13) becomes

$$\frac{4(\gamma - i\Delta\beta)^2}{\kappa^2} e^{2SL} = -1 \tag{16.3-14}$$

Equating the phases on both sides of (16.3-14) results in

$$2 \tan^{-1} \frac{(\Delta\beta)_m}{\gamma_m} - 2(\Delta\beta)_m L + \frac{(\Delta\beta)_m L\kappa^2}{\gamma_m^2 + (\Delta\beta)_m^2} = (2m + 1)\pi \tag{16.3-15}$$

$$m = 0, \pm 1, \pm 2, \ldots$$

In the limit $\gamma_m \gg (\Delta\beta)_m$, κ, the oscillating mode frequencies are given by

$$(\Delta\beta_m)L \cong -\left(m + \frac{1}{2}\right)\pi \tag{16.3-16}$$

and since $\Delta\beta \equiv \beta - \beta_0 \approx (\omega - \omega_0)n_{\text{eff}}/c$

$$\omega_m = \omega_0 - \left(m + \frac{1}{2}\right)\frac{\pi c}{n_{\text{eff}}L} \tag{16.3-17}$$

We note that no oscillation can take place exactly at the Bragg frequency ω_0. The mode frequency spacing is

$$\omega_{m-1} - \omega_m \simeq \frac{\pi c}{n_{\text{eff}}L} \tag{16.3-18}$$

which is approximately the same as in a two-reflector resonator of length L. (See 4.1-10.)

The threshold gain value γ_m is obtained from equating the amplitudes in (16.3-14):

$$\frac{e^{2\gamma_m L}}{\gamma_m^2 + (\Delta\beta)_m^2} = \frac{4}{\kappa^2} \tag{16.3-19}$$

indicating an increase in threshold with increasing mode number m. This is also evident from the numerical gain plots (Figures 16-7 and 16-8). An important feature that follows from (16.3-19) is that the threshold gain for modes with the same $|\omega - \omega_0|$, or equivalently the same $|\Delta\beta|$, is the same. Thus two modes will exist with the lowest threshold, one on each side of ω_0. This property of DFB lasers is usually undesirable, and methods for obtaining single-mode operation are discussed in the last part of this section.[3] The periodic perturbation in semiconductor DFB lasers is achieved by incorporating a grating, usually in the form of a rippled interface, in the laser structure. This is achieved by interrupting the crystal growth at the appropriate stage and wet-chemical etching a corrugation into the topmost layer by

[3] High-speed (data rate) optical communication in fibers requires that the optical source put out a single frequency in order to minimize the temporal spread of the optical pulses with distance, which is caused by group velocity dispersion.

using an interferometrically produced photoresist mask [10]. Growth of a layer with a different index of refraction, or optical absorption on top of the rippled surface, results in the desired spatial modulation.

A diagram of a distributed feedback laser using a GaAs-GaAlAs structure is shown in Figure 16-9. The waveguiding layer, as well as that providing the gain (active layer), is that of *p*-GaAs. The feedback is provided by corrugating the interface between the *p*-$Ga_{.93}Al_{.07}As$ and *p*-$Ga_{.7}Al_{.3}As$, where the main index discontinuity responsible for the guiding occurs. Figure 16-12 shows an example of a periodic gain grating. The laser in this example is based on the quaternary $GA_{1-x}In_xAs_{1-y}P_y$ as the active region and InP as the high-energy gap, low index-cladding layer. The feedback is achieved by growing an extra-absorbing, i.e., low-energy gap, layer and then etching through a mask to leave behind a periodic array of absorbing islands.

The increase in threshold gain with the longitudinal mode index m predicted by (13.6-19) and by the plots of Figures 16-7 and 16-8 manifests itself in a high degree of mode discrimination in the distributed feedback laser.

It follows from (16.3-17) and (16.3-19) that the two lowest threshold modes are those with $m = 0$ and $m = -1$ and that they are situated symmetrically on either side of the Bragg frequency ω_0 just outside the bandgap.

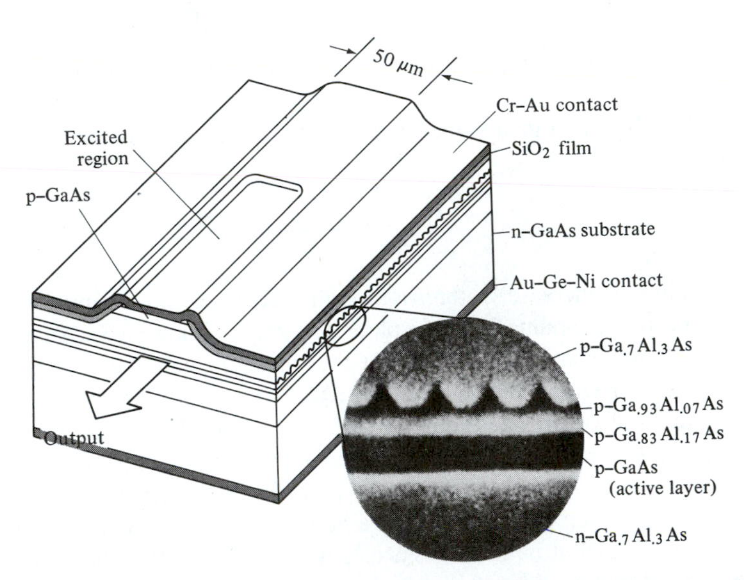

Figure 16-9 A GaAs–GaAlAs *cw* injection laser with a corrugated inferface. The insert shows a scanning electron microscope photograph of the layered structure. The feedback is in third order ($\ell = 3$) and is provided by a corrugation with a period $\Lambda = 3\lambda_g/2 \doteq 0.345$ μm. The thin (0.2 μm) *p*-$Ga_{.83}A_{.17}As$ layer provides a potential barrier which confines the injected electrons to the active (*p*-GaAs) layer, thus increasing the gain. (After Reference [11].)

To understand why the basic DFB laser of Figure 16-6, in which the index of refraction is spatially periodic, does not oscillate at the Bragg frequency, consider Figure 16-10(a). Let the reflection coefficient of a wave (at ω) incident from the left on the plane $z = 0$ be r_2, and for a wave incident from the right, r_1. The reflectivity r_2 is given according to (16.3-11) by

$$r_1 = \frac{-\kappa \sinh SL_1}{(\gamma - i\Delta\beta) \sinh SL_1 - S \cosh SL_1} \tag{16.3-20}$$

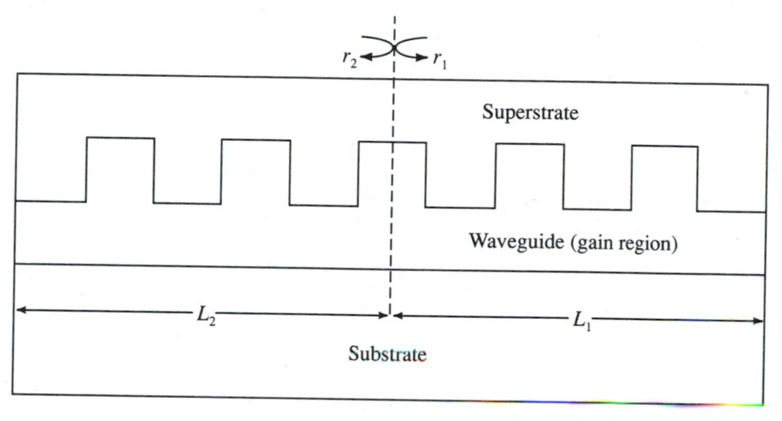

(a)

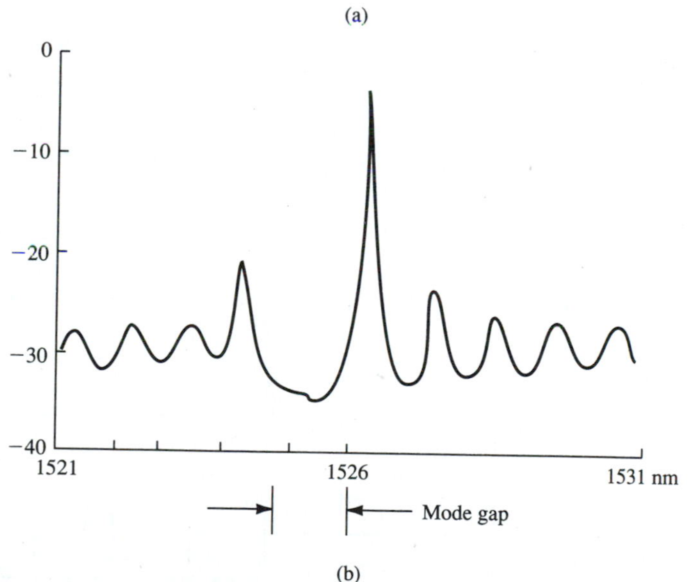

(b)

Figure 16-10 A periodic waveguide model used to derive Equation (16.3-12). (a) A periodic (DFB) GaInAsP waveguide laser. (b) The spontaneous emission spectrum below, but near, threshold of a DFB laser showing the mode gap. (c) A DFB laser with a phase shift section. (d) a "quarter wavelength shifted" DFB laser. (e) The spontaneous emission spectrum below threshold of a $\lambda/4$-shifted DFB laser. (Courtesy of P. C. Chen, ORTEL Corporation).

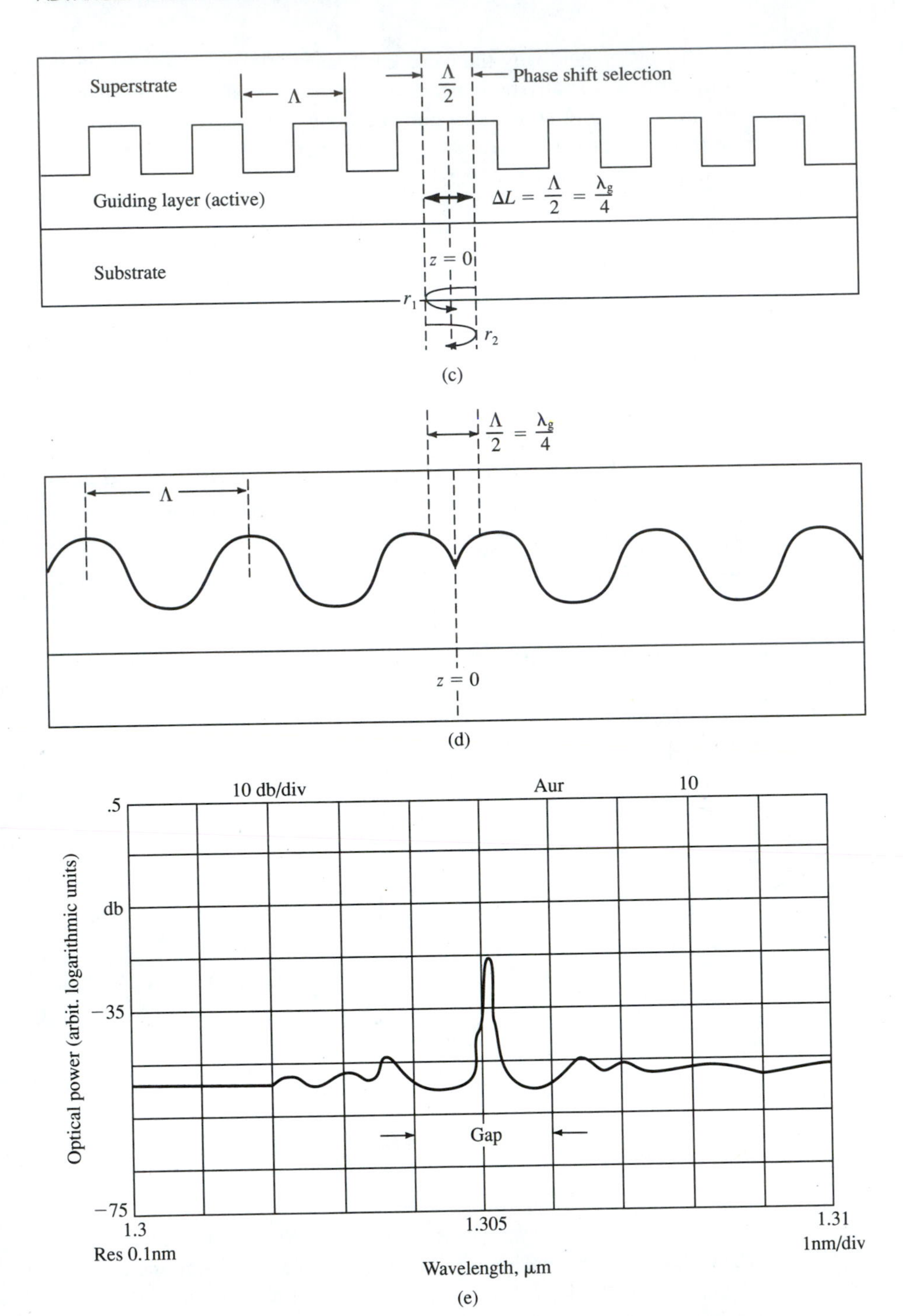

Figure 16-10 *(Continued)*

where

$$S = \sqrt{\kappa^2 + (\gamma - i\Delta\beta)^2}$$

$$\Delta\beta = \frac{\omega}{c} n_{\text{mode}} - \frac{\pi}{\Lambda} = (\omega - \omega_0) \frac{n_{\text{mode}}}{c} \tag{16.3-21}$$

and we used the fact that for an index perturbation of odd symmetry (in z), $\Delta n^2(z) \propto \sin(\eta z)$, the coupling coefficient κ given by (13.4-17) is a *real number*,[4] i.e., $\kappa_{ab} = \kappa_{ba} = \kappa$. The reflectivity r_1, "looking" to the left, is

$$r_1 = \frac{-\kappa \sinh SL_1}{(\gamma - i\Delta\beta) \sinh SL_1 - S \cosh SL_1}$$

The reason for the difference in sign between r_2 and r_1 is due to the fact that we chose, in (13.4-10), the index perturbation to have odd symmetry. An observer "looking" to the right sees $\Delta n(z) \propto \sin \eta z$, while an observer "looking" to the left will see a perturbation in index $\Delta n(z) \propto -\sin \eta z$. It follows that on resonance ($\Delta\beta = 0$), $r_1 r_2$ = negative number. The oscillation condition for a laser, on the other hand, is[5]

$$r_1(\omega) r_2(\omega) = 1 \tag{16.3-22}$$

It follows that the periodic index DFB laser cannot oscillate at the Bragg frequency ω_0 where $\Delta\beta = 0$. Oscillation thus takes place at the symmetrically situated frequencies shown in Figure 16-8. The two oscillation frequencies nearest the Bragg frequency require the lowest gain and are given, according to (16.3-17), by

$$\omega_0 \pm \frac{\pi c}{2 n_{\text{eff}} L} \tag{16.3-23}$$

The threshold gain for oscillation at these two frequencies, is, according to (16.3-24) and Figure 16-8, equal, so that they are in practice equally likely to oscillate. This situation is highly undesirable in practice, since it results in wavelength instabilities and spectral broadening. This is unacceptable, for example, in long-haul, high-data-rate fiber links where the increased spectral width due to multiwavelength oscillation was shown in Chapter 3 to limit the data rate due to pulse broadening by group velocity dispersion.

The existence of two such oscillating wavelengths is shown in the spectrum of Figure 16-10(b) as the two peaks on either side of the "gap."

A widely employed method [12] for forcing the DFB laser to oscillate preferentially at a single midgap frequency is shown in Figure 16-10(c, d). An extra section

---

[4]Had we chosen a reference plane other than $z = 0$, κ_{ab} will not be real, but since $\kappa_{ab} = \kappa_{ba}^*$, all the results remain the same.

[5]This is just a sophisticated way of saying that at steady state, the oscillation condition is equivalent to demanding that a wave launched, say, to the right, returns after one round trip with the same amplitude and the same phase (modulo $m2\pi$).

of length $\lambda_g/4$ is inserted at the center of the laser (λ_g is the "guide" wavelength). The reflectivities r_1 and r_2 "looking" to the left and right, respectively, from the midplane are now given by their previous values [i.e., those corresponding to Figure 16-10(a)], each multiplied by $\exp[-i(\pi/2)]$ to account for the added propagation delay in the $\lambda_g/4$ section. At $\omega = \omega_0(\Delta\beta = 0)$, r_1 becomes $r_1 \exp[-i(\pi/2)]$, and r_2 becomes $r_2 \exp[-i(\pi/2)]$. The product of the reflectivities $r_1' r_2'$ is now $-r_1 r_2$, which is a *positive* number, so that oscillation at ω_0 is possible. This is illustrated in Figure 16-10(e).

Gain-Coupled Distributed Feedback Lasers (13)

Another type of distributed feedback laser is one where the periodic modulation is not of the index of refraction but of the gain or losses of the medium. To analyze this situation, we remind ourselves that gain or losses can be represented by taking the dielectric constant ϵ of a medium as complex. It is a straightforward matter to show (see Problem 16.6) that ϵ can be expressed as

$$\epsilon = \epsilon_0 n^2 \left(1 + i \frac{2\gamma}{k_0 n}\right)$$

$$k_o = \frac{2\pi}{\lambda} = \omega\sqrt{\mu\epsilon_0} \qquad (16.3\text{-}24)$$

and $\gamma(\ll k_0)$ is the exponential gain constant of the field amplitude. In a lossy medium, $\gamma < 0$.

In the case where n and γ are periodic, we can write

$$n(z) = n_0 + n_1 \cos \frac{2\pi}{\Lambda} z$$

$$\gamma(z) = \gamma_0 + \gamma_1 \cos \frac{2\pi}{\Lambda} z \qquad (16.3\text{-}25)$$

Limiting ourselves to the case $n_1 \ll n_0$, $\gamma_1 \ll \gamma_0$, we can write

$$\omega^2 \mu\epsilon(z)\, E = [k_0^2 n_0^2 + i2k_0 n_0 \gamma_0 + 4k_0 n_0 \left(\frac{\pi n_1}{\lambda} + i\frac{\gamma_1}{2}\right) \cos \frac{2\pi}{\Lambda} z]\, E \qquad (16.3\text{-}26)$$

This last result shows that we can use the coupled mode equations (16.3-5, 6) in the general case of both index and gain modulation, provided we generalize the definition of the coupling constant to

$$\kappa = \frac{\pi n_1}{\lambda} + i \frac{\gamma_1}{2} \qquad (16.3\text{-}27)$$

Not surprising: The coupling constant due to gain modulation differs by a factor $\exp(i\,\pi/2)$ from that due to index modulation. We can use, for example, the expression for the reflectivity r_2 in (16.3-20) but replace κ by $i\kappa$. This renders the product

$r_1 r_2$ at ω_0 a *positive* number, so that laser oscillation can now take place at the exact Bragg frequency (ω_0). If we plot $|r^2|$ vs. $\Delta\beta$ for this case, we obtain the result shown in Figure 16-11(a). For comparison, we show in Figure 16-11(b) a plot of the reflectivity in the case of index modulation, which shows two modes situated symmetrically about ω_0. The experimental oscillation spectrum of a gain-coupled laser

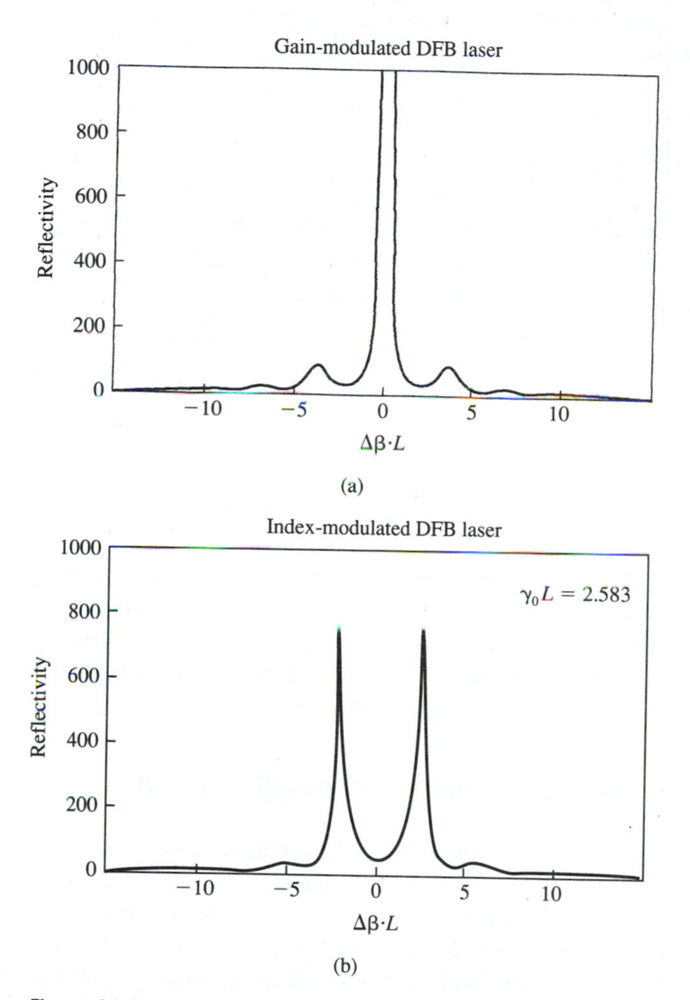

Figure 16-11 (a) A theoretical plot of the reflectivity $|E_r(0)/E_i(0)|^2$ of a waveguide with a gain $\gamma = \gamma_0 + \gamma_1 \sin(2\pi/\Lambda)z$. (b) A similar plot for an index modulated waveguide, $\gamma = \gamma_0$, $n = n_0 + n_1 \sin(2\pi/\Lambda)z$. (c) The measured oscillation spectrum of a GaInAsP distributed feedback laser with gain coupling. $\lambda = 1.5427$ μm. A single oscillating mode is present at the Bragg wavelength. Higher-order modes have output powers that are down by a factor of >45 db (i.e., >32,000), compared to that of the fundamental mode. (a) and (b) Courtesy of M. McAdam—Caltech. (c) Courtesy of Dr. P. C. Chen, ORTEL Corporation.

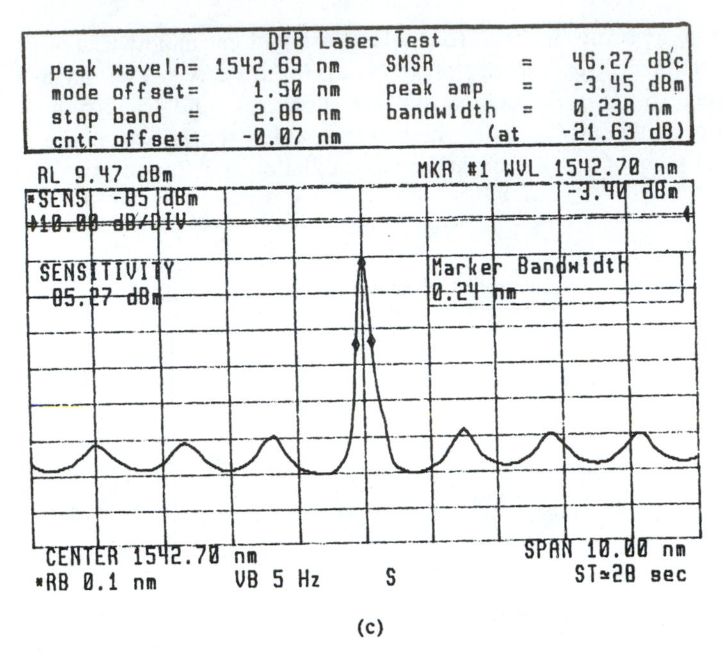

(c)

Figure 16-11 *(Continued)*

is shown in Figure 16-11(c). It demonstrates the strong suppression of higher-order modes.

A cross section of a commercial gain-coupled DFB laser is shown in Figure 16-12. The periodic modulation of the gain is achieved by photolithographic corrugation [13] of an absorbing layer near the active region. The layer is incorporated for this purpose in the epitaxially grown laser structure. Additional layers grown epitaxially result in "burying" the periodic loss layer.

16.4 VERTICAL CAVITY SURFACE EMITTING SEMICONDUCTOR LASERS (14, 15, 16)

Vertical cavity surface emitting semiconductor lasers (VCSELs) differ from their more conventional relatives in that the optical beam travels at right angles to the active region instead of in the plane of the active regions. A typical VCSEL structure is illustrated in Figure 16-13. The top and bottom reflectors consist, each, of alternating layers of semiconductor GaInAlAs with different *x* and *y* compositions. The difference in the index of refraction between adjacent layers gives rise to a high reflection (>99 percent) at the vicinity of the Bragg wavelength from each such "stack." The mirror layers are grown epitaxially along with the rest of the laser layers. The laser biasing current flows through the mirrors so that they are highly doped to reduce the series resistance. The gain is provided by a small number, typically 1 to 4, of quantum wells that are placed near a maximum of the standing wave pattern to maximize the stimulated emission rate into the oscillation field. The

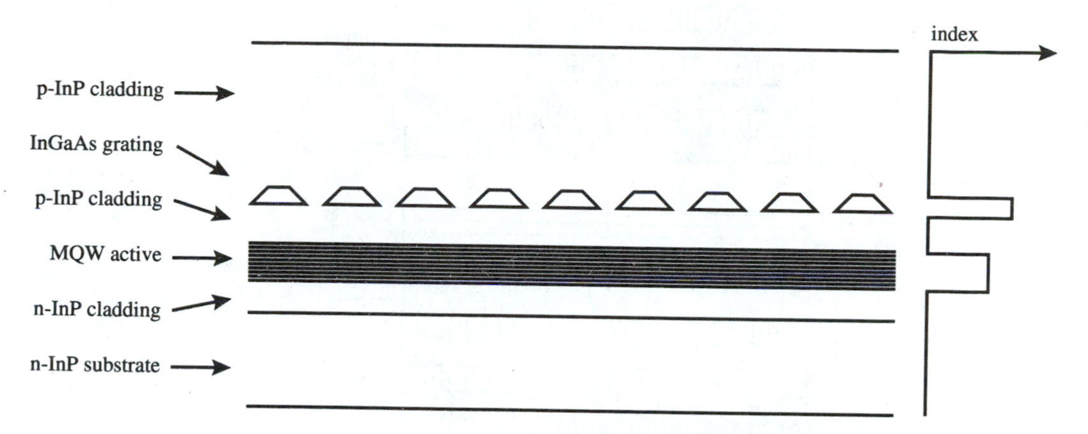

Figure 16-12 A periodic lossy layer, i.e., a layer with an energy gap smaller than $\hbar\omega_{\text{oscill}}$, provides the periodic gain coupling in a semiconductor DFB laser. (Courtesy of Dr. P. C. Chen, ORTEL Corporation.)

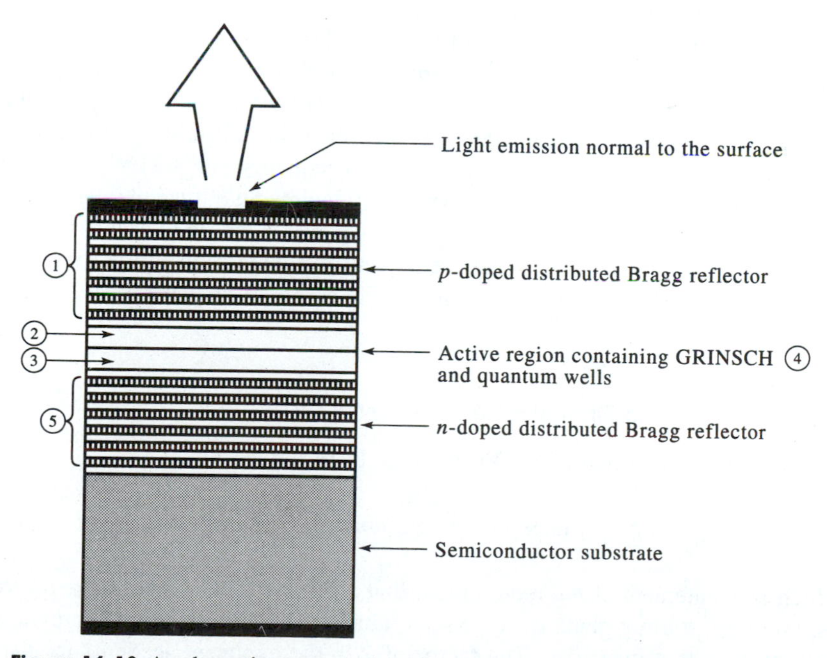

Figure 16-13 A schematic cross section of a vertical cavity surface emitting semiconductor laser based on the GaInAlAs alloy system.

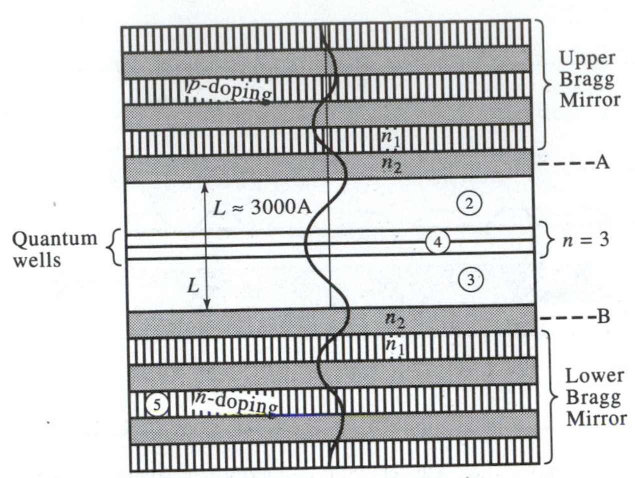

Figure 16-14 The field distribution of the laser mode inside a vertical cavity laser with $L \approx \lambda/n$ with three quantum wells. Notice the evanescent decay of the field envelope inside the Bragg mirrors and the constant amplitude standing wave between the mirrors.

total length of the spacer region, 2 and 3, that straddles the active region is typically $L = \lambda$, where λ is the wavelength in the medium. This translates, near $\lambda = 1\ \mu$m, to $L \approx 0.3\ \mu$m. Typical mode diameters are in the range of 3 to 10 μm. A typical Bragg stack consisting of, say, 15 $\lambda/4$ layers is 2 μm thick.

The field distribution inside a vertical cavity laser is shown in Figure 16-14. We note that inside the Bragg mirror the optical wave amplitude undergoes exponential evanescence. This is in agreement with Equation (13.5-6) and Figures 13-7 and 18-8, which describe the evanescent decay of an optical wave inside a periodic medium for optical frequencies within the "forbidden" frequency gap [17].

Since the distance L_z traveled in the amplifying medium is small (approximately 100 Å per quantum well), the gain per pass is very small, and laser oscillation is made possible by the extremely high reflectance (>99 percent) of the Bragg mirror and the very low losses in regions 2 and 3. Figure 16-14 conveys the relative scale of the key layer thicknesses.

The Oscillation Condition of a Vertical Cavity Laser

The oscillation condition of the VCSEL can be written as

$$r_1(\omega)r_2(\omega)\exp\left[2\sum_{m=1}^{N} \gamma_m(\omega)L_z - i2\frac{\omega}{c}nL\right] = 1 \qquad \textbf{(16.4-1)}$$

which is a statement of the requirement that after one round trip a wave returns to its, arbitrary, starting plane with the same amplitude and, to within an integral multiple of 2π, the same phase. The factor of 2 in the exponent accounts for the fact that the quantum wells are placed at the peak of the standing wave pattern where they are exposed to an intensity that is twice the spatially averaged value. The number of quantum wells is N. In what follows, we will assume that each quantum well contributes equally to the gain so that $\sum_{m=1}^{N} \gamma_m(\omega)L_z \equiv N\gamma(\omega)L_z$. The average index of refraction of the path is n. The Bragg mirrors' (amplitude) reflectances $r_1(\omega)$ and $r_2(\omega)$ refer to their respective input planes A and B in Figure 16-14.

The amplitude condition of (16.4-1) is

$$|r(\omega)|^2 = \exp(-2N\gamma(\omega)L_z) \tag{16.4-2}$$

Since the optical wave travels at right angles to the plane of the quantum wells, the gain γ is not the modal gain g_{model} of (16.2-13) but the bulk gain γ of a quantum well medium. We note that according to (16.2-9), the product $\gamma(\omega)L_z$ is independent of L_z. (This is strictly true when L_z is sufficiently small so that contributions to the gain from excited states ($\ell > 1$) in the quantum well are negligible. In practice, this is satisfied at room temperature for $L_z < 70$Å.) From experimental data of edge-emitting quantum well lasers, we determine that for $L_z \approx 70$ Å the maximum gain due to the $\ell = 1$ quantized well level with a fully inverted population is $\gamma(\omega_0) \cong 5 \times 10^3$ cm^{-1}. Using this value in (16.4-2), taking $L_z = 70$Å, leads to

$$|r(\omega)|^2 = \exp[-2N \times 5 \times 10^3 \times 7 \times 10^{-7}]$$

for the reflectivity needed for oscillation.

$$N = 1, \; |r(\omega)|^2 = 0.993$$

$$N = 2, \; |r(\omega)|^2 = 0.986$$

$$N = 3, \; |r(\omega)|^2 = 0.979$$

$$N = 4, \; |r(\omega)|^2 = 0.972$$

where N is the number of quantum wells so that reflectivities around $R(\equiv |r(\omega)|^2)$ = 98 percent are required of the Bragg reflectors. We will next make a small detour to study these reflectors.

The Bragg Mirror

The analysis of the Bragg mirror is an excellent example of the power of the coupled-mode formalism developed in Chapter 13. The periodic perturbation of the index of refraction couples, exactly as in the case of the DFB laser, two waves propagating in opposite directions. The coupling is strongest when the propagation constants $\pm\beta$ of the two coupled waves

$$B(z)\mathscr{E}_y(x, y)\exp[i(\omega t - \beta z)] \quad \text{(forward wave)}$$

$$A(z)\mathscr{E}_y(x, y)\exp[i(\omega t + \beta z)] \quad \text{(backward wave)} \tag{16.4-3}$$

obey very nearly the Bragg condition

$$\beta = \ell\frac{\pi}{\Lambda} \qquad \ell = 1, 2, \ldots \tag{16.4-4}$$

for some integer ℓ. If we retain in Equation (13.4-3) only the two Bragg-coupled waves $A_s^{(-)} \to A$, $A_s^{(+)} \to B$, we obtain

$$\frac{dA}{dz} = \frac{i\omega\epsilon_0}{4} B \exp(-i2\beta z) \int_{-\infty}^{\infty} \Delta n^2(x, y, z)\mathscr{E}_y^2(x, y)\, dxdy \tag{16.4-5}$$

We also assume that the modes A and B are both y-polarized and have a normalized transverse distribution, $\mathscr{E}_y(x, y)$. The index of refraction of the Bragg mirror can be represented by

$$n^2(x, y, z) = \frac{1}{2}(n_1^2 + n_2^2) + \frac{1}{2}(n_1^2 - n_2^2)\, f(z)$$

where n_1, n_2 are the indices of refraction of the two alternating layers, and $f(z)$ is a square wave of unity amplitude as shown in Figure 16-14.

$$f(z) = \sum_\ell a_\ell e^{i\ell\frac{2\pi}{\Lambda}z}, \quad a_\ell = i\frac{(e^{-i\pi\ell} - 1)}{\ell\pi}$$

$$\Delta n^2(x, y, z) = \left(\frac{n_1^2 - n_2^2}{2}\right) f(z) \tag{16.4-6}$$

Assuming that the Bragg condition (16.4-4) is satisfied by the ℓth term in the Fourier series expansion of $f(z)$, we can rewrite (16.4-5) as

$$\frac{dA}{dz} = \frac{i\omega\epsilon_0(n_1^2 - n_2^2)a_\ell}{8} \int_{-\infty}^{\infty} \mathscr{E}^2(x, y)\, dx\, dy\, B \exp\left[i\left(\ell\frac{2\pi}{\Lambda} - 2\beta\right)z\right] \tag{16.4-7}$$

when $\ell = 1$ we have

$$\frac{dA}{dz} = \kappa B \exp(i\Delta\beta z)$$

$$\frac{dB}{dz} = \kappa A \exp(-i\Delta\beta z) \tag{16.4-8}$$

$$\kappa = \frac{\omega\epsilon_0}{4\pi}(n_1^2 - n_2^2)\int_{-\infty}^{\infty} \mathscr{E}_y^2(x, y)\, dx\, dy \approx \frac{2\Delta n}{\lambda}$$

$$\Delta\beta(\omega) = 2\left(\frac{\pi}{\Lambda} - \beta(\omega)\right) \tag{16.4-9}$$

In the second approximate equality of (16.4-9), we assumed $|\Delta n| = |n_1 - n_2| \ll n_1$, n_2, $\beta \approx \omega\sqrt{\mu\epsilon_0}n$, $n^2 \equiv (1/2)(n_1^2 + n_2^2)$, and used the normalization integral, (13.2-8). Equations (16.4-8) constitute a pair of first-order, linear-coupled differential equations. Their solution requires that we specify two boundary conditions. Our chief interest is in the operation of the Bragg stack as a reflector. The incident amplitude $B(0)$ thus becomes one of the given conditions. Since the backward-going wave A is due completely to internal reflections, we take $A(L) = 0$. The solution is thus given by Equations (13.5-2) so that the amplitude reflectance is

$$r(\omega) = \frac{A(0)}{B(0)} = \frac{-i\kappa \sinh(SL)}{-\Delta\beta(\omega)\sinh(SL) + iS\cosh(SL)}$$
$$S(\omega) = \sqrt{\kappa^2 - \Delta\beta(\omega)^2} \tag{16.4-10}$$

where $\omega_0 = \pi c/\Lambda n$ is the Bragg frequency.

To obtain an appreciation for the magnitude of reflectivities that we may expect in a typical Bragg mirror, we will design a Bragg mirror to operate at a center wavelength of $\lambda_0 = 0.875\ \mu$m. The unit cell consists of a pair of epitaxially grown $Ga_{.8}Al_{.2}As$ and AlAs layers. The index of refraction difference is as $\Delta n = n_{Ga_{0.8}Al_{0.2}As} - n_{AlAs} = 0.55$. The average index is $n = 3.3$. The peak reflectivity is obtained from (16.4-10) with $\Delta\beta = 0$. Since the thickness of a unit cell is $\lambda_0/2n$, the length of the Bragg mirror with N_m periods is $L = N_m\lambda_0/2n$. The result in the case of $N_m = 15$ is $R(\omega_0) = |r(\omega_0)|^2 = \tanh^2\left(N_m \frac{\Delta n}{n}\right) = \tanh^2\left(\frac{15 \times 0.55}{3.3}\right) = 0.973$. This value is sufficient, according to the discussion following (16.4-2), to satisfy the oscillation conditions in vertical cavity lasers with more than four inverted ($N \geq 4$) quantum wells.

A plot of the reflectivity $|r(\omega)|^2$ based on (16.4-10) and the experimental parameters of the above example is shown in Figure 16-15(a). An experimental plot of a Bragg mirror with the same parameters is shown in Fig. 16-15(b). The phase shift $\phi(\omega)$ of the complex reflectance $r(\omega) = |r(\omega)| \exp(-i\phi(\omega))$ is shown in Fig. 16-15(c). For a more detailed treatment of Bragg mirrors and light propagation in stratified media, the reader is referred to Reference [17].

The Oscillation Frequencies

The phase part of (16.4-1) is used to obtain an expression for the oscillation frequencies of a surface-emitting Bragg mirror laser. If, for simplicity's sake, we take two identical $r_1(\omega) = r_2(\omega) = |r(\omega)|e^{i\phi(\omega)}$, the phase condition is

$$-\phi(\omega) + \frac{\omega}{c} nL = m\pi$$
$$m = 1, 2, \ldots \tag{16.4-11}$$

Let us denote the two neighboring oscillation frequencies corresponding to m and $m+1$ as ω_m and ω_{m+1}, respectively:

$$-\phi(\omega_m) + \frac{\omega_m}{c} nL = m\pi$$

$$-\phi(\omega_{m+1}) + \frac{\omega_{m+1}}{c} nL = (m+1)\pi \tag{16.4-12}$$

so that

$$\left[-\phi(\omega_{m+1}) + \phi(\omega_m) + \frac{\omega_{m+1} - \omega_m}{c} nL\right] = \pi \tag{16.4-13}$$

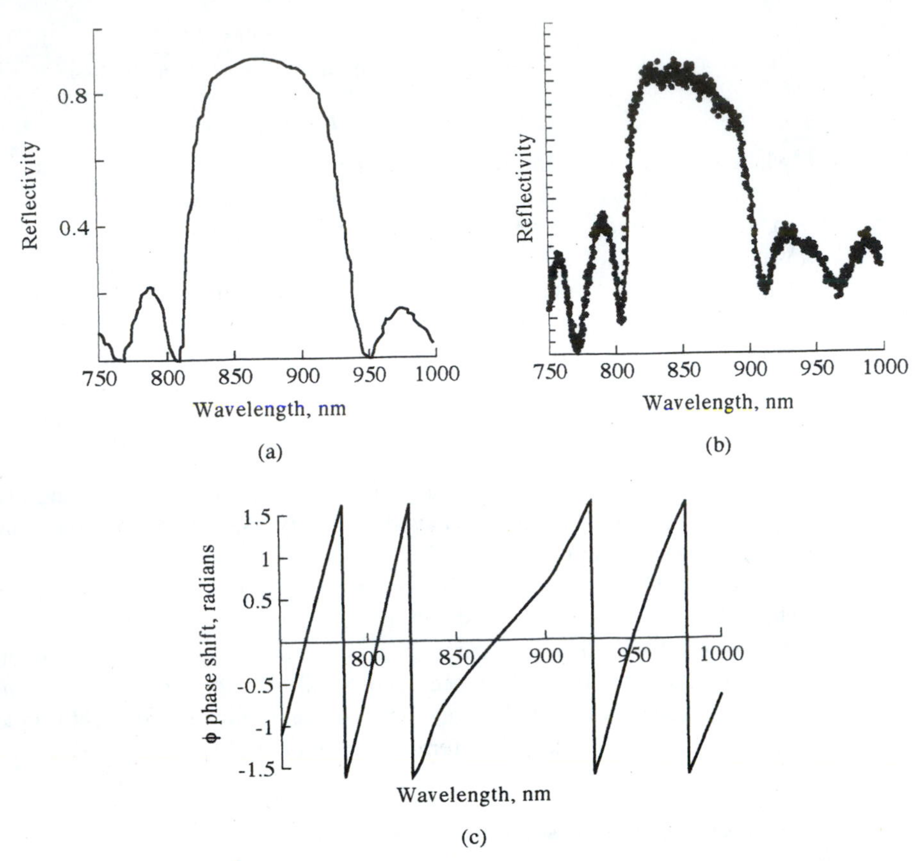

Figure 16-15 Calculated (a) and measured (b) reflectivity of a 15-period $Al_{0.2}GA_{0.8}AS/AlAs$ distributed Bragg reflector. The calculated phase shift $\phi(\omega)$ is plotted in (c). (Courtesy of J. Obrien, Caltech.)

According to Figure 16-15(c), we can approximate $\phi(\omega)$ in the region of high reflectivity by

$$\phi(\omega) \cong -a(\omega - \omega_0)$$

$$a \approx \frac{\pi n}{2\kappa c} \tag{16.4-14}$$

> The expression for the slope a is obtained by dividing the maximum phase deviation of π in Figure 16-15(c) by the corresponding (horizontal) frequency interval that, according to Equation (13.5-7), is $(\Delta\omega)_{\text{gap}} = 2\kappa c/n$.

which when applied to (16.4-13) results in

$$2\pi\Delta\nu = (\omega_{m+1} - \omega_m) = \frac{\pi c}{n\left(L + \dfrac{\pi}{2\kappa}\right)} \tag{16.4-15}$$

for the intermode frequency interval.

The effective length of the Bragg mirror resonator is thus not the mirror spacing L but

$$L_{\text{eff}} = L + \frac{\pi}{2\kappa} \tag{16.4-16}$$

The contribution $\pi/2\kappa$ is due to the evanescent penetration of the oscillating laser field into the Bragg mirrors, as illustrated in Figure 16-14. Since two Bragg mirrors are assumed in the analysis, the Bragg penetration distance into a single mirror is $\pi/4\kappa$.

We recall that the field behavior inside the periodic Bragg mirror (at the Bragg frequency ω_0) is given by (13.5-6) as

$$\exp(-i\beta' z) = \exp(-i\frac{\pi}{\Lambda} z)\exp(-\kappa z)$$

which corresponds to an effective penetration distance of $\sim\kappa^{-1}$ to be compared to the value of $\pi/4\kappa$ of (16.4-16)

Numerical example—intermode frequency separation. To obtain an appreciation for the intermode frequency spacing of (16.4-15), we will consider the laser depicted in Figure 16-14. The data for the Bragg mirror is the same as used in the example following (16.4-10). The basic parameters are:

$$\lambda = 1\ \mu\text{m} \qquad L = \lambda = 1\ \mu\text{m}$$

$$\kappa = \frac{2\Delta n}{\lambda} = \frac{2 \times 0.55}{1} = 1.1\ \mu\text{m}^{-1}$$

$$L_{\text{eff}} = L + \frac{\pi}{2\kappa} = (1 + 1.427)\ \mu\text{m} = 2.427\ \mu\text{m}$$

(Note that the penetration depth, 1.427 μm, is larger than the intermirror spacing, L.)

$$\Delta\nu \equiv \frac{\omega_{m+1} - \omega_m}{2\pi} = \frac{c}{2nL_{\text{eff}}} = \frac{3 \times 10^{10}}{2 \times 3.3 \times 2.427 \times 10^{-4}} = 1.873 \times 10^{13}\ \text{Hz} = 624.3\ \text{cm}^{-1}$$

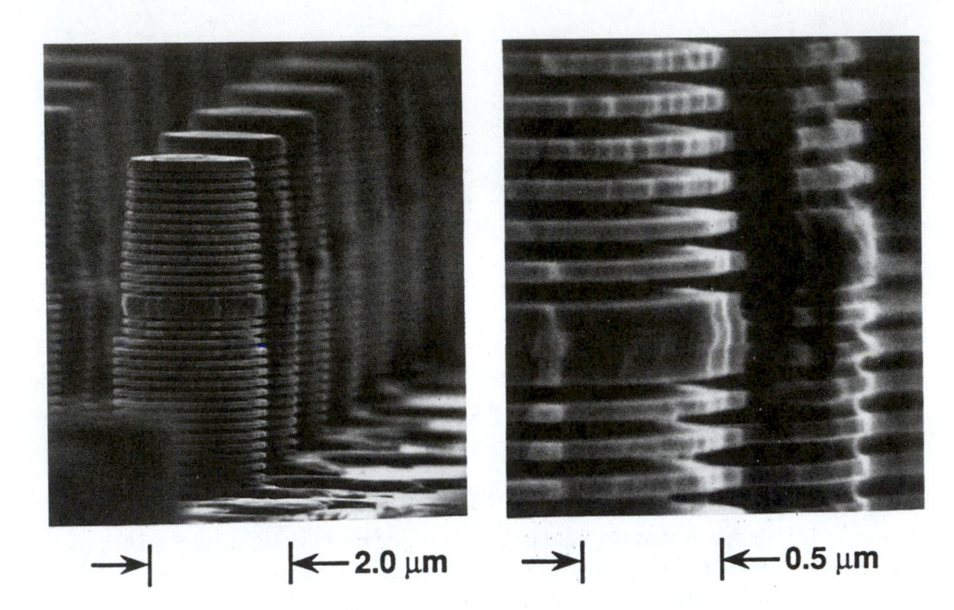

Figure 16-16 A scanning electron micrograph of a two-dimensional array of vertical-cavity, surface-emitting lasers. The multilayered structure of alternating GaAs and AlAs layers is demonstrated by partial preferential etch of the AlAs layers. (Courtesy of Scientific American magazine and of A. Scherer, Caltech.)

The high-reflectivity region of the Bragg mirror is given by (see box following (16.4-14))

$$(\Delta\nu)_{\text{Bragg}} \approx \frac{\kappa c}{\pi n} = \frac{1.1 \times 10^4(\text{cm}^{-1}) \times 3 \times 10^{10}}{\pi \times 3.3} = 1.486 \times 10^{13} \text{ Hz}$$

This number is comparable to the intermode spacing $\Delta\nu = 1.873 \times 10^{13}$ Hz, so that only one mode at a time will experience high reflectivity and will satisfy the oscillation condition. This leads in most cases to a single-mode oscillation. This is to be contrasted with more conventional cleaved-facet-reflectors, edge-emitting semiconductor lasers, where $L \approx 300$ μm and the mode spacing is correspondingly shorter.

We conclude by showing in Figure 16-16 a photograph of a two-dimensional array of surface emitting lasers.

Problems

16.1 Solve the one-dimensional Schrödinger equation (16.1-3) in the case of a simple square potential well where

$$V(z) = -V_0, \quad \frac{L}{2} < z < \frac{L}{2}, \; V(z) = 0 \text{ elsewhere}$$

16.2 Assume that as we scale the length L of a quantum well laser we maintain the differential quantum efficiency η_{ex} constant by increasing R.

a. Derive the expression relating R (mirror reflectivity) to L.
b. Show that $I_{threshold}$ is proportional to L.

16.3 Show qualitatively that for a given m_c and injection current the maximum gain obtains when $m_v = m_c$.

16.4 Estimate the coupling constant κ of the DFB laser whose spontaneous emission spectrum is given in Figure 16-10(e).

16.5

a. Using a computer program, plot the magnitude of the reflection coefficient [see Equation (16.3-11)] of a periodic amplifying waveguide as a function of $\Delta\beta L$, assuming $\kappa L = 0.4$. Let γL be the parameter, and generate plots with $\gamma L = 2, 2.9, 3.5$, and 3.8.
b. Plot the equigain contours in the $\gamma L - \Delta\beta L$ plane as in Figure 16-12.

16.6

a. Derive the coupled-mode equations of a DFB laser with a periodic modulation of its losses. The spatial periodic loss can be accounted for according to Maxwell's equations by taking the dielectric constant of the waveguide as $\epsilon'(\mathbf{r}) = \epsilon(\mathbf{r}\left[1 - i\dfrac{\sigma(\mathbf{r})}{\omega\epsilon(\mathbf{r})}\right]$, with $\sigma(\mathbf{r}) = \sigma_0 + \sigma_1(x)\cos\dfrac{2\pi}{\Lambda}z$ where $\sigma(\mathbf{r})$ is the medium conductivity.
b. Compare the coupling coefficient κ in this case to that of index modulation [See (13.4-7).]
c. Estimate the magnitude of κ in the case of a loss-modulated waveguide where the effective index is $n_{eff} = 3.5$, $\Lambda = 0.22\ \mu m$, $\lambda \cong 1.55\ \mu m$. The lossy layer has an absorption coefficient of $\alpha = 300\ cm^{-1}$ and a thickness of 1000 Å. It is situated at the center of guiding layer. Assume that the waveguide mode is highly confined to the inner layer ($n = 3.51$).

References

1. van der Ziel, J. P., R. Dingle, R. C. Miller, W. Wiegmann, and W. A. Nordland, Jr., "Laser oscillation from quantum states in very thin GaAs-$Al_{0.2}Ga_{0.8}As$ multilayer structures," *Appl. Phys. Lett.* 26:463, 1975.
Dupuis, R. D., P. D. Dapkus, *IEEE J. Quant. Elec.* QE-16:170, 1980.
2. Dingle, R., W. Wiegmann, and C. H. Henry, "Quantum states of confined carriers in very thin $Al_xGa_{1-x}As$-GaAs-$Al_xGA_{1-x}As$ heterostructures," *Phys. Rev. Lett.* 33:827, 1974. Also see G. Bastard and J. A. Brum, "Electronic states in semiconductor heterostructures," *IEEE J. Quant. Elec.* QE-22:1625, 1986.

3. Arakawa, Y., and A. Yariv, ''Theory of gain, modulation response and spectral linewidth in AlGaAs quantum-well lasers,'' *IEEE J. Quantum Elec.* QE-21:1666, 1985.
4. M. Mittelstein, ''Theory and experiments on unstable resonators and quantum well GaAs/GaAlAs lasers,'' Ph.D. thesis in applied physics, California Institute of Technology, Pasadena, CA, p. 54, 1989.
5. Tsang, W. T., ''Extremely low threshold (AlGa)As modified multiquantum well heterostructure laser grown by MBE,'' *Appl. Phys. Lett.* 39:786, 1981.
6. Mehuys, D., ''Linear, nonlinear and tunable guided wave modes for high power (GaAl)As semiconductor lasers,'' Ph.D. thesis, California Institute of Technology, Pasadena, CA, June 1989.
7. Derry, P., et al., ''Ultra low threshold graded-index separate confinement single quantum well buried heterostructure (Al, Ga) as lasers with high reflectivity coatings,'' *Appl. Phys. Lett.* 50:1773, 1987.
8. Eng, L. E., et al., ''Sub milliampere threshold current pseudomorphic InGaAs/AlGaAs buried heterostructure quantum well lasers grown by molecular beam epitaxy,'' *Appl. Phys. Lett.* 55:Oct. 1989.
9. Kogelnik, H., and C. V. Shank, ''Coupled wave theory of distributed feedback lasers,'' *J. Appl. Phys.* 43:2328, 1972.
10. Nakamura, M., A. Yariv, H. W. Yen, S. Somekh, and H. L. Garvin, ''Optically pumped GaAs surface laser with corrugation feedback,'' *Appl. Phys. Lett.* 22:515, 1973.
11. K. Aiki, M. Nakmura, J. Umeda, A. Yariv, A. Katzir, and H. W. Yen, ''GaAs-GaAlAs distributed feedback laser with separate optical and carrier confinement,'' *Appl. Phys. Lett.* 27:145, 1975.
12. Haus, H. A., and C. V. Shank, ''Antisymmetric taper of distributed feedback lasers,'' *IEEE J. Quant. Elec.* QE-12:532, 1976.
13. Nakano, Y., Y. Luo, and K. Tada, ''Facet reflection independent, single longitudinal mode oscillation in a GaAlAs/GaAs distributed feedback laser equipped with a gain-coupling mechanism,'' *Appl. Phys. Lett.* 55:16:1606, 1989.
14. K. Iga, S. Ishikawa, S. Ohkouchi, and T. Nishimura, ''Room-temperature pulsed oscillation of GaAlAs/GaAs surface emitting injection laser,'' *Appl. Phys. Lett.* 45:348, 1984.
15. Jewell, J. L., J. P. Harbison, A. Scherer, Y. H. Lee, and L. T. Florez, ''Vertical cavity surface emitting lasers,'' *IEEE J. Quant. Elec.* 27:1332, 1991.
16. Jewell, J. L., J. P. Harbison, and A. Scherer, ''Microlasers,'' *Scientific American* 86:1991.
17. Yariv, A., and P. Yeh, *Optical Waves in Crystals*, New York: *Wiley Interscience*, 1984.
18. A good general reference to vertical cavity lasers is L. A. Coldren, and S. W. Corzine, *Diode Lasers and Photonic Integrated Circuits,* New York: Wiley, 1995.

Phase Conjugate Optics—Theory and Applications

17.0 INTRODUCTION AND BACKGROUND

The field of phase conjugate optics can be traced to early (1967–1971) experiments [1–4] in transient holography. Application, starting in 1976 [5–11], of the theoretical and experimental tools of nonlinear optics caused a renewed interest and an ever-increasing activity in the field. Most of this interest can be traced to potential applications in image processing and in dynamic (real-time) compensation for distortion inside laser resonators.

Phase conjugate optics in its most basic form deals with situations in which some input monochromatic optical field

$$E_1(\mathbf{r}, t) = \mathrm{Re}[\psi(\mathbf{r})e^{i(\omega t - kz)}] \tag{17.0-1}$$

is converted in *real time*, by means to be discussed later, to a new field that is proportional to

$$E_2(\mathbf{r}, t) = \mathrm{Re}[\psi^*(\mathbf{r})e^{i(\omega t + kz)}] \tag{17.0-2}$$

In most situations of interest, the spatial dependence of the envelope function $\psi(\mathbf{r})$ is "slow"[1] compared to $\exp(ikz)$ so (17.0-1) represents a nearly plane wave propagating in the z direction. Amplitude and phase deviations from the plane wave are represented by the (complex) function $\psi(\mathbf{r})$. These may be due, for example, to distortions in the wave path, to finite aperture effects (diffraction), or to spatial information impressed on the beam. We note that E_2 can be obtained from E_1 by replacing the spatial part of the analytic function (the part inside the brackets) by its

[1]That is, the change of $\psi(\mathbf{r})$ in one optical wavelength is negligibly small.

complex conjugate but leaving the factor $\exp(i\omega t)$ as is.[2] We refer to the field E_2 defined by (17.0-2) as the *phase conjugate* replica of E_1.

Let us consider the implication of the relationship between E_1 and its phase conjugate replica E_2. If we took a picture of the wavefronts of E_1 and E_2, we would not be able to tell them apart. The two sets of *wavefronts coincide* everywhere. To differentiate between them, we will need to go back to the time domain. We will then find that the family of wavefronts of E_1 moves to the right ($+z$), while those of E_2 move to the left, all the while evolving so that at any one moment the two families of wavefronts are identical. This situation is depicted in Figure 17-1.

17.1 THE DISTORTION CORRECTION THEOREM

The wave picture of Figure 17-1 suggests an important theorem, the *distortion correction theorem* [12], which may be stated as follows: "If a (scalar) wave $E_1(\mathbf{r})$ propagates from left to right through an arbitrary dielectric (but lossless) medium, then if we generate in some region of space (say, near $z = 0$) its phase conjugate replica $E_2(\mathbf{r})$, then E_2 will propagate backward from right to left through the dielectric medium remaining *everywhere* the phase conjugate of E_1." An immediate consequence of this theorem can be appreciated with reference to Figure 17-1. Let the "blob" represent some optical distortion that causes the near spherical Gaussian-beam wavefronts incident from the left to be distorted as shown. The phase conjugate wave generated to the right of the distortion traverses it in reverse, regaining to the left of the distorting "blob" its original waveform. This healing property of phase conjugation can be utilized in numerous ways, some of which will be described further on.

[2]We can also write (17.0-2) as $E_2 = \mathrm{Re}[\psi(\mathbf{r})e^{i[\omega(-t)-kz]}]$ so that E_2 can be viewed as the "time-reversed" ($t \rightarrow -t$) replica of E_1.

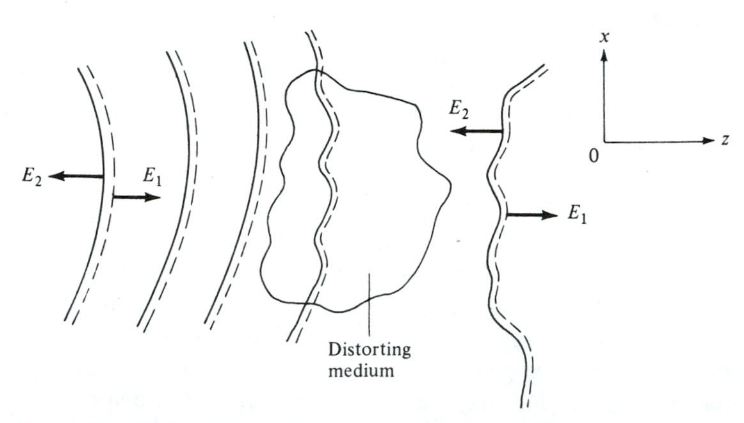

Figure 17-1 The dashed curves represent the wavefronts of a monochromatic optical beam E_1 propagating to the right, while the solid wavefronts are those of the phase conjugate replica of E_1. For generality we include some lossless distorting "blob" in the path.

The proof of the distortion correction theorem follows. We take the (scalar) right-going wave E_1 as

$$E_1 = \psi_1(r)\, e^{i(\omega t - kz)} \tag{17.1-1}$$

where k is taken as a real constant. In the paraxial limit E_1 obeys the wave equation (3.1-1)

$$\nabla^2 E_1 + \omega^2 \mu \varepsilon(\mathbf{r}) E_1 = 0 \tag{17.1-2}$$

where $\varepsilon(\mathbf{r})$ represents the spatial dependence of the dielectric constant of the medium including distortions. Using (17.1-1) the paraxial[3] wave equation (17.1-2) becomes

$$\nabla^2 \psi_1 + [\omega^2 \mu \varepsilon(\mathbf{r}) - k^2]\psi_1 - 2ik \frac{\partial \psi_1}{\partial z} = 0 \tag{17.1-3}$$

Taking the complex conjugate of (17.1-3) leads to

$$\nabla^2 \psi_1^* + [\omega^2 \mu \varepsilon^*(\mathbf{r}) - k^2]\psi_1^* + 2ik \frac{\partial \psi_1^*}{\partial z} = 0 \tag{17.1-4}$$

Had we started, instead of with E_1, with a backward-propagating wave

$$E_2 = \psi_2(\mathbf{r}) e^{i(\omega t + kz)} \tag{17.1-5}$$

then instead of (17.1-3) we would obtain

$$\nabla^2 \psi_2 + [\omega^2 \mu \varepsilon(\mathbf{r}) - k^2]\psi_2 + 2ik \frac{\partial \psi_2}{\partial z} = 0 \tag{17.1-6}$$

But when $\varepsilon(r) = \varepsilon^*(r)$, i.e., a lossless (and gainless) medium, Equations (17.1-6) and (17.1-4) are identical in that the waves ψ_2 and ψ_1^* obey the *same* differential equation. It follows immediately that if $\psi_2 = a\psi_1^*$ (a being an arbitrary constant) over some plane (say, $z = 0$), then, due to the uniqueness property of the solutions to second-order linear differential equations, $\psi_2(x, y, z) = a\psi_1^*(x, y, z)$ at all $x, y, z < 0$. This completes the proof. All that is required now is to find some means for rendering the phase conjugate of an electromagnetic wave.

17.2 THE GENERATION OF PHASE CONJUGATE WAVES

In the last section we have shown that if a phase conjugate replica of an input wave can be generated, this new wave will propagate in reverse through the dielectric medium, regaining everywhere the original form of the input wave. In this section we will show how a phase conjugate wave is generated by means of nonlinear optical techniques.

In Chapter 8, and specifically in the treatment involving relations (8.1-14) and (8.1-17), we discussed the second-order optical nonlinearity $P \propto E^2$ that gives rise

[3]See Section 3.1 for the limitation imposed by the paraxial approximation.

to the phenomena of second harmonic generation, parametric amplification, and frequency addition and subtraction. Symmetry considerations restricted these phenomena to noncentrosymmetric crystals.

The next order of optical nonlinearity involves the third power of the electric field $P \propto E^3$. This leads to the phenomena of optical third-harmonic generation, optical Kerr effect, and the phenomenon whereby a material medium that is subjected to waves at frequencies ω_1, ω_2, and ω_3 radiates a fourth wave at $\omega_1 + \omega_2 - \omega_3$. The latter effect, called *four-wave mixing*, is used in phase conjugation.

To be specific, we assume that a material medium is irradiated simultaneously by three optical fields:

$$\mathbf{E}_1(\mathbf{r}, t) = \tfrac{1}{2}\mathbf{A}_1(r)e^{i(\omega_1 t - \mathbf{k}_1 \cdot \mathbf{r})} + \text{c.c.}$$
$$\mathbf{E}_2(\mathbf{r}, t) = \tfrac{1}{2}\mathbf{A}_2(r)e^{i(\omega_2 t - \mathbf{k}_2 \cdot \mathbf{r})} + \text{c.c.}$$
$$\mathbf{E}_3(\mathbf{r}, t) = \tfrac{1}{2}\mathbf{A}_3(r)e^{i(\omega_3 t - \mathbf{k}_3 \cdot \mathbf{r})} + \text{c.c.} \qquad (17.2\text{-}1)$$

There is induced in the medium a nonlinear (NL) polarization

$$P_i^{(\mathrm{NL})}(\mathbf{r}, t) = 6\chi_{ijkl}^{(3)} A_{1j} A_{2k} A_{3l}^{*}\, e^{i[(\omega_1+\omega_2-\omega_3)t - (\mathbf{k}_1+\mathbf{k}_2-\mathbf{k}_3)\cdot \mathbf{r}]} + \text{c.c.} \qquad (17.2\text{-}2)$$

where i, j, k, l refer to Cartesian coordinates. $\chi_{ijkl}^{(3)}$ is a fourth-rank tensor characteristic of the medium [13] that depends on the input frequencies ω_1, ω_2, ω_3. If we apply to (17.2-2) the argument leading to (8.1-19), we can convince ourselves that, unlike the phenomenon of second-harmonic generation, the third-order optical effects considered here exist in all media, including noncentrosymmetric crystals. The form of χ_{ijkl} but not its magnitude is determined by the symmetry properties of the medium that are discussed and tabulated in Reference [13].

We can, equivalently, express (17.2-2) as a relationship between the complex amplitude of the induced polarization and the complex amplitudes of the inducing fields

$$P_i^{(\mathrm{NL})}(\omega_1 + \omega_2 - \omega_3) = D\chi_{ijkl}^{(3)}(\omega_4 = \omega_1 + \omega_2 - \omega_3) A_{1j}^{(\omega_1)} A_{2k}^{(\omega_2)} [A_{3l}^{(\omega_3)}]^{*} \qquad (17.2\text{-}2a)$$

$$D = 6 \text{ when } \omega_1 \neq \omega_2 \neq \omega_3 \qquad D = 3 \text{ when } \omega_1 = \omega_2 = \omega_3$$

Table 17-1 Nonlinear Constant χ_{ijkl} for Some Common Materials

Material	Index of Refraction	Nonlinear Constant (esu units)*	n_2' (esu units)
CS_2	1.5	$\chi_{1221} \simeq 2.7 \times 10^{-13}$	
		$\chi_{1111} \simeq 3.9 \times 10^{-13}$	0.48×10^{-11}
2-Methyl-4-nitroaniline (MNA)	1.8	$\chi_{1111} \simeq 1.19 \times 10^{-11}$	12×10^{-11}
PTS Polydiacetylene	1.88	$\chi_{1111} \cong 4.0 \times 10^{-11}$	40×10^{-11}

*See Equation (17.3-23) for conversion from esu to MKS units.

A list of the nonlinear coefficients of some optical materials is included in Table 17-1. We also include in the table a listing of the Kerr constant [13] n_2'. This constant, which is often tabulated, describes the dependence of the index of refraction of an isotropic medium on the peak optical field E or the optical intensity I according to

$$n = n_0 + n_2'E^2$$

$$n = n_0 + n_2 I \qquad n_2 = 2\sqrt{\frac{\mu}{\epsilon}}\, n_2'$$

It follows directly from (17.2-2) that

$$n_2' = \frac{6\pi}{n_0}\chi_{1111} \text{ in esu units}$$

$$n_2' = \frac{3}{2\varepsilon_0 n_0}\chi_{1111} \text{ in MKS units}$$

$$n_2 = 2\sqrt{\frac{\mu}{\epsilon}}\, n_2' \text{ (MKS)} = \frac{3}{n_0^2}\sqrt{\frac{\mu}{\varepsilon_0^{3/2}}}\,\chi_{1111} \text{ (MKS)} \tag{17.2-3}$$

The relatively large value of χ_{1111} in MNA is due to large charge separation (of the order of 30 Å) and hence large induced dipoles that can obtain in certain organic molecules.

17.3 THE COUPLED-MODE FORMULATION OF PHASE CONJUGATE OPTICS (8)

In this section we will consider the wave mixing shown in Figure 17-2. We will find that the result is a new wave generated in and radiated by the nonlinear medium. This new wave 3 will be shown to be the phase conjugate of input wave 4. The nonlinear medium is traversed simultaneously by four beams of the same frequency.

$$\mathbf{E}_1(\mathbf{r}, t) = \tfrac{1}{2}\mathbf{A}_1'(r_1)e^{i(\omega t - \mathbf{k}_1\cdot\mathbf{r})} + \text{c.c.}$$

$$\mathbf{E}_2(\mathbf{r}, t) = \tfrac{1}{2}\mathbf{A}_2'(r_2)e^{i(\omega t - \mathbf{k}_2\cdot\mathbf{r})} + \text{c.c.}$$

$$\mathbf{E}_3(\mathbf{r}, t) = \tfrac{1}{2}\mathbf{A}_3'(z)e^{i(\omega t + kz)} + \text{c.c.}$$

$$\mathbf{E}_4(\mathbf{r}, t) = \tfrac{1}{2}\mathbf{A}_4'(z)e^{i(\omega t - kz)} + \text{c.c} \tag{17.3-1}$$

$$k^2 \equiv \omega^2\mu\varepsilon$$

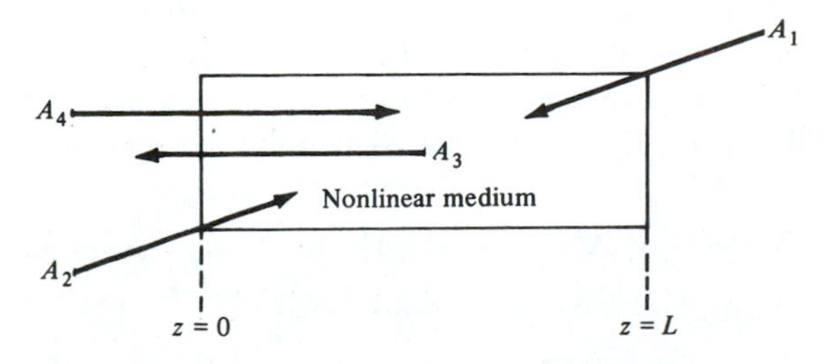

Figure 17-2 The "canonical" geometry of phase conjugation by four-wave mixing.

Waves 1 and 2 propagate along the directions $\mathbf{k}_1$ and $\mathbf{k}_2$, respectively. In the analysis that follows, these two waves correspond to the pump beams and their amplitudes $|A'_1|$ and $|A'_2|$ will be taken as much larger than $|A'_3|$ and $|A'_4|$ and thus will be scarcely affected by the interaction so that they will be taken as constants throughout the interaction volume. We will further take $|A'_1| = |A'_2|$. This causes the effect of each one of these two waves on the phase velocity of the other to be the same so that $k_2 = k_1$ [14]. Furthermore, the pump waves 1 and 2 are made to propagate through the nonlinear medium in opposite directions so that

$$\mathbf{k}_1 + \mathbf{k}_2 = 0 \tag{17.3-2}$$

Wave 4 corresponds to the input beam[4] and in what follows wave 3, which is generated by the interaction of beams 1, 2, and 4, is the (desired) phase conjugate replica of wave 4.

We start with the wave equation

$$\nabla^2 \mathbf{E} - \mu\varepsilon \frac{\partial^2 \mathbf{E}}{\partial t^2} = \mu \frac{\partial^2}{\partial t^2} \mathbf{P}^{(\mathrm{NL})} \tag{17.3-3}$$

Since the physical situation considered here is one where only four beams with well-separated spatial directions are involved, we can apply (17.3-3) to each beam separately. Starting with E_4 and using

$$\nabla^2 \mathbf{E}_4 = \frac{1}{2}\left[-k^2 \mathbf{A}'_4 - 2ik\frac{d\mathbf{A}'_4}{dz} + \frac{d^2\mathbf{A}'_4}{dz^2}\right] e^{i(\omega t - kz)} + \text{c.c.} \tag{17.3-4}$$

in (17.3-3) leads to

$$\frac{1}{2}\left[(\omega^2\mu\varepsilon - k^2)A'_{4i} - 2ik\frac{dA'_{4i}}{dz} + \frac{d^2A'_{4i}}{dz^2}\right] e^{i(\omega t - kz)} + \text{c.c.} = \mu\frac{\partial^2}{\partial t^2} P_i^{(\mathrm{NL})} \tag{17.3-5}$$

Using $\omega^2\mu\varepsilon \equiv k^2$ and the assumed "slow" variation $d^2A'_{4i}/dz^2 \ll k\, dA'_{4i}/dz$, the last equation becomes

$$-ik\frac{dA'_{4i}}{dz} e^{i(\omega t - kz)} + \text{c.c.} = \mu\frac{\partial^2}{\partial t^2} P_i^{(\mathrm{NL})} \tag{17.3-6}$$

The choice of $P_i^{(\mathrm{NL})}$ in (17.3-6) requires some judicious reasoning. First, since we assumed that A'_{4i} is not time dependent, $P_i^{(\mathrm{NL})}$ must contain the exponential time factor $\exp(i\omega t)$ to match that of the left side of the equation. Second, in order that $A'_{4i}(z)$ not vary significantly on the scale of one optical wavelength (i.e., "slow" variation), $P_i^{(\mathrm{NL})}$ need also include the factor $\exp(-ikz)$ so that we must look for polarization terms that contain the wave factor $\exp[i(\omega t - kz)]$.

Recalling that $\mathbf{k}_1 + \mathbf{k}_2 = 0$, it follows from (17.2-2) that the only third-order products of the fields $\mathbf{E}_1$, $\mathbf{E}_2$, $\mathbf{E}_3$, and $\mathbf{E}_4$ that contain the factors $\exp[\pm i(\omega t - kz)]$ are

$$\begin{aligned} P_i^{(\mathrm{NL})} = 3(&\chi_{ijkl}A'_{1j}A'_{2k}A'^{*}_{3l} + \chi_{ijji}A'_{1j}A'^{*}_{1j}A'_{4i} + \chi_{ikki}A'_{2k}A'^{*}_{2k}A'_{4i} \\ &+ \chi_{iiii}A'_{4i}A'^{*}_{4i}A'_{4i} + \chi_{illi}A'_{3l}A'^{*}_{3l}A'_{4i})\, e^{i(\omega t - kz)} + \text{c.c.} \end{aligned} \tag{17.3-7}$$

---

[4]In the applications this is the wave that is to be conjugated.

In what follows we will drop the tensorial notation and limit ourselves to cases where a single χ_{ijkl} is involved. This will correspond in practice to situations where all the fields have the same polarization in which case only χ_{iiii} is involved, or when waves 1 and 2 have one polarization while 3 and 4 possess the orthogonal polarization. In this case $\chi_{ijji}(j \neq i)$ is used. We will refer to the first as case (a) and to the second as case (b). Next we define

$$\chi^{(3)} \equiv 6\chi_{iiii} \qquad \text{case (a)}$$

$$\chi^{(3)} \equiv 6\chi_{ijji} \qquad \text{case (b)} \tag{17.3-8}$$

Furthermore, we will neglect the last two terms on the right side of (17.3-7) since $|A_3'|, |A_4'| \ll |A_1'|, |A_2'|$. Using (17.3-7) in the wave equation (17.3-6) results in

$$\frac{dA_4'}{dz} = -i\frac{\omega}{2}\sqrt{\frac{\mu}{\varepsilon}}\,\chi^{(3)}(|A_1'|^2 + |A_2'|^2)A_4' - i\frac{\omega}{2}\sqrt{\frac{\mu}{\varepsilon}}\,\chi^{(3)}\,A_1'A_2'A_3'^* \tag{17.3-9}$$

We note that the first term on the right side of (17.3-9), acting alone, merely modifies the propagation phase constant of wave 4 from k to $k + (\omega/2)\sqrt{\mu/\varepsilon}\chi^{(3)}\,(|A_1'|^2 + |A_2'|^2)$. (This is the optical Kerr effect [13].) We can thus simplify the analysis by introducing a new set of field amplitudes A_s that are defined by

$$A_s' = A_s e^{-i(\omega/2)\sqrt{\mu/\varepsilon}\chi^{(3)}(|A_1'|^2+|A_2'|^2)z} \tag{17.3-10}$$

$s = 1, 2, 3, 4$. The new amplitudes A_s are thus related to the set A_s' by a mere phase factor. The wave equation (17.3-10) becomes

$$\frac{dA_4}{dz} = -i\frac{\omega}{2}\sqrt{\frac{\mu}{\varepsilon}}\,\chi^{(3)}A_1A_2A_3^* \tag{17.3-11}$$

A similar derivation for wave 3 would lead to

$$\frac{dA_3^*}{dz} = -i\frac{\omega}{2}\sqrt{\frac{\mu}{\varepsilon}}\,\chi^{(3)*}A_1^*A_2^*A_4 \tag{17.3-12}$$

Defining

$$\kappa^* \equiv \frac{\omega}{2}\sqrt{\frac{\mu}{\varepsilon}}\,\chi^{(3)}A_1A_2 \tag{17.3-13}$$

and taking the complex conjugate of (17.3-11) and (17.3-12) results in our final form of the coupled-mode equations for phase conjugate optics [8]

$$\frac{dA_4^*}{dz} = i\kappa A_3$$

$$\frac{dA_3}{dz} = i\kappa^* A_4^* \tag{17.3-14}$$

The student is urged to ponder at this point how a relatively complex physical experiment involving four optical beams interacting through the nonlinear electronic response of a material medium can be described by equations as simple as (17.3-14). This is possible through a "ruthless," but justifiable, elimination of mathematical terms whose effects are physically negligible but whose inclusion will have rendered the analysis intractable. This is an *essential* difference between mathematics and physics.

Since wave 4 propagates in the $+z$ direction, while wave 3 propagates in the $-z$ direction, we can specify their complex amplitudes at their respective input planes $z = 0$ and $z = L$ (see Figure 17-2). These are taken as $A_4(0)$ and $A_3(L)$. Subject to these boundary conditions, the solution of (17.3-14) is

$$A_3(z) = \frac{\cos |\kappa| z}{\cos |\kappa| L} A_3(L) + i \frac{\kappa^* \sin |\kappa|(z - L)}{|\kappa| \cos |\kappa| L} A_4^*(0)$$
$$A_4(z) = -i \frac{|\kappa| \sin |\kappa| z}{\kappa \cos |\kappa| L} A_3^*(L) + \frac{\cos |\kappa|(z - L)}{\cos |\kappa| L} A_4(0) \tag{17.3-15}$$

In the basic phase conjugate experiments, there is but a single input, $A_4(0)$ (the "pump" beams A_1 and A_2 are considered here as part of the "apparatus" and are lumped in our analysis into the coupling constant κ). Putting $A_3(L) = 0$, we obtain from (17.3-15) for the reflected wave at the input,

$$A_3(0) = -i\left(\frac{\kappa^*}{|\kappa|} \tan |\kappa| L\right) A_4^*(0) \tag{17.3-16}$$

while at the output ($z = L$)

$$A_4(L) = \frac{A_4(0)}{\cos |\kappa| L} \tag{17.3-17}$$

We note that $|A_4(L)| > |A_4(0)|$, i.e., the device acts as a phase *coherent optical amplifier* with a gain of $|(\cos |\kappa| L)^{-1}|$. When

$$\frac{\pi}{4} \leq |\kappa| \, L \leq \frac{3\pi}{4} \tag{17.3-18}$$

the result $|A_3(0)| > |A_4(0)|$ obtains, so that *the reflectivity of the phase conjugate mirror exceeds unity.* The intensity distribution of the two waves inside the nonlinear medium for a value of $|\kappa| L$ satisfying (17.3-18) is shown in Figure 17-3.

Of particular interest is the condition $|\kappa| L = \pi/2$. In this case,

$$\frac{A_3(0)}{A_4(0)} = \infty \qquad \frac{A_4(L)}{A_4(0)} = \infty \tag{17.3-19}$$

that is, both the transmission gain $[A_4(L)/A_4(0)]$ and the reflection gain $[A_3(0)/A_4(0)]$ become infinite so that finite outputs $A_3(0)$ and $A_4(L)$ can result even when the input

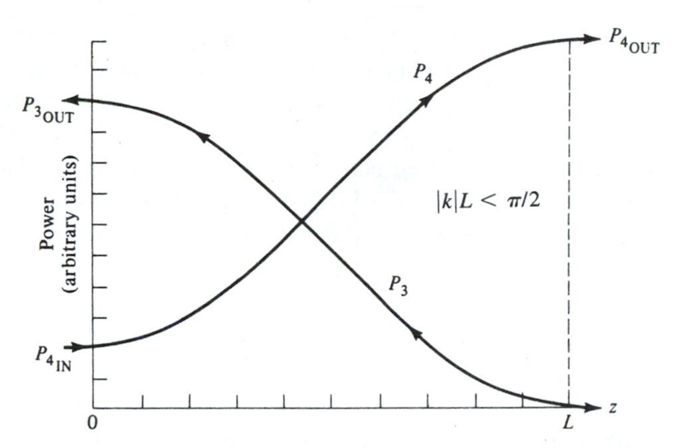

Figure 17-3 The intensity distribution inside the interaction region corresponding to the amplifier case $\pi/4 < |\kappa|L < \pi/2$.

$A_4(0)$ is zero. This corresponds to *oscillation.* This oscillation takes place *without the benefit of mirror feedback.* The feedback process that is essential to oscillation is provided by the fact that waves 3 and 4 propagate in opposite directions, so that $A_4(z_1)$, for example, is influenced by $A_4(z_2)$ even when $z_2 > z_1$, the information being carried from z_2 to z_1 by the backward-going wave 3. The intensity distribution corresponding to the oscillation condition is shown in Figure 17-4.

Another point of physical interest is that of the source of the power. Since energy is conserved, it follows that the increase in the output powers of beams 3 and 4 relative to their input values must come at the expense of the "pump" beams 1 and 2. A more exact analysis that does not neglect the spatial dependence of beams 1 and 2 shows that this indeed is the case. A quantum mechanical description of this

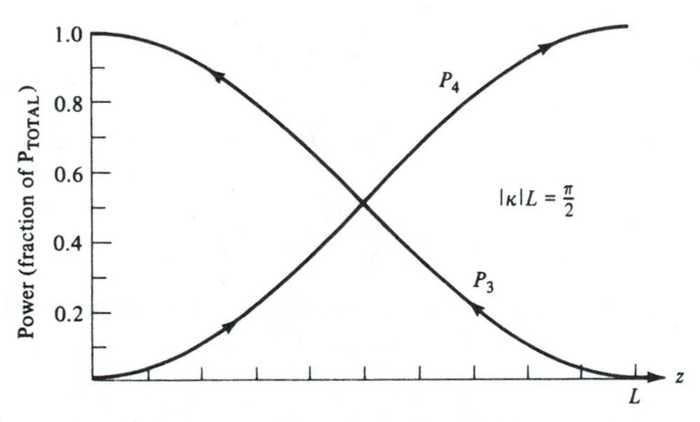

Figure 17-4 The intensity distribution inside the interaction region when the oscillation condition $|\kappa|L = \pi/2$ is satisfied.

process [14] shows that on the atomic scale the basic process is one where, simultaneously, two photons, one from beam 1 and one from beam 2, are annihilated while two photons are created—one of these photons is added to beam 3 and the other to beam 4.

We have strayed somewhat from our main purpose, which is to show that the four-wave mixing geometry of Figure 17-2 is capable of rendering in real time the phase conjugate replica of an input beam. Returning to our basic plane wave result (17.3-16),

$$A_3(0) = -i\left(\frac{\kappa^*}{|\kappa|}\tan|\kappa|L\right)A_4^*(0) \tag{17.3-20}$$

it follows that the amplitude A_3 of the backward wave 3 at the input plane $z = 0$ is proportional to the complex amplitude $A_4^*(0)$ of the input wave 4 at the same plane. It follows directly that since an input wave with an arbitrarily complex wave front $A_4(x, y, z)$ can be expanded in terms of plane wave components (which in the paraxial limit adopted here will span a small solid angle centered about the $+z$ axis), we can extend (17.3-20) to each plane wave component individually [8] to obtain

$$A_3(x, y, z < 0) = -i\left(\frac{\kappa^*}{|\kappa|}\tan|\kappa|L\right)A_4^*(x, y, z < 0) \tag{17.3-21}$$

This is the basic result of phase conjugation by four-wave mixing. It shows that the reflected beam $A_3(\mathbf{r})$ to the left of the nonlinear medium ($z < 0$) is the phase conjugate of the input beam $A_4(\mathbf{r})$.

Some Consideration of Units

As in the rest of this book, our analysis of phase conjugation employs the MKS system of units. Much of the research literature, unfortunately, uses the esu system. The relations that follow should facilitate the translation from one system to another. These are

$$\begin{aligned}\kappa^*_{\text{esu}} &= \frac{2\pi\omega}{cn}\,\chi^{(3)}_{\text{esu}}A_1A_2\\ \kappa^*_{\text{MKS}} &= \frac{\omega}{2cn\varepsilon_0}\,\chi^{(3)}_{\text{MKS}}A_1A_2\\ &= \frac{\omega}{2}\sqrt{\frac{\mu_0}{\varepsilon}}\,\chi^{(3)}_{\text{MKS}}A_1A_2\end{aligned} \tag{17.3-22}$$

where $n = \sqrt{\varepsilon/\varepsilon_0}$ is the index of refraction and where for nonmagnetic material we put $\mu = \mu_0$.

Another useful relation is

$$\chi^{(3)}_{\text{MKS}} = \frac{\chi^{(3)}_{\text{cgs}}}{8.1 \times 10^{18}} \tag{17.3-23}$$

Numerical Example: Phase Conjugation in CS_2

Many of the first experiments in phase conjugate optics were performed in carbon disulfide, CS_2. In the case when the waves A_1 and A_2 are polarized along one direction that we will call y, while A_3 and A_4 are polarized along x, the relevant nonlinear coefficient is χ_{1221}, which in MKS units has the value [obtained from Table 17-1 and Equation (17.3-23)]

$$\chi_{1221} = 1.7 \times 10^{-32}$$

The coefficient $\chi^{(3)}$ used in our analysis, we recall, is according to (17.3-8) equal to $6\chi_{1221}$. We will further assume that the experiment is carried out with waves at $\lambda = 10^{-6}$ m and that the two pump beams are of equal intensities with $I_1 = I_2 = 5 \times 10^{10}$ watts/m^2 (i.e., 5×10^6 watts/cm^2). The index of refraction of CS_2 is $n \simeq 1.5$. Using these data we obtain

$$A_1 = A_2 = \sqrt{\frac{2I_1}{\varepsilon_0 cn}} = 5 \times 10^6 \text{ V/m}$$

and from the last of (17.3-22), $|\kappa| = 0.6 \text{ m}^{-1}$.

It follows that in CS_2 we need to use pump intensities of the order of megawatts per square centimeter with path lengths of the order of magnitude of 1 m in order to satisfy the condition $|\kappa|L \sim 1$ needed according to (17.3-17) for appreciable ($\sim$1) phase conjugate reflectivities. Experiments in CS_2, which are discussed in detail in Section 17.4, verify the basic feature with the value calculated above.

17.4 SOME EXPERIMENTS INVOLVING PHASE CONJUGATION

Probably the two main features of interest to an experimentalist embarking for the first time on the field of phase conjugate optics would be (1) the verification of the basic phase conjugation equations (17.3-16), and (2) the demonstration that the backward-going wave A_3 is indeed the phase conjugate replica of the incident A_4. The latter can be achieved by introducing some distortion or passive optical element, say a lens, and seeing if A_3 in its reverse propagation assumes everywhere the same wavefronts as A_4, i.e., is "healed" of the distortion.

An experiment verifying both of the above points is illustrated in Figure 17-5. The collimated pump beam A_1 ($\lambda = 0.694$ μm) is polarized in the plane of the figure ($\uparrow$) and originates in a Q-switched ruby laser. The counter-propagating pump beam A_2 is generated by reflection from a mirror on the extreme left. A small part of the incident ($\uparrow$) energy is converted to the orthogonal $\odot$ polarization by the birefringent plate P. This radiation is separated spatially from the main beam by polarizing prism A to yield the input-probing beam A_4. A spherical mirror C focuses the collimated

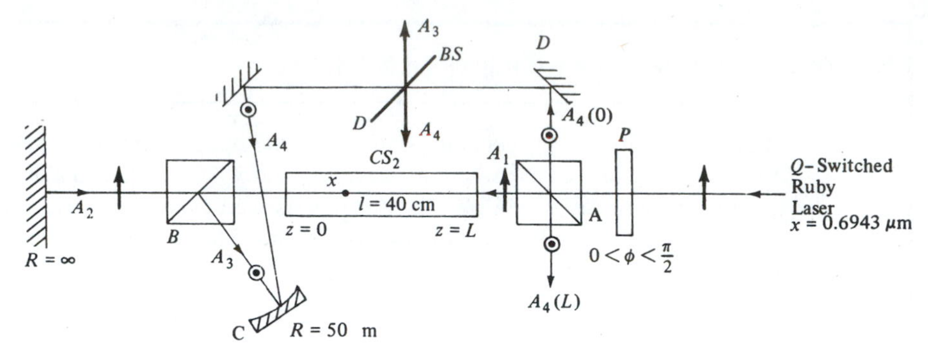

Figure 17-5 The experimental arrangement of a basic phase conjugation experiment. (After Reference [8].)

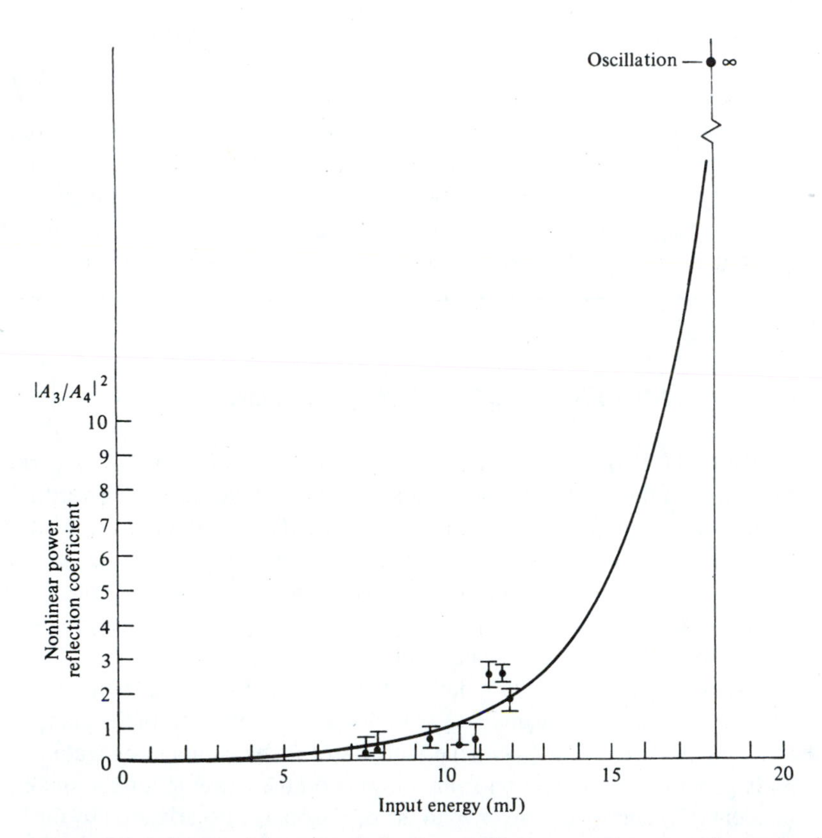

Figure 17-6 A plot of the power reflection coefficient versus pump pulse energy in (millijoules) for the experimental setup of Figure 17-5. Data points [·] include oscillation ($|A_3/A_4|^2 = \infty$). The solid curve is least square fit to $R = \tan^2(CI)$. (After Reference [8].)

beam A_4 to point x in the CS_2 cell. This focusing constitutes the "distortion." (One person's focusing is another person's distortion.)

The reflected beam A_3 generated by the mixing of A_1, A_2, and A_4 thus corresponds to (17.3-8) case (b) and consequently is ⊙ polarized and can be photographed and/or read off beam splitter D.

The measurement of the reflection coefficient $|A_3/A_4|^2$ for various pump intensities is shown in Figure 17-6. The solid curve is a plot of the theoretical expression (17.3-16)

$$\left|\frac{A_3(0)}{A_4(0)}\right|^2 = \tan^2|\kappa|L$$

The verification of the phase conjugate relationship between A_3 and A_4 is provided by the fact that, after reflection from C, beam 3 is collimated. Beam 3 thus emanates from point source x, which is the very point on which beam 4 is converging. It follows that the wavefronts of these two beams are identical.

The prediction of oscillation for $|\kappa|L = \pi/2$ is verified by eliminating the input beam $A_4(0)$, say by removing mirror D, and observing the simultaneous emergence of ⊙ polarized beams 3 and 4 with further increase of the pump intensity.

17.5 OPTICAL RESONATORS WITH PHASE CONJUGATE REFLECTORS

One of the more interesting consequences of phase conjugate optics involves optical resonators where one of the two conventional reflectors is replaced by a phase conjugate mirror [14], henceforth referred to as a PCM. The situation is demonstrated in Figure 17-7.

Consider some arbitrary transverse Gaussian beam with transverse quantum numbers m, n (of the type considered in Section 4.3). Let the phase shift of this

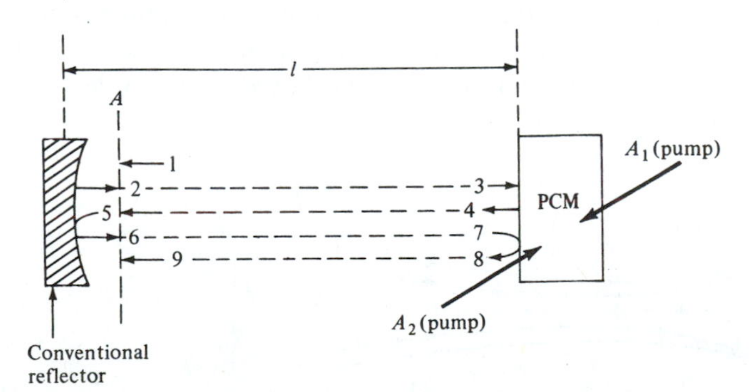

Figure 17-7 An optical resonator formed between a conventional reflector and a phase conjugate mirror (PCM).

beam due to propagation between the two mirrors (spacing l) be $\phi_l(m, n)$. The phase shift upon reflection from the conventional reflector is taken as ϕ_R while that of the PCM is α. We shall now derive the resonance condition for the resonator in the vein of the reasoning used in Section 4.6 by requiring that the phase of the internally shuttling beam reproduce itself after a given number of round trips to within an integer multiple of 2π. We designate the arbitrary starting phase of, say, a left-propagating beam at some plane A, as ϕ_1. Without loss of generality we take A to lie just to the right of the reflector.

The phases of the beam at the various stages are given by

$$\phi_2 = \phi_1 + \phi_R$$

$$\phi_3 = \phi_2 + \phi_l(m, n) = \phi_1 + \phi_R + \phi_l(m, n)$$

$$\phi_4 = -\phi_3 + \alpha = -(\phi_1 + \phi_R + \phi_l(m, n)) + \alpha$$

Notice the sign inversion of ϕ_3 due to phase conjugation

$$\phi_5 = \phi_4 + \phi_l(m, n) = -\phi_1 - \phi_R + \alpha$$

$$\phi_6 = \phi_5 + \phi_R = -\phi_1 + \alpha$$

$$\phi_7 = \phi_6 + \phi_l(m, n) = -\phi_1 + \alpha + \phi_l(m, n)$$

$$\phi_8 = -\phi_7 + \alpha = \phi_1 - \phi_1(m, n)$$

$$\phi_9 = \phi_8 + \phi_l(m, n) = \phi_1 \tag{17.5-1}$$

The self-consistent condition $\phi_9 = \phi_1$ is thus satisfied automatically. The phase conjugate resonator has a resonance at the frequency of the pump beams. (This follows since no allowance was made for a frequency shift upon reflection from the PCM, which would be the case if the resonant mode frequency did not equal that of the pump) and *the resonance condition is satisfied independently of the length l of the resonator or the transverse order (m, n) of the Gaussian beam.* This requires *two* complete round trips. By tracing the arrows of Figure 17-8, we can verify that the radius of curvature of the Gaussian beam will also reproduce itself after two round trips. It follows that *the phase conjugate resonator is stable* (in the sense defined in Sections 4.4 and 4.5) *regardless of the radius of curvature R of the mirror and the spacing l.*

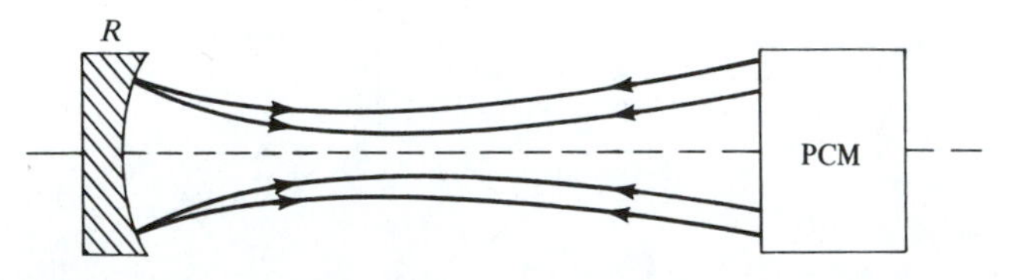

Figure 17-8 A self-consistent beam solution inside a phase conjugate resonator reproduces its wavefront curvature after two round trips.

17.6 THE *ABCD* FORMALISM OF PHASE CONJUGATE OPTICAL RESONATORS

In this section we extend the *ABCD* Gaussian mode formalism that we employed in Section 4.5 to the case of a phase conjugate resonator. The analysis follows closely that of Reference [15].

The *ABCD* Matrix of a Phase Conjugate Mirror

Consider a Gaussian field E_i propagating along the z axis, to be incident upon the PCM. In this case, using (4.3-1),

$$E_i = \mathscr{E}_i(\mathbf{r}) \exp\left[i\left(\omega t - kz - \frac{kr^2}{2\rho}\right) - \frac{r^2}{w^2}\right] \tag{17.6-1}$$

where $\mathscr{E}_i(\mathbf{r})$ is the complex amplitude of E_i, and ρ and w are the radius of curvature and the spot size of the incident field, respectively. This field can also be written as

$$E_i = \mathscr{E}_i(\mathbf{r}) \exp\left[i\left(\omega t - kz - \frac{kr^2}{2q_i}\right)\right] \tag{17.6-2}$$

The complex radius of curvature q_i is defined by (2.4-9) and (2.6-5)

$$\frac{1}{q_i} = \frac{1}{\rho} - \frac{i\lambda}{\pi w^2} \tag{17.6-3}$$

The effect of the PCM is to "reflect" such an incident field as to yield its conjugate replica, leaving the wavefront and the spot size unchanged. The reflected field is thus

$$E_r \propto \mathscr{E}_i^*(\mathbf{r}) \exp\left[i\left(\omega t + kz + \frac{kr^2}{2\rho}\right) - \frac{r^2}{w^2}\right] \tag{17.6-4}$$

which can also be expressed as

$$E_r \propto \mathscr{E}_i^*(\mathbf{r}) \exp\left[i\left(\omega t + kz - \frac{kr^2}{2q_r}\right)\right] \tag{17.6-5}$$

The reflected field complex radius of curvature subject to (17.6-3) and (17.6-4) is given by

$$\frac{1}{q_r} = -\frac{1}{\rho} - \frac{i\lambda}{\pi w^2} = -\frac{1}{q_i^*} \tag{17.6-6}$$

An observer traveling with the beam will find the spot size unchanged after phase conjugate reflection but will see an opposite sign for the curvature of the wavefront.

If we introduce the ray matrix formalism of Section 2.1, the effect of the PCM can thus be represented by the matrix

$$\mathbf{M} = \begin{pmatrix} A & B \\ C & D \end{pmatrix} = \begin{pmatrix} 1 & 0 \\ 0 & -1 \end{pmatrix} \tag{17.6-7}$$

with the output and input q parameters related by

$$q_r = \frac{Aq_i^* + B}{Cq_i^* + D} \tag{17.6-8}$$

Note the *conjugation* operation upon q_i, as opposed to the conventional formalism (Section 2.1), where the input field is *not* conjugated. We note that this matrix also describes the reflection of rays from the conjugate mirror.

It follows directly that the ordinary *ABCD* formalism for treating the propagation of Gaussian beams through a sequence of lenslike media (Section 2.7) can be applied also in the case when one of the elements is a PCM. The matrix representing the PCM is given by (17.6-7). The q parameter at any plane *following* the PCM is related to the input q by

$$q_{\text{out}} = \frac{A_T q_i^* + B_T}{C_T q_i^* + D_T} \tag{17.6-9}$$

where the subscript T implies that the matrix elements correspond to that of the resultant matrix for the given sequence of optical elements, including that of the PCM. Since all the matrices are assumed to be real, the conjugation operation imposed by (17.6-9) can be performed at any plane.

Consider next the situation sketched in Figure 17-9. The resonator is bounded on one end by a mirror having a radius of curvature R, containing arbitrary intracavity optical components described collectively by an $A'B'C'D'$ matrix $\mathbf{M}'$ for optical propagation from left to right and again by an $A''B''C''D''$ matrix $\mathbf{M}''$ for propagation from right to left. The resonator is bounded on the other end by a PCM. In order to investigate the stability criterion for such a cavity, we apply the standard self-consistent condition whereby we require that the complex radius of curvature of the beam reproduce itself after two round trips. Choosing a plane to the immediate right

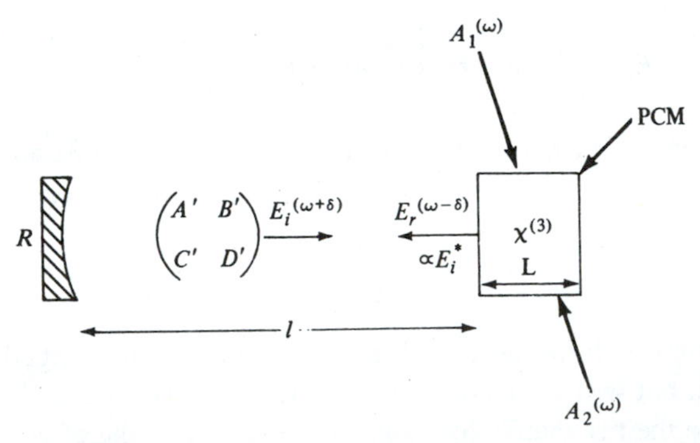

Figure 17-9 The phase conjugate resonator (PCR). This general resonator is formed by placing some arbitrary optical components, represented collectively by an equivalent $A'B'C'D'$ ray matrix, between a "real" mirror (of radius R) on one end, and a phase conjugate mirror (PCM) on the other end. In the case of degenerate modes, which is considered here, $\delta = 0$.

of the real mirror, we trace a beam that propagates to the right and get, after one round trip, the following matrix product:

$$\mathbf{M}_1 = \begin{pmatrix} A_1 & B_1 \\ C_1 & D_1 \end{pmatrix} = \begin{pmatrix} 1 & 0 \\ \dfrac{-2}{R} & 1 \end{pmatrix} \begin{pmatrix} A'' & B'' \\ C'' & D'' \end{pmatrix} \begin{pmatrix} 1 & 0 \\ 0 & -1 \end{pmatrix} \begin{pmatrix} A' & B' \\ C' & D' \end{pmatrix}$$
$$= \begin{pmatrix} 1 & 0 \\ \dfrac{-2}{R} & 1 \end{pmatrix} \begin{pmatrix} 1 & 0 \\ 0 & -1 \end{pmatrix} \tag{17.6-10}$$

where we have used the relation

$$\mathbf{M}''\mathbf{M} = \mathbf{M}(\mathbf{M}')^{-1} \tag{17.6-11}$$

which can be shown straightforwardly, using the reciprocity property of the group of optical elements represented by $\mathbf{M}'$ (or $\mathbf{M}''$), where $\mathbf{M}$ is given by (17.6-7). Equation (17.6-10) is merely a reaffirmation of the fact that an arbitrary sequence of passive and lossless optical elements followed by a PCM is equivalent to the PCM alone. This is due to the ''time reversal'' occurring at the PCM and the reciprocity of the passive components.

We have already established in Section 17.5 that the self-consistent phase condition of a mode in a phase conjugate resonator is satisfied automatically after two round trips. The *ABCD* matrix describing the effect of two round trips on the complex beam radius is $\mathbf{M}_2 = (\mathbf{M}_1)^2$, where $\mathbf{M}_1$, given by (17.6-10), is the single round-trip Gaussian beam evolution matrix. Using (17.6-10) we obtain

$$\mathbf{M}_2 = (\mathbf{M}_1)^2 = \mathbf{I} \tag{17.6-12}$$

where $\mathbf{I}$ is the identity matrix. It follows that *any* Gaussian beam (i.e., one with an arbitrary waist location, waist size, and transverse mode order m, n) is a proper mode solution as far as shape reproducibility (after two round trips) is concerned. This, coupled with the above demonstration (17.5-1) concerning the phase condition, completes the proof that any arbitrary Gaussian beam with a frequency equal to that of the pump beams is a proper mode solution of a phase conjugate resonator independent of the resonator length and the radius of curvature of its one spherical mirror.

For a discussion of this topic including the problem of modes whose resonant frequencies differ from that of the pump waves, one should consult References [15–17].

17.7 DYNAMIC DISTORTION CORRECTION WITHIN A LASER RESONATOR

One of the more interesting practical applications of phase conjugate optics is in dynamic real-time correction of distortion in optical resonators. The situation is depicted in Figure 17-10. A laser oscillator consists of a gain medium, a mirror, a phase conjugate mirror, and a distortion. The distortion may be due to the gain medium itself or to ''bad'' optics. Let us assume, for a moment, that the wave

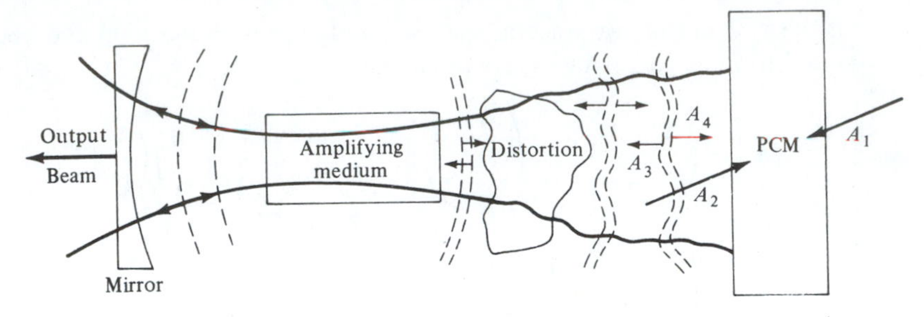

Figure 17-10 A phase conjugate reflector compensates in real time for a time-varying distortion inside an optical resonator.

(a)

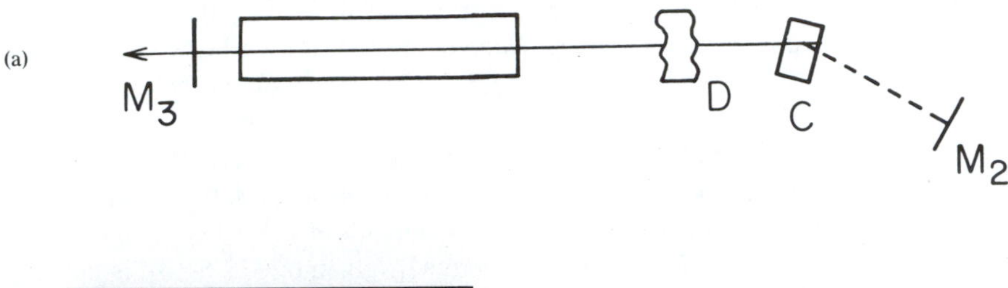

(b)

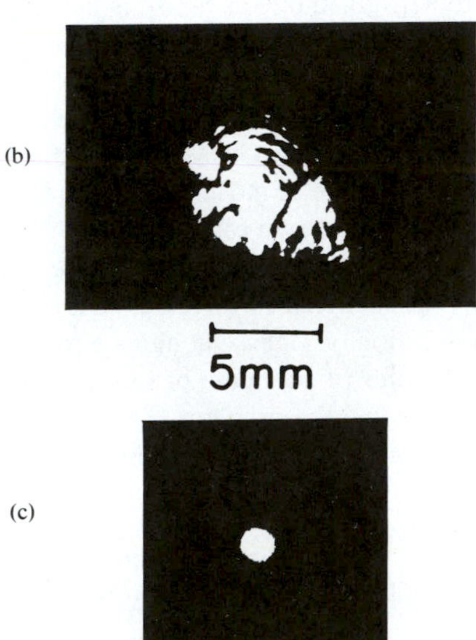

(c)

Figure 17-11 (a) An argon laser gain tube with a distortion D, a (photorefractive) phase conjugating crystal C and a feedback mirror M_2. (b) The highly degraded output beam from M_3 when C is replaced by a conventional mirror. (c) The beam regains its diffraction limited shape in the presence of the distortion when the configuration (a) is used.

incident on the distortion from the left corresponds to a perfect Gaussian beam whose radius of curvature at the left mirror matches that of the mirror. The beam is distorted in passage through the distortion, but after reflection from the PCM and the reverse propagation through the distortion it regains, according to the distortion correction theorem of Section 17.1, its original undistorted form with the reflected (left-going) wavefronts coinciding in space with those of the right-going beam. It follows immediately by repeating the above scenario that the situation depicted in Figure 17-10 is self-reproducing and self-consistent (if not necessarily unique). It should thus be possible to extract the full available power of a laser oscillator in the form of a near ideal Gaussian beam, i.e., the output on the left side of Figure 17-10, in the presence of considerable and even time-varying distortion inside the resonator, corresponds to that of a Gaussian beam.

An experimental demonstration [18] of a laser oscillator with dynamic phase conjugate distortion correction is illustrated in Figure 17-11. The phase conjugate mirror utilizes a crystal of barium titanate. The gain medium is a commercial argon laser tube and the distortion is an acid-etched glass flat.

17.8 HOLOGRAPHIC ANALOGS OF PHASE CONJUGATE OPTICS

The analogy between phase conjugate optics and holography is interesting both from the formal and the practical points of view and suggests that nearly all of the applications envisaged or demonstrated with conventional holography can be performed using phase conjugate optics. The main attraction in the use of phase conjugate optics to replace conventional holography is the real-time aspect of the former that obviates the need to develop the hologram (see Chapter 14). To appreciate this analogy we use the expression (17.3-16) for the reflectivity of a phase conjugate mirror

$$A_3(0) = -i\left(\frac{\kappa^*}{|\kappa|}\tan|\kappa|L\right)A_4^*(0)$$

To simplify the discussion, consider the case of small reflectivity $|A_3(0)/A_4(0)|^2 \ll 1$. The last relation simplifies to

$$A_3(0) = -i\kappa^* A_4^*(0)L$$

where L is the thickness of the phase conjugating medium. Using (17.3-13) we obtain

$$A_3(0) = -i\frac{\omega}{2}\sqrt{\frac{\mu}{\varepsilon}}\chi^{(3)}LA_1(A_2A_4^*) \tag{17.8-1}$$

$$= -i\frac{\omega}{2}\sqrt{\frac{\mu}{\varepsilon}}\chi^{(3)}L(A_1A_4^*)A_2 \tag{17.8-2}$$

The placement of the parentheses in (17.8-1) is to suggest that we may view the process of phase conjugation as the reflection of beam A_1 from the *stationary* holographic grating formed by the interference of A_2 and A_4. This situation is depicted in Figure 17-12(b). We may, likewise, using the grouping of (17.8-2), view

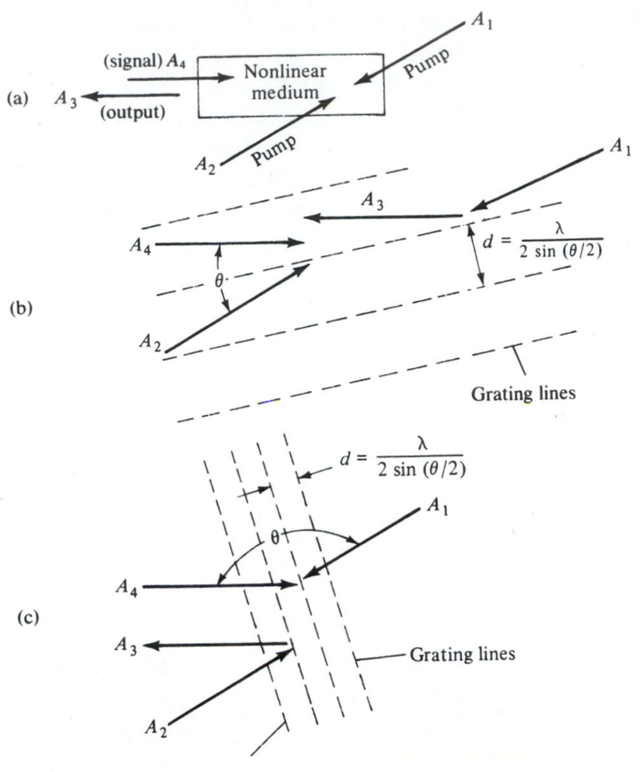

Figure 17-12 (a) The conventional geometry of phase conjugate optics. (b) Beams 2 and 4 interfere to form a grating $A_2A_4^*$. Beam 1 is Bragg diffracted from the grating to yield the output phase conjugate (to A_4) beam $A_3 \propto (A_2A_4^*)A_1$. (c) Beams 1 and 4 interfere to form a grating $A_1A_4^*$. Beam 2 is Bragg diffracted from the grating to yield the phase conjugate beam $A_3 \propto (A_1A_4^*)A_2$.

the process as the reflection of A_2 from the grating formed by beams A_1 and A_4. This situation is depicted in Figure 17-12(c).

It should be emphasized here that the grating point of view used above is employed mostly for pedagogic reasons and contains no new physics. Both sets of "gratings" [Figures 17-12(b) and (c)] are accounted for automatically in the electromagnetic formulation of phase conjugation in Section 17.3.

The multiplication property

$$A_3 \propto A_1A_2A_4^*$$

of the phase conjugate optical configuration of Figure 17-12(a) is the basis for numerous "real-time holographic applications" [22–25]. We will describe in what follows two generic applications: (1) image transmission through a distortion and

(2) real-time image processing with emphasis on the operations of correlation and convolution.

This task may become easier if we refer back to our discussion of the grating formation in conventional holography leading up to Equation (14.2-4). If we simply relabel $E_1 \rightarrow A_1$ (= "reference" wave) and $E_2 \rightarrow A_4$ (= "picture" wave) and take the total field at the hologram plane as in (14.2-4) as the sum of both these waves

$$\begin{aligned} E_{(x,z,t)} &= \mathrm{Re}[(A_1 e^{ik(+x\sin\theta + z\cos\theta)} + A_4 e^{ik(-x\sin\theta + z\cos\theta)})e^{i\omega t}] \\ &\equiv \mathrm{Re}[V(x, z, t)] \end{aligned} \tag{17.8-3}$$

then the *stationary* hologram, say the modulation of the index of refraction, is proportional to the temporal average of E^2:

$$\begin{aligned} \Delta n \propto \overline{E^2(x, z, t)} &= \tfrac{1}{2}(VV^*) \\ &= \tfrac{1}{2}|A_1|^2 + \tfrac{1}{2}|A_4|^2 + \tfrac{1}{2}(A_1 A_4^* e^{i2kX\sin\theta} + \text{c.c.}) \end{aligned} \tag{17.8-4}$$

The holographic grating thus consists of the term containing the factor $A_1 A_4^*$ as in (17.8-2). To complete the analogy we "illuminate" the grating (17.8-4) with a field E_2 traveling in the opposite sense to E_1, i.e.,

$$E_2(x, z, t) = \mathrm{Re}[(A_2 e^{-ik(x\sin\theta + z\cos\theta)})e^{i\omega t}] \tag{17.8-5}$$

The result is a new (diffracted) field that is proportional to the product of the grating function Δn and E_2[5]

$$E_3(x, z, t) \propto \Delta n(x, z) E_2(x, z, t) = \tfrac{1}{2}\mathrm{Re}[(A_1 A_4^*) A_2 e^{ik(x\sin\theta - z\cos\theta) + i\omega t}]$$

17.9 IMAGING THROUGH A DISTORTED MEDIUM

To illustrate how real-time holography, or more fundamentally, the three-wave multiplication (17.8-1) and (17.8-2) can be used for distortion correction, we refer to the experimental configuration of Figure 17-13. The object here is to transmit the transparency image $f(\mathbf{x}'')$ from plane (3) to plane (1) passing in the process a "thin" (but not necessarily weak) phase distorter that is characterized by the added (distorting) phase shift $\phi(\mathbf{x}')$.

Let us start with an intuitive approach. Wave A_2 starts as a plane wave, passes through the transparency $f(\mathbf{x})$ and continues on to the nonlinear medium. At the same time a "spy" plane wave A_1 that is temporally coherent with A_2 passes through and samples the distortion and is then imaged (here we use the precise optics definition of imaging) on the nonlinear medium that, for the sake of this discussion, will be taken to have negligible thickness in the z direction. A third wave A_3 (plane or spherical) arrives from the left (we recall here that three input waves are needed in four-wave mixing). These three waves "mix" by multiplication in the medium to

---

[5]In the case of a "thin" index grating Δn of thickness t, a field E_2 passing through it will emerge as $E_2 e^{-i(\omega/c)(\Delta n)t} \approx E_2(1 - i(\omega/c)\,\Delta n t)$ so that $E_4 \equiv \Delta E_2 \approx -iE_2(\omega/c)\,\Delta n$.

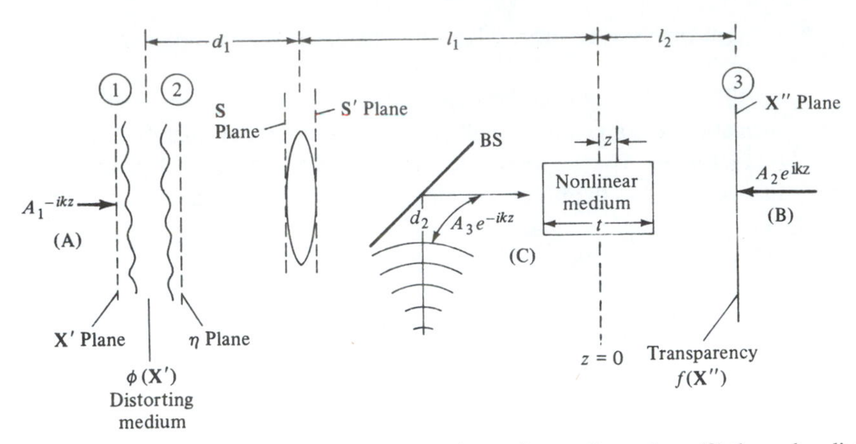

Figure 17-13 The configuration for propagating an image from plane (3) through a distorting medium to plane (1) with no distortion. (After References [18, 19].)

generate a product wave that contains both the transparency information $f(\mathbf{x})$ as well as the distortion information. The phase of the distortion, however, is reversed due to complex conjugation. This new wave with the negative distortion passes in reverse (from right to left) through the distortion and is healed.

The more rigorous mathematical treatment of this is quite formal and complicated. A somewhat simplified analysis can, however, capture much of the relevant physics.

The "spy" wave A_1 just to the right of the distortion is of the form

$$E_1(\boldsymbol{\eta}) = A_1 e^{i\phi(\boldsymbol{\eta})} \tag{17.9-1}$$

(We leave out everywhere the propagation factor $e^{\pm ikz}$.) It is, next, imaged by the lens onto the nonlinear medium at $z = 0$ where it regains, according to the law of imaging [21], its complex field distribution

$$E_1(z = 0, x) = A_1 e^{i\phi(\mathbf{x})}$$

Let the picture wave E_2 at $z = 0$ have the form $E_2(\mathbf{x})$ while the third wave is taken as a plane wave with amplitude $E_3(z = 0, \mathbf{x}) = A_3$. We note that apart from spatial modulation of the wavefronts by the information $f(\mathbf{x})$ and the distortion, the three waves incident on the nonlinear medium are arrayed as in the canonical four-wave mixing geometry of Figure 17-12(a). The result of the mixing is a new, left-traveling wave[6]

$$E_4 \propto E_1^*(\mathbf{x})E_2(\mathbf{x})A_3 = A_3 A_1^* e^{-i\phi(\mathbf{x})}E_2(\mathbf{x}) \tag{17.9-2}$$

Notice the change from $\phi(\mathbf{x})$ in E_1 to $-\phi(\mathbf{x})$ in E_4 due to complex conjugation. This wave is next imaged on plane (2) where it merely replicates the distribution

[6]To satisfy ourselves that this wave travels to the left, we need merely reinsert the propagation factors $\exp(\pm ikz)$ of the input waves E_1^*, E_2, and A_3.

(17.9-2). The passage from right to left through the distortion causes the wave to be multiplied by the factor $\exp[i\phi(\mathbf{x})]$ so that it emerges in plane (1) with the form

$$\begin{aligned} E_4(\mathbf{x}') &= A_3 A_1^* e^{-i\phi(\mathbf{x}')} E_2(\mathbf{x}') e^{i\phi(\mathbf{x}')} \\ &= A_3 A_1^* E_2(\mathbf{x}') \end{aligned}$$

i.e., beam 4 at the $\mathbf{x}'$ plane to the left of the distortion is, apart from the constant factor $A_3 A_1^*$, identical to the image bearing field E_2 at $z = 0$. We have thus, in effect, imaged the information $f(\mathbf{x})$ through the distortion while compensating perfectly for the latter. Figure 17-14 shows the results of an experimental demonstration of the one-way imaging through a distortion using the scheme of Figure 17-13 with a crystal of $BaTiO_3$ as the nonlinear medium.

17.10 IMAGE PROCESSING BY FOUR-WAVE MIXING

Before embarking on the topic of this section, we will provide some needed background results. The first involves the Fourier transform of a product of Fourier transforms. If we denote the Fourier transform of $g(x, y)$ by

$$F\{g(x, y)\} \equiv \tilde{g}(p, q) = \left(\frac{1}{2\pi}\right)^2 \int_{-\infty}^{\infty}\int_{-\infty}^{\infty} g(x, y) e^{-i(px+qy)}\, dx\, dy \qquad (17.10\text{-}1)$$

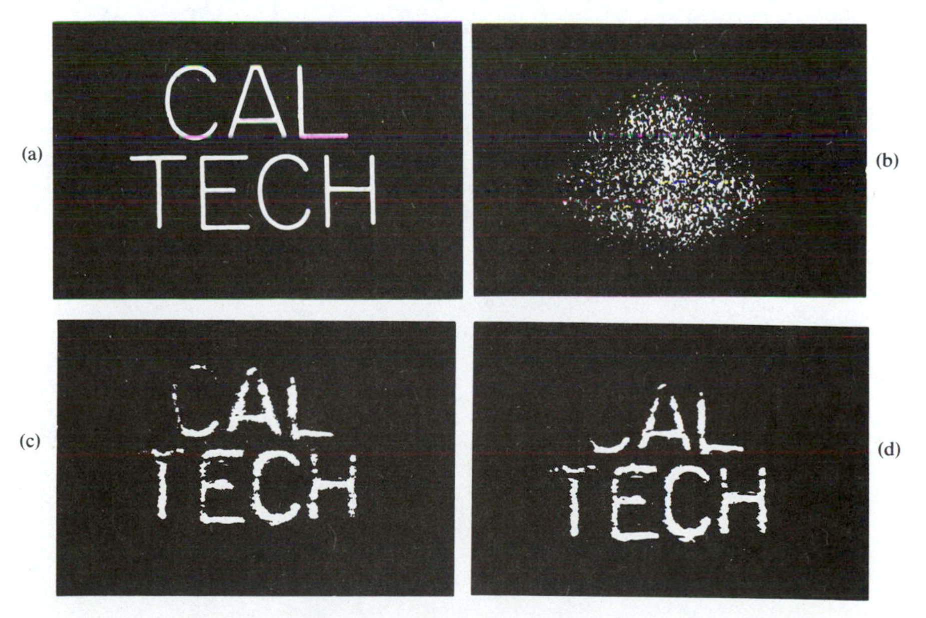

Figure 17-14 (a) Original transparency, (b) seen through distortion (no correction), (c) seen through distortion–phase conjugate window combination using configuration of Figure 17-13, and (d) seen through phase conjugate window (no distortion).

so that

$$g(x, y) = \int_{-\infty}^{\infty}\int_{-\infty}^{\infty} \tilde{g}(p, q)e^{i(px+qy)}\, dp\, dq$$

then it follows straightforwardly that the Fourier transform of the product

$$\begin{aligned} F[\tilde{f}(p, q)\tilde{g}^*(p, q)] &= \left(\frac{1}{2\pi}\right)^2 \int_{-\infty}^{\infty}\int_{-\infty}^{\infty} [\tilde{f}(p, q)\tilde{g}^*(p, q)]e^{-i(ps+qt)}\, dp\, dq \\ &= \left(\frac{1}{2\pi}\right)^4 \int_{-\infty}^{\infty}\int_{-\infty}^{\infty} f(x, y)g^*(x + s, y + t)\, dx\, dy \\ &\equiv \left(\frac{1}{2\pi}\right)^4 f(x, y)\star g(x, y) \end{aligned} \tag{17.10-2}$$

where $\star$ stands for the spatial correlation integral. Stating the last result in words, "The Fourier transform of the product of the transforms $\tilde{f}$ and $\tilde{g}^*$ is the correlation of the original functions f and g."

The second result to be stated here (see Appendix D) concerns the spatial Fourier transformation property of lenses. It is a basic result of the theory of optical [20] diffraction that the relation between the coherent (complex) field distributions in plane (2) distance f to the right of a lens and plane (1) a focal distance f to the left is one of Fourier transformation. Referring to Figure 17-15, the "output" field at plane (2) is related to the incident field at (1) by

$$u_2(x, y, 2f) = i\,\frac{4\pi^2 e^{-i2kf}}{\lambda f}\,\tilde{u}_1\left(p = \frac{kx}{f}, q = \frac{ky}{f}\right) \tag{17.10-3}$$

We can now use (17.10-2) and (17.10-3) to explain the operation of a real-time image processor based on four-wave mixing. Referring to Figure 17-16, collimated laser beams 1 and 4 from the left are modulated spatially by transparencies (or other means) placed at distance f(=focal length) to the left of lens L1. The resulting

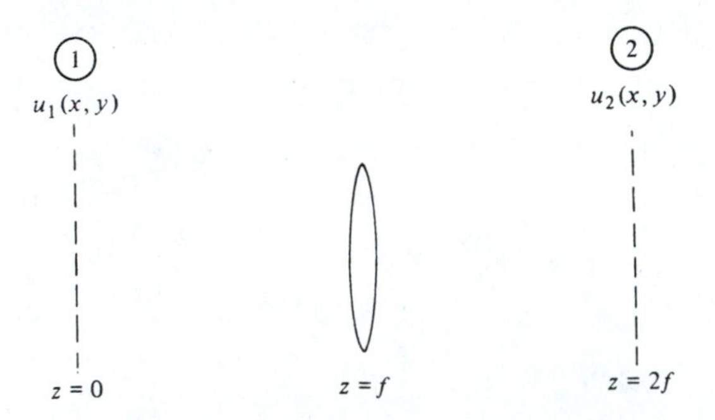

Figure 17-15 A transformation of a coherent field between the front and back focal planes of a thin lens.

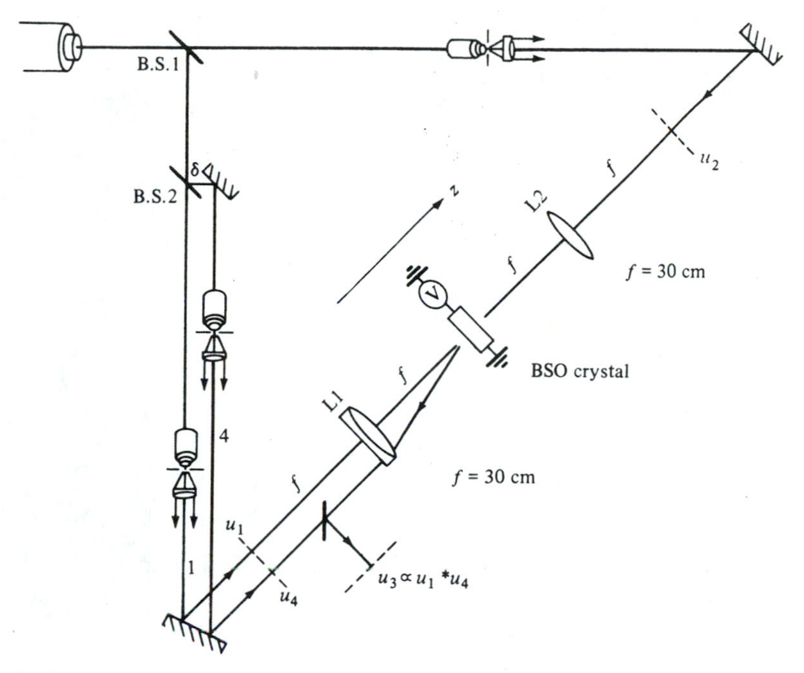

Figure 17-16 A setup for real-time image processing. In the experiment described above $u_2(x) = \delta(x)$, i.e., a pinhole so that $\tilde{u}_2$ = constant and $u_3 = u_1 \star u_4$.

spatially modulated fields are denoted by u_1 and u_4, respectively. Similarly, beam 2 is collimated by passing through a pinhole in plane z and then through lens L2 resulting in a plane wave $\tilde{u}_2$. The three beams are then incident on a nonlinear medium (a crystal of BSO in the original experiment). Using the Fourier transformation property of a lens [see Equation (17.10-3)], the three fields incident on the nonlinear medium are $\tilde{u}_1$, $\tilde{u}_4$, and $\tilde{u}_2$ (=const) where $\tilde{u}_1$ and $\tilde{u}_4$ are, according to (17.10-3), the spatial Fourier transforms of u_1 and u_4, respectively.

The multiplication property (17.8-2) of the nonlinear medium causes a field

$$E_3 \propto A_2 \tilde{u}_1 \tilde{u}_4^* \tag{17.10-4}$$

to be radiated to the left by the nonlinear medium. This field is thus proportional as indicated to the product of the Fourier transforms of u_1 and u_4^*. A distance f to the left of L1 in plane (3), one obtains according to (17.10-3) the Fourier transform of E_3 or according to (17.10-2)

$$u_3(x, y) \propto u_1 \star u_4 \tag{17.10-5}$$

The field u_3 is thus proportional to the two-dimensional correlation of the images u_1 and u_4. Note that this complicated mathematical operation is performed in real time. The result of a correlation experiment based on the geometry of Figure 17-16 is shown in Figure 17-17.

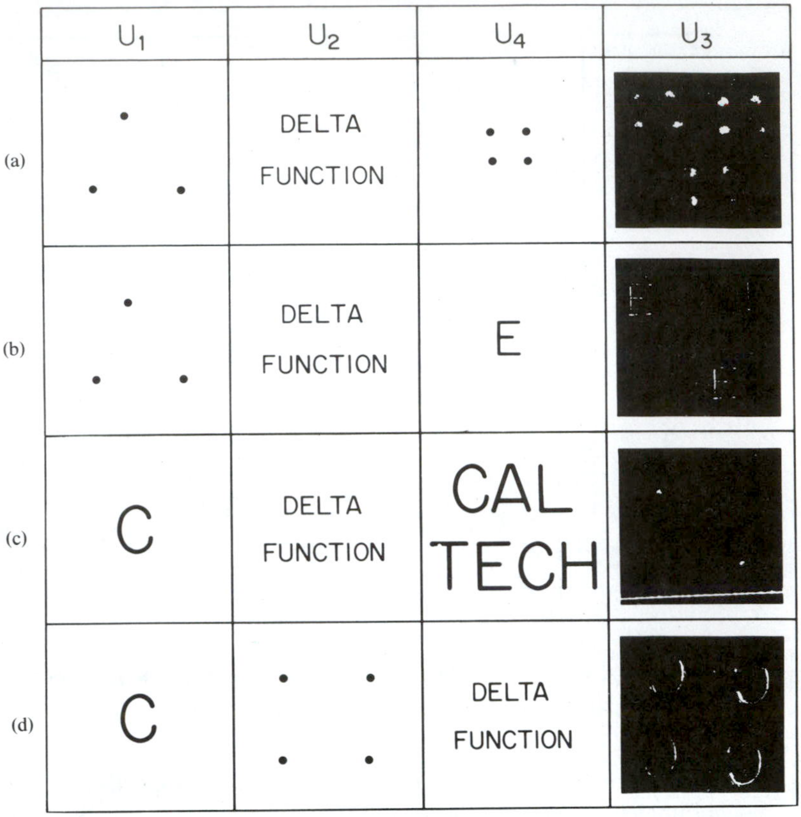

	U_1	U_2	U_4	U_3
(a)		DELTA FUNCTION		
(b)		DELTA FUNCTION	E	
(c)	C	DELTA FUNCTION	CAL TECH	
(d)	C		DELTA FUNCTION	

Figure 17-17 The results of the four-wave image processing experiment sketched in Figure 17-16. The input fields are transparencies u_1, u_2, u_4, while the output is u_3. Rows (a) through (c) illustrate correlation while row (d) illustrates convolution. (After Reference [22].)

Much of the present research in phase conjugate optics centers on the use of noncentrosymmetric photorefractive crystals such as barium titanate and strontium barium niobate [23–25]. In these crystals a standing optical wave pattern, such as that produced by the interference of two beams, A_1 and A_4, generates, by excitation and retrapping of impurity atom electrons, a corresponding spatially alternating electric field grating. This leads, via the electrooptic effect, to a spatial grating of the indices of refraction. Because of the diffusion and drift of the electrons responsible for the electric field, the index grating is displaced spatially with respect to the optical intensity pattern. This can cause power exchange between the very two beams, A_1 and A_4, that "write" the grating. This possibility does not exist in the conventional four-wave mixing, treated in this chapter, where no corresponding spatial shift exists.

This phenomenon of power exchange has been used to perform phase conjugation experiments without the need to supply externally the pump waves A_1 and A_2 [26]. It forms the main subject of Chapter 18.

17.11 COMPENSATION OF FIBER DISPERSION

One of the most important new applications of optical phase conjugation is the compensation for pulse spreading due to group velocity dispersion in optical fibers [26, 27]. This topic is discussed in Sec. 3.5.

Problems

17.1 Show, using the arguments of Section 8.1, that third-order optical effects as defined by (17.3-2) can exist in all homogeneous media.

17.2 Show that the reflection of the holograms (b) and (c) in Figure 17-12 each satisfy the Bragg condition.

17.3 Derive the coupled-mode equations in a manner similar to that leading to (17.3-12, 17.3-13) for the case where the frequency of the incident wave ω_4 is related to that of the pump beams (ω) by

$$\omega_4 = \omega - \delta$$

a. Show that the reflected wave frequency is $\omega_3 = \omega + \delta$.
b. Solve the coupled-mode equations for the reflection coefficient $|A_3(0)/A_4(0)|^2$. Plot it as a function of the frequency offset δ.

17.4

a. Solve the degenerate ($\omega_1 = \omega_2 = \omega_3 = \omega_4$) coupled-mode equations (17.3-14) as modified for (ordinary) optical losses. The new equations are

$$\frac{dA_4^*}{dz} = i\kappa e^{-\alpha L} A_3 - \alpha A_4^*$$

$$\frac{dA_3}{dz} = i\kappa e^{-\alpha L} A_4^* + \alpha A_3$$

$$\kappa = \frac{\omega}{2}\sqrt{\frac{\mu}{\varepsilon}}\,\chi^{(3)} A_1(L) A_2(0)$$

where α is the optical amplitude loss coefficient (assumed the same for all four beams).

b. Plot the reflection coefficient $|A_3(0)/A_4(0)|^2$ as a function of (κL) for $\alpha L =$ 0.1, 0.5, 1, 2. Discuss qualitatively the effect of the losses.

17.5 Invent an optical AND gate using phase conjugate optics.

17.6 Justify the image processing demonstrated in rows (a), (b), and (d) of Figure 17-17.

References

1. Gerritsen, H. J., "Nonlinear effects in image formation," *Appl. Phys. Lett.* 10:237, 1967.

2. Amodei, J. J., *RCA Rev.* 32:185, 1971.
3. Woerdman, J. P., ''Formation of a transient free carrier hologram in Si,'' *Opt. Commun.* 2:212, 1971.
4. Stepanov, B. I., E. V. Ivakin, and A. S. Rubanov, ''Recording two-dimensional and three-dimensional dynamic holograms in bleachable substances,'' *Doklady Akademii Nauk SSSR.* 196:567 (Translation: *Sov. Phys.-Doklady-Tech. Phys.* 16:46, 1971.)
5. Yariv, A., ''Three-dimensional pictorial transmission in optical fibers,'' *Appl. Phys. Lett.* 28:88, 1976.
6. Yariv, A., ''On Transmission and recovery of three-dimensional image information in optical waveguides,'' *J. Opt. Soc. Am.* 66:301, 1976.
7. Hellwarth, R. W., ''Generation of time-reversed wave fronts by nonlinear refraction,'' *J. Opt. Soc. Am.* 67:1, 1977.
8. Yariv, A., and D. M. Pepper, ''Amplified reflection, phase conjugation, and oscillation in degenerate four-wave mixing,'' *Opt. Lett.* 1:16, 1977.
9. Pepper, D. M., D. Fekete, and A. Yariv, *Appl. Phys. Lett.* 33:41, 1978.
10. Abrams, R. L., and R. C. Lind, ''Degenerate four-wave mixing in absorbing media,'' *Opt. Lett.* 2:94, 1978; Erratum, *Opt. Lett.* 3:205, 1978.
11. Liao, P. F., D. M. Bloom, and N. P. Economou, ''CW optical wave-front conjugation by saturated absorption in atomic sodium vapor,'' *Appl. Phys. Lett.* 32:813, 1978; ''Continuous-wave backward-wave generation by degenerate four-wave mixing in ruby,'' *Opt. Lett.* 3:4, 1978.
12. Yariv, A., ''Compensation for atmospheric degradation of optical beam transmission,'' *Opt. Commun.* 21:49, 1977.
13. Hellwarth, R. W., ''Third order susceptibilities of liquids and gases,'' *Progr. Quant. Elec.* 5:1, 1977.
14. Fisher, R. A., ed., *Optical Phase Conjugation.* New York: Academic Press, 1983. A comprehensive collection of material covering many aspects of phase conjugation.
15. Auyeung, J., D. Fekete, D. M. Pepper, and A. Yariv, ''A theoretical and experimental investigation of the modes of optical resonators with phase-conjugate mirrors,'' *IEEE J. Quant. Elec.* QE15:1180, 1979.
16. Beldyugin, I. M., and E. M. Zemskov, ''Theory of resonators with wavefront reversing mirrors,'' *Sov. J. Quant. Elec.* 9:1198, 1979.
17. Siegman, A. E., P. Belanger, and A. Hardy, chapter in Reference [14] entitled, ''Optical resonators using phase-conjugate mirrors.''
18. Cronin-Golomb, M., B. Fischer, J. Nilsen, J. O. White, and A. Yariv, ''Laser with dynamic holographic intracavity distortion correction capability,'' *Appl. Phys. Lett.* 41:220, 1982.
19. Yariv, A., and T. L. Koch, ''One way coherent imaging through a distorting medium using four-wave mixing,'' *Opt. Lett.* 7:113, 1982. Also G. J. Dunning, and R. C. Lind, ''Demonstration of image transmission through fibers by optical phase conjugation,'' *Opt. Lett.* 558, 1983.
20. Goodman, J. W., *Introduction to Fourier Optics.* San Francisco: McGraw-Hill, 1968.

21. White, J. O., and A. Yariv, "Real time image processing via four-wave-mixing in a photorefractive medium," *Appl. Phys. Lett.* 37:5, 1980.
22. Kukhtarev, N. V., V. B. Markov, S. G. Odulov, M. S. Soskin, and V. L. Vinetskii, "Holographic storage in electrooptic crystals," *Ferroelectrics* 22:949, 1979.
23. Cronin-Golomb, M., B. Fischer, J. O. White, and A. Yariv, "Theory and applications of four-wave mixing in photorefractive media," *IEEE J. Quant. Elec.* 20:12, 1984.
24. Huignard, J. P., J. P. Herriot, and G. Rivet, "Phase conjugation and spatial frequency dependence of wavefront reflectivity in $Bi_{12}SiO_{20}$(BSO) crystals," *Opt. Lett.* 5:102, 1980.
25. White, J. O., M. Cronin-Golomb, B. Fischer, and A. Yariv, "Coherent oscillation by self-induced gratings in the photorefractive crystal $BaTiO_3$," *Appl. Phys. Lett.* 40:450, 1982.
26. Yariv, A., D. Fekete, and D. M. Pepper, "Compensation for channel dispersion by nonlinear optical phase conjugation," *Opt. Lett.* 4:52, 1979.
27. Watanabe, S., T. Naito, and T. Chikama, "Compensation of chromatic dispersion in a single mode fiber by optical phase conjugation," *IEEE Photonics Tech. Lett.* 5:92, 1993.

Two-Beam Coupling and Phase Conjugation in Photorefractive Media

18.0 INTRODUCTION

In Chapter 17 we investigated the exchange of power among four optical waves at frequencies ω_1, ω_2, ω_3 and $\omega_4 = \omega_1 + \omega_2 - \omega_3$ that is mediated via the nonlinear optical response of the medium [see (17.2-2)]

$$P_i^{\mathrm{NL}}(\mathbf{r}) = 6\chi_{ijk\ell}^{(3)} A_{1j}(\mathbf{r}) A_{2k}(\mathbf{r}) A_{3\ell}^*(\mathbf{r}) e^{i[(\omega_1+\omega_2-\omega_3)t-(\mathbf{k}_1+\mathbf{k}_2-\mathbf{k}_3)\cdot\mathbf{r}} + \text{c.c.} \qquad \textbf{(18.0-1)}$$

where $A_{sm}(s = 1, 2, 3)$ is the complex amplitude of the mth Cartesian component of the sth wave. The coefficient $\chi_{ijk\ell}^{(3)}$ characterizes the nonlinear polarization response of the material medium (atoms, molecules). Relation (18.0-1) is local (it involves field quantities at the point $\mathbf{r}$ only), and the coefficient $\chi_{ijk\ell}^{(3)}$ can, in principle, be obtained by solving for the (local) nonlinear response of the atoms or molecules [2]. There exists a very important class of nonlinear interactions in which the response is nonlocal. Among these the photorefractive effect and stimulated Brillouin scattering are the most important since they both lead to large effects. Both of these cases can be described by a scenario in which two of the incident waves, say 1 and 3, "write" an index of refraction grating in the medium that is proportional to their (spatially and temporally varying) intensity interference pattern.

$$\Delta n \propto A_1 A_3^* e^{i[(\omega_1-\omega_3)t-(\mathbf{k}_2-\mathbf{k}_3)\cdot\mathbf{r}]} + \text{c.c.}$$

The third wave at ω_2 is Bragg-scattered from the grating, resulting in the fourth wave

$$A_4 \propto \Delta n A_3 \propto A_1 A_2 A_3^* \exp\{i[(\omega_1 + \omega_2 - \omega_3)t - (\mathbf{k}_1 + \mathbf{k}_2 - \mathbf{k}_3) \cdot \mathbf{r}]\}$$

which has the same form as (18.0-1).

Before considering this dynamic case, we take up the simpler situation of a fixed index grating.

18.1 TWO-WAVE COUPLING IN A FIXED GRATING

We consider a two-wave optical field at a radian frequency ω

$$\mathbf{E}(\mathbf{r}) = \left[\frac{1}{2}\,\mathbf{A}_1(r)e^{-i\mathbf{k}_1\cdot\mathbf{r}} + \frac{1}{2}\,\mathbf{A}_2(r)e^{-i\mathbf{k}_2\cdot\mathbf{r}}\right] e^{i\omega t} + \text{c.c.} \tag{18.1-1}$$

so that the complex amplitudes of the beams are $A_{1,2}$. The beam polarization is taken, for simplicity, to be perpendicular to the plane of the paper.

The two waves are propagating in a medium with a spatially periodic stationary index distribution ("grating")

$$n(\mathbf{r}) = n_0 + n_1 \cos(\mathbf{K} \cdot \mathbf{r} + \phi) \tag{18.1-2}$$

as shown in Figure 18-1.

The paraxial wave equation that is obeyed by the field (see 17.1-2) is

$$\nabla^2\mathbf{E} + \omega^2\mu\epsilon(\mathbf{r})\mathbf{E} = 0 \tag{18.1-3}$$

where the dielectric constant now has a contribution from the grating

$$\epsilon(r) = \epsilon_0 n^2(r) \cong \epsilon_0\,[n_0^2 + (n_0 n_1 e^{-i(\mathbf{K}\cdot\mathbf{r}+\phi)} + \text{c.c.})] \tag{18.1-4}$$

Substituting (18.1-1) and (18.1-4) in (18.1-3) leads to

$$\begin{aligned}
&\frac{1}{2}\left(-2ik_1\frac{dA_1}{dz} - k_1^2 A_1\right)e^{-i\mathbf{k}_1\cdot\mathbf{r}} + \text{c.c.}\\
&\quad + \frac{1}{2}\left(-2ik_2\frac{dA_2}{dz} - k_2^2 A_2\right)e^{-i\mathbf{k}_2\cdot\mathbf{r}} + \text{c.c.}\\
&\quad + \omega^2\mu\epsilon_0\,[n_0^2 + (n_0 n_1 e^{-i\phi}e^{-i\mathbf{K}\cdot\mathbf{r}} + \text{c.c.})]\\
&\quad \times \left[\frac{A_1}{2}\,e^{-i\mathbf{k}_1\cdot\mathbf{r}} + \frac{A_2}{2}\,e^{-i\mathbf{k}_2\cdot\mathbf{r}}\right] = 0
\end{aligned} \tag{18.1-5}$$

where we neglected

$$\frac{d^2A}{dz^2} \ll k\,\frac{dA}{dz}$$

We observe by inspection that spatially cumulative exchange of power takes place when the (Bragg) condition

$$\mathbf{k}_2 - \mathbf{k}_1 = \mathbf{K} \tag{18.1-6}$$

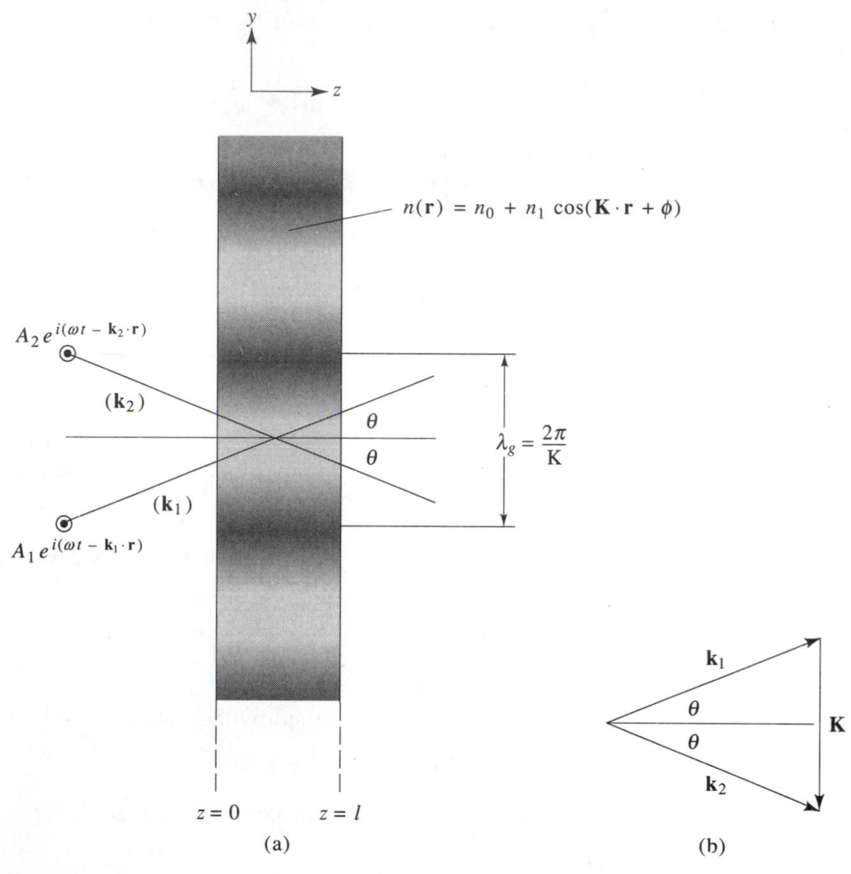

Figure 18-1 (a) Coupling between beam 1 propagating along $\mathbf{k}_1$ and beam 2 ($\mathbf{k}_2$) caused by a fixed index grating with a grating vector $\mathbf{K}$. (b) The Bragg condition diagram $\mathbf{k} = \mathbf{k}_2 - \mathbf{k}_1$.

is satisfied.[1] Keeping only synchronous terms (terms with similar exponents) and recalling that in an isotropic medium $k_1 = k_2 = \omega\sqrt{\mu\epsilon_0}n_0$ helps us simplify (18.1-5)

$$\cos\theta \frac{dA_1}{dz} = -\frac{\alpha}{2} A_1 - i \frac{\pi n_1}{\lambda_0} e^{i\phi} A_2 e^{i(\mathbf{k}_1-\mathbf{k}_2+\mathbf{K})\cdot\mathbf{r}}$$

$$\cos\theta \frac{dA_2}{dz} = -\frac{\alpha}{2} A_2 - i \frac{\pi n_1}{\lambda_0} e^{-i\phi} A_1 e^{-i(\mathbf{k}_1-\mathbf{k}_2+\mathbf{K})\cdot\mathbf{r}} \qquad \textbf{(18.1-7)}$$

where loss terms $-(\alpha/2)\, A_{1,2}$ were added phenomenologically to account for absorption and $\lambda_0 = 2\pi/(\omega\sqrt{\mu\epsilon_0})$ is the wavelength in free space. 2θ is the angle

[1]When condition (18.1-6) is not satisfied the power exchange reverses sign every $\Delta\ell = \pi/(|\mathbf{k}_2 - \mathbf{k}_1 - \mathbf{K}|)$. When (18.1-6) is satisfied the product term in (18.1-5) contains synchronous terms with factors $\exp(-i\mathbf{k}_1 \cdot \mathbf{r})$ and $\exp(-i\mathbf{k}_2 \cdot \mathbf{r})$.

between $\mathbf{k}_1$ and $\mathbf{k}_2$, and z is the distance measured along the bisector, so that $z = r_{1,2} \cos\theta$. Expressing the amplitudes in terms of magnitudes and phases by using the definition $A_j \equiv \sqrt{I_j} \exp(-i\phi_j)$ leads to (in what follows we take $\mathbf{k}_1 - \mathbf{k}_2 + \mathbf{K} = 0$, i.e., the Bragg condition is satisfied)

$$\cos\theta \frac{dI_1}{dz} = -\alpha I_1 + \frac{2\pi n_1}{\lambda_0} \sqrt{I_1 I_2} \sin(\phi_1 - \phi_2 + \phi)$$

$$\cos\theta \frac{dI_2}{dz} = -\alpha I_2 - \frac{2\pi n_1}{\lambda_0} \sqrt{I_1 I_2} \sin(\phi_1 - \phi_2 + \phi) \qquad \textbf{(18.1-8)}$$

Note that the coupling at a point $\mathbf{r}$ depends on the local phase $\psi \equiv (\phi_1 - \phi_2 + \phi)$. If the phase ψ is zero, no power exchange takes place. If $\psi = \pm\pi/2$, the exchange is maximum. The case $\psi = \pm\pi/2$ according to Equations (18.1-1) and (18.1-2) corresponds to a grating that is displaced by a quarter period with respect to the intensity interference pattern of waves 1 and 2. In the most common scenario, a single wave, say 1, is incident on the grating and wave 2 is the diffracted wave. In this case it follows from the second equation of (18.1-7) that wave 2 is generated with a phase $\phi_2 = \phi_1 + \phi + \pi/2$, i.e. $\psi = -\pi/2$, which results, according to the second equation of (18.1-8), in a *maximum* positive value for the power exchange dI_2/dz.

The solution of (18.1-8) in the case of $\psi = -\pi/2$ becomes

$$I_1(z) = I_1(0)e^{-\alpha z} \cos^2\left(\frac{\pi n_1 z}{\lambda_0 \cos\theta}\right)$$

$$I_2(z) = I_1(0)e^{-\alpha z} \sin^2\left(\frac{\pi n_1 z}{\lambda_0 \cos\theta}\right) \qquad \textbf{(18.1-9)}$$

so that in a grating of length ℓ the diffraction efficiency is

$$\eta = \frac{I_2(\ell)}{I_1(0)} = \exp\left(-\frac{\alpha\ell}{\cos\theta}\right) \sin^2\left(\frac{\pi n_1 \ell}{\lambda_0 \cos\theta}\right) \qquad \textbf{(18.1-10)}$$

This formula, first obtained by Kogelnik [1], is very useful in interpreting a large variety of experimental data involving fixed volume gratings and holograms.

We are now ready to consider the more interesting and varied case of dynamic scattering where the grating is not fixed but is generated in "real time" by the very two waves that scatter from it.

18.2 THE PHOTOREFRACTIVE EFFECT—TWO-BEAM COUPLING

In Section 18.1 we discussed the phenomenon of two-beam coupling (or diffraction) in a *fixed* hologram. In this section we will discuss the two-beam coupling by a hologram that is formed by the intensity interference pattern of the two (coupled) beams themselves. Under such circumstances the ''writing'' of the hologram by the two beams and the coupling of the (same) beams by the hologram cannot be considered separately, as was done in Section 18.1, and need be treated self-consistently. The phenomenon is known variously as two-beam coupling, dynamic holography, and real-time holography. The two most important classes of interactions that give rise to two-beam coupling are stimulated Brillouin scattering and the photorefractive coupling.

The photorefractive effect can be defined as a change in the index of refraction (n) of a material medium that is proportional to the intensity pattern of the light. It thus follows that since every material possesses a nonlinear optical response of the type given by Equation (17.3-7)

$$P_i^{(\mathrm{NL})} = 3\chi_{ijk\ell}^{(3)} E_j E_k^* E_\ell \qquad (18.2\text{-}1)$$

the material also displays a photorefractive effect since for the case of a single beam, i.e., putting $i = j = k = \ell$, we can obtain from (18.2-1) the relation $\Delta n = (1/2) \chi^{(3)}|E|^2/(n\epsilon_0)$ where n is the index of refraction when the field amplitude is zero. The nonlinear coefficient $\chi^{(3)}$ (see Table 17-1) in most materials is very small so that one requires very large optical intensities such as are available from pulsed lasers or from guided waves in small cross-sectional optical waveguides to affect appreciable ($\Delta n > 10^{-5}$) changes of the index of refraction. The cases of stimulated Brillouin scattering and that of the photorefractive effect are exceptions to the rule and lead to very strong effects. In each of these two cases, however, the interaction is mediated by a nonlocal effect—a traveling hypersonic wave in the case of Brillouin scattering, and of a traveling, or stationary, spatially periodic charge distribution in the photorefractive case. This gives rise to index changes that are orders of magnitude larger than those due to local atomic (or molecular) nonlinear response of the type described by (18.2-1). We will start with a description of the photorefractive effect.

This effect takes place in impurity-doped electrooptic crystals (i.e., crystals lacking inversion symmetry as discussed in Section 8.1). Let such a crystal be subject to a sinusoidal intensity distribution

$$I(x) = I_0 + I_1 e^{i(\Omega t - Kx)} + \text{c.c.} \qquad (18.2\text{-}2)$$

caused by the interference of two, mutually coherent, optical beams as shown in Figure 18-2. The optical intensity causes carriers, say electrons, to be excited from occupied donor states (the N_D^0 states of Figure 18-3) to the conduction band. Once excited the now highly mobile electrons will migrate away under the influence of diffusion and any internal or external electric fields until captured by a trapping center N_D^+, usually an empty donor (i.e., a donor that has lost its outer valence electron either by excitation to the conduction band or to a deep acceptor N_A). It follows that at steady state the high intensity regions will lose electrons while those of low intensity will acquire an excess of electrons.

The resultant space charge distribution ρ_{sc} is shown in Figure 18-2. Also shown is the electric field $E_{sc} = \dfrac{1}{\varepsilon} \int \rho_{sc}\, dx$ that results from the charge separation. Since the

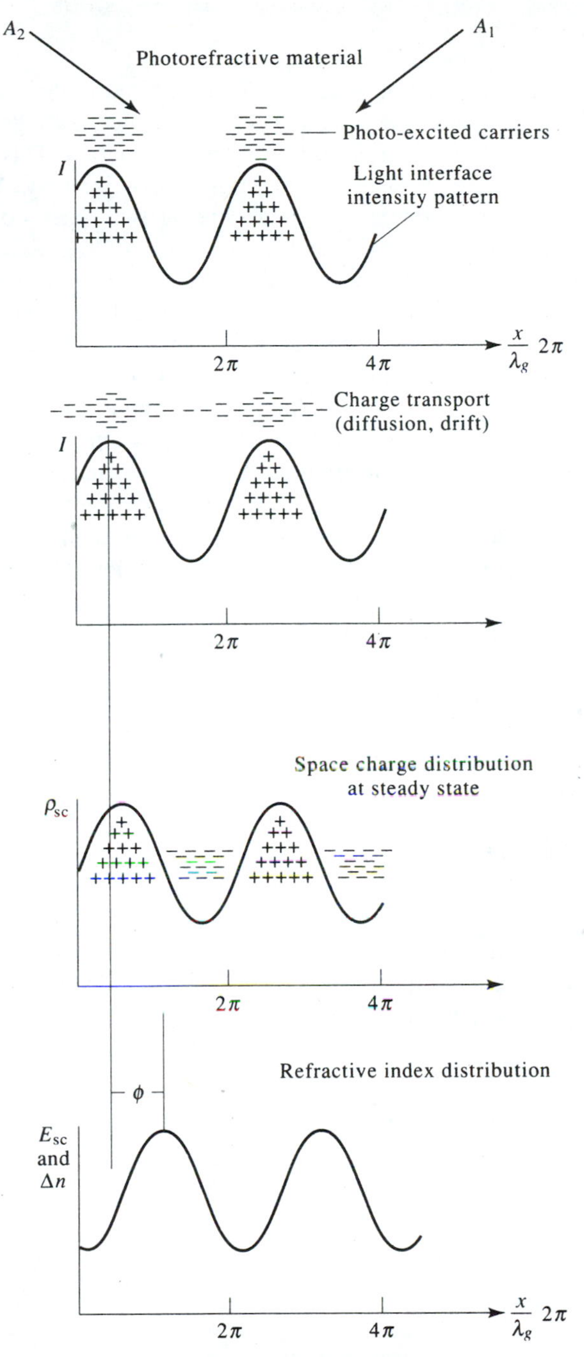

Figure 18-2 The photorefractive mechanism. Two coherent light beams intersect in an electrooptic crystal, forming an interference pattern. Electrons are excited where the intensity is large and migrate to regions of low intensity. The electric field associated with the resultant space charge operates through the electrooptic effect to produce a refractive index grating. ϕ is the phase shift (in radians) between the light interference pattern and the index grating.

crystal is electrooptic, an index grating $\Delta n \propto rE_{sc}$ is induced in the crystal by this field where r is the appropriate electrooptic coefficient. This index grating is displaced by a quarter period with respect to the charge distribution. This shift is due to the relation $\nabla \cdot E_{sc} = \rho_{sc}/\epsilon$. To analyze the physics of the grating formation we refer to Figure 18-3. The photorefractive crystal contains two species of atoms, the donor atoms whose density is $N_D(\text{cm}^{-3})$ and acceptor atoms (N_A). Since the energy of a valence electron in the acceptor atom state is lower than that of the donor, each acceptor atom has deprived a donor atom of an electron. This leaves behind a density $\langle N_D^+ \rangle = N_A$ of ionized donors (the $\langle\rangle$ brackets represent a spatial average). The remainder $(N_D - N_D^+)$ of the donor atoms are candidates for excitation by the optical field. Each such excitation generates a free (mobile) electron in the conduction band while, simultaneously, converting a unionized donor atom whose density is $(N_D - N_D^+)$ into an ionized N_D^+ site. Electrons can be trapped by the N_D^+ ions, returning them in the process into the unionized state. While in the conduction band the electrons are free to drift under the influence of the local electric field and diffuse.

In the above discussions we took the mobile charge carrier to be an electron. In some crystals it could be a hole. Simultaneous existence of electrons and holes is also possible. Also, the charge designation, N_D^+ and $N_D^0(=N_D - N_D^+)$, for example, are meant to represent the change in the charge state of the atom and not its true state. In BaTiO_3 for example, N_D^+ stands for Fe^{3+} ion, in which case N_D^0 will represent the Fe^{2+} state. This will be discussed further on in this section.

The density of electrons in the conduction band is denoted by n_e. The three species listed above coexist in the presence of the interaction with the optical field as well as the diffusion, drift, and trapping processes. The process indicated in Figure 18-3 converts the donor (N_D^0) atom into an electron trap (N_D^+) while the trap, having gained an electron, becomes an N_D^0 atom so that the spatial average of each species remains constant while the local concentration may vary. The acceptor atoms N_A are fully occupied by electrons at all times so that in the dark $\langle N_D^+ \rangle = N_A$ where $\langle\rangle$ stands for spatial averaging. The main role of the deep acceptors N_A is to ensure that there exists everywhere a large population of traps $\langle N_D^+ \rangle \cong N_A$ that can readily capture

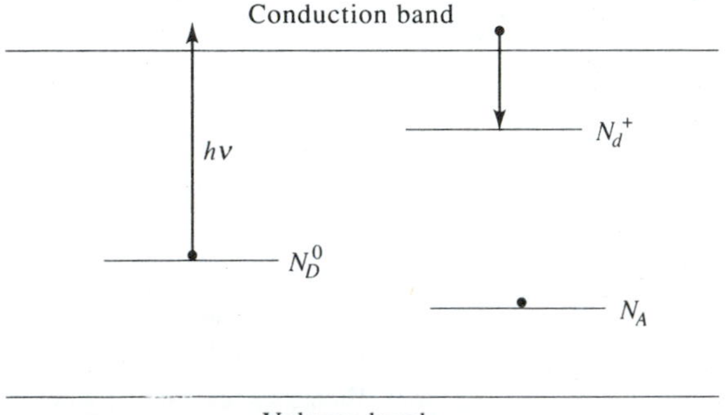

Figure 18-3 The deep impurity levels involved in the charge migration and trapping of a photorefractive crystal.

mobile electrons. Otherwise electrons could be trapped only at the sites from which they were excited, which would not give rise to the desired charge separation (i.e., grating).

The rate equation for the donor atom density is

$$\frac{\partial N_D^+(x, t)}{\partial t} = \left(\frac{\alpha_D}{h\nu}\right) I(x)(N_D - N_D^+) - \gamma_D n_e N_D^+ \tag{18.2-3}$$

where α_D is the absorption cross section of the $N_D^0 = N_D - N_D^+$ donor state atoms. γ_D is the recombination coefficient of a free electron at an N_D^+ site. γ_D is related to the commonly used recombination cross section σ_D by $\gamma_D = \sigma_d v_{\text{th}}$ where v_{th} is the mean thermal velocity of the free electrons.

The current density $J_x(A/m^2)$ is the sum of a drift and a diffusion term

$$J_x = \mu e n_e E_x + eD \frac{\partial n_e}{\partial x} = \mu e n_e E_x + k_B T \mu \frac{\partial n_e}{\partial x} \tag{18.2-4}$$

where we use $e \equiv |e|$ and the Einstein relation $eD = k_B T \mu$ relating the electron diffusion coefficient D to the Boltzmann constant k_B, the mobility μ, and the temperature T. The current continuity relation $\nabla \cdot \mathbf{J} = -\partial \rho / \partial t$ becomes

$$\frac{\partial J_x}{\partial x} = -e \frac{\partial}{\partial t}(N_D^+ - n_e) \tag{18.2-5}$$

and the Gauss relation

$$\frac{\partial E_x}{\partial x} = \rho/\varepsilon = \frac{e(N_D^+ - n_e - N_A)}{\varepsilon} \tag{18.2-6}$$

We will first solve for $n(x, t)$, $\rho(x, t)$, and $E_x(x, t)$ by assuming that the spatial modulation is small so that it can be represented by the first two harmonics ($n = 0$, $n = 1$) of the spatial Fourier amplitudes

$$N_D^+(x, t) = D_0 + [D_1 e^{-iKx} + \text{c.c.}] \tag{18.2-7}$$

$$n_e(x, t) = n_{e0} + [n_{e1} e^{-iKx} + \text{c.c.}] \tag{18.2-8}$$

$$E_x(x, t) = E_0 + [E_1^{\text{sc}} e^{-iKx} + \text{c.c.}] \tag{18.2-9}$$

Since the crystal is charge-balanced, it follows that

$$\langle \rho \rangle = e\langle N_D^+ - n_e - N_A \rangle \equiv (D_0 - n_{e0} - N_A) = 0$$

where $\langle\rangle$ denotes averaging over x. In addition, E_0, the average value of the internal field E_x, is equal to the externally applied electric field, if one exists.

The following approximations that are justified by the actual numerical values in real crystals are made

$$\begin{array}{ccccccc} N_D & \gg & N_A & \gg & n_{e0} & \quad & D_1 \gg n_{e1} \\ \downarrow & & \downarrow & & \downarrow & & \\ \sim 10^{19}\ \text{cm}^{-3} & & \sim 10^{16}\ \text{cm}^{-3} & & 10^{13}\ \text{cm}^{-3} & & \end{array}$$

The electric fields of the two interfering beams are

$$\mathbf{E}_1(\mathbf{r}, t) = \hat{\mathbf{e}}_1 A_1(\mathbf{r}) e^{i(\omega_1 t - \mathbf{k}_1 \cdot \mathbf{r})} + \text{c.c.}$$
$$\mathbf{E}_2(\mathbf{r}, t) = \hat{\mathbf{e}}_2 A_2(\mathbf{r}) e^{i(\omega_2 t - \mathbf{k}_2 \cdot \mathbf{r})} + \text{c.c.} \tag{18.2-10}$$

The squared magnitude of the field is thus[2]

$$\begin{aligned} I(x) &\equiv \tfrac{1}{2}\langle(\mathbf{E}_1 + \mathbf{E}_2) \cdot (\mathbf{E}_1 + \mathbf{E}_2)\rangle_{\text{space-time}} \\ &= |A_1|^2 + |A_2|^2 + \hat{\mathbf{e}}_1 \cdot \hat{\mathbf{e}}_2 (A_1 A_2^* e^{i[(\omega_1 - \omega_2)t - (\mathbf{k}_1 - \mathbf{k}_2)\cdot \mathbf{r}]} + \text{c.c.}) \end{aligned} \tag{18.2-11}$$

Comparing the last result to (18.2-2) leads to

$$\mathbf{K} = \mathbf{k}_1 - \mathbf{k}_2 = \hat{\mathbf{e}}_x |\mathbf{k}_1 - \mathbf{k}_2| \qquad I_0 \equiv |A_1|^2 + |A_2|^2$$
$$\Omega = \omega_2 - \omega_1 \qquad I_1 \equiv \hat{\mathbf{e}}_1 \cdot \hat{\mathbf{e}}_2 A_1 A_2^* \tag{18.2-11a}$$

We note that the interference term I_1 is zero when the two beams are mutually orthogonal. We have also assumed that in general $\omega_1 \neq \omega_2$.

We substitute (18.2-4) into (18.2-5) and eliminate n_{e0}, E_0, n_{e1}, D_0, and D_1 using (18.2-3) and (18.2-6). We take advantage of the inequalities and neglect the product of second-order terms, and after a good deal of algebra, obtain [3]

$$E_1^{\text{sc}} = -i \frac{I_1}{I_0} \frac{E_N(E_0 + iE_D)(e^{i\Omega t} - e^{-t/\tau})}{[E_0 - \Omega t_0(E_D + E_\mu)] + i(E_N + E_D + \Omega t_0 E_0)}$$

$$\Omega \equiv \omega_2 - \omega_1 \qquad E_N = \frac{eN_A}{\varepsilon K} \qquad E_\mu = \frac{\gamma_D N_A}{\mu K} \tag{18.2-12}$$

$$E_0 = \text{externally applied field}, \quad E_D = \frac{k_B T K}{e}$$

$$t_0 = \frac{N_A h\nu}{\alpha_D N_D I_0} \qquad \tau = t_0 \frac{E_0 + i(E_D + E_\mu)}{E_0 + i(E_N + E_D)}$$

The steady-state response is obtained at $t \gg \tau$, at which time the transient term $\exp(-t/\tau)$ can be neglected. Under typical conditions using a grating period $2\pi K^{-1} = 2\ \mu\text{m}$ and the above approximate values for N_D, N_D^+, and N_A, we estimate in $BaTiO_3$

$$E_D \sim 800 \text{ V/cm} \qquad E_N \sim 118 \text{ V/cm}$$

If no applied field is present $E_0 = 0$, and the steady-state internal field (18.2-12) is

$$E_1^{\text{sc}} \simeq -i\left(\frac{I_1}{I_0}\right) \frac{1}{1 + i\Omega\tau} \frac{E_D}{(1 + E_D/E_N)} e^{i\Omega t}$$

$$\tau = t_0 \frac{1 + E_\mu/E_D}{1 + E_N/E_D} e^{i\Omega t} \tag{18.2-13}$$

[2]The temporal averaging $\langle\rangle_{\text{time}}$ is over a few optical periods so that "slow" variations such as $\exp[i(\omega_1 - \omega_2)t]$ survive. The spatial averaging is over a few optical wavelengths.

If we take the steady-state limit of Equation (18.2-12) in the case $\omega_1 = \omega_2$ $(\Omega = 0)$ and no external field $(E_0 = 0)$, we obtain

$$E_1^{sc} \approx -i \frac{I_1}{I_0} \frac{E_D}{1 + \dfrac{E_D}{E_N}} \tag{18.2-14}$$

A few basic features of fundamental importance for practical applications stand out:

1. The factor $(-i)$ represents a quarter period shift of the index grating with respect to the intensity pattern.
2. E_1^{sc} (and Δn) depend not on the total intensity but on the fractional modulation I_1/I_0.
3. The space charge field tends toward the smaller of the E_N and E_D (multiplied by I_1/I_0).

These last two conclusions merit some further discussion: The total amount of separable charge is limited. The maximum separation would result when approximately *all* the traps in the low intensity regions are full, while all the traps in the high intensity regions are emptied. Since the initial density of such traps is $N_D^+ = N_A$, the resulting charge density can be approximated by

$$\rho_{sc} = N_A e \cos Kx = Re(N_A e e^{iKx})$$

From the Gauss law $\nabla \cdot \mathbf{E} = \rho/\epsilon$, we obtain, using complex notation,

$$E_1^{sc} = -\frac{ieN_A}{\epsilon K} = -iE_N \tag{18.2-15}$$

We thus identify E_N with the *maximum* space charge field that results from *full separation* (by half the grating period $\lambda/2 = \pi/K$) of the available charge (eN_A per unit volume).

The question then arises as to why the internal field is prevented from reaching a value $\sim E_N$ and is limited, instead, to a value of $\sim E_D = k_B TK/e$ when $E_N > E_D$. To answer this question, consider the electron current resulting from the excitation of some initial distribution of mobile carriers into the conduction band in the *presence* of a space charge field E_1^{sc} (due to earlier charge separation). From (18.2-4)

$$J_e = \mu e n_e E_1^{sc} + k_B T \mu \frac{\partial n_e}{\partial x}$$

Since the initial distribution n_e mimics the (negative) of the intensity (18.2-2), we can write

$$n_e(x) = n_{e0} - (n_{e1} e^{-iKx} + \text{c.c.}) \qquad \text{and}$$

$$E_{sc}(x) = E_1^{sc} e^{-iKx} + \text{c.c.}$$

The fundamental component of the current density becomes

$$J_e = (\mu e n_{e0} E_1^{sc} + iK k_B T \mu n_{e1}) e^{-iKx} + \text{c.c.}$$

It follows that the electron current vanishes when the space charge field E_1^{sc} reaches a value

$$E_1^{sc} = -i\frac{k_BTK}{e}\left(\frac{n_{e1}}{n_{e0}}\right) \equiv -iE_D\left(\frac{n_{e1}}{n_{e0}}\right)$$

and no further charge separation takes place. The internal space periodic field E_1^{sc} thus gets arrested at a value $\sim E_D$ and cannot achieve the charge-limited value of E_N. This situation changes in the presence of an external field E_0 that can "overpower" the internal space charge field so that E_1^{sc} tends, according to (18.2-12), to a value of $\sim -i(I_1/I_0)E_N$, corresponding to full charge separation. Figure 18-4 shows the theoretical dependence of the space charge field E_1^{sc} on the external field E_0. Of special interest is the change of E_1^{sc} from an initial value smaller than the smaller of E_N and E_D to a final value approaching E_N. In this particular example a large trap concentration N_A leading to a limiting field of $E_{N2} = 2 \times 10^3$ V/cm should give, according to curve (b), a fourfold increase in the internal space charge field with the application of an external field $E_0 \sim 10^4$ V/cm. Figure 18-5 shows experimental data from two strontium barium niobate $Sr_xBa_{1-x}Nb_2O_6$ (SBN:x = 60%) crystals doped with Cr. An increase in the space charge field by about a factor of three is seen for $E_0 \sim 10^4$ V/cm. Experimental data of the dependence of E_1^{sc} on E_0 is shown in Figure 18-5. The values of E_1^{sc} are deduced from two-beam coupling experiments that are discussed later.

Two of the most important features of the analysis are the dependence of the

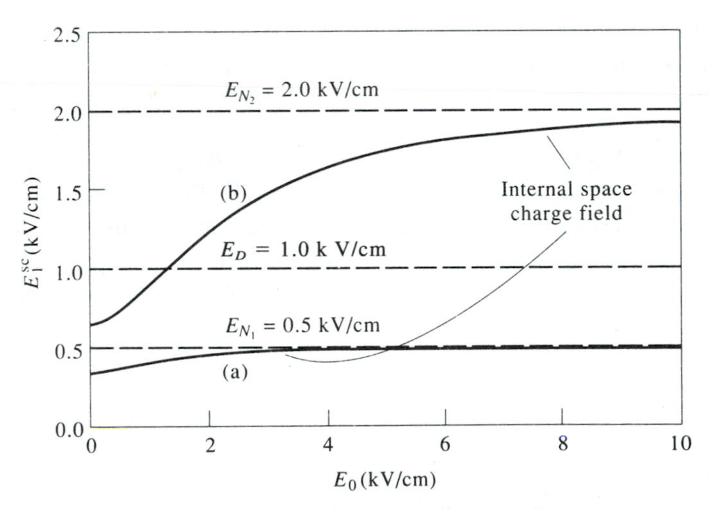

Figure 18-4 A theoretical plot of the amplitude E_1^{sc} of the spatially periodic internal electric field in photorefractive crystals [see Equation (18.2-12)] with $\Omega = 0$, $t \rightarrow \infty$ as a function of the externally applied field E_0. The characteristic fields are: $E_D = 10^3$ V/cm and (a) $E_N = 5 \times 10^2$ V/cm, (b) $E_N = 2 \times 10^3$ V/cm. (Private communication K. Sayano, The California Institute of Technology and R. R. Neurgaonkar, Rockwell International Corp.)

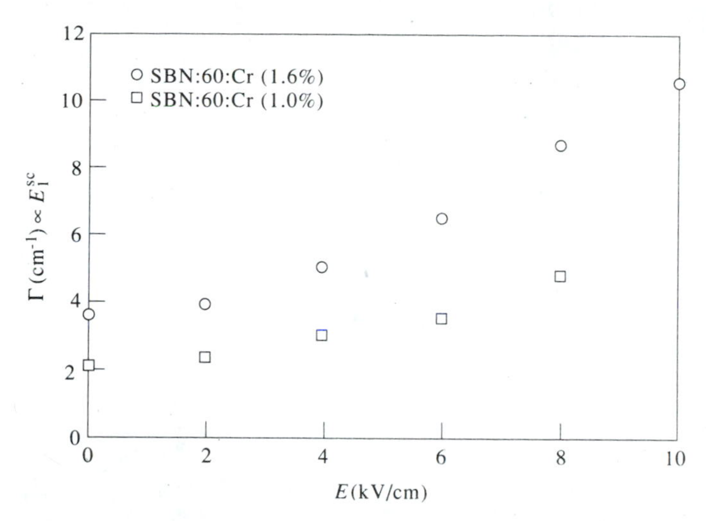

Figure 18-5 Experimental data of the two-beam coupling gain $\Gamma \propto E_1^{sc}$ as a function of applied electric field E_0 in two doped SBN:60 crystals. (After Reference [4].)

internal field E_1^{sc} on the grating period λ_g and the dependence of τ on intensity [6]. If we use the above definitions of E_D and E_N, we can rewrite E_1^{sc} of (18.2-12) as

$$E_1^{sc} = -i\left(\frac{I_1}{I_0}\right)\left(\frac{1}{1+i\Omega\tau}\right)\frac{(k_BTK/e)}{1+\dfrac{\varepsilon k_BTK^2}{e^2N_A}} \tag{18.2-16}$$

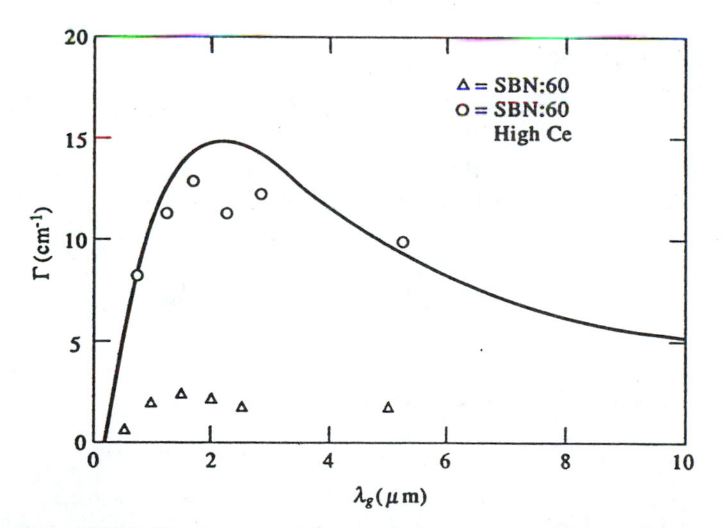

Figure 18-6 Two-beam coupling coefficient versus grating wavelength for $E_0 = 0$. The coupling coefficient Γ is proportional to the internal field E_1^{sc} of Equation (18.2-16). (After Reference [5].)

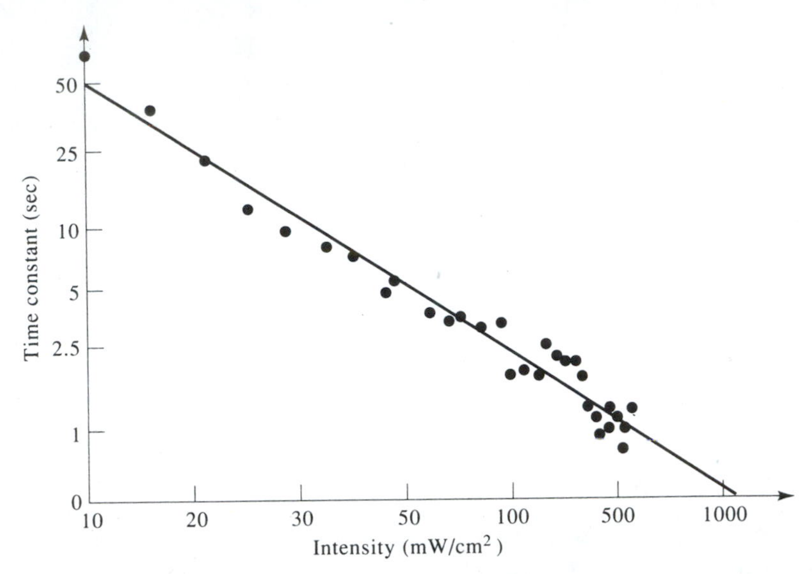

Figure 18-7 Photorefractive response time of the $BaTiO_3$ crystal versus intensity for $\lambda = .605$ μm, $\lambda_g = 1.4$ μm. (After Reference [5].)

so that the limits of $K \to 0$ ($\lambda_g \to \infty$) and $K \to \infty$ ($\lambda_g \to 0$), $E_1^{sc} \to 0$. Experimental evidence illustrating the dependence of E_1 on the grating period is shown in Figure 18-6, which shows an experimental plot of a quantity proportional to the internal field as a function of the grating period $\lambda_g = 2\pi K^{-1}$. Figure 18-7 shows the measured dependence of τ on the intensity I_0.

The Grating Formation

Now that we have determined the space-periodic electric field created by two interfering optical beams (18.2-12), we will derive an expression for the resulting refractive index grating induced by this field.

The change in the optical indicatrix due to a (low frequency) electric field E_k is given according to (9.2-1) by

$$\Delta\left(\frac{1}{n^2}\right)_{ij} = r_{ijk}E_k$$

where r_{ijk} is the electrooptic tensor element. This corresponds to a change in the index of refraction experienced by a propagating wave

$$\Delta n \equiv -\frac{1}{2}\, n_0^3 r_{\text{eff}} E \tag{18.2-17a}$$

where E is the low-frequency electric field, and r_{eff} is some linear combination of the electrooptic tensor elements r_{ijk} that depends on the crystal orientation and the field direction. We describe the spatial dependence of the index of refraction as

$$n(x, t) \equiv n_0 + \frac{1}{2}\left[\frac{n_1 e^{-i\phi} I_1}{I_0} e^{i(\Omega t - Kx)} + \text{c.c.}\right] \tag{18.2-17b}$$

where $I_0 = |A_1|^2 + |A_2|^2$, $I_1 = \hat{\mathbf{e}}_1 \cdot \hat{\mathbf{e}}_2 A_1 A_2^*$ so that, using (18.2-17a), n_1 is defined by

$$\frac{n_1 e^{-i\phi} I_1}{I_0} = \frac{n_1 e^{-i\phi} \hat{\mathbf{e}}_1 \cdot \hat{\mathbf{e}}_2 A_1 A_2^*}{I_0} = -n_0^3 r_{\text{eff}} E_1^{\text{sc}} \tag{18.2-18}$$

We take n_1 as real so that ϕ is the phase shift between the intensity pattern and that of the index $n(x)$. The quantity n_1 is thus the index modulation amplitude (within a factor I_1/I_0). From (18.2-12) in the case $\omega_1 = \omega_2$ $(\Omega = 0)$

$$n_1 e^{-i\phi} \simeq r_{\text{eff}} n_0^3 \frac{iE_N(E_0 + iE_D)}{E_0 + i(E_N + E_D)} \tag{18.2-19}$$

We note from (18.2-17) and (18.2-19) that the index modulation does not depend on the absolute intensity but only on the spatial modulation index $|I_1/I_0| \equiv |\hat{\mathbf{e}}_1 \cdot \hat{\mathbf{e}}_2 A_1 A_2^*|/I_0$. This reflects the fact that the role of the optical field is only to redistribute the electronic charge so that the maximum space charge field and Δn are limited, as discussed above, only by drift diffusion processes and by the total available charge, but not by the intensity.

Refractive Two-Beam Coupling

The index grating registered by beams E_1 and E_2 in a photorefractive medium causes power transfer by coupling E_1 and E_2 to each other. The coupling is due to simultaneous Bragg scattering of both beams from the grating into each other. Since the grating is due to the interference of the two beams, the Bragg condition $\mathbf{K} = \mathbf{k}_1 - \mathbf{k}_2$ is automatically satisfied and ensures that beam 1 is diffracted exactly in the direction of beam 2 and vice versa, hence the coupling between the two beams. What's more, in the case where $\omega_1 \neq \omega_2$, the grating moves with just the right velocity $((\omega_2 - \omega_1)/K)$ so that the Doppler-shifted frequency of the incident beam 1, i.e., $\omega_1 + \Delta\omega_{\text{Doppler}}$, is equal to that of beam 2 (ω_2) and vice versa. The analysis starts with the wave equation (6.5-3).

$$\nabla^2 \mathbf{E} + \omega^2 \mu \varepsilon(\mathbf{r}) \mathbf{E} = 0, \qquad \varepsilon(\mathbf{r}) = \varepsilon_0 n^2(\mathbf{r}) \tag{18.2-20}$$

Using Equation (18.2-18) for $n(\mathbf{r})$, assuming $n_1 \ll n_0$, and putting $\omega^2 \mu \varepsilon_0 = \omega^2/c^2$ leads to

$$\nabla^2 \mathbf{E} + \frac{\omega^2}{c^2}\left[n_0^2 + \frac{n_0 n_1 e^{-i\phi} \hat{\mathbf{e}}_1 \cdot \hat{\mathbf{e}}_2 A_1 A_2^*}{I_0} e^{i(\Omega t - Kx)} + \text{c.c.}\right]\mathbf{E} = 0 \tag{18.2-21}$$

The field $\mathbf{E}$ is the sum of the two fields $\mathbf{E}_1$ and $\mathbf{E}_2$ (18.2-10) that "write" the grating. If we substitute $\mathbf{E}$ into (18.2-21) and for simplicity replace the vector $\mathbf{E}$ by a scalar E [this requires the use of the proper electrooptic coefficient, or combination of coefficients, r_{eff} in (18.2-18)], the result is

$$\left[\left(-2ik_1\frac{dA_1}{dr_1}+\frac{d^2}{dr_1^2}A_1(r_1)-k_1^2A_1(r_1)\right)e^{i(\omega_1t-k_1r_1)}+\text{c.c.}\right]$$
$$+\left[\left(-2ik_2\frac{dA_2}{dr_2}+\frac{d^2}{dr_2^2}A_2(r_2)-k_2^2A_2(r_2)\right)e^{i(\omega_2t-k_2r_2)}+\text{c.c.}\right]$$
$$+\frac{\omega^2}{c^2}\left[n_0^2+n_0n_1e^{-i\phi}\frac{\hat{\mathbf{e}}_1\cdot\hat{\mathbf{e}}_2A_1A_2^*}{I_0}e^{i[(\omega_1-\omega_2)t-(k_1r_1-k_2r_2)]}+\text{c.c.}\right]$$
$$\times\,[A_1(r_1)\varepsilon^{i(\omega_1t-k_1r_1)}+A_2(r_2)e^{i(\omega_2t-k_2r_2)}+\text{c.c.}]=0 \qquad (18.2\text{-}22)$$

Recognizing that $k_{1,2}=\dfrac{\omega_{1,2}n_0}{c}$, neglecting the second derivative terms compared to those involved in the first derivatives (this is the slowly varying amplitude approximation), and equating separately terms with the same exponential factors lead to the coupled wave equations

$$\cos\theta_1\frac{dA_1}{dz}=-\frac{\alpha}{2}A_1-i\frac{\pi n_1}{\lambda_0}e^{-i\phi}\frac{|A_2|^2}{I_0}\hat{\mathbf{e}}_1\cdot\hat{\mathbf{e}}_2A_1$$
$$\cos\theta_2\frac{dA_2}{dz}=-\frac{\alpha}{2}A_2-i\frac{\pi n_1^*}{\lambda_0}e^{+i\phi}\frac{|A_1|^2}{I_0}\hat{\mathbf{e}}_1\cdot\hat{\mathbf{e}}_2A_2 \qquad (18.2\text{-}23)$$

where θ_1 and θ_2 are the angles between $\mathbf{k}_1$ and $\mathbf{k}_2$ and the normal to the crystal input face, taken as $z=0$. The loss term α was added phenomenologically to account for absorption in the crystal.

Before considering some exact consequences of (18.2-23), we might contemplate some qualitative features. Using (18.2-19) in the limit $E_N \gg E_D$, $E_0=0$ and $\phi=-\pi/2$, we can recast (18.2-23) in the form of

$$\cos\theta_1\frac{dA_1}{dz}=-\frac{\alpha}{2}A_1+\left[\frac{\pi n_0^3}{\lambda_0}r_{\text{eff}}(\hat{\mathbf{e}}_1\cdot\hat{\mathbf{e}}_2)\frac{E_D|A_2|^2}{I_0}\right]A_1$$
$$\cos\theta_2\frac{dA_2}{dz}=-\frac{\alpha}{2}A_2-\left[\frac{\pi n_0^3}{\lambda_0}r_{\text{eff}}(\hat{\mathbf{e}}_1\cdot\hat{\mathbf{e}}_2)\frac{E_D|A_1|^2}{I_0}\right]A_2 \qquad (18.2\text{-}24)$$

which for $r_{\text{eff}}>0$ indicates the growth of A_1 at the expense of A_2 with an initial exponential growth constant

$$\gamma=\left[\frac{\pi n_0^3}{\lambda}r_{\text{eff}}(\hat{\mathbf{e}}_1\cdot\hat{\mathbf{e}}_2)E_D\frac{|A_2|^2}{I_0}-\frac{\alpha}{2}\right] \qquad (18.2\text{-}25)$$

The direction of power flow depends on the sign of r_{eff} and thus can be reversed by inverting the crystal orientation. Defining normalized intensities as $\mathcal{I}_1=|A_1|^2$, $\mathcal{I}_2=|A_2|^2$, we obtain directly from (18.2-23)

$$\cos\theta_1\frac{d\mathcal{I}_1}{dz}=-\alpha\mathcal{I}_1-\frac{2\pi n_1}{\lambda_0}\hat{\mathbf{e}}_1\cdot\hat{\mathbf{e}}_2\sin(\phi)\mathcal{I}_1\mathcal{I}_2/(\mathcal{I}_1+\mathcal{I}_2)$$
$$\cos\theta_2\frac{d\mathcal{I}_2}{dz}=-\alpha\mathcal{I}_2+\frac{2\pi n_1}{\lambda_0}\hat{\mathbf{e}}_1\cdot\hat{\mathbf{e}}_2\sin(\phi)\mathcal{I}_1\mathcal{I}_2/(\mathcal{I}_1+\mathcal{I}_2) \qquad (18.2\text{-}26)$$

A similar analysis for ordinary nonlinear transparent materials that are characterized by a real $\chi^{(3)}$ will lead to $\phi = 0$ ($I(x)$ and $n(x)$ "in step") so that, according to (18.2-26), no power exchange takes place. In a photorefractive material, on the other hand, where $n(x)$ is given by Equation (18.2-17, 18.2-19), $\phi = \pm\pi/2$ is possible and the power transfer is maximum. By adding the last two equations, we obtain

$$\frac{d}{dz}(\mathcal{I}_1 \cos\theta_1 + \mathcal{I}_2 \cos\theta_2) = -\alpha(\mathcal{I}_1 + \mathcal{I}_2)$$

which in the case of $\alpha = 0$ amounts to the conservation of total power.

Two-Beam Coupling—Symmetric Geometry

In the case of $\theta_1 = -\theta_2 \equiv \theta$, illustrated in Figure 18.8, Equation (18.2-26) can be solved exactly. If we define $J_{1,2} \equiv \mathcal{I}_{1,2}e^{\alpha r}$, $r_1 = r_2 \equiv r = z/\cos\theta$ is the distance measured along the beams' propagation directions

$$\frac{dJ_1}{dr} = -2\Gamma\frac{J_1J_2}{J_t} = -2\Gamma\frac{J_1(J_t - J_1)}{J_t}$$

$$\frac{dJ_2}{dr} = 2\Gamma\frac{J_1J_2}{J_t} = 2\Gamma\frac{J_2(J_t - J_2)}{J_t} \tag{18.2-27}$$

$$J_t \equiv J_1 + J_2$$

$$2\Gamma = \frac{2\pi n_1}{\lambda}\hat{\mathbf{e}}_1 \cdot \hat{\mathbf{e}}_2 \sin\phi \tag{18.2-28}$$

These equations can be integrated directly. The result, when expressed in terms of the original intensity variables, is

$$\mathcal{I}_1(r) = \mathcal{I}_1(O)e^{-\alpha r}\frac{\mathcal{I}_1(O) + \mathcal{I}_2(O)}{\mathcal{I}_1(O) + \mathcal{I}_2(O)e^{2\Gamma r}}$$

$$\mathcal{I}_2(r) = \mathcal{I}_2(O)e^{-\alpha r}\frac{\mathcal{I}_1(O) + \mathcal{I}_2(O)}{\mathcal{I}_1(O)e^{-2\Gamma r} + \mathcal{I}_2(O)} \tag{18.2-29}$$

In the case of $\mathcal{I}_2(O) \ll \mathcal{I}_1(O)e^{-2\Gamma r}$, the last equation becomes

$$\mathcal{I}_2(r) = \mathcal{I}_2(O)e^{(2\Gamma - \alpha)r} \tag{18.2-30}$$

This predicted power exchange has been observed first by Staebler and Amodei in 1972 (Reference [7]).

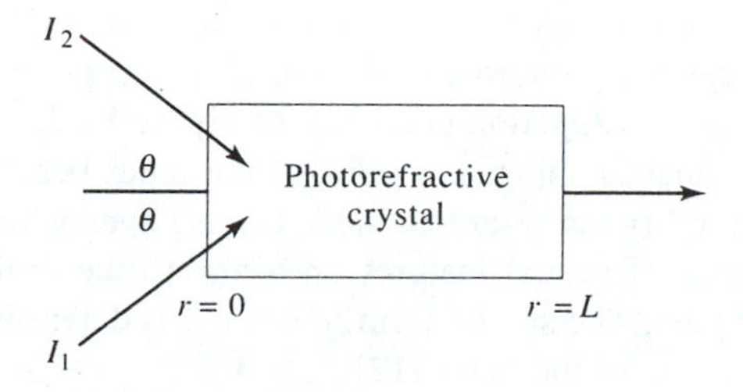

Figure 18-8 The symmetric two-beam coupling configuration.

Numerical Example: Two-Beam Coupling in $BaTiO_3$

$BaTiO_3$ is an electrooptic crystal with a perovskite structure that is ferroelectric at room temperature (its electrooptic constants are listed in Table 9-2), which possesses an extremely large electrooptic coefficient. The particular coefficient that comes into play in the two-beam coupling is $r_{51} = r_{42} = 16.4 \times 10^{-10}$ m/V. Using the following data $n_0 = 2.5$, $r_{\text{eff}} \approx r_{42}/2$, $E_1^{sc} \sim E_D \simeq 5 \times 10^2$ V/cm, $\lambda = 0.5$ μm, $\phi = \pi/2$, $\hat{\mathbf{e}}_1 = \hat{\mathbf{e}}_2$. $\alpha = 0$, $I_1 = I_2 = I_0/2$, and using Equations (18.2-28) and (18.2-18)

$$2\Gamma = \frac{2\pi n_1}{\lambda}\sin\phi \approx \frac{\pi}{\lambda} n_0^3 r_{\text{eff}} E_D \approx 40 \text{ cm}^{-1}$$

This is a very large gain constant. To put it in perspective we may recall that most laser media provide gain of good deal less than 1 cm^{-1}. As a matter of fact, only in semiconductor lasers do we encounter similar gains.

The analogy of photorefractively induced gain to ordinary laser gain is quite fundamental since in the presence of gain all we need do to obtain oscillation is to provide optical feedback. A whole new class of optical devices has sprung over the last few years that depends on photorefractively pumped oscillators [11, 12]. Some of these devices will be described in the remainder of this chapter. Reference [15] contains a number of review articles on photorefractive topics.

18.3 PHOTOREFRACTIVE SELF-PUMPED PHASE CONJUGATION (10, 11, 12)

The large optical amplification by two-beam coupling in photorefractive crystals can be used to "pump" a new class of optical oscillators, and these in turn can perform a variety of tasks. Most of these tasks involve passive phase-conjugate reflectors, that is, phase-conjugating mirrors that do not require externally supplied pump beams [9, 10]. To illustrate this principle, consider the configuration of Figure 18-9.

An input beam 4 provides gain, by two-beam coupling, to beam 1 in a photorefractive crystal placed inside a two-mirror (R_1, R_2) optical resonator. If this gain (see Eq. 18.2-30) is sufficient to overcome the crystal and mirror losses, an oscillating optical field builds up inside the resonator. The two traveling beams 1 and 2 that make up the oscillating field then play the role of the conventional pump beams as in the canonical four-wave phase-conjugation geometry of Figure 17-2, resulting in a reflected beam 3 that is the phase-conjugate replica of the input beam 4. A key point here is that beam 4 can amplify the resonator mode beam 1 even when it (beam 4) bears an image. The presence of spatial features on beam 4 (due to the image) only influences the amount of gain. The spatial form of beam 1 is determined solely by the resonator, since it is a mode of the latter [12].

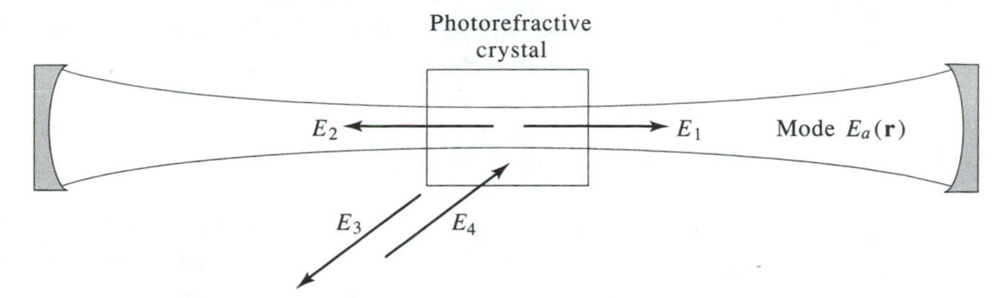

Figure 18-9 A photorefractively pumped optical oscillator.

The possibility of phase conjugation without externally provided pump beams, the so-called "self-phase conjugation," opens up a new area of practical applications involving image processing and distortion correction. Before we move on to discuss these applications, we show in Figure 18-10 an impressive demonstration of correction for propagation distortion following passive (self-pumped) phase conjugation by J. Feinberg [10]. The feedback for the photorefractive oscillation in this case is provided not by external reflectors, as in Figure 18-9, but by total internal reflection in the photorefractive ($BaTiO_3$ crystals). The point to appreciate here is the com-

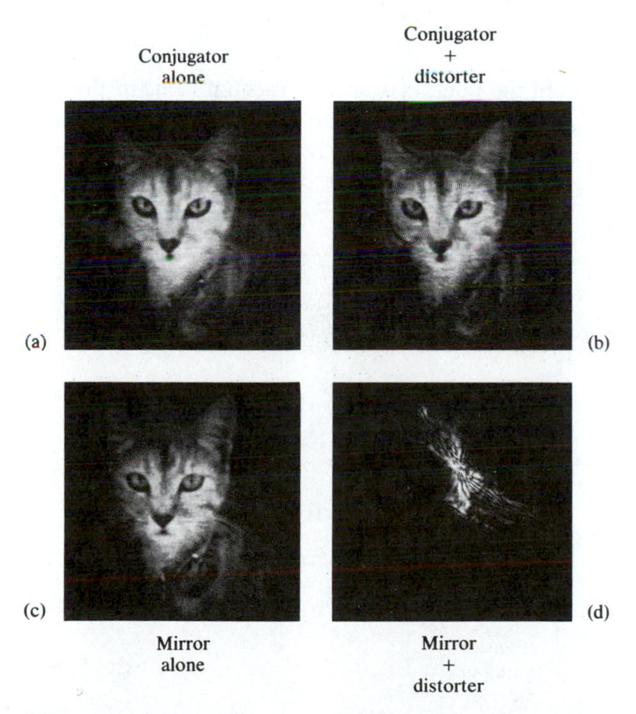

Figure 18-10 Images of a cat reflected from a self- pumped phase conjugate mirror (SPPCM) of $BaTiO_3$, a combination of a distorter and a SPPCM, an ordinary mirror, and an ordinary mirror plus a distorter. (Courtesy of J. Feinberg, University of Southern California [10].)

parison of the distorted image (image + distorter) to the distorted and (phase) conjugated image (conjugator + distorter).

18.4 APPLICATIONS OF PHOTOREFRACTIVE OSCILLATORS

In this section we will describe some generic applications of photorefractive oscillators. In principle almost any application that uses "conventional" phase conjugate reflectors, i.e., ones that depend on externally supplied pump waves (A_1 and A_2 in Figure 17-10) can also use a self-pumped phase conjugate mirror [16]. As a matter of fact, the dynamic distortion correction of a laser mode demonstrated in Figure 17-11 employs a self-pumped phase conjugate mirror consisting of the crystal c and the *single* mirror M_2.

Rotation Sensing

As a representative example of the applications of phase conjugate mirrors we take up the case of fiber rotation sensors [17, 18]. A rotation sensor based on the Sagnac effect is shown in Figure 18-11. The basic principle of such a sensor is that light propagating in a coiled fiber (coil radius = R) will undergo an extra phase shift

$$\phi = \frac{2\pi RL}{\lambda c}\Omega \tag{18.4-1}$$

where L is the length of the fiber, c, λ are the vacuum value of the velocity of light and the wavelength, respectively, and Ω is the angular rotation rate (rad/s) about an axis normal to the plane of the loop. $\Omega > 0$ when the light propagates in the same

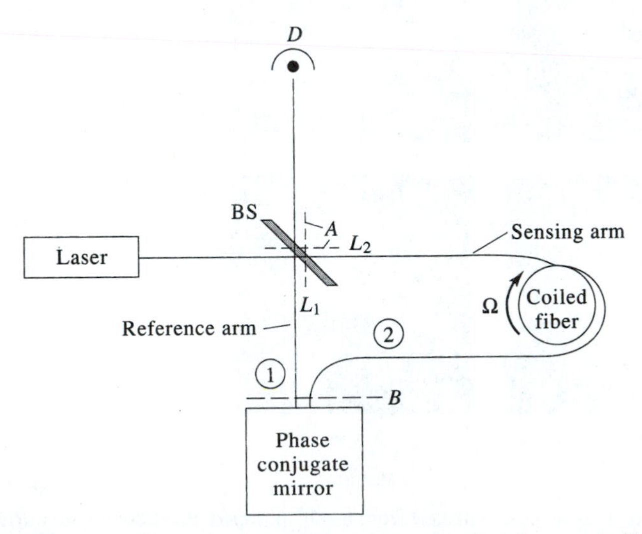

Figure 18-11 The generic configuration of an interferometric fiber rotation sensor employing the Sagnac effect and a self-pumped phase conjugate mirror.

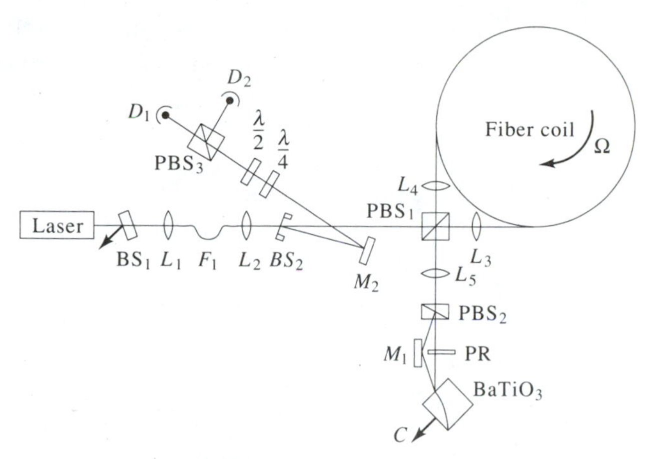

Figure 18-12 Experimental fiber-optic gyroscope setup using a self-pumped phase conjugate mirror. Instead of the two fibers in Figure 18-13, the experimental setup of this figure uses the two polarization modes of a polarization preserving fiber. (After Reference [18].)

sense as the rotation and is negative otherwise. Note that ϕ is independent of the refractive index n of the fiber material [17].

The derivation of ϕ is based on relativistic considerations. The correct result, however, can be obtained almost rigorously, from a consideration of the effective elongation (shortening) of a fiber experienced by light propagating in the same (opposite) sense as the rotation.

In the interferometric arrangement of Figure 18-12 a beam splitter sends one part of the input beam through a coiled fiber (sensing arm) while the other portion of the split beam enters the "reference" uncoiled arm. Both beams are then reflected from a self-pumped phase conjugate mirror, retraverse their original path and recombine interferometrically at detector D. Consider first the beam in the sensing arm. On its forward path between planes A and B it undergoes a phase shift

$$\phi_1 = kL_2 + \frac{2\pi R L_2}{c\lambda}\Omega + \phi_{R2}(t) \tag{18.4-2}$$

where $k = 2\pi n/\lambda$, and $\phi_{R2}(t)$ accounts for random ("noise") phase fluctuations in the sensing leg that are "slow" on the time scale of a round trip (these could be due, for example, to strain or temperature variations). The phase of the beam after reflection from the conjugator at point B is the reverse of the input phase plus an additional phase ϕ_c

$$\phi_2 = -kL_2 - \frac{2\pi R L_2}{c\lambda}\Omega - \phi_{R2}(t) + \phi_c[\Omega, kL_1, kL_2, \phi_{R1}(t), \phi_{R2}(t)] \tag{18.4-3}$$

The last term ϕ_c in (18.4-3) is of fundamental importance. In the case of a single input (i.e., with the reference arm blocked-off) $\phi_c = 2kL_2$ and the overall phase delay $\phi_1 + \phi_2$ due to L_2 is $2kL_2$, as required by causality. In the case of more than one input beam, the phase delay term ϕ_c is the same for all of them[3] [19], so that the relative phases of the beams are reversed, as required for phase conjugation, while the absolute phases are not. After retracing the original path, the sensing beam returns to A with a phase

$$\begin{aligned}\phi_{\text{sense}} &= \left[-kL_2 - \frac{2\pi RL_2}{c\lambda}\Omega - \phi_{R2}(t) + \phi_c[\Omega, kL_1, kL_2, \phi_{R1}(t), \phi_{R2}(t)]\right] \\ &\quad + \left(kL_2 - \frac{2\pi RL_2}{c\lambda}\Omega + \phi_{R2}(t)\right) \\ &= -\frac{4\pi RL_2}{c\lambda}\Omega + \phi_c[\Omega, kL_1, kL_2, \phi_{R1}(t), \phi_{R2}(t)]\end{aligned} \tag{18.4-4}$$

note that the phase shift due to Ω has the opposite sign on the return trip since the sense of rotation relative to the beam is now reversed.

Repeating the same procedure in the case of the reference beam leads to a phase of the reflected beam at plane A

$$\begin{aligned}\phi_{\text{ref}} &= -[kL_1 + \phi_{R1}(t)] + \phi_c(\Omega, kL_1, kL_2, \phi_{R1}(t), \phi_{R2}(t) \\ &\quad + [kL_1 + \phi_{R1}(t)] \\ &= \phi_c(\Omega, kL_1, kL_2, \phi_{R1}(t), \phi_{R2}(t))\end{aligned} \tag{18.4-5}$$

The interference output signal at D thus involves the phase difference

$$\phi_{\text{ref}} - \phi_{\text{sense}} = \frac{4\pi RL_2}{c\lambda}\Omega \tag{18.4-6}$$

from which the extraneous effects of phase fluctuations ϕ_{R1}, ϕ_{R2}, that usually limit the sensitivity of such sensors, disappeared.

A basic point that should have emerged from the preceding discussion is that although a passive (self-pumped) phase conjugate mirror does not reverse the absolute[4] phase of an incoming beam, it does reverse the *relative* phases of the Fourier components (partial plane waves) that make up the beam. This property is sufficient to guarantee wavefront reversal and enable various sensor applications [19]. Further discussion of this point is included in Problem 5 at the end of this chapter.

Mathematical and Logic Operations on Images

We have already discussed in Chapter 17 how nonlinear optical techniques can be used to perform spatial correlation, convolution, and other operations. In the following we will discuss some new mathematical operations that can be performed with passive (or externally pumped) phase conjugate mirrors.

---

[3]The crystal "regards" the multiplicity of input beams, which are coherent relative to each other, as a single, albeit complex, beam. The phase ϕ_c is thus due to the complex but *single* grating "written" in the crystal by this composite beam.

[4]"Absolute phase" here is taken to mean the phase relative to some coherent reference wave.

Consider the configuration shown in Figure 18-13. A plane wave with amplitude E_{in} is split by a lossless beam splitter BS_1 whose reflection and transmission coefficients are equal to r and t, respectively. Let r′ and t′ be the amplitude reflection and transmission coefficients for waves incident from the opposite side of the beam splitter. Each of the two resulting (split) waves passes through a transparency with amplitude transmittance T_1 for beam 1 and T_2 for beam 2. The two beams are then reflected by a self-pumped phase conjugate mirror (SPCM) with phase conjugate amplitude reflectivity R. The phase conjugate beams recombine interferometrically at beam splitter BS_1 to form an output field. The total field at D_1 is

$$E_{out} \propto E_{in}^* [(rT_2)^*R\, T_2 t + (tT_1)^*RT_1 r'] = E_{in}^* R(r^*t)(|T_2|^2 - |T_1|^2) \quad (18.4\text{-}7)$$

where we took advantage of the time reversibility condition

$$r't^* + r^*t = 0 \quad (18.4\text{-}8)$$

The intensity at D_1 becomes

$$\begin{aligned} I_{out} &\equiv |E_{out}|^2 \propto |R|^2 \| T_1|^2 - |T_2|^2|^2 |r^*t|^2 I_{in} \\ &\propto |T_1|^2 \mathbf{O} |T_2|^2 \end{aligned} \quad (18.4\text{-}9)$$

where $\mathbf{O}$ represents the Boolean ''exclusive or'' operation. The field at D_1 is thus the difference (squared) of the intensity pattern of the two transparencies. Similarly the field intensity I' at D_2 is

$$\begin{aligned} I' &\propto \| T_1|^2 + |r|^2(|T_2|^2 - |T_1|^2)|^2 |R|^2 I_{in} \\ &\propto |T_1|^2 + |T_2|^2 \qquad \text{when } |r|^2 = .5 \end{aligned} \quad (18.4\text{-}10)$$

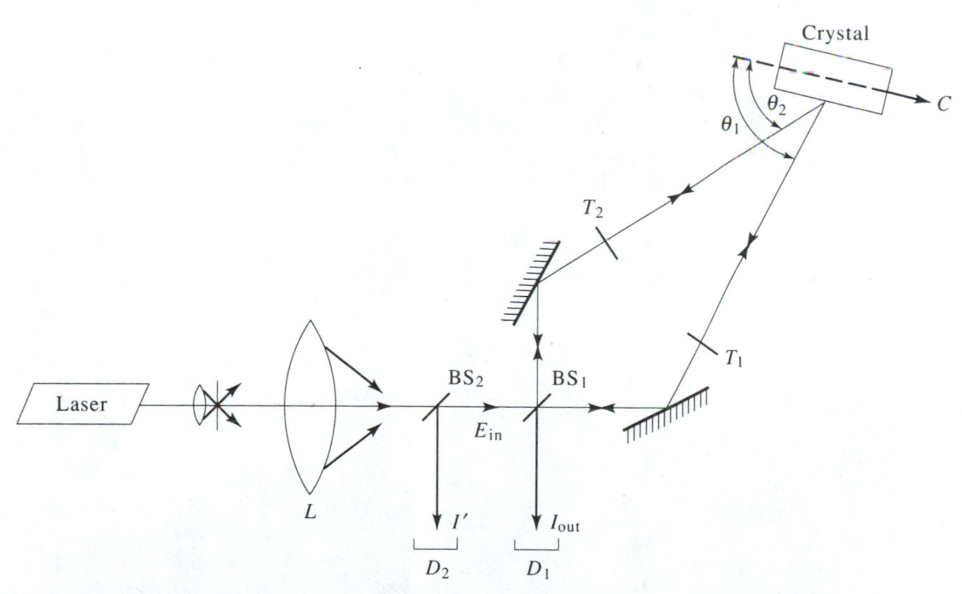

Figure 18-13 An experimental arrangement to demonstrate image addition and subtraction (the ''exclusive or'' operation). (After Reference [20].)

Note that when $T_1 = e^{i\theta_1(\mathbf{x}_t)}$ and $T_2 = e^{i\theta_2(\mathbf{x}_t)}$, i.e., pure phase modulation, $I_{\text{out}} = 0$. This is at first sight, an amazing result. A little thought, however, will convince us that this is as it should be. When the transparencies are lossless (which is the case for $T_{1,2} = \exp[i\theta_{1,2}(\mathbf{x}_t)]$) a "time reversal" by the conjugator should result in a return field at the beam splitter BS_1 that is a time-reversed (phase conjugate) replica of the incoming field. Since no field entered BS_1 from the direction of D_1, none can emerge that propagate toward BS_1. Since in this case $|T_1| = |T_2| =$ constant, the vanishing of the field propagating toward D_1 comes about by destructive interference, point by point, of the two fields as expressed by (18.4-7). When T_1 and T_2 are, each, spatially intensity modulated the point by point destructive interference is no longer operative. A dark area in T_1 cannot, obviously, interfere with a bright spot on T_2. More globally, the path becomes lossy (or amplifying) and, except for the trivial case of uniform and equal, $|T_1| = |T_2|$, the time reversibility property of phase conjugation (see Section 17-1) does not hold. The striking advantage of the phase-conjugating interferometer described above is that the system works without any need to balance L_1 and L_2 or to maintain their path difference constant to within a small fraction of λ as is usually required with conventional interferometers.

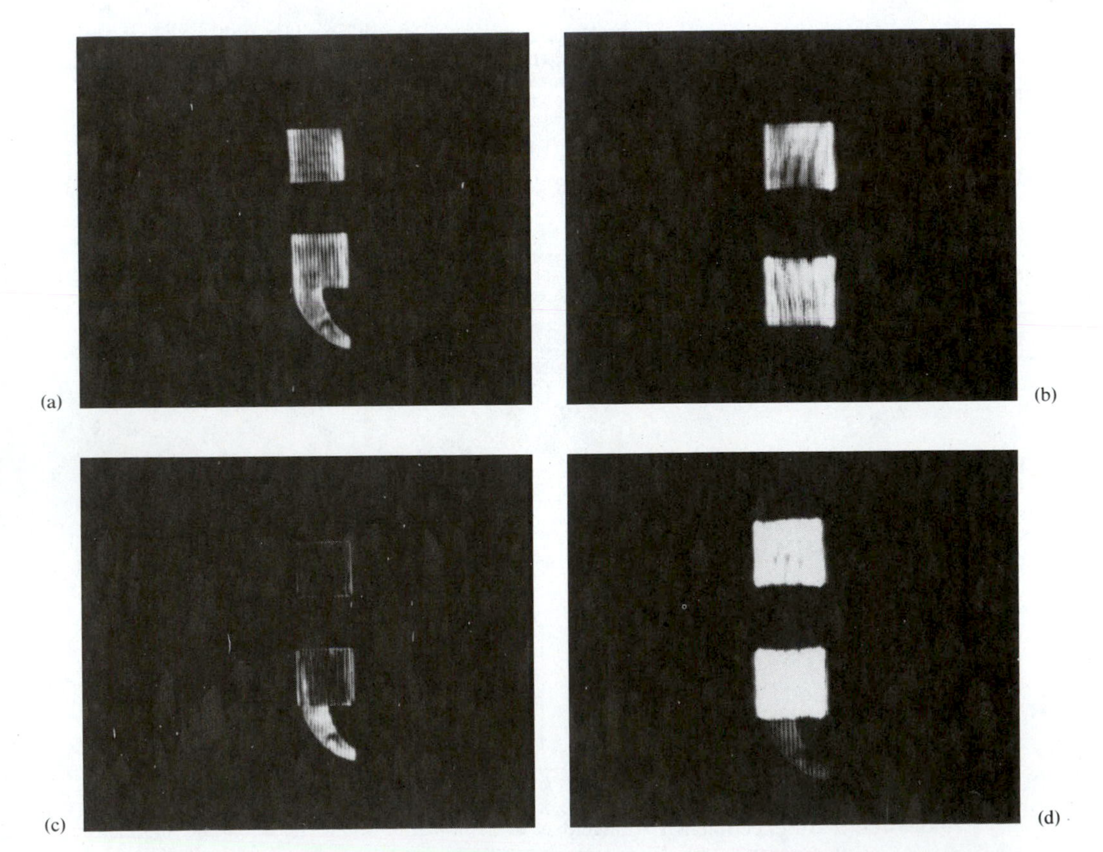

Figure 18-14 The results of image subtraction using the setup of Figure 18-15 (see text for details). (After Reference [20].)

Figure 18-14 shows the result of experiments using the setup of Figure 18-13. Figure 18-16(a) shows the image at D_2 of a semicolon transparency place at T_1 with beam 2 blocked and in (b) the image of a colon placed T_2 with path 1 blocked. Figure 18-14(c) shows the result of "subtraction" as observed at D_1, while (d) shows the image at D_2.

Problems

18.1 Obtain the expression for the diffraction efficiency of a fixed hologram as in (18.1-9) when a slight deviation from the Bragg condition exists, i.e.,

$$\Delta \equiv \mathbf{k}_2 - \mathbf{k}_1 - \mathbf{K} \neq 0$$

Consider the effects on the diffraction efficiency of:

a. A small angular departure (δ) of the incident beam from that of the Bragg condition.
b. A small deviation of the wavelength. [*Hint:* Using Figure 18-1(b) show that for $\delta \ll 1$, $(\mathbf{k}_2 - \mathbf{k}_1 - \mathbf{K}) \cdot \mathbf{r} \approx k\delta(\sin\theta)z$.]

18.2

a. Obtain an expression for the diffraction efficiency of a transmission hologram (Figure 18-1) as a function of $\Delta\ell$ where Δ is defined by

$$\Delta \equiv |\mathbf{k}_1 - \mathbf{k}_2 + \mathbf{K}|$$

and ℓ is the length of the hologram (in the z direction).
b. Plot the diffraction efficiency of a fixed hologram as a function of λ for a fixed incidence angle. Assume $\Delta = 0$ at some nominal λ_0.

18.3 Derive an expression for the diffraction efficiency $|A_1(0)/A_2(0)|^2$, of a reflection hologram as shown in the accompanying figure.

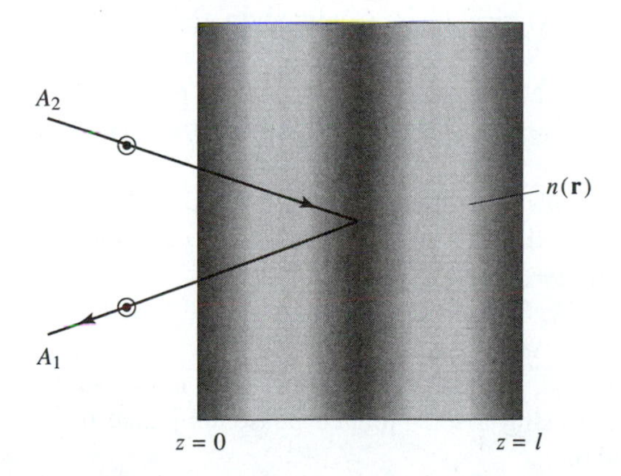

References

1. Kogelnik, H., "Coupled wave theory for thick hologram gratings," *Bell Syst. Tech. J.* 48:2909, 1969.

2. A. Yariv, *Quantum Electronics,* 3d ed. New York: Wiley, Chs. 16, 17.
3. Kukhtaerev, N. V., M. B. Markov, S. G. Odulov, M. S. Soskin, and V. L. Vinetsky, ''Holographic storage in electrooptic crystals,'' Part I: Steady state, Part II: Beam coupling light amplification,'' *Ferroelectrics* 22:949, 1979.
4. Sayano, K., A. Yariv, and R. R. Neurgaonkar, ''Reduction of the photorefractive response time in rhodium doped $Sr_{0.6}Ba_{0.4}Nb_2O_6$ with a dc field,'' *Opt. Lett.* 15:9–11, 1990.
5. Rakuljic, G., ''Photorefractive properties and applications of BaTiO3 and tungsten bronze ferroelectrics,'' Ph.D. thesis, California Institute of Technology, Pasadena, California, 1987.
6. Huignard, J. P., J. P. Herriau, and G. Rivet, ''Progress in photorefractive tungsten bronze crystals,'' *Opt. Lett.* 5:102, 1980.
7. Staebler, J. I., and D. L. Amodei, ''Coupled-wave analysis of holographic storage in LiNbO3,'' *J. Appl. Phys.* 46:3510, 1975.
8. J. O. White, S. Kwong, M. C. Golomb, B. Fischer, and A. Yariv, ''Wave propagation in photorefractive media,'' in *Topics in Applied Physics* 62:101, Berlin: Springer-Verlag, 1989. This whole volume, edited by P. Guenther and J. P. Huignard consists of authoritative review papers, seven on various aspects of the photorefractive effect and its applications.
9. M. Cronin-Golomb, B. Fischer, J. O. White, and A. Yariv, ''Passive (self-pumped) phase conjugate mirror,'' *Appl. Phys. Lett.* 41:689, 1982.
10. Feinberg, J., ''Self pumped continuous wave phase conjugator using internal reflection,'' *Opt. Lett.* 7:486, 1982.
11. Kwong, S. K., M. Cronin-Golomb, and A. Yariv, ''Oscillation with photorefractive gain,'' *IEEE J. Quant. Elec.* QE-22:1508, 1986.
12. Yariv, A., and S. K. Kwong, ''Theory of Laser Oscillators with Photorefractive Gain,'' *IEEE J. Quant. Elec.* QE-22:1508, 1986. 13. Slater, J. C., *Microwave Electronics.* Princeton, N. J.: Van Nostrand, 1950.
14. Yariv, A., *Quantum Electronics,* 2d ed. (Wiley, New York 1975).
15. Special issue on Nonlinear Optical Phase Conjugation, *IEEE J. Quant. Elec.* 25, D. M. Pepper, ed., March 1989.
16. Kyuma, K., S. K. Kwong, M. Cronin-Golomb, and A. Yariv, ''Principles and applications of optical phase conjugation in photorefractive materials,'' *Optoelectronics,* 2:1, 1987.
17. G. Joos, *Theoretical Physics* New York: Hefner, 1950, p. 471.
18. P. Yeh, I. McMichael, and M. Khroshevisan, ''Phase conjugate fiber-optic gyro,'' *Appl. Opt.* 25:1029, 1986.
19. Y. Tomita, R. Yahalom, A. Yariv, ''On the phase shift of a self-pumped phase conjugate mirror,'' *J. Opt. Soc. Am, B.*
20. Kwong, S. K., G. Rakuljic, and A. Yariv, ''Real time image subtraction and 'exclusive or' operation using a self-pumped phase conjugate mirror,'' *Appl. Phys. Lett.* 48:201, 1986.
21. Kwong, S. K., and A. Yariv, ''One-way real time wavefront converters,'' Appl. Phys. Lett. 48:9, 1986.

19 Optical Solitons

19.0 INTRODUCTION

Optical solitons are pulses that propagate without changing their shape in media optical. This is due to a balance between two effects: (1) The group velocity dispersion of the medium. This effect characterized by the parameter β'' $[\beta'' \equiv (d^2\beta/d\omega^2)]$, and as shown in Section 3.4, causes transform-limited pulses to broaden. (2) The Kerr effect described by the relation $n = n_0 + n_2 I$ and treated in Section 17.2. It is interesting to note that solitons are not limited to optics. As a matter of historical interest, a solitary water wave was observed and first described by John Scott Russel in a barge canal in Great Britain in 1834 [1].

19.1 THE MATHEMATICAL DESCRIPTION OF SOLITONS

To illustrate qualitatively the basic physics of this phenomenon, we consider first in Figure 19-1 what happens to a transform-limited optical pulse as it propagates in a fiber whose index of refraction depends on the field intensity I.

$$n = n_0 + n_2 I \tag{19.1-1}$$

To be specific, we will assume a Gaussian pulse

$$E_{\text{in}} = C\sqrt{I}\exp\left(i\omega_0 t - \frac{\alpha t^2}{2}\right) \tag{19.1-2}$$

where C is a constant that does not play a role in this discussion. After a short propagation distance, ΔL, the pulse will emerge as

$$E_{\text{out}} = C\sqrt{I}\exp\left(i(\omega_0 t - \frac{\omega_0 \Delta L}{C} n(I))\right) \exp\,(-\alpha t^2/2)$$

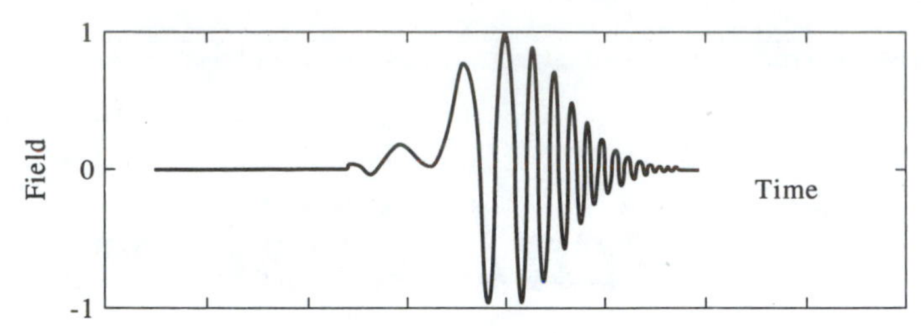

Figure 19-1 Phase modulation (chirping) of an optical pulse propagating in a medium with $n_2 > 0$.

Its output phase is thus

$$\phi(\Delta L, t) = \omega_0 t - \frac{\omega_0 \Delta L}{C}[n_0 + n_2 I_0 \exp(-\alpha t^2)]$$

while the instantaneous frequency is

$$\omega(\Delta L, t) = d\phi/dt = \omega_0 + 2\frac{\omega_0 \alpha \Delta L n_2 I_0}{C} t \exp(-\alpha t^2) \tag{19.1-3}$$

and is plotted in Figure 19-1 for the case $n_2 > 0$. Over the central, and most important portion of the pulse, it is chirped positively, i.e., $d\omega/dt > 0$. We note that the sign of the chirp in Figure 19-1 is the opposite of that of Figure 3-12. It follows directly that if a pulse with, say, a positive chirp, $d\omega/dt > 0$, enters a linear dispersive fiber with $\beta'' \equiv \dfrac{d^2\beta}{d\omega^2} = \left(-\dfrac{1}{v_g^2}\dfrac{dv_g}{d\omega}\right) < 0$, it will narrow with propagation distance since the late arriving high frequencies will be sped up during their transit relative to the lower frequencies. The position along the fiber where all frequency components ''catch up'' with each other is where the minimum pulse width occurs. Beyond this point the pulse rebroadens.

Imagine a hypothetical fiber made up of alternating short segments, half of which have $\beta'' < 0$ and $n_2 = 0$, while in the remainder $\beta'' = 0$ and $n_2 > 0$. A section of such a fiber with three segments is shown in Figure 19-2. We start with a transform-limited pulse in plane 1, which by plane 2 becomes broadened and chirped due to the dispersion $\beta'' \neq 0$, as discussed in Section 3.4. Propagation in the nonlinear fiber ($n_2 > 0$) results in a reversal of the sign of the chirp at plane 3 without a change in the pulse length. The relative slowing down of the high frequencies occurring in the dispersive ($\beta'' < 0$) last segment gives rise to a pulse with the original shape and zero chirp at 4. In an actual fiber, both effects (i.e., n_2 and β'') exist simultaneously. This results in steady-state nonspreading pulses—solitons.

To find the properties of the soliton, we will derive first the wave equation governing propagation in a dispersive nonlinear channel.

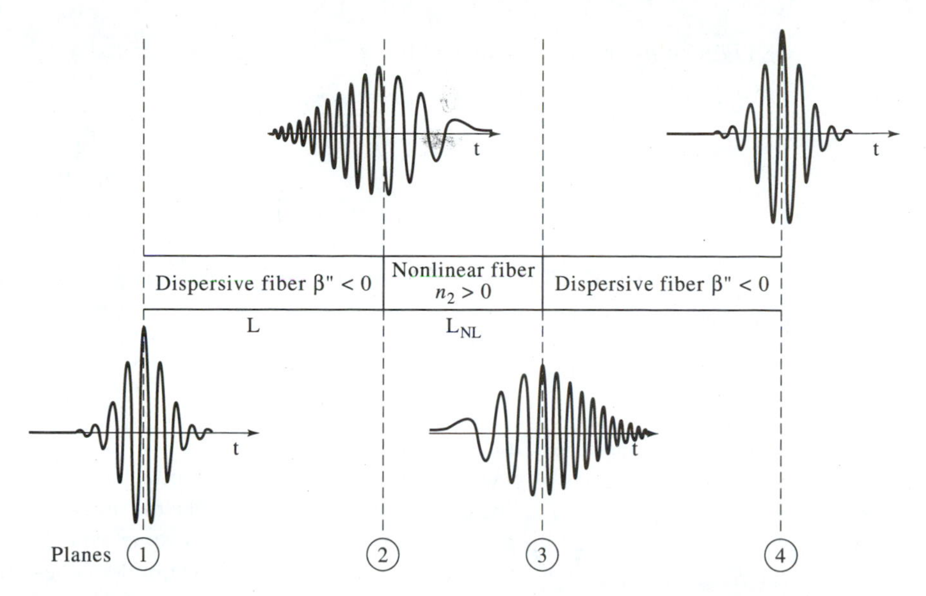

Figure 19-2 A representation of the soliton physics in which the simultaneous effects of linear dispersion due to β'' and nonlinear chirping due to n_2 are shown, for the sake of illustration as operating on the propagating pulse in succession. One unit cell of the periodic fiber is shown.

The Wave Equation

The physical nonlinearity that plays a role is that described above where the index of refraction depends on the wave intensity:

$$n = n_0 + n_2 I \tag{19.1-4}$$

This last relation can be described alternatively using the language of nonlinear optics—see (17.3-7)—as

$$P_{\rm NL}(\omega) = 3\chi^{(3)}|E(\omega)|^2 \, E(\omega) \tag{19.1-5}$$

where $\chi^{(3)} = \chi^{(3)}(-\omega, \omega, \omega, -\omega)$ is the third-order nonlinear optical constant of the medium. Using the relations

$$\begin{aligned} D &= \epsilon_0 E + P = \epsilon_0(1 + \chi^{(1)})E + P_{\rm NL} \\ &= \epsilon E \equiv \epsilon_0 n^2 E \end{aligned}$$

and assuming that the nonlinear polarization is small compared to the linear one, i.e.,

$$P_{\rm NL} \ll \epsilon_0(1 + \chi^{(1)})E$$

we find that n_2 and $\chi^{(3)}$ are related by

$$\chi^{(3)} = \frac{\epsilon_0 n_0^2}{3\ \eta_0} n_2$$

$$\left(\eta_0 = \sqrt{\frac{\mu}{\epsilon_0}}\right) \tag{19.1-6}$$

With relations (19.1-1) to (19.1-6) at hand, we are ready to solve the scalar Hemlholtz wave equation (17.3-3):

$$\nabla^2 E(x, y, z, t) - \mu\epsilon \frac{\partial^2 E}{\partial t^2} = \mu \frac{\partial^2}{\partial t^2} P_{\mathrm{NL}} \tag{19.1-7}$$

where E is the dominant transverse field component.

We are interested in fiber solitons where the mode confinement in the transverse dimensions renders the problem one-dimensional. We take the propagation direction as z and replace the field $E(x, y, z, t)$ by $E(z, t)$, which is some average of $E(x, y, z, t)$ over the transverse cross section (x, y). This renders the problem one-dimensional.

$$E(z, t) = Re[\phi(z, t) \exp[i(\omega_0 t - \beta_0 z)]] \tag{19.1-8}$$

where $\beta_0(\omega)$ is the propagation constant of the mode.

In a typical case, the temporal duration of ϕ might be $\sim 10^{-11}$ s, while the carrier period $2\pi/\omega_0$ is $\sim 10^{-14}$ s. Substituting (19.1-8) in (19.1-7) and using (19.1-5) leads to

$$\begin{aligned}[-2i\beta_0\phi' - 2i\omega_0\mu\epsilon\dot{\phi} - \beta_0^2\phi + \omega_0^2\mu\epsilon\phi] \exp[i\omega_0 t - \beta_0 z)] \\ = 3\ \mu\chi^{(3)} \frac{\partial^2}{\partial t^2} [|\phi|^2\ \phi \exp(i(\omega_0 t - \beta_0 z))]\end{aligned} \tag{19.1-9}$$

where following (19.1-5) we took

$$P_{\mathrm{NL}}(z, t) = Re[3\ \chi^{(3)} |\phi(z, t)|^2\ \phi(z, t) \exp(i(\omega_0 t - \beta_0 z))]$$

which is valid for the case where ϕ varies slowly compared to the physical processes responsible for $\chi^{(3)}$. We used the definitions $\phi' \equiv \partial\phi/\partial z$, $\dot{\phi} \equiv \partial\phi/\partial t$ and neglected "small" terms involving $\ddot{\phi}$ and ϕ''. If, on the right of (19.1-9), we take advantage of the inequality $\omega_0^2\phi \gg \omega_0\dot{\phi} \gg \ddot{\phi}$, we obtain

$$\begin{aligned}-2i\beta_0\phi' - 2i\omega_0\mu\epsilon\dot{\phi} - \beta_0^2\phi + \omega_0^2\mu\epsilon\phi = -3\ \chi^{(3)}\mu\omega_0^2|\phi|^2 \\ \phi\ \beta_0 \equiv \omega_0\sqrt{\mu\epsilon} \qquad \epsilon \equiv \epsilon(\omega_0)\end{aligned} \tag{19.1-10}$$

Next we take the double (space and time) Fourier transforms (FT) of (19.1-10). Recalling that

$$\mathrm{FT}\{\phi(z, t)\} = \frac{1}{4\pi^2} \iint \phi(z, t) \exp[-i(pz + \Omega t]\ dz\ dt \equiv \tilde{\tilde{\phi}}(p, \Omega) \tag{19.1-11}$$

(19.1-10) transforms to

$$[-2i\beta_0(ip) - 2i\omega_0\mu\epsilon(i\Omega) - \beta_0^2 + \omega_0^2\mu\epsilon]\tilde{\tilde{\phi}}(p, \Omega)$$
$$= -3\chi^{(3)}\mu\omega_0^2\mathrm{FT}\{|\phi(z, t)|^2 \phi\} \quad (19.1\text{-}12)$$

From Equations (19.1-8) and (19.1-11) it follow that

$$E(z, t) \equiv \iint \tilde{\tilde{\phi}}(p, \Omega) \exp[i((\omega_0 + \Omega)t - (\beta_0 - p)z)]\, dp\, d\Omega$$

$\tilde{\tilde{\phi}}(p, \Omega)$ is thus associated with the optical frequency $(\omega_0 + \Omega)$, so that in (19.1-12) we replace ϵ everywhere by

$$\epsilon(\omega_0 + \Omega) = \epsilon + \epsilon'\Omega + \frac{1}{2}\epsilon''\Omega^2 + \ldots$$

$$\epsilon' = \left.\frac{d\epsilon}{d\omega}\right|_{\omega_0} \qquad \epsilon'' = \left.\frac{d^2\epsilon}{d\omega^2}\right|_{\omega_0} \quad (19.1\text{-}13)$$

With the substitution of (19.1-13), Equation (19.1-12) becomes

$$\left[-2i\beta_0(ip) - \beta_0^2 + \omega_0^2\mu\epsilon + \left(2\omega_0\mu\epsilon + \frac{\beta_0^2}{\epsilon}\epsilon'\right)\Omega + \left(2\omega_0\mu\epsilon' + \frac{\beta_0^2\epsilon''}{2\epsilon}\right)\Omega^2\right]\tilde{\tilde{\phi}} = -3\chi^{(3)}\mu\omega_0^2\mathrm{FT}\{|\phi|^2\phi\} \quad (19.1\text{-}14)$$

Expanding

$$\beta(\omega_0 + \Omega) = \beta_0 + \beta'\Omega + \frac{1}{2}\beta''\Omega^2 \ldots$$
$$= (\omega_0 + \Omega)\sqrt{\mu\epsilon(\omega_0 + \Omega)} \quad (19.1\text{-}14a)$$

If we use the expansion (19.1-13) for $\epsilon(\omega_0 + \Omega)$ in the second equality of the last expression and compare the result to the first equality, we find, after equating equal powers of Ω,

$$\beta'\left(= \frac{1}{v_g}\right) = \frac{\beta_0\epsilon'}{2\epsilon} + \sqrt{\mu\epsilon} \qquad \beta'' = \frac{\beta_0\epsilon''}{2\epsilon} + \sqrt{\mu\epsilon}\,\frac{\epsilon'}{\epsilon}$$

With this last result, (19.1-14) can be rewritten as

$$\beta_0(ip\tilde{\tilde{\phi}}) + \frac{\beta_0}{v_g}(i\Omega\tilde{\tilde{\phi}}) + i\,\frac{\beta_0\beta''}{2}\,\Omega^2\tilde{\tilde{\phi}} = -i\,\frac{3}{2}\,\chi^{(3)}\mu\omega_0^2\mathrm{FT}\{|\phi|^2\phi\}$$

where we assumed $\omega_0\mu\epsilon' \ll \beta_0^2\epsilon''/\epsilon$.

We now return to the (z, t) domain by taking the (double) inverse Fourier (IFT) transform of the last equation. We use the following relations:

$$\mathrm{IFT}\{ip\tilde{\tilde{\phi}}\} = \frac{\partial}{\partial z}\phi(z, t) \qquad \mathrm{IFT}\{i\Omega\tilde{\tilde{\phi}}\} = \frac{\partial}{\partial t}\phi(z, t) \qquad \mathrm{IFT}\{\mathrm{FT}[|\phi|^2\phi]\} = |\phi|^2\phi$$

This results in the envelope wave equation for $\phi(z, t)$

$$\frac{\partial\phi}{\partial z} + \frac{1}{v_g}\frac{\partial\phi}{\partial t} - i\,\frac{\beta''}{2}\frac{\partial^2\phi}{\partial t^2} = -i\,\frac{3\chi^{(3)}}{2}\,\omega_0\eta|\phi|^2\phi \quad (19.1\text{-}15)$$

We find it useful to transform to a coordinate system moving at the group velocity, v_g.

$$T = t - \frac{z}{v_g}$$

In addition we define

$$\gamma \equiv \frac{\omega_0 n_2 n_0}{2c\eta_0} = \frac{3\chi^{(3)}}{2} \omega_0 \eta$$

and rewrite (19.1-15) as

$$-i \frac{\partial \phi}{\partial z} - \frac{\text{sgn}(\beta'')}{2} |\beta''| \frac{\partial^2 \phi}{\partial T^2} + \gamma |\phi|^2 \phi = 0 \tag{19.1-16}$$

which is the main result. To solve it, it is convenient to define the following dimensionless variables:

$$\tau = \frac{T}{\tau_0} \qquad z_0 = \frac{\tau_0^2}{2|\beta''|} \qquad \xi = \frac{z}{2z_0} = \frac{|\beta''|z}{\tau_0^2}$$

$$\psi = \frac{\phi}{\phi_0} = \tau_0 \left(\frac{\gamma}{|\beta''|}\right)^{1/2} \phi \qquad \phi_0 = \left(\frac{|\beta''|}{\gamma \tau_0^2}\right)^{1/2} \tag{19.1-17}$$

which transform Equation (19.1-16) to the soliton equation

$$\boxed{-i \frac{\partial \psi}{\partial \xi} - \text{sgn}(\beta'') \left(\frac{1}{2}\right) \frac{\partial^2 \psi}{\partial \tau^2} + |\psi|^2 \psi = 0} \tag{19.1-18}$$

This last equation is formally the same as the Schrödinger time-dependent equation, except for the nonlinear term $|\psi|^2\psi$. It is often referred to as the nonlinear Schrödinger equation. A solution (the fundamental solution) of (19.1-18) in the case of $\beta'' < 0$ is

$$\psi(\xi, \tau) = \text{Sech}(\tau) \exp \frac{-i\xi}{2} \tag{19.1-19}$$

In terms of the original (z, t) variables, the fundamental soliton solution is

$$\phi(z, t) = \phi_0 \text{Sech}\left(\frac{t - z/v_g}{\tau_0}\right) \exp\left(-i \frac{z}{2z_0}\right) \tag{19.1-20}$$

and is illustrated in Figure 19-3. We may view the soliton width τ_0 as an independent parameter that uniquely characterizes the soliton in a fiber with given n_2 and β''. It follows from (19.1-17) that for a given τ_0 the two remaining constants of $\phi(z, t)$, ϕ_0, and z_0, are determined. Recalling that the pulse energy is proportional to $A_{\text{fiber}} \times \phi_0^2 \tau_0$, we obtain from (19.1-17)

$$\frac{|\beta''/\gamma|}{\text{pulse energy}} \propto \tau_0 \tag{19.1-21}$$

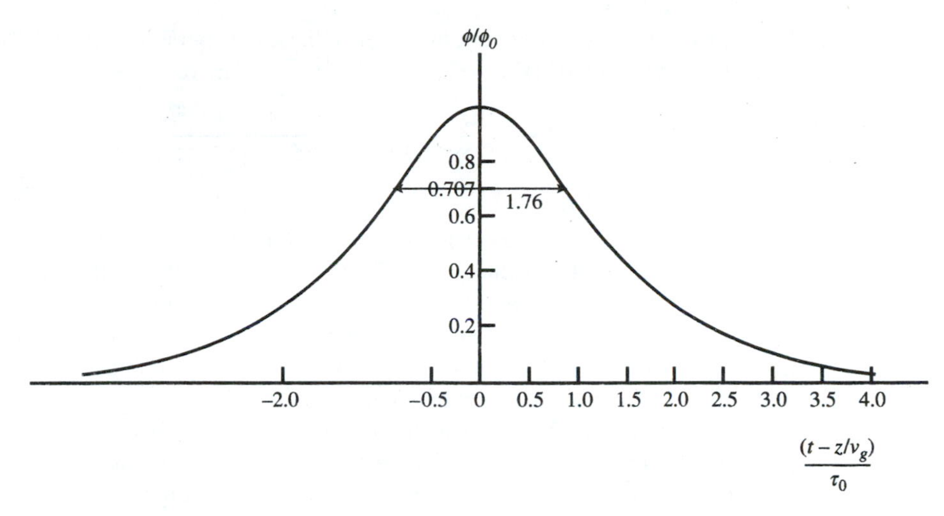

Figure 19-3 The fundamental soliton.

so that lower-energy pulses are broader. This causes pulse broadening due to attenuation in long fibers, a problem that can be remedied with the use of optical amplifiers. We note that in a given fiber (fixed β'' and n_2), the peak electric field ϕ_0 of the fundamental soliton is fixed by the relation $\phi_0^2 = (|\beta''|/\gamma\tau_0^2)$. Since the field is limited essentially to the fiber core area σ_f, the total power in a soliton is

$$P_{\text{soliton}} = \frac{\sigma_f n \beta''}{2\eta_0 \gamma \tau_0^2}$$

$$P_{\text{soliton}} \sim \sigma_f \frac{|\beta''|}{2\gamma\tau_0^2\eta_0}$$

The numerical example that follows will help us get some appreciation of the soliton properties and the range of soliton powers.

Numerical Example—Optical Solitons in Silica Fibers

We will use the following data: $\lambda = 1.55 \times 10^{-6}$ m, $n = 1.45$, $n_2 = 3 \times 10^{-20}$ m^2/W, $\gamma = \pi n_2 n/\lambda\eta_0 = 2.39 \times 10^{-16}$ m/W ohm, $D_{\text{fiber}} = 10^{-5}$ m. We will assume a data transmission rate $f_{\text{bit}} = 10^{10}$ bits/s and a corresponding $\tau_{\text{FWHM}} = 3 \times 10^{-11}$ s, $\tau_0 = \tau_{\text{FWHM}}/1.76 = 1.76 \times 10^{-11}$ s.

The peak pulse power in the fiber soliton mode is

$$P_{\text{soliton}} \sim \frac{\sigma_f \phi_0^2 n}{2\eta_0} = \frac{\sigma_f n|\beta''|}{2\eta_0\gamma\tau_0^2}$$

(we used $\phi_0^2 = |\beta''|/\gamma\tau_0^2$), where σ_f = effective cross-sectional area of fiber ~100 (μm)^2. We use expression (2.9-32) for β'', $\beta'' = (-\lambda^2/2\pi c)D$, and use a value of $D \approx 18 \times 10^{-6}$ s/m^2. (This corresponds in Figure 3-10 to a value of $D = 18$ ps/nm-km appropriate to $\lambda = 1.55\ \mu$m in a fiber with zero dispersion at $\lambda = 1.33\ \mu$m. The corresponding value of β'' is -2.31×10^{-26} s^2/m. The result is

$$P_{\text{soliton}} = \frac{\sigma_f n |\beta''|}{2\eta_0 \gamma \tau_0^2} = \frac{10^{-10} \times 1.45 \times 7.8 \times 10^{-27}}{2 \times 377 \times 2.79 \times 10^{-16} \times (1.7 \times 10^{-11})^2}$$
$$= 0.056 \text{ watt (peak power)}$$

We note that if we wish to communicate at a higher bit rate, i.e. $f_{\text{bit}} > 10^{10}$ bit/s, we need to use smaller values of τ_0. This will increase the peak power, which is proportional to τ_0^{-2}. At power levels, exceeding, say 100 mw in our fiber, the corresponding optical intensity, in a typical fiber, will be near 10^5 watts/cm^2. This intensity is sufficiently elevated so that nonlinear effects such as Raman and four-wave mixing become important and lead to degradation [2].

Much of the present interest in solitons for optical fiber communication is a result of a series of experimental demonstrations by L. Mollenauer [3, 4] which have established its viability for long distance (thousands of kilometers transmission).

Problems

19.1 Rewriting Equation (19.1-15) as

$$\frac{\partial \phi}{\partial z} + \frac{1}{v_g} - i\,\frac{\beta''}{2}\frac{\partial^2 \phi}{\partial t^2} = -i\,\frac{\omega_0 n_2}{2c\eta}\,|\phi|^2\phi$$

where we used

$$\chi^{(3)} = \frac{\epsilon_0 n_0^2 n_2}{3\eta_0}, \qquad v_g = \frac{1}{\beta'}$$

Show that for short distances, where the dispersive (β'') effects can be neglected, a pulse with an input envelope

$$\phi(t) = \phi_0 \exp(-\alpha t^2)$$

propagating in a nondispersive ($\beta'' = 0$) fiber becomes

$$\phi(z,t) = \phi\left(t - \frac{z}{v_g}\right) \exp\left[-i\,\frac{\omega_0 n_2 z}{2c\eta}\,|\phi(z,t)|^2\right] \tag{1}$$

19.2 Show that for a Gaussian optical pulse with an envelope $\phi(t) = \phi_0 \exp(-\alpha t^2)$ and input intensity (in watts/m^2)

$$I(t) = \frac{|\phi(t)|^2}{2\eta} = I_0 \exp(-2\alpha t^2), \qquad I_0 = \frac{\phi_0^2}{2\eta} \tag{2}$$

Propagation in a dispersive ($\beta'' \neq 0$) and nonlinear ($n_2 \neq 0$) fiber results in a chirped output pulse whose envelope near the peak can be approximated by

$$\phi(z,t) \simeq \phi\left(t - \frac{z}{v_g}\right) \exp\left[i \frac{2\omega_0 n_2 \alpha I_0 z}{c} \left(t - \frac{z}{v_g}\right)^2 \right]$$

where an, uninteresting time-independent phase shift was left out.

19.3 Consider the propagation of a pulse through a periodic fiber such as shown in Figure 19-2. Assume that in plane 1 the pulse envelope is a chirpless Gaussian pulse $\phi(t) = \exp(-\alpha t^2)$. Show, using the results of Section 3.4 (Eq. 3.4-26 is especially relevant) and problem 19.2, that the chirping of the pulse due to propagation between planes 1 and 2 can be reversed between planes 2 and 3 so that the pulse returns to its original envelope at 4 as shown in the figure. Derive the conditions necessary for the pulse envelope to repeat itself between 1 and 4. [*Hint:* You should find that the condition for phase reversal and pulse restoration between planes 1 and 4 is

$$I_0 = \frac{\lambda_0 |\beta''| \alpha}{\pi |n_2|} \left(\frac{L}{L_{NL}}\right) \sqrt{1 + (2\beta'' L \alpha)^2} \tag{3}$$

where $2L$ is the length of the dispersive fiber segments (Figure 19-2 shows a unit cell which contains two dispersive sections of length L straddling a nonlinear fiber of length L_{NL}].

19.4 Show that in the case of

$$L = (L_{NL})$$

and in the limit of

$$4|\beta''|^2 \alpha^2 L^2 \ll 1$$

expression (3) agrees with the results of Section 19.1 for the soliton power $P_{soliton}$. (Do not worry about factors of the order of magnitude of unity.)

References

1. Dodd, R. K., J. C. Eilbeck, J. D. Gibbon, and H. C. Morris, *Solitons and Nonlinear Wave Equations*, London: Academic Press, 1982.
2. Agrawal, G. P., *Nonlinear Fiber Optics*, Boston: Academic Press, 1989.
3. Mollenauer, L. F., R. H. Stolen, J. P. Gordon, *Phys. Rev. Lett.* 45:1095, 1980.
4. Mollenauer, L. F. and R. H. Stolen, *Opt. Lett.* 9:13, 1984.
5. Mollenauer, L. F. and J. P. Gordon, "Long-distance, high-bit-rate transmission using solitons in optical fibers," *The Froehlic/Kent Encyclopedia of Telecommunication,* New York: M. Dekker, 1995, p. 329. Excellent semipopular review article.
6. Gordon, J. P. and H. A. Haus, "Random walk of coherently amplified solitons in optical fiber," *Opt. Lett.* 11665, 1986.

7. Mecozzi, A., J. D. Morres, H. A. Haus, and Y. Lai, ''Soliton transmission control,'' *Opt. Lett.,* 16:1841–1843, 1991.
8. Kodama, Y. and A. Hasegawa, ''Generation of asymptotically stable optical solitons and suppression of the Gordon Haus effect,'' *Opt. Lett.* 17:33, 1992.
9. Mollenauer, L. F., J. P. Gordon, and S. G. Evangelides, ''The sliding frequency guiding filter: An improved form of soliton jitter control,'' *Opt. Lett.* 17:1675, 1992.
10. L. F. Mollenauer, E. Lichtman, M. J. Neubelt, and Harvey, G. T., ''Demonstration using sliding-frequency guiding filters of error-free soliton transmission over more than 20 Mm at 10Gbit/s single channel and over more than 13Mm at 20 Gbit/s in a two channel WDM,'' *Elect. Letts.* 29:910, 1993.

20 A Classical Treatment of Quantum Optics, Quantum Noise, and Squeezing

20.0 INTRODUCTION

Some of the most important and elegant phenomena related to optical waves and their detection can only be explained using the extension of the formalism of quantum mechanics to optics, i.e., "quantum optics." Important topics that fit this category involve amplitude and phase noise (fluctuations), the statistics of photo-generated electrons, and the new field of nonlinear squeezing. These areas are just too important to forego. Somewhat to my surprise, I found that by asking the student to accept just *one* result from quantum mechanics, it is possible to treat all the above-mentioned phenomena classically and obtain results that agree with those of quantum optics.

20.1 THE QUANTUM UNCERTAINTY GOES CLASSICAL

One of the better-known uncertainties of quantum mechanics relates to the simultaneous measurement of the position (x) and momentum (p) of a particle and decrees that the product of the uncertainties Δp and Δx must obey [1]

$$\Delta p \, \Delta x \geqq \hbar/2$$

where $\hbar \equiv h/2\pi$ and $h = 6.62377 \times 10^{-34}$ joule-sec is the Planck's constant. These fundamental uncertainties extend to optical measurements such as measurements of the amplitudes and phases of optical fields. Their proper study involves the elegant formalism of quantum optics [1, 2, 3]. Since we have foresworn quantum mechanics in this book, we cannot approach this subject from first principles. We can, however, appreciate many of the consequences and even obtain numerically correct results for the important scenarios by accepting just *one* basic consequence of quantum mechanics: that of the *uncertainty principle*.

The Uncertainty Principle

Let us represent the classical monochromatic electric field of some mode oscillating in a resonator as

$$e(t) = |E| \cos(\omega t + \beta) = Re[E \exp(i\omega t)] \tag{20.1-1}$$

where

$$E = |E| \exp(i\beta) = E_1 + iE_2 \tag{20.1-2}$$

is the complex phasor representing the field. It is shown in Figure 20-1(a) as a vector in the complex **E** plane of length $|E|$ and projections E_1 and E_2 along the real and imaginary axes, respectively.

According to quantum mechanics [1, 2], the complex amplitude E in (20.1-2) *cannot be* specified exactly. This uncertainty is represented in Figure 20-1(b) by means of the ''uncertainty circle.'' The most probable position of the tip of the phasor E, on measurement, will be found near the center of the circle. The field phasor corresponding to the center of this circle is denoted as $\langle E \rangle$, the ''expectation value'' of E. The expectation value corresponds to the quantum mechanical ensemble average, that is, to the average of a large number of independent field determinations (measurements) under *identical* conditions. This expectation value obeys in all respects the same (Maxwell's) equations as its classical counterpart. There is even a theorem in quantum mechanics, Ehrenfest's theorem [2], to prove it.

Repeated measurements of the projections of E, $E_1(t)$, and $E_2(t)$ will yield different results, and the results for $E(t) = E_1(t) + iE_2(t)$ will tend to cluster about the center of the circle, which is a graphical way to describe the most probable region in which the tip of $E(t)$ will fall. The uncertainty that results from this inherent quantum-imposed spread in the values of $E(t)$ can be thought of as *quantum noise*. We shall devote the rest of this chapter to a consideration of the consequences of this noise in optical measurements.

A classical *approximation* of the quantum physics is to write the basic monochromatic field of an electromagnetic mode as

$$e(t) = Re[E(t) \exp(i\omega t)] = E_1(t) \cos \omega t - E_2(t) \sin \omega t \tag{20.1-2a}$$

where the complex amplitude $E(t)$ is

$$E(t) = E_1(t) + iE_2(t)$$

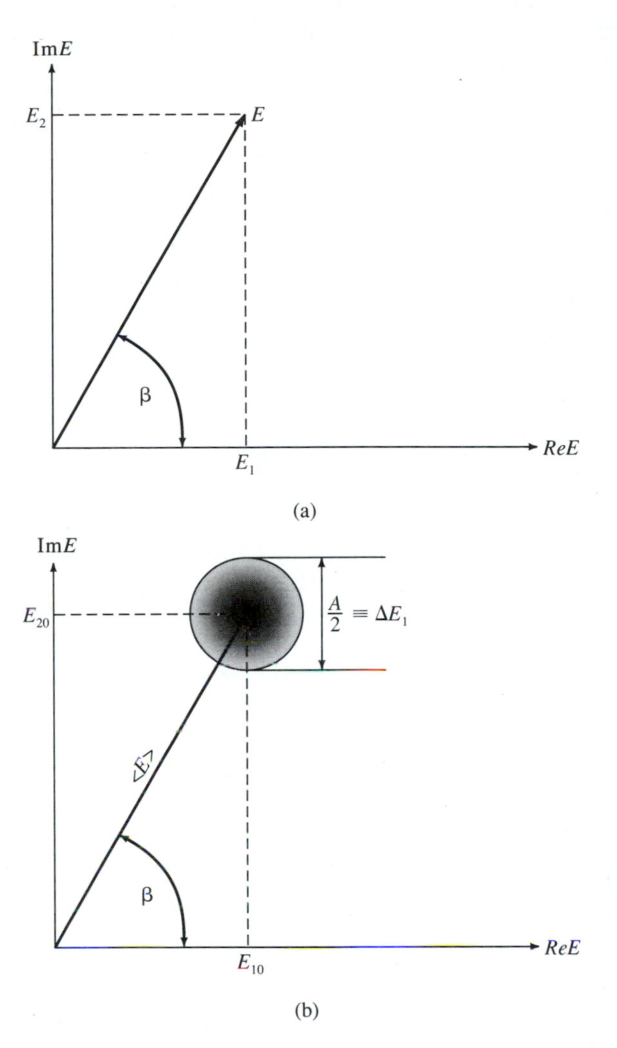

Figure 20-1 (a) A classical phasor representation of the optical field. (b) A representation of a coherent field that is consistent with quantum optics. In this special case, $\Delta E_1 = \Delta E_2$. (c) The electromagnetic field representation of an unexcited vacuum ($n = 0$) optical mode. (d) An equivalent representation of the field with a random phasor added, vectorially, to the tip of the classical phasor. (e) A ''squeezed'' field. (f) A squeezed vacuum ($n = 0$) mode. The squeezing factor is $s > 1$.

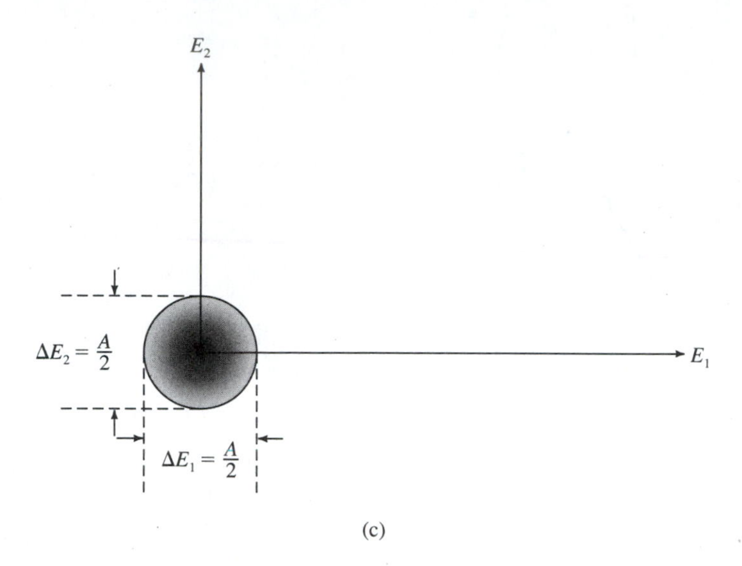

(c)

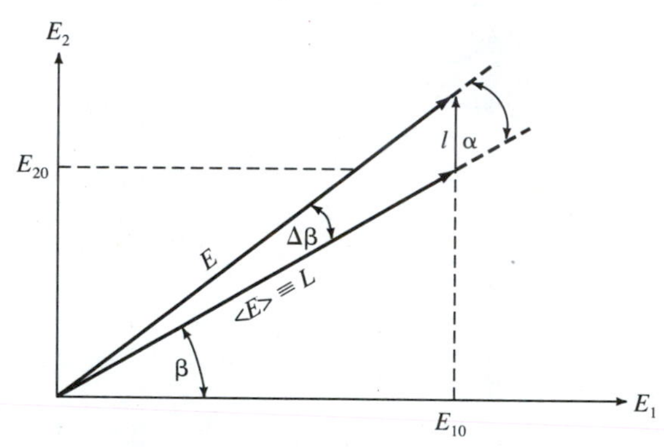

(d)

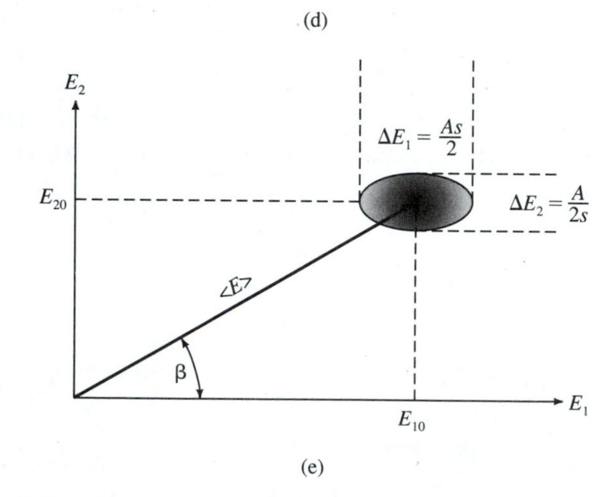

(e)

Figure 20-1 (*Continued*)

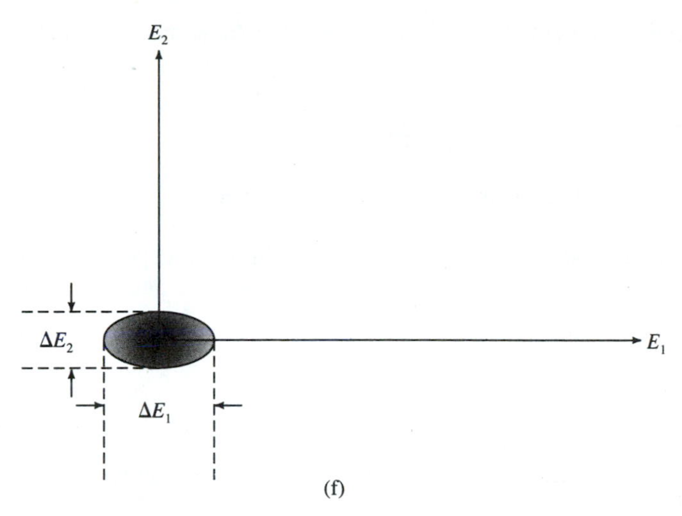

Figure 20-1 (*Continued*)

and

$$
\begin{aligned}
E_1(t) &= \langle E_1(t)\rangle + \Delta E_1(t) = E_{10} + \Delta E_1(t) \\
E_2(t) &= \langle E_2(t)\rangle + \Delta E_2(t) = E_{20} + \Delta E_2(t)
\end{aligned}
\tag{20.1-3}
$$

where $\langle E_{1,2}(t)\rangle \equiv E_{10,20}$ is the expectation value of $E_{1,2}(t)$. $\Delta E_1(t)$ and $\Delta E_2(t)$ represent the fundamental quantum mechanical uncertainties. They have zero mean $\langle \Delta E_{1,2}(t)\rangle = 0$, and are uncorrelated $\langle \Delta E_1 \Delta E_2\rangle = 0$. $\langle\ \rangle$ everywhere indicates ensemble (or temporal) averaging.

Before proceeding with a description of these fluctuations, we will, as is the practice in quantum optics, find it useful to relate the *mean* electric field of a mode to the mean number n of optical photons (quanta) in the mode. Taking the mode volume as V, the dielectric constant of the medium as ϵ, and using Equation (1.3-22) yields

$$
\frac{\epsilon}{2}\,|\langle E\rangle|^2 V = n\hbar\omega = \text{Field Energy within volume } V
$$

so that the mean field amplitude is expressible as

$$
\begin{aligned}
|\langle E\rangle| &= \left(\frac{2\hbar\omega}{\epsilon V}\right)^{1/2} \sqrt{n} \equiv A\sqrt{n} \\
A &\equiv \left(\frac{2\hbar\omega}{\epsilon V}\right)^{1/2}
\end{aligned}
\tag{20.1-4}
$$

If we define the measure of the fundamental quantum mechanical uncertainties as

$$
\Delta E_1 = \langle (E_1 - E_{10})^2\rangle^{1/2} = \langle (\Delta E_1(t))^2\rangle^{1/2}
$$

with a similar expression in which $1 \rightarrow 2$ for ΔE_2, then according to quantum mechanics [2]

$$\Delta E_1 \Delta E_2 \geq \frac{A^2}{4} \tag{20.1-5}$$

This is the only quantum mechanical result that we will use.

The output field of most laser oscillators is in the so-called *coherent state* [1, 2] in which the uncertainty is divided equally between two quadrature components E_1 and E_2:

$$\Delta E_1 = \Delta E_2 = \frac{A}{2} \tag{20.1-6}$$

or using the normalized dimensionless field, $x \equiv \frac{E}{A}$, we can rewrite (20.1-5) as

$$\Delta x_1 \, \Delta x_2 \geq \frac{1}{4} \tag{20.1-7}$$

and for the *coherent state* field

$$\Delta x_1 = \Delta x_2 = \frac{1}{2} \tag{20.1-8}$$

A profound consequence of the uncertainty relation, Equation (20.1-5), is that it applies even to a mode that, classically, is *not excited*, that is, $n = 0$. This so-called *vacuum state*, which corresponds to (20.1-3) with $E_{10} = E_{20} = 0$, is illustrated by Figure 20.1(c).

Electromagnetic fields in which the uncertainties of the two quadrature components are unequal, i.e., $\Delta x_1 \neq \Delta x_2$, are called *squeezed states*. Such states have been produced recently using nonlinear optical techniques [4, 5]. A squeezed electromagnetic field is illustrated in Figure 20.1(e). A squeezed vacuum field is shown in Figure 20.1(f). Instead of representing the field in terms of the quadrature amplitudes E_1 and E_2, we can use the description of Figure 20.1(d) in which a random phasor

$$\ell(t) = |\ell| e^{i\alpha} = \Delta E_1(t) + i\Delta E_2(t) \tag{20.1-9}$$

is added vectorially to the average phasor $\langle E \rangle$.

The phase angle α of this fluctuation phasor measured from $\langle E \rangle$ is uniformly distributed (is equally likely to occur) between 0 and 2π while

$$\langle |\ell|^2 \rangle = 2\langle \Delta E_1^2(t) \rangle = \frac{A^2}{2} \tag{20.1-10}$$

Some of the consequences of the uncertainty relation (20.1-5) are explored in what follows.

The Energy of an Electromagnetic Mode

This energy is given classically by

$$\mathscr{E} = \frac{1}{2}\epsilon V(EE^*) = \frac{1}{2}\epsilon V[(L + |\ell|\cos\alpha)^2 + |\ell|^2\sin^2\alpha] \qquad \textbf{(20.1-11)}$$

and choosing, without loss of generality, the direction of $\langle E\rangle$ as the real axis so that

$$E = L + |\ell|\cos\alpha + i|\ell|\sin\alpha \qquad L \equiv \langle E\rangle$$

$$\langle\mathscr{E}\rangle = \frac{\epsilon V}{2}\langle(L^2 + 2L|\ell|\cos\alpha + |\ell|^2\cos^2\alpha + |\ell|^2\sin^2\alpha\rangle$$

$$= \frac{\epsilon V}{2}(L^2 + \langle|\ell|^2\rangle) = \frac{\epsilon V}{2}\left(A^2 n + \frac{1}{2}A^2\right) = \hbar\omega\left(n + \frac{1}{2}\right) \qquad \textbf{(20.1-12)}$$

where we used the definition of A in (20.1-4) and $\langle\sin^2\alpha\rangle = \langle\cos^2\alpha\rangle = \frac{1}{2}$, as well as the fact that $\langle\cos\alpha\rangle = 0$, since α is distributed uniformly between 0 and 2π. It follows that in the case when the classical field is zero, i.e., $\langle E\rangle$ and n, according to (20.1-4), are zero, the mode energy is $\hbar\omega/2$. This is the so-called *zero point vibration energy of the mode*. It is one of the main consequences of quantum mechanics and it does *not* have a classical counterpart.

Uncertainty in Energy

The uncertainty of the mode energy $\mathscr{E}$ can be defined by

$$\langle(\Delta\mathscr{E})^2\rangle = \langle\mathscr{E} - \langle\mathscr{E}\rangle\rangle^2 = \langle\mathscr{E}^2\rangle - \langle\mathscr{E}\rangle^2$$

Using (20.1-11) for $\mathscr{E}$, taking $\langle\cos^n\alpha\rangle = 0$ for n odd, and neglecting terms $0(\ell^4/L^4)$, we obtain

$$\Delta\mathscr{E} \equiv \langle(\Delta\mathscr{E})^2\rangle^{1/2} = \hbar\omega\sqrt{n}$$

or in terms of the number of photons, N, in the resonator

$$\Delta N = \frac{\Delta\mathscr{E}}{\hbar\omega} = \sqrt{n} \equiv \sqrt{\langle N\rangle} \qquad (\text{recall } n \equiv \langle N\rangle)$$

$$(\Delta N)^2 = \langle N\rangle = n \qquad \textbf{(20.1-13)}$$

where $\Delta N \equiv \langle(N - n)^2\rangle^{1/2}$.

The relation (20.1-13) between the mean square uncertainty in the number of photons N and the average number $\langle N\rangle$ *applies to a* **Poissonian** *probability distribution of the photon number* **N** *[2, 3]*

$$p(N) = \frac{\langle N\rangle^N e^{-\langle N\rangle}}{N!} \qquad \textbf{(20.1-14)}$$

for the number of photons N in the resonator.

Phase Uncertainty

The uncertainty $\Delta\beta$ in the value of the field phase is obtainable from Figure 20.1(d)

$$(\Delta\beta)^2 \equiv \langle(\Delta\beta(t)^2\rangle = \left\langle\left(\frac{|\ell| \sin\alpha}{L}\right)^2\right\rangle$$

$$= \frac{\langle|\ell|^2\rangle\langle\sin^2\alpha\rangle}{L^2} = \frac{\frac{1}{2}A^2\frac{1}{2}}{A^2 n}$$

$$= \frac{1}{4n}$$

where we assumed $|\ell| \ll L$ and used the fact that $\ell(t)$ and $\alpha(t)$ are not correlated.

Using (20.1-13) we obtain

$$\Delta N \Delta\beta = \frac{1}{2} \tag{20.1-15}$$

where $\Delta\beta \equiv \langle(\Delta\beta(t)^2\rangle^{1/2}$. This is a most important result and states the fundamental quantum mechanical limit on the simultaneous measurement of the phase (β) and excitation level N of an electromagnetic field.

Fluctuation of Photoelectron Number

If an optical wave is incident on a perfect photodetector whose area is A_{rea}, then for each incident and absorbed photon, ideally, one photoelectron is emitted. The resulting current is thus

$$i = \frac{ec\epsilon|E|^2 A_{\text{rea}}}{2\hbar\omega} \tag{20.1-16}$$

where e is the absolute value of the electronic charge, and the power incident on the detector is $c\epsilon|E|^2 A_{\text{rea}}$. If we use the dimensionless field $x \equiv E/A$ (A is defined in Eq. 20.1-4), the expression for the current becomes

$$i = \frac{ecA_{\text{rea}}}{V}[x_1^2 + x_2^2]$$

$$\cong \frac{ecA_{\text{rea}}}{V}(x_{10}^2 + x_{20}^2 + 2x_{10}\Delta x_1 + 2x_{20}\Delta x_2) \tag{20.1-17}$$

$$= i_0 + \Delta i$$

where $x_1 = x_{10} + \Delta x_1$, $x_2 = x_{20} + \Delta x_2$,

$$i_0 = \frac{ecA_{\text{rea}}}{V}(x_{10}^2 + x_{20}^2) = \frac{ec}{L_R} n \tag{20.1-18}$$

$$\Delta i = 2\frac{ec}{L_R}(x_{10}\Delta x_1 + x_{20}\Delta x_2) \tag{20.1-19}$$

where we used $x_{10}^2 + x_{20}^2 = |\langle E \rangle|^2/A^2 = n$:

$$L_R = \frac{V}{A_{\text{rea}}} = \text{length of quantization volume used in Eq. (20.1-4)}$$

The total number of photoelectrons emitted during a time interval $\tau = L_R/c$ (corresponding to a bandwidth $B = 1/\tau$) is

$$N_e = \frac{i\tau}{e} = \frac{[i_0 + \Delta i(t)]\tau}{e}$$

$$\langle N_e \rangle = \frac{i_0 \tau}{e} = n$$

$$\begin{aligned}\langle \Delta N_e^2 \rangle &= \frac{\langle (\Delta i)^2 \rangle \, \tau^2}{e^2} = 4\langle (x_{10}\Delta x_1 + x_{20}\Delta x_2)^2 \rangle \\ &= 4[x_{10}^2 \langle (\Delta x_1)^2 \rangle + x_{20}^2 \langle (\Delta x_2)^2 \rangle] \\ &= 4(x_{10}^2 + x_{20}^2)\langle (\Delta x_1)^2 \rangle = n\end{aligned}$$

where we used $\langle \Delta x_1 \Delta x_2 \rangle = 0$, Equation (20.1-18), and $\langle \Delta i \rangle = 0$ as well as

$$\langle (\Delta x_1)^2 \rangle = \langle (\Delta x_2)^2 \rangle \equiv \langle (\Delta x)^2 \rangle = \frac{1}{4}$$

It follows that

$$\langle \Delta N_e^2 \rangle = \langle N_e \rangle \tag{20.1-20}$$

so that the photoelectrons number N_e obeys Poisson's statistics. This was also shown in (10.10-18) to be true for the photons. Poissonian statistics apply to the case where each event (electron emission in this case) is completely random and independent so that there exist no correlations between individual emission events. This is exactly the scenario shown in Sections 10.3 and 10.4, which leads to shot noise in the current spectrum

$$S_i(\nu) = \frac{\overline{i_N^2(\nu)}}{\Delta \nu} = 2ei_0 \tag{20.1-21}$$

where $S_i(\nu)$ is the spectral density of the photocurrent at the radio frequency ν. We have thus *demonstrated that the electronic shot noise in the photocurrent can be attributed to the quantum field fluctuations.*

Minimum Detectable Optical Power Increment

Most of the methods used to measure the power of an electromagnetic wave employ detectors that convert absorbed optical power to a proportional output current. In a perfect detector, which releases one electron into the external circuit for each absorbed photon, we have

$$i = \frac{Pe}{\hbar\omega} \tag{20.1-22}$$

where P is the optical power to be measured. Let our measurement of the current consist of accumulating it for T seconds, which results in a total number of collected electrons (or holes)

$$N_e(T) = \frac{iT}{e} = \frac{PT}{\hbar\omega} \quad \Rightarrow P = \frac{\hbar\omega}{T} N_e(T)$$

The mean-squared uncertainty in the power measurement is thus given by

$$\langle(\Delta P)^2\rangle = \left(\frac{\hbar\omega}{T}\right)^2 \langle(\Delta N_e)^2\rangle = \left(\frac{\hbar\omega}{T}\right)^2 \langle N_e(T)\rangle$$

where, in the last equality, we used (20.1-20). The average number of collected electrons during the observation interval T is given by

$$\langle N_e(T)\rangle = \frac{\langle P\rangle}{\hbar\omega} T$$

so that

$$\langle(\Delta P)^2\rangle = \frac{\hbar\omega}{T} \langle P\rangle$$

Defining, arbitrarily, the minimum detectable power as that power at which the root-mean-squared fluctuation is equal to the average, i.e.,

$$\langle P\rangle_{\min} = \langle(\Delta P)^2\rangle^{1/2}$$

we obtain

$$P_{\min} = \hbar\omega B \tag{20.1-23}$$

$B = 1/T$ is the bandwidth of the current integrating system. One often refers to the quantity $\hbar\omega B$ as the minimum detectable power. It is treated in some detail in Section 11.4.

20.2 SQUEEZING OF OPTICAL FIELDS

It is possible to take a coherent optical field, such as the output of a laser, and reduce the fluctuation of one of its quadrature components, say, ΔE_1, at the expense of ΔE_2, or vice versa. The resulting uncertainty diagram becomes elliptical, while the product $\Delta E_1 \Delta E_2$ retains its initial value of $A^2/4$. This is referred to as *squeezing* [4, 5]. Squeezing is accomplished, usually, by a nonlinear optical operation on the field. One of the most common methods of achieving squeezing employs degenerate optical parametric amplification, which is discussed in Section 8.6. In this, degenerate, case, the "pump" frequency is twice that of the "signal," $\omega_{\text{pump}} = 2\omega_{\text{signal}}$. To demonstrate how squeezing is accomplished in this situation, we will need, first, to revisit the topic of parametric amplification.

Optical parametric amplification is described by Equations (8.7-2). In the case of a degenerate phase-matched parametric amplifier, we have

$$\omega_1 = \omega_2 \equiv \omega \qquad \omega_3 = 2\omega$$

If we designate the complex amplitude of the field at ω as E, the amplifier equations become

$$\frac{dE}{dz} = -i\frac{g}{2}E^* \qquad (20.2\text{-}1)$$

$$g = \frac{\omega d}{n_0}\sqrt{\mu/\epsilon_0}\,E_3 \qquad (20.2\text{-}2)$$

where n_0 is the index of refraction.

The coupling constant g is complex, since it is proportional to the complex pump amplitude E_3. Without loss of generality, we can take the pump field at $z = 0$ as

$$E_3(t) = -|E_3|\sin 2\omega t = \frac{i|E_3|}{2}(e^{i2\omega t} - e^{-i2\omega t}) \qquad (20.2\text{-}3)$$

This choice determines the time reference. In this case, $E_3 = +i|E_3|$, the coupling constant g is imaginary, and (20.2-1) assumes the form

$$\frac{dE}{dz} = \frac{1}{2}|g|E^*$$

$$|g| = \frac{\omega d}{n_0}\sqrt{\mu/\epsilon_0}\,|E_3| \qquad (20.2\text{-}4)$$

It is convenient to express the "signal" field at ω as in Equation (20.1-2) in terms of its quadrature amplitudes, E_1 and E_2,

$$E = (E_1 + iE_2) \qquad E_1 = \frac{1}{2}(E + E^*) \qquad E_2 = \frac{-i}{2}(E - E^*) \qquad (20.2\text{-}5)$$

so that the time dependent ("signal") field is given by

$$e(t) = Re[E\exp(i\omega t)] = Re[(E_1 + iE_2)\exp(i\omega t)]$$
$$= E_1\cos\omega t - E_2\sin\omega t \qquad (20.2\text{-}6)$$

If we substitute the first of Equations (20.2-5) in (20.2-4), we obtain

$$\frac{dE_1}{dz} = \frac{|g|}{2}E_1$$

$$\frac{dE_2}{dz} = -\frac{|g|}{2}E_2 \qquad (20.2\text{-}7)$$

so that at output of the parametric amplifier $z = L$

$$E_1(L) = E_1(0)\exp\left(\frac{|g|}{2}L\right) = E_1(0)s$$

$$E_2(L) = E_2(0)\exp\left(-\frac{|g|}{2}L\right) = \frac{E_1(0)}{s} \tag{20.2-8}$$

where the squeezing factor s is defined, in accordance with Figure 20-1(e), by

$$s = \exp(|g|L/2)$$

Degenerate parametric amplification is thus seen to lead to amplification of one quadrature component (E_1) and to the attenuation of the other component (E_2). The choice of which component is amplified is determined by the phase of the pump E_3. This is illustrated by Figure 20-2. If we now express the field amplitudes as in (20.1-3), including their quasi quantum mechanical fluctuations, the last two equations become

$$E_1(L, t) = (E_{10}(0) + \Delta E_1(0, t))\exp\left(\frac{|g|L}{2}\right)$$

$$E_2(L, t) = (E_{20}(0) + \Delta E_2(0, t))\exp\left(-\frac{|g|L}{2}\right) \tag{20.2-9}$$

The mean fields E_{10} and E_{20} *as well as the fluctuations* $\Delta E_1(t)$ and $\Delta E_2(t)$ are thus found to be amplified (attenuated) by the nonlinear parametric interaction. The output ($z = L$) fluctuations are

$$\Delta E_1(L) \equiv \langle(\Delta E_1(L, t))^2\rangle^{1/2} = \Delta E_1(0)\exp\left(\frac{|g|L}{2}\right) = \frac{A}{2}\exp\left(\frac{|g|L}{2}\right) \tag{20.2-10}$$

$$\Delta E_2(L) = \Delta E_2(0)\exp\left(-\frac{|g|L}{2}\right) = \frac{A}{2}\exp\left(-\frac{|g|L}{2}\right)$$

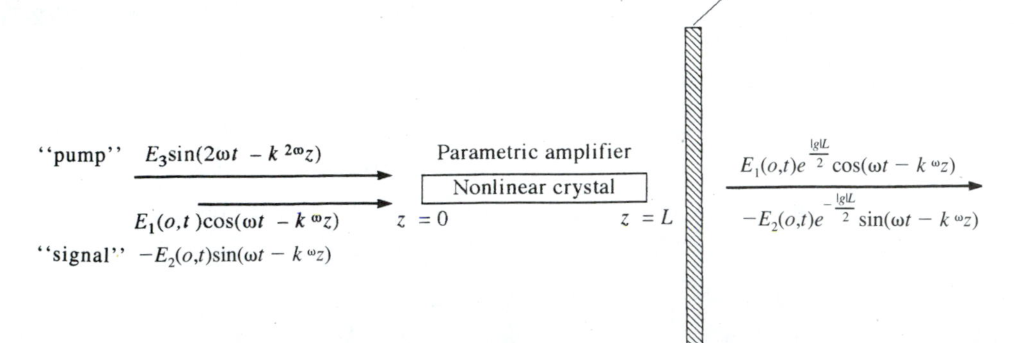

Figure 20-2 A degenerate parametric amplifier (pump frequency equal to twice the signal frequency) used in generating squeezed fields.

The uncertainty product

$$\Delta E_1(L)\Delta E_2(L) = \Delta E_1(0)\Delta E_2(0) = \frac{A^2}{4} \tag{20.2-11}$$

remains unchanged, although the uncertainty area is now elliptical rather than circular. This is illustrated in Figure 20-3(a) for the case of $\exp\left(\frac{|g|L}{2}\right) = 2$.

The case of a parametric amplifier with no input is particularly interesting. Classically we expect no output. Quantum mechanically, however, there exists an

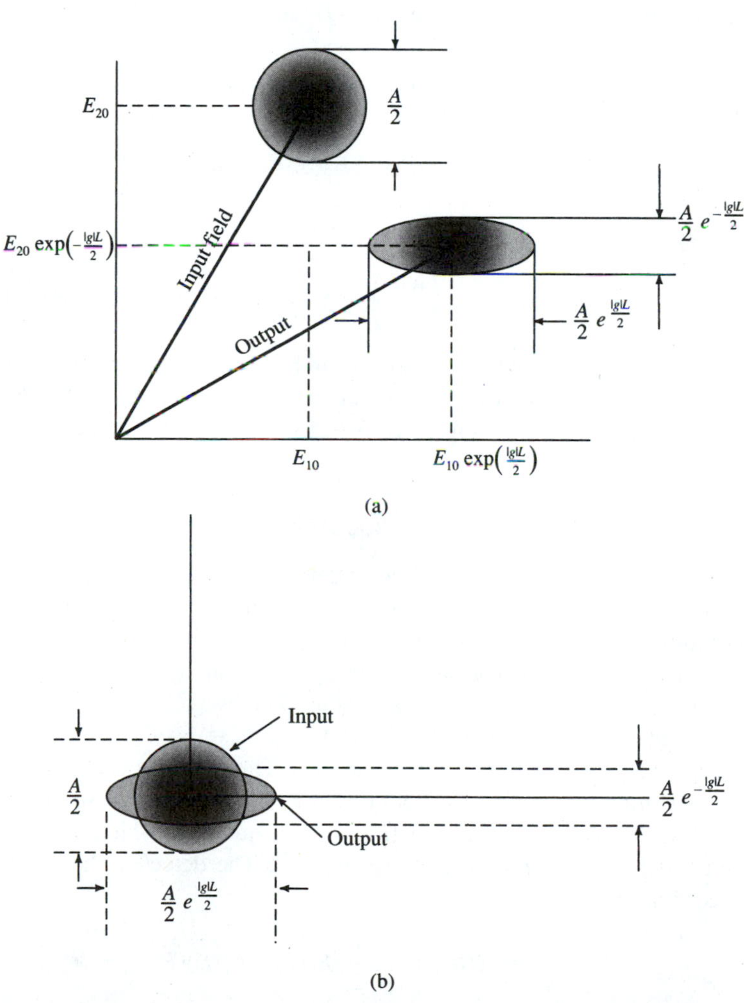

Figure 20-3 (a) The input field to a degenerate parametric amplifier shown in Figure 20-2 and the squeezed output fields for the case $\exp(|g|L/2) = 2$ (6 db squeezing). (b) Same as (a) except that the input field is zero.

input field, the so-called *vacuum field*, represented by the origin-centered circle of Figure 20-1(c). This field can be represented classically by Figure 20-1(a) with $\langle E_1 \rangle = \langle E_2 \rangle = 0$.

$$e_{\text{in}}(t) = Re\{[\Delta E_1(0, t) + i\Delta E_2(0, t)] \exp(i\omega t)\} = \Delta E_1(0, t) \cos \omega t - \Delta E_2(0, t) \sin \omega t$$

$$\langle \Delta E_1(0, t) \rangle = \langle \Delta E_2(0, t) \rangle = 0 \tag{20.2-12}$$

$$\langle (\Delta E_1(0, t)^2 \rangle = \langle (\Delta E_2(0, t)^2 \rangle = \frac{A^2}{4}$$

The resulting output is given by (20.2-9) with $E_{10}(0) = E_{20}(0) = 0$ and is

$$E_1(L, t) \equiv \Delta E_1(0, t) \exp\left(\frac{|g|L}{2}\right)$$

$$E_2(L, t) = \Delta E_2(0, t) \exp\left(-\frac{|g|L}{2}\right) \tag{20.2-13}$$

so that

$$e_{\text{out}}(t) = \Delta E_1(0, t) \exp\left(\frac{|g|L}{2}\right) \cos \omega t - \Delta E_2(0, t) \exp\left(\frac{-|g|L}{2}\right) \sin \omega t$$

corresponding to a field with a zero mean but with squeezed vacuum fluctuations—the so-called *squeezed vacuum*. The input (circle) and output (ellipse) fluctuation in this case are depicted by Figure 20-3(b) for a parametric gain $\exp(|g|L) = 4$.

Experimental Demonstration of Squeezing

The experimental setup used often to demonstrate squeezing [5] is shown in Figure 20-4. The output of an optical parametric amplifier at ω is combined in a balanced homodyne receiver with the strong local oscillator, also of frequency ω, which is *coherent* with that of the pump field at 2ω. (Usually these two fields are derived from the same master laser oscillator at ω). The two combined fields, whose complex amplitudes are $\mathcal{E}_1 = 1/\sqrt{2}[E(t) - \mathcal{E}_L(t)]$ and $\mathcal{E}_2 = 1/\sqrt{2}[E(t) + \mathcal{E}_L(t)]$ are detected, respectively, by photodetectors D_1 and D_2. The resulting currents i_1 and i_2 are subtracted from each other. The net current $i_2 - i_1$ is fed to a spectrum analyzer that displays the spectral density of $(i_2 - i_1)$, a quantity, which according to (10.2-5) and (10.2-7), is proportional to $\langle (i_2 - i_1)^2 \rangle$. The detected photocurrents are given according to (11.1-2) by

$$i_1 = t\mathcal{E}_1(t)\mathcal{E}_1^*(t) = \frac{t}{2}(E(t) - \mathcal{E}_L(t))(E^*(t) - \mathcal{E}_L^*(t))$$

$$i_2 = \frac{t}{2}(E(t) + \mathcal{E}_L(t))(E^*(t) + \mathcal{E}_L^*(t)) \tag{20.2-14}$$

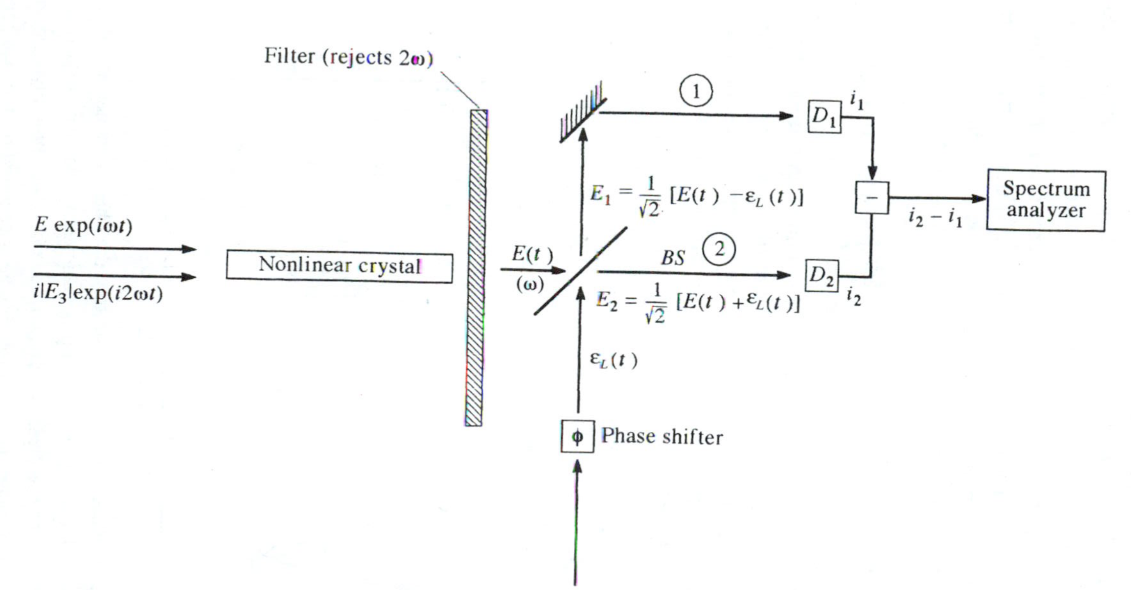

Figure 20-4 A balanced homodyne receiver for measuring the squeezing of the vacuum fluctuations of an electromagnetic field. The squeezing is caused by degenerate parametric amplification. Only the complex amplitudes are marked in the homodyne receiver section since the factor $\exp(i\omega t)$ has been dropped.

where t is a detector constant. The result of the electronic subtraction is thus

$$i_2 - i_1 = t[\mathscr{E}_L^*(t)E(t) + \text{c.c}] \tag{20.2-15}$$

This result illustrates the raison d'être for the balanced homodyne receiver. Since the output $(i_2 - i_1)$ contains only mixed (''signal'' × local oscillator) product terms, fluctuations $\Delta\mathscr{E}_L(t)$ of the local oscillator field, which lead to ''large'' terms $\mathscr{E}_L\Delta\mathscr{E}_L^*(t)$ in the photocurrents, i_1 and i_2, cancel out in the current subtraction. These terms would, in the case of a single detector receiver, mask the signal term $\mathscr{E}_L^*E(t)$.

In most of the recent experiments [4, 5] demonstrating squeezing, there exists no input to the parametric amplifier. In this case, the output of the parametric amplifier is given by Equation (20.2-9) with $E_{10}(0) = E_{20}(0) = 0$

$$E(t) = \Delta E_1(0, t)\exp\left(\frac{|g|L}{2}\right) + i\Delta E_2(0, t)\exp\left(-\frac{|g|L}{2}\right) \tag{20.2-16}$$

i.e., the squeezed vacuum field. The complex amplitude of the local oscillator field at the beam splitter is taken as the sum of the average field plus a fluctuation term. The fluctuation may be due to basic quantum causes or any other cause. A phase factor $\exp(i\phi)$ accounts for the phase shifter

$$\mathscr{E}_L(t) = [\mathscr{E}_{LO} + \Delta\mathscr{E}_L(t)]\exp(i\phi) \tag{20.2-17}$$

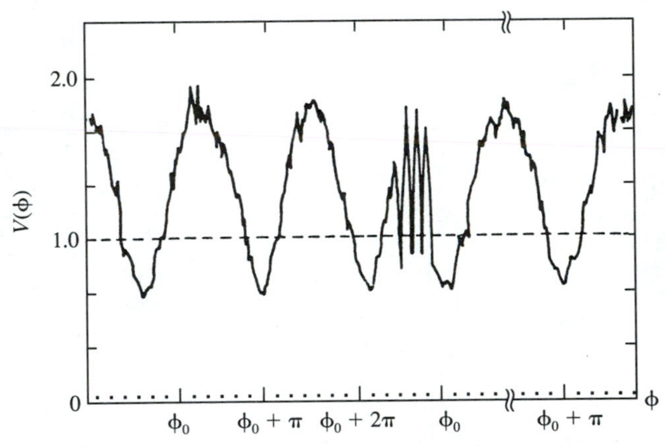

Figure 20-5 Measurement demonstrating the phase dependence of the quantum fluctuations in a squeezed state. The squeezing is achieved by degenerate parametric amplification. The phase dependence of the rms voltage from a balanced homodyne receiver is displayed vs. the local oscillator phase ϕ. The noise voltage is centered on $\nu = 1.8$ MHz. With the parametric amplifier blocked, $|g| = 0$, the output is given by the dashed horizontal line with no ϕ dependence. The dips represent 50 percent of the electronic noise power relative to that of unsqueezed vacuum (i.e., $|g| = 0$) input [5].

$\mathscr{E}_{LO}$ can, without any loss of generality, be taken as a real number. Substituting the last two equations for $\mathscr{E}_L(t)$ and $E(t)$ in (20.2-15) and neglecting the terms involving $\Delta\mathscr{E}_L(t)$, since $\Delta\mathscr{E}_L(t) \ll \mathscr{E}_{LO}$ results in

$$(i_2 - i_1) = 2t\mathscr{E}_{LO}\left(\Delta E_1(0, t)\exp\left(\frac{|g|L}{2}\right)\cos\phi - \Delta E_2(0, t)\exp\left(\frac{-|g|L}{2}\right)\sin\phi\right) \quad \text{(20.2-18)}$$

Since both $\langle \Delta E_1(t) \rangle$ and $\langle \Delta E_1(t) \rangle$ are zero, the time-averaged $(i_2 - i_1)$, the quantity that normally will be registered by a sensitive ammeter, is zero. This problem is avoided by squaring $(i_2 - i_1)$. This is accomplished usually [5] by the spectrum analyzer, which displays the spectral density $S_f(\Omega)$ of the input $f(t)$ where

$$\langle f^2(t) \rangle = \int_0^\infty S_f(\Omega)d\Omega$$

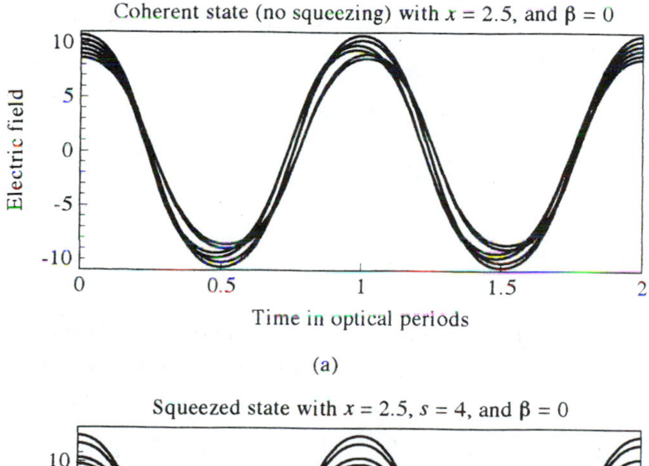

(a)

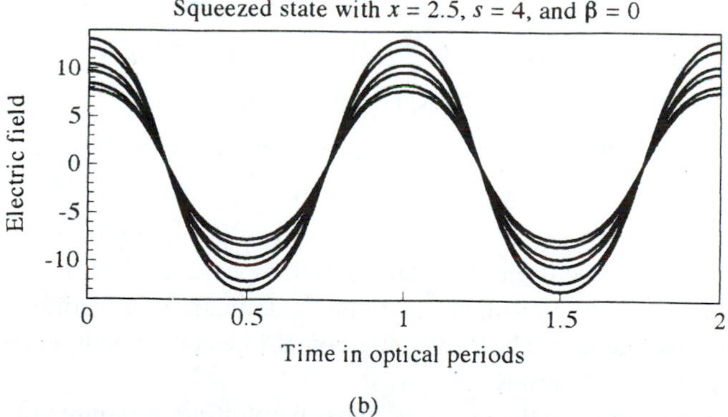

(b)

Figure 20-6 Representation of: (a) unsqueezed electric field, (b) squeezed ($s = 4$) state, (c) unsqueezed "vacuum" field, and (d) squeezed ($s = 4$) vacuum field.

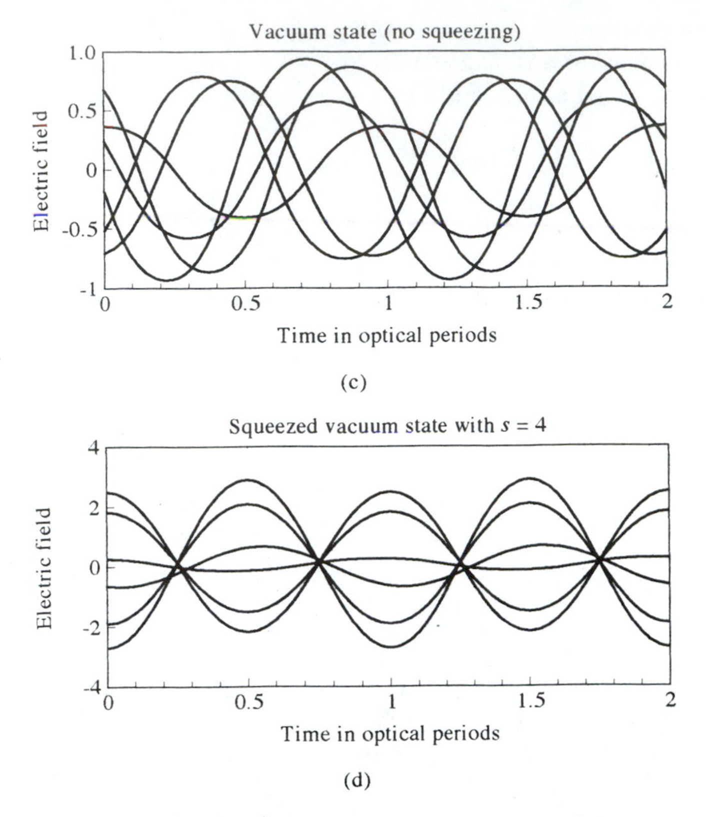

Figure 20-6 *(Continued)*

as discussed in Section 10.2. In our case, the input is $f(t) = i_2(t) - i_1(t)$, so that the output of the spectrum analyzer is proportional, in the case of a constant $S_{(i_2-i_1)}(\Omega)$, to

$$\begin{aligned} S_{i_2-i_1}(\Omega) \propto \langle (i_2 - i_1)^2 \rangle &= 4t^2\mathscr{E}_{LO}^2[\langle (\Delta E_1(0, t))^2 \rangle \exp(|g|L)\cos^2\phi \\ &\quad + \langle (\Delta E_2(0, t)^2 \rangle \exp(-|g|L)\sin^2\phi] \\ &= t^2\mathscr{E}_{\mathrm{LO}}^2 A^2[\exp(|g|L)\cos^2\phi + \exp(-|g|L)\sin^2\phi] \end{aligned} \tag{20.2-19}$$

where we used (20.1-6) and $\langle \Delta E_1(t)\Delta E_2(t) \rangle = 0$.

A typical result of such an experiment is shown in Figure 20-5. The observed dependence of the photocurrent fluctuations on the phase ϕ of local oscillator field is in agreement with (20.2-19) and constitutes a dramatic verification of squeezing of the vacuum fluctuations. The dashed horizontal line is the result when the optical parametric amplifier is blocked.

It is instructive to view the squeezed states by plotting in Figure 20-6, the actual sinusoidal optical fields corresponding to six representative points inside the uncertainty ellipse of Figure 20-1(e). We recall that each such point represents a possible realization of the field (complex) amplitude. The case of no squeezing ($s = 0$) is

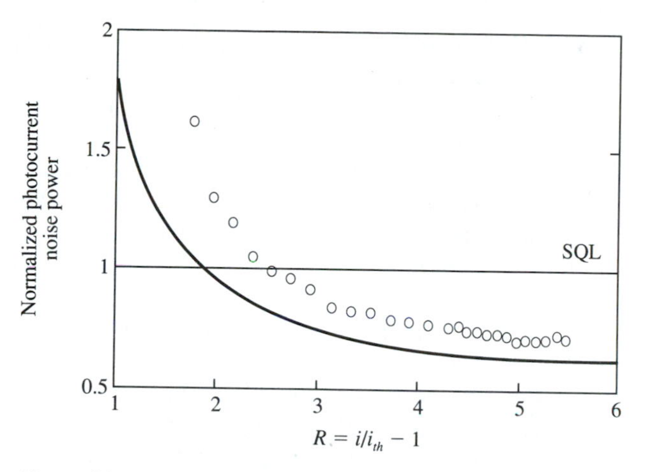

Figure 20-7 The normalized photocurrent noise current spectral density at $\nu = 29$ MHz as a function of the injection current to a semiconductor laser. i_{th} is the oscillation threshold current. SQL—the standard quantum limit—is the level corresponding to shot noise. (After Reference [7].)

shown in Figure 20-6(a), while a squeezed state with $x = 2.5$, $s = 4$, $\beta = 0$ is shown in (b). We note that in the squeezed state we trade an increase in the accuracy of measuring the frequency (or phase) for an increased amplitude fluctuation in qualitative agreement with (20.1-15). The vacuum state ($\langle E \rangle = 0$) is shown in (c), which contains plots of representative points from Figure 20-1(c), while the squeezed vacuum, Figure 20-1(f), is shown in (d).

Another type of squeezing, ''number squeezing,'' which results in photocurrent noise level below that of shot noise, can be exhibited in semiconductor diode lasers [6]. This squeezing results when the injection current to the lasers is highly constant and/or when proper, frequency dependent feedback is employed [7]. Such lasers may find practical uses in communication [8] and atomic measurements [9], for example. Experimental data showing such squeezing is shown in Figure 20-7.

In conclusion it is worthwhile to remind ourselves that the classical treatment of this chapter seems to do a good job in representing the results of the rigorous quantum treatment—but only up to second-order electric field products. If we were to ask some more difficult questions, say those involving expectation values of the electric field raised to third power or higher, the classical approach fails.

A comprehensive review of the topic quantum noise in optics is found in Reference [10].

References

1. See, for example, A. Yariv, *Quantum Electronics*, 4th ed. Wiley, 1988, p. 13.
2. Glauber, R. J., ''Coherent and incoherent states of radiation fields,'' *Phys. Rev.* 131:2776, 1963.
3. Louisell, W. H., *Radiation and Noise in Quantum Electronics*, New York: McGraw-Hill, 1964.

4. Yuen, H. P., and J. H. Shapiro, "Generation and detection of two-photon coherent states in degenerate four wave mixing," *Opt. Lett.* 4:334, 1979.

5. Wu, L., J. H. Kimble, J. L. Hall, and H. Wu, "Generation of squeezed states by parametric down conversion," *Phys. Rev. Lett.* 57:2520, 1986.

6. Yamamoto, Y., S. Machida, and O. Nilsson, "Amplitude squeezing in a pump-noise-suppressed laser oscillator," *Phys. Rev. A* 34:4025–4042, 1986.

7. Kitching, J., A. Yariv, and Y. Shevy, "Room temperature generation of squeezed light from a semiconductor laser with weak optical feedback," *Phys. Rev. Lett.* 74, 1995.

8. Saleh, B. E. A., and M. C. Teich, "Information transmission with number-squeezed light," *Proc. IEEE* 80:451–460, 1992. Also by the same authors, *Fundamentals of Photonics*, New York: Wiley, pp. 414–416.

9. Wieman, C. E., and L. Hollberg, "Using diode lasers for atomic physics," *Rev. Sci. Instrum.* 62:1–20, 1991.

The Kramers–Kronig Relations

Here we present the derivation of the Kramers–Kronig relations

$$
\begin{aligned}
\chi'(\omega) &= \frac{1}{\pi}\,\text{P.V.}\int_{-\infty}^{+\infty} \frac{\chi''(\omega')}{\omega' - \omega}\,d\omega' \\
\chi''(\omega) &= -\frac{1}{\pi}\,\text{P.V.}\int_{-\infty}^{+\infty} \frac{\chi'(\omega')}{\omega' - \omega}\,d\omega'
\end{aligned}
\tag{A-1}
$$

that were given without proof in Equation (5.4-4). These relations are valid when $\chi(\omega) = \chi'(\omega) - i\chi''(\omega)$ has no poles in the lower half-complex-ω-plane [when $\chi(\omega)$ has no singularities in the upper half-plane, similar relations, but with opposite signs, result]. For this case, we integrate the function $\chi(\omega')/(\omega' - \omega)$ over the contour shown in Figure A-1.

$$
\int_{c'} \frac{\chi(\omega')}{\omega' - \omega}\,d\omega' + \int_{-R}^{\omega-\varepsilon} \frac{\chi(\omega')}{\omega' - \omega}\,d\omega' + \int_{\omega+\varepsilon}^{R} \frac{\chi(\omega')}{\omega' - \omega}\,d\omega' + \int_{c} \frac{\chi(\omega')}{\omega' - \omega}\,d\omega' = 0 \tag{A-2}
$$

where c' is the semicircle extending from $-R$ to R, whereas c is the semicircle around ω. The right side is zero since $\chi(\omega')/(\omega' - \omega)$ has no singularities inside the contour. We next take the limit of (A-2) as $R \to \infty$ and $\varepsilon \to 0$. The integral over c' vanishes for $\chi(\infty) = 0$ while the integral over c becomes

$$
\lim_{\varepsilon\to 0} \int_{c} \frac{\chi(\omega')}{\omega' - \omega}\,d\omega' = \lim_{\varepsilon\to 0} \int_{\pi}^{2\pi} \frac{\chi(\omega + \varepsilon e^{i\phi})\; i\varepsilon e^{i\phi}}{\varepsilon e^{i\phi}}\,d\phi = \pi i \chi(\omega)
$$

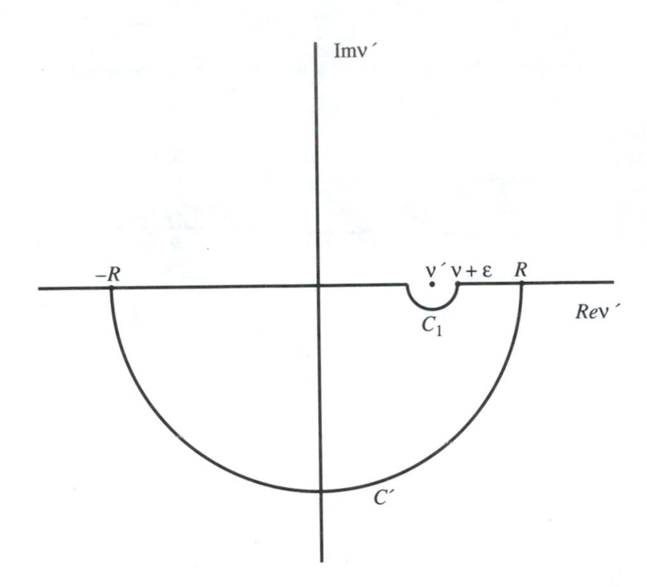

Figure A-1 The integration contour of (A-2) used to derive the Kramers–Kronig relations.

where we took $\omega' = \omega + \varepsilon e^{i\phi}$ over c. The sum of the second and third integrals of (A-2) in the limit $\varepsilon \to 0$, $R \to \infty$, is, by definition, the principal value of the integral between $-\infty$ and ∞. The final result is

$$\chi(\omega) = \frac{i}{\pi}\,\text{P.V.}\int_{-\infty}^{+\infty} \frac{\chi(\omega')}{\omega' - \omega}\,d\omega' \tag{A-3}$$

Taking $\chi(\omega) = \chi'(\omega) - i\chi''(\omega)$ and equating the real and imaginary parts of both sides of (A-3) yield Eqs. (A-1).

Another useful form results from the requirement that $\chi(-\omega) = \chi^*(\omega)$ so that $\chi''(\omega')$ is an odd and $\chi'(\omega')$ is an even function of ω'. We can multiply (A-1) by $(\omega' + \omega)/(\omega' + \omega)$ and obtain

$$\chi'(\omega) = \frac{2}{\pi}\,\text{P.V.}\int_{0}^{+\infty} \frac{\chi''(\omega')\omega'}{\omega'^2 - \omega^2}\,d\omega' \tag{A-4}$$

where the integral involving $\chi''(\omega')$ is zero, since $\chi''(\omega')$ is odd. In a similar fashion, we derive

$$\chi''(\omega) = \frac{2\omega}{\pi}\,\text{P.V.}\int_{0}^{+\infty} \frac{\chi'(\omega')}{\omega'^2 - \omega^2}\,d\omega' \tag{A-5}$$

The requirement that $\chi(\omega)$ have no poles in the lower half-plane is satisfied by *passive* linear systems. This is so because the Fourier transform $P(\omega)$ of the polarization $P(t)$ is equal in a linear system to the product $\chi(\omega)E(\omega)$ where $E(\omega)$ is the Fourier transform of $E(t)$. One consequence of this relation is that the natural frequencies of vibration are the poles of $\chi(\omega)$ and a pole $\omega_0 - i\omega_i$ in the lower half-

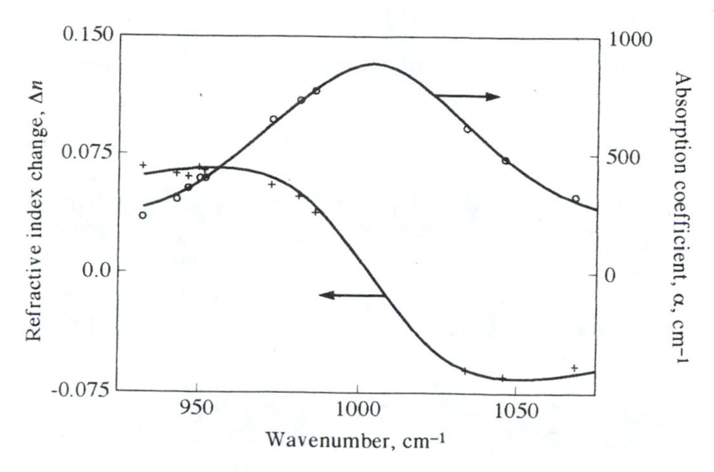

Figure A-2 The contribution to the refractive index and absorption coefficient of 30 periods of 70 Å GaAs quantum wells doped to a surface density of 2×10^{12} cm^{-1} and separated by 440 Å GaAl$_{0.4}$As barriers. (From: G. Almogy, A. Shakouri, and A. Yariv, "Observation of birefringence induced by intersubband transitions in quantum wells," *Appl. Phys. Lett.* 63:20, 2720, 1993.)

plane corresponds to a solution $e^{i\omega_0 t}e^{\omega_i t}$. This represents an indefinite increase in energy with time, which is not possible in a passive linear system. The presence of poles in the lower half-plane can also be shown to violate the causality relation since it corresponds to a response of the polarization $P(t)$, which precedes the driving "force" $E(t)$.

Figure A-2 shows plots of the absorption coefficient $\alpha(\nu)$, which is proportional to $\chi''(\nu)$, in semiconductor quantum wells (see Chapter 16), and of the change in the index of refraction $\Delta n(\nu)$, which is proportional to $\chi'(\nu)$. The crosses and dots are experimental data. The solid Δn curve is derived using the Kramers–Kronig relations from the solid curve fit to the $\alpha(\nu)$ data.

The Electrooptic Effect in Cubic $\bar{4}3m$ Crystals

As an example of transverse modulation[1] and of the application of the electrooptic effect we consider the case of crystals of the $\bar{4}3m$ symmetry group. Examples of this group are: InAs, CuCl, GaAs, and CdTe. The last two are used for modulation in the infrared since they remain transparent beyond 10 μm. These crystals are cubic and have axes of fourfold symmetry along the cube edges (⟨100⟩ directions) and threefold axes of symmetry along the cube diagonals ⟨111⟩.

To be specific, we apply the field in the ⟨111⟩ direction—that is, along a threefold-symmetry axis. Taking the field magnitude as E, we have

$$\mathbf{E} = \frac{E}{\sqrt{3}} (\mathbf{e}_1 + \mathbf{e}_2 + \mathbf{e}_3) \tag{B-1}$$

where $\mathbf{e}_1$, $\mathbf{e}_2$, and $\mathbf{e}_3$ are unit vectors directed along the cube edges x, y, and z, respectively. The three nonvanishing electrooptic tensor elements are, according to Table 9-1 [see $\bar{4}3m$ tensor], r_{41}, $r_{52} = r_{41}$, and $r_{63} = r_{41}$. Thus, using Equations (9.1-2) through (9.1-4), with

$$\left(\frac{1}{n^2}\right)_1 = \left(\frac{1}{n^2}\right)_2 = \left(\frac{1}{n^2}\right)_3 \equiv \frac{1}{n_o^2}$$

we obtain

$$\frac{x^2 + y^2 + z^2}{n_o^2} + \frac{2r_{41}E}{\sqrt{3}} (xy + yz + xz) = 1 \tag{B-2}$$

[1] *Transverse modulation* is the term applied to the case when the field is applied normal to the direction of propagation.

as the equation of the index ellipsoid. One can proceed formally at this point to derive the new directions x', y', and z' of the principal axes of the ellipsoid. A little thought, however, will show that the $\langle 111 \rangle$ direction along which the field is applied will continue to remain a threefold-symmetry axis, whereas the remaining two orthogonal axes can be chosen *anywhere* in the plane normal to $\langle 111 \rangle$. Thus (B-2) is an equation of an ellipsoid of revolution about $\langle 111 \rangle$. To prove this we choose $\langle 111 \rangle$ as the z' axis, so

$$z' = \frac{1}{\sqrt{3}} x + \frac{1}{\sqrt{3}} y + \frac{1}{\sqrt{3}} z \tag{B-3}$$

and take

$$x' = \frac{1}{\sqrt{2}} y - \frac{1}{\sqrt{2}} z$$

$$y' = \frac{-2}{\sqrt{6}} x + \frac{1}{\sqrt{6}} y + \frac{1}{\sqrt{6}} z \tag{B-4}$$

Therefore

$$x = -\frac{2}{\sqrt{6}} y' + \frac{1}{\sqrt{3}} z'$$

$$y = \frac{1}{\sqrt{2}} x' + \frac{1}{\sqrt{6}} y' + \frac{1}{\sqrt{3}} z' \tag{B-5}$$

$$z = -\frac{1}{\sqrt{2}} x' + \frac{1}{\sqrt{6}} y' + \frac{1}{\sqrt{3}} z'$$

Substituting (B-5) in (B-2), we obtain the equation of the index ellipsoid in the x', y', z' coordinate system as

$$(x'^2 + y'^2)\left(\frac{1}{n_o^2} - \frac{r_{41}E}{\sqrt{3}}\right) + z'^2\left(\frac{1}{n_o^2} + \frac{2r_{41}}{\sqrt{3}} E\right) = 1 \tag{B-6}$$

so the principal indices of refraction become

$$n_{y'} = n_{x'} = n_o + \frac{n_o^3 r_{41} E}{2\sqrt{3}}$$

$$n_{z'} = n_o - \frac{n_o^3 r_{41} E}{\sqrt{3}} \tag{B-7}$$

It is clear from (B-6) that other choices of x' and y', as long as they are normal to z' and to each other, will work as well since x' and y' enter (B-6) as the combination $x'^2 + y'^2$, which is invariant to rotations about the z' axis. The principal axes of the index ellipsoid (B-6) are shown in Figure B-1.

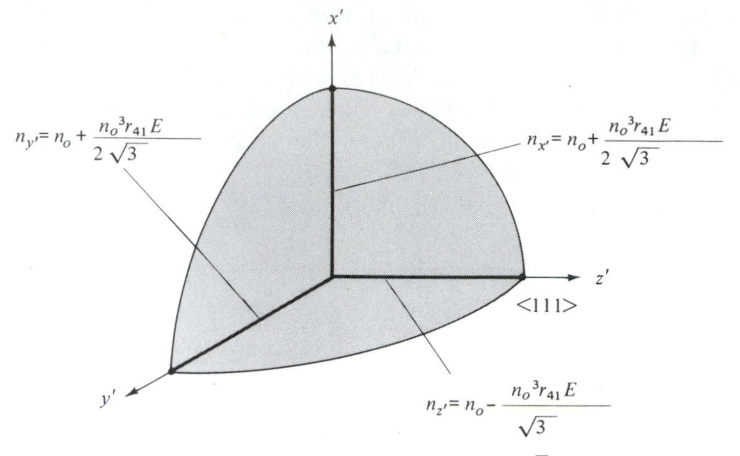

Figure B-1 The intersection of the index ellipsoid of $\bar{4}3m$ crystals (with **E** parallel to $\langle 111 \rangle$) with the planes $x' = 0$, $y' = 0$, $z' = 0$. The principal indices of refraction for this case are $n_{x'}$, $n_{y'}$, and $n_{z'}$.

An amplitude modulator based on the foregoing situation is shown in Figure B-2. The fractional intensity transmission is given by (9.3-4) as

$$\frac{I_o}{I_i} = \sin^2 \frac{\Gamma}{2}$$

where the retardation, using (B-7), is

$$\Gamma = \phi_{z'} - \phi_{y'} = \frac{(\sqrt{3}\pi) n_o^3 r_{41}}{\lambda_0} \left(\frac{Vl}{d} \right) \tag{B-8}$$

An important difference between this case where the electric field is applied normal to the direction of propagation and the longitudinal case (9.2-4) is that here Γ is proportional to the crystal length l.

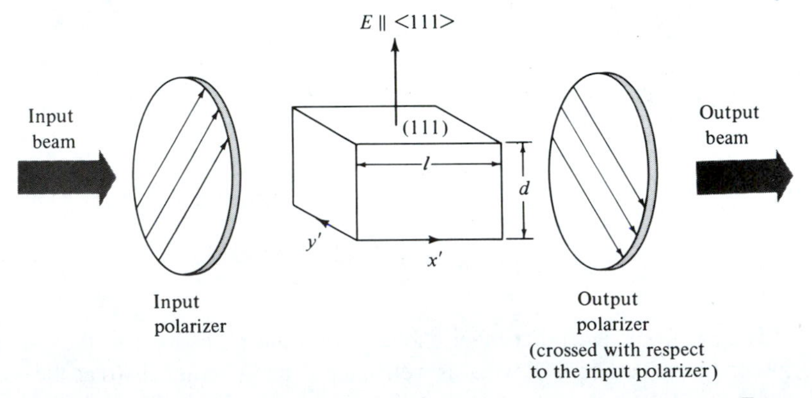

Figure B-2 A transverse electrooptic modulator using a zinc-blende-type ($\bar{4}3m$) crystal with **E** parallel to the cube diagonal $\langle 111 \rangle$ direction.

A complete discussion of the electrooptic effect in $\bar{4}3m$ crystals is given in C. S. Namba, *J. Opt. Soc. Am.*, vol. 51, p. 76, 1961. A summary of his analysis is included in Table B-1.

Table B-1 Electrooptical Properties and Retardation in $\bar{4}3m$ (zinc-blende structure) Crystals for Three Directions of Applied Field (After C. S. Namba, *J. Opt. Soc. Am.*, vol. 51, p. 76, 1961.)

	$E_\perp(001)$ plane	$E_\perp(110)$ plane	$E_\perp(111)$ plane
	$E_x = E_y = 0, E_z = E$	$E_x = E_y = \frac{E}{\sqrt{2}}, E_z = 0$	$E_x = E_y = E_z = \frac{E}{\sqrt{3}}$
Index ellipsoid	$\frac{x^2 + y^2 + z^2}{n_o^2} + 2r_{41}Exy = 1$	$\frac{x^2 + y^2 + z^2}{n_o^2} + \sqrt{2}r_{41}E(yz + zx) = 1$	$\frac{x^2 + y^2 + z^2}{n_o^2} + \frac{2}{\sqrt{3}} r_{41}E(yz + zx + xy) = 1$
n'_x	$n_o + \frac{1}{2}n_o^3 r_{41}E$	$n_o + \frac{1}{2}n_o^3 r_{41}E$	$n_o + \frac{1}{2\sqrt{3}} n_o^3 r_{41}E$
n'_y	$n_o - \frac{1}{2}n_o^3 r_{41}E$	$n_o - \frac{1}{2}n_o^3 r_{41}E$	$n_o + \frac{1}{2\sqrt{3}} n_o^3 r_{41}E$
n'_z	n_o	n_o	$n_o - \frac{1}{\sqrt{3}} n_o^3 r_{41}E$
x' y' z' coordinates	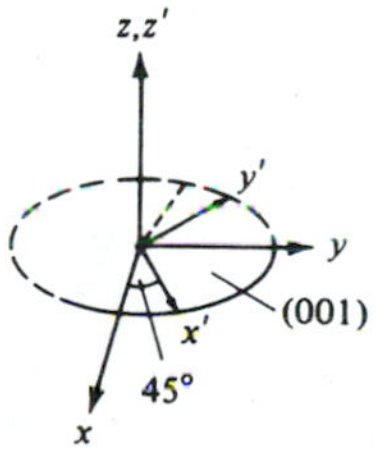 z,z′; y′; y; (001); x′; 45°; x	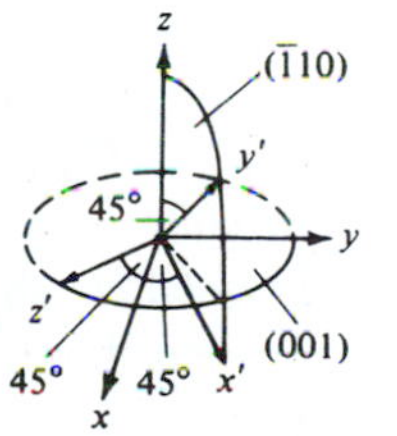 z; ($\bar{1}$10); y′; 45°; y; z′; (001); 45°; 45°; x′; x	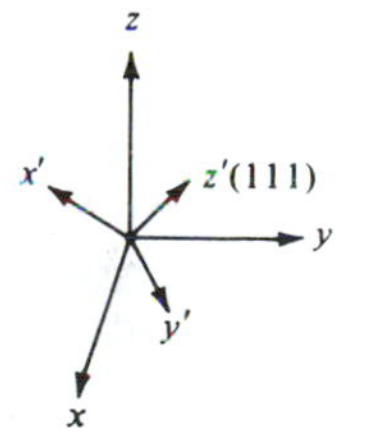 z; x′; z′(111); y; y′; x
Directions of optical path and axes of crossed polarizer	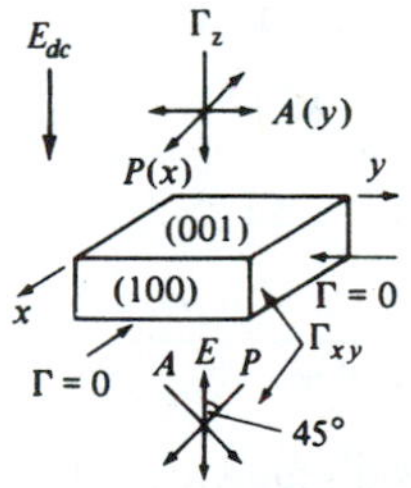 E_{dc}; Γ_z; A(y); P(x); y; (001); (100); x; Γ = 0; Γ_{xy}; Γ = 0; A; E; P; 45°	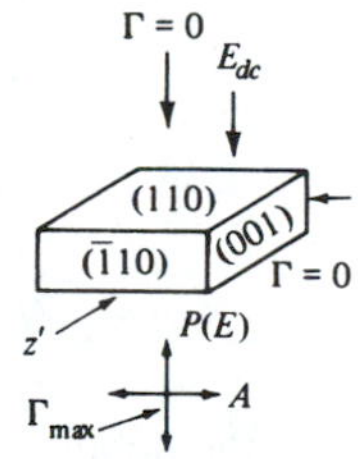 Γ = 0; E_{dc}; (110); ($\bar{1}$10); (001); Γ = 0; z′; P(E); Γ_{max}; A	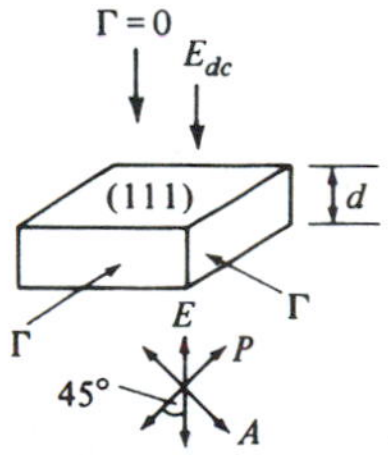 Γ = 0; E_{dc}; (111); d; Γ; E; Γ; P; 45°; A
Retardation phase difference $\Gamma(V = Ed)$	$\Gamma = \frac{2\pi}{\lambda} n_o^3 r_{41} V$ $\Gamma_{xy} = \frac{\pi}{\lambda}\frac{l}{d} n_o^3 r_{41} V$	$\Gamma_{max} = \frac{2\pi}{\lambda}\frac{l}{d} n_o^3 r_{41} V$	$\Gamma = \frac{\sqrt{3}\pi}{\lambda}\frac{l}{d} n_o^3 r_{41} V$

Noise in Traveling Wave Laser Amplifiers

In Section 10.6 we discussed the effect of spontaneous emission power on the spectral width of the laser output. In this appendix we will derive the effect of spontaneous emission noise on a laser amplifier in which the gain medium, with no mirrors, is used to amplify a weak input field. The basic engineering problem is to find the degradation of the signal-to-noise power that is caused by the (inevitable) addition of some spontaneous emission (noise) power to the amplified signal. A typical experimental situation is shown in Figure C-1.

An inverted atomic medium with population densities N_2 and N_1 in the upper and lower transition levels occupies the space between $z = 0$ and $z = L$. An optical beam with power P is focused through an aperture with an area A_1 into the gain medium and exits through an aperture A_2. The coherent amplification of the input beam power P due to stimulated emission is given by

$$dP = \gamma P \, dz \tag{C-1}$$

where γ, the exponential gain constant, is given by (5.3-3) as

$$\gamma(\nu) = (N_2 - N_1)\frac{c^2 g(\nu)}{8\pi n^2 \nu^2 t_{\text{spont}}} \tag{C-2}$$

Let us consider next the details of how spontaneous radiation (noise) is emitted, amplified, and mixes with the signal beam to degrade its signal-to-noise ratio.

An element dz at z with area A emits *spontaneously*

$$P_N = \frac{N_2 h\nu A \, dz}{t_{\text{spont}}} \tag{C-3}$$

watts of power. Since this power is emitted isotropically over the 4π solid angle, only a fraction $d\Omega/4\pi$ of the total is fed into the solid angle $d\Omega$ subtended by the

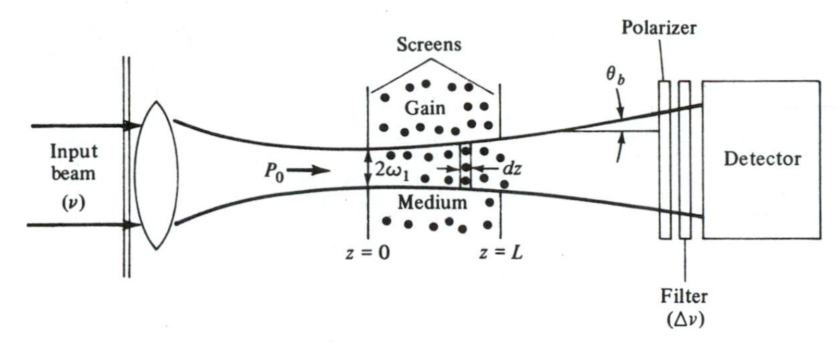

Figure C-1 A laser amplifier consisting of an inverted atomic medium contained between two screens. The signal beam is injected so that its waist coincides with the front screen.

laser beam and ultimately intercepted by the detector. Similarly, it follows from the definition in Section 5.1 of the line-shape function $g(\nu)$ that only a fraction $g(\nu)\Delta\nu$ of the total spectrum of the spontaneous radiation falls within the transmission bandpass $\Delta\nu$ of the filter. The total noise power emitted by the elemental volume $A\ dz$ within the optical spectral region $\Delta\nu$ and solid angle $d\Omega$ allowed into the detector is thus

$$(dP)' = \frac{1}{2} \frac{N_2 h\nu g(\nu)\Delta\nu A}{t_{\text{spont}}} \frac{d\Omega}{4\pi} dz \tag{C-4}$$

where the factor $\frac{1}{2}$ in front accounts for the polarizer that can remove half of the (isotropically polarized) noise without affecting the (linearly polarized) signal power. The smallest solid angle $d\Omega$ that we can use without sacrificing signal power is that subtended by the beam

$$(d\Omega)_{\min} = (d\Omega)_b = \pi\theta_b^2 = \frac{\lambda^2}{n^2 A} \tag{C-5}$$

where $A = \pi\omega_1^2$ and $\theta_b = \lambda/\pi\omega_1 n$ is the far-field diffraction angle of the signal beam as shown in Figure C-1. The value λ^2/n^2A is often referred to as the *solid angle per mode.*

Using (C-1) and (C-5), we rewrite (C-4) as

$$(dP)' = \frac{N_2 \gamma h\nu}{N_2 - N_1} \Delta\nu \tag{C-6}$$

The total evolution of beam power including the induced and spontaneous transitions is thus given by the sum of the emitted powers. (If the two contributions were coherent we would add their fields.)

$$\frac{dP}{dz} = \gamma P + \frac{N_2}{N_2 - N_1} \gamma h\nu\ \Delta\nu \tag{C-7}$$

The solution of (C-7) subject to the boundary condition $P(0) = P_0$ is

$$\mathrm{P}(z) = \underbrace{P_0 e^{\gamma z}}_{\text{amplified signal}} + \underbrace{\mu h\nu\, \Delta\nu(e^{\gamma z} - 1)}_{\text{amplified noise}} \tag{C-8}$$

where

$$\mu \equiv \frac{N_2}{N_2 - N_1} \tag{C-9}$$

is the population inversion factor. The signal-to-noise power ratio at the output of the amplifier is

$$\left(\frac{\mathrm{S}}{\mathrm{N}}\right)_{\text{output}} = \frac{P_0}{\mu h\nu\, \Delta\nu} \frac{G}{G - 1} \tag{C-10}$$

$G \equiv \exp(\gamma L)$ is the one-pass gain. From the point of view of power bookkeeping, the effect of spontaneous emission is seen to be equivalent to a noise input power

$$N_{\text{eff}} = \mu h\nu\, d\nu \left(1 - \frac{1}{G}\right) \tag{C-11}$$

which for an ideal four-level gain medium ($\mu = 1$) and high gain ($G \gg 1$) becomes

$$N_{\text{eff}} = h\nu\, \Delta\nu$$

If the laser amplifier were to be employed as a preamplifier in an optical receiver, then the minimum detectable power in the sense defined in Section 11.4 is given by

$$(P_s)_{\min} \sim h\nu\, \Delta\nu \tag{C-12}$$

which is the same as that obtained in (11.4-10) in the case of a heterodyne detection scheme with unity quantum efficiency ($\eta = 1$). The laser preamplifier is thus an "ideal" quantum limited receiver.[1]

The approach leading to (C-7) is quite general and should apply also to an atomic medium that is in thermal equilibrium (at T) and hence is absorbing. We can use (C-7) in this case, provided we put $\gamma(\nu) \rightarrow -\alpha(\nu)$, α (being the medium absorption coefficient) and $(N_2/N_1) = \exp(-h\nu/kT)$ as appropriate to a medium in thermal equilibrium.

The result is

$$\frac{dP}{dz} = -\alpha P + \frac{\alpha h\nu\, \Delta\nu}{e^{h\nu/kT} - 1} \tag{C-13}$$

---

[1]A practical note: To achieve the minimum detectable signal power, we need, according to (C-12), to reduce the bandwidth $\Delta\nu$ as much as possible. This is more easily done at the radio frequencies of the heterodyne signal than at optical frequencies. The practical advantage thus lies with heterodyne reception.

whose solution is

$$P(z) = P(0)\, e^{-\alpha z} + \frac{h\nu\, \Delta\nu}{e^{h\nu/kT} - 1}(1 - e^{-\alpha z}) \tag{C-14}$$

If the medium is "black" $e^{-\alpha L} \ll 1$ (i.e., all incident radiation is absorbed), the output power is

$$P(L) = \frac{h\nu}{e^{h\nu/kT} - 1}\, \Delta\nu \tag{C-15}$$

independent of L and the input. This result is the same as the Johnson noise formula (10.5-6) which was obtained using quite a different point of view.

If the laser medium contains a transition, other than that responsible for the gain, that causes an absorption coefficient α and is characterized by a temperature T (this would be the temperature appearing in the Boltzmann ratio of the populations involved), then we must add the spontaneous emission from the upper to lower level of this transition to that from the upper, amplifying, laser level. Using (C-7) and (C-13) we obtain

$$\frac{dP}{dz} = (\gamma - \alpha)P + \frac{N_2}{N_2 - N_1}\gamma h\nu\, \Delta\nu + \frac{\alpha h\nu \Delta\nu}{e^{h\nu/kT} - 1} \tag{C-16}$$

whose solution is

$$P(z) = P(0)\, e^{[\gamma(\nu) - \alpha(\nu)]z} + \left(\frac{\alpha h\nu\, \Delta\nu}{e^{h\nu/kT} - 1} + \frac{\gamma h\nu\, d\nu}{1 - N_1/N_2}\right)\frac{e^{(\gamma - \alpha)z} - 1}{\gamma - \alpha} \tag{C-17}$$

At optical frequencies where $h\nu \gg kT$, the contribution of the first term in the brackets, which represents spontaneous emission due to atomic levels involved in the absorbing transitions, is, in most cases, negligible compared to the second term, which is due to spontaneous emission in the lasing transition. At lower, say microwave, frequencies the loss contribution may become appreciable.

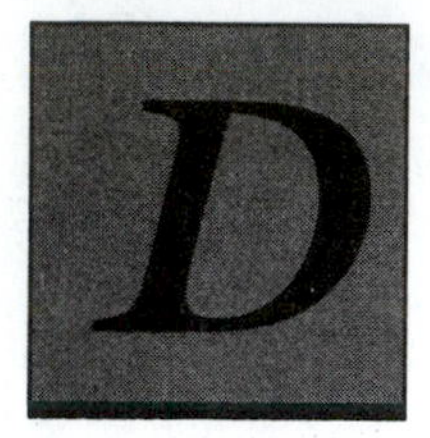

D Transformation of a Coherent Electromagnetic Field by a Thin Lens

In this appendix we will derive one of the most important results of the theory of coherent optics, which deals with the transformation of a coherent monochromatic field by a lens. A special case of this derivation was stated as Equation (17.10-3).

Consider the propagation of an optical beam

$$E = \mathrm{Re}[u(x, y, z)e^{i\omega t}]$$

from an "input" plane $z = 0$ through a lens at z and then to the back focal "output" plane at $z + f$ as shown in Figure D-1. $u(\mathbf{x})$ is thus the complex amplitude of the field, and f is the focal length of the lens.

We use Equation (4.8-2) to transform the input beam at $z = 0$ to the plane 1.

$$u_1(x_1, y_1) = \frac{i}{\lambda z} \iint_{\Sigma_0} u(x, y)e^{-ikr}\, dxdy \tag{D-1}$$

We used $u(x, y) \equiv u(x, y, z = 0)$ and

$$r = \sqrt{(x_1 - x)^2 + (y_1 - y)^2 + z^2} \approx z + \frac{(x_1 - x)^2}{2z} + \frac{(y_1 - y)^2}{2z} \tag{D-2}$$

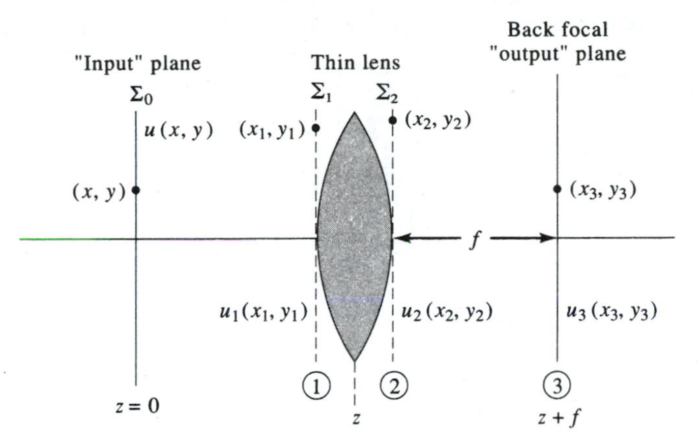

Figure D-1 A lens at z transforms an "input" beam $u(x, y)$ at $z = 0$ to an "output" at $z + f$. The lens plane is designated as Σ_1 and assumed infinite in its transverse dimensions so that truncation effects are neglected.

to rewrite (D-1) as

$$u_1(x_1, y_1) = \frac{ie^{-ikz}}{\lambda z} \iint_{\Sigma_0} u(x, y) e^{-\frac{ik}{2z}[(x_1-x)^2 + \text{s.y.}]}\, dxdy \tag{D-3}$$

In (D-2) and the rest of this appendix, s.y. stands for "similar terms with x → y." As an example $[(x_1 - x)^2 + \text{s.y.}] \equiv [(x_1 - x)^2 + (y_1 - y)^2]$. Relation (D-3), which results from using the approximate form of (D-2), is called the Kirchhoff diffraction integral. The approximation is valid when the neglected terms in (D-2) multiplied by $k(=2\pi/\lambda)$ are small compared to 2π. The field at plane Σ_2 is obtained by multiplying the field at u_1 at 1 by the lens transfer function (2.3-1)

$$\begin{aligned} u_2(x_2, y_2) &= u_1(x_2, y_2) e^{\frac{ik}{2f}(x_2^2 + y_2^2)} \\ &= \frac{ie^{-ikz}}{\lambda z} \iint_{\Sigma_0} dxdyu(x, y) e^{-i[\frac{k}{2z}(x_2-x)^2 - \frac{k}{2f}x_2^2 + \text{s.y.}]} \end{aligned} \tag{D-4}$$

Next we apply (D-3) again to "propagate" from plane 2 to 3

$$u_3(x_3, y_3, z + f) = -\frac{e^{-ik(z+f)}}{\lambda^2 zf} \iint_{\Sigma_2} dx_2 dy_2 e^{-i\frac{k}{2f}[(x_3-x_2)^2 + \text{s.y.}]} \left[\iint_{\Sigma_0} dxdyu(x, y) e^{-i[\frac{k}{2z}(x_2-x)^2 - \frac{k}{2f}x_2^2 + \text{s.y.}]} \right] \tag{D-5}$$

We rearrange the sum of *all* of the terms in the exponents of (D-5). The result is

$$k(z+f) + \frac{k}{2f}[(x_3 - x_2)^2 + \text{s.y.}] + \left[\frac{k}{2z}(x_2 - x)^2 - \frac{k}{2f}x_2^2 + \text{s.y.}\right]$$

$$= k(z+f) + k\frac{(x_3^2 + y_3^2)}{2f}\left(1 - \frac{z}{f}\right) - \frac{k}{f}(xx_3 + yy_3)$$
$$+ \frac{k}{2z}\left\{\left[x_2 - \left(x + \frac{z}{f}x_3\right)\right]^2 + \left[y_2 - \left(y + \frac{z}{f}y_3\right)\right]^2\right\}$$

Changing the order of integration, we rewrite (D-5) as

$$u_3(x_3, y_3, z+f) = -\frac{e^{-ik\left[(z+f)+\frac{x_3^2+y_3^2}{2f}\left(1-\frac{z}{f}\right)\right]}}{\lambda^2 zf} \cdot \iint_{\Sigma_0} dxdyu(x, y)\, e^{i\left(\frac{kxx_3}{f} + \frac{kyy_3}{f}\right)}$$
$$\iint_{\Sigma_2} dx_2 dy_2 e^{-\frac{ik}{2z}\left\{\left[x_2 - \left(x + \frac{z}{f}x_3\right)\right]^2 + \left[y_2 - \left(y + \frac{z}{f}y_3\right)\right]^2\right\}} \quad \text{(D-6)}$$

Considering the apertures Σ_0, Σ_2 as infinite and using the definite integral

$$\int_{-\infty}^{\infty} e^{-ia^2x^2}\, dx = \frac{\sqrt{\pi}}{|a|}\, e^{-i\pi/4}$$

the integral over Σ_2 is equal to $-i2\pi z/k$ so that recalling that $k\lambda = 2\pi$, (D-6) becomes

$$u_3(x_3, y_3, z+f) = \frac{ie^{-ik(z+f)}}{\lambda f}\, e^{i\frac{k}{2f}\left(\frac{z}{f} - 1\right)\left(x_3^2 + y_3^2\right)} \iint_{\Sigma_0} u(x,y) e^{ik\left(\frac{xx_3}{f} + \frac{yy_3}{f}\right)}\, dxdy \quad \text{(D-7)}$$

Recalling the definition (17.10-1) of the Fourier transform, we can rewrite (D-7) as

$$u_3(x_3, y_3, z+f)$$
$$= i\frac{e^{-ik(z+f)+\frac{ik}{2f}\left(\frac{z}{f} - 1\right)\left(x_3^2 + y_3^2\right)}}{\lambda f}(2\pi)^2 F\{u(x, y)\}_{p=-kx_3/f, q=-ky_3/f} \quad \text{(D-8)}$$

$F\{u(x, y)\}$ is the double (x, y) Fourier transform of $u(x, y)$ and is a function of the variables p and q. An especially simple form results if the plane Σ_0 is the front focal plane, i.e., $z = f$. In this case

$$u_3(x_3, y_3, 2f) = i\frac{4\pi^2 e^{-i2kf}}{\lambda f} F\{u(x, y)\}_{p=-kx_3/f, q=-ky_3/f} \quad \text{(D-9)}$$

The output field $u_3(x_3, y_3, 2f)$ is thus the (scaled) Fourier transform of the input field $u(x, y)$. Equation (D-9) was stated in the book as Equation (17.10-3).

Index